Handbook of
Plant
Nutrition

BOOKS IN SOILS, PLANTS, AND THE ENVIRONMENT

Soil Biochemistry, Volume 1, edited by A. D. McLaren and G. H. Peterson

Soil Biochemistry, Volume 2, edited by A. D. McLaren and J. Skujins

Soil Biochemistry, Volume 3, edited by E. A. Paul and A. D. McLaren

Soil Biochemistry, Volume 4, edited by E. A. Paul and A. D. McLaren

Soil Biochemistry, Volume 5, edited by E. A. Paul and J. N. Ladd

Soil Biochemistry, Volume 6, edited by Jean-Marc Bollag and G. Stotzky

Soil Biochemistry, Volume 7, edited by G. Stotzky and Jean-Marc Bollag

Soil Biochemistry, Volume 8, edited by Jean-Marc Bollag and G. Stotzky

Soil Biochemistry, Volume 9, edited by G. Stotzky and Jean-Marc Bollag

Organic Chemicals in the Soil Environment, Volumes 1 and 2,
   edited by C. A. I. Goring and J. W. Hamaker

Humic Substances in the Environment, M. Schnitzer and S. U. Khan

Microbial Life in the Soil: An Introduction, T. Hattori

Principles of Soil Chemistry, Kim H. Tan

Soil Analysis: Instrumental Techniques and Related Procedures,
   edited by Keith A. Smith

*Soil Reclamation Processes: Microbiological Analyses and
      Applications*, edited by Robert L. Tate III and Donald A. Klein

Symbiotic Nitrogen Fixation Technology, edited by Gerald H. Elkan

Soil--Water Interactions: Mechanisms and Applications, Shingo Iwata
   and Toshio Tabuchi with Benno P. Warkentin

Soil Analysis: Modern Instrumental Techniques, Second Edition,
   edited by Keith A. Smith

Soil Analysis: Physical Methods, edited by Keith A. Smith
   and Chris E. Mullins

Growth and Mineral Nutrition of Field Crops, N. K. Fageria,
   V. C. Baligar, and Charles Allan Jones

Semiarid Lands and Deserts: Soil Resource and Reclamation,
   edited by J. Skujins

Plant Roots: The Hidden Half, edited by Yoav Waisel, Amram Eshel,
   and Uzi Kafkafi

Plant Biochemical Regulators, edited by Harold W. Gausman

Maximizing Crop Yields, N. K. Fageria

Transgenic Plants: Fundamentals and Applications, edited by
   Andrew Hiatt

*Soil Microbial Ecology: Applications in Agricultural and Environmental
      Management*, edited by F. Blaine Metting, Jr.

Principles of Soil Chemistry: Second Edition, Kim H. Tan

Water Flow in Soils, edited by Tsuyoshi Miyazaki

Handbook of Plant and Crop Stress, edited by Mohammad Pessarakli

Genetic Improvement of Field Crops, edited by Gustavo A. Slafer

Agricultural Field Experiments: Design and Analysis,
   Roger G. Petersen

Environmental Soil Science, Kim H. Tan

*Mechanisms of Plant Growth and Improved Productivity: Modern
      Approaches*, edited by Amarjit S. Basra

Selenium in the Environment, edited by W. T. Frankenberger, Jr.
   and Sally Benson

Plant–Environment Interactions, edited by Robert E. Wilkinson

Handbook of Plant and Crop Physiology, edited by
   Mohammad Pessarakli

Handbook of Phytoalexin Metabolism and Action, edited by M. Daniel and R. P. Purkayastha

Soil–Water Interactions: Mechanisms and Applications, Second Edition, Revised and Expanded, Shingo Iwata, Toshio Tabuchi, and Benno P. Warkentin

Stored-Grain Ecosystems, edited by Digvir S. Jayas, Noel D. G. White, and William E. Muir

Agrochemicals from Natural Products, edited by C. R. A. Godfrey

Seed Development and Germination, edited by Jaime Kigel and Gad Galili

Nitrogen Fertilization in the Environment, edited by Peter Edward Bacon

Phytohormones in Soils: Microbial Production and Function, William T. Frankenberger, Jr., and Muhammad Arshad

Handbook of Weed Management Systems, edited by Albert E. Smith

Soil Sampling, Preparation, and Analysis, Kim H. Tan

Soil Erosion, Conservation, and Rehabilitation, edited by Menachem Agassi

Plant Roots: The Hidden Half, Second Edition, Revised and Expanded, edited by Yoav Waisel, Amram Eshel, and Uzi Kafkafi

Photoassimilate Distribution in Plants and Crops: Source–Sink Relationships, edited by Eli Zamski and Arthur A. Schaffer

Mass Spectrometry of Soils, edited by Thomas W. Boutton and Shinichi Yamasaki

Handbook of Photosynthesis, edited by Mohammad Pessarakli

Chemical and Isotopic Groundwater Hydrology: The Applied Approach, Second Edition, Revised and Expanded, Emanuel Mazor

Fauna in Soil Ecosystems: Recycling Processes, Nutrient Fluxes, and Agricultural Production, edited by Gero Benckiser

Soil and Plant Analysis in Sustainable Agriculture and Environment, edited by Teresa Hood and J. Benton Jones, Jr.

Seeds Handbook: Biology, Production, Processing, and Storage, B. B. Desai, P. M. Kotecha, and D. K. Salunkhe

Modern Soil Microbiology, edited by J. D. van Elsas, J. T. Trevors, and E. M. H. Wellington

Growth and Mineral Nutrition of Field Crops: Second Edition, N. K. Fageria, V. C. Baligar, and Charles Allan Jones

Fungal Pathogenesis in Plants and Crops: Molecular Biology and Host Defense Mechanisms, P. Vidhyasekaran

Plant Pathogen Detection and Disease Diagnosis, P. Narayanasamy

Agricultural Systems Modeling and Simulation, edited by
    Robert M. Peart and R. Bruce Curry

Agricultural Biotechnology, edited by Arie Altman

Plant–Microbe Interactions and Biological Control, edited by
    Greg J. Boland and L. David Kuykendall

*Handbook of Soil Conditioners: Substances That Enhance
    the Physical Properties of Soil*, edited by Arthur Wallace
    and Richard E. Terry

Environmental Chemistry of Selenium, edited by
    William T. Frankenberger, Jr., and Richard A. Engberg

Principles of Soil Chemistry: Third Edition, Revised and Expanded,
    Kim H. Tan

Sulfur in the Environment, edited by Douglas G. Maynard

Soil–Machine Interactions: A Finite Element Perspective, edited by
    Jie Shen and Radhey Lal Kushwaha

Mycotoxins in Agriculture and Food Safety, edited by Kaushal K.
    Sinha and Deepak Bhatnagar

Plant Amino Acids: Biochemistry and Biotechnology, edited by
    Bijay K. Singh

Handbook of Functional Plant Ecology, edited by Francisco I.
    Pugnaire and Fernando Valladares

*Handbook of Plant and Crop Stress: Second Edition, Revised
    and Expanded*, edited by Mohammad Pessarakli

*Plant Responses to Environmental Stresses: From Phytohormones
    to Genome Reorganization*, edited by H. R. Lerner

Handbook of Pest Management, edited by John R. Ruberson

Environmental Soil Science: Second Edition, Revised and Expanded,
    Kim H. Tan

Microbial Endophytes, edited by Charles W. Bacon
    and James F. White, Jr.

Plant–Environment Interactions: Second Edition, edited by
    Robert E. Wilkinson

Microbial Pest Control, Sushil K. Khetan

*Soil and Environmental Analysis: Physical Methods, Second Edition,
    Revised and Expanded*, edited by Keith A. Smith
    and Chris E. Mullins

*The Rhizosphere: Biochemistry and Organic Substances at the
    Soil–Plant Interface*, Roberto Pinton, Zeno Varanini,
    and Paolo Nannipieri

*Woody Plants and Woody Plant Management: Ecology, Safety,
    and Environmental Impact*, Rodney W. Bovey

Metals in the Environment, M. N. V. Prasad

Plant Pathogen Detection and Disease Diagnosis: Second Edition, Revised and Expanded, P. Narayanasamy

Handbook of Plant and Crop Physiology: Second Edition, Revised and Expanded, edited by Mohammad Pessarakli

Environmental Chemistry of Arsenic, edited by William T. Frankenberger, Jr.

Enzymes in the Environment: Activity, Ecology, and Applications, edited by Richard G. Burns and Richard P. Dick

Plant Roots: The Hidden Half, Third Edition, Revised and Expanded, edited by Yoav Waisel, Amram Eshel, and Uzi Kafkafi

Handbook of Plant Growth: pH as the Master Variable, edited by Zdenko Rengel

Biological Control of Major Crop Plant Diseases edited by Samuel S. Gnanamanickam

Pesticides in Agriculture and the Environment, edited by Willis B. Wheeler

Mathematical Models of Crop Growth and Yield, , Allen R. Overman and Richard Scholtz

Plant Biotechnology and Transgenic Plants, edited by Kirsi-Marja Oksman Caldentey and Wolfgang Barz

Handbook of Postharvest Technology: Cereals, Fruits, Vegetables, Tea, and Spices, edited by Amalendu Chakraverty, Arun S. Mujumdar, G. S. Vijaya Raghavan, and Hosahalli S. Ramaswamy

Handbook of Soil Acidity, edited by Zdenko Rengel

Humic Matter in Soil and the Environment: Principles and Controversies, edited by Kim H. Tan

Molecular Host Plant Resistance to Pests, edited by S. Sadasivam and B. Thayumanayan

Soil and Environmental Analysis: Modern Instrumental Techniques, Third Edition, edited by Keith A. Smith and Malcolm S. Cresser

Chemical and Isotopic Groundwater Hydrology, Third Edition, edited by Emanuel Mazor

Agricultural Systems Management: Optimizing Efficiency and Performance, edited by Robert M. Peart and W. David Shoup

Physiology and Biotechnology Integration for Plant Breeding, edited by Henry T. Nguyen and Abraham Blum

Global Water Dynamics: Shallow and Deep Groundwater: Petroleum Hydrology: Hydrothermal Fluids, and Landscaping, , edited by Emanuel Mazor

Principles of Soil Physics, edited by Rattan Lal

Seeds Handbook: Biology, Production, Processing, and Storage, Second Edition, Babasaheb B. Desai

Field Sampling: Principles and Practices in Environmental Analysis, edited by Alfred R. Conklin

Sustainable Agriculture and the International Rice-Wheat System, edited by Rattan Lal, Peter R. Hobbs, Norman Uphoff, and David O. Hansen

Plant Toxicology, Fourth Edition, edited by Bertold Hock and Erich F. Elstner

Drought and Water Crises: Science, Technology, and Management Issues, edited by Donald A. Wilhite

Soil Sampling, Preparation, and Analysis, Second Edition, Kim H. Tan

Climate Change and Global Food Security, edited by Rattan Lal, Norman Uphoff, B. A. Stewart, and David O. Hansen

Handbook of Photosynthesis, Second Edition, edited by Mohammad Pessarakli

Environmental Soil-Landscape Modeling: Geographic Information Technologies and Pedometrics, edited by Sabine Grunwald

Water Flow In Soils, Second Edition, Tsuyoshi Miyazaki

Biological Approaches to Sustainable Soil Systems, edited by Norman Uphoff, Andrew S. Ball, Erick Fernandes, Hans Herren, Olivier Husson, Mark Laing, Cheryl Palm, Jules Pretty, Pedro Sanchez, Nteranya Sanginga, and Janice Thies

Plant–Environment Interactions, Third Edition, edited by Bingru Huang

Biodiversity In Agricultural Production Systems, edited by Gero Benckiser and Sylvia Schnell

Organic Production and Use of Alternative Crops, Franc Bavec and Martina Bavec

Handbook of Plant Nutrition, edited by Allen V. Barker and David J. Pilbeam

Handbook of
Plant
Nutrition

Edited by

Allen V. Barker
David J. Pilbeam

Taylor & Francis
Taylor & Francis Group
Boca Raton   London   New York

CRC is an imprint of the Taylor & Francis Group,
an informa business

Cover photo by Allen V. Barker.

CRC Press
Taylor & Francis Group
6000 Broken Sound Parkway NW, Suite 300
Boca Raton, FL 33487-2742

© 2007 by Taylor & Francis Group, LLC
CRC Press is an imprint of Taylor & Francis Group, an Informa business

No claim to original U.S. Government works
Printed in the United States of America on acid-free paper
10 9 8 7 6 5 4 3

International Standard Book Number-10: 0-8247-5904-4 (Hardcover)
International Standard Book Number-13: 978-0-8247-5904-9 (Hardcover)

Library of Congress Cataloging-in-Publication Data

Barker, Allen V., 1937-
　　Handbook of plant nutrition / Allen V. Barker, David J. Pilbeam.
　　　　p. cm. -- (Books in soils, plants, and the environment ; v. 117)
　　Includes bibliographical references.
　　ISBN 0-8247-5904-4
　　1. Plants--Nutrition--Handbooks, manuals, etc. I. Pilbeam, D. J. II. Title. III. Series.

QK867.B29 2006
631.8--dc22   2006044539

**Visit the Taylor & Francis Web site at
http://www.taylorandfrancis.com**

**and the CRC Press Web site at
http://www.crcpress.com**

Preface

For over 150 years, scientists have studied plant nutrition with goals of understanding the acquisition, accumulation, transport, and functions of chemical elements in plants. From these studies, much information has been obtained about the growth and composition of plants in response to soil-borne elements and to fertilization of crops in the soil or in soil-less media, as in hydroponic culture of plants. A compilation of elements known as *plant nutrients* and *beneficial elements* has also been developed from this work.

Plant nutrients are chemical elements that are essential for plant growth. For an element to be essential, it must be required for a plant to complete its life cycle, it must be required by all plants, and no other nutrient can replace this requirement fully. If an element does not meet all of these requirements, for example, being required by some plants or only enhancing the growth of plants, the element may be a *beneficial element*. Much interest in plant nutrition lies in the development and use of diagnostic techniques for assessment of the status of plants with respect to plant nutrients and beneficial elements.

Soil testing is a common approach to assessments of soil fertility and plant nutrition. With correlation to plant growth, development, and yield, soil testing indicates the capacity of soils to supply plant nutrients and suggests appropriate corrective measures. Plant analysis, used in conjunction with plant symptoms and soil testing, is another common tool for assessment of the nutritional status of plants.

This handbook covers principles of plant nutrition from a historical standpoint to current knowledge of the requirements of crops for certain elements and the beneficial effects of others. Its layout owes much to Homer D. Chapman's 1966 book *Diagnostic Criteria for Plants and Soils* and, as with that book, presents contributions from eminent plant and soil scientists from around the world. The purpose of this handbook is to provide a current, readily available source of information on the nutritional requirements of world crops.

In the Introduction, the editors provide an overview of plant nutrients and beneficial elements and note diagnostic criteria and research approaches used by current investigators who are interested in plant nutrition.

Each of the chapters dealing with plant nutrients starts with historical information of each nutrient, including the demonstration of essentiality and functions in plants. Each of these chapters will include diagnosis of the nutritional status of plants through assessments of plant appearance and composition. Tabulated data will help correlate plant appearance and composition with regard to nutritional needs. A discussion of the value of soil tests for assessment of the nutritional status of plants will be provided in each chapter. Each chapter will conclude with fertilizers that can be applied to remedy nutritional deficiencies in plants.

Chapters concerning beneficial elements will discuss the history of the relation of the beneficial effects of these elements to crop growth and yield and will relate the benefits to growth stimulation and plant metabolism for particular plant species.

A separate CD-ROM containing all the photographs and some line drawings in color is included with the book, because color versions of the illustrations offer details not obvious in black-and-white pictures.

With the world population increasing rapidly, and projected to do so for some time, and with improved plant nutrition remaining as one of the major factors increasing crop yields, use of our knowledge of plant nutrition to maximize agricultural yields grows in importance. However, public interest in minimizing the use of chemical inputs in agriculture also is increasing with emphasis on less use of chemical fertilizers and more use of alternative fertilizers. Attention to precision agriculture, in which plant nutrition is controlled or monitored carefully, has grown in research and practice. All of these situations require knowledge of plant nutrition.

The handbook is intended to be a practical reference work for anyone who needs to know the requirements of the world's major crops for essential or beneficial elements. It will also give information on how to assess and govern the nutritional status of crops. It should be of use to farmers, agricultural advisers, soil scientists, and plant scientists.

Contributors

Allen V. Barker
Department of Plant, Soil, and Insect Sciences
University of Massachusetts
Amherst, Massachusetts

Elke Bloem
Institute of Plant Nutrition and
    Soil Science
Federal Agricultural Research Centre (FAL)
Braunschweig, Germany

Patrick H. Brown
Department of Plant Sciences
University of California
Davis, California

Gretchen M. Bryson
Department of Plant, Soil, and Insect Sciences
University of Massachusetts
Amherst, Massachusetts

Lawrence E. Datnoff
Plant Pathology Department
University of Florida/IFAS
Gainesville, Florida

Luit J. de Kok
Laboratory of Plant Physiology
University of Groningen
Haren, The Netherlands

Khaled Drihem
School of Biology
University of Leeds
Leeds, United Kingdom

Michael A. Dunn
Department of Human Nutrition, Food and
    Animal Sciences
University of Hawaii at Manoa
Honolulu, Hawaii

John Gorham
Arid Land Research Center
Tottori University
Tottori, Japan

and
Centre for Arid Zone Studies and School of
    Biological Sciences
University of Wales
Bangor, United Kingdom

Robin D. Graham
Discipline of Plant and Food Science
School of Agriculture, Food and Wine
University of Adelaide
Adelaide, Australia

Umesh C. Gupta
Crops and Livestock Research Centre
Agriculture and Agri-Food Canada
Charlottetown, Prince Edward Island, Canada

Russell L. Hamlin
Coggins Farms and Produce
Lake Park, Georgia

Silvia Haneklaus
Institute of Plant Nutrition and Soil
    Science
Federal Agricultural Research
    Centre (FAL)
Braunschweig, Germany

Joseph R. Heckman
Plant Biology and Pathology Department
Rutgers University
New Brunswick, New Jersey

N.V. Hue
Department of Tropical Plant and Soil
    Sciences
University of Hawaii at Manoa
Honolulu, Hawaii

Julia M. Humphries
Discipline of Plant and Food Science
School of Agriculture, Food and Wine
University of Adelaide
Adelaide, Australia

David E. Kopsell
School of Agriculture
University of Wisconsin
Platteville, Wisconsin

Dean A. Kopsell
Plant Sciences Department
University of Tennessee
Knoxville, Tennessee

Vladimir V. Matichenkov
Institute of Basic Biological Problems
Russian Academy of Sciences
Pushchino, Russia

Konrad Mengel
Institute of Plant Nutrition
Justus Liebig University
Giessen, Germany

Donald J. Merhaut
Department of Botany and Plant Sciences
University of California
Riverside, California

Susan C. Miyasaka
Department of Tropical Plant and Soil
  Sciences
University of Hawaii at Manoa
Hilo, Hawaii

Philip S. Morley
Wight Salads Ltd.
Arreton, United Kingdom

Miroslav Nikolic
Centre for Multidisciplinary Studies
University of Belgrade
Belgrade, Serbia

David J. Pilbeam
Institute of Intergrative and Comparative
  Biology
University of Leeds
Leeds, United Kingdom

Volker Römheld
Institute of Plant Nutrition
University of Hohenheim
Stuttgart, Germany

Charles A. Sanchez
Department of Soil, Water, and Environmental
  Sciences
Yuma Agricultural Center
Yuma, Arizona

Ewald Schnug
Institute of Plant Nutrition and
  Soil Science
Federal Agricultural Research Centre (FAL)
Braunschweig, Germany

Archana Sharma
Department of Botany
University of Calcutta
Kolkata, India

George H. Snyder
University of Florida/IFAS
Everglades Research and Education Center
Belle Glade, Florida

James C.R. Stangoulis
Discipline of Plant and Food Science
School of Agriculture, Food and Wine
University of Adelaide
Adelaide, Australia

J. Benton Storey
Department of Horticultural Science
Texas A&M University
College Station, Texas

Ineke Stulen
Laboratory of Plant Physiology
University of Groningen
Haren, The Netherlands

Geeta Talukder
Vivekananda Institute of Medical Sciences
Kolkata, India

Contents

Section I
Introduction 1

Chapter 1 Introduction..3
Allen V. Barker and David J. Pilbeam

Section II
Essential Elements—Macronutrients..19

Chapter 2 Nitrogen 21
Allen V. Barker and Gretchen M. Bryson

Chapter 3 Phosphorus 51
Charles A. Sanchez

Chapter 4 Potassium 91
Konrad Mengel

Chapter 5 Calcium 121
David J. Pilbeam and Philip S. Morley

Chapter 6 Magnesium 145
Donald J. Merhaut

Chapter 7 Sulfur 183
Silvia Haneklaus, Elke Bloem, Ewald Schnug, Luit J. de Kok, and Ineke Stulen

Section III
Essential Elements—Micronutrients..239

Chapter 8 Boron 241
Umesh C. Gupta

Chapter 9 Chlorine 279
Joseph R. Heckman

Chapter 10 Copper 293
David E. Kopsell and Dean A. Kopsell

Chapter 11 Iron 329
Volker Römheld and Miroslav Nikolic

Chapter 12 Manganese 351
Julia M. Humphries, James C.R. Stangoulis, and Robin D. Graham

Chapter 13 Molybdenum 375
Russell L. Hamlin

Chapter 14 Nickel 395
Patrick H. Brown

Chapter 15 Zinc 411
J. Benton Storey

Section IV
Beneficial Elements **437**

Chapter 16 Aluminum 439
Susan C. Miyasaka, N.V. Hue, and Michael A. Dunn

Chapter 17 Cobalt 499
Geeta Talukder and Archana Sharma

Chapter 18 Selenium 515
Dean A. Kopsell and David E. Kopsell

Chapter 19 Silicon 551
George H. Snyder, Vladimir V. Matichenkov, and Lawrence E. Datnoff

Chapter 20 Sodium 569
John Gorham

Chapter 21 Vanadium 585
David J. Pilbeam and Khaled Drihem

Section V
Conclusion **597**

Chapter 22 Conclusion 599
Allen V. Barker and David J. Pilbeam

Index 605

Section I

Introduction

1 Introduction

Allen V. Barker
University of Massachusetts, Amherst, Massachusetts

David J. Pilbeam
University of Leeds, Leeds, United Kingdom

CONTENTS

1.1 Definitions 3
    1.1.1 Plant Nutrient 3
1.2 Diagnostic Criteria 5
    1.2.1 Visual Diagnosis 5
    1.2.2 Plant Analysis 8
    1.2.3 Quantitative Analysis 8
    1.2.4 Tissue Testing 9
    1.2.5 Biochemical Tests 10
    1.2.6 Soil Tests 11
1.3 Approaches in Research 12
References 13

1.1 DEFINITIONS

1.1.1 PLANT NUTRIENT

A *plant nutrient* is a chemical element that is essential for plant growth and reproduction. *Essential element* is a term often used to identify a plant nutrient. The term *nutrient* implies essentiality, so it is redundant to call these elements essential nutrients. Commonly, for an element to be a nutrient, it must fit certain criteria. The principal criterion is that the element must be required for a plant to complete its life cycle. The second criterion is that no other element substitutes fully for the element being considered as a nutrient. The third criterion is that all plants require the element. All the elements that have been identified as plant nutrients, however, do not fully meet these criteria, so, some debate occurs regarding the standards for classifying an element as a plant nutrient. Issues related to the identification of new nutrients are addressed in some of the chapters in this handbook.

The first criterion, that the element is essential for a plant to complete its life cycle, has historically been the one with which essentiality is established (1). This criterion includes the property that the element has a direct effect on plant growth and reproduction. In the absence of the essential element or with severe deficiency, the plant will die before it completes the cycle from seed to seed. This requirement acknowledges that the element has a function in plant metabolism; that with short supply of the nutrient, abnormal growth or symptoms of deficiency will develop as a result of the disrupted metabolism; and that the plant may be able to complete its life cycle with restricted

3

growth and abnormal appearance. This criterion also notes that the occurrence of an element in a plant is not evidence of essentiality. Plants will accumulate elements that are in solution without regard to the elements having any essential role in plant metabolism or physiology.

The second criterion states that the role of the element must be unique in plant metabolism or physiology, meaning that no other element will substitute fully for this function. A partial substitution might be possible. For example, a substitution of manganese for magnesium in enzymatic reactions may occur, but no other element will substitute for magnesium in its role as a constituent of chlorophyll (2). Some scientists believe that this criterion is included in the context of the first criterion (3).

The third criterion requires that the essentiality is universal among plants. Elements can affect plant growth without being considered as essential elements (3,4). Enhancement of growth is not a defining characteristic of a plant nutrient, since although growth might be stimulated by an element, the element is not absolutely required for the plant to complete its life cycle. Some plants may respond to certain elements by exhibiting enhanced growth or higher yields, such as that which occurs with the supply of sodium to some crops (5,6). Also, some elements may appear to be required by some plants because the elements have functions in metabolic processes in the plants, such as in the case of cobalt being required for nitrogen-fixing plants (7). Nitrogen fixation, however, is not vital for these plants since they will grow well on mineral or inorganic supplies of nitrogen. Also, plants that do not fix nitrogen do not have any known need for cobalt (3). Elements that might enhance growth or that have a function in some plants but not in all plants are referred to as *beneficial elements*.

Seventeen elements are considered to have met the criteria for designation as plant nutrients. Carbon, hydrogen, and oxygen are derived from air or water. The other 14 are obtained from soil or nutrient solutions (Table 1.1). It is difficult to assign a precise date or a specific researcher to the discovery of the essentiality of an element. For all the nutrients, their roles in agriculture were the subjects of careful investigations long before the elements were accepted as nutrients. Many

TABLE 1.1
Listing of Essential Elements, Their Date of Acceptance as Essential, and Discoverers of Essentiality

Element	Date of Essentiality[a]	Researcher[a]
Nitrogen	1804	de Saussure[b]
	1851–1855	Boussingault[b]
Phosphorus	1839	Liebig[c]
	1861	Ville[b]
Potassium	1866	Birner & Lucanus[b]
Calcium	1862	Stohmann[b]
Magnesium	1875	Boehm[b]
Sulfur	1866	Birner & Lucanus[b]
Iron	1843	Gris[c]
Manganese	1922	McHargue[c]
Copper	1925	McHargue[c]
Boron	1926	Sommer & Lipman[c]
Zinc	1926	Sommer & Lipman[c]
Molybdenum	1939	Arnon & Stout[c]
Chlorine	1954	Broyer, Carlton, Johnson, & Stout[c]
Nickel	1987	Brown, Welch, & Cary (11)

[a]The dates and researchers that are listed are those on which published articles amassed enough information to convince other researchers that the elements were plant nutrients. Earlier work preceding the dates and other researchers may have suggested that the elements were nutrients.
[b]Cited by Reed (22).
[c]Cited by Chapman (13).

individuals contributed to the discovery of the essentiality of elements in plant nutrition. Much of the early research focused on the beneficial effects or sometimes on the toxic effects of the elements. Generally, an element was accepted as a plant nutrient after the body of evidence suggested that the element was essential for plant growth and reproduction, leading to the assignment of certain times and individuals to the discovery of its essentiality (Table 1.1).

Techniques of hydroponics (8,9) initiated in the mid-1800s and improved in the 1900s enabled experimenters to grow plants in defined media purged of elements. Elements that are required in considerable quantities (*macronutrients*), generally accumulating to 0.1% and upward of the dry mass in plant tissues, were shown to be nutrients in the mid-1800s. Most of the elements required in small quantities in plants (*micronutrients*), generally accumulating to amounts less than 0.01% of the dry mass of plant tissues, were shown to be essential only after techniques were improved to ensure that the water, reagents, media, atmosphere, and seeds did not contain sufficient amounts of nutrients to meet the needs of the plants. Except for iron, the essentiality of micronutrients was demonstrated in the 1900s.

Beneficial elements may stimulate growth or may be required by only certain plants. Silicon, cobalt, and sodium are notable beneficial elements. Selenium, aluminum, vanadium, and other elements have been suggested to enhance growth of plants (3,10). Some of the beneficial elements may be classified in the future as essential elements as developments in chemical analysis and methods of minimizing contamination during growth show that plants will not complete their life cycles if the concentrations of elements in plant tissues are diminished sufficiently. Nickel is an example of an element that was classified as beneficial but recently has been shown to be essential (11).

Studies of the roles of nutrients in plants have involved several diagnostic criteria that address the accumulation of nutrients and their roles in plants. These criteria include visual diagnosis, plant analysis, biochemical tests, and soil tests.

1.2 DIAGNOSTIC CRITERIA

1.2.1 VISUAL DIAGNOSIS

Careful observations of the growth of plants can furnish direct evidence of their nutritional conditions. Metabolic disruptions resulting from nutrient deficiencies provide links between the function of an element and the appearance of a specific visible abnormality. Symptoms of disorders, therefore, provide a guide to identify nutritional deficiencies in plants. Careful experimental work and observations are needed to characterize symptoms. For example, nitrogen is needed for protein synthesis and for chlorophyll synthesis, and symptoms appear as a result of the disruption of these processes. Symptoms of nitrogen deficiency appear as pale-green or yellow leaves starting from the bottom and extending upward or sometimes covering the entire plant. Magnesium deficiency also affects protein synthesis and chlorophyll synthesis, but the symptoms may not resemble those of nitrogen deficiency, which affects the same processes. Experience is necessary to distinguish the symptoms of nitrogen deficiency from symptoms of magnesium deficiency or in the identification of the deficiency of any nutrient.

Symptoms on foliage have been classified into five types (12): (a) chlorosis, which may be uniform or interveinal (Figure 1.1); (b) necrosis, which may be at leaf tips or margins, or be interveinal (Figure 1.2); (c) lack of new growth, which may result in death of terminal or axillary buds and leaves, dieback, or rosetting (Figure 1.3); (d) accumulation of anthocyanin, which results in an overall red color (Figure 1.4); and (e) stunting with normal green color or an off-green or yellow color (Figure 1.5). Symptoms of deficiency can be quite specific according to nutrient, especially if the diagnosis is made early in the development of the symptoms. Symptoms may become similar among deficiencies as the intensities of the symptoms progress.

Generalities of development of deficiency symptoms can be made among species. Many references are available with descriptions, plates, or keys that enable identification of nutrient deficiencies (12–20). As mentioned above, for example, nitrogen deficiency appears across plant species as chlorosis of lower or of all leaves on plants. Advanced stages of nitrogen deficiency can lead to leaf death and leaf drop. Nitrogen-deficient plants generally are stunted and spindly in addition to

FIGURE 1.1 Interveinal chlorosis of iron-deficient borage (*Borago officinalis* L.). (Photograph by Allen V. Barker.) (For a color presentation of this figure, see the accompanying compact disc.)

FIGURE 1.2 Deficiency symptoms showing necrosis of leaf margins, as in this case of potassium deficiency on cucumber (*Cucumis sativus* L.) leaf. (Photograph by Allen V. Barker.) (For a color presentation of this figure, see the accompanying compact disc.)

showing the discoloration that is imparted by chlorosis. Potassium-deficient plants have marginal and tip necrosis of lower leaves. On the other hand, for elements that are immobile (not transported in phloem) or slowly mobile in plants, the deficiency symptoms will appear on the young leaves first. The symptoms might appear as chlorosis, as with sulfur, iron, manganese, zinc, or copper deficiency, or the symptoms might be necrosis of entire plant tips, as occurs with boron or calcium deficiency. Brooms or rosetting may occur in cases where deficiencies (e.g., copper or zinc) have caused death of the terminal bud and lateral buds have grown or where internode elongation has been restricted by

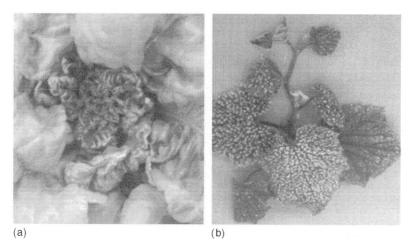

(a)   (b)

FIGURE 1.3 Deficiency symptoms showing necrosis on young leaves of (a) calcium-deficient lettuce (*Lactuca sativa* L.) and necrosis on young and old leaves of (b) calcium-deficient cucumber (*Cucumis sativus* L.). With cucumber the necrosis has extended to all leaves that have not expanded to the potential size of full maturity. (Photographs by Allen V. Barker.) (For a color presentation of this figure, see the accompanying compact disc.)

FIGURE 1.4 Stunting and development of red color and loss of green color of phosphorus-deficient tomato (*Lycopersicon esculentum* Mill.). (Photograph by Allen V. Barker.) (For a color presentation of this figure, see the accompanying compact disc.)

nutrient (e.g., zinc) deficiencies. Accumulation of anthocyanin, exhibited by reddening of leaves, may indicate phosphorus deficiency, although nitrogen deficiency can lead to a similar development. Some people try to distinguish the two deficiencies by noting whether the symptoms of reddening develop between the veins (phosphorus deficiency) or along the veins (nitrogen deficiency). Stunting is a good indication of nutrient deficiency, but often stunting cannot be recognized unless a well-nourished plant is available as a standard of comparison. A stunted plant may have normal color and not be recognized as being deficient until abnormal coloration develops with advanced stages of deficiency. In some cases, symptoms may not develop during the growth cycle of crops, but yields may be suppressed relative to plants that have optimum nutrition. *Hidden hunger* is a term applied to cases where yield suppression occurred but symptoms did not develop.

Deficiency symptoms can occur at any stage of growth of a plant. The most typical symptoms are those that appear early in the cycle of deficiency. Early diagnosis of deficiencies may also allow

FIGURE 1.5   Cabbage (*Brassica oleracea* var. *capitata* L.) plants showing symptoms of stunting. Left: stunting and dark green color diagnosed as being caused by salinity in nutrient solution. Middle: stunting and mottling of foliage due to condition diagnosed as magnesium deficiency. Right: stunting and discoloration of foliage due to condition diagnosed as phosphorus deficiency. (Photographs by Allen V. Barker.) (For a color presentation of this figure, see the accompanying compact disc.)

time for remedial action to take place. Generally, however, if symptoms have appeared, irreparable damage has occurred, with quantity or quality of yields being suppressed or diminished with annual crops or with slowing or damaging of growth and development of perennial crops. Also, symptoms that resemble nutrient deficiency can develop on plants as a result of conditions that are not related to nutrient deficiencies, for example, drought, wet soils, cold soils, insect or disease infestations, herbicide damage, wind, mechanical damage, salinity, or elemental toxicities. Deficiency symptoms are only one of several diagnostic criteria that can be used to assess the nutritional status of plants. Plant analysis, biological tests, soil analysis, and application of fertilizers containing the nutrient in question are additional tools used in diagnosis of the status of plant nutrition.

1.2.2   PLANT ANALYSIS

Plant analysis as a means of understanding plant physiology perhaps started with de Saussure (21). With plant analysis, de Saussure corrected the misunderstanding at the time that the mineral matter of plants had no importance. He showed that the mineral matter in plants came from the soil and not from the air and that little growth of plants occurred if they were grown in distilled water. Through plant analysis, he also demonstrated that plants absorbed minerals in ratios that differed from the proportions existing in solution or in soil and that plants absorbed substances from solution, whether the substances were beneficial to the plants or not.

Plant analysis was one of the means used by scientists in the 1800s to determine the essentiality of chemical elements as plant nutrients (22). Further refinements and applications of plant analysis led to studies of the relationship between crop growth or yield and nutrient concentrations in plants (23–26). Elemental analysis of leaves is commonly used as a basis for crop fertilizer recommendations (27,28).

Plants can be tested for sufficiency of nutrition by analytical tests, which employ quantitative analysis (total or specific components) in laboratories, or by *tissue tests* (semiquantitative analysis), often applied in the field. With proper means of separation of constituents, quantitative tests may measure nutrients that have been incorporated into plant structures or that are present as soluble constituents in the plant sap. The tissue tests generally deal with soluble constituents.

1.2.3   QUANTITATIVE ANALYSIS

Quantitative plant analysis has several functions in assessing the nutrient status of plants (29). Among these functions, plant analysis can be used to confirm a visual diagnosis. Plant analysis also can help in identifying hidden hunger or incipient deficiencies. In confirming diagnoses or in identifying incipient deficiencies, comparisons are made between laboratory results and critical values or ranges that assess the nutritional status as deficient, low, sufficient, or high, or in other applicable terms. The *critical concentration* of a nutrient is defined as the concentration of the nutrient below which yields are suppressed (26,30). In the determination of critical concentration, analysis of a specific tissue of a specific organ at a designated state of development is required. Because of the amount of work involved, critical concentrations are rarely determined; consequently, *ranges of sufficiency* are most commonly used in assessment of plant nutrition (27). For each nutrient or beneficial element mentioned in this handbook, ranges of sufficiency are reported.

For any plant, it could be that only one nutrient is deficient or in excess, but it is also possible that more than one nutrient may be out of its range of sufficiency. Furthermore, the actual requirement for an individual nutrient may be different if other nutrients are not present in the plant above their own critical concentrations. For this reason, it is becoming common to consider concentrations of nutrients in relation to the concentrations of other nutrients within the plant. Forms of multivariate analysis such as *principal component analysis* and *canonical discriminant analysis* have been used to investigate relationships between the internal concentrations of many nutrients together and plant growth (31). Currently, a commonly used application of plant analysis is the Diagnosis and Recommendation Integrated System (DRIS), which compares ratios of concentrations of all the possible pairs of elements analyzed to establish values that help to identify nutrients that are most likely to be deficient (32,33).

Plant analysis is also used to determine if an element entered a plant. Fertilization is employed to correct deficiencies, often in response to a visual diagnosis. It is important to know that nutrients actually entered plants after the application of the nutrients to the soil or foliage. No response to the application of a nutrient may be understood as meaning that the element was not lacking, when in fact, it might not have been absorbed by the plant being treated. Plant analysis can also indicate the effects of application of plant nutrients on plant composition with regard to elements other than the one being studied. Interactions may occur to enhance or to suppress the absorption of other nutrients. In some cases, growth may be stimulated by a nutrient to the point that other nutrients become deficient, and further growth cannot occur. Plant analysis can help to detect changes in plant composition or growth that are synergistic or antagonistic with crop fertilization.

Collecting samples of plant organs or tissues is important in assessing nutrition by plant analysis. Comparable leaves or other organs or tissues from the same plant or from similar plants should be collected as samples that show symptoms and samples that do not. Samples of abnormal and normal material from the same plant or similar plants allow for development of standards of comparison for deficient, optimum, or excessive nutrition. The composition of plants varies with time (diurnal and stage of growth) and with parts of plants as well as with nutrition (34). It is wise to take samples from plant parts that have been studied widely and for which published standards of comparisons for deficient, sufficient, and optimum concentrations of nutrients are available. Jones and Steyn (35) discuss methods of sampling and sample preparation prior to analysis, along with methods of extracting nutrients for analysis and methods of analysis of plant tissues. A handbook edited by Kalra (36) also addresses sampling and analysis of plant tissues.

1.2.4 TISSUE TESTING

Plant tissue testing is a technique for rapid determination of the nutritional status of a crop and is often conducted on the field sites where crops are grown. The test generally assesses the nutrient status by direct measurements of the unassimilated fraction of the nutrient in question in the plant. For example, determination of nitrate in leaf petioles, midribs, or blades or in roots is often a chosen tissue test for assessment of the nitrogen status of a plant (37–40). Nitrate in these plant parts represents an unassimilated form of nitrogen that is in transit to the leaves and often shows greater variations in response to soil nutrient relations than determinations of total nitrogen in plant parts, although some research indicates that total nitrogen concentration in the whole plant gives the best index of plant nitrogen nutrition (41). Generally, in a tissue test, the sap of the tissues is extracted by processes such as crushing or grinding along with filtering to collect liquid for testing (34). Testing of a component, such as nitrate in the sap, is often done by semiquantitative determinations with nitrate-sensitive test strips (37,40,42,43), by hand-held nitrate-testing meters (44), or by quantitative laboratory measurements (45). In tissue testing, ammonium determinations are used less often than nitrate determinations because accumulation of ammonium can be an artifact of sampling and analysis (46).

An exception to the direct determination of an element to assess deficiency was the corn (*Zea mays* L.) stalk test of Hoffer (47). This test was based on the observation that insoluble iron compounds appeared at the nodes of corn plants under stress of potassium deficiency (48). The corn stalk test provided only a rough indication of the potassium nutrition of the plant but had a fair agreement with other tests for potassium deficiency and had some application to crops other than corn (34). Similarly, Leeper (49) noted that manganese-deficient oats (*Avena sativa* L.) accumulated nitrate in stems.

Selection of the plant part for testing varies with the nutrient being assessed. With nitrate, it may be important that conductive tissue be selected so that the sampling represents the nutrient in transit to a site of assimilation and before metabolic conversions occur. However, potassium is not assimilated into organic combinations in plants; hence, selection of a plant part is of lesser importance than with determination of nitrate, and leaf petioles, midribs, blades, or other tissues can be used for potassium determination by quick tests or by laboratory measurements (50,51).

Color of leaves can be used as a visual assessment of the nutrient status of plants. This assessment can also be quantitative in a quick test, and chlorophyll-measuring meters have been used to nondestructively evaluate the nitrogen status of plants (52). The meters have to be used in reference to predetermined readings for plants receiving adequate nutrition and at selected stages of development, which are usually before flowering and maturation. Correlations of readings with needs for nitrogen fertilization may not be good as the plant matures and flowers and as materials are transported from leaves to fruits.

Leaf canopy reflectance (near-infrared or red), as employed in remote sensing techniques, can be used to assess the nutrient status of fields. Reflectance has been shown to be related to chlorophyll concentrations and to indicate the nitrogen status of crops in a field (53).

1.2.5 BIOCHEMICAL TESTS

Activities of specific enzymes can provide rapid and sensitive indicators of nutrient deficiencies in plants (54). Deficiencies of micronutrients can lead to inhibited activities of enzymes for which the nutrient is part of the specific enzyme molecule. Assays of enzymatic activity can help identify deficiencies when visual diagnosis does not distinguish between deficiencies that produce similar symptoms (55), when soil analysis does not determine if nutrients enter plants, or when plant analysis does not reflect the concentration of a nutrient needed for physiological functions (56). The enzymatic assays do not give concentrations of nutrients in plants, but the enzyme activity gives an indication of sufficiency or deficiency of a nutrient. The assay can be run on deficient tissue or on tissue into which the suspected element has been infiltrated to reactivate the enzymatic system. The assays are run on crude extracts or leaf disks to provide quick tests (57).

Peroxidase assays have been used to distinguish iron deficiency from manganese deficiency in citrus (*Citrus* spp. L.) (55,58). Peroxidases are heme-containing enzymes that use hydrogen peroxide as the electron acceptor to catalyze a number of oxidative reactions. In this application, during iron deficiency, peroxidase activity is inhibited, whereas during manganese deficiency peroxidase activity may be increased. Iron is a constituent of peroxidase, but manganese is not. Kaur et al. (59) reported associations of limited catalase and peroxidase activities with iron deficiency in chickpeas (*Cicer arietinum* L.). Leidi et al. (60) evaluated catalase and peroxidase activities as indicators of iron and manganese nutrition for soybeans (*Glycine max* Merr.). Nenova and Stoyanov (61) reported that intense iron deficiency resulted in low activities of peroxidase, catalase, and nitrate reductase in corn (*Zea mays* L.). Ranieri et al. (62) observed a suppression of peroxidase activity in iron-deficient sunflower (*Helianthus annuus* L.). On the other hand, carbonic anhydrase has been employed to identify zinc deficiency in citrus (63), sugarcane (*Saccharum officinarum* L.) (64), black gram (*Vigna mungo* L.) (65), and pecan (*Carya illinoinensis* Koch) (66). Zinc deficiency was associated with a decrease in messenger RNA for carbonic anhydrase along with a decrease in carbonic anhydrase activity in rice (*Oryza sativa* L.) (67). In another assay, alcohol dehydrogenase was twice as high in roots of zinc-sufficient rice as in zinc-deficient rice, and activity of alcohol dehydrogenase in roots was correlated with zinc concentration in leaves (68). Ascorbic acid oxidase assays have been used in the identification of copper deficiency in citrus (69). Molybdenum deficiency has been associated with low levels of nitrate reductase activity in citrus (70). Polle et al. (71) reported that the activities of superoxide dismutase and some other protective enzymes increased in manganese-deficient leaves of Norway spruce (*Picea abies* L.).

Applications of enzymatic assays for the micronutrient status of plants have not been adopted widely in agronomic or horticultural practice, although interest in usage may be increasing as is shown by the number of investigations associating enzymatic activity with plant nutrients. The peroxidase test in the assessment of iron deficiency has perhaps been employed more than other assays (57,72). Macronutrients have numerous functions in plants, and association of specific enzymatic activity with deficiencies of macronutrients is difficult. However, some assays have been developed, such as nitrate reductase activity for assessment of nitrogen deficiency, glutamate-oxaloacetate aminotransferase for phosphorus deficiency, and pyruvic kinase for potassium deficiency (54). Measurement of pyruvic kinase activity may also be useful for establishing the optimum balance between potassium, calcium, and magnesium concentrations in tissues (73).

1.2.6 SOIL TESTS

A soil test is a chemical or physical measurement of soil properties based on a sample of soil (74). Commonly, however, a soil test is considered as a rapid chemical analysis or quick test to assess the readily extractable chemical elements of a soil. Interpretations of soil tests provide assessments of the amount of *available nutrients*, which plants may absorb from a soil. Recommendations for fertilization may be based on the results of soil tests. Chemical soil tests may also measure salinity, pH, and presence of elements that may have inhibitory effects on plant growth.

A basic principle of soil testing is that an area can be sampled so that chemical analysis of the samples will assess the nutrient status of the entire sampled area. Methods of sampling may differ with the variability of the area being sampled and with the nutrients being tested. A larger number of samples may need to be taken from a nonuniform area than from a uniform area. Movement of nutrients into the soil, as with nitrate leaching downward, may cause the need for sampling of soil to be at a greater depth than with nutrients that do not move far from the site of application. Wide differences in test results across a field bring into question whether a single recommendation for fertilization can be made for the entire field (74,75). Fertilization of fields can increase the variability of nutrients of a field, and the assessment of the fertility level with respect to nutrients will become more difficult. Variations in patterns of applications of fertilizers, such as placement of fertilizers in bands in contrast to broadcasting of fertilizers, can affect soil samples. The proceedings of an international conference on precision agriculture addressed variability in fields, variable lime and fertilizer applications in fields, and other factors involved in site-specific collection of data, such as soil samples (76).

Results of soil tests must be calibrated to crop responses in the soil. Crop responses, such as growth and yields, are obtained through experimentation. In the calibrations, the results of soil tests are treated as independent variables affecting crop growth and yields; otherwise, all other variables such as weather, season, diseases, soil types, weeds, and other environmental factors must be known and interpreted. The consideration of results of soil test as independent variables may impart difficulties in interpreting the results, especially if the environmental factors have marked effects on crop yields.

Results of soil analysis, sometimes called *total analysis*, in which soil mineral and organic matter are destroyed with strong mineral acids, heat, or other agents do not correlate well with crop responses (77). Generally, soil tests involve determination of a form of a plant nutrient with which a variation in amount is correlated with crop growth and yield. These forms of nutrients are commonly called *available plant nutrients*. The different forms of nutrients are extracted from the soil with some solvent. Many different methods of extraction of soil samples are being used for measurement of available nutrients in soils. Extractants are various combinations of water, acids, bases, salts, and chelating agents at different strengths. The extractants are designed to extract specific nutrients or are universal extractants (77–83). Much discussion has occurred as to whether one method of extraction is better than another. Morgan (77) noted that any chemical method of soil extraction is empirical and that the results give only an approximate quantitative expression of the various chemical constituents in soil. Morgan stated further that no one solvent acting on the soil for a period of minutes or hours will duplicate the conditions involved in provision of nutrients from soil to plants. Researchers may choose to continue to test soils with extraction procedures with which they have experience and for which they have compilations of results. Researchers who analyze only a relatively few samples may choose to use procedures for which published results are readily and commonly available. Methods of extraction and analysis for specific elements are addressed in several monographs and handbooks (84–86). Chemical analyses are the most accurate part of soil testing since they are chemically reproducible or precise measurements of the amounts of nutrients extracted from soils. Selection of the method of analysis depends largely on the facilities that are available to scientists.

1.3 APPROACHES IN RESEARCH

Research in plant nutrition is a continuing program. The development of new crop varieties and the introduction of new management practices to increase crop yields impart changes in nutrient requirements of plants. The increasing application of genomics is providing more understanding of the genetic basis for the efficiency with which different plants utilize nutrients. For example, a study of induction of *Arabidopsis* genes by nitrate confirmed that genes encoding nitrate reductase, the nitrate transporter NRT1 (but not the nitrate transporter NRT2), and glutamate synthase were all highly induced, and this work also demonstrated induction of a further 15 genes that had not previously been shown to be induced (87). Nitrate influences root architecture through induction of genes that control lateral root growth (88).

Research is conducted, and will continue to be conducted, to ensure that soil tests correlate with use of nutrients by plants and that fertilizer recommendations are calibrated for crops (89). These correlations must be developed for individual crops and different land areas. Some research is directed toward development of systems for evaluation of soil and crop conditions through methods other than traditional soil and plant analysis. Much of the past and current research addresses chemical, physical, and biological properties of soils (90,91). Some researchers have studied the interaction of these quantitative aspects to determine *soil quality* and to develop a *soil quality index* that correlates with crop productivity and environmental and health goals (92). Soil quality has been defined to include productivity, sustainability, environmental quality, and effects on human nutrition (93). To quantify soil quality, specific soil indicators are measured and integrated to form a soil quality index.

Research in plant nutrition addresses methods of economically and environmentally sound methods of fertilization. Worldwide, large increases have occurred in the use of fertilizers because of their effects on yields and availability. Traditionally, fertilizer use has followed Sprengel's law of the minimum, made famous by Liebig (94), and the application of the law of diminishing returns by Mitscherlich (95). Applying these two laws has given us fertilizers with the nutrients blended in the correct proportions for the world's major crops and rates of fertilizer use that lead to maximum yields commensurate with the cost of the fertilizer.

More recently, interest has turned to issues related to the impact of this intensified agriculture and fertilizer use on the environment and to greater interest in fertilizer use efficiency to help avoid pollution of land and water resources (96). Research is conducted on dairy manure management to protect water quality from nutrient pollution from the large amounts of nitrogen and phosphorus that may be added to heavily manured land (97,98). In its most extreme manifestation, this interest in avoiding excessive fertilization of farmland has given rise to increased practice of organic farming, where synthetic inorganic fertilizers are eschewed in favor of organic sources of nutrients. Regardless of whether nutrients are supplied from organic or synthetic sources, it is still the same inorganic elements that plants are absorbing.

Research is conducted on the use of plants to clean metal-polluted land. Phytoextraction is a plant-based technology to remove metals from contaminated sites through the use of metal-accumulating plants (99,100). Research interests have focused on identifying plants that will accumulate metals and on methods of enhancing accumulation of metals in plants (101–103). Another suggested use of knowledge about the uptake of mineral elements by plants is in the identification of geographical origin of foodstuffs. Analysis of 18 elements in potato tubers has been shown to give a distinctive signature that allows a sample to be correctly assigned to its place of origin, something that could be of great use in tracing of foodstuffs (104).

Research also gives attention to the accumulation of elements that are beneficial in plant, animal, and human nutrition. Accumulation of selenium is addressed in research and in this handbook (105,106). Chapters on aluminum, cobalt, and silicon discuss research on these elements.

Traditional soil testing provides information on patterns in soil fertility and management, and plant vigor provides an indication of plant response to soil properties and management often based on soil testing. Shortcomings of current soil testing methodology are the inability to predict yields, large soil test spatial and temporal variability, inability to reflect dynamics of field parameters that affect nutrient availability, lack of accurate tests for nutrient mineralization, and lack of accurate nutrient response functions (107).

Precision agriculture considers spatial variability across a field to optimize application of fertilizer and other inputs on a site-specific basis (76,90,108–110). Precision agriculture employs technologies of global positioning and geographic information systems and remote sensing. These technologies permit decisions to be made in the management of crop-yield-limiting biotic and abiotic factors and their interactions on a site-specific basis rather than on a whole-field basis (111–114). Remote sensing is a term applied to research that assesses soil fertility and plant responses through means other than on-the-ground sampling and analysis (115). Research has applied video image analysis in monitoring plant growth to assess soil fertility and management (116). Spectral reflection and digital processing of aerial photographs have been researched to assess soil fertility (117). In precision agriculture, it is possible for the fertilizer spreader on the back of a tractor to operate at different speeds in different parts of a field in response to data obtained on the growth of the crop underneath and stored in a geographic information system. These data may have been obtained by remote sensing, or even by continuous measurement of yields by the harvesting equipment operating in the same field at the previous harvest. The precise location of the fertilizer spreader at any moment of time is monitored by global positioning.

REFERENCES

1. D.I. Arnon, P.R. Stout. The essentiality of certain elements in minute quantity for plants with special reference to copper. *Plant Physiol.* 14:371–375, 1939.
2. H. Marschner. *Mineral Nutrition of Higher Plants*, 2nd ed. London: Academic Press, 1995, p. 889.
3. E. Epstein, A.J. Bloom. *Mineral Nutrition of Plants: Principles and Perspectives*, 2nd ed. Sunderland, Mass.: Sinauer, 2005, p. 400.
4. J.B. Jones, Jr. *Hydroponics. A Practical Guide for the Soilless Grower.* Boca Raton, Fla.: St. Lucie, 2000, p. 230.

5. P.M. Harmer, E.J. Benne. Sodium as a crop nutrient. *Soil Sci.* 60:137–148, 1945.

6. M. Johnston, C.P.L. Grof, P.F. Brownell. Responses to ambient CO_2 concentration by sodium-deficient C_4 plants. *Aus. J. Plant Physiol.* 11:137–141, 1984.

7. S. Ahmed, H.J. Evans. Cobalt: a micronutrient for the growth of soybean plants under symbiotic conditions. *Soil Sci.* 90:205–210, 1960.

8. D.R. Hoagland. *Lectures on the Inorganic Nutrition of Plants.* Waltham, Mass.: Chronica Botanica Co., 1948, p. 226.

9. J. von Sachs. *Lectures on the Physiology of Plants* (transl. by H.M. Ward), Oxford: Clarendon Press, 1887, p. 836.

10. C.J. Asher. Beneficial elements, functional nutrients, and possible new essential elements. In: J.J. Mortvedt, F.R. Cox, L.M. Shuman, R.M. Welch, eds. *Micronutrients in Agriculture.* Madison, Wis.: Soil Science Society of America, Book Series No. 4, 1991, pp. 703–723.

11. P.H. Brown, R.M. Welch, E.E. Cary. Nickel: a micronutrient essential for higher plants. *Plant Physiol.* 85:801–803, 1987.

12. W.F. Bennett. Plant nutrient utilization and diagnostic plant symptoms. In: W.F. Bennett, ed. *Nutrient Deficiencies and Toxicities in Plants.* St. Paul, Minn.: APS Press, 1993, pp. 1–7.

13. H.D. Chapman, ed. *Diagnostic Criteria for Plants and Soils.* Riverside, Cal.: H.D. Chapman, 1966, p. 793.

14. G.C. Cresswell, R.G. Weir. *Plant Nutrient Disorders 2. Tropical Fruit and Nut Crops.* Melbourne: Inkata Press, 1995a, pp. 1–112.

15. G.C. Cresswell, R.G. Weir. *Plant Nutrient Disorders 4. Pastures and Field Crops.* Melbourne: Inkata Press, 1995b, pp. 1–126.

16. G.C. Cresswell, R.G. Weir. *Plant Nutrient Disorders 5. Ornamental Plants and Shrubs.* Melbourne: Inkata Press, 1998, pp. 1–200.

17. J.E. English, D.N. Maynard. A key to nutrient disorders of vegetable plants. *HortScience* 13:28–29, 1978.

18. H.B. Sprague. *Hunger Signs in Crops. A Symposium.* New York: McKay, 1964, pp. 1–461.

19. R.G. Weir, G.C. Cresswell. *Plant Nutrient Disorders 1. Temperate and Subtropical Fruit and Nut Crops.* Melbourne: Inkata Press, 1993a, pp. 1–93.

20. R.G. Weir, G.C. Cresswell. *Plant Nutrient Disorders 3. Vegetable Crops.* Melbourne: Inkata Press, 1993b, pp. 1–104.

21. N.T. de Saussure. *Chemical Research on Plants* (French). Paris: Nyon, 1804, p. 328.

22. H.S. Reed. *A Short History of the Plant Sciences.* Waltham, Mass.: Chronica Botanica Co., 1942, pp. 241–265.

23. D.I. Arnon. Growth and function as criteria in determining the essential nature of inorganic nutrients. In: E. Truog, ed. *Mineral Nutrition of Plants.* Madison, Wis.: University of Wisconsin, Press, 1951, pp. 313–341.

24. R.D. Munson, W.L. Nelson. Principles and practices in plant analysis. In: L.M. Walsh, J.D. Beaton, eds. *Soil Testing and Plant Analysis.* Madison, Wis.: Soil Science Society of America, Inc., 1973, pp. 223–248.

25. A Ulrich. Critical nitrate levels of sugar beets estimated from analysis of petiole and blades, with special reference to yields and sucrose concentrations. *Soil Sci.* 69:291–309, 1949.

26. A Ulrich. Plant tissue analysis as a guide in fertilizing crops. In: H.M. Reisenhauer, ed. *Soil and Plant Tissue Testing in California.* Riverside: University of California Bulletin 1976, 1879, pp. 1–4.

27. H. Mills, J.B. Jones, Jr. *Plant Analysis Handbook II.* Athens, Ga.: Micro Macro Publishing, Inc., 1996, p. 422.

28. L.M. Walsh, J.D. Beaton, eds. *Soil Testing and Plant Analysis.* Revised edition. Madison, Wis.: Soil Science Society of America, Inc., 1973, p. 491.

29. S.R. Aldrich. Plant analysis: problems and opportunities. In: L.M. Walsh, J.D. Beaton, eds. *Soil Testing and Plant Analysis.* Revised edition. Madison, Wis.: Soil Science Society of America, Inc., 1973, pp. 213–221.

30. T.E. Bates. Factors affecting critical nutrient concentrations in plants and their evaluation: a review. *Soil Sci.* 112:116–130, 1971.

31. J.G. Cruz-Castillo, S. Ganeshanandam, B.R. McKay, G.S. Lawes, C.R.O. Lawoko, D.J. Woolley. Applications of canonical discriminant analysis in horticultural research. *HortScience* 29:1115–1119, 1994.

32. R.B. Beverly. *A Practical Guide to the Diagnosis and Recommendation Integrated System.* Athens, Ga.: Micro Macro Publishing, Inc., 1991, pp. 1–70.

33. J.L. Walworth, M.E. Sumner. Foliar diagnosis: a review. In: B. Tinker, A. Läuchli, eds. *Advances in Plant Nutrition,* Vol. 3. New York: Praeger, 1988, pp. 193–245.

34. D.W. Goodall, F.G. Gregory. *Chemical Composition of Plants as an Index of Their Nutritional Status.* Technical Communication No. 17. East Malling, Kent, England: Imperial Bureau of Horticulture and Plantation Crops, 1947, pp. 1–167.

35. J.B. Jones, Jr, W.J.A. Steyn. Sampling, handling, and analyzing plant tissue samples. In: L.M. Walsh, J.D. Beaton, eds. *Soil Testing and Plant Analysis.* Revised edition, Madison, Wis.: Soil Science Society of America, Inc., 1973, pp. 249–270.

36. Y.P. Kalra, ed. *Handbook of Reference Methods for Plant Analysis.* Boca Raton, Fla.: CRC Press, 1998, p. 300.

37. T.K. Hartz, W.E. Bendixen, L. Wierdsma. The value of presidedress soil nitrate as a nitrogen management tool in irrigated vegetable production. *HortScience* 35:651–656, 2000.

38. A. Ulrich. Nitrate content of grape leaf petioles as an indicator of the nitrogen status of the plant. *Proc. Am. Soc. Hortic. Sci.* 41:213–218, 1942.

39. D.D. Warncke. Soil and plant tissue testing for nitrogen management in carrots. *Commun. Soil Sci. Plant Anal.* 27:597–605, 1996.

40. C.M.J. Williams, N.A. Maier. Determination of the nitrogen status of irrigated potato crops. 2. A simple on farm quick test for nitrate-nitrogen in petiole sap. *J. Plant Nutr.* 13:985–993, 1990.

41. B. Vaughan, K.A. Barbarick, D.G. Westfall, P.L. Chapman. Tissue nitrogen levels for dryland hard red winter wheat. *Agron. J.* 82:561–565, 1990.

42. J.M. Jemison, R.H. Fox. A quick-test procedure for soil and plant-tissue nitrates using test strips and a hand-held reflectometer. *Commun. Soil Sci. Plant Anal.* 19:1569–1582, 1988.

43. A. Scaife, K.L. Stevens. Monitoring sap nitrate in vegetable crops—comparison of test strips with electrode methods, and effects of time of day and leaf position. *Commun. Soil Sci. Plant Anal.* 14:761–771, 1983.

44. M.P. Westcott, C.J. Rosen, W.P. Inskeep. Direct measurement of petiole sap nitrate in potato to determine crop nitrogen status. *J. Plant Nutr.* 16:515–521, 1993.

45. H.D. Sunderman, A.B. Onken, L.R. Hossner. Nitrate concentration of cotton petioles as influenced by cultivar, row spacing, and N application rate. *Agron. J.* 71:731–737, 1979.

46. U. Kafkafi, R. Ganmore-Neumann. Ammonium in plant tissue: real or artifact? *J. Plant Nutr.* 20:107–118, 1997.

47. G.N. Hoffer. Testing corn stalks chemically to aid in determining their food needs. *Indiana Agric. Exp. Sta. Bull.* 298, 1930, p. 31.

48. G.N. Hoffer, J.F. Trost. The accumulation of iron and aluminum compounds in corn plants and its probable relation to root rots. *J. Am. Soc. Agron.* 15:323–331, 1923.

49. G.W. Leeper. Manganese deficiency and accumulation of nitrates in plants. *J. Aus. Inst. Agric. Sci.* 7:161–162, 1941.

50. W.Z. Huang, X.Y. Liang, X.J. Lun. Diagnosis of potassium deficiency in bananas using the method of different values. *Commun. Soil Sci. Plant Anal.* 23:75–84, 1992.

51. A. Ulrich. Potassium content of grape leaf petioles and blades contrasted with soil analysis as an indicator of the potassium status of the plant. *Proc. Am. Soc. Hortic. Sci.* 41:204–212, 1942.

52. F.T. Turner, M.F. Jund. Chlorophyll meter to predict nitrogen topdress requirement for semidwarf rice. *Agron. J.* 83:926–928, 1991.

53. C.S.T. Daughtry, C.L. Walthall, M.S. Kim, E.B. deColstoun, J.E. McMurtrey III. Estimating corn leaf chlorophyll concentration from leaf and canopy reflectance. *Remote Sensing Environ.* 74:229–239, 2000.

54. R. Lavon, E.E. Goldschmidt. Enzymatic methods for detection of mineral element deficiencies in citrus leaves: a mini-review. *J. Plant Nutr.* 22:139–150, 1999.

55. A. Bar Akiva. Biochemical indications as a means of distinguishing between iron and manganese deficiency symptoms in citrus plants. *Nature,* 190:647–648, 1961.

56. A. Bar Akiva. Leaf analysis: possible limitations. *Proc. 18th Int. Hort. Congr.* 4:333–345, 1972.

57. A. Bar Akiva. Substitutes for benzidine as H-donors in the peroxidase assay for rapid diagnosis of iron deficiency in plants. *Commun. Soil Sci. Plant Anal.* 15:929–934, 1984.

58. A. Bar Akiva, M. Kaplan, R. Lavon. The use of biochemical indicator for diagnosing micronutrient deficiencies of grapefruit trees under field conditions. *Agrochimica* 11:283–288, 1967.

59. N.P. Kaur, P.N. Takkar, V.K. Nayyar. Catalase, peroxidase, and chlorophyll relationship to yield and iron deficiency chlorosis in *Cicer* genotypes. *J. Plant Nutr.* 7:1213–1220, 1984.

60. E.O. Leidi, M. Gomez, M.D. de la Guardia. Evaluation of catalase and peroxidase activity as indicators of Fe and Mn nutrition for soybean. *J. Plant Nutr.* 9:1239–1249, 1986.

61. V. Nenova, I. Stoyanov. Physiological and biochemical changes in young maize plants under iron deficiency. 2. Catalase, peroxidase, and nitrate reductase activities in leaves. *J. Plant Nutr.* 18:2081–2091, 1995.

62. A. Ranieri, A. Castagna, B. Baldan, G.F. Soldatini. Iron deficiency differently affects peroxidase isoforms in sunflower. *J. Exp. Bot.* 52:25–35, 2001.

63. A. Bar Akiva, R. Lavon. Carbonic anhydrase activity as an indicator of zinc deficiency in citrus leaves. *J. Hortic. Sci.* 44:359–362, 1969.

64. C. Chatterjee, R. Jain, B.K. Dube, N. Nautiyal. Use of carbonic anhydrase for determining zinc status of sugar cane. *Trop. Agric.* 75:480–483, 1998.

65. N. Pandey, G.C. Pathak, A.K. Singh, C.P. Sharma. Enzymic changes in response to zinc nutrition. *J. Plant Physiol.* 159:1151–1153, 2002.

66. I. Snir. Carbonic anhydrase activity as an indicator of zinc deficiency in pecan leaves. *Plant Soil* 74:287–289, 1983.

67. H. Sasaki, T. Hirose, Y. Watanabe, R. Ohsugi. Carbonic anhydrase activity and CO_2-transfer resistance in Zn-deficient rice leaves. *Plant Physiol.* 118:929–934, 1998.

68. P.A. Moore, Jr., W.H. Patrick, Jr. Effect of zinc deficiency on alcohol dehydrogenase activity and nutrient uptake in rice. *Agron. J.* 80:882–885, 1988.

69. A. Bar Akiva, R. Lavon, J. Sagiv. Ascorbic acid oxidase activity as a measure of the copper nutrition requirements of citrus trees. *Agrochimica* 14:47–54, 1969.

70. A. Shaked, A. Bar Akiva. Nitrate reductase activity as an indication of molybdenum requirement in citrus plants. *Phyochemistry* 6:347–350, 1967.

71. A. Polle, K. Chakrabarti, S. Chakrabarti, F. Seifert, P. Schramel, H. Rennenberg. Antioxidants and manganese deficiency in needles of Norway spruce (*Picea abies* L.) trees. *Plant Physiol.* 99:1084–1089, 1992.

72. A. Bar Akiva, D.N. Maynard, J.E. English. Rapid tissue test for diagnosing iron deficiencies in vegetable crops. *HortScience* 13:284–285, 1978.

73. J.M. Ruiz, I. López-Cantero, L. Romero. Relationship between calcium and pyruvate kinase. *Biol. Plant* 43:359–362, 2000.

74. S.W. Melsted, T.R. Peck. The principles of soil testing. In: L.M. Walsh, J.D. Beaton, eds. *Soil Testing and Plant Analysis*. Revised edition, Madison, Wis.: Soil Science Society of America, Inc., 1973, pp. 13–21.

75. T.R. Peck, S.W. Melsted. Field sampling for soil testing. In: L.M. Walsh, J.D. Beaton, eds. *Soil Testing and Plant Analysis*. Revised edition, Madison, Wis.: Soil Science Society of America, Inc., 1973, pp. 67–75.

76. P.C. Robert. Precision agriculture: a challenge for crop nutrition management. *Plant Soil*, 247:143–149, 2002.

77. M.F. Morgan. Chemical Soil Diagnosis by the Universal Soil Testing System. New Haven: Connecticut Agric. Exp. Sta. Bull. 450, 1941, pp. 579–628.

78. A.V. Barker. Nitrate Determinations in Soil, Water and Plants. Massachusetts Agric. Exp. Sta. Bull. 611, 1974, p. 35.

79. R.H. Bray, L.T. Kurtz. Determination of total, organic, and available forms of phosphorus in soils. *Soil Sci.* 59:39–45, 1945.

80. A. Mehlich. Mehlich-3 soil test extractant—a modification of Mehlich-2 extractant. *Commun. Soil Sci. Plant Anal.* 15:1409–1416, 1984.

81. *Northeast Coordinating Committee on Soil Testing. Recommended Soil Testing Procedures for the Northeastern United States*. 2nd ed. Newark, Del.: Northeastern Regional Publication No. 493. College of Agriculture and Natural Resources, University of Delaware, 1995, pp. 1–15.

82. S.R. Olsen, L.A. Dean. Phosphorus. In: C.A. Black, ed-in-chief, *Methods of Soil Analysis. Part 2. Chemical and Microbiological Properties, Agronomy 9*. Madison, Wis.: American Society of Agronomy, 1965, pp. 1035–1049.

83. F.S. Watanabe, S.R. Olsen. Colorimetric determination of phosphorus in water extracts of soils. *Soil Sci.* 93:183–188, 1962.
84. C.A. Black, ed-in-chief. *Methods of Soil Analysis. Part 2. Chemical and Microbiological Properties.* *Agronomy 9.* Madison, Wis.: American Society of Agronomy, 1965, p. 1572.
85. A.L. Page, R.H. Miller, D.R. Keeney, eds. *Methods of Soil Analysis, Part 2. Chemical and Microbiological Properties, 2nd Ed., Agronomy 9.* Madison, Wis.: American Society of Agronomy, 1982, p. 1159.
86. Soil and Plant Analysis Council, Inc. *Handbook on Reference Methods for Soil Analysis.* Athens, Ga.: Council on Soil Testing and Plant Analysis, 1992, p. 202.
87. R. Wang, K. Guegler, S.T. LaBrie, N.M. Crawford. Genomic analysis of a nutrient response in *Arabidopsis* reveals diverse expression patterns and novel metabolic and potential regulatory genes induced by nitrate. *Plant Cell* 12:1491–1509, 2000.
88. H. Zhang, B.G. Forde. Regulation of *Arabidopsis* root development by nitrate availability. *J. Exp. Bot.* 51:51–59, 2000.
89. P.N. Soltanpour. 1999. Soil testing. Colorado State University Cooperative Extension Fact Sheet 501. Fort Collins, Colo. http://www.ext.colostate.edu/pubs/crops/00501.html
90. R.H. Beck. Applications in sustainable production. *Commun. Soil Sci. Plant Anal.* 31:1621–1625, 2000.
91. P.N. Soltanpour, J.A. Delgado. Profitable and sustainable soil test-based nutrient management. *Commun. Soil Sci. Plant Anal.* 33:2557–2583, 2002.
92. D. Granatstein, D.F. Bezdicek. The need for a soil quality index: local and regional perspectives. *Am. J. Altern. Agric.* 7:12–16, 1992.
93. J.L. Smith, R.I. Papendick, J.J. Halvorson. Using multiple-variable indicator kriging for evaluating soil quality. *Soil Sci. Soc. Am. J.* 57:743–749, 1993.
94. J.F. von Liebig. *Principles of Agricultural Chemistry, With Special Reference to the Late Researches Made in England.* London: Walton & Maberly, 1855, p. 136.
95. G.O. Ware, K. Ohki, L.C. Moon. The Mitscherlich plant growth model for determining critical nutrient deficiency level. *Agron. J.* 74:88–91, 1982.
96. A.E. Johnston. Efficient use of nutrients in agricultural production systems. *Commun. Soil Sci. Plant Anal.* 31:1599–1620, 2000.
97. L.E. Lanyon. Dairy manure and plant nutrient management issues affecting water quality and the dairy industry. *J. Dairy Sci.* 77:1999–2007, 1994.
98. G.M. Pierzynski, G.F. Vance, J.T. Sims. *Soils and Environmental Quality.* Boca Raton, Fla.: CRC Press, 2000, 459 p.
99. R.L. Hamlin, C. Schatz, A.V. Barker. Zinc accumulation in *Brassica juncea* as influenced by nitrogen and phosphorus nutrition. *J. Plant Nutr.* 26:177–190, 2003.
100. P.B.A.N. Kumar, V. Dushenkov, H. Motto, I. Raskin. Phytoextraction: the use of plants to remove heavy metals from soils. *Environ. Sci. Tech.* 29:1232–1238, 1995.
101. M.J. Blaylock, J.W. Huang. Phytoextraction of metals. In: I. Raskin, B.D. Ensley, eds. *Phytoremediation of Toxic Metals—Using Plants to Clean Up the Environment.* New York: Wiley, 2000, pp. 53–70.
102. S.D. Cunningham, D.W. Ow. Promises and prospects of phytoremediation. *Plant Physiol.* 110:715–719, 1996.
103. J.W. Huang, J. Chen, W.R. Berti, S.D. Cunningham. Phytoremediation of lead-contaminated soils: role of synthetic chelates in lead phytoextraction. *Environ. Sci. Tech.* 31:800–805, 1997.
104. K.A. Anderson, B.A. Magnuson, M.L. Tschirgi, B. Smith. Determining the geographic origin of potatoes with trace metal analysis using statistical and neural network classifiers. *J. Agric. Food Chem.* 47:1568–1575, 1999.
105. D.A. Kopsell, W.M. Randle. Genetic variances and selection potential for selenium accumulation in a rapid-cycling *Brassica oleracea* population. *J. Am. Soc. Hortic. Sci.* 126:329–335, 2001.
106. N. Terry, A.M. Zayed, M.P. deSouza, A.S. Tarun. Selenium in higher plants. *Annu. Rev. Plant Physiol. Plant Mol. Biol.* 51:401–432, 2000.
107. S. Rahman, L.C. Munn, G.F. Vance. Detecting salinity and soil nutrient deficiencies using spot satellite data. *Soil Sci.* 158:31–39, 1994.
108. J.A. Delgado, R.R. Riggenbach, M.J. Shaffer, A. Thompson, R.T. Sparks, R.F. Follett, M.A. Dillon, R.J. Ristau, A. Stuebe, H.R. Duke. Use of innovative tools to increase nitrogen use efficiency and protect environmental quality in crop rotations. *Commun. Soil Sci. Plant Anal.* 32:1321–1354, 2001.

109. G.W. Hergert. A futuristic view of soil and plant analysis and nutrient recommendations. *Commun. Soil Sci. Plant Anal.* 29:1441–1454, 1998.

110. P.C. Robert, R.H. Rust, W.E. Larson. Proceedings of the 4th International Conference on Precision Agriculture, 19–22 July 1998, Part A and Part B. St. Paul, Minne. Madison, Wis.: American Society of Agronomy, 1999, p. 1938.

111. R.W. Heiniger. Understanding geographic information systems and global positioning systems in horticultural applications. *HortTechnology* 9:539–547, 1999.

112. H. Melakeberhan. Embracing the emerging precision agriculture technologies for site-specific management of yield-limiting factors. *J. Nematol.* 34:185–188, 2002.

113. J.T. Moraghan, L. Smith, A. Sims. Remote sensing of sugarbeet canopies for improved nitrogen fertilizer recommendations for a subsequent wheat crop. *Commun. Soil Sci. Plant Anal.* 31:827–836, 2000.

114. A.L. Sims, L.J. Smith, J.T. Moraghan. Spring wheat response to fertilizer nitrogen following a sugar beet crop varying in canopy color. *Precision Agric.* 3:283–295, 2002.

115. E. Schnug, K. Panten, S. Haneklaus. Sampling and nutrient recommendations—the future. *Commun. Soil Sci. Plant Anal.* 29:1456–1462, 1998.

116. R.B. Beverly. Video image analysis as a nondestructive measure of plant vigor for precision agriculture. *Commun. Soil Sci. Plant Anal.* 27:607–614, 1996.

117. F. Zheng, H. Schreier. Quantification of soil patterns and field soil fertility using spectral reflection and digital processing of aerial photographs. *Fert. Res.* 16:15–30, 1988.

Section II

Essential Elements—Macronutrients

2 Nitrogen

Allen V. Barker
University of Massachusetts, Amherst, Massachusetts

Gretchen M. Bryson
University of Massachusetts, Amherst, Massachusetts

CONTENTS

2.1 Determination of Essentiality 22
2.2 Nitrogen Metabolism and Nitrogenous Constituents in Plants.............................22
   2.2.1 Nitrate Assimilation 23
      2.2.1.1 Nitrate Reductase 23
      2.2.1.2 Nitrite Reductase 23
   2.2.2 Ammonium Assimilation 23
      2.2.2.1 Glutamine Synthetase..24
      2.2.2.2 Glutamate Synthase 24
      2.2.2.3 Glutamic Acid Dehydrogenase 24
      2.2.2.4 Transamination...24
      2.2.2.5 Amidation..24
   2.2.3 Proteins and Other Nitrogenous Compounds 25
2.3 Diagnosis of Nitrogen Status in Plants 26
   2.3.1 Symptoms of Deficiency and Excess 26
   2.3.2 Concentrations of Nitrogen in Plants 28
      2.3.2.1 Concentrations of Nitrogen in Plant Parts 29
      2.3.2.2 Ratios of Concentrations of Nitrogen to Other Nutrients in Plants...........31
2.4 Nitrogen in Soils 32
   2.4.1 Forms of Nitrogen in Soils 32
      2.4.1.1 Organic Nitrogen in Soil 33
      2.4.1.2 Inorganic Nitrogen in Soil 35
2.5 Soil Testing for Nitrogen...35
   2.5.1 Determinations of Total Nitrogen 36
   2.5.2 Biological Determinations of Availability Indexes 36
      2.5.2.1 Determination of Inorganic Nitrogen..36
         2.5.2.1.1 Ammonium...36
         2.5.2.1.2 Nitrate..37
         2.5.2.1.3 Amino Sugars 38
2.6 Nitrogen Fertilizers 39
   2.6.1 Properties and Use of Nitrogen Fertilizers 40
      2.6.1.1 Anhydrous Ammonia (82% N) 40
      2.6.1.2 Aqua Ammonia (21% N) 40
      2.6.1.3 Urea (46% N)..40

2.6.1.4   Ammonium Nitrate (34% N) 41
2.6.1.5   Ammonium Sulfate (21% N) 41
2.6.1.6   Nitrogen Solutions (28–32% N) 41
2.6.1.7   Ammonium Phosphates (10–21% N) 42
2.6.1.8   Other Inorganic Nitrogen Fertilizers 42
2.6.1.9   Organic Nitrogen Fertilizers (0.2–15% N) 42
References 43

2.1   DETERMINATION OF ESSENTIALITY

Discovery of the essentiality of nitrogen is often credited to de Saussure (1–3), who in 1804 recognized that nitrogen was a vital constituent of plants, and that nitrogen was obtained mainly from the soil. De Saussure noted that plants absorb nitrates and other mineral matter from solution, but not in the proportions in which they were present in solution, and that plants absorbed substances that were not required for plant growth, even poisonous substances (2). Other scientists of the time believed that nitrogen in plant nutrition came from the air. The scientists reasoned that if it was possible for plants to obtain carbon from the air, which is a mere 0.03% carbon dioxide (by volume), then it would be easy for plants to obtain nitrogen from the air, which is almost 80% nitrogen gas. Greening was observed in plants that were exposed to low levels of ammonia in air, further suggesting that nitrogen nutrition came from the air. Liebig (1–3) wrote in the 1840s, at the time when he killed the humus theory (the concept that plants obtain carbon from humus in soil rather than from the air), that plants require water, carbon dioxide, ammonia, and ash as constituents. Liebig supported the theory that plants obtained nitrogen as ammonium from the air, and his failure to include nitrogen in his "patent manure" was a weakness of the product. Plants will absorb ammonia at low concentrations from the air, but most air contains unsubstantial amounts of ammonia relative to that which is needed for plant nutrition.

The concept that nitrogen was acquired from the air or from soil organic matter was dismissed in the mid-1800s, as it was shown that crop yields rose as a result of fertilization of soil. Using laboratory methods of de Saussure, Boussingault (1), in field research of 1838, developed balances of carbon, dry matter, and mineral matter in crops. Boussingault established a special position for legumes in nitrogen nutrition, a position that Liebig did not support (1). Other research also showed that different nitrogen fertilizers varied in their effectiveness for supporting crop production, with potassium nitrate often being a better fertilizer than ammonium salts (1). Microbial transformations of nitrogen in the soil made it doubtful as to which source was actually the best and which form of nitrogen entered into plants. Studies made with sterile media and in water culture demonstrated that plants may utilize nitrate or ammonium and that one or the other might be superior depending on the species and other conditions. At the time when much of this research was performed, organic fertilizers (farm manures) and gas-water (ammonia derived from coal gases) were the only ones that were cost-effective, considering the value of farm crops and the cost of the fertilizers. With the development of the Haber process in 1909 for the synthesis of ammonia from hydrogen and nitrogen gases, ammonia could be made cheaply, leading to the development of the nitrogen fertilizer industry.

The recognition of the importance of nitrogen in plants predates much of the relatively modern-day research of de Saussure and others. It was written as early as the 1660s and 1670s (1,3) that plants benefitted from nitre or saltpeter (potassium nitrate), that plants accumulated nitre, and that the fertility of the land with respect to nitre affected the quality of crops for storage and yields of sugar.

2.2   NITROGEN METABOLISM AND NITROGENOUS
CONSTITUENTS IN PLANTS

Nitrogen has a wide range of valence states in compounds, which may be used in plant metabolism. Although some compounds have oxidation–reduction states of +7, as in pernitric acid, plant metabolites have oxidation–reduction states ranging from $+5$ (nitric acid, nitrate) to -3 (ammonia, ammonium) (4). Organic, nitrogen-containing compounds are at the oxidation–reduction state of nitrogen in ammonium (-3). Biologically important organic molecules in plants include proteins, nucleic acids, purines, pyrimidines, and coenzymes (vitamins), among many other compounds.

2.2.1 NITRATE ASSIMILATION

Nitrate and ammonium are the major sources of nitrogen for plants. Under normal, aerated conditions in soils, nitrate is the main source of nitrogen. Nitrate is readily mobile in plants and can be stored in vacuoles, but for nitrate to be used in the synthesis of proteins and other organic compounds in plants, it must be reduced to ammonium. Nitrate reductase converts nitrate into nitrite in the nonorganelle portions of the cytoplasm (5,6). All living plant cells have the capacity to reduce nitrate to nitrite, using the energy and reductant (NADH, NADPH) of photosynthesis and respiration in green tissues and of respiration in roots and nongreen tissues (5). Nitrite reductase, which is located in the chloroplasts, reduces nitrite into ammonium, utilizing the energy and reductant of photosynthesis (reduced ferredoxin).

2.2.1.1 Nitrate Reductase

> Nitrate + reduced pyridine nucleotides (NADH, NADPH)
> $\rightarrow$ nitrite + oxidized pyridine nucleotides (NAD$^+$, NADP$^+$)

Nitrate reduction requires molybdenum as a cofactor. A two-electron transfer takes place to reduce nitrate (N oxidation state, $+5$) to nitrite (N oxidation state, $+3$). Respiration is the likely source of reduced pyridine nucleotides in roots and also, along with photosynthesis, can be a source in shoots.

The conversion of nitrite into ammonia is mediated by nitrite reductase, which is located in the chloroplasts of green tissues and in the proplastids of roots and nongreen tissues (5,7,8).

2.2.1.2 Nitrite Reductase

> Nitrite + reduced ferredoxin $\rightarrow$ ammonium + oxidized ferredoxin

In leaves, nitrite reduction involves the transfer of six electrons in the transformation of nitrite to ammonium. No intermediates, such as hyponitrous acid ($H_2N_2O_2$) or hydroxylamine ($HONH_2$), are released, and the reduction takes place in one transfer. The large transfer of energy and reducing power required for this reaction is facilitated by the process being located in the chloroplasts (8). In roots, a ferredoxin-like protein may function, and the energy for producing the reducing potential is provided by glycolysis or respiration (9,10).

In plants, roots and shoots are capable of nitrate metabolism, and the proportion of nitrate reduced in roots or shoots depends on plant species and age, nitrogen supply, temperature, and other environmental factors (11–15).

The assimilation of nitrate is an energy-consuming process, using the equivalent of 15 mol of adenosine triphosphate (ATP) for each mole of nitrate reduced (16). The assimilation of ammonia requires an additional five ATP per mole. In roots, as much as 23% of the respiratory energy may be used in nitrate assimilation compared with 14% for ammonium assimilation (17). However, nitrate can be stored in cells without toxic effects, but ammonium is toxic at even low concentrations and must be metabolized into organic combination. Consequently, ammonium metabolism for detoxification may deplete carbon reserves of plants much more than nitrate accumulation.

2.2.2 AMMONIUM ASSIMILATION

The metabolism of ammonium into amino acids and amides is the main mechanism of assimilation and detoxification of ammonium. Glutamic acid formation is a port of entry of nitrogen into organic compounds and occurs in the chloroplasts or mitochondria. Ammonium assimilation in root mitochondria probably uses ammonium absorbed in high concentrations from nutrient solutions. One enzyme is involved in ammonium assimilation in mitochondria: glutamic acid dehydrogenase. Ammonium assimilation in chloroplasts utilizes the ammonium that is formed from the reduction of nitrite by nitrite reductase and that which is released in photorespiration. Two enzymes are involved in chloroplasts, glutamine synthetase and glutamate synthase. Glutamine synthetase forms glutamine from ammonium and glutamate (glutamic acid). Glutamate synthase forms glutamate from glutamine and α-oxoglutarate (α-ketoglutaric acid). These enzymes are also active in roots and nodules (N_2 fixation). These enzymes assimilate most of the ammonium derived from absorption from dilute solutions, reduction of nitrate, N_2 fixation, or photorespiration (18–25). Further discussions of glutamine synthetase, glutamate synthase, and glutamic acid dehydrogenase follow.

2.2.2.1 Glutamine Synthetase

Ammonium + glutamate + ATP + reduced ferredoxin $\rightarrow$ glutamine + oxidized ferredoxin

2.2.2.2 Glutamate Synthase

Glutamine + α-oxoglutarate $\rightarrow$ 2 glutamate

Sum (or net): Ammonium + α-oxoglutarate + ATP + reduced ferredoxin
$\rightarrow$ glutamate + oxidized ferredoxin

Glutamine synthetase has a high affinity for ammonium and thus can assimilate ammonium at low concentrations, such as those that occur from the reduction of nitrate. If this enzyme is inhibited, however, ammonium may accumulate to phytotoxic levels. Ammonium accumulation to toxic levels from the inhibition of glutamine synthetase is the mode of action of the herbicide glufosinate ammonium (26,27).

2.2.2.3 Glutamic Acid Dehydrogenase

Ammonium + α-oxoglutarate + ATP + reduced pyridine nucleotide (NADH, NADPH)
$\rightarrow$ glutamate + oxidized pyridine nucleotide (NAD$^+$, NADP$^+$)

Another pathway for ammonium assimilation into organic compounds is by glutamic acid dehydrogenase, which is located in the mitochondria (28). Glutamic acid dehydrogenase has a low affinity for ammonium and becomes important in ammonium assimilation at high concentrations of ammonium and at low pH in growth media (15).

2.2.2.4 Transamination

Glutamate + α-oxyacid $\rightarrow$ α-oxoglutarate + α-amino acid

Ammonium that is assimilated into glutamate from mitochondrial or chloroplastic assimilation can be transferred by aminotransferases (transaminases) to an appropriate α-oxyacid (α-ketoacid) to form an α-amino acid. The transfer can also be to other keto-groups on carbon chains to form, for example, γ- or δ-amino acids. The keto acids for the synthesis of amino acids are derived from photosynthesis, glycolysis, and the tricarboxylic acid cycle, among other processes.

2.2.2.5 Amidation

Glutamate + ammonium + ATP $\rightarrow$ glutamine + ADP

Amides are formed by the amidation of carboxyl groups. Amides are nitrogen-rich compounds that can store or transport nitrogen. Common amides are glutamine (5C, 2N) and asparagine

(4C, 2N). Glutamine is formed from amidation of glutamic acid (glutamate), and asparagine is formed by amidation of aspartic acid (aspartate). Often, when the external supply of ammonium is high, asparagine, a metabolite unique to plants, will dominate among the amides, as plants respond to conserve carbon in the detoxification of ammonium.

2.2.3 PROTEINS AND OTHER NITROGENOUS COMPOUNDS

Unlike animals, plants do not eliminate nitrogen from their bodies but reuse nitrogen from the cycling of proteins and other nitrogenous constituents. Nitrogen losses from plants occur mainly by leaching of foliage by rain or mist and by leaf drop (29). Nitrogen in plants is recycled as ammonium. In the case of hydrolysis (breakdown) of proteins, the amino acids of proteins do not accumulate, but rather nitrogen-rich storage compounds (amides, arginine, and others) accumulate as reserves of nitrogen at the oxidation–reduction level of ammonium. These compounds are formed from the catabolism of proteins. The carbon and hydrogen of proteins are released as carbon dioxide and water. These nitrogen-rich products also accumulate if accumulation of nitrogenous compounds occurs in excess of their conversion into proteins. The amino acids that enter into proteins are not mingled with the storage reserves or translocated products but are made at the same site where protein synthesis occurs. The carbon framework (carbon skeletons) remaining after the donation of nitrogen (ammonium) for amino acid synthesis for incorporation into proteins is metabolized into carbon dioxide and water. Thus, the products of protein catabolism are ammonium, carbon dioxide, and water. Protein turnover (breakdown and resynthesis) may occur in plants in a diurnal cycle, with synthesis occurring in the light and breakdown occurring in the dark, or anabolism and catabolism of proteins may proceed in different compartments of the same cell at the same time (29–31). In a 24-h period, one quarter of the protein in a healthy leaf may be newly synthesized as a result of protein turnover. Most authors indicate a protein turnover of 0.1 to 2% per hour (32,33). With *Lemma minor*, Trewavas (34,35) measured turnover rates of 7% per day. In an excised leaf, protein synthesis does not proceed after protein hydrolysis, and soluble nitrogenous compounds accumulate. In a nitrogen-deficient plant, the nitrogen will be translocated to a site of need. Also, under normal conditions, leaves will donate some of their nitrogen in leaf proteins to fruits and seeds.

Amino acids are assimilated into proteins or other polypeptides (28). Although plants contain more than 100 amino acids (1,29), only about 20 enter into proteins (Table 2.1). Hydroxyproline may be formed after incorporation of proline into proteins. Cystine is the dimer of cysteine and is formed after incorporation of cysteine into protein. Animal proteins occasionally contain amino acids other than those listed in Table 2.1.

TABLE 2.1
Amino Acids Occurring Regularly in Plant Proteins

Alanine	Glutamic acid	Leucine	Serine
Arginine	Glutamine	Lysine	Threonine
Asparagine	Glycine	Methionine	Tryptophan
Aspartic acid	Histidine	Phenylalanine	Tyrosine
Cysteine	Isoleucine	Proline	Valine

Source: From McKee, H.S., *Nitrogen Metabolism in Plants*, Oxford University Press, London, 1962, pp. 1–18 and Steward, F.C. and Durzan, D.J., in *Plant Physiology: A Treatise. Vol IVA: Metabolism: Organic Nutrition and Nitrogen Metabolism*, Academic Press, New York, 1965, pp. 379–686.

TABLE 2.2
Approximate Fractions and Common Ranges of Concentrations of Nitrogen-Containing Compounds in Plants

Compound	Fraction of Total Nitrogen (%)	Concentration (μg/g Dry Weight)
Proteins	85	10,000 to 40,000
Nucleic acids	5	1000 to 3000
Soluble organic	<5	1000 to 3000
Nitrate	<1	10 to 5000
Ammonium	<0.1	1 to 40

The major portion of nitrogen in plants is in proteins, which contain about 85% of the total nitrogen in plants (Table 2.2). Nucleic acids (DNA, RNA) contain about 5% of the total nitrogen, and 5 to 10% of the total nitrogen is in low-molecular-weight, water-soluble, organic compounds of various kinds (36).

Some of the low-molecular-weight, water-soluble, organic compounds are intermediates in the metabolism of nitrogen. Some have specific roles in processes other than intermediary metabolism. Amides and amino acids have roles in transport and storage of nitrogen in addition to their occurrence in proteins. Ureides (allantoin and allantoic acid) are prominent in xylem sap and transport nitrogen fixed in root nodules of legumes (15,29). Amines (ethanolamine) and polyamines (putrescine, spermine, spermidine) have been assigned roles or have putative roles in the lipid fraction of membranes, as protectants, and in processes involved in plant growth and development (15,37–43). Putrescine accumulation in plants may be a physiological response to stresses such as the form of nitrogen supplied and the nutrient status of plants (39,44–46). Simple nitrogen bases, such as choline, are related to alkaloids in plants and to lipids (29). Analogs of purines and pyrimidines have functions in growth regulation (29). Various amino acids other than those in proteins exist in plants. Often, the nonprotein amino acids are related to those occurring in proteins. β-Alanine, homoserine, and γ-aminobutyric acid are common examples of these amino acids (1,29). Accumulation of amino acids such as ornithine and citrulline is generally rare in plants, but they may be the major soluble nitrogenous constituents of some species (1). Nonprotein amino acids may be natural products or metabolites, but their functions are generally unclear.

2.3 DIAGNOSIS OF NITROGEN STATUS IN PLANTS

2.3.1 SYMPTOMS OF DEFICIENCY AND EXCESS

A shortage of nitrogen restricts the growth of all plant organs, roots, stems, leaves, flowers, and fruits (including seeds). A nitrogen-deficient plant appears stunted because of the restricted growth of the vegetative organs. Nitrogen-deficient foliage is a pale color of light green or yellow (Figure 2.1). Loss of green color is uniform across the leaf blade. If a plant has been deficient throughout its life cycle, the entire plant is pale and stunted or spindly. If the deficiency develops during the growth cycle, the nitrogen will be mobilized from the lower leaves and translocated to young leaves causing the lower leaves to become pale colored and, in the case of severe deficiency, to become brown (firing) and abscise. Until the 1940s crops received little nitrogen fertilizer (a typical application of N was 2 or 3 kg/ha), and when the light green color and firing appeared, farmers assumed that the soil was droughty (47). Sometimes under conditions of sufficiency of nitrogen, leaves, especially the lower ones, will provide nitrogen to fruits and seeds, and symptoms of deficiency may develop on the leaves. These symptoms, which develop late in the growing season, may not be evidence of yield-limiting deficiencies but are expressions of transport of nitrogen from old leaves to

(a)   (b)

(c)

FIGURE 2.1 Photographs of nitrogen deficiency symptoms on (a) corn (*Zea mays* L.), (b) tomato (*Lycopersicon esculentum* Mill.), and (c) parsley (*Petroselinum crispum* Nym.). (Photographs by Allen V. Barker.) (For a color presentation of this figure, see the accompanying compact disc.)

other portions of the plant. For additional information on nitrogen-deficiency symptoms, readers should consult Cresswell and Weir (48–50), Weir and Cresswell (51,52) or Sprague (53).

At least 25%, more commonly more than 75%, of the nitrogen in leaves is contained in the chloroplasts (29,54). Most of the nitrogen of chloroplasts is in enzymatic proteins in the stroma and lamellae. Chlorophyll and proteins exist in lamellae as complexes referred to as chlorophyll proteins or holochromes (55–59). Nitrogen-deficient chloroplasts may be circular in profile rather than elliptical and may appear swollen. Nitrogen deficiency generally brings about a decrease in protein in chloroplasts and a degradation of chloroplast fine (lamellar) structure (60). Almost all membranous structure may be disrupted. Grana are often reduced in number or are indistinguishable. The loss of membranous structures is associated with the loss of proteins (61). A loss of chlorophyll occurs simultaneously with the loss of membranes and proteins, leading to the loss of green color from nitrogen-deficient leaves.

The loss of fine structure in chloroplasts during nutrient deficiency is not unique to nitrogen deficiency. Association of chloroplast aberrations with specific nutritional disorders has been difficult because of similarities in appearance of nutrient-deficient chloroplasts (62,63). The similarities are due to the effects that the deficiencies have on protein or chlorophyll synthesis (64,65). Elemental toxicities can also impart structural changes that resemble elemental deficiencies in chloroplasts (66).

2.3.2 CONCENTRATIONS OF NITROGEN IN PLANTS

Many attempts have been made to relate yields of crops to nutrient supply in media and to accumulation in plants. Deficiency of nitrogen or another nutrient is associated with suboptimum development of a plant, as reflected by the appearance of symptoms of deficiency, the suppression of yields, or to the response of plants after the accumulation of the deficient nutrient following its application as a fertilizer. Plant analysis (tissue testing) is used in the diagnosis of nutritional deficiency, sufficiency, or excess. Generally, the concentrations of nitrogen in plants reflect the supply of nitrogen in the root medium, and yields increase as internal concentration of nitrogen in plants increases. The use of information on internal concentrations of nitrogen in plants should not be directed toward forecasting of yields as much as it should be used in assessing how yields can be improved by fertilization.

Various models have been developed to describe the response of plants to nutrient supply and accumulation (67). Pfeiffer et al. (68) proposed a hyperbolic model in which plants approached an asymptote or maximum value as nutrient accumulation increased. Linear models have been proposed to describe growth responses to nutrient accumulation (67). Other researchers identified a three-phase model (69–71) (Figure 2.2). In this model, growth curves describe a deficient level of nutrient accumulation, region of poverty adjustment, or minimum percentage where yields rise with increasing internal concentrations of nitrogen. In the second zone of the growth curve, a transition from deficiency to sufficiency occurs followed by a region known as luxury consumption in which internal concentration of nitrogen rises but yield does not rise. The concentration of nitrogen at the transition from deficiency to sufficiency is known as the *critical concentration*. Eventually, nitrogen accumulation will rise to excessive or toxic levels.

Nitrogen concentrations in plants vary with species and with varieties within species (72,73). Nitrogen accumulation in plants also varies among families. Herbaceous crops from fertilized fields commonly have concentrations of nitrogen that exceed 3% of the dry mass of mature leaves. Leaves of grasses (Gramineae, Poaceae) (1.5 to 3.5% N) are typically lower in total nitrogen concentrations

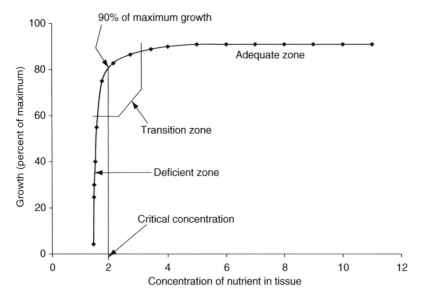

FIGURE 2.2 Model of plant growth response to concentration of nutrients in plant tissue. Units of concentration of nutrient in tissue are arbitrary. The model shows the critical concentration of nutrient at a response that is 90% of the maximum growth obtained by nutrient accumulation in the tissue. Deficient zone, transition zone, and adequate zone indicate concentrations at which nutrients may be lacking, marginal, or sufficient for crop yields.

TABLE 2.3
Concentrations of Total Nitrogen in Plant Parts

Plant Part	Concentration of Total Nitrogen (% Dry Weight)	
	Range	Optimum
Leaves (blades)	1 to 6	>3
Stems	1 to 4	>2
Roots	1 to 3	>1
Fruits	1 to 6	>3
Seeds	2 to 7	>2

than those of legumes (Leguminosae, Fabaceae) (>3% N). Leaves of trees and woody ornamentals may have <1.5% N in mature leaves. Genetic differences attributable to species or families are due to many factors affecting absorption and metabolism of nitrogen and plant growth in general.

The concentrations of nitrogen in leaves, stems, and roots changes during the growing season. In the early stages of growth, concentrations will be high throughout the plant. As plants mature the concentrations of nitrogen in these organs fall, and is usually independent of the initial external supply of nitrogen. Mobilization of nitrogen from old leaves to meristems, young leaves, and fruits leads to a diminished concentration of nitrogen in old, bottom leaves of plants. Whether a plant is annual, biennial, or perennial affects considerations of yield relations and the state of nutrient accumulation in organs (leaves) during the season. If the development of a plant is restricted by low levels of external factors, such as other nutrients, water, or temperature, internal concentration of nitrogen may rise. Root structure and metabolism can lead to differential accumulation of nitrogen. Assimilation and transport of nitrogenous compounds in plants can lead to differential accumulation among species and within the plants. Nitrogen sources can have large effects on total nitrogen concentrations in plants. Plants grown on ammonium nutrition can have twice the nitrogen concentrations in vegetative parts as plants grown on nitrate nutrition.

The choice of tissue for plant analysis is important in plant diagnosis (Table 2.3). Generally, leaves are the most satisfactory plant part to use for diagnosis (69,72,74). Blades are used more frequently than leaf petioles or whole leaves. Blades are chosen as the diagnostic part if total nitrogen is to be assessed, whereas petioles may be selected if the nitrogenous component is soluble, such as nitrate. Total nitrogen quantity in tissues is the most commonly measured fraction, although some researchers believe that nitrate contents reflect the nutritional status better than total nitrogen.

2.3.2.1 Concentrations of Nitrogen in Plant Parts

With a nutrient supply in which all elements except nitrogen are held at a constant high level, the concentration of nitrogen in a plant will be expected to rise, along with growth and yields, with increases in nitrogen supply. Nitrogen concentrations in leaves are often not correlated with increased growth and yields. Shortages of other nutrients or stresses imposed by growth-limiting temperatures or water supply can cause concentrations of total nitrogen or nitrate to increase, along with a suppression of yield (75). The age of plant tissues is important in diagnosis of nitrogen sufficiency. In the early stages of plant growth, the concentration of nitrogen in plants will be higher than at the later stages. Increased external concentrations of nitrogen will increase the concentration of nitrogen in plant organs, but the trend is for nitrogen concentrations to fall in leaves, stems, and roots as plants mature. These changes will vary with whether the plant is annual, biennial, or perennial (67). It is important to sample plants for nitrogen determinations at a given time of the year or stage of plant development. Some researchers recommend that samples be taken at a certain time of the day, since light intensity and duration can affect the amount of nitrate in tissues (76). Nutrient concentrations in leaves can vary by as much as 40% during a diurnal period (67). Nitrate can vary with time of day, with lower concentrations occurring in the afternoon than in the morning.

Analysis of whole shoots may be the best index of the nutritional status of plants even though each organ of a plant will vary in nitrogen concentrations. Since organs of plants vary in composition and since the proportions of organs vary with the nitrogen status of plants, a particular organ of a plant is usually chosen for analysis. Conducting tissue, such as that of stems or petioles, may provide the best index of the response of plants to nutrient applications or the best index of the nutrient status at a given time in growth. Nitrate concentrations in corn (*Zea mays* L.) stalks are usually several times higher than those of leaves (77). Measurement of nitrate in the lower stalk of corn is valuable in the diagnosis of the nitrogen status of the crop (78–80). Brouder et al. (79) noted that analysis of grain for total nitrogen was as good as the stalk test in determining sufficiency or deficiency of corn. Leaf petioles as conducting tissues are often analyzed to assess the nutritional status of vegetable crops (81). Leaves are often taken as samples for nitrogen determinations since they are the organs of active assimilation and hence likely to be the best for analysis to reflect the nutrient status of the whole plant. Leaf samples can be taken conveniently in nondestructive harvests of plants, and leaves can be identified by position or stage of development on plants. Random sampling of leaves is not as good a technique as sampling based on position on plant, size, and age. Nitrogen is a mobile element in plants; hence, it moves from lower leaves to upper leaves, and analysis of lower leaves might be a better index of deficiency than analysis of upper leaves. Sometimes, young leaves or the first-fully expanded leaves are chosen for analysis because of convenience in identifying the sample and because the lower leaves might be dead or contaminated with soil. Deficient, sufficient, and high concentrations of nitrogen in the leaves of plants are reported in Table 2.4.

TABLE 2.4
Concentrations of Nitrogen in Leaves of Various Crops Under Cultivated Conditions

Type of Crop	Diagnostic Range (% Dry Mass of Leaves)		
	Low	Sufficiency[a]	High
Agronomic Crops			
Grass grains	<1.5	1.8 to 3.6	>3.6
Legume grains	<3.6	3.8 to 5.0	>5.0
Cotton	<3.0	3.0 to 4.5	>5.0
Tobacco		4.1 to 5.7	>5.7
Rapeseed		2.0 to 4.5	>4.5
Sugarbeet		4.3 to 5.0	>5.0
Sugarcane	<1 to 1.5	1.5 to 2.7	>2.7
Bedding Plants		2.8 to 5.6	
Trees			
Conifers	<1.0	1.0 to 2.3	>3.0
Broadleaf	<1.7	1.9 to 2.6	>3.0
Cut Flowers	<3.0	3.1 to 4.7	>5
Ferns		1.8 to 2.9	
Potted Floral		2.5 to 4.2	
Forage Crops			
Grasses	<1.5	2.0 to 3.2	>3.6
Legumes	<3.8	3.8 to 4.5	5 to 7
Tree Fruits and Nuts			
Nuts	<1.7	2.0 to 2.9	>3.9

TABLE 2.4 (*Continued*)

| Type of Crop | Diagnostic Range (% Dry Mass of Leaves) | | |
	Low	Sufficiency[a]	High
Citrus	<2.0 to 2.2	2.3 to 2.9	>3.3
Pome	<1.5 to 1.8	2.1 to 2.9	>3.3
Stone	<1.7 to 2.4	2.5 to 3.0	>3.8
Small Woody	<1.5	1.5 to 2.3	>4.5
Strawberry	<2.1	2.1 to 4.3	>4.3
Banana		3.0 to 3.8	
Pineapple		1.5 to 2.5	
Foliage Plants		2.2 to 3.8	
Herbaceous Perennials	<2.2	2.2 to 3.2	>4.0
Ornamental Grasses	<1.6	1.6 to 2.5	>3.0
Ground Covers			
Herbaceous-broadleaf	<2.0	2.0 to 3.9	>4.0
Herbaceous-monocot	<1.5	1.6 to 2.4	>4.0
Woody		1.5 to 2.5	
Turfgrasses		2.6 to 3.8	
Vegetables			
Broadleaf	<2.6	3.5 to 5.1	
Sweet corn		2.5 to 3.2	
Forest and Landscape Trees	<1.9	1.9 to 2.6	
Woody Shrubs			
Palms		2.1 to 3.2	

Note: Values with few exceptions are mean concentrations in mature leaves. 'Low' is value where symptoms of deficiency are showing. 'Sufficiency' is mean range of lower and upper concentrations commonly reported in healthy plants showing no deficiencies. 'High' is a concentration that might represent excessive accumulation of nitrogen.

[a]Optimum or sufficient values for maximum yield or for healthy growth of plants will vary with species, age, and nutrition of plant, position of organ on plant, portion of plant part sampled, and other factors.

Source: Adapted from Chapman, H.D., *Diagnostic Criteria for Plants and Soils*, HD Chapman, Riverside, Cal., 1965, pp. 1–793; Mills, H.A. and Jones, J.B. Jr., *Plant Analysis Handbook II*, MicroMacro Publishing, Athens, Ga., 1996, pp. 155–414; Goodall, D.W. and Gregory, F.G., Chemical composition of plants as an index of their nutritional status, Technical Communication No. 17, Imperial Bureau of Horticulture and Plantation Crops, East Malling, Kent, England, 1947, pp. 1–167; Weir, R.G. and Cresswell, G.C., *Plant Nutrient Disorders 1. Temperate and Subtropical Fruit and Nut Crops*, Inkata Press, Melbourne, 1993, pp. 1–93; Weir, R.G. and Cresswell, G.C., *Plant Nutrient Disorders 3. Vegetable Crops*, Inkata Press, Melbourne, 1993, pp. 1–104; Walsh, L.M. and Beaton, J.D., *Soil Testing and Plant Analysis*, revised edition, Soil Science Society of America, Madison, Wis., 1973, pp. 1–491; and from other sources cited in references.

2.3.2.2   Ratios of Concentrations of Nitrogen to Other Nutrients in Plants

The *critical concentration* (see Section 2.3.2) of nitrogen is the value in a particular plant part sampled at a given growth stage below which plant growth and yield are suppressed by 5 or 10% (82). The responses of plants to nutrient additions are essentially independent of the source of nutrients; hence, the symptoms and nutrient concentrations of affected tissues, and relationships to growth and yields, are identical regardless of the growth medium or location. Therefore, the critical concentration is proposed to have universal application to media and geographic locations (82). However, since leaf (tissue) composition varies with age, the critical concentration can vary and be insensitive or inflexible to diagnosis of nutrient deficiency (83). For example, if a leaf sample is taken at an early plant-growth stage, the concentration of nitrogen may exceed the critical concentration that was determined for tissue at a later stage of growth. Likewise, a sample taken at a late stage of growth might mistakenly be diagnosed as indicating a deficiency of nitrogen. To deal with the problem of variable critical concentrations with plant age, several sets of critical values are needed, one for each growth stage. Determinations of critical concentrations are difficult because of the many observations that must be made of growth and yield in response to nutrient concentrations in leaves. Hence, few critical concentrations have been determined at one growth stage, not considering that multiple stages should be assessed. Applications of sufficiency ranges, such as those reported (Table 2.4), are often too wide to be used for precise diagnoses.

The Diagnostic and Recommendations Integrated System (DRIS) was developed to assess plant nutrition without regard to variety, age, or position of leaves on plants (83,84). The DRIS method considers nutrient balance and utilizes ratios of nutrient concentrations in leaves to determine the relative sufficiency of nutrients (85). The DRIS method differs from standard diagnostic methods in the interpretation of analytical results based on the concentrations of individual elements. Instead of considering each nutrient concentration independently, DRIS evaluates nutrient relationships that involve ratios between pairs of nutrients and evaluates the adequacy of a nutrient in relation to others. Generation of the DRIS index yields positive and negative numbers, which are deviations from a norm and which sum to zero for all nutrients considered. DRIS norms are standard values suggested to have universal application to a crop. Norms are determined by research and have been published for several crops (86).

The optimum range for plant DRIS indices is −15 to 15. If the index is below −15, that element is considered to be deficient. If the index is above 15, that element is considered to be in excess. DRIS indices must be interpreted in comparison with other nutrients. A negative number does not indicate that a nutrient is deficient, but it may be used to compare relative deficiencies among nutrients. DRIS may be useful in identifying hidden hunger or imbalances. For example, if nitrogen had an index of −12, phosphorus an index of −8, and potassium an index of 6, the order of likely growth-limiting effects would be nitrogen > phosphorus > potassium. Variations in DRIS (M-DRIS or modified DRIS) consider dry matter in generation of indices (87,88).

2.4 NITROGEN IN SOILS

2.4.1 FORMS OF NITROGEN IN SOILS

The total nitrogen of the Earth is about 1.67×10^{23} g (89,90). Stevenson (89,90) reported that about 98% of the nitrogen of the Earth is in the lithosphere (rocks, soil, coal, sediments, core, sea bottom). About 2% of the nitrogen is in the atmosphere, with the portions in the hydrosphere and biosphere being insignificant relative to that in the lithosphere and atmosphere. Most of the nitrogen of the Earth, including the nitrogen in the rocks and in the atmosphere, is not available for plant nutrition. The nitrogen in soils, lakes, streams, sea bottoms, and living organisms is only about 0.02% of the total nitrogen of the Earth (89,90). Plants obtain most of their nitrogen nutrition from the soil. The nitrogen in the soil is about 2.22×10^{17} g, most of which is in soil organic matter and which is a negligible component of the total nitrogen content of the world (89,90). Living organisms (biosphere) contain about 2.8×10^{17} g of nitrogen. The nitrogen of living organisms and of the soil is in a constant state of flux, with some forms of nitrogen being readily transformed in this group and some forms being inactive over a long time (91). Transformations are insignificant in the lithosphere and atmosphere. The amount of interchange of nitrogen among the lithosphere (not including soil), atmosphere, and living organisms is very small.

The total amount of nitrogen in the soil to the depth of plowing is considerable relative to the amounts required for crop production, often above 3000 kg/ha but ranging from 1600 kg/ha in sands through 8100 kg/ha in black clay loams to 39,000 kg/ha in deep peats (Table 2.5) (92). Note that the nitrogen in the atmosphere above a hectare of land exceeds 100 million kg at sea level. When land is

TABLE 2.5
Estimated Content and Release of Nitrogen from Various Soils

Type of Soil	Nitrogen in Soil (kg/ha)	
	Total[a]	Annual Release[b]
Sands	1400	28
Yellow sandy loam	2200	44
Brown sandy loam	3100	62
Yellow silt loam	2000	40
Grey silt loam	3600	72
Brown silt loam	5000	100
Black clay loam	7200	144
Deep peats	39,000	780

[a]From Schreiner O. and Brown B.E., in *United States Department of Agriculture, Soils and Men, Yearbook of Agriculture, 1938*, United States Government Printing Office, Washington, DC, 1938, pp. 361–376.
[b]Estimated at 2% annual mineralization rate of soil organic matter.

put for crop production, the nitrogen content of soils declines to a new equilibrium value (90,92). Crop production that relies on the reserves of nitrogen cannot be effective for long, as the reserves become exhausted. Most plants cannot tap into the large reserve of nitrogen in the atmosphere, although biological nitrogen fixation is a means of enhancing the nitrogen content of soils. Biological nitrogen fixation is the principal means of adding nitrogen to the soil from the atmosphere (89). More than 70% of the atmospheric nitrogen added or returned to soils is by biological fixation, and can exceed 100 kg of nitrogen addition per year by nitrogen-fixing legumes. Most of this nitrogen enters into the organic fraction of the soils. Unless nitrogen-fixing legumes are grown, the addition of nitrogen to soils by biological fixation, averaging about 9.2 kg/ha annually, is too small to support crop production. The remainder is from atmospheric precipitation of ammonium, nitrate, nitrite, and organically bound nitrogen (terrestrial dust). The amount of nitrogen precipitated is normally too small to support crop production but might be of significance in natural landscapes (90). Virtually no interchange of nitrogen occurs between rocks and soils.

2.4.1.1 Organic Nitrogen in Soil

The concentrations of nitrogen range from 0.02% in subsoils to 2.5% in peats (93). Nitrogen concentrations in soils generally fall sharply with depth, with most of the nitrogen being in the top one-meter layer of soils (89). Surface layers (A-horizon, plow-depth zone) of cultivated soils have between 0.08 and 0.4% nitrogen. Well over 90%, perhaps over 98%, of the nitrogen in the surface layers (A-horizon, plow-depth zone) of soil is in organic matter (93,94). Since most of the nitrogen in soil is organic, determination of total nitrogen has been a common method of estimating organic nitrogen. The Kjeldahl method, a wet digestion procedure (93,95,96), provides a good estimate of organic, soil nitrogen in surface soils, even though some forms of nitrogen (fixed ammonium, nitrates, nitrites, some organic forms) are not determined by this analysis. In depths below the A-horizon or plow zone, although the amounts of total nitrogen are small, inorganic nitrogen, particularly fixed ammonium, is a high proportion of the total, perhaps 40%, and results from Kjeldahl analysis should be treated with some caution as this fraction would not be determined (93). The Dumas method, a dry digestion procedure, is seldom used for determination of nitrogen in soils but generally gives results in close agreement with Kjeldahl determinations, if certain precautions are taken in the analysis (93).

Soil organic matter is a complex mixture of compounds in various states of decay or stability (97). Soil organic matter may be classified into humic and nonhumic fractions, with no sharp demarcation between the two fractions. The partially decayed or nonhumic portion is the major source of energy for soil organisms. Depending on the nature of the plant materials, about half of fresh plant residues added to soil decompose in a few weeks or months (98,99). Humus, or humic substances, are the degradation products or residues of microbial action on organic matter and are more stable than the nonhumic substances. Humus is classified into three fractions, humin, humic acids, and fulvic acids, based on their solubilities. Humin is the highest molecular weight material and is virtually insoluble in dilute alkali or in acid. Humic acids are alkali-soluble and acid-insoluble. Fulvic acids are alkali- or acid-soluble. The humic and fulvic fractions are the major portions, perhaps 90%, of the humic soil organic matter and are the most chemically reactive substances in humus (100). Humus is slow to mineralize, and unless present in large quantities may contribute little to plant nitrogen nutrition in most soils. About 60 to 75% of the mineralized nitrogen may be obtained by a crop (99). The turnover rate of nitrogen in humus may be about 1 to 3% of the total nitrogen of the soil, varying with type of soil, climate, cultivation, and other factors (93,99). The mineralization rate is likely to be closer to 1% than to 3%. Bremner (96) and Stanford (101) discussed several methods to assess availability of organic nitrogen in soils. Among these procedures were biochemical methods (estimation of microbial growth, mineral nitrogen formed, or carbon dioxide released) and chemical methods (estimation of soil total nitrogen, mineral nitrogen, and organic matter and application of various extraction procedures). The chemical methods are applied more commonly than the biological methods in the estimation of mineralization. Correlation of crop yields to estimations of mineralization generally have not been satisfactory in the assessment of the potential for soils to supply nitrogen for crop growth.

Most studies on the fractionation of total soil organic matter have dealt with the hydrolysis of nitrogenous components with hot acids (3 or 6 M hydrochloric acid for 12 to 24 h) (Table 2.6). The fraction that is not hydrolyzed is called the *acid-insoluble nitrogen*. The acid-soluble nitrogen is fractionated into *ammonium*, *amino acid*, *amino sugar*, and *unidentified* components. The origins and composition of each of the named fractions are not clear. The absolute values vary with soil type and with cultivation (94). All of these forms of nitrogen, including the acid-stable form, appear to be biodegradable and, hence, to contribute to plant nutrition (94,102). Organic matter that is held to clays is recalcitrant to biodegradation and increases in relative abundance in heavily cropped soils (94,103,104). This fraction may have little importance in nitrogen nutrition of plants.

TABLE 2.6
Fractions of Nitrogen in Soil Organic Matter Following Acid Hydrolysis

Nitrogen Component	Fraction of Total Organic Nitrogen (%)
Acid insoluble	20 to 35
Ammonium	20 to 35
Amino acid	30 to 45
Amino sugar	5 to 10
Unidentified	10 to 20

Source: From Bremner, J.M., in *Soil Nitrogen*, American Society of Agronomy, Madison, Wis., 1965, pp. 1324–1345 and Stevenson, F.J., *Nitrogen in Agricultural Soils*, American Society of Agronomy, Madison, Wis., 1982, pp. 67–122.

Cultivation reduces the total amount of organic matter in soils but has little effect on the relative distribution of the organic fractions in soils, suggesting that the results of acid hydrolysis are of little value as soil tests for available nitrogen or for predicting crop yields (94). Humic substances contain about the same forms of nitrogen that are obtained from the acid hydrolysis of soils but perhaps in different distribution patterns (94). Agricultural systems that depend on soil reserves do not remain productive without the input of fertilizer nitrogen.

2.4.1.2 Inorganic Nitrogen in Soil

Soil inorganic nitrogen is commonly less than 2% of the total nitrogen of surface soils and undergoes rapid changes in composition and quantity. Inorganic nitrogen varies widely among soils, with climate, and with weather. In humid, temperate zones, soil inorganic nitrogen in surface soil is expected to be low in winter, to increase in spring and summer, and to decrease with fall rains, which move the soluble nitrogen into the depths of the soil (105). Despite being small in magnitude, the inorganic fraction is the source of nitrogen nutrition for plants. Unless supplied by fertilizers, inorganic nitrogen in soil is derived from the soil organic matter, which serves as a reserve of nitrogen for plant nutrition. Plant-available nitrogen is released from organic matter by mineralization and is transformed back into organic matter (microbial cells) by immobilization. Absorption by plants is the chief means of removal of inorganic nitrogen from soils, although nitrate leaching and denitrification, ammonium volatilization and fixation, and nitrogen immobilization lead to losses of inorganic nitrogen from soils or from the soil solution (105).

Detectable inorganic nitrogen forms in soil are nitrate, nitrite, exchangeable and fixed ammonium, nitrogen (N_2) gas, and nitrous oxide (N_2O gas) (106). Nitrate and exchangeable ammonium are important in plant nutrition. The other forms are generally not available for plant nutrition. Fixed ammonium, entrapped in clays, is a principal nitrogenous constituent of subsoils and is probably derived from parent rock materials; however, the fixed ammonium in surface soils may be of recent origin from organic matter (106). Fixed ammonium is resistant to removal from clay lattices and has little importance in plant nutrition. The gaseous constituents diffuse from the atmosphere or arise from denitrification and have no role in plant nutrition, other than in considerations of losses of nitrogen from soils (107).

Exchangeable or dissolved ammonium is available to plants, but ammonium concentrations in soils are low, usually in a magnitude of a few mg/kg or kg/ha. In well-aerated soils, ammonium is oxidized rapidly to nitrate by nitrification, so that nitrate is the major source of plant-available nitrogen in soil (108,109). Nitrite, an intermediate in nitrification, is oxidized more rapidly than ammonium (109). Hence, little ammonium or nitrite accumulates in most soils. Ammonium and nitrite are toxic to most plants (110). Toxicity of ammonium or nitrite might occur if the concentration of either rises above 50 mg N/kg in soil or in other media, especially if either is the principal source of nitrogen for plant nutrition (110,111). Nitrification is sensitive to soil acidity and is likely to be inhibited in soils under pH 5; this acidity may lead to ammonium accumulation (108).

2.5 SOIL TESTING FOR NITROGEN

Testing for plant-available soil nitrogen is difficult. This difficulty arises in part because most of the nitrogen in soil is in organic forms, which have varying rates of microbial transformation into available forms. Also, nitrate, the main form of plant-available nitrogen, is subject to leaching, denitrification, and immobilization. Many attempts have been made to develop availability indexes for release of nitrogen from organic matter and to correlate yields with tests for inorganic nitrogen in soils (93,101,112–114). Biological tests are time consuming and may give variable results if the methodology is not standardized among researchers. Chemical tests for estimating plant-available nitrogen have been empirical in approach and have had low correlations with production of mineral nitrogen and crop accumulation of nitrogen.

2.5.1 DETERMINATIONS OF TOTAL NITROGEN

The determination of nitrogen by the Kjeldahl method gives an estimation of the total nitrogen in soils (93,113). This test, often considered a chemical index, is essentially a test for total soil organic matter, since the nitrogen concentration of soil organic matter is relatively constant. This measurement does not estimate the rates of transformations of organic nitrogen into inorganic forms that are available for plants; hence, many irregularities in predicting available nitrogen occur in its use. However, considering that transformations depend on the type of organic matter, temperature, aeration, water supply, acidity, and other factors, total nitrogen is likely as informative as determination of other availability indexes. Nevertheless, determinations of availability indexes have been investigated extensively (96).

2.5.2 BIOLOGICAL DETERMINATIONS OF AVAILABILITY INDEXES

Aerobic incubation of soil samples for 2 to 4 weeks under nearly optimum conditions of microbial decomposition of organic matter and measurements of nitrogen mineralization is an extensively employed *biological procedure* for the development of an availability index (96,101,112–114). Incubated samples are tested for the amounts of nitrate, ammonium, or both forms released. Since determinations are run under nearly optimum conditions, only an estimate of the potential for mineralization is provided. Results may differ from mineralization in a field in a particular year. Determinations of indexes by anaerobic incubation involve estimations of ammonium released (115). Other biological tests involve bioassays of microbial growth or pigment production (116), chlorophyll production by algae (117), and carbon dioxide production (118).

2.5.2.1 Determination of Inorganic Nitrogen

These determinations are considered to be chemical indexes of availability of nitrogen soil organic matter. The utility of chemical indexes depends on their correlation for a broad range of soils with biological criteria, such as crop yields, nitrogen accumulation in plants, and biological indexes (101). Inorganic nitrogen is determined in an extraction of soil with water or solutions of acids, bases, chelating agents, or salts at differing concentrations and temperatures (101). Severe extractants, such as moderately concentrated (4.5 to 6 M) boiling mineral acids or bases, generally give nitrogen releases that correlate well with total soil nitrogen. However, total soil nitrogen as such is not a reliable index of nitrogen availability in soils. Also, release of nitrogen by moderate extraction procedures, such as alkaline permanganate, sodium carbonate, and molar solutions of mineral acids and bases, generally are poorly correlated with biological measurements (96,101). Relatively mild extractions with cold, hot, or boiling water or solutions of cold dilute (0.01 M) acids, bases, or salts have been used with the premise that these methods determine nitrogen of which a high proportion is derived from microbial action on the soils (101). Ammonium or nitrate may be determined in the extracts (96,105,106).

2.5.2.1.1 Ammonium

The rate-controlling step in nitrogen mineralization is the conversion of organic nitrogen into ammonium. The conversion of ammonium into nitrate is a rapid step, as a result ammonium generally does not accumulate in well-drained mineral soils. Ammonium in soil, initially present in soils at sampling, is correlated weakly with nitrogen accumulation in plants (113). Temperatures in handling and storage of soil samples are important in judging the correlation between ammonium in soils and accumulation in plants (119). Waterlogging, high acidity (pH < 5.0) or alkalinity (pH > 8.0), or use of nitrification inhibitors can lead to mineralization that stops with the formation of ammonium and hence to accumulation beyond that occurring in well-drained, mineral soils. Determination of ammonium present in soil without any manipulation generally gives better correlations with biological processes than the correlation of ammonium that accumulates with manipulation of processes that lead to ammonium accumulation.

2.5.2.1.2 Nitrate

Nitrate is the form of nitrogen that is used most commonly by plants and that may accumulate in agricultural soils. In combination with other factors, such as soil water, nitrate concentrations in soils have been used in assessments of soil fertility since the early 1900s (113,120–122). Ozus and Hanway (123) reported that nitrogen accumulation in crops during early growth was related to nitrate content in soils. Early workers related nitrate in soils to crop yields. Nitrate in soil was shown to be a reliable evaluation of soil nitrogen that is residual from previous fertilization (124–126). Recent work has related tests for nitrate in soils to prediction of the needs of crops for nitrogen fertilization. These tests are commonly called *preplant nitrate tests* and are conducted in the early spring to a soil-sampling depth of 60, 90, 120 cm, or greater.

Nitrate is a soluble form of nitrogen that is subject to downward movement in soils in humid temperate climates (105). Sometimes, soil tests for nitrate in the top 15 or 30 cm of soils have not been well correlated with crop yields because of depletion of nitrate in these zones by leaching in humid regions (113). Good correlations between soil nitrate tests and crop yields have been noted with soil samples taken from 120- to 180-cm depth in the profile. Roth and Fox (125) reported nitrate concentrations that ranged from 36 to 295 kg N/ha in the 120-cm profile following the harvest of corn. Soils fertilized with nitrogen applied at economiclly optimum amounts had nitrate concentrations ranging from 41 to 138 kg N/ha. Soils with more than 169 kg nitrate-N/ha in the 120-cm profile did not show an increase in corn yields in response to nitrogen fertilization. Jokela and Randall (124) reported that nitrate concentrations in a 150-cm profile ranged from 150 to 500 kg N/ha over a range of fertilizer treatments after corn harvest in the fall but fell by 50 to 70% by the following spring.

Nitrate concentrations vary among soils and among seasons of the year for a given soil and climate (105,127). In humid temperate climatic areas, nitrate in soils is low in the cold of winter, rises in spring and through the summer with warming of soils and falls in the fall with the rains. In unfertilized fields in the winter, nitrate in topsoil (top 30 to 60 cm) is less than 5 or 10 mg N/kg (105). The concentration can rise to 40 to 60 mg nitrate-N/kg in spring and summer. Depending on the permeability of soil, the depletion of nitrate from topsoil can be rapid with fall rains. Tillage of land can bring about an increase of nitrate, as mineralization and nitrification are increased by aeration of the soil due to tillage. Generally, the more intensive the tillage, the greater the nitrate concentrations in the soil (128–130). For example, in the 120-cm-deep soil profile, following a crop of corn, the nitrate in conventionally tilled soils (100 to 120 kg N/ha) was twice that in the profile of soils cropped in a no-tillage system (129). In dry seasons, soil nitrate can be very low due to low microbiological activity, perhaps less than 10 mg N/kg, but increases as rain falls and mineralization and nitrification result in the wetted soil. In some cases, if the subsoil contains nitrate, nitrate may rise with capillary action and accumulate in dry surface soils. Absorption by plants is a principal path of removal of nitrate from soils. Removal is unique with various soils and crops (105). Perennial crops having a developed root system can absorb nitrate as soon as conditions are favorable for plant growth. Grassland soils generally are low in nitrate throughout the year. However, annual crops do not absorb much nitrate from soils until the root systems are developed.

Many soil test recommendations for correlation of soil nitrate with crop yields require soil sampling to a minimum depth of 60 cm (113). Sampling to this depth involves considerable costs, and attempts have been made to develop a test based on shallower sampling. Alvarez et al. (131) developed prediction equations that related nitrate in the top 30 cm stratum to that in the top 60 cm stratum. Recent research has shown good correlations between crop yields and concentrations of nitrate in the surface 30 cm layer of soils early in the growing season (132–135). Determination of the amount of nitrate in the upper stratum of soil early in the season has led to the development of a test called the *early season nitrate test* or *pre-sidedress soil nitrate test* (PSNT).

The basis of the PSNT is the concentration of nitrate in the surface 30 cm of soils at the time that a crop starts rapid growth, for example, when corn is 30 cm tall (133,134). The amount of nitrate in the soil at this depth at this time is an assessment of the amount of nitrogen available for crop growth for the remainder of the season and of the need for nitrogen fertilization. The critical concentration of soil nitrate for the PSNT is the concentration above which yields are not expected to increase with additional nitrogen fertilization. For corn production, Sims et al. (135) in Delaware reported that the PSNT test identified nitrogen-deficient or nitrogen-sufficient sites with about 70% success. Binford et al. (132) in Iowa determined that the critical concentration of nitrate for corn was 23 to 26 mg N/kg for a 30 cm depth. Sampling 60 cm deep improved correlations between corn grain yields and soil nitrate, but it was felt that the improvement did not justify the additional costs of deep sampling. The critical concentration for the 60 cm depth was 16 to 19 mg N/kg soil. Other research has given similar results. Meisinger et al. (136) in Maryland determined a critical nitrate concentration of 22 mg N/kg with the PSNT successfully identifying nitrogen-sufficient sites across a range of textures, drainage classes, and years. Including ammonium in the analysis slightly improved the predictive use of the test (136). Heckman et al. (137) in New Jersey reported a critical nitrate concentration at the 30 cm depth to be 22 mg N/kg for corn. Evanylo and Alley (138) in Virginia reported critical nitrate concentrations of 18 mg N/kg for corn and noted that the PSNT was applicable to soils without regard to texture or physiographic region. Also for corn, Sainz-Rozas et al. (139) in Argentina reported a critical nitrate concentration of 17 to 27 mg N/kg at the 30 cm depth. They also reported that there was no improvement in reliability if the test was done on samples to 60 cm depth or with the inclusion of ammonium in the determinations. Critical concentrations, similar in magnitude to those for corn have been reported for sweet corn (*Zea mays rugosa* Bonaf.) (140), lettuce (*Lactuca sativa* L.), celery (*Apium graveolens dulce* Pers.) (141), cabbage (*Brassica oleracea capitata* L.) (142), and tomato (*Lycopersicon esculentum* Mill.) (143).

If the concentration of nitrate is below the critical concentration, fertilization of the crops is necessary. However, the need to collect soil samples during the growing season has limited the usage of the PSNT. Fertilization is delayed until the results of the PSNT are obtained, and bad weather can delay applications of nitrogen.

2.5.2.1.3 Amino Sugars

Fractionation of soil hydrolysates has been used to determine a labile pool of organic nitrogen in soil and to relate this fraction to crop responses to nitrogen fertilizers (102,144). The results of most of these studies have shown little variation among soil types or cultivation patterns in the partitioning of hydrolyzable soil nitrogen into various nitrogenous components and the capacity of soil organic matter to form nitrate. The uniformity among soils was attributed in part to errors in analysis (145,146). Mulvaney and Khan (147) developed a diffusion method for accurately determining amino sugar nitrogen in soil hydrolysates. Mulvaney et al. (145) noted that hydrolysates (6 M HCl) of soils in which crops were nonresponsive to nitrogen fertilization had higher concentrations of amino sugars (e.g., glucosamine, galactosamine, mannosamine, muramic acid) than did hydrolysates of soils in which crops responded to nitrogen fertilization. They reported no consistent differences among the total nitrogen, the ammonium nitrogen, or the amino acid nitrogen fraction of the soil hydrolysate. The amounts of amino sugars were related to mineralization of soil organic nitrogen, since production of inorganic nitrogen upon aerobic incubation of the nonresponsive soils was much greater than that in the responsive soils (145). Concentrations of amino sugars were correlated with response to fertilizer nitrogen applied. Mulvaney et al. (145) classified soils with more than 250 mg amino sugar nitrogen per kg as being nonresponsive and those with less than 200 mg amino sugar nitrogen per kg as being responsive to nitrogen fertilization. Khan et al. (146) developed a simpler test for determining amino sugar nitrogen than the processes involving soil hydrolysis. The simpler test involved soil being treated with base (2 M NaOH), followed by heating (50°C) to release ammonia, and then determining the amount of ammonia releases by volumetric methods. This method determined ammonium and amino sugar nitrogen without liberating substantial nitrogen from amino acids and none from nitrate or nitrite. Test values for soils nonresponsive to nitrogen fertilization were 237 to 435 mg N/kg and for responsive soils were 72 to 223 mg N/kg soil.

Amino sugars may constitute 5 or 6% of the humic substances in soils (148). Variations in kind and amount of amino sugars have been noted with climate and with cultivation of soils (149,150).

2.6   NITROGEN FERTILIZERS

Soils have little capacity to retain oxidized forms of nitrogen, and ammonium accumulation in soils is small; consequently, most of the soil nitrogen is associated with organic matter. Release of nitrogen from organic matter is slow and unpredictable. If soil organic matter is depleted, as occurs in cultivated soils, nitrogen for plant growth is limited. Nitrogen is usually the most deficient nutrient in cultivated soils of the world, and fertilization of these soils with nitrogen is required. To maintain or increase productivity of soils, worldwide consumption of nitrogen fertilizers continues to increase with time (Figure 2.3). However, the consumption of phosphorus and potassium fertilizers has leveled.

Anhydrous ammonia (NH_3 gas) is the starting product for manufacture of most nitrogen fertilizers. Anhydrous ammonia is manufactured from hydrogen and nitrogen gases by the Haber process (Haber–Bosch process). The reaction is performed at high temperature (400 to 500°C) and high pressure (300 to 1000 atm) in the presence of a catalyst (iron or other metal) (151–153). The nitrogen gas is obtained from the air, which is about 79% nitrogen by volume, and the hydrogen is obtained from natural gas (methane), oil, coal, water, or other sources.

Jones (152) and Moldovan et al. (154) describe the production of other nitrogen fertilizers from ammonia. A brief summary of these processes follows. Nitric acid, produced from ammonia, is another basic material in the manufacture of nitrogen fertilizers. To produce nitric acid, compressed ammonia and air are heated in the presence of a catalyst and steam. The nitric acid can be reacted with ammonia to produce ammonium nitrate. Sodium nitrate is the product of the reaction of nitric acid with sodium bicarbonate. Sodium nitrate also is produced from caliche (Chilean saltpeter), which is a mineral that contains sodium nitrate and various salts of sodium, calcium, potassium, and magnesium. Sodium nitrate, sometimes called Chilean nitrate, is one of the earliest commercial nitrogen fertilizers marketed. Until 1929, all of the sodium nitrate marketed was extracted from Chilean saltpeter (154). Urea is manufactured chiefly by combining ammonia with carbon dioxide under high pressure. Ammonium sulfate is manufactured by the reaction of ammonia with sulfuric acid, gypsum, or sulfur dioxide.

The merits of nitrate and ammonium fertilizers have been researched and reviewed extensively (155–166). Many manufactured fertilizers and most organic fertilizers are ammonical; however, the ammonium that is inherent in the fertilizer or that is released upon contact with soils is soon oxidized to nitrate, unless nitrification is inhibited (167–171). Nitrification inhibitors may be employed with ammoniacal fertilizers to restrict losses of nitrogen from soils by leaching or denitrification.

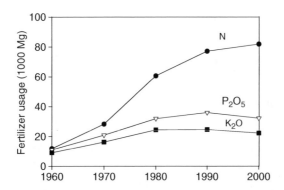

FIGURE 2.3   Worldwide consumption of nitrogen, phosphorus, and potassium in fertilizers for the period 1960–2000. Units of Mg are 1000 kg or one metric ton. (Adapted from http://www.fertilizer.org/ifa/statistics/indicators/tablen.asp.)

2.6.1 PROPERTIES AND USE OF NITROGEN FERTILIZERS

The nitrogen concentrations of the following fertilizers have been rounded to values of commonly marketed grades.

2.6.1.1 Anhydrous Ammonia (82% N)

Anhydrous ammonia is the most-used nitrogen-containing fertilizer for direct application to land in the United States (152). Worldwide, consumption of anhydrous ammonia is ranked fourth or fifth among nitrogen fertilizers (Table 2.7). In agriculture, anhydrous gaseous ammonia is compressed into a liquid and is applied under high pressure with a special implement by injection at least 15 cm deep into a moist soil. The ammonia gas reacts with water to form ammonium ions, which can be held to clay or organic matter. If the ammonia is not injected deeply enough or soil is too wet or dry, ammonia can be lost by volatilization. Anhydrous ammonia is usually the cheapest source of nitrogen, but equipment and power requirements of the methods of application are specific and high.

2.6.1.2 Aqua Ammonia (21% N)

Aqua ammonia is ammonia dissolved in water under low pressure. Aqua ammonia must be incorporated into land to avoid losses of nitrogen by ammonia volatilization; however, it needs not be incorporated as deeply as anhydrous ammonia.

2.6.1.3 Urea (46% N)

Urea is the most widely used dry nitrogen fertilizer in the world (Table 2.7). After application to soils, urea is converted into ammonia, which can be held in the soil or converted into nitrate. Ammonia volatilization following fertilization with urea can be substantial, and if urea is applied to the surface

---

TABLE 2.7
Worldwide Nitrogen Fertilizer Consumption in the Year 2000

Nitrogen Fertilizer	Nitrogen Fertilizer Usage (Metric Tons)
Straight N Fertilizer	
Urea	41042
Ammonium nitrate	5319
Calcium ammonium nitrate	4768
N solutions	3812
Anhydrous ammonia	3581
Ammonium sulfate	2738
Other	7907
Total straight	69168
Mixed N Fertilizer	
NPK-N	6347
Ammonium phosphate	4631
Other NP-N	1656
NK-N	74
Total mixed	12708
Total N fertilizer	81880

Source: Compiled from http://www.fertilizer.org/ifa/

---

of the land, considerable loss of nitrogen can occur (172,173). Hydrolysis of urea by urease produces ammonium carbonate. With surface-applied urea, alkalinity of pH 9 or higher can develop under the urea granule or pellet, and ammonia will volatilize into the air. Volatilization occurs on bare ground, on debris, or on plant leaves. Urea is readily soluble in water, and rainfall or irrigation after its application move it into the soil and lessens volatilization losses. Use of urease inhibitors has been suggested to lessen the volatilization losses of ammonia from surface-applied urea (174). Manufactured urea is identical to urea in animal urine.

Calcium nitrate urea (calurea, 34% N, 10% Ca) is a double-compound fertilizer of calcium nitrate and urea to supply calcium and nitrogen (152).

Several derivatives of urea are marketed as slow-release fertilizers (175,176). Urea formaldehyde (ureaform, 38% N) is a slow-release fertilizer manufactured from urea and formaldehyde and is used for fertilization of lawns, turf, container-grown plants, and field crops (177–180). Urea formaldehyde is also a glue and is used for the manufacture of plywood and particle board (181,182). Dicyandiamide (cyanoguanidine) (66% N) is a nitrogen fertilizer but is used most commonly as an additive (2% of the total N fertilizer) as a nitrification inhibitor with urea (153,183–185). Sulfur-coated urea (186,187) is a slow-release formulation (30–40% N) used as a fertilizer for field crops, orchards, and turfgrass (175,177,188–191).

Isobutylidene diurea (IBDU) is similar to urea formaldehyde, but contains 32% nitrogen. However, utilization of IBDU is less dependent on microbial activity than urea formaldehyde, as hydrolysis proceeds rapidly following dissolution of IBDU in water (175). Nitrogen is released when soil moisture is adequate. IBDU is used most widely as a lawn fertilizer (176,192). Its field use is to restrict leaching of nitrogen (181).

Methylene ureas are a class of sparingly soluble products, which were developed during the 1960s and 1970s. These products contain predominantly intermediate chain-length polymers. The total nitrogen content of these polymers is 39 to 40%, with between 25 and 60% of the nitrogen present as cold-water-insoluble nitrogen. This fertilizer is used primarily in fertilization of turfgrass, although it has been used with other crops on sandy soils or where leaching of nitrate is an environmental concern (176,191,193).

2.6.1.4 Ammonium Nitrate (34% N)

Ammonium nitrate is a dry material sold in granular or prilled form. It can be broadcasted or sidedressed to crops and can be left on the surface or incorporated. It does not give an alkaline reaction with soils; hence, it does not volatilize readily. However, incorporation is recommended with calcareous soils. Ammonium nitrate is decreasing in popularity because of storage problems, e.g., with fire and explosion.

Calcium ammonium nitrate (ammonium nitrate limestone, about 20% N and 6% Ca) is a mixture of ammonium nitrate and limestone. This fertilizer is not acid-forming and is used to supply nitrogen and calcium to crops (152).

2.6.1.5 Ammonium Sulfate (21% N)

Ammonium sulfate is marketed as a dry crystalline material. It is recommended for use on alkaline soils where it may be desirable to lower soil pH. Nitrification of ammonium is an acidifying process. Ammonium sulfate can be broadcasted or sidedressed. It can left on surfaces or incorporated, although on calcareous soils watering in or incorporating is recommended to avoid ammonia volatilization (176).

2.6.1.6 Nitrogen Solutions (28–32% N)

These fertilizers are mixtures of ammonium nitrate and urea dissolved in water. In the solutions, half of the nitrogen is supplied as urea, and half is supplied as ammonium nitrate. Because of the difficulties in handling, urea and ammonium nitrate should not be mixed together in dry form. The solution acts once the dry materials are applied to the soil. Ammonia volatilization may be substantial during warm weather, especially with surface application. The solutions should be watered into the soil and should not be applied to foliage.

2.6.1.7 Ammonium Phosphates (10–21% N)

Ammonium phosphates are important phosphorus-containing fertilizers because of their high concentrations of phosphorus and water solubility. Diammonium phosphate (commonly 18% N, 46% P_2O_5) is a dry granular or crystalline material. It is a soil-acidifying fertilizer and is useful on calcareous soils. It should be incorporated into the soil. It is a common starter fertilizer and is a common component of greenhouse and household fertilizers. Monoammonium phosphate (commonly 11% N, 48% P_2O_5) has uses similar to those of diammonium phosphate. Ammonium polyphosphate (10% N, 34% P_2O_5) is marketed as a solution. Its use is similar to that of monoammonium phosphate and diammonium phosphate. Ammonium phosphates are made by reaction of ammonia with orthophosphoric acid (mono- and diammonium salts) or with superphosphoric (pyrophosphoric) acid (152).

2.6.1.8 Other Inorganic Nitrogen Fertilizers

Many other nitrogen-containing fertilizers include double-salt mixtures such as ammonium nitrate sulfate (30% N), ammonium phosphate nitrate (25% N), urea ammonium phosphate (25–34% N), nitric phosphate, and ammoniated superphosphate (8% N) (152). These materials are used in the manufacture of mixed N-P-K fertilizers or for special needs in soil fertility.

2.6.1.9 Organic Nitrogen Fertilizers (0.2–15% N)

Although naturally occurring, sodium nitrate may not be recognized as an organic fertilizer. Most organic fertilizers are derived from plant and animal sources and are proteinaceous

TABLE 2.8
Representative Nitrogen Concentrations and Mineralization of Some Organic Fertilizers

Fertilizer	% N (Dry Mass)[a]	Mineralization[b]
Feather meal, hair, wool, silk	15	Moderate–Rapid
Dried blood, blood meal	12	Rapid
Fish scrap (dry)	9	Moderate–Rapid
Tankage, animal	8	Moderate–Rapid
Seed meals[c]	6	Rapid
Poultry manure	2–3	Moderate–Rapid
Livestock manure	1–2	Slow
Sewage biosolids	1–4	Slow
Bone meal, steamed	1	Moderate–Rapid
Kelp	0.7	Slow
Compost	0.5–1	Slow

[a]Concentrations will vary from these representative values, depending on the handling of the products, nutrition of livestock, and source of materials.

[b]Mineralization rate will vary with the products. Rapid mineralization is more than 70% of the organic N expected to be mineralized in a growing season; moderate is 50 to 70% mineralization; and slow is less than 50% mineralization.

[c]Includes by-products such as cottonseed meal, soybean meal, linseed meal, corn gluten meal, and castor pomace.

materials. The fertilizer industry started with meat and other food processors, who wanted to dispose of and find a use for wastes and by-products (152,194). Around 1900, about 90% of nitrogen fertilizer was derived from proteinaceous wastes and by-products, but today usage has declined to less than 1%. Organic materials range from less than 1 to about 15% N compared with the chemical sources described above, which range upward to over 80% N. Costs of handling, shipping, and spreading of the bulky, low-analysis organic materials have led to their decline in usage with time. Also, many of the proteinaceous by-products of food processing have higher value as feeds for poultry and livestock than as fertilizers (194,152). Nevertheless, demand for organic fertilizers remains, as organic farmers require these products in the maintenance of soil fertility on their cropland (195).

The value of organic nitrogen fertilizers depends on their rate of mineralization, which is closely related to their nitrogen concentration (152,195,196). Generally, the more nitrogen in the fertilizer, the faster the rate of mineralization. Some common organic fertilizers are listed in Table 2.8.

REFERENCES

1. H.S. McKee. *Nitrogen Metabolism in Plants.* London: Oxford University Press, 1962, pp. 1–18.
2. H.S. Reed. *A Short History of the Plant Sciences.* Waltham, Mass.: Chronica Botanica Co., 1942, pp. 241–254.
3. E.W. Russell. *Soil Conditions and Plant Growth.* 9th ed. New York: Wiley, 1973, pp. 1–23.
4. D.J.D. Nicholas. Inorganic nutrition of microorganisms. In: F.C. Steward, ed. *Plant Physiology: A Treatise Vol. III.* New York: Academic Press, 1963, pp. 363–447.
5. L.H. Beevers, R.H. Hageman. Nitrate reduction in higher plants. *Annu. Rev. Plant Physiol.* 20:495–522, 1969.
6. H.J. Evans, A. Nason. Pyridine nucleotide nitrate reduction from extracts of higher plants. *Plant Physiol.* 28:233–254, 1953.
7. B.J. Miflin. The location of nitrate reductase and other enzymes related to amino acid biosynthesis in the plastids of roots and leaves. *Plant Physiol.* 54:550–555, 1974.
8. A. Oaks, B. Hirel. Nitrogen metabolism in roots. *Annu. Rev. Plant Physiol.* 36:345–365, 1985.
9. C.G. Bowsher, D.P. Hucklesby, M.J. Emes. Nitrite reduction and carbohydrate metabolism in plastids purified from roots. *Planta* 177:359–366, 1989.
10. L.P. Solomonson, M.J. Barber. Assimilatory nitrate reductase: functional properties and regulation. *Annu. Rev. Plant Physiol. Plant Mol. Biol.* 41:225–253, 1990.
11. M. Andrews. Partitioning of nitrate assimilation between root and shoot of higher plants. *Plant Cell Environ.* 9:511–519, 1986.
12. M. Andrews. Nitrate and reduced-N concentrations in the xylem sap of *Stellaria media, Xanthium strumarium* and six legume species. *Plant Cell Environ.* 9:605–608, 1986.
13. M. Andrews, J.D. Morton, M. Lieffering, L. Bisset. The partitioning of nitrate assimilation between root and shoot of a range of temperature cereals and pasture grasses. *Ann. Bot. London* 70:271–276, 1992.
14. W.J. Hunter, C.J. Fahring, S.R. Olsen, L.K. Porter. Location of nitrate reduction in different soybean cultivars. *Crop Sci.* 22:944–948, 1982.
15. H. Marschner. *Mineral Nutrition of Higher Plants.* 2nd ed. San Diego: Academic Press, 1995, pp. 229–265.
16. L. Salsac, S. Chaillou, J.F. Morot-Gaudry, C. Lesaint, E. Jolivet. Nitrate and ammonium nutrition in plants. *Plant Physiol. Biochem.* 25:805–812, 1987.
17. A. Bloom, S.S. Sukrapanna, R.L. Warner. Root respiration associated with ammonium and nitrate absorption and assimilation by barley. *Plant Physiol.* 99:1294–1301, 1992.
18. A.C. Baron, T.H. Tobin, R.M. Wallsgrove, A.K. Tobin. A metabolic control analysis of the glutamine synthetase/glutamate synthase cycle in isolated barley (*Hordeum vulgare* L.) chloroplasts. *Plant Physiol.* 105:415–424, 1994.
19. D.G. Blevins. An overview of nitrogen metabolism in higher plants. In: J.E. Poulton, J.T. Romeo, E.E. Conn, eds. *Plant Nitrogen Metabolism.* New York: Plenum Press, 1989, pp. 1–41.

20. A. Bravo, J. Mora. Ammonium assimilation in *Rhizobium phaseoli* by the glutamine synthetase-glutamate synthase pathway. *J. Bacteriol.* 170(2):980–984, 1988.

21. B. Hirel, C. Perrot-Rechenmann, A. Suzuki, J. Vidal, P. Gadal. Glutamine synthetase in spinach leaves. Immunological studies and immunocytochemical localization [*Spinacia oleracea*]. *Plant Physiol.* 69:983–987, 1982.

22. O.A.M. Lewis, S. Chadwick, J. Withers. The assimilation of ammonium by barley roots. *Planta* 159:483–486, 1983.

23. J.M. Ngambi, P. Amblard, E. Bismuth, M.L. Champigny. Study of enzymatic activities of nitrate reductase and glutamine synthetase related to assimilation of nitrates in millet *Pennisetum americanum* 23 DB. *Can. J. Bot.* 59:1050–1055, 1981.

24. R.M. Wallsgrove. The roles of glutamine synthetase and glutamate synthase in nitrogen metabolism of higher plants. In: W.R. Ulrich, ed. *Inorganic Nitrogen Metabolism*. Berlin: Springer Verlag, 1987, pp. 137–141.

25. K.C. Woo, J.F. Morot-Gaudry, R.E. Summons, C.B. Osmond. Evidence for the glutamine synthetase/glutamate synthase pathway during the photorespiratory nitrogen cycle in spinach leaves [*Spinacia oleracea*]. *Plant Physiol.* 70:1514–1517, 1982.

26. J.F. Seelye, W.M. Borst, G.A. King, P.J. Hannan, D. Maddocks. Glutamine synthetase activity, NH_4^+ accumulation and growth of callus cultures of *Asparagus officinalis* L. exposed to high NH_4^+ or phosphinothricin. *J. Plant Physiol.* 146:686–692, 1995.

27. A. Wild, H. Sauer, W. Ruhle. The effects of phosphinothricin (glufosinate) on photosynthesis. I. Inhibition of photosynthesis and ammonia accumulation. *Zeitschrift fur Naturforschung-Section C-Biosciences* 42c:263–269, 1987.

28. A.L. Lehninger. *Biochemistry*. New York: Worth Publishers, 1975, pp. 1–1104.

29. F.C. Steward, D.J. Durzan. Metabolism of nitrogen compounds. In: F.C. Steward, ed. *Plant Physiology: A Treatise. Vol IVA: Metabolism: Organic Nutrition and Nitrogen Metabolism*. New York: Academic Press, 1965, pp. 379–686.

30. F.G. Gregory, P.K. Sen. Physiological studies in plant nutrition. VI. The relation of respiration rate to the carbohydrate and nitrogen metabolism of the barley leaf, as determined by nitrogen and potassium deficiency. *Ann. Bot. N.S.* 1:521–561, 1937.

31. E.W. Yemm. The respiration of plants and their organs. In: F.C. Steward, ed. *Plant Physiology: A Treatise*. New York: Academic Press, 1965, pp. 231–310.

32. J.A. Helleburst , R.G.S. Bidwell. Protein turnover in wheat and snapdragon leaves. *Can. J. Bot.* 41:961–983,1963.

33. D. Racusen, M. Foote. Protein turnover rate in bean leaf discs. *Plant Physiol.* 37:640–642, 1960.

34. A. Trewavas. Determination of the rates of protein synthesis and degradation in *Lemma minor*. *Plant Physiol.* 49:40–46, 1972.

35. A. Trewavas. Control of the protein turnover rates in *Lemma minor*. *Plant Physiol.* 49:47–51, 1972.

36. W.H. Pearsall. The distribution of the insoluble nitrogen in *Beta* leaves of different ages. *J. Exp. Biol.* 8:279–285, 1931.

37. W. Bors, C. Langebartels, C. Michel, H. Sandermann, Jr. Polyamines as radical scavengers and protectants against ozone damage. *Phytochemistry* 28:1589–1595, 1989.

38. P.T. Evans, R.L. Malmberg. Do polyamines have roles in plant development? *Annu. Rev. Plant Physiol. Plant Mol. Biol.* 40:235–269, 1989.

39. A.W. Galston, R. Kaur-Sawhney. Polyamines as endogenous growth regulators. In: P.J. Davies, ed. *Plant Hormones and Their Role in Plant Growth and Development*. Boston: Martinus Nijhoff, 1987, pp. 280–295.

40. A.W. Galston, R.K. Sawhney. Polyamines in plant physiology. *Plant Physiol.* 94:406–410,1990.

41. R. Krishnamurthy, K.A. Bhagwat. Polyamines as modulators of salt tolerance in rice cultivars. *Plant Physiol.* 91:500–504, 1989.

42. K.A. Nielsen. Polyamine content in relation to embryo growth and dedifferentiation in barley (*Hordeum vulgare* L.). *J. Exp. Bot.* 41:849–854, 1990.

43. R.A. Saftner, B.G. Baldi. Polyamine levels and tomato fruit development: possible interaction with ethylene. *Plant Physiol.* 92:547–550, 1990.

44. K.A. Corey, A.V. Barker. Ethylene evolution and polyamine accumulation by tomato subjected to interactive stresses of ammonium toxicity and potassium deficiency. *J. Am. Soc. Hortic. Sci.* 114:651–655, 1989.

45. T.A. Smith, C. Sinclair. The effect of acid feeding on amine formation in barley. *Ann. Bot. (London) [N.S.]* 31:103–111, 1967.

46. H.A. Zaidan, F. Broetto, E.T. De Oliveira, L.A. Gallo, O.J. Crocomo. Influence of potassium nutrition and the nitrate/ammonium ratio on the putrescine and spermidine contents in banana vitroplants. *J. Plant Nutr.* 22:1123–1140, 1999.

47. T.C. Tucker. Diagnosis of nitrogen deficiency in plants. In: R.D. Hauck, ed. *Nitrogen in Crop Production*. Madison, Wis.: American Society of Agronomy, 1984, pp. 249–262.

48. G.C. Cresswell, R.G. Weir. *Plant Nutrient Disorders 2. Tropical Fruit and Nut Crops*. Melbourne: Inkata Press, 1995, pp. 1–112.

49. G.C. Cresswell, R.G. Weir. *Plant Nutrient Disorders 4. Pastures and Field Crops*. Melbourne: Inkata Press, 1995, pp. 1–126.

50. G.C. Cresswell. R.G. Weir. *Plant Nutrient Disorders 5. Ornamental Plants and Shrubs*. Melbourne: Inkata Press, 1998, pp. 1–200.

51. R.G. Weir, G.C. Cresswell. *Plant Nutrient Disorders 1. Temperate and Subtropical Fruit and Nut Crops*. Melbourne: Inkata Press, 1993, pp. 1–93.

52. R.G. Weir, G.C. Cresswell. *Plant Nutrient Disorders 3. Vegetable Crops*. Melbourne: Inkata Press, 1993, pp. 1–104.

53. H.B. Sprague. *Hunger Signs in Crops. A Symposium*. New York: McKay, 1964, pp. 1–461.

54. D. Spencer, J.V. Possingham. The effect of nutrient deficiencies on the Hill reaction of isolated chloroplasts from tomato. *Aus. J. Biol. Sci.* 13:441–445, 1960.

55. H. Ji, Tae J.L. Hess, A.A. Benson. Chloroplast membrane structure. I. Association of pigments with chloroplast lamellar protein. *Biochim. Biophys. Acta* 1504:676–685, 1968.

56. T. Oku, G. Tomita. Protochlorophyllide holochrome 1. Plastoquinone attached to a Protochlorophyllide holochrome. *Photosynthetica* 4:295–301, 1970.

57. K. Shinashi, H. Satoh, A. Uchida, K. Nakayama, M. Okada, I. Oonishi. Molecular characterization of a water-soluble chlorophyll protein from main veins of Japanese radish. *J. Plant Physiol.* 157:255–262, 2000.

58. J.P. Thornber, J.C. Stewart, M.W.C. Hatton, J.C. Bailey. Nature of chloroplast lamellae. II. Chemical composition and further physical properties of two chlorophyll-protein complexes. *Biochemistry* 6:2006–2014, 1967.

59. C. Tietz, F. Jelezko, U. Gerken, S. Schuler, A. Schubert, H. Rogl, J. Wrachtrup. Single molecule spectroscopy on the light-harvesting complex II of higher plants. *Biophys. J.* 81:556–562, 2001.

60. A.V. Barker. Nutritional factors in photosynthesis of higher plants. *J. Plant Nutr.* 1:309–342, 1979.

61. I. Terashima, J.R. Evans. Effects of light and nitrogen nutrition on the organization of the photosynthetic apparatus of spinach. *Plant Cell Physiol.* 29:143–155, 1988.

62. W.W. Thomson, T.W. Weier. The fine structure of chloroplasts from mineral-deficient leaves of *Phaseolus vulgaris. Am. J. Bot.* 49:1047–1055, 1962.

63. M. Vesk, J.V. Possingham, F.V. Mercer. The effect of mineral nutrient deficiencies on structure of the leaf cells of tomato, spinach, and maize. *Aus. J. Bot.* 14:1–18, 1966.

64. R.J. Deshaies, L.E. Fish, A.T. Jagendorf. Permeability of chloroplast envelopes to Mg^{2+}. Effects on protein synthesis. *Plant Physiol.* 74:775–782, 1984.

65. J.D. Hall, R. Barr, A.H. Al-Abbas, F.L. Crane. The ultrastructure of chloroplants in mineral-deficient maize leaves. *Plant Physiol.* 50:404–409, 1972.

66. G.S. Puritch, A.V. Barker. Structure and function of tomato leaf chloroplasts during ammonium toxicity. *Plant Physiol.* 42:1229–1238, 1967.

67. D.W. Goodall, F.G. Gregory. Chemical composition of plants as an index of their nutritional status. Technical Communication No. 17. Imperial Bureau of Horticulture and Plantation Crops, East Malling, Kent, England. 1947, pp. 1–167.

68. T. Pfeiffer, W. Simmermacher, A. Rippel. The content of nitrogen, phosphorus, and potassium in oat plants under differing conditions and their relationships to the nutrient supply for obtaining high yields (in German). *J. Fur Landwirtschaft* 67:1–57, 1942.

69. T.E. Bates. Factors affecting critical nutrient concentrations in plants and their evaluation: a review. *Soil Sci.* 112:116–130, 1971.

70. P. Macy. The quantitative mineral nutrient requirements of plants. *Plant Physiol.* 11:749–764, 1936.

71. A. Ulrich. Plant tissue analysis as a guide in fertilizing crops. In: H.M. Reisenhauer, ed. *Soil and Plant Tissue Testing in California*. University of California Bulletin 1976, 1879, pp. 1–4.

72. H.A. Mills, J.B. Jones, Jr. *Plant Analysis Handbook II*. Athens, Ga.: MicroMacro Publishing, 1996, pp. 155–414.

73. P.B. Vose. Varietal differences in plant nutrition. *Herbage Abstr.* 33(1):1–13, 1963.

74. J.B. Jones, Jr., W.J.A. Steyn. Sampling, handling, and analyzing plant tissue samples. In: L.M. Walsh, J.D. Beaton, eds. *Soil Testing and Plant Analysis*. Madison, Wis.: Soil Science Society of America, 1973, pp. 249–270.

75. J.J. Hanway, J.B. Herrick, T.L. Willrich, P.C. Bennett, J.T. McCall. *The Nitrate Problem*. Iowa Agric. Exp. Stn. Special Report No. 1963, 34:1–20.

76. D.N. Maynard, A.V. Barker, P.L. Minotti, N.H. Peck. Nitrate accumulation in vegetables. *Adv. Agron.* 28:71–118, 1976.

77. L.E. Schrader. Uptake, accumulation, assimilation, and transport of nitrogen in higher plants. In: D.R. Nielsen, J.G. MacDonald, eds. *Nitrogen in the Environment. Vol. 2. Soil-Plant-Nitrogen Relationships*. New York: Academic Press, 1978, pp. 101–141.

78. G.D. Binford, A.M. Blackmer, N.M. El-Hout. Optimal concentrations of nitrate in corn stalks at maturity. *Agron. J.* 84:881–887, 1992.

79. S.M. Brouder, D.B. Mengel, B.S. Hoffman. Diagnostic efficiency of the blacklayer stalk nitrate and grain nitrogen tests for corn. *Agron. J.* 92:1236–1247, 2000.

80. B.A. Hooker, T.F. Morris. End-of-season corn stalk test for excess nitrogen in silage corn. *J. Prod. Agric.* 12:282–288,1999.

81. G.M. Geraldson, G.R. Klacan, O.A. Lorenz. Plant analysis as an aid in fertilizing vegetable crops. In: L.M. Walsh, J.D. Beaton, eds. *Soil Testing and Plant Analysis*, revised edition. Madison, Wis.: Soil Science Society of America, 1973, pp. 365–379.

82. A. Ulrich, F.J. Hills. Plant analysis as an aid in fertilizing sugar crops: Part I. Sugar beets. In: L.M. Walsh, J.D. Beaton, eds. *Soil Testing and Plant Analysis*, revised edition. Madison, Wis.: Soil Science Society of America, 1973, pp. 271–288.

83. M.B. Sumner, Interpretation of foliar analysis for diagnostic purposes. *Agron. J.* 71:343–348, 1979.

84. M.B. Sumner. Preliminary N, P, and K foliar diagnostic norms for soybeans. *Agron. J.* 69:226–230, 1977.

85. J.B. Jones, Jr., B. Wolf, H.A. Mills. *Plant Analysis Handbook*. Athens, Ga.: Micro-Macro Publishing, Inc., 1991, pp. 205–213.

86. R.B. *Beverly. A Practical Guide to the Diagnosis and Recommendation Integrated System*. Athens, Ga: Micro Macro Publishing, Inc., 1991, pp. 1–70.

87. W.B. Hallmark, J.L. Walworth, M.E. Sumner, C.J. DeMooy, J. Pesek, K.P. Shao. Separating limiting from non-limiting nutrients. *J. Plant Nutr.* 10:1381–139, 1987.

88. J.L. Walworth, M.E. Sumner, R.A. Isaac, C.O. Plank, Preliminary DRIS norms for alfalfa in the Southeastern United States and a comparison with Midwestern norms. *Agron. J.* 78:1046–1052, 1986.

89. F.J. Stevenson. Origin and distribution of nitrogen in the soil. In: W.V. Bartholomew, F.E. Clark, eds. *Soil Nitrogen*. Madison, Wis.: American Society of Agronomy, 1965, pp. 1–42.

90. F.J. Stevenson. Origin and distribution of nitrogen in the soil. In: F.J. Stevenson, ed. *Nitrogen in Agricultural Soils*. Madison, Wis.: American Society of Agronomy, 1982, pp. 1–42.

91. R.D. Hauck, K.K. Tanji. Nitrogen transfers and mass balances. In: F.J. Stevenson, ed. *Soil Nitrogen*. Madison, Wis.: American Society of Agronomy, 1985, pp. 891–925.

92. O. Schreiner, B.E. Brown. Soil nitrogen. In: *United States Department of Agriculture, Soils and Men, Yearbook of Agriculture, 1938*. Washington, DC: United States Government Printing Office, 1938, pp. 361–376.

93. J.M. Bremner. Organic nitrogen in soils. In: C.A. Black, ed. *Methods of Soil Analysis*. Madison, Wis.: American Society of Agronomy, 1965, pp. 93–149.

94. F.J. Stevenson. Organic forms of soil nitrogen, In: F.J. Stevenson, ed. *Nitrogen in Agricultural Soils*. Madison, Wis.: American Society of Agronomy, 1982, pp. 67–122.

95. R.B. Bradstreet. *The Kjeldahl Method for Organic Nitrogen*. New York: Academic Press, 1965, pp. 1–166.

96. J.M. Bremner. Nitrogen availability indexes. In: W.V. Bartholomew, F.E. Clark, eds. *Soil Nitrogen*. Madison, Wis.: American Society of Agronomy, 1965, pp. 1324–1345.

97. A.V. Barker, M.L. Stratton, J.E. Rechcigl. Soil and by-product characteristics that impact the beneficial use of by-products. In: W.A. Dick, ed. *Land Application of Agricultural, Industrial, and Municipal By-Products*. Madison, Wis.: Soil Science Society of America, 2000, pp. 169–213.

98. H.A. Ajwa, M.A. Tabatabai. Decomposition of different organic materials in soils. *Bio. Fertile Soils* 18:175–182, 1994.

99. E.A. Paul, F.E. Clark. *Soil Microbiology and Biochemistry*. 2nd ed. San Diego: Academic Press, 1996, pp. 1–340.

100. S. Waksman. *Humus*. Baltimore: Williams and Wilkins, 1936, pp. 1–194.

101. G. Stanford. Assessment of soil nitrogen availability. In: F.J. Stevenson, ed. *Nitrogen in Agricultural Soils*. Madison, Wis.: American Society of Agronomy, 1982, pp. 651–688.

102. D.R. Keeney, J.M. Bremner. Effect of cultivation on the nitrogen distribution in soils. *Soil Sci. Soc. Am. Proc.* 28:653–656, 1964.

103. D.A. Laird, D.A. Martens, W.L. Kingery. Nature of clay-humic complexes in an agricultural soil. I. Chemical, biochemical, and spectroscopic analyses. *Soil Sci. Soc. Am. J.* 65:1413–1418, 2001.

104. A.D. McLaren, G.H. Peterson. Physical chemistry and biological chemistry of clay mineral-organic nitrogen complexes. In: W.V. Bartholomew, F.E. Clark, eds. *Soil Nitrogen*. Madison, Wis.: American Society of Agronomy, 1965, pp. 259–284.

105. G.W. Harmsen, G.J. Kolenbrander. Soil inorganic nitrogen. In: W.V. Bartholomew, F.E. Clark, eds. *Soil Nitrogen*. Madison, Wis.: American Society of Agronomy, 1965, pp. 43–92.

106. J.L. Young, R.W. Aldag. Inorganic forms of nitrogen in soil. In: F.J. Stevenson, ed. *Soil Nitrogen*. Madison, Wis.: American Society of Agronomy, 1982, pp. 43–66.

107. M.K. Firestone. Biological denitrification. In: F.J. Stevenson, ed. *Soil Nitrogen*. Madison, Wis.: American Society of Agronomy, 1982, pp. 289–326.

108. M. Alexander. Nitrification. In: W.V. Bartholomew, F.E. Clark, eds. *Soil Nitrogen*. Madison, Wis.: American Society of Agronomy, 1965, pp. 307–343.

109. E.L. Schmidt. Nitrification in soil. In: F.J. Stevenson, ed. *Soil Nitrogen*. Madison, Wis.: American Society of Agronomy, 1982, pp. 253–288.

110. S.S. Goyal, R.C. Huffaker. Nitrogen toxicity in plants. In: R.D. Hauck, ed. *Nitrogen in Crop Production*. Madison, Wis.: American Society of Agronomy, 1984, pp. 97–118.

111. M.L. Stratton, A.V. Barker. Growth and mineral composition of radish in response to nitrification inhibitors. *J. Am. Soc. Hortic. Sci.* 112:13–17, 1987.

112. L.G. Bundy, J.J. Meisinger. Nitrogen availability indices. In: R.W. Weaver, ed. *Methods of Soil Analysis, Part 2*. Madison, Wis.: Soil Science Society of America, 1994, pp. 951–984.

113. W.C. Dahnke, E.H. Vasey. Testing for soil nitrogen. In: L.M. Walsh, J.D. Beaton, eds. *Soil Testing and Plant Analysis*. Madison, Wis.: Soil Science Society of America, 1973, pp. 97–114.

114. D.R. Keeney. Nitrogen—Availability indices. In: A.L. Page, ed. *Methods of Soil Analysis*, Part 2, 2nd ed. Madison, Wis.: Agronomy 9, 1982, pp. 711–733.

115. S.A. Waring, J.M. Bremner. Ammonium production in soil under water-logged conditions as an index of nitrogen availability. *Nature* 201:951–952, 1964.

116. F.C. Boswell, A.C. Richer, L.E. Casida, Jr. Available soil nitrogen measurements by microbiological techniques and chemical methods. *Soil Sci. Soc. Am. Proc.* 26:254–257, 1962.

117. D.R. Cullimore. A qualative method of assessing the available nitrogen, potassium and phosphorus in the soil. *J. Sci. Food Agr.* 17:321–323, 1966.

118. A.H. Cornfield. Carbon dioxide production during incubation of soils treated with cellulose as a possible index of the nitrogen status of soils. *J. Sci. Food Agric.* 12:763–765, 1961.

119. D.S. Jenkinson. Studies of methods of measuring forms of available soil nitrogen. *J. Sci. Food Agric.* 19:160–168, 1968.

120. H.O. Buckman. Moisture and nitrate relations in dry-land agriculture. *J. Am. Soc. Agron.* 2:121–138, 1910.

121. L.E. Call. The effect of different methods of preparing a seed bed for winter wheat upon yield, soil moisture, and nitrates. *J. Am. Soc. Agron.* 6:249–259, 1914.

122. L.G. Bundy, E.S. Malone. Effect of residual profile nitrate on corn response to applied nitrogen. *Soil Sci. Soc. Am. J.* 52:1377–1383, 1988.

123. T. Ozus, J.J. Hanway. Comparisons of laboratory and greenhouse tests for nitrogen and phosphorus availability in soils. *Soil Sci. Soc. Am. Proc.* 30:224–228, 1966.

124. W.E. Jokela, G.W. Randall. Corn yield and residual soil nitrate as affected by time and rate of nitrogen application. *Agron. J.* 81:720–726, 1989.

125. G.W. Roth, R.H. Fox. Soil nitrate accumulations following nitrogen-fertilized corn in Pennsylvania. *J. Environ. Qual.* 19:243–248, 1990.

126. W.C. White, J. Pesek. Nature of residual nitrogen in Iowa soils. *Soil Sci. Soc. Am. Proc.* 23:39–42, 1959.

127. J. Daliparthy, S.J. Herbert, P.L.M. Veneman. Dairy manure applications to alfalfa: crop response, soil nitrate, and nitrate in soil water. *Agron. J.* 86:927–933, 1994.

128. J.S. Angle, C.M. Gross, R.L. Hill, M.S. McIntosh. Soil nitrate concentrations under corn as affected by tillage, manure, and fertilizer applications. *J. Environ. Qual.* 22:141–147, 1993.

129. Z. Dou, R.H. Fox, J.D. Toth. Seasonal soil nitrate dynamics in corn as affected by tillage and nitrogen source. *Soil Sci. Soc. Am. J.* 59:858–864, 1995.

130. A. Katupitiya, D.E. Eisenhauer, R.B. Ferguson, R.F. Spalding, F.W. Roeth. Long-term tillage and crop rotation effects on residual nitrate in the crop root zone and nitrate accumulation in the intermediate vadose zone. *Trans. ASAE* 40:1321–1327, 1997.

131. C.R. Alvarez, R. Alvarez, H.S. Steinbach. Predictions of available nitrogen in soil profile depth using available nitrogen concentration in the surface layer. *Commun. Soil Sci. Plant Anal.* 32:759–769, 2001.

132. G.D. Binford, A.M. Blackmer, M.E. Cerrato. Relationship between corn yields and soil nitrate in late spring. *Agron. J.* 84:53–59, 1992.

133. F.R. Magdoff. Understanding the Magdoff pre-sidedress nitrate test for corn. *J. Prod. Agric.* 4:297–305, 1991.

134. F.R. Magdoff, W.E. Jokela, R.H. Fox, G.F. Griffith. A soil test for nitrogen availability in the northeastern United States. *Commun. Soil Sci. Plant Anal.* 21:1103–1115, 1990.

135. J.T. Sims, B.L. Vasilas, K.L. Gartley, B. Milliken, V. Green. Evaluation of soil and plant nitrogen tests for maize on manured soils of the Atlantic Coastal Plain. *Agron. J.* 87:213–222, 1995.

136. J.J. Meisinger, V.A. Bandel, J.S. Angle, B.E. O'Keefe, C.M. Reynolds. Preside dress soil nitrate test in Maryland. *Soil Sci. Soc. Am. J.* 56:1527–1532, 1992.

137. J.R. Heckman, R. Govindasamy, D.J. Prostak, E.A. Chamberlain, W.T. Hlubik, R.C. Mickel, E.P. Prostko. Corn response to side dress nitrogen in relation to soil nitrate concentration. *Commun. Soil Sci. Plant Anal.* 27:575–583, 1996.

138. G.K. Evanylo, M.M. Alley. Presidedress soil nitrogen test for corn in Virginia. *Commun. Soil Sci. Plant Anal.* 28:1285–1301, 1997.

139. H. Sainz-Rozas, H.E. Echeverria, G.A. Studdert, G. Dominguez. Evaluation of the presidedress soil nitrogen test for no-tillage maize fertilized at planting. *Agron. J.* 92:1176–1183, 2000.

140. J.R. Heckman, W.T. Hublik, D.J. Prostak, J.W. Paterson. Pre-sidedress soil nitrate test for sweet corn. *HortScience* 30:1033–1036, 1995.

141. T.K. Hartz, W.E. Bendixen, L. Wierdsma. The value of the presidedress soil nitrate testing as a nitrogen management tool in irrigated vegetable production. *HortScience* 35:651–656, 2000.

142. J.R. Heckman, T. Morris, J.T. Sims, J.B. Sieczka, U. Krogmann, P. Nitzsche, R. Ashley. Pre-sidedress soil nitrate test is effective for fall cabbage. *HortScience* 37:113–117, 2002.

143. H.H. Krusekopf, J.P. Mitchell, T.K. Hartz, E.M. May, E.M. Miyao, M.D. Cahn. Pre-sidedress soil nitrate test identifies processing tomato fields not requiring sidedress N fertilizer. *HortScience* 37:520–524, 2002.

144. L.K. Porter, B.A. Stewart, H.J. Haas. Effects of long-term cropping on hydrolyzable organic nitrogen fractions in some Great Plains soils. *Soil Sci. Soc. Am. Proc.* 28:368–370, 1964.

145. R.L. Mulvaney, S.A. Khan, R.G. Hoeft, H.M. Brown. A soil nitrogen fraction that reduces the need for nitrogen fertilization. *Soil Sci. Soc. Am. J.* 65:1164–1172, 2001.

146. S.A. Khan, R.L. Mulvaney, R.G. Hoeft. A simple soil test for detecting sites that are nonresponsive to nitrogen fertilizer. *Soil Sci. Soc. Am. J.* 65:1751–1760, 2001.

147. R.L. Mulvaney, S.A. Khan. Diffusion methods to determine different forms of nitrogen in soil hydrolysates. *Soil Sci. Soc. Am. J.* 65:1284–1292, 2001.

148. H.R. Schulten, M. Schnitzer. The chemistry of soil organic nitrogen: a review. *Biol. Fert. Soils* 26:1–15, 1998.

149. W. Amelung, X. Shang, K.W. Flach, W. Zech. Amino sugars in native grassland soils along a climosequence in North America. *Soil Sci. Soc. Am. J.* 63:86–92, 1999.

150. D. Solomon, J. Lehmann, W. Zech. Land use effects on amino sugar signature of chromic luvison in the semi-arid part of northern Tanzania. *Biol. Fert. Soils* 33:33–40, 2001.

151. L.C. Axelrod, T.E. O'Hare. Production of synthetic ammonia. In: V. Sauchelli, ed. *Fertilizer Nitrogen—its Chemistry and Technology*. New York: Reinhold Publishing Corp, 1964, pp. 58–88.

152. U.S. Jones. *Fertilizers and Soil Fertility.* Reston, Va.: Reston Publishing Co., 1979, pp. 29–103.

153. J. Pesek, G. Stanford, N.L. Case. Nitrogen production and use. In: R.A. Olson, ed. *Fertilizer Technology & Use.* Madison, Wis.: Soil Science Society of America, 1971, pp. 217–269.

154. I. Moldovan, M. Popovici, G. Chivu, *The Technology of Mineral Fertilizers.* London: The British Sulfur Corporation Ltd., 1969, pp. 1–793.

155. A.V. Barker, H.A. Mills. Ammonium and nitrate nutrition of horticultural crops. *Hortic. Rev.* 2:395–423, 1980.

156. M.S. Colgrove, Jr., A.N. Roberts. Growth of the azalea as influenced by ammonium and nitrate nitrogen. *Proc. Am. Soc. Hortic. Sci.* 68:522–536, 1956.

157. J.C. Cain. A comparison of ammonia and nitrate nitrogen on blueberries. *Proc. Am. Soc. Hortic. Sci.* 59:161–166, 1952.

158. D.A. Cox, J.G. Seeley. Ammonium injury to poinsettia: effects of NH_4 - N: NO_3 - N ratio and pH control in solution culture on growth, N absorption and N utilization. *J. Am. Soc. Hortic. Sci.* 109:57–62, 1984.

159. R.H. Hageman. Ammonium versus nitrate nutrition of higher plants. In: R.D. Hauck, ed. *Nitrogen in Crop Production.* Madison, Wis.: American Society of Agronomy, 1984, pp. 67–85.

160. R.J. Haynes. Uptake and assimiulation of mineral nitrogen by plants. In: R.J. Haynes, ed. *Mineral Nitrogen in the Plant-Soil System.* Orlando, Fla.: Academic Press, 1986, pp. 303–378.

161. H. Matsumoto, K. Tamura. Respiratory stress in cucumber roots treated with ammonium or nitrate nitrogen. *Plant Soil* 60:195–204, 1981.

162. D.N. Maynard, A.V. Barker. Studies on the tolerance of plants to ammonium nutrition. *J. Am. Soc. Hortic. Sci.* 94:235–239, 1969.

163. H.M. Reisenauer. Absorption and utilization of ammonium nitrogen by plants. In: D.R. Nielsen, J.G. MacDonald, eds. *Nitrogen in the Environment*, Vol. 2. New York: Academic Press, 1978, pp. 157–189.

164. H.E. Street, D.E.G. Sheat. The absorbtion and availability of nitrate and ammonium. In: W. Ruhland, ed., *Encyclopedia of Plant Physiology*, Vol. 8. Berlin: Springer Verlag, 1958, pp. 150–166.

165. A.H. Uljee. Ammonium nitrogen accumulation and root injury to tomato plants. *New Zealand J. Agric. Res.* 7:343–356, 1964.

166. H.M. Vines, T.D. Wedding. Some effects of ammonia on plant metabolism and possible mechanisms for ammonia toxicity. *Plant Physiol.* 35:820–825, 1960.

167. J.K.R. Gasser. Nitrification inhibitors—their occurrence, production and effects of their use on crop yields and composition. *Soils Fert.* 33:547–554, 1970.

168. J. Glasscock, A. Shaviv, J. Hagin. Nitrification inhibitors—interaction with applied ammonium concentration. *J. Plant Nutr.* 18:105–116, 1995.

169. C.A.I. Goring. Control of nitrification of ammonium fertilizers and urea by 2-chloro-6-trichloromethylpyridine. *Soil Sci.* 93:431–439, 1962.

170. R. Prasad, G.B. Rajale, B.A. Lakhdive. Nitrification retarders and slow-release nitrogen fertilizers. *Adv. Agron.* 23:337–383, 1971.

171. S.C. Rao. Evaluation of nitrification inhibitors and urea placement in no-tillage winter wheat, *Agron. J.* 88:904–908, 1996.

172. R.H. Fox, W.P. Piekielek, K.E. Macneal. Estimating ammonia volatilization losses from urea fertilizers using a simplified micrometeorological sampler. *Soil Sci. Soc. Am. J.* 60:596–601, 1996.

173. M.A. Gameh, J.S. Angle, J.H. Axley. Effects of urea-potassium chloride and nitrogen transformations on ammonia volatilization from urea. *Soil Sci. Soc. Am. J.* 54:1768–1772, 1990.

174. J.M. Bremner. Recent research on problems in the use of urea as a nitrogen fertilizer. *Fert. Res.* 42:321–329, 1995.

175. S.E. Allen. Slow-release nitrogen fertilizers. In: R.D. Hauck, ed. *Nitrogen in Crop Production.* Madison, Wis.: American Society of Agronomy, 1984, pp. 195–206.

176. J.B. Sartain, J.K. Kruse. Selected Fertilizers Used in Turfgrass Fertilization. Gainesville, Fla: University of Florida Cooperative Extension Service Circular CIR 1262, 2001.

177. R.N. Carrow. Turfgrass response to slow-release nitrogen fertilizers. *Agron. J.* 89:491–496, 1997.

178. M.F. Carter, P.L.G. Vlek, J.T. Touchton. Agronomic evaluation of new urea forms for flooded rice. *Soil Sci. Soc. Am. J.* 50:1055–1060, 1986.

179. R.W. Moore, N.E. Christians, M.L. Agnew. Response of three Kentucky bluegrass cultivars to sprayable nitrogen fertilizer programs. *Crop Sci.* 36:1296–1301, 1996.

180. R.L. Mikkelsen, H.M. Williams, A.D. Behel, Jr. Nitrogen leaching and plant uptake from controlled-release fertilizers. *Fert. Res.* 37: 43–50, 1994.

181. F.L. Wang, K. Alva. Leaching of nitrogen from slow-release urea sources in sandy soils. *Soil Sci. Soc. Am. J.* 60:1454–1458, 1996.

182. H.M. Keener, W.A. Dick, C. Marugg, R.C. Hansen. Composting spent press-molded, wood fiber pallets bonded with urea formaldehyde: a pilot scale evaluation. *Compost Sci. Util.* 2(3):73–82, 1994.

183. D.E. Clay, G.L. Malzer, J.L. Anderson. Ammonia volatilization from urea as influenced by soil temperature, soil water content, and nitrification and hydrolysis inhibitors. *Soil Sci. Soc. Am. J.* 54:263–266, 1990.

184. G.A. Cowie. Decomposition of cyanamide and dicyandiamide. *J. Agric. Sci.* 9:113–136, 1918.

185. M.D. Serna, F. Legaz, E. Primo-Millo. Efficacy of dicyandiamide as a soil nitrification inhibitor in citrus production. *Soil Sci. Soc. Am. J.* 58:1817–1824, 1994.

186. G.R. McVey, D.P. Horn, E.L. Scheiderer, A.D. Davidson. Correlation of analytical methods and biological activity of various sulfur-coated urea fertilizer products. *J. Assoc. Off. Anal. Chem.* 68:785–788, 1985.

187. R. Puchades, E.P. Yufera, J.L. Rubio. The release, diffusion and nitrification of nitrogen in soils surrounding sulfur-coated urea granules. *Plant Soil* 78:345–356, 1984.

188. B.D. Brown, A.J. Hornbacher, D.V. Naylor. Sulfur-coated urea as a slow-release nitrogen source for onions. *J. Am. Soc. Hortic. Sci.* 113:864–869, 1988.

189. E.A. Guertal. Preplant slow-release nitrogen fertilizers produces similar bell pepper yields as split applications of soluble fertilizer. *Agron. J.* 92:388–393, 2000.

190. W.R. Raun, D.H. Sander, R.A. Olson. Nitrogen fertilizer carriers and their placement for minimum till corn under sprinkler irrigation. *Agron. J.* 81:280–285, 1989.

191. M. Zekri, R.C.J. Koo. Evaluation of controlled-release fertilizers for young citrus trees. *J. Am. Soc. Hortic. Sci.* 116:987–999, 1991.

192. J. Halevy. Efficiency of isobutylidene diurea, sulphur-coated urea and urea plus nitrapyrin, compared with divided dressings of urea, for dry matter production and nitrogen uptake of ryegrass. *Exp. Agric.* 23:167–179, 1987.

193. L.B. Owens, W.M. Edwards, R.W. Van Keuren. Nitrate leaching from grassed lysimeters treated with ammonium nitrate or slow-release nitrogen fertilizer. *J. Environ. Qual.* 28:1810–1816, 1999.

194. A.V. Barker, T.A. O'Brien, M.L. Stratton. Description of food processing by-products. In: W.A. Dick, ed. *Land Application of Agricultural, Industrial, and Municipal By-Products.* Madison, Wis.: Soil Science Society of America, 2000, pp. 63–106.

195. E.E. Huntley, A.V. Barker, M.L. Stratton. Composition and uses of organic fertilizers. In: J.E. Rechcigl, H.C. MacKinnon, eds. *Agricultural Uses of By-Products and Wastes.* Washington, DC: American Chemical Society, 1997, pp. 120–139.

196. A.L. Mehring. *Dictionary of Plant Foods.* Philadelphia, Pa.: Ware Bros. Co., 1958, pp. 1–51.

197. H.D. Chapman. *Diagnostic Criteria for Plants and Soils.* Riverside, Cal.: HD Chapman, 1965, pp. 1–793.

198. L.M. Walsh, J.D. Beaton. *Soil Testing and Plant Analysis,* revised edition. Madison, Wis.: Soil Science Society of America, 1973, pp. 1–491.

3 Phosphorus

Charles A. Sanchez
Yuma Agricultural Center, Yuma, Arizona

CONTENTS

3.1 Background Information 51
    3.1.1 Historical Information 51
    3.1.2 Phosphorus Functions in Plants 52
    3.1.3 Nature and Transformations of Soil Phosphorus 53
3.2 Diagnosing Phosphorus Deficiency 54
    3.2.1 Visual Symptoms of Deficiency and Excess 54
    3.2.2 Tissue Testing for Phosphorus 55
    3.2.3 Soil Testing for Phosphorus 71
3.3 Factors Affecting Management of Phosphorus Fertilization 75
    3.3.1 Crop Response to Phosphorus 75
    3.3.2 Soil Water 76
    3.3.3 Soil Temperature 78
    3.3.4 Sources of Phosphorus 79
    3.3.5 Timing of Application of Phosphorus Fertilizers 79
    3.3.6 Placement of Phosphorus Fertilizers 79
    3.3.7 Foliar-Applied Phosphorus Fertilization 81
    3.3.8 Fertilization in Irrigation Water 81
References 82

3.1 BACKGROUND INFORMATION

3.1.1 HISTORICAL INFORMATION

Incidental phosphorus fertilization in the form of manures, plant and animal biomass, and other natural materials, such as bones, probably has been practiced since agriculture began. Although specific nutritional benefits were unknown, Arthur Young in the *Annuals of Agriculture* in the mid-nineteenth century describes experiments evaluating a wide range of products including poultry dung, gunpowder, charcoal, ashes, and various salts. The results showed positive crop responses to certain materials. Benefiting from recent developments in chemistry by Antoine Lavoisier (1743–1794) and others, Theodore de Saussure (1767–1845) was perhaps the first to advance the concept that plants absorb specific mineral elements from the soil.

The science of plant nutrition advanced considerably in the nineteenth century owing to contributions by Carl Sprengel (1787–1859), A.F. Wiegmann (1771–1853), Jean-Baptiste Boussingault (1802–1887), and Justus von Liebig (1803–1873). Based on the ubiquitous presence of phosphorus in soil and plant materials, and crop responses to phosphorus-containing products, it became apparent that phosphorus was essential for plant growth.

Liebig observed that dissolving bones in sulfuric acid enhanced phosphorus availability to plants. Familiar with Liebig's work, John Lawes in collaboration with others, evaluated several apatite-containing products as phosphorus nutritional sources for plants. Lawes performed these experiments in what ultimately became the world's most famous agricultural experiment station—his estate in Rothamsted. The limited supply of bones prompted developments in the utilization of rock phosphates where Lawes obtained the first patent concerning the utilization of acid-treated rock phosphate in 1842, The first commercial production of rock phosphate began in Suffolk, England, in 1847. Mining phosphate in the United States began in 1867. Thus began the phosphorus fertilizer industry.

Crop responses to phosphorus fertilization were widespread. For many years phosphorus fertilization practices were based on grower experience often augmented with empirical data from experiment station field tests. Although researchers and growers realized that customized phosphorus fertilizer recommendations would be invaluable, early work often focused on total element content of soils and produced disappointing results. The productivity of soil essentially showed no correlation to total content of nutrients in them.

It was during the twentieth century that the recognition that the plant itself was an excellent indicator of nutrient deficiency coupled with considerable advances in analytical methodology gave way to significant advances in the use of tissue testing. Hall (1) proposed plant analysis as a means of determining the normal nutrient contents of plants. Macy (2) proposed the basic theory that there was a critical concentration of nutrient in a plant above which there was luxury consumption and below which there was poverty adjustment, which was proportional to the deficiency until a minimum percentage was reached.

Also during the twentieth century, a greater understanding of soil chemistry of phosphorus and the observation that dilute acids seem to correlate to plant-available phosphorus in the soil gave way to the development of successful soil-testing methodologies. The early contributions of Dyer (3), Truog (4), Morgon (5), and Bray and Kutrz (6) are noteworthy. Plant tissue testing and soil testing for phosphorus are discussed in greater detail in the subsequent sections. For more detailed history on plant nutrition and soil–plant relationships, readers are referred to Kitchen (7) and Russell (8).

3.1.2 PHOSPHORUS FUNCTIONS IN PLANTS

Phosphorus is utilized in the fully oxidized and hydrated form as orthophosphate. Plants typically absorb either $H_2PO_4^-$ or HPO_4^{2-}, depending on the pH of the growing medium. However, under certain conditions plants might absorb soluble organic phosphates, including nucleic acids. A portion of absorbed inorganic phosphorus is quickly combined into organic molecules upon entry into the roots or after it is transported into the shoot.

Phosphate is a trivalent resonating tetraoxyanion that serves as a linkage or binding site and is generally resistant to polarization and nucleophilic attack except in metal-enzyme complexes (9). Orthophosphate can be condensed to form oxygen-linked polyphosphates. These unique properties of phosphate produce water-stable anhydrides and esters that are important in energy storage and transfer in plant biochemical processes. Most notable are adenosine diphosphate and triphosphate (ADP and ATP). Energy is released when a terminal phosphate is split from ADP or ATP. The transfer of phosphate molecules to ATP from energy-transforming processes and from ATP to energy-requiring processes in the plants is known as phosphorylation. A portion of the energy derived from photosynthesis is conserved by phosphorylation of ADP to yield ATP in a process called photophosphorylation. Energy released during respiration is similarly harnessed in a process called oxidative phosphorylation.

Beyond their role in energy-transferring processes, phosphate bonds serve as important linkage groups. Phosphate is a structural component of phospholipids, nucleic acids, nucleotides, coenzymes, and phosphoproteins. Phospholipids are important in membrane structure. Nucleic acids of genes and chromosomes carry genetic material from cell to cell. As a monoester, phosphorus provides an essential ligand in enzymatic catalysis. Phytic acid, the hexaphosphate ester of *myo*-inositol phosphate, is the most common phosphorus reserve in seeds. Inorganic and organic phosphates in plants also serve as buffers in the maintenance of cellular pH.

Total phosphorus in plant tissue ranges from about 0.1 to 1%. Bieleski (10) suggests that a typical plant might contain approximately 0.004% P as deoxyribonucleic acid (DNA), 0.04% P as ribonucleic acid (RNA), 0.03% as lipid P, 0.02 % as ester P, and 0.13% as inorganic P.

3.1.3 NATURE AND TRANSFORMATIONS OF SOIL PHOSPHORUS

Soils contain organic and inorganic phosphorus compounds. Because organic compounds are largely derived from plant residues, microbial cells, and metabolic products, components of soil organic matter are often similar to these source materials. Approximately 1% of the organic phosphorus is in the phospholipid fraction; 5 to 10% is in nucleic acids or degradation products, and up to 60% is in an inositol polyphosphate fraction (11). A significant portion of the soil organic fraction is unidentified.

Phospholipids and nucleic acids that enter the soil are degraded rapidly by soil microorganisms (12,13). The more stable, and therefore more abundant, constituents of the organic phosphorus fraction are the inositol phosphates. Inositol polyphosphates are usually associated with high-molecular-weight molecules extracted from the soil, suggesting that they are an important component of humus (14,15).

Soils normally contain a wide range of microorganisms capable of releasing inorganic orthophosphate from organic phosphates of plant and microbial origin (16,17). Conditions that favor the activities of these organisms, such as warm temperatures and near-neutral pH values also favor mineralization of organic phosphorus in soils (16,18). The enzymes involved in the cleavage of phosphate from organic substrates are collectively called phosphatases. Microorganisms produce a variety of phosphatases that mineralize organic phosphate (19).

Phosphorus released to the soil solution from the mineralization of organic matter might be taken up by the microbial population, taken up by growing plants, transferred to the soil inorganic pool, or less likely lost by leaching and runoff (Figure 3.1). Phosphorus, like nitrogen, undergoes mineralization and immobilization. The net phosphorus release depends on the phosphorus concentration of the residues undergoing decay and the phosphorus requirements of the active microbial population (16).

In addition to phosphorus mineralization and immobilization, it appears that organic matter has indirect, but sometimes inconsistent, effects on soil phosphorus reactions. Lopez-Hernandez and Burnham (20) reported a positive correlation between humification and phosphate-sorption capacity. Wild (21) concluded that the phosphorus-sorption capacity of organic matter is negligible. It is observed more commonly that organic matter hinders phosphorus sorption, thereby enhancing availability. Humic acids and other organic acids often reduce phosphorus fixation through the formation of complexes (chelates) with Fe, Al, Ca, and other cations that react with phosphorus (22–24). Studies have shown that organic phosphorus is much more mobile in soils than inorganic sources (25). The

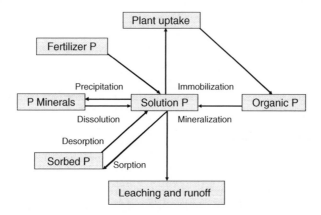

FIGURE 3.1 Phosphorus cycle in agricultural soils.

interaction between the organic and inorganic phosphorus fractions is understood poorly. It is generally presumed that phosphorus availability to plants is controlled by the inorganic phosphorus fraction, although the contribution of organic phosphorus to plant nutrition should not be dismissed.

Inorganic phosphorus entering the soil solution, by mineralization or fertilizer additions, is rapidly converted into less available forms. Sorption and precipitation reactions are involved. The sorption of inorganic phosphorus from solution is closely related to the presence of amorphous iron and aluminum oxides and hydrous oxides (26–30) and the amounts of calcium carbonate ($CaCO_3$) (24,31,32).

Hydrous oxides and oxides of aluminum and iron often occur as coatings on clay mineral surfaces (27,28,33), and these coatings may account for a large portion of the phosphorus sorption associated with the clay fraction of soils. Even in calcareous soils, hydrous oxides have been demonstrated as being important in phosphorus sorption, as was demonstrated by Shukla (34) for calcareous lake sediments, Holford and Mattingly (24) for calcareous mineral soils, and Porter and Sanchez (35) for calcareous Histosols.

In calcareous soils, phosphorus (or phosphate) sorption to $CaCO_3$ may be of equal or greater importance than sorption to aluminum and iron oxides (35). In a laboratory investigation with pure calcite, Cole (31) concluded that the reaction of phosphorus with $CaCO_3$ consisted of initial sorption reactions followed by precipitation with increasing concentrations of phosphorus. Phosphorus sorption may occur in part as a multilayer phenomenon on specific sites of the calcite surface (24,32). As sorption proceeds, lateral interactions occur between sorbed phosphorus, eventually resulting in clusters. These clusters in turn serve as centers for the heterogeneous nucleation of calcium phosphate crystallites on the calcite surface.

Phosphorus sorption is probably limited to relatively low initial phosphorus solution concentrations and precipitation is likely a more important mechanism of phosphorus removal from the soil solutions at higher concentrations (31). Lindsay (36) identified, by x-ray crystallography, what he considered to be an incomplete list of 32 forms of phosphate compounds as reaction products from phosphorus fertilizers. The nature of the reaction products formed when phosphorus fertilizer is added to soil depends primarily on the coexisting cation, the pH of the saturated solution, the quantity of phosphorus fertilizer added, and the chemical characteristics of the soil (37). In acidic soils, aluminum and iron will generally precipitate phosphorus. In calcareous soils, an acidic fertilizer solution would dissolve calcium, and it is anticipated that most of the added phosphorus fertilizer would precipitate initially as dicalcium phosphate dihydrate (DCPD) and dicalcium phosphate (DCP) (38,39). These products are only moderately stable and undergo a slow conversion into compounds such as octacalcium phosphate, tricalcium phosphate, or one of the apatites.

As discussed above, soil transformations of phosphorus are complex and often ambiguous. Phosphorus availability has often been characterized in general terms (a) as solution phosphorus, often known as the intensity factor, (b) as readily available or labile phosphorus, often known as the quantity factor, and (c) as nonlabile phosphorus. The labile fraction might include easily mineralizable organic phosphorus, low-energy sorbed phosphorus, and soluble mineral phosphorus. The nonlabile fraction might include resistant organic phosphorus, high-energy sorbed phosphorus, and relatively insoluble phosphate minerals. As plants take up phosphorus from the solution, it is replenished from the labile fraction, which in turn is more slowly replenished by the nonlabile fraction. The soil buffer capacity, known as the capacity factor, governs the distribution of phosphorus among these pools. As will be shown in a subsequent section, although some soil tests aim to characterize only the intensity factor, most aim to characterize quantity and capacity factors as indices of phosphorus availability.

3.2 DIAGNOSING PHOSPHORUS DEFICIENCY

3.2.1 Visual Symptoms of Deficiency and Excess

Phosphorus deficiency suppresses or delays growth and maturity. Although phosphorus- deficient plants are generally stunted in appearance, they seldom exhibit the conspicuous foliar symptoms characteristic of some of the other nutrient deficiencies. Furthermore, appreciable overlap often occurs with the symptoms of other nutrient deficiencies. Plant stems or leaves are sometimes dark green, often developing red and purple colors. However, when weather is cool purpling of leaves can also be associated with nitrogen deficiency, as is often observed in *Brassica* species, or with phosphorus deficiency. Plants stunted by phosphorus deficiency often have small, dark-green leaves and short and slender stems. Sustained phosphorus deficiency will probably produce smaller-sized fruit and limited harvestable vegetable mass. Because phosphorus is mobile in plants, it is translocated readily from old to young leaves as deficiency occurs, and chlorosis and necrosis on older leaves is sometimes observed. Readers are referred to tables of phosphorus deficiency symptoms specific to individual crops and compiled by other authors (40–43).

Most soils readily buffer phosphorus additions, and phosphorus is seldom present in the soil solution at levels that cause direct toxicity. Perhaps the most common symptoms of phosphorus excess are phosphate-induced micronutrient deficiencies, particularly Zn or Cu deficiencies (43,44).

3.2.2 TISSUE TESTING FOR PHOSPHORUS

As noted previously, visual indications of phosphorus deficiency are seldom conclusive; consequently, accurate diagnosis typically requires a tissue test. Most diagnostic standards are generated using the theory of Macy (2), as noted previously concerning critical levels, sufficiency ranges, and poverty adjustment. In practice, critical levels or sufficiency ranges are usually determined by plotting final relative yield against phosphorus concentration in plant tissues and interpreting the resulting curvilinear function at some specified level of maximum yield. For many agronomic crops, values of 90 to 95% maximum yield are frequently used. However, for vegetable crops, which have a higher market value and an economic optimum closer to maximum yield, values of 98% have been used (Figure 3.2). Sometimes researchers use discontinuous functions such as the "linear response and plateau" or "quadratic response and plateau" and define adequacy by the plateau line (Figure 3.3). Yet, other researchers have suggested that the correlation to final yield is less than ideal and have proposed the use of incremental growth-rate analysis in developing critical concentrations (45).

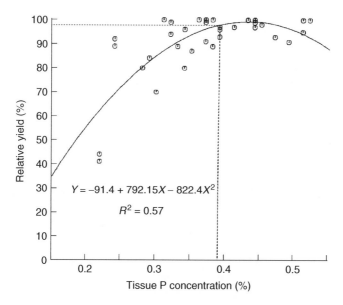

FIGURE 3.2 Calculated critical phosphorus concentration in the midribs of endive at the eight-leaf stage using curvilinear model. (Adapted from C.A. Sanchez and H.W. Burdine, *Soil Crop Sci. Soc. Fla. Proc.* 48:37–40, 1989.)

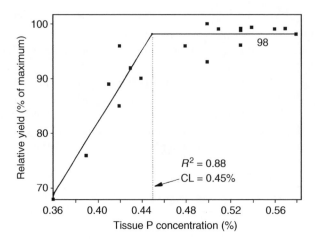

FIGURE 3.3 Calculated critical phosphorus concentration (CL) of radish leaves using linear-response and plateau model. Plateau is at 98%. (Adapted from C.A. Sanchez et al., *HortScience* 26:30–32, 1991.)

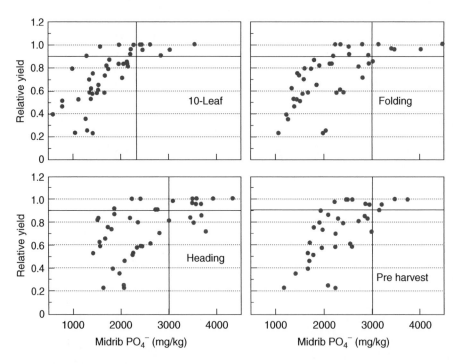

FIGURE 3.4 Calculated critical acetic acid extractable phosphate-P concentrations at four growth stages for lettuce. (Gardner and Sanchez, unpublished data.)

Levels of deficiency, sufficiency, and excess have been determined in solution culture and in greenhouse and field experiments. Total phosphorus content of a selected plant part at a certain growth stage is used for most crops. However, many standards developed for vegetable crops are based on a 2% acetic acid extraction (Figure 3.4). Diagnostic standards for various plant species are summarized in Table 3.1. This compilation includes data from other compilations and from research studies. When data from other compilations were used, priority was given to research that cited original source of data (46–48) so that potential users can scrutinize how the values were determined. However, when

TABLE 3.1
Diagnostic Ranges for Phosphorus Concentrations in Crop and Ornamental Plants

A. Field Crops

Species	Growth Stage	Plant Part	Deficient	Low	Sufficient	High	Reference
Barley	GS 2	WP	<0.30				130
(*Hordeum*	GS 6	WP	<0.30	0.30–0.40	>4.0		130
vulgare L.)	GS 9	WP	<0.15	0.15–0.20	>0.20		130
	GS 10.1	WP	<0.15	0.15–0.20	0.20–0.50	>0.5	131
Cassava	Veg.	YML	<0.20	0.40	0.30–0.50		132
(*Manihot*							
esculentum							
Crantz)							
Chickpea (*Cicer*	45 DAP	WP	0.09–0.25		0.29–0.33		133
arietinum L.)	77 DAP	WP	0.15–0.20		>0.26		133
Dent corn (*Zea*	<30 cm tall	WP			0.30–0.50		134
mays var.	40–60 cm tall	WP		0.22–0.26			135
indentata	Tassel	Ear L		0.25			136
L.H. Bailey)	Silking	Ear L		0.28–0.32			137
	Silking	Ear L	<0.20		>0.29		138
	Silking	Ear L	0.22–0.32		0.27–0.62		139
	Silking	6th L from base		<0.32			140
	Silking	6th L from base	<0.21	<0.30	<0.33		141
	Silking	Ear L	0.16–0.24		0.25–0.40	0.41–0.50	142
	Silking	Ear L			0.25–0.40		143
	Silking	Ear L			0.22–0.23		135
	Silking	Ear L			0.26–0.35		144
	Silking	Ear L		0.27			145
Cotton	<1st Fl	YML			0.30–0.50		134
(*Gossypium*	July–August	L			0.30–0.64		146
hirsutum L.)	Early fruit	YML		0.31			147
	Late fruit	YML		0.33			147
	Late Mat	YML		0.24			147
	1st Fl	PYML PO$_4$-P		0.15		0.20	148
	Peak Fl	PYML PO$_4$-P		0.12		0.15	148
	1st bolls open	PYML PO$_4$-P		0.10		0.12	148
	Mat	PYML PO$_4$-P		0.08		0.10	148
Cowpea (*Vigna*	56 DAP	WP			0.28		149
unguiculata	30 cm	WP	0.28		0.27–0.35		150
Walp.)	Early Fl	WP	0.19–0.24		0.23–0.30		150
Faba or field bean	Fl	L 3rd node from A			0.32–0.41		151
(*Vicia faba* L.)							
Field pea	36 DAS	WP	<0.06		>0.92		152
(*Pisum*	51 DAS	WP	<0.53		>0.71		152
sativum L.)	66 DAS	WP	<0.46		>0.64		152
	81 DAS	WP	<0.40		>0.55		152
	96 DAS	WP	<0.43		>0.60		152

Continued

TABLE 3.1 (*Continued*)

Species	Growth Stage	Plant Part	Deficient	Low	Sufficient	High	Reference
	8–9 nodes	L 3rd node from A			0.36–0.51		151
	Pre-Fl	WP			0.16		153
Dry beans (*Phaseolus vulgaris* L.)	10% Fl	YML			0.40		154
	50–55 DAE	WP	0.22		0.33		155
Oats (*Avena sativa* L.)	GS 10.1	WP	<0.15	0.15–0.19	0.20–0.50	>0.50	131
	Pre-head	Upper L			0.20–0.40		134
Peanuts (*Arachis hypogaea* L.)	Early pegging	Upper L+S			0.20–0.35		156
	Pre Fl or Fl	YML			0.25–0.50		134
Pigeon pea (*Cajanus cajan* Huth.)	91 DAP	L	0.08		0.24		157
	30 DAP	L			0.35–0.38		158
	60 DAP	L			0.30–0.33		158
	90–100 DAP	L			0.19–0.28		158
	120–130 DAP	L			0.15–0.20		158
	160–165 DAP	L			0.15–0.18		158
Rice (*Oryza sativa* L.)	25 DAS	WP	<0.70	0.70–0.80	0.80–0.86		159
	50DAS	WP	<0.18	0.18–0.26	0.26–0.40		159
	75 DAS	WP	<0.26	0.26–0.36	0.36–0.48		159
	35 DAS	WP		0.25			160
	Mid till	Y blade			0.14–0.27		131
	Pan init	Y blade			0.18–0.29		131
PO_4-P	Mid till	Y blade		0.1	0.1–0.18		161
PO_4-P	Max till	Y blade		0.08	0.1–0.18		161
PO_4-P	Pan init	Y blade		0.08	0.1–0.18		161
PO_4-P	Flagleaf	Y blade		0.1	0.08–0.18		161
Sorghum (*Sorghum bicolor* Moench.)	23–29 DAP	WP	<0.25	0.25–0.30	0.30–0.60	>0.60	162
	37–56 DAP	YML	<0.13	0.13–0.25	0.20–0.60		162
	66–70 DAP (Bloom)	3L below head	<0.18	0.18–0.22	0.20–0.35	>0.35	162
	82–97 DAP (Dough)	3 L below head	<0.13	0.13–0.15	0.15–0.25	>0.25	162
	NS	YML			0.25–0.40		163
Soybean (*Glycine max* Merr.)	Pre-pod	YML			0.26–0.50		156
	Early pod	YML		0.35			136
	Early pod	YML			0.30–0.50		134
	Pod	Upper L		0.37			164
	August	L			0.25–0.60		165
Sugar beet (*Beta vulgaris* L.)	25 DAP	Cotyledon PO_4-P	0.02–0.15		0.16–1.30		166
	25 DAS	Oldest P PO_4-P	0.05–0.15		0.16–0.50		166
	25 DAS	Oldest L PO_4-P	0.05–0.32		0.35–1.40		166
	NS	PYML PO_4-P	0.15–0.075		0.075–0.40		167
	NS	YML PO_4-P	0.025–0.070		0.10–.80		167

TABLE 3.1 (*Continued*)

Species	Growth Stage	Plant Part	Deficient	Low	Sufficient	High	Reference
Sugarcane	5 month	3rd LB		0.21			168
(*Saccharum*	ratoon	below A					
officinarum L.)	4th mo.	3rd & 4th			0.24–0.30		
		LB below A			0.24–0.30		169
	3 mo.	Leaves	0.15–0.18		0.18–0.24	0.24–0.30	170
	Early rapid	Sheath 3–6	<0.05	0.08	0.05–0.20		171
	growth						
Tobacco	Fl	YML			0.27–0.50		134
(*Nicotiana*	Mat	L	0.12–0.17		0.22–0.40		172
tabacum L.)							
Wheat (*Triticum*	GS 3–5	WP			0.4–0.70		173
aestivum L.)	GS 6–10	WP			0.2–0.40		173
	GS 10	Flag L			0.30–0.50		173
	GS 10	WP		030			136
	GS 10.1	WP	0.15–0.20		0.21–0.50	>0.50	131
	Pre-head	Upper LB			0.20–0.40		134

B. Forages and Pastures

Species	Growth Stage	Plant Part	Deficient	Low	Sufficient	High	Reference
Alfalfa	Early Fl	WP		<0.20			174
(*Medicago*	Early Fl	WP		<0.30			174
sativa L.)	Early Fl	WP	<0.18		0.25–0.50		174
	Early Fl	WP	<0.20	0.21–0.22	0.23–0.30	>0.30	174
	Early Fl	WP		<0.25			174
	Early Fl	WP		<0.25			174
	Early Fl	WP		<0.25			174
	Early Fl	Top 15 cm	<0.20	0.20–0.25	0.26–0.70	>0.70	174
	Early Fl	Upper stem		0.35			174
	Early Fl	Midstem PO$_4$P	<0.05	0.05–0.08	0.08–0.20	>0.20	174
Bermuda grass,	4–5 weeks	WP	<0.16	0.18–0.24	0.24–0.30	>0.40	174
Coastal	between						
(*Cynodon*	clippings						
dactylon Pers.)							
Bermuda grass,	4–5 weeks	WP	<0.22	0.24–0.28	0.28–0.34	>0.40	174
Common and	between						
Midland	clippings						
(*Cynodon*							
dactylon Pers.)							
Birdsfoot trefoil	Growth	WP		<0.24			174
(*Lotus*							
corniculatus L.)							
Clover, Bur	Growth	WP			2.5		174
(*Medicago*							
hispida Gaertn.)							
Clover, Ladino	Growth	WP		<0.23			174
or White	Growth	WP		<0.30			174
(*Trifolium*	Growth	WP		0.10–0.20	0.30		174
repens L.)	Growth	WP		<0.25	0.25–0.30		174

Continued

TABLE 3.1 (*Continued*)

Species	Growth Stage	Plant Part	Deficient	Low	Sufficient	High	Reference
	Growth	WP		0.15–0.25	0.30–0.35		174
	Growth	WP PO$_4$P		0.06	0.06–0.12		174
Clover, Red	Growth	WP		<0.25	0.25–0.80		174
(*Trifolium*	Growth	WP			0.20–0.40		174
pratense L.)	Growth	WP		<0.27			174
Clover, Rose	Growth	WP	0.10–0.14	0.14–0.18	0.19–0.24		174
(*Trifolium*	Growth	WP			0.20–0.25		174
hirtum All.)	Growth	WP	0.07	<0.19			174
Clover,	Growth	WP		0.30–0.31			174
Subterranean	Growth	WP			0.20–0.28		174
(*Trifolium*	Growth	WP			0.26–0.32		174
subterraneum L.)	Growth	WP		<0.25			174
	Growth	WP		<0.14			174
	Growth	WP		0.08–0.13			174
	Growth	L	0.07		0.20–0.26		175
Dallisgrass (*Paspalum dilatatum* Poir.)	3–5 weeks	WP	<0.24	<0.26	0.28–0.30		174
Johnsongrass (*Sorghum halepense* Pers.)	4–5 weeks after clipping	WP	<0.14	0.16–0.20	0.20–0.25		174
Kentucky bluegrass (*Poa pratensis* L.)	4–6 weeks between clippings	WP	<0.18	0.24–0.30	0.28–0.36	>0.40	174
Millet (*Pennisetum glaucum* R. Br.)	4–5 wks after clipping	WP	<0.16	0.16–0.20	0.22–0.30	>0.40	174
Orchardgrass (*Dactylis glomerata* L.)	3–4 weeks between clippings	WP	<0.18	0.22–0.24	0.23–0.28	>0.35	174
Pangolagrass (*Digitaria decumbens* Stent.)	4–5 weeks between clippings	WP	<0.10	0.12–0.16	0.16–0.24	>0.28	174
Ryegrasses, perennial (*Lolium perenne* L.)	4–5 weeks between clippings	WP	<0.28	0.28–0.34	0.36–0.44	>0.50	174
Sudangrass (*Sorghum sudanese* Stapf.) and Sorghum sudan hybrids	4 to 5 weeks after clipping	WP	<0.14	0.14–0.18	0.20–0.30	>0.35	174
Stylo, Capica (*Stylosanthes capitata* Vog.)	56 DAP	WP		0.11–0.18			176

TABLE 3.1 *(Continued)*

Species	Growth Stage	Plant Part	Deficient	Low	Sufficient	High	Reference
Stylo, Macrocephala (*Stylosanthes macrocephala* M.B. Ferr. & Sousa Costa)	56 DAP	WP		0.10			176
Tall fescue (*Festuca arundinacea* Schreb.)	5–6 weeks	WP	<0.24	0.26–0.32	0.24–0.40	>0.45	174

C. Fruits and Nuts

Species	Growth Stage	Plant Part	Deficient	Low	Sufficient	High	Reference
Almond	July–August	L			0.09–0.19		177
(*Prunus amygdalus* Batsch.)	July–August	L		0.08	0.12	>0.30	178
Apple	July–August	L	<0.11	0.11–0.13	0.13–0.20		179
(*Malus domestica* Borkh.)	July–August	L			0.11–0.30		177
	Harvest	L			0.21		43
	July–August	L		0.15–0.19	0.20–0.30		43
	June–Sept.	L/tips of shoots			0.19–0.32		43
	20 DAfl	L			0.28		43
	200 DAfl	L			0.10		43
	July–August	L		0.08	0.12	>0.30	178
	July–August	L			0.23		180
	110 DAfl	L/mid shoot			0.20		181
Apricot	August	L			0.09		177
(*Prunus armeniaca* L.)	110 Dafl	L/mid shoot			0.1		181
Avocado	Mature	L		0.065	0.065–0.20		43
(*Persea americana* Mill.)	December–January	YML			0.10–0.15		43
	August–October	YML/nonfruiting terminals	0.05		0.08–0.25	0.3	182
Banana	NS	L	<0.20		0.45		183
(*Musa* spp.)	5th L Stage	L			0.20		177
	8th L Stage	L			0.18		177
	15th L stage	L			0.15		177
Blueberry, High Bush	Mid-season	L/mature shoots	0.02–0.03	<0.07	0.10–0.32		184
(*Vaccinium corymbosum* L.)	July–August	L			0.10–0.12		177
	July–August	YML/fruiting shoot	<0.10		0.12–0.40	>0.41	185
Cacao (*Theobroma* spp.)	NS	L	<0.13	0.13–0.20	>0.20		186

Continued

TABLE 3.1   (*Continued*)

Species	Growth Stage	Plant Part	Deficient	Low	Sufficient	High	Reference
Cherry	July–August	L			0.13–0.67		177
(*Prunus* spp.)	July–August	L			0.25		180
	110 Dafl	L/midshoot			0.30		181
	July–August	L			0.13–0.30		187
Citrus,	February	L			0.05–0.11		177
Grapefruit	July	L			0.12		177
(*Citrus xparadisi*	October	L			0.07–0.11		177
Macfady)							
Citrus, Lemon	July	L			0.13–0.22		177
(*Citrus limon*							
Burm. f.)							
Citrus, Orange	4–7 mo.	L	<0.09	0.09–0.11	0.12–0.16	>0.30	188
(*Citrus sinensis*	spring flush						
Osbeck.)			0.09–0.11		0.12–0.16	0.17–0.25	189
Currants	NS	L		<0.17	0.25–0.30		190
(*Ribes nigrum* L.)							
Coffee (*Coffea*		L	<0.10		0.11–0.20	>0.20	191
arabica L.)							
Fig (*Ficus*	April	Basal L			0.42		43
carica L.)	May	Basal L			0.15		43
	July	Basal L			0.10		43
	September	Basal L			0.08		43
Grapevine	May–July	P/YML	<0.10		0.10–0.40		177
(*Vitis labrusca* L.)							
Grapevine	Fl	YML			0.20–0.40		192
(*Vitis vinifera* L.)							
Mango	NS				0.08–0.20		193
(*Mangifera*							
indica L.)							
Coconut palm	NS	YML		<0.10			43
(*Cocos*							
nucifera L.)							
Date palm	NS	YML			0.1–0.14		43
(*Phoenix*							
dactyifera L.)							
Oil palm	NS	YML			0.21–0.23		43
(*Elaeis*	NS	YML			0.23		43
guineensis Jacq.)							
Olive (*Olea*	July–August	L			0.10–0.30		177
europea L.)							
Papaya (*Carica*	NS	P/YML			0.22–0.40		49
papaya L.)							
Peach (*Prunus*	Midsummer	L			0.19–0.25		177
persica Batsch.)	July–August	L			0.26		180
	July–August	L		0.080	0.12	>0.30	178
	110 DAfl	L/mid shoot			0.3		181

TABLE 3.1   (*Continued*)

Species	Growth Stage	Plant Part	Deficient	Low	Sufficient	High	Reference
Pear (*Pyrus*	Midsummer	L			0.11–0.25		194
communis L.)	Midsummer	L			0.14–0.16		179
	Sept.	L	0.07		0.11–0.16		177
	110 DAfl	L/mid-shoot			0.20		181
Pecan (*Carya*	September	L			0.11–0.16		177
illinoinensis							
K. Koch)							
Pineapple	3–12 mo.	L	0.08		0.20–0.25		177
(*Ananas*							
comosus Merr.)							
Pistachio	September	L			0.14–0.17		195
(*Pistacia vera* L.)							
Plum	NS	L		<0.14			196
(*Prunus* spp.)	August	L			0.14–0.25		177
	110 DAfl	L/mid-shoot			0.20		181
Raspberry, Red	NS	YML		<0.30			190
(*Rubus idaeus* L.)		nonbearing					
		canes					
	Before Fl	YML			0.30–0.50		49
Strawberry	Pre-Fl	YML	0.10–0.30	0.10	0.30–0.50		197
(*Fragaria* spp.)	NS	YML			0.18–0.24		178
Walnut (*Juglans*	July	L	0.05–0.12		0.12–0.30		177
regia L.)	July–August	L		0.08	0.12	<0.30	178

D. Ornamentals

Species	Growth Stage	Plant Part	Deficient	Low	Sufficient	High	Reference
Chinese evergreen	NS	YML			0.20–0.40		49
(*Aglaonema*							
commutatum							
Schott.)							
Allamanda	NS	YML			0.25–1.0		49
(*Allamanda* spp.)							
Amancay or	NS	YML			0.30–0.75		49
Inca lily							
(*Alstroemeria*							
aurantiaca)							
Anthurium spp.	NS	B+MR+P/			0.20–0.75		49
		YML					
Asparagus fern	NS	YMCL			0.20–0.30		49
(*Asparagus*							
densiflorus							
Jessop)							
Asparagus Myers	NS	YMCL			0.30–0.70		49
(*Asparagus*							
densiflorus							
Jessop)							

Continued

TABLE 3.1   (*Continued*)

Species	Growth Stage	Plant Part	Deficient	Low	Sufficient	High	Reference
Azalea (*Rhododendron indicum* Sweet)	Fl	YML on Fl shoot	<0.20		0.29–0.50		198
Baby's breath (*Gypsophila paniculata* L.)	NS	YML			0.30–0.70		49
Begonia spp.	NS	YML			0.30–0.75		49
Bird of paradise (*Caesalpinia gilliesii* Benth.)	NS	B+MR+P/ YML			0.20–0.40		49
Bougainvillea spp.	NS	YML			0.25–0.75		49
Boxwood, Japanese (*Buxus japonica* Mull. Arg.)	NS	YML			0.30–0.50		49
Bromeliad Aechmea (*Aechmea* spp.)	Before FL				0.30–0.70		49
Caladium (*Caladium* spp.)	NS	B+MR			0.30–0.70		49
Calathea (*Calathea* spp.)	NS	YML			0.20–0.50		49
	5 mo	5th pr L from A of Lat	<0.1–0.15				199
Carnation (*Dianthus caryophyllus* L.)	17 mo	5th pr L from A of Lat			0.25–0.30		199
	1.5–2 mo	Unpinched plants	<0.05		0.20–0.30		198
Chrysanthemum (*Chrysanthemum xmorifolium* Ramat.)	Veg.&Fl	Upper L on Fl stem	<0.21		0.26–1.15		200
Christmas cactus (*Opuntia leptocaulis* DC)	NS	YML			0.60–1.0		49
Dieffenbachia (*Dieffenbachia exotica*)	Near Maturity	YML			0.20–0.35		201
Dracaena (*Dracaena* spp.)	NS	YML			0.20–0.50		49
Eugenia (*Eugenia* spp.)	NS	YML			0.40–0.80		49
Fern, Birdsnest (*Asplenium nidus* L.)	NS	YML			0.30–0.50		49

TABLE 3.1 *(Continued)*

Species	Growth Stage	Plant Part	Deficient	Low	Sufficient	High	Reference
Fern, Boston (*Nephrolepis exaltata* Schott.)	5–10 mo after planting	YMF			0.50–0.70		202
Fern, Leather-leaf (*Rumohra adaintiformis* G. Forst.)	NS	YMF			0.25–0.50		49
Fern, Maiden-hair (*Adiantum* spp.)	NS	YMF			0.30–0.60		49
Fern, Table (*Pteris* spp.)	NS	YMF			0.21–0.30		49
Fern, Pine (*Podocarpus* spp.)	NS	YML			0.25–1.0		49
Ficus spp.	NS	YML			0.10–0.50		49
Gardenia (*Gardenia jasminoides* Ellis)	NS	YML			0.16–0.40		49
Geranium (*Pelargonium zonale* L. Her.)	Fl	YML	<0.28		0.40–0.67		198
Gladiolus (*Gladiolus tristis* L.)	NS	YML			0.25–1.0		49
Gloxinia (*Gloxinia* spp.)	NS	YML			0.25–0.70		49
Hibiscus (H*ibiscus syriacus* L.)	NS	YML			0.25–1.0		49
Holly (*Ilex aquifolium* L.)	NS	YML			0.10–0.20		49
Hydrangea, Garden (*Hydrangea macrophylla* Ser.)	NS	YML			0.25–0.70		49
Ixora, Jungle Flame (*Ixora coccinea* L.)	NS				0.15–1.0		49
Jasmine (*Jasminum* spp.)	NS	YML			0.18–0.50		49
Juniper (*Juniperus* spp.)	Mature shoots	Tips/Stem			0.20–0.75		49
Kalanchoe (*Kalanchoe* spp.)	NS	4 L from tip			0.25–1.0		49

Continued

TABLE 3.1 *(Continued)*

Species	Growth Stage	Plant Part	Deficient	Low	Sufficient	High	Reference
Japanese privet (*Ligustrum japonicum* Thunb.)	NS	YML			0.20–0.50		49
Lilac (*Syringa xpersica* L.)	NS	YML			0.25–0.40		49
Lipstick plant (*Bixa orellana* L.)	NS	YML			0.20–0.40		49
Liriope (*Liriope muscari* L.H. Bailey)	NS	YML			0.25–0.35		49
Mandevilla (*Mandevilla* spp.)	NS	YML			0.20–0.50		49
Nepthytis (*Syngonium podophyllum* Schott.)	NS	YML			0.20–0.50		49
Natal plum (*Carissa macrocarpa* A. DC)	NS				0.18–0.6		49
Norfolk Island pine (*Araucaria hetrophylla* Franco)	NS	YML			0.20–0.30		49
Orchid, Cattleya (*Cattleya* spp.)	NS	5 cm tips / YML		0.07	0.11–0.17		49
Orchid, Cymbidium (*Cymbidium* spp.)	NS	5 cm tips / YML		0.07	0.11–0.17		49
Orchid, Phalaenopsis (*Phalaenopsis* spp.)	NS	5 cm tips LYML		0.10	0.30–0.17		49
Philodendron, Monstera (*Monstera deliciosa* Liebm.)	NS	B+MR+P/ YML			0.20–0.40		49
Philodendron, Split leaf (*Philodendron selloum* C. Koch)	NS	B+MR+P/ YML			0.25–0.40		49
Pittosporum, Japanese (*Pittosporum tobira* Ait.)	NS	YML			0.25–1.0		49

TABLE 3.1 *(Continued)*

Species	Growth Stage	Plant Part	Deficient	Low	Sufficient	High	Reference
Poinsettia (*Euphorbia pulcherrima* Willd.)	Before Fl 70 DAE	YML WP	<0.20		0.30–0.70 0.30–0.37		198 203
Pothos (*Epipremnum aureum* Bunt.)	NS	YML			0.20–0.50		49
Rose, Floribunda (*Rosa floribunda* Groep.)	Harvest	2nd & 3rd 5-leaflet L from Fl shoots	0.14		0.28–0.36		204
Rose, Hybrid Tea (*Rosa* spp.)	Harvest	2nd & 3rd 5-leaflet L from Fl shoot			0.28–0.36		204
Salvia (*Salvia* spp.)	NS	YML			0.30–0.70		49
Sanservieria (*Sansevieria* spp.)	NS	YML			0.15–0.40		49
Snapdragon (*Antirrhinum majus* L.)	NS	YML			0.30–0.50		49
Spathiphyllum (*Spathiphyllum wallisi* Regel)	< 4 mo > 4 mo	B+MR+P/ YML B+MR+P/ YML			0.25–1.0 0.20–0.80		49 49
Spider plant (*Chlorophytum comosum* Jacques)	NS	YML			0.15–0.40		49
PStatice (*Limonium perezii* F.T. Hubb)	NS	YMCL			0.30–0.70		
Umbrella plant (*Schefflera* spp.)	NS	Central L			0.20–0.35		205
Viburnum (*Viburnum* spp.)	NS	YML			0.15–0.40		49
Violet, African (*Saintpaulia ionantha* H. Wendl.)	NS	YML			0.30–0.70		49
Yucca (*Yucca* spp.)	NS	YML			0.15–0.80		49
Zebra plant (*Aphelandra squarrosa* Nees)	NS	YML			0.20–0.40		49

Continued

TABLE 3.1 (*Continued*)

Species	Growth Stage	Plant Part	Deficient	Low	Sufficient	High	Reference
E. Vegetable Crops							
Asparagus	Mid-growth	Fern needles from top 30 cm		0.17	0.20–0.23		43
(*Asparagus officinalis* L.) YP	Mid-growth	New fern from 10 cm tip PO$_4$-P	0.08		0.16		206
Garden bean	Harvest	L			0.24		207
(*Phaseolus*	Harvest	Pods			0.30		207
vulgaris L.)	Harvest	Seeds			0.36		207
	Mid-growth	P/4th L from tip PO$_4$-P	0.10		0.30		206
	Early Fl	P/4th L from tip PO$_4$-P	0.08		0.20		206
	Mature	L			0.30		43
Beets	Harvest	L		0.15	0.28	0.56	43
(*Beta*	Harvest	R		0.10	0.27	0.62	43
vulgaris L.)	NS	YML			0.25–0.50		49
Broccoli	Harvest	Head			0.79–1.07		43
(*Brassica*	Mid-growth	MR/YML PO$_4$-P	0.25		0.50		206
oleracea var.							
italica Plenck	Budding	MR/YML PO$_4$-P	0.20		0.40		206)
Brussels sprouts	Mid-growth	MR/YML PO$_4$-P	0.20		0.35		206
(*Brassica*							
oleracea var.	Late-growth	MR/YML PO$_4$-P	0.10		0.30		206
gemmifera Zenk.)							
Cabbage	Harvest	Head		0.13	0.38	0.77	43
(*Brassica*	Heading	MR/WL PO$_4$-P	0.25		0.35		206
oleracea var.							
capitata L.)							
Carrot	Harvest	L			0.26		43
(*Dacus carota*	Harvest	R		0.14	0.33	0.65	43
var. *sativus*	Mid-growth	PYML PO$_4$-P	0.20		0.40		206
Hoffm.)							
Cauliflower	Harvest	L (immature 4 cm)			0.62–0.70		43
(*Brassica*							
oleracea var.	Harvest	Heads		0.51	0.76	0.88	43
botrytis L.)	Buttoning	MR/YML PO$_4$-P	0.25		0.35		206
Celery	Mid-season	YML			0.30–0.50		208
(*Apium*	Mid-season	Outer P		<0.55			209
graveolens var.	Mid-season	Outer P		<0.46			210
dulce Pers.)	Harvest	Stalks		0.43	0.64	0.90	43
	Mid-season	P PO$_4$-P			0.28–0.34		43
	Mid-season	PYML PO$_4$-P	0.20		0.40		206
	Near maturity	PYML PO$_4$-P	0.20		0.40		206

TABLE 3.1 (*Continued*)

Growth Species	Plant Stage	Part	Deficient	Low	Sufficient	High	Reference
Cucumber (*Cucumis sativus* L.)	Budding	L/5th L from tip		0.28–0.34	0.34–1.25	>1.25	49
	Fruiting	L/5th L from tip		0.22–0.24	0.25–1.0	>1.0	49
	Early fruiting	P/6th L from tip PO_4-P	0.15		0.25		206
Eggplant (*Solanum melongena* L)	Mature leaves	PYML		0.25–0.29	0.30–0.12	>1.2	49
Endive (*Cichorium endiva* L.)	8-L	YML			0.45–0.80		211
	Maturity	YML			0.40–0.60		211
	8-L	YML			0.54		212
Escarole (*Cichorium endiva* L.)	8-L	YML			0.45–0.60		211
	Maturity	YML			0.35–0.45		211
	6-L	YML			0.50		212
Lettuce (*Lactuca sativa* L.)	28 DAP	L			0.55–0.76		213
	8-L stage	MR/YML		<0.43			214
	Mid-growth	MR/YML		<0.40			215
	Mid-growth	MR/YML			0.35–0.60		216
	Heading	MR/YML PO_4-P	0.20		0.40		206
	Harvest	MR/YML PO_4-P	0.15		0.25		206
Melons (*Cucumis melo* L.)	Harvest	B			0.25–0.40		208
	Early growth	P/6th L from GT PO_4-P	0.20		0.40		206
	Early fruit	P/6th L from GT PO_4-P	0.15		0.25		206
	1st Mature fruit	P/6th L from GT PO_4-P	0.10		0.20		206
Onion (*Allium cepa* L.)	2-leaf				0.44		216
	4-leaf				0.31		216
	6-leaf				0.34		216
Peas (*Pisum sativum* L.)	Mid-growth	YML			0.25–0.35		208
	Early flowering	L			0.33		207
	Flowering	Entire Tops			0.30–0.35		208
		Entire Tops		0.19	0.29		43
	Early flowering	Pods			0.20		207
	Harvest	Seeds			0.35		207
	Early flowering	Pods		0.23	0.57	0.78	43
Pepper (*Capsicum annuum* L.)	Mid-growth	YML			0.30–0.70		208
	Early-growth	PYML PO_4-P	0.20		0.30		206
	Early fruit set	PYML PO_4-P	0.15		0.25		206
Potato (*Solanum tuberosum* L.)	Mid-growth	PYML			0.20–0.40		208
	Tuber initiation				0.38–0.45		217
	Tubers mature				0.14–0.17		217

Continued

TABLE 3.1 (*Continued*)

Species	Growth Stage	Plant Part	Deficient	Low	Sufficient	High	Reference
	Early season	P/4th L from growing tip PO$_4$-P	0.12		0.20		206
	Mid-season	P/4th L from growing tip PO$_4$-P	0.08		0.16		206
	Late-season	P/4th L from growing tip PO$_4$-P	0.05		0.10		206
Radish	Maturity	L		<0.40			215
(*Raphanus sativus* L.)	Maturity	L		<0.45			219
Spinach	48 DAP	L		0.10	0.25–0.35		43
(*Spinacia	40–50 DAP	YML			0.48–0.58		208
oleracea* L.)	Mature	YML			0.30–0.50		208
	Mature	WP		0.27	0.72	1.17	43
	Mid-growth	PYML PO$_4$-P	0.20		0.40		206
Sweet corn	Silking	Ear-leaf		<0.25			136
(*Zea mays* var.	Silking	Ear-leaf			0.20–0.30		208
rugosa Bonaf.)	8-L stage	Ear-leaf		<0.31			220
	8-L stage	Ear-leaf		<0.38			221
	Tasseling	MR of 1st L above ear PO$_4$-P	0.05		0.10		206
Sweet potato	4th L	L		0.20	0.23		43
(*Ipomoea	Mid-growth	ML			0.20–0.30		208
batatas* Lam.)	Harvest	Tubers		0.06	0.12	0.22	43
	Mid-growth	P/6th L from GT PO$_4$-P	0.10		0.20		206
Tomato	Early fruiting	L	0.24–0.35		0.42–0.72		43
(*Lycoperscion	Harvest	YML		<0.13	0.40		222
esculentum* Mill.)	Early bloom	P/4th L from GT PO$_4$-P	0.20		0.30		206
	Fruit 2.5 cm	P/4th L from GT PO$_4$-P	0.20		0.30		206
	Fruit color	P/4th L from GT PO$_4$-P	0.20		0.30		207
Watermelon	Flowering	L/5th L from tip			0.30–0.80		49
(*Citrullus lanatus							
Matsum. & Nakai)	Fruiting	L/5th L from tip			0.25–0.70		49

TABLE 3.1 (*Continued*)

Species	Growth Stage	Plant Part	Deficient	Low	Sufficient	High	Reference
	P/6th L from tip	P/6th L from GT PO$_4$-P	0.15		0.25		206

Note: Phosphorus is reported in units of percent total phosphorus on a dry mass basis except where designated otherwise under plant part. Units of PO$_4$-P are phosphorus in sap of petioles or leaf midribs.

Abbreviations used for plant parts:

A = apex	LB = leaf blade
B = blades	MR = midrib
DAP = days after planting	NS = not specified (pertaining to growth stage)
DAE = days after emergence	P = petiole
DAfl = days after flowering	PYML = petiole from young mature leaf
F = fern	R = roots
Fl = flowers or flowering	WP = whole aboveground plant
GT = growing tip	YML = young mature leaves synonymous with recently mature leaf and most recently
L = leaves	developed leaf

no other values were available, some values were drawn from sources that did not cite original research (49). Generally, crops require a preplant application of phosphorus fertilizer in the case of annual crops or before the fruiting cycle begins in the case of perennial crops. Diagnosis of a phosphorus deficiency by tissue analysis for annual crops is often postmortem for the existing crop.

3.2.3 SOIL TESTING FOR PHOSPHORUS

As noted in a previous section, crop response to phosphorus is correlated poorly to the total amount of phosphorus in a soil. Therefore, a successful soil test should represent some index of phosphorus availability. The development of a soil test requires selection of an extractant, development of studies that correlate the amount of nutrient extracted with phosphorus accumulation by crops, and calibration studies that determine a relationship between soil test results and amount of fertilizer required for optimal production.

Over the past century, a number of soil-testing procedures have been proposed, and several excellent reviews on soil testing for phosphorus have been published (50–53).

This chapter focuses on historical developments, mode of action, and generalized interpretations of the major phosphorus soil tests utilized in the United States.

The major soil tests that have been used or proposed in the United States are summarized in Table 3.2. Most early soil tests were developed empirically and were based on simple correlations between extractant and some measure of crop response to fertilization with phosphorus. However, based on the phosphorus-fractionation method developed by Chang and Jackson (54), inferences have been made concerning the mode of action, or the forms of phosphorus extracted by various solutions. The inferred modes of action for various chemical extractant components are presented in Table 3.3. Generally, water or dilute salt solutions characterize phosphorus in the soil solution or the intensity factor, whereas acids, complexing solutions, or alkaline buffer solutions generally characterize the quantity factor. Tests based on water extraction often correlate well with phosphorus accumulation in shallow-rooted, fast- growing vegetable crops. However, soil tests capable of better characterizing the labile fraction and capacity factor generally produce more reliable results for field and orchard crops.

An early soil test for phosphorus aimed at characterizing available phosphorus was the 1% citric acid test developed by Dyer (3). This test was adapted in England but was not used widely in the

TABLE 3.2
Some Historical and Commonly Used Soil Test and Extracting Solutions for Determining Available Soil Phosphorus

Name of Test	Extractant	Reference
AB-DPTA	1M NH_4HCO_3 + 0.005 M DPTA, pH 5	59
Bray I	0.025 N HCl + 0.03 N NH_4F	6
Bray II	0.1 N HCL + 0.03 N NH_4F	6
Citric acid	1% Citric acid	3
EDTA	0.02 M Na_2-EDTA	61
Mehlich 1	0.05 M HCl + 0.0125 M H_2SO_4	224
Mehlich 3	0.015 M NH_4F + 0.2 M CH_3COOH	56
	+ 0.25 M NH_4NO_3 + 0.013 M HNO_3	
Morgan[a]	0.54 N HOAc + 0.7 N NaOAc, pH4	5
Olsen	0.5 M $NaHCO_3$, pH 8.5	58
Truog	0.001 M H_2SO_4 + $(NH4)_2SO_4$, pH 3	4
Water[b]	Water	225

[a]A modification of the Morgan by Wolf to include 0.18 g/L DPTA gives better correlations for micronutrients.
[b]From: C.A. Sanchez. Soil Testing and Fertilizer Recommendations for Crop Production on Organic Soils in Florida. University of Florida Agricultural Experiment Station Bulletin 876, Gainesville, 1990.

TABLE 3.3
Forms of Phosphorus Extracted by Constituent Components of Commonly Used Soil Test Extractants[a]

Chemical	Form of Phosphorus Extracted
Acid (H^+)	Solubilizes all chemical P in the following order Ca-P>Al-P>Fe-P
Bases (OH^-)	Solubilizes Fe-P and Al-P in respective order. Also results in release of some organic P
Fluoride ion	Forms complexes with Al thus releasing Al-P. Also precipitates Ca as CaF_2 and thus will extract more Ca-P as $CaHPO_4$. No effect on basic Ca-P and Fe-P
Bicarbonate ions	Precipitate Ca as $CaCO_3$ thus increasing solubility of Ca-P. Also remove Al-bound P
Acetate ions	Form weak complexes with polyvalent metal ions. Possibly prevents readsorption of P removed by other ions
Sulfate ions	Appear to reduce readsorption of P replaced by H ions

[a]Adapted from G.W. Thomas and D.E. Peaslee, in *Soil Testing and Plant Analysis*. Madison, WI: Soil Sci. Soc. Am. Inc., 1973 and E.J. Kamprath and M.E. Watson, in *The Role of Phosphorus In Agriculture*. American Society of Agronomy Inc. 677 South Segoe Road, Madison WI 53711, 1980.

United States. A dilute acid test proposed by Truog (4) and a test based on a universal soil extracting solution proposed by Morgan (5) were among the earliest soil tests used in the United States.

The test based on the Bray-I extractant was perhaps the first to be implemented widely in soil-testing laboratories in the United States, and it is still extensively used in the midwestern United States. This mild-acid solution has been shown reliably to predict crop response to phosphorus fertilization on neutral to acidic soils. However, the test is much less effective in basic soils, where the acid is neutralized quickly by the soil bases present and fluoride ions are precipitated by calcium (55).

In the southeastern United States, the Mehlich 1 (M-I) soil-test extractant is used commonly for simultaneous extraction of P, K, Ca, Mg, Cu, Mn, Fe, and Zn. The M-I soil test does not correlate with crop response on calcareous soils probably for the same reasons the Bray-I test does not. Consequently, the Mehlich 2 (M-II) test was introduced as an extractant that would allow simultaneous determinations of the same nutrients over a wide range of soil properties. However, the corrosive properties of the M-II in instruments discouraged wide acceptance of this extractant and prompted modifications that ultimately became the Mehlich 3 (M-III) extraction. The M-III has been shown to be reliable across a wide range of soil–crop production circumstances (56,57).

The sodium bicarbonate (NaHCO$_3$) (58) soil test for phosphorus generally correlates well with crop response on calcareous soils in the western United States. The NH$_4$HCO$_3$-DPTA (diethylenetriaminepentaacetic acid) soil test also has been used for the simultaneous determination of P, K, Zn, Fe, Cu, and Mn (59,60) and performs similar to the NaHCO$_3$ test with respect to phosphorus. Another test that shows good correlations on calcareous soils is the EDTA (ethylenediaminetetraacetic acid) soil test (61).

Isotopic dilution techniques (53) and phosphorus sorption isotherms (62) have been used not only to characterize the labile phosphorus fraction but also the phosphorus-buffering capacity of soils. However, these approaches are too tedious and costly to be used as routine soil tests.

Ultimately, soil-test phosphorus levels must be converted into phosphorus fertilizer recommendations for crops. A useful starting point is the determination of critical soil-test levels, that is the soil-test phosphorus level above which there is no response to phosphorus fertilizer. An example of a critical phosphorus soil-test level based on water extraction for celery is shown in Figure 3.5. Using the double calibration approach described by Thomas and Peaslee (50) information on how much fertilizer is required to achieve the critical concentration would result in a fertilizer recommendation. This approach is used for Histosols by the Soil Testing Laboratory at the University of Florida. An example of resulting fertilizer recommendations for several commodities is shown in Figure 3.6.

The laboratory mentioned above makes recommendations for Histosols over a limited geographical location. However, most soil-testing laboratories make recommendations over large geographical area and across more diverse soil types. Under most situations, quantitative information on how phosphorus fertilizer additions change with soil-test phosphorus levels across a range of soil types rarely exist. Owing to this uncertainty, most soil-testing laboratories make phosphorus fertilizer recommendations based on probability of response using class interval grouping such as low, medium, and high.

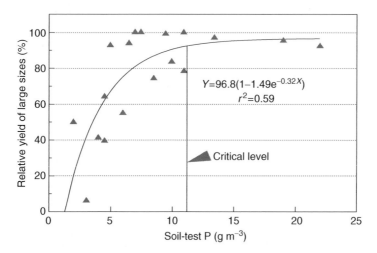

FIGURE 3.5 Critical soil-test phosphorus levels for large, harvest-size celery on Florida Histosols. (Adapted from C.A. Sanchez et al., *Soil Crop Sci. Soc. Fla. Proc.* 29:69–72, 1989.)

Crops produced on a soil scoring very low or low have a very high probability of responding to moderate to high rates of fertilization. Crops produced on soils classified as medium frequently respond to moderate rates of fertilization, and typically, crops produced on soils testing high for phosphorus would not respond to fertilization (Table 3.4). General soil-test phosphorus interpretations for mineral soils in California and Florida are shown in Tables 3.5 and 3.6 for comparative purposes. In California, only the probability of response to $NaHCO_3$-phosphorus is indicated, and it is presumed that specific fertilizer recommendations are left to service laboratories, crop consultants, or the grower. In Florida, specific fertilizer recommendations for phosphorus are made for each level of M-I-extractable phosphorus. Furthermore, research aimed at validating and calibrating soil-test fertilizer recommendations for phosphorus in Florida is ongoing (63–65). It must be stressed that all fertilizer recommendations must be calibrated locally, and readers are advised to consult the cooperative extension service for recommendation guidelines specific to their region.

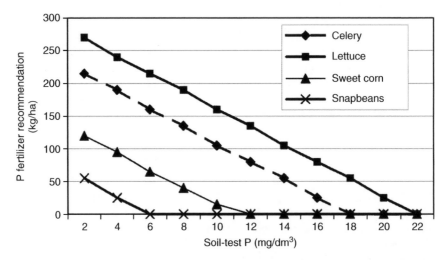

FIGURE 3.6 Fertilizer phosphorus recommendations for selected crops on Everglades Histosols. (Adapted from C.A. Sanchez, Soil Testing and Fertilizer Recommendations for Crop Production on Organic Soils in Florida. University of Florida Agricultural Experiment Station Bulletin 876, Gainesville, 1990.)

TABLE 3.4
Classifications for Soil Nutrient Tests and Yield Potential and Crop Response to Application of Phosphorus-Containing Fertilizers

Classification	Yield Potential and Need for Fertilizer
Very low	Very high probability of response to fertilizer. Crop-yield potential less than 50% of maximum. Deficiency symptoms possible. Highest recommended rate of fertilizer required
Low or poor	High probability of response to fertilizer. Crop yield potential 50 to 75%. No pronounced deficiency symptoms. Needs modest to high fertilizer application
Medium	Crop yield potential >75% without fertilizer addition. Low to modest rates of fertilizer may be required for economic maximum yield when yield potential high or for quality for high value crops
High	Very low probability of yield increase due to added fertilizer
Very High	No positive response to fertilizer. Crop may be affected adversely by fertilizer addition

Source: Adapted from B. Wolf, *Diagnostic Techniques for Improving Crop Production*. Binghampton, New York: The Hayworth Press Inc., 1996.

TABLE 3.5
General Guidelines for Interpreting the NaHCO3 Phosphorus Test for Fertilizing Vegetable Crops in California

Vegetable	Response Likely (mg/kg)	Response Unlikely (mg/kg)
Lettuce	<20	>40
Muskmelon	<8	>12
Onion	<8	>12
Potato (mineral soils)	<12	>25
Tomato	<6	>12
Warm-season vegetables	<5	>9
Cool-season vegetables	<10	>20

Source: Adapted from Soil and Plant Testing in California, University of California, Division of Agricultural Science Bulletin 1879 (1983). Modified based on personal communication with Husien Ajwa, University of California, Davis.

3.3 FACTORS AFFECTING MANAGEMENT OF PHOSPHORUS FERTILIZATION

3.3.1 CROP RESPONSE TO PHOSPHORUS

As noted in the previous section, the amounts of phosphorus applied to crops should be based ideally on a well-calibrated soil test. However, even at a given soil-test phosphorus level, the amount of phosphorus fertilizer required for economic-optimum yield often will vary with crop. Generally, fast-growing, short-season vegetable crops have higher phosphorus requirements than field and orchard crops. Many deciduous fruit crops infrequently respond to phosphorus fertilization even if soil tests are low (47). It is presumed often that surface soil tests fail to characterize the full soil volume where trees take up nutrients or the fact that trees take up nutrients over a considerable time period.

There is considerable variability in phosphorus response among species of vegetable crops (66–70). For example, lettuce generally shows larger responses to phosphorus than most other vegetable crops including cucurbit and brassica species. Furthermore, genetic variation in response to phosphorus within species also exists. For example, Buso and Bliss (71), in sand culture experiments found that some butterhead types of lettuce (*Lactuca sativa* L.) were less efficient than other types under phosphorus-deficient regimes. However, the magnitude of this variation is usually small compared to the uncertainties and natural variation in soil-test-based phosphorus fertilizer recommendations. Generally, field experiments show that lettuce has a similar response to phosphorus regardless of cultivar or morphological type (72,73). As shown by the data presented in Figure 3.7, a similar soil-test phosphorus index level of 22 mg dm^3 was required for maximum yield regardless of lettuce type (73).

Mechanisms of phosphorus-utilization efficiency have been classified into three broad classes including (a) secretion or exudation of chemical compounds into the rhizosphere, (b) variation in the geometry or architecture of the root system, and (c) association with microorganisms (74). Future opportunities for improving phosphorus-utilization efficiency in crops through genetic manipulation of traits exist (75).

In conclusion, as available data permit, soil-test recommendations for phosphorus should be customized by crop. However, at present, soil-test-based recommendations are generally not sufficiently sensitive to allow recommendations to accommodate the more subtle genetic variation among cultivars within crop species.

TABLE 3.6
Phosphorus Fertilizer Recommendations for Various Vegetable Crops on Sandy Soils in Florida Based on the Mehlich 1 Soil Test

Soil Test P (mg/kg)	<10	10–15	16–30	31–60	>60
Classification	Very Low	Low	Medium	High	Very High
Crop	P Fertilizer Recommendation (kg/ha)				
Bean	60	50	40	0	0
Beet	60	50	40	0	0
Broccoli	75	60	50	0	0
Brussel sprouts	75	60	50	0	0
Cabbage	75	60	50	0	0
Carrot	75	60	50	0	0
Cauliflower	75	60	50	0	0
Celery	100	75	50	0	0
Corn, sweet	75	60	50	0	0
Cucumber	60	50	40	0	0
Eggplant	75	60	50	0	0
Endive	75	60	50	0	0
Escarole	75	60	50	0	0
Kale	75	60	50	0	0
Lettuce	75	60	50	0	0
Muskmelon	75	60	50	0	0
Mustard	75	60	50	0	0
Okra	75	60	50	0	0
Onion/bulb	75	60	50	0	0
Onion/leek	60	50	40	0	0
Onion/bunching	60	50	40	0	0
Parsley	75	60	50	0	0
Pea	40	40	30	0	0
Pepper, bell	75	60	50	0	0
Potato	60	60	30	0	0
Potato, sweet	60	50	40	0	0
Pumpkin	60	50	40	0	0
Radish	60	50	40	0	0
Spinach	60	50	40	0	0
Squash	60	50	40	0	0
Strawberry	75	60	50	0	0
Tomato	75	60	50	0	0
Turnip	75	60	50	0	0
Watermelon	75	60	50	0	0

Source: Adapted from G. Hochmuth and E. Hanlon, IFAS Standarized Fertilization Recommendations for Vegetable Crops. Fla. Coop. Ext. Serv. Circ. 1152, 1995.

3.3.2 SOIL WATER

Phosphorus availability is affected by soil water conditions. Soil water affects soil reactions governing the release and diffusion of phosphorus in the soil solution and ultimately the positional availability of phosphorus relative to root growth. Generally, maximum availability of phosphorus for most crops has been associated with a soil water tension of about 1/3 bar (76).

The dissolution of fertilizer phosphorus and all amorphous and mineral phosphorus compounds in the soil depends on soil water. Furthermore, under anaerobic conditions, the reduction of ferric

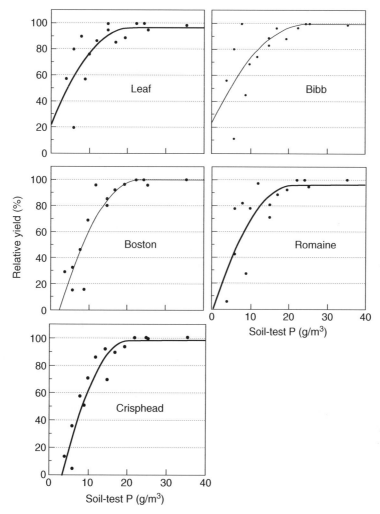

FIGURE 3.7 Response of five lettuce types to soil-test phosphorus. (Adapted from C.A. Sanchez and N.M. El-Hout, *HortScience* 30:528–531, 1995.)

phosphates to ferrous phosphates might result in additional increased phosphorus solubility (77,78). Nevertheless, it is the general view that with the exception of aquatic crops, excessive water resulting in poor aeration would actually restrict phosphorus uptake by crops in spite of this enhanced solubility. However, Bacon and Davey (79), using trickle irrigation in an orchard, noted increased phosphorus availability during and immediately after each irrigation and noted that available phosphorus decreased rapidly as soil moisture declined below field capacity. These authors attributed this increased phosphorus availability to the reduction of amorphous iron phosphates in anaerobic micro-sites.

The volume of soil that is occupied by water affects the cross-sectional area through which phosphorus can diffuse (80). Thus, the lower the soil moisture, the more tortuous the path of diffusion and the greater the likelihood of contact with soil constituents that render phosphorus insoluble.

Under most conditions, phosphorus is applied near the soil surface. Thus, during dry periods in nonirrigated production systems, crops largely draw soil moisture from lower soil depths, and phosphorus deficiencies can arise (81). This condition is generally not a problem in irrigated production systems where root growth extends to near the soil surface.

3.3.3 SOIL TEMPERATURE

Soil temperature affects reactions that govern the dissolution, adsorption and diffusion of phosphorus. Although sorption and desorption generally occur concurrently, an increase in soil temperature increases kinetics of reactions (82) and enables more rapid equilibration among nonlabile, labile, and solution phosphorus pools, resulting in more rapid replenishment of solution phosphorus as phosphorus is taken up by crops. Sutton (83) concluded that most of the effect of temperature on available phosphorus was due to inorganic reactions, since the effect occurred too rapidly to be explained by microbial mineralization.

Soil temperature also has the potential to affect root uptake of phosphorus. With excised corn roots in solution culture experiments, Carter and Lathwell (84) reported that absorption increased as temperature was increased from 20 to 40°C. The effects of temperature on soil reactions may be more important than effects on plant physiology. Singh and Jones (85) noted that changes in temperature had a more pronounced effect on the phosphorus nutrition of Boston lettuce in soil culture than in solution culture.

In production systems where crops are seeded and harvested over the same time interval each year, soil temperature is unlikely to substantially confound soil-test-based fertilizer recommendations for phosphorus. However, in crop production situations where planting and harvesting are extended over seasonal changes, such as many vegetable production systems, temperature changes can affect the amount of fertilizer required for maximum production. Lingle and Davis (86) reported that tomatoes seeded in cool soils showed a larger growth (dry mass) response to phosphorus than those seeded in warm soils. Locascio and Warren (87) noted that tomato (*Lycopersicon esculentum* Mill.) growth increased with applications up to 550 kg P/ha at 13°C but only to 140 kg P/ha at 21 or 30°C. Research has shown that the phosphorus rate required for maximum production of lettuce in deserts increased as temperatures during the growing season decreased (88,89). Lettuce produced in the desert of southwestern United States is planted every day from September through January and is harvested daily from November through April with mean soil temperatures ranging from 4 to 18°C. As illustrated in Figure 3.8, soil-test levels for phosphorus requirement for maximum lettuce yield decreased as mean soil temperature during the growing season increased.

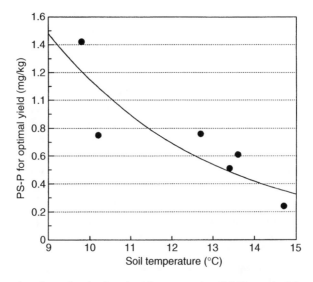

FIGURE 3.8 Soil test phosphorus level using phosphorus sorption (PS-P) required for maximum yield of lettuce as affected by soil temperature. (Adapted from Gardner and Sanchez, unpublished data.)

3.3.4 SOURCES OF PHOSPHORUS

Most phosphorus-containing fertilizers are derived from mined phosphate rock. In some unique production situations on acidic soils, phosphate rock can be used directly as a phosphorus source. Most cropping systems show the best response to water-soluble phosphorus fertilizers. Water-soluble phosphorus fertilizers are produced by reacting phosphate rock with sulfuric or phosphoric acid (90). Ammonium phosphates are made by passing anhydrous ammonia through phosphoric acid. This production includes diammonium phosphate and monoammonium phosphate.

The agronomic effectiveness of phosphorus fertilizers was reviewed by Engelstad and Terman (91). Most crops require readily available phosphorus, and most soluble sources perform similarly. However, in some situations the ammonium phosphates produce phytotoxicity (92), and their use is often discouraged when high amounts of phosphorus are required. For example, for economic reasons, diammonium phosphate typically is broadcast applied for lettuce production in the southwestern desert, but its use is discouraged when broadcast rates are high or when phosphorus fertilizer is banded near the plants.

Soluble, dry fertilizers and solution fertilizers perform similarly under many production systems. However, there are some unique production situations where solution sources may present logistical advantages. Often solution sources are easier to use in band placement or point-injection technologies. Generally, solution sources would be utilized in application with irrigation water.

In conclusion, under most conditions, cost considerations, available application technologies, and the potential for phytotoxicity are the major determining factors influencing the selection of sources of phosphorus fertilizers.

3.3.5 TIMING OF APPLICATION OF PHOSPHORUS FERTILIZERS

Overwhelming evidence indicates that for annual crops, phosphorus fertilizers should largely be applied preplant. Phosphorus moves to plant roots primarily by diffusion, and young seedlings of most annual crops are very sensitive to phosphorus deficits. Furthermore, yields of some crops often fail to recover fully from transitory phosphorus deficits (93).

Grunes et al. (94) showed that the proportion of fertilizer phosphorus absorbed by sugar beets (*Beta vulgaris* L.) decreased as the time of application was delayed. Lingle and Wright (95) reported that muskmelons (*Cucumis melo* L.), which showed large responses to phosphorus at seeding, showed no response to sidedressed phosphorus fertilization. Sanchez et al. (96) reported that a preplant phosphorus deficit in lettuce could not be corrected by sidedressed fertilization. Preplant broadcast or band applications are usually recommended for annual crops.

3.3.6 PLACEMENT OF PHOSPHORUS FERTILIZERS

The literature contains many accounts recording the positive effects of applying phosphorus fertilizer to a localized area, usually near the plant roots, as opposed to a general soil broadcast application. Reviews on the subject of fertilizer placement should be consulted for detailed information (97,98). Localized placement of phosphorus fertilizers might include row, band, or strip placement.

It is generally presumed that a localized or band application reduces fertilizer contact with the soil thereby resulting in less phosphorus sorption and precipitation reactions and, thus, enhanced availability to crops. However, for soils with a high phosphorus-fixing capacity, where phosphorus is relatively immobile, placement of the fertilizer where root contact is enhanced may be an equally or more important mechanism than restricting fixation (99–101).

The relative benefits of localized placement of phosphorus fertilizers are neither constant nor universal across crop production situations. This fact is illustrated by a series of experiments that the author conducted to improve phosphorus fertilizer use for vegetable crops produced on Histosols (102,103). The amount of phosphorus required for lettuce production could be reduced by at least 50% if phosphorus was banded instead of broadcast (Figure 3.9). However, band placement was not a viable strategy for improving phosphorus-use efficiency for celery under the

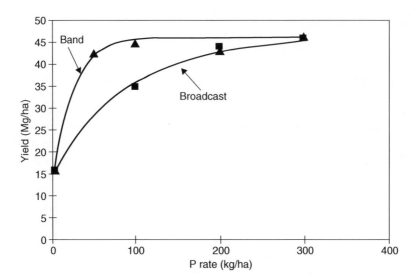

FIGURE 3.9   Marketable yield of lettuce as affected by phosphorus rate and placement. (Adapted from C.A. Sanchez et al. *J. Am. Soc. Hortic. Sci.* 115:581–584, 1990.)

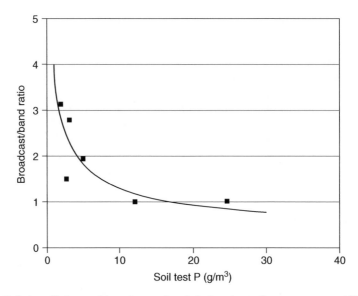

FIGURE 3.10   Relative efficiency of broadcast to banded phosphorus for sweet corn as affected by soil-test phosphorus level.

existing production system. For sweet corn (*Zea mays rugosa* Bonaf.), the relative efficiency of banded to broadcast phosphorus depended on soil-test level (Figure 3.10). The relative efficiency was greater than 3:1 (band:broadcast) at low soil-test phosphorus levels but approached 1:1 as soil-test phosphorus approached the critical value. Others have reported a relationship between the relative efficiency of the localized placement of phosphorus and soil-test levels (105–107). Many factors including crop root morphology, length of crop growing season, soil chemical and physical characteristics, and crop cultural practices interact to influence the relative crop response to broadcast or band fertilization.

3.3.7 Foliar-Applied Phosphorus Fertilization

Foliar fertilization with phosphorus is generally not practiced to the extent that it is done with nitrogen and micronutrient fertilizers although a limited amount of fertilizer phosphorus can be absorbed by plant foliage. Silberstein and Witwer (108) tested various organic and inorganic phosphorus-containing compounds on vegetable crops. They generally observed small responses in plant growth, but some compounds caused injury at phosphorus concentrations as low as 0.16%. They concluded that orthophosphoric acid was the most effective foliar phosphorus fertilizer evaluated. Barrel and Black (109,110) reported that several condensed phosphates and some phosphate fertilizers containing phosphorus and nitrogen could be applied at 2.5 to 3 times the quantity of orthophosphate without causing leaf damage. Yields of corn and soybeans (*Glycine max* Merr.) were higher with tri-polyphosphate and tetra-polyphosphate than with orthophosphate.

Teubner (111) reported that although about 12% of the phosphorus in the harvested plant parts of some field-grown vegetable crops could be supplied through multiple foliar sprays, foliar phosphorus fertilization did not increase total phosphorus absorbed or crop yields. Upadhyay (112) reported that the yield of soybeans were highest when all fertilizer phosphorus was soil-applied, intermediate where 50% of the phosphorus was soil-applied and 50% foliar-applied, and lowest where all the phosphorus was foliar-applied.

Some research suggests that phosphorus in combination with other nutrients might delay senescence and increase yields, but results are inconsistent. Garcia and Hanway (113) reported that foliar applications of N, P, K, and S mixtures during seed filling seemed to delay senescence and increase yield in soybean and the complete mixture produced greater yields than foliar sprays where the mixture was incomplete. Subsequent work with soybeans by others ranged from no-yield response (114) to yield reduction (115) for foliar mixtures containing phosphorus. Similar negative responses have been obtained with other crops. Harder et al. (116,117) observed temporary decrease in photosynthesis and a decrease in grain yield of corn (*Zea mays* L.) receiving foliar N, P, K, and S. Batten and Wardlaw (118) reported that applying monobasic ammonium phosphate to the flag-leaf of phosphate-deficient wheat (*Triticum aestivum* L.) delayed senescence but failed to increase grain yield.

Because only a modest portion of the crop's total phosphorus requirement can be met by foliar application and foliar fertilization does not produce consistent positive responses where residual soil phosphorus or soil-applied fertilizer phosphorus is sufficient, foliar fertilization with phosphorus is seldom recommended as a substitute for soil fertilization practices.

3.3.8 Fertilization in Irrigation Water

Although application of fertilizer in irrigation water (fertigation) is a common practice with mobile nutrients such as nitrogen, it is less common with phosphorus because of concerns about efficiency of utilization. Owing to the soil reactions discussed in a previous section, it is often presumed that much of the phosphorus applied with water will be tied up at its point of contact with the soil. Nevertheless, there are some situations where fertigation is a viable and economical means of delivering phosphorus for crop production.

The downward movement of phosphorus in soil is influenced strongly by soil texture as shown in the laboratory (119,120) and field experiments (121,122). In one study, sprinkler-applied phosphorus moved to a depth of approximately 5 cm in a clay loam soil and to approximately 18 cm in a loamy sand (121). On a basin surface-irrigated Superstition sand that received 91 cm of water, phosphorus moved to a depth of 45 cm (123).

Phosphorus source seems to be another important factor affecting phosphorus movement in soils and thus the efficacy of fertigation. Stanberry et al. (124), using radioautographs to trace P32 movement in Superstition sand, noted that phosphorus from phosphoric acid and monocalcium phosphate moved vertically across the length of the photographic film (20 cm) compared to dicalcium phosphate and tricalcium phosphate, which showed negligible movement. Lauer (122) reported that sprinkler-applied monoammonium phosphate, urea phosphate, and phosphoric acid showed similar movement in soils. However, ammonium polyphosphate penetrated only to 60 to 70% of the depth of the other sources. Rauschkolb (125) reported that glycerophosphate moved slightly farther than orthophosphate when injected through a trickle-irrigation system but phosphorus from both sources moved a sufficient distance into the root zone such that phosphorus availability was adequate for tomatoes. O'Neill (126) reported that orthophosphoric acid applied in the irrigation water for trickle-irrigated citrus (*Citrus* spp. L.) was delivered to a greater soil volume than triple superphosphate applied directly below the emitter. The phosphoric acid also lowered the pH of the irrigation water sufficiently to eliminate clogging problems associated with the precipitation of phosphorus in the irrigation lines.

In established perennial crops such as citrus or deciduous fruits, fertigation is often a viable means of phosphorus delivery, regardless of the method of irrigation, because tractor application and incorporation would likely cause root damage and broadcast application would not necessarily be more efficient than fertigation. For fast-growing annual crops, where most phosphorus should be applied preplant, fertigation might not result consistently in production benefits compared to band application but might be economical or even necessary depending on the opportunities and constraints of the irrigation delivery system. Bar-Yosef et al. (127) noted no difference between broadcast and drip-injected phosphorus for sweet corn on a sandy soil. Carrijo et al. (128) reported that phosphorus applied through the irrigation system was more efficient than preplant incorporation for tomato produced on sandy soils testing low in phosphorus. Reports that phosphorus fertigation sometimes produced positive responses have been attributed to band-like effects where phosphorus is delivered in or close to the root zone and not widely mixed with the soil (128,129). Overall, the efficacy of phosphorus fertigation depends on soil texture, phosphorus source, irrigation method and amount, and cropping system utilized.

REFERENCES

1. A.D. Hall. The analysis of the soil by means of the plant. *J. Agric. Sci.* 1:65–88, 1905.
2. P. Macy. The quantitative mineral nutrient requirements of plants. *Plant Physiol.* 11:749–764, 1936.
3. B. Dyer. Analytical determination of probably available mineral plant food in soils. *Trans. Chem. Soc.* 65:115–167, 1894.
4. E. Truog. Determination of the readily available phosphorus of soils. *Agron. J.* 22:874–882, 1930.
5. M.F. Morgan. Chemical Soil Diagnosis by the Universal Soil Testing System. Connecticut Agricultural Experiment Station Bulletin 45, 1941.
6. R.H. Bray, L.T. Kurtz. Determination of total, organic, and available forms of phosphorus in soils. *Soil Sci.* 59:39–45, 1945.
7. H.B. Kitchen (ed.). *Diagnostic Techniques for Soil and Crops.* The American Potash Institute. Washington DC, 1948.
8. E.J. Russell. *Soil Conditions and Plant Growth.* 9th ed. New York: Wiley, 1961.
9. D.T. Clarkson, J.B. Hanson. The mineral nutrition of higher plants. *Ann. Rev. Plant Physiol.* 32:239–298, 1980.
10. R.L. Bieleski. Phosphate pools, phosphate transport, and phosphate availability. *Ann. Rev. Plant Physiol.* 24:225–252, 1973.
11. R.L. Halstead, R.B. McKercher. Biochemistry and cycling of phosphorus. In: A. Paul, A.D. McLaren, eds. *Soil Biochemistry.* Vol. 4. New York: Marcel Dekker, 1975.
12. W.H. Ko, F.K. Hora. Production of phospholipases by soil microorganisms. *Soil Sci.* 10:355–358, 1970.
13. G. Anderson. Nucleic acids, derivatives, and organic phosphorus. In: A.D. McLaren, G.H. Peterson, eds. *Soil Biochemistry.* Vol. 1. New York: Marcel Dekker, 1967.
14. T.I. Omotoso, A. Wild. Content of inositol phosphates in some English and Nigerian soils. *J. Soil Sci.* 21:216, 1970.
15. J.H. Steward, M.E. Tate. Gel chromatography of soil organic phosphorus. *J. Chromat.* 60:75–78, 1971.
16. M. Alexander. Microbial transformations of phosphorus. In: *Introduction to Soil Microbiology.* New York: Wiley, 1977.

17. D.J. Cosgrove. Microbial transformations in the phosphorus cycle. In: M. Alexander, ed. *Advances in Microbial Ecology*. New York: Plenum Press, 1977.

18. G. Anderson. Other organic phosphorus compounds. In: *Soil Organic Components*. Vol 1. New York: Springer Verlag, 1975.

19. J. Feder. The phosphatases. In: E.J. Griffith, A. Beeton, J.M. Spencer, D.T. Mitchell, eds. *Environmental Phosphorus Handbook*. New York: Wiley, 1973.

20. D. Lopez-Hernandez, C.P. Burnham. Phosphate sorption by organic soils in Britain. *Transactions of the 10th International Soil Science Congress*, 1974, pp. 73–80.

21. A. Wild. The retention of phosphate by soil—A review. *J. Soil Sci.* 1:221–238, 1950.

22. D.B. Bradely, D.H. Sieling. Effect of organic anions and sugars on phosphate precipitation by iron and aluminum as influenced by pH. *Soil Sci.* 76:175–179, 1953.

23. S. Nagarajah, A.M. Posner, J.P. Quirk. Competitive adsorption of phosphate with polygalacturonate and other organic acids on kaolinite and oxide surfaces. *Nature* 228:83–85, 1970.

24. I.C.R. Holford, G.E.G. Mattingly. Phosphate sorption by Jurassic Oolitic limestones. *Geoderma* 13:257–264, 1975.

25. R.J. Hannapel, W.H. Fuller, S. Bosma, J.S. Bullock. Phosphorus movement in a calcareous soil. I. Predominance of organic forms of phosphorus in phosphorus movement. *Soil Sci.* 97:350–357, 1964.

26. M. Fried, L.A. Dean. Phosphate retention by iron and aluminum in cation exchange systems. *Soil Sci. Soc. Am. Proc.* 19:142–147, 1955.

27. E.G. Williams, N.M. Scott, M.J. McDonalds. Soil properties and phosphate sorption. *J. Sci. Food Agric.* 9:551–559, 1958.

28. D.J. Greenland, J.M. Oades, T.W. Sherwin. Electron microscope observations of iron oxides in some red soils. *J. Soil Sci.* 19:123–126, 1968.

29. R.L. Fox, E.J. Kamprath. Adsorption and leaching of P in acid organic soils and high organic matter sand. *Soil Sci. Soc. Am. Proc.* 35:154–156, 1970.

30. C. Cogger, J.M. Duxbury. Factors affecting phosphorus losses from cultivated organic soils. *J. Environ. Qual.* 13:111–114, 1984.

31. C.V. Cole, S.R. Olsen, C.O. Scott. The nature of phosphate sorption by calcium carbonate. *Soil Sci. Soc. Am. Proc.* 17:352–356, 1953.

32. R.A. Griffin, J.J. Jurinak. The interaction of phosphate with calcite. *Soil Sci. Soc. Am. Proc.* 37:847–850, 1973.

33. M.J. Shen, C.I. Rich. Aluminum fixation in montmorillonite. *Soil Sci. Soc. Am. Proc.* 26:33–36, 1962.

34. S.S. Shukla, J.K. Syers, J.D.H. Williams, D.E. Armstrong, R.F. Harris. Sorption of inorganic phosphorus by lake sediments. *Soil Sci. Soc. Am. Proc.* 35:244–249, 1971.

35. P.S. Porter, C.A. Sanchez. The effect of soil properties on phosphorus sorption by Everglades Histosols. *Soil Sci.* 154:387–398, 1992.

36. W.L. Lindsay, A.W. Frazier, H.F. Stephenson. Identification of reaction products from phosphate fertilizers in soils. *Soil Sci. Soc. Am. Proc.* 26:446–452, 1962.

37. W.L. Lindsay. *Chemical Equilibrium in Soil*. New York: Wiley Interscience, 1979.

38. G.L. Terman, D.R. Bouldin, J.R. Lehr. Calcium phosphate fertilizers: I. Availability to plants and solubility in soils varying in pH. *Soil Sci. Soc. Am. Proc.* 22:25–29, 1958.

39. W.L. Lindsay, H.F. Stephenson. Nature of the reactions of monocalcium phosphate monohydrate in soils: II. Dissolution and precipitation reactions involving iron, aluminum, manganese and calcium. *Soil Sci. Soc. Am. Proc.* 26:446–452, 1959.

40. G. Hambidge. *Hunger Sign in Crops*. Washington DC: American Society of Agronomy and The National Fertilizer Council, 1941.

41. J.E. McMurtrey Jr, Visual symptoms of malnutrition in plants. In: H.B. Kitchen, ed. *Diagnostic Techniques for Soils and Crops*. Washington DC: American Potash Insititute, 1948.

42. T. Wallace. *The Diagnosis of Mineral Deficiencies in Plants by Visual Diagnosis*. New York: Chemical Publishing Co Inc., 1961.

43. F.T. Bingham. Phosphorus. In: H.D. Chapman, ed. *Diagnostic Criteria for Plant and Soils*. Division of Agricultural Sciences, University of California, 1966.

44. W.T. Forsee Jr, R.V. Allison. Evidence of phosphorus interference in the assimilation of copper by citrus on the organic soils of the lower east coast of Florida. *Soil Sci. Soc. Fla. Proc.* 6:162–165, 1944.

45. A. Scaife. Derivation of critical nutrient concentrations for growth rate from data from field experiments. *Plant Soil* 109:159–169, 1988.

46. H.D. Chapman (ed.). *Diagnostic Criteria for Plants and Soils*. Berkeley, CA: Division of Agricultural Sciences, University of California, 1966.

47. N.F. Childers (ed.). *Temperate to Tropical Fruit Nutrition*. New Brunswick, NJ: Rutgers–The State University, 1966.

48. R.L. Westerman (ed.). *Soil Testing and Plant Analysis*. Madison, WI: Soil Science Society of America, 1990.

49. B. Wolf. *Diagnostic Techniques for Improving Crop Production*. Binghampton, New York: The Hayworth Press Inc., 1996.

50. G.W. Thomas, D.E. Peaslee. Testing soils for phosphorus. In: L.M. Walsh, J.D. Beaton, eds. *Soil Testing and Plant Analysis*. Madison, WI: Soil Sci. Soc. Am. Inc., 1973.

51. E.J. Kamprath, M.E. Watson. Conventional soil and tissue tests for assessing the phosphorus status of soils. In: F.E. Khasawneh, E.C. Sample, E.J. Kamprath, eds. *The Role of Phosphorus In Agriculture*. American Society of Agronomy Inc. 677 South Segoe Road, Madison WI 53711, 1980.

52. P.E. Fixen, J.H. Grove. In: R.L. Westerman, ed. *Soil Testing and plant analysis*. Soil Science Society of America Book Series 3:141–180, 1990.

53. S. Kuo. Phosphorus. In: D.L. Spark, A.L. Page, P.A. Helmke, R.H. Loeppert, P.N. Soltanpour, M.A. Tabatabai, C.T. Johnston, M.E. Sumner, eds. *Methods of Soil Analysis Part 3 Chemical Methods*. Soil Science Society of America 5:869–920, 1996.

54. S.C. Chang, M.L. Jackson. Fractionation of soil phosphorus. *Soil Sci.* 84:113–144, 1957.

55. A. Mehlich. New extractant for soil test evaluation of phosphorus, potassium, magnesium, calcium, sodium, manganese, and zinc. *Commun. Soil Sci. Plant Anal.* 9:477–492, 1978.

56. A. Mehlich. Mehlich 3 soil test extractant: a modification of the Mehlich 2 extractant. *Commun. Soil Sci. Plant Anal.* 15:1409–1416, 1984.

57. E.A. Hanlon, G.V. Johnson. Bray/Kurtz, Mehlich III, AB/D, and ammonium acetate extractions of P, K, and Mg in four Oklahoma soils. *Commun. Soil Sci. Plant Anal.* 15:277–294, 1984.

58. S.R. Olsen, C.V. Cole, F.S. Watanabe, L.A. Dean. Estimation of Available P in Soils by Extraction with NaHCO3. USDA Cir. 939. US Government Printing Office, Washington DC, 1954.

59. P.N. Soltanpour, A.P. Schwab. A new soil test for the simultaneous extraction of macro and micro nutrients in alkaline soils. *Commun. Soil Sci. Plant Anal.* 8:195–207, 1977.

60. P.N. Soltanpour, S. Workman. Modification of the NH4HCO3-DPTA soil tests to omit carbon black. *Commun. Soil Sci. Plant Anal.* 10:1411–1420, 1979.

61. B.A. Ahmed, A. Islam. The use of sodium EDTA as an extractant for determining available phosphorus in soil. *Geoderma* 14:261–265, 1972.

62. R.L. Fox, E.J. Kamprath. Phosphate sorption isotherms for evaluating the phosphate requirements of soils. *Soil Sci. Soc. Am. Proc.* 34:902–907, 1970.

63. E.A. Hanlon, G. Hochmuth. Recent changes in phosphorus and potassium fertilizer recommendations for tomato, pepper, muskmelon, watermelon, and snapbean in Florida. *Commun. Soil Sci. Plant Anal.* 23:2651–2665, 1992.

64. G. Hochmuth, P. Weingartner, C. Hutchinson, A. Tilton, D. Jesseman. Potato yield and tuber quality did not respond to phosphorus fertilization of soils testing high in phosphorus content. *Hortech* 12(3) 2002.

65. G. Hochmuth, O. Carrujo, K. Shuler. Tomato yield and fruit size did not respond to P fertilization of a sandy soil testing very high in Mehlich-1 P. *HortScience* 34:653–656, 1999.

66. R.K. Nishomoto, R.L. Fox, P.E. Parvin. Response of vegetable crops to phosphorus concentrations in soil solution. *J. Am. Soc. Hort. Sci.* 102:705–709, 1977.

67. S. Itoh, S.A. Barber. Phosphorus uptake by six plant species as related to root hairs. *Agron. J.* 75:457–461, 1983.

68. D.J. Greenwood, T.J. Cleaver, M.K. Turner, J. Hunt, K.B. Niendorf, S.M.H. Loquens. Comparison of the effects of phosphate fertilizer on the yield, phosphate content and quality of 22 different vegetable and agricultural crops. *J. Agric. Sci. Camp.* 95:457–469, 1980.

69. D. Alt. Influence of P- and K- fertilization on yield of different vegetable species. *J. Plant Nutr.* 10:1429–1435, 1987.

70. C.A. Sanchez. Soil Testing and Fertilizer Recommendations for Crop Production on Organic Soils in Florida. University of Florida Agricultural Experiment Station Bulletin 876, Gainesville, 1990.

71. G.S.C. Buso, F.A. Bliss. Variability among lettuce cultivars grown at two levels of available phosphorus. *Plant Soil* 111:67–93, 1988.

72. R.T. Nagata, C.A. Sanchez, F.J. Coale. Crisphead lettuce cultivar response to fertilizer phosphorus. *J. Am. Soc. Hort. Sci.* 117:721–724, 1992.

73. C.A. Sanchez, N.M. El-Hout. Response of diverse lettuce types to fertilizer phosphorus. *HortScience* 30:528–531, 1995.

74. N.J. Pearson, Z. Rengel. Mechanisms of plant resistance to nutrient deficiency stresses. In: A.S. Basra, R.K. Basra, eds. *Mechanisms of Environmental Stress Resistance in Plants.* Amsterdam, Netherlands: Harwood Academic Publishers, 1997, pp. 213–240.

75. J. Lynch. The role of nutrient efficient crops in modern agriculture. In: Z. Rengel, ed. *Nutrient Use in Crop Production.* Binghamton, NY: The Haworth Press Inc., 1998, pp. 241–264.

76. F.S. Watanabe, S.R. Olsen, R.E. Danielson. Phosphorus Availability as Related to Soil Moisture. *Int. Congress Soil Sci. Trans.* 7th, Madison, WI, III: 1960.

77. F.N. Ponnamperuma. The chemistry of submerged soils. *Adv. Agron.* 24:29–96, 1972.

78. I.C.R. Holford, W.H. Patrick Jr. Effect of reduction and pH changes on phosphate sorption and mobility in an acid soil. *Soil Sci. Soc. Am. J.* 43:292–297, 1979.

79. P.E. Bacon, B.G. Davey. Nutrient availability under trickle irrigation: distribution of water and Bray No 1 phosphate. *Soil Sci. Soc. Am. J.* 46:981–987, 1982.

80. S.S. Barber. Soil-plant interaction in the phosphorus nutrition of plants. In: Khasawneh, Sample, Kamprath, eds. *The Role of Phosphorus in Agriculture.* 1980.

81 J.J. Hanway, R.A. Olson. Phosphate nutrition of corn, sorghum, soybeans, and small grains. In: Khasawneh, Sample, Kamprath, eds. *The Role of Phosphorus in Agriculture.* 1980.

82. B.R. Gardner, J. Preston Jones. Effects of temperature on phosphate sorption isotherms and phosphate desorption. *Commun. Soil Sci. Plant Anal.* 4:83–93, 1973.

83. C.D. Sutton. Effects of low temperature on phosphate nutrition of plants–a review. *J. Sci. Food Agric.* 20:1–3, 1969.

84. O.G. Carter, D.J. Lathwell. Effect of temperature on orthophosphate absorption by excised corn roots. *Plant Physiol.* 42:1407–1412, 1967.

85. B.B. Singh, J.P. Jones. Phosphorus sorption isotherm for evaluating phosphorus temperature requirements of lettuce at five temperature regimes. *Plant Soil* 46:31–44, 1977.

86. J.C. Lingle, R.M. Davis. The influence of soil temperature and phosphorus fertilization on the growth and mineral absorption of tomato seedling. *Proc. Am. Soc. Hortic. Sci.* 73:312–322, 1959.

87. S.J. Locascio, G.F. Warren, G.E. Wilcox. The effect of phosphorus placement on uptake of phosphorus and growth of direct seeded tomatoes. *Proc. Am. Soc. Hortic. Sci.* 76:503–514, 1960.

88. O.A. Lorenz, K.B. Tyler, O.D. McCoy. Phosphate sources and rates for winter lettuce on a calcareous soil. *Proc. Am. Soc. Hortic. Sci.* 84:348–355, 1964.

89. B.R. Gardner. Effects of soil P levels on yields of head lettuce. *Agron. Abst. P* 204, 1984.

90. W.H. Wagganman. *Phosphoric Acid, Phosphates, and Phosphatic Fertilizers.* 2nd ed. New York: Hafner Publications Company, 1969.

91. O.P. Engelstad, G.L. Teramn. Agronomic effectiveness of phosphate fertilizers. In: Khasawneh, Sample, Kamprath, eds. *The Role of Phosphorus In Agriculture.* 1980.

92. A.C. Bennett, F. Adams. Calcium deficiency and ammonium toxicity versus separate causal factors of (NH4)2HPO4-injury to seedling. *Soil Sci. Soc. Am. Proc.* 34:255–259, 1973.

93. I.G. Burns. Effects of interruptions in N, P, or K supply on the growth and development of lettuce. *J. Plant Nutr.* 10:1571–1578, 1987.

94. D.L. Grunes, H.R. Haise, L.O. Fine. Proportional uptake of soil and fertilizer phosphorus by plants as affected by nitrogen fertilization: field experiments with sugarbeets and potatoes. *Soil Sci. Soc. Am. Proc.* 22:49–52, 1958.

95. J.C. Lingle, J.R. Wright. Fertilizer Experiments with Cantaloupes. California Agricultural Exp. Bull. 807, Berkeley, 1964.

96. C.A. Sanchez, V.L. Guzman, R.T. Nagata. Evaluation of sidedress fertilization for correcting nutritional deficits in crisphead lettuce on Histosols. *Proc. Fla. State Hortic. Soc.* 103:110–113, 1990.

97. R.E. Lucas, M.T. Vittum. Fertilizer placement for vegetables. In: G.E. Richards, ed. *Phosphorus Fertilization-Principles and Practices of Band Application.* St Louis, MO: Olin Corp., 1976.

98. G.W. Randall, R.G. Hoeft. Placement methods for improved efficiency of P and K fertilizers: a review. *J. Prod. Agri.* 1:70–78, 1988.

99. I. Anghinoni, S.A. Barber. Phosphorus influx and growth characteristics of corn roots as influenced by phosphorus supply. *J. Agron.* 72:685–688, 1980.

100. I. Anghinoni, S.A. Barber. Predicting the most efficient phosphorus placement for corn. *Soil Sci. Soc. Am. J.* 44:1016–1020, 1980.

101. D.M. Sleight, D.H. Sander, G.A. Peterson. Effect of fertilizer phosphorus placement on the availability of phosphorus. *Soil Sci. Soc. Am. J.* 48:336–340, 1984.

102. C.A. Sanchez, S. Swanson, P.S. Porter. Banding P to improve fertilizer use efficiency in lettuce. *J. Am. Soc. Hortic. Sci.* 115:581–584, 1990.

103. C.A. Sanchez, P.S. Porter, M.F. Ulloa. Relative efficiency of broadcast and banded phosphorus for sweetcorn produced on Histosols. *Soil Sci. Soc. Am. J.* 55:871–875, 1991.

104. L. Espinoza, C.A. Sanchez, T.J. Schueneman. Celery yield responds to phosphorus rate but not placement on Histosols. *HortScience* 28:1168–1170, 1993.

105. S.A. Barber. Relation of fertilizer placement to nutrient uptake and crop yield. I. Interaction of row phosphorus and the soil level of phosphorus. *Agron. J.* 50:535–539, 1958.

106. L.F. Welch, D.L. Mulvaney, L.V. Boone, G.E. McKibben, J.W. Pendleton. Relative efficiency of broadcast versus banded phosphorus for corn. *Agron. J.* 58:283–287, 1966.

107. G.A. Peterson, D.H. Sanders, P.H. Grabouski, M.L. Hooker. A new look at row and broadcast phosphate fertilizer recommendations for winter wheat. *Agron. J.* 73:13–17, 1981.

108. O. Silberstein, S.H. Wittwer. Foliar application of phosphatic fertilizers to vegetable crops. *Proc. Am. Soc. Hortic. Sci.* 58:179–180, 1951.

109. D. Barel, C.A. Black. Foliar application of P. I. Screening of various inorganic and organic P compounds. *Agron. J.* 71:15–21, 1979.

110. D. Barel, C.A. Black. Foliar application of P. II. Yield response of crin and soybeans sprayed with various condensed phosphates and P-N compounds in greenhouse and field experiments. *Agron. J.* 71:21–24, 1979.

111. F.G. Teubner, M.J. Bukovac, S.H. Wittwer, B.K. Guar. The utilization of foliar-applied radiophosphorus by several vegetable crops and tree fruits under field conditions. *Michigan Agric. Exp. Stn. Quar. Bull.* 44:455–465, 1962.

112. A.P. Upadhyay, M.R. Deshmukh, R.P. Rajput, S.C. Deshmukh. Effect of sources, levels and methods of phosphorus application on plant productivity and yield of soybean. *Indian J. Agron.* 33:14–18, 1988.

113. R. Garcia, J.J. Hanway. Foliar fertilization of soybeans during the seed-filling period. *Agron. J.* 68:653–657, 1976.

114. W.K. Robertson, K. Hinson, L.C. Hammond. Foliar fertilization of soybeans (*Glycine max* L) Merr in Florida. *Soil Crop Sci. Soc. Fla. Proc.* 36:77–79, 1977.

115. M.B. Parker, F.C. Boswell. Foliage injury, nutrient uptake, and yield of soybeans as influenced by foliar fertilization. *Agron. J.* 72:110–113, 1980.

116. H.J. Harder, R.E. Carlson, R.H. Shaw. Corn grain yield and nutrient response to foliar fertilization applied during grain fill. *Agron. J.* 74:106–110, 1982.

117. H.J. Harder, R.E. Carlson, R.H. Shaw. Leaf photosynthetic response to foliar fertilizer applied to corn during grain fill. *Agron. J.* 74:759–761, 1982.

118. G.D. Batten, I.F. Wardlaw. Senescence of the flag leaf and grain yield following late foliar and root applications of phosphate on plants of differing phosphate status. *J. Plant Nutr.* 10:735–748, 1987.

119. T.J. Logan, E.O. McLean. Effects of phosphorus application rate, soil properties and leaching mode on P movement in soil columns. *Soil Sci. Soc. Am. Proc.* 37:371–374, 1973.

120. T.J. Logan, E.O. McLean. Nature of phosphorus retention and absorption with depth in soil columns. *Soil Sci. Soc. Am. Proc.* 37:351–355, 1973.

121. G.W. Hergert, J.O. Reuss. Sprinkler application of P and Zn fertilizer. *Agron. J.* 68:5–8, 1976.

122. D.A. Lauer. Vertical distribution in soil of sprinkler-applied phosphorus. *Soil Sci. Soc. Am. J.* 52:862–868, 1988.

123. C.O. Stanberry, C.D. Converse, H.R. Haise, A.J. Kelly. Effect of moisture and phosphate variables on alfalfa hay production on the Yuma mesa. *Soil Sci. Soc. Am. Proc.* 19:303–310, 1955.

124. C.O. Stanberry, H.A. Schreiber, L.R. Cooper, S.D. Mitchell. Vertical Movement of Phosphorus in Some Calcareous Soils of Arizona. Unpublished data. University of Arizona Agric. Exp. Stn. Tech. Paper, 1965.

125. R.S. Rauschkolb, D.E. Rolston, R.J. Miller, A.B. Carlton, R.G. Burau. Phosphorus fertilization with drip irrigation. *Soil Sci. Soc. Am. J.* 40:68–72, 1979.

126. M.K. O'Neill, B.R. Gardner, R.L. Roth. Orthophosphoric acid as a phosphorus fertilizer in trickle irrigation. *Soil Sci. Soc. Am. J.* 43:283–286, 1979.

127. B. Bar-Yosef, B. Sagiv, T. Markovitch, I. Levkovitch. Phosphorus placement effects on sweet corn growth, uptake, and yield. *Proceedings of Dahlia Greidinger International Symposium on Fertigation.* Haifa, Israel, 1995, pp. 141–154.

128. O.A. Carrijo, G. Hochmuth. Tomato responses to preplant incorporated or fertigated phosphorus on soils varying in Mehlich-1 extractable phosphorus. *HortScience* 35:67–72, 2000.

129. R.L. Mikklelsen. Phosphorus fertilization through drip irrigation. *J. Prod. Agric.* 2:279–286, 1989.

130. A.J. Dow. Critical Nutrient Ranges in Northwest Crops. Western Reg. Publ. 43 Washington State Univ. Irrigated Agric. Res. Ext. Ctr. Prosser, WA, 1980.

131. R.C. Ward, D.A. Whitney, D.G. Westfall. Plant analysis as an aid in fertilizing small grains. In: L.M. Walsh, J.D. Beaton, eds. *Soil Testing and Plant Analysis.* Madison, WI: SSSA, 1973, pp. 329–348.

132. R.H. Howeler. The mineral nutrition and fertilization of cassava. In: *Cassava Production Course.* Cali, Colombia: CIAT, 1978, pp. 247–292.

133. D. Satinder, R.D. Kaushik, V.K. Gupta. Relationship between P and Mn in chickpea (*Cicer arietinum* L). *Plant Soil* 72:85–90, 1983.

134. J.B. Jones. *Plant Analysis Handbook for Georgia.* Univ. Georgia Coll. Agric. Bull. 735, 1974.

135. G.W. Rehm, R.C. Sorensen, R.A. Wiese. Application of phosphorus, potassium and zinc to corn grown for grain or silage: nutrient concentration and uptake. *Soil Sci. Soc. Am. J.* 47:697–700, 1983.

136. S.W. Melsted, H.L. Motto, T.R. Peck. Critical plant nutrient composition values useful in interpreting plant analysis data. *Agron. J.* 61:7–20, 1969.

137. L.C. Dumenil. Nitrogen and phosphorus composition of corn leaves and corn yields in relation to critical levels and nutrient balance. *Soil Sci. Soc. Am. Proc.* 25:295–298, 1961.

138. J.J. Hanway. Corn growth and composition in relation to soil fertility: III. Percentages of N, P and K in different plant parts in relation to stage of growth. *Agron. J.* 54:222–229, 1962.

139. R.D. Powell, J.R. Webb. Effect of high rates of fertilizer N, P, and K on corn (*Zea mays* L), leaf nutrient concentrations. *Commun. Soil Sci. Plant Anal.* 5:93–104, 1974.

140. E.H. Tyner. The relation of corn yields to leaf nitrogen, phosphorus, and potassium content. *Soil Sci. Soc. Am. Proc.* 11:317–323, 1946.

141. W.F. Bennett, G. Stanford, L. Dumenil. Nitrogen, phosphorus and potassium content of the corn leaf and grain related to nitrogen fertilization and yield. *Soil Sci. Soc. Am. Proc.* 17:252–258, 1953.

142. J.B. Jones. Interpretation of plant analysis for several agronomic crops. In: G.W. Hardy, ed. *Soil Testing and Plant Analysis: Part 2.* Madison, WI: Soil Sci. Soc. Am. Special Public No. 2. 1967, pp. 49–85.

143. P.W. Neubert, W. Wrazidlo, N.P. Vielemeyer, I. Hundt, F. Gullmick, W. Bergmann. *Tabellen Zur planzenanalyze-Erste orientierende Ubersicht.* Berlin: Inst fur Planzenernahrung Jena, 1969.

144. C.R. Escano, C.A. Jones, G. Uehara. Nutrient diagnosis in corn grown on hydric dystrandepts: I. Optimum tissue concentrations. *Soil Sci. Soc. Am. J.* 45:1135–1139, 1981a.

145. J.P. Jones, J.A. Benson. Phosphate sorption isotherms for fertilizer P needs of sweet corn (*Zea mays*) grown on a high phosphorus fixing soil. *Commun. Soil Sci. Plant Anal.* 6:465–477, 1975.

146. W.E. Sabbe, A.J. MacKenzie. Plant analysis as an aid to cotton fertilization. In: *Soil Testing and Plant Analysis.* Madison, WI: SSSA, 1973, pp. 299–313.

147. R. Maples, J.L. Keogh. Phosphorus Fertilization Experiments with Cotton on Delta Soils of Arkansas. Arkansas Agric. Exp. Stn. Bull. 781, 1973.

148. D.M. Bassett, A.J. MacKenzie. Plant Analysis as a Guide to Cotton Fertilization. University of California Bulletin 1879, pp. 6–17, 1978.

149. A. Kashirad, A. Bassiri, M. Kheradnam. Responses of cowpea to applications of P and Fe in calcareous soils. *Agron. J.* 70:67–70, 1978.

150. C.L. Godfrey, F.L. Fisher, M.J. Norris. A comparison of ammonium metaphosphate and ammonium orthophosphate with superphosphate on the yield and chemical composition of crops grown under field conditions. *Soil Sci. Soc. Am. Proc.* 23:43–46, 1959.

151. R.F. Bishop, G.G. Smeltzer, C.R. MacEachern. Effect of nitrogen, phosphorus and potassium on yields, protein contents and nutrient levels in soybean, field peas and faba beans. *Commun. Soil Sci. Plant Anal.* 7:387–404, 1976.

152. N.K. Fageria. Effect of phosphatic fertilization on growth and mineral composition of pea plants (*Pisum sativum* L). *Agrochimica* 21:75–78, 1977.

153. J.L. Haddock, D.C. Linton. Yield and phosphorus content of canning peas as affected by fertilization, irrigation regime and sodium bicarbonate-soluble soil phosphorus. *Soil Sci. Soc. Am. Proc.* 21:167–171, 1957.

154. D.C. MacKay, J.S. Leefe. Optimum leaf levels of nitrogen, phosphorus and potassium in sweet corn and snap beans. *Can. J. Plant Sci.* 42:238–246, 1962.

155. J.T. Moraghan. Differential responses of five species to phosphorus and zinc fertilizers. *Commun. Soil Sci. Plant Anal.* 15:437–447, 1984.

156. H.G. Small, A.J. Ohlrogge. Plant analysis as an aid in fertilizing soybeans and peanuts. In: L.M. Walsh, J.D. Beaton, eds. *Soil Testing and Plant Analysis.* Madison, WI: SSSA, 1973, pp. 315–327.

157. R. Nichols. Studies on the major-element deficiencies of the pigeon pea (*Cajanus cajan*) in sand culture. II. The effects of major-element deficiencies on nodulation, growth and mineral composition. *Plant Soil* 22:12–126, 1965.

158. A.R. Sheldrake, A Narayanan. Growth, development and nutrient uptake of pigeon peas (*Cajanus cajan*). *J. Agric. Sci.* 92:513–526, 1979.

159. N.K. Fageria. Critical P, K, Ca and Mg contents in the tops of rice and peanut plants. *Plant Soil* 45:421–431, 1976.

160. L.R. Hossner, J.A. Freeouf, B.L. Folsom. Solution phosphorus concentration and growth of rice (*Oryza sativa* L.) in flooded soils. *Soil Sci. Soc. Am. Proc.* 37:405–408, 1973.

161. D.S. Mickkelsen, R.R. Hunziker. A plant analysis survey of California rice. *Agrichem. Age* 14:8–22, 1971.

162. R.B. Lockman. Mineral composition of grain sorghum plant samples. III. Suggested nutrient sufficiency limits at various stages of growth. *Commun. Soil Sci. Plant Anal.* 3:295–303, 1972.

163. D. Whitney. A Soil and Plant Analysis for Corn and Sorghum Survey. Kansas State Univ. Mimeo 3a-162-1-300, 1970.

164. R.J. Miller, J.T. Pesck, J.J. Hanway. Relationships between soybean yield and concentrations of phosphorus and potassium in plant parts. *Agron. J.* 53:393–396, 1961.

165. W.E. Sabbe, J.L. Keogh, R. Maples, L.H. Hileman. Nutrient analysis of Arkansas cotton and soybean leaf tissue. *Arkansas Farm Res.* 21:2, 1972.

166. K.M. Sipitanos, A Ulrich. Phosphorus nutrition of sugarbeet seedlings. *J. Am. Soc. Sugar Beet Technol.* 15:332–346, 1969.

167. A. Ulrich. Plant and analysis as a guide to the nutrition of the sugar beets in California. *Proceedings of the 5th General Meeting of American Society of Sugar Beet Technolnogy,* 1948, pp. 364–377.

168. P. Halais. The detection of N P K deficiency trends in sugarcane crops means of foliar diagnosis run from year to year on a follow-up basis. *Proceedings of 11th International Society of Sugar Cane Technology,* 1962, pp. 214–221.

169. E. Malavolta, F.P. Gomes. Foliar diagnosis in Brazil. In: W. Reuther, ed. *Plant Analysis and Fertilizer Problems.* Washington DC: Am. Inst. Biol. Sci., 1961, pp. 180–189.

170. G. Samuels. The influence of the age of sugar cane on the leaf nutrient (N-P-K) content. *Proc. Int. Soc. Sugar Cane Technol.* 10:508–514, 1959.

171. W.R. Schmehl, R.P. Humbert. Nutrient deficiencies in sugar crops. In: H.W. Sprague, ed. *Hunger Signs in Crops. A Symposium.* David McKay Company, New York, 1964, pp. 415–450.

172. W.W. Garner. In: *The Production of Tobacco.* 1st ed. New York: Blakiston Co, 1951.

173. C.O. Plank. *Plant Analysis Handbook for Georgia.* Athens: Coop. Ext. Serv. University of Georgia, 1988.

174. K.A. Kelling, J.E. Matocha. Plant analysis as an aid in the fertilization of forage crops. In: *Soil Testing and Plant Analysis.* Madison, WI: SSSA Book Series: 3. 1990, pp. 603–636.

175. W.E. Martin, J.E. Matocha. Plant analysis as an aid in the fertilization of forage crops. In: *Soil Testing and Plant Analysis.* Revised edition. Madison, WI, 1973, pp. 394–423.

176. CIAT. Tropical Pastures Program Report 1981. CIAT Series 02SETP(1)82. pp. 167–191, 1982.

177. A.L. Kenworthy, L. Martin. Mineral content of important fruit plants. XXIV. In: N.F. Childers, ed. *Temperate to Tropical Fruit Nutrition.* New Brunswick, NJ: Rutgers–The State University, 1966, pp. 813–870.

178. M.N. Westwood. *Temperate–Zone Pomology.* W.H. Freeman and Co., San Francisco, 1978.

179. C. Bould. Leaf analysis of deciduous fruits. XXI. In: N.F. Childers, ed. *Temperate to Tropical Fruit Nutrition.* New Brunswick, NJ: Rutgers–The State University, 1966, pp. 651–684.

180. A.L. Kenworthy. Leaf analysis as an aid in fertilizing orchards. In: L.M. Walsh, J.D. Benton, eds. *Soil Testing and Plant Analysis*. Revised edition. Madison, WI: SSSA. 1973.
181. M. Faust. *Physiology of Temperate Zone Fruit Trees*. John Wiley and Sons, New York, 1989.
182. T.W. Embleton, W.W. Jones. Avocado and mango. II. In: N.F. Childers, ed. *Temperate to Tropical Fruit Nutrition*. New Brunswick, NJ: Rutgers–The State University, 1966, pp. 51–76.
183. S.R. Freiberg. Banana nutrition. III. In: N.F. Childers, ed. *Temperate to Tropical Fruit Nutrition*. New Brunswick, NJ: Rutgers–The State University, 1966, pp. 77–100.
184. J.C. Cain, P. Eck. Blueberry and cranberry. IV. In: N.F. Childers, ed. *Temperate to Tropical Fruit Nutrition*. New Brunswick, NJ: Rutgers–The State University, 1966, pp. 101–129.
185. C.C. Doughty, E.B. Adams, L.W. Martin. High Bush Blueberry Production in Washington and Oregon Cooperative Extension, Washington and Oregon State Universities and University of Idaho Bull. No. PNW215, 1981.
186. D.B. Murray. Cacao nutrition. IX. In: N.F. Childers, ed. *Temperate to Tropical Fruit Nutrition*. New Brunswick, NJ: Rutgers–The State University, 1966, pp. 229–251.
187. T.L. Reghetti, K.L. Wilder, G.A. Cummings. *Plant Analysis as an Aid in Fertilizing Orchards*. 3rd ed. Madison, WI: SSSA Book Series 3, 1990, pp. 563–602.
188. P. Smith. Citrus nutrition. VII. In: N.F. Childers, ed. *Temperate to Tropical Fruit Nutrition*. Rutgers–The State University, 1966, pp. 174–207.
189. T.W. Embleton, W.W. Jones, C. Pallares, R.G. Platt. Effect of fertilization of citrus on fruit quality and ground water nitrate–pollution potential. *Proc. Int. Soc. Citriculture*, pp. 280–285, 1978.
190. B. Ljones. Bush fruits nutrition. V. In: N.F. Childers, ed. *Temperate to Tropical Fruit Nutrition*. Rutgers–The State University. New Brunswick, NJ, 1966, pp. 130–157.
191. L. M?ller. Coffee nutrition. XXII. In: N.F. Childers, ed. *Temperate to Tropical Fruit Nutrition*. New Brunswick, NJ: Rutgers–The State University. New Brunswick, NJ, 1966, pp. 685–776.
192. J.A. Cook. Grape nutrition. XXIII. In: N.F. Childers, ed. *Temperate to Tropical Fruit Nutrition*. New Brunswick, NJ: Rutgers–The State University, 1966, pp. 777–812.
193. T.W. Young, R.C.J. Koo. Increasing yield of Parwin and Kent mangos on Lakewood sand by increased nitrogen and potassium fertilization. *Proc. Fla. Sta. Hortic. Soc.* 87:380–384, 1974.
194. D. Boynton, G.H. Oberly. Pear nutrition. XIII. In: N.F. Childers, ed. *Temperate to Tropical Fruit Nutrition*. New Brunswick, NJ: Rutgers–The State University, 1966, pp. 489–503.
195. K. Uriu, J.C. Crane. Mineral element changes in pistachio leaves. *J. Am. Soc. Hortic. Sci.* 102:155–158, 1977.
196. N.R. Benson, R.C. Lindner, R.M. Bullock. Plum, prune and apricot. XIV. In: N.F. Childers, ed. *Temperate to Tropical Fruit Nutrition*. New Brunswick, NJ: Rutgers–The State University, 1966, pp. 504–517.
197. A. Ulrich, M.A.E. Mostafa, W.W. Allen. Strawberry Deficiency Symptoms: A Visual and Plant Analysis Guide to Fertilization. University of California Division of Agricultural Science Publication No. 4098, 1980.
198. J.W. Mastarlerz. *The Greenhouse Environment*. New York: Wiley, 1977, pp. 510–516.
199. G.W. Winsor, M.I.E. Long, B. Hart. The nutrition of the glasshouse carnation. *J. Hortic. Sci.* 45:401–413, 1970.
200. O.R. Lunt, A.M. Kofranek, J.J. Oertli. Some critical nutrient levels in *Chyrsanthemum morifolium* cv Good News. *Plant Anal. Fert. Problems* 4:398–413, 1964.
201. J.N. Joiner, W.E. Waters. Influence of cultural conditions on the chemical composition of six tropical foliage plants. *Proc. Trop. Reg. Am. Soc. Hortic. Sci.* 14:254–267, 1969.
202. G.H. Gilliam, C.E. Evans, R.L. Schumack, C.O. Plank. Foliar sampling of Boston fern. *J. Am. Soc. Hortic. Sci.* 108(1):90–93, 1983.
203. Anon. *Mineral Analysis Interpretation Key for Ornamental Plants* (Manual). California: Soil and Plant Laboratory Inc., 1974.
204. M. Prasad, R.E. Widme, R.R. Marshall. Soil testing of horticultural substrates for cyclamen and poinsettia. *Commun. Soil Sci. Plant Anal.* 14(7):553–573, 1983.
205. J. Johanson. Effects of nutrient levels on growth, flowering and leaf nutrient content of greenhouse roses. *Acta Agric. Scand.* 28:363–386, 1963.
206. R.T. Poole, C.A. Conover, J.N. Joiner. Chemical composition of good quality foliage plants. *Proc. Fla. Sta. Hortic. Soc.* 89:307–308, 1976.

207. O.A. Lorenz, K.B. Tyler. Plant Tissue Analysis of Vegetable Crops; Soil and Plant Tissue Testing in California. H. Reisenauer, ed. University of California Bulletin No. 1879. pp. 22–23, 1978.

208. N.H. Peck, D.L. Grunes, R.M. Welch, G.E. MacDonald. Nutritional quality of vegetable crops as affected by phosphorus and zinc fertilizers. *Agron. J.* 74:583–585, 1982.

209. M.C. Geraldson, G.R. Klacan, O.A. Lorenz. Plant analysis as an aid in fertilizing vegetable crops. In: L.M. Walsh, J.D. Beaton, eds. *Soil Testing and Plant Analysis.* Madison, WI: Am. Soc. Agron., 1973, pp. 365–579.

210. C.A. Sanchez, H.W. Burdine, V.L. Guzman. Soil-testing and plant analysis as guides for the fertilization of celery on Histosols. *Soil Crop Sci. Soc. Fla. Proc.* 29:69–72, 1989.

211. L. Espinoza, C.A. Sanchez, T.J. Schueneman. Celery yield responds to phosphorus rate but not phosphorus placement on Histosols. *HortScience* 28:1168–1170, 1993.

212. H.W. Burdine. Some Winter Grown Leafy Crop Responses to Varying Levels of Nitrogen Phosphorus, and Potassium On Everglades Organic Soil. Belle Glade AREC Research Report EV-1976-5, April 1976.

213. C.A. Sanchez, H.W. Burdine. Soil testing and plant analysis as guides for the fertilization of Escarole and Endive on Histosols. *Soil Crop Sci. Soc. Fla. Proc.* 48:37–40, 1989.

214. W.L. Berry, D.T. Krizek, D.P. Ormord, J.C. McFarlane, R.W. Langhans, T.W. Tibbits. Variation in elemental content of lettuce grown under baseline conditions in five controlled-environment facilities. *J. Am. Soc. Hortic. Sci.* 106:661–666, 1981.

215. C.A. Sanchez, H.W. Burdine, V.L. Guzman. Yield, quality, and leaf nutrient composition of crisphead lettuce as affected by N, P, and K on Histosols. Vegetable Section. *Proc. Fla. State Hortic. Soc.* 101:346–350, 1988.

216. R.B. Beverly. Nutritional survey of the Everglades vegetable industry. *Proc. Fla. State Hortic. Soc.* 97:201–205, 1984.

217. C.A. Sanchez, G.H. Snyder, H.W. Burdine, C.B. Hall. DRIS evaluation of the nutritional status of crisphead lettuce. *HortScience* 26:274–276, 1991.

218. F.W. Zink . Growth and nutrient absorption of green bunching onions. *Proc. Am. Soc. Hortic. Sci.* 180:430–435, 1962.

219. S. Roberts, A.I. Dow. Critical nutrient ranges for petiole phosphorous levels of sprinkler-irrigated Russet Burbank potatoes. *Agron. J.* 74:583–585, 1982.

220. C.A. Sanchez, M. Lockhart, P.S. Porter. Response of radish to phosphorus and potassium fertilization of Histosols. *HortScience* 26:30–32, 1991.

221. C.A. Sanchez, H.W. Burdine, F.S. Martin. Yield and quality responses of three sweet corn hybrids as affected by fertilizer phosphorus. *J. Fert. Issues* 6:17–24, 1989a.

222. C.A. Sanchez, P.S. Porter, M.F. Ulloa. Relative efficiency of broadcast and banded phosphorus for sweet corn produced on Histosols. *Soil Sci. Soc. Am. J.* 55:871–875, 1991.

223. R.T. Besford. Uptake and distribution of phosphorous in tomato plants. *Plant Soil* 51:331–340, 1979.

224. A. Mehlich. Determination of P, Ca, Mg, K, Na and NH_4^-. North Carolina Soil Testing Div. Mimeo, Raleigh, 1953.

225. W.T. Forsee Jr. Development and evaluation of methods for the determination of phosphorus in Everglades peat under various conditions of treatment. *Soil Sci. Soc. Fla. Proc.* 4:50–54, 1942.

226. G. Hochmuth, E. Hanlon. IFAS Standarized Fertilization Recommendations for Vegetable Crops. Fla. Coop. Ext. Serv. Circ. 1152, 1995.

4 Potassium

Konrad Mengel
Justus Liebig University, Giessen, Germany

CONTENTS

4.1 Historical Information 91
4.2 Determination of Essentiality 92
    4.2.1 Function in Plants 93
        4.2.1.1 Enzyme Activation 93
        4.2.1.2 Protein Synthesis 93
        4.2.1.3 Ion Absorption and Transport 94
            4.2.1.3.1 Potassium Absorption 94
            4.2.1.3.2 Potassium Transport within Tissues 95
            4.2.1.3.3 Osmotic Function 95
        4.2.1.4 Photosynthesis and Respiration 96
        4.2.1.5 Long-Distance Transport 97
4.3 Diagnosis of Potassium Status in Plants..99
    4.3.1 Symptoms of Deficiency 99
    4.3.2 Symptoms of Excess 100
4.4 Concentrations of Potassium in Plants 101
4.5 Assessment of Potassium Status in Soils 105
    4.5.1 Potassium-Bearing Minerals 105
    4.5.2 Potassium Fractions in Soils 107
    4.5.3 Plant-Available Potassium..109
    4.5.4 Soil Tests for Potassium Fertilizer Recommendations 111
4.6 Potassium Fertilizers 112
    4.6.1 Kinds of Fertilizers 112
    4.6.2 Application of Potassium Fertilizers...113
References 116

4.1 HISTORICAL INFORMATION

Ever since ancient classical times, materials that contained potassium have been used as fertilizers, such as excrement, bird manure, and ashes (1), and these materials certainly contributed to crop growth and soil fertility. However, in those days people did not think in terms of modern chemical elements. Even an excellent pioneer of modern chemistry, Antoine Laurent de Lavoisier (1743–1794), assumed that the favorable effect of animal excrement was due to the humus present in it (2). Humphry Davy (1778–1827) discovered the chemical element potassium and Martin Heinrich Klaproth (1743–1817) was the first person to identify potassium in plant sap (3). Home (1762, quoted in 4) noted in pot experiments that potassium promoted plant growth. Carl Sprengel (1787–1859) was the first to propagate the idea that plants feed from inorganic nutrients and thus also from potassium (5). Justus Liebig (1803–1873) emphasized the importance of inorganic plant nutrients as cycling between the living nature and the inorganic nature, mediated by plants (6). He quoted that farmers in the area of Giessen fertilized their fields with charcoal burners' ash and prophesied that future farmers would fertilize their fields with potassium salts and with the ash of burned straw. The first potash mines for the production of potash fertilizer were sunk at Stassfurt, Germany in 1860.

4.2 DETERMINATION OF ESSENTIALITY

Numerous solution culture and pot experiments with K^+-free substrates have shown that plants do not grow without K^+. As soon as the potassium reserves of the seed are exhausted, plants die. This condition may also occur on strongly K^+-fixing soils. In contrast to other plant nutrients such as N, S, and P, there are hardly any organic constituents known with K^+ as a building element. Potassium ions activate various enzymes, which may also be activated by other univalent cationic species with a similar size and water mantle such as NH_4^+, Rb^+, and Cs^+ (7). These other species, however, play no major role under natural conditions as the concentrations of Cs^+, Rb^+, and also NH_4^+ in the tissues are low and will not reach the activation concentration required. In vitro experiments have shown that maximum activation is obtained within a concentration range of 0.050 to 0.080 M K^+. Ammonium may attain high concentrations in the soil solution of flooded soils, and ammonium uptake rates of plant species such as rice (*Oryza sativa* L.) are very high. In the cytosol, however, no high NH_4^+ concentrations build up because NH_4^+ is assimilated rapidly, as was shown for rice (8). Activation of enzymes in vivo may occur at the same high K^+ concentration as seen in in vitro experiments, as was shown for ribulose bisphosphate carboxylase (9).

It is assumed that K^+ binds to the enzyme surface, changing the enzymic conformation and thus leading to enzyme activation. Recent research has shown that in the enzyme dialkyl-glycine carboxylase, K^+ is centered in an octahedron with O atoms at the six corners. As shown in Figure 4.1, these O atoms are provided by three amino acyls, one water molecule, and the O of hydroxyl groups of each of serine and aspartate (10). As compared with Na^+, the K^+ binding is very selective because the dehydration energy required for K^+ is much lower than for Na^+. If the latter binds to the enzyme, the natural conformation of the enzyme is distorted, and the access of the substrate to the binding site is blocked. Lithium ions (Li^+) inactivate the enzyme in an analogous way. It is supposed that in most K^+-activated enzymes, the required conformation change is brought about by the central position of K^+ in the octahedron, where its positive charge attracts the negative site of the O atom located at each corner of the octahedron. This conformation is a unique structure that gives evidence of the unique function of K^+. In this context, it is of interest that the difference between K^+ and Na^+ binding to the enzyme is analogous to the adsorption of the cationic species to the

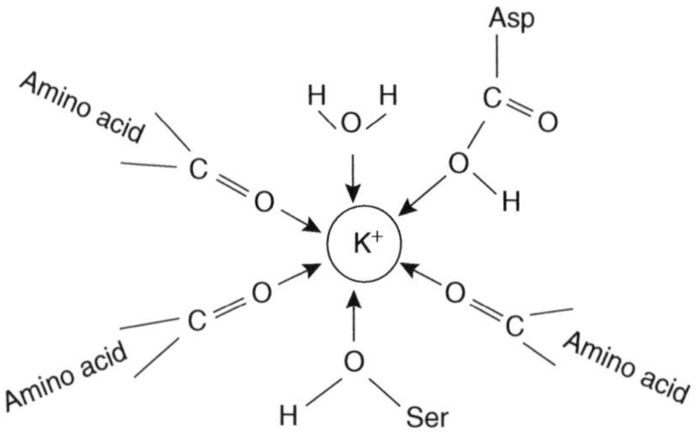

FIGURE 4.1 Potassium complexed by organic molecules of which the oxygen atoms are orientated to the positive charge of K^+. (Adapted from K. Mengel and E.A. Kirkby, *Principles of Plant Nutrition*. 5th ed. Dordrecht: Kluwer Academic Publishers, 2001.)

TABLE 4.1
Effect of Metal Chlorides on the H$^+$ Release by Roots of Intact Maize Plants

	Treatment of Water or Chloride Salt				
Outer medium	H$_2$O	K$^+$	Na$^+$	Ca^{2+}	Mg^{2+}
H$^+$ release (μmol/pot)	29.5	128***	46.5*	58.1*	78**

Significant difference from the control (H$_2$O) at *$P \leq$ 0.05, **$P \leq 0.01$, and ***$P \leq 0.001$, respectively.

Source: From K. Mengel and S. Schubert, *Plant Physiol.* 79:344–348, 1985.

interlayer of some 2:1 clay minerals, where the adsorption of K$^+$ is associated with the dehydration of the K$^+$, thus leading to a shrinkage of the mineral; Na$^+$ is not dehydrated and if it is adsorbed to the interlayer, the mineral is expanded.

It is not yet known how many different enzymes activated by K$^+$ possess this octahedron as the active site. There is another enzyme of paramount importance in which the activity is increased by K$^+$, namely the plasmalemma H$^+$-ATPase. This enzyme is responsible for excreting H$^+$ from the cell. As can be seen from Table 4.1 the rate of H$^+$ excretion by young corn (*Zea mays* L.) roots depends on the cationic species in the outer solution, with the lowest rate seen in the control treatment, which was free of ions. The highest H$^+$ release rate was in the treatment with K$^+$. Since the other cationic species had a promoting effect on the H$^+$ release relative to pure water, the influence of K$^+$ is not specific. However, a quantitative superiority of K$^+$ relative to other cations may have a beneficial impact on plant metabolism since the H$^+$ concentration in the apoplast of root cells is of importance for nutrients and metabolites taken up by H$^+$ cotransport as well as for the retrieval of such metabolites (11). The beneficial effect of cations in the outer solution is thought to originate from cation uptake, which leads to depolarization of the plasma membrane so that H$^+$ pumping out of the cytosol requires less energy. This depolarizing effect was highest with K$^+$, which is taken up at high rates relative to other cationic species. High K$^+$ uptake rates and a relatively high permeability of the plasmalemma for K$^+$ are further characteristics of K$^+$, which may also diffuse out of the cytosol across the plasma membrane back into the outer solution.

4.2.1 FUNCTION IN PLANTS

4.2.1.1 Enzyme Activation

The function of potassium in enzyme activation was considered in the preceding section.

4.2.1.2 Protein Synthesis

A probable function of potassium is in polypeptide synthesis in the ribosomes, since that process requires a high K$^+$ concentration (12). Up to now, however, it is not clear which particular enzyme or ribosomal site is activated by K$^+$. There is indirect evidence that protein synthesis requires K$^+$ (13). Salinity from Na$^+$ may affect protein synthesis because of an insufficient K$^+$ concentration in leaves and roots, as shown in Table 4.2 (14). Sodium chloride salinity had no major impact on the uptake of ^{15}N-labelled inorganic N but severely depressed its assimilation and the synthesis of labelled protein. In the treatment with additional K$^+$ in the nutrient solution, particularly in the treatment with 10 mM K$^+$, assimilation of inorganic N and protein synthesis were at least as good as in the control treatment (no salinity). In the salinity treatment without additional K$^+$, the K$^+$ concentrations in roots and shoots were greatly depressed. Additional K$^+$ raised the K$^+$ concentrations in roots and shoots to levels that were even higher than the K$^+$ concentration in the control treatment, and at this high cytosolic K$^+$ level, protein synthesis was not depressed.

TABLE 4.2
Effect of Na$^+$ Salinity on the K$^+$ Concentration in Barley Shoots and on ^{15}N Incorporation in Shoots

Treatment	K (mmol/kg fresh weight)	Total ^{15}N (mg/kg fresh weight)	% of Total ^{15}N in Protein	% of Total ^{15}N in Soluble Amino N	% Total ^{15}N in Inorganic N Compounds
Control	1260	54.4	43.9	53.1	3.0
80 mM NaCl	800	55.4	28.7	51.3	20.0
80 mM NaCl + 5 mM KCl	1050	74.2	39.9	53.8	6.3
80 mM NaCl + 10 mM KCl	1360	74.5	49.0	50.1	0.9

Note: ^{15}N solution was applied to roots of intact plants for 24 h. After pre-growth of plants in a standard nutrient solution for 5 weeks, plants were exposed to nutrient solutions for 20 days differing in Na$^+$ and K$^+$ concentrations.

Source: From H.M. Helal and K. Mengel, *Plant Soil* 51:457–462, 1979.

4.2.1.3 Ion Absorption and Transport

4.2.1.3.1 Potassium Absorption

Plant membranes are relatively permeable to K$^+$ due to various selective K$^+$ channels across the membrane. Basically, one distinguishes between low-affinity K$^+$ channels and high-affinity channels. For the function of the low-affinity channels, the electrochemical difference between the cytosol and the outer medium (liquid in root or leaf apoplast) is of decisive importance. The K$^+$ is imported into the cell for as long as the electrochemical potential in the cytosol is lower than in the outer solution. With the import of the positive charge (K$^+$) the electrochemical potential increases (decrease of the negative charge of the cytosol) and finally attains that of the outer medium, equilibrium is attained, and there is no further driving force for the uptake of K$^+$ (15). The negative charge of the cytosol is maintained by the activity of the plasmalemma H$^+$ pump permanently excreting H$^+$ from the cytosol into the apoplast and thus maintaining the high negative charge of the cytosol and building up an electropotential difference between the cytosol and the apoplast in the range of 120 to 200 mV. If the plasmalemma H$^+$ pumping is affected (e.g., by an insufficient ATP supply), the negative charge of the cytosol drops, and with it the capacity to retain K$^+$, which then streams down the electrochemical gradient through the low-affinity channel, from the cytosol and into the apoplast. Thus in roots, K$^+$ may be lost to the soil, which is, for example, the case under anaerobic conditions. This movement along the electrochemical gradient is also called *facilitated diffusion*, and the channels mediating facilitated diffusion are known as *rectifying channels* (16). Inwardly and outwardly directed K$^+$ channels occur, by which uptake and retention of K$^+$ are regulated (17). Their 'gating' (opening and closure) are controlled by the electropotential difference between the cytosol and the apoplast. If this difference is below the electrochemical equilibrium, which means that the negative charge of the cytosol is relatively low, outwardly directed channels are opened and vice versa. The plasmalemma H$^+$-ATPase activity controls the negative charge of the cytosol to a high degree since each H$^+$ pumped out of the cytosol into the apoplast results in an increase of the negative charge of the cytosol. Accordingly, hampering the ATPase (e.g., by low temperature) results in an outwardly directed diffusion of K$^+$ (18). Also, in growing plants, darkness leads to a remarkable efflux of K$^+$ into the outer solution, as shown in Figure 4.2. Within a period of 4 days, the K$^+$ concentration in the nutrient solution in which maize seedlings were grown increased steadily under dark conditions, whereas in light it remained at a low level of <10 μM (19). The outwardly directed channels may be blocked by Ca^{2+} (20). The blocking may be responsible for the so-called *Viets effect* (21), which results in an enhanced net uptake of potassium through a decrease in K$^+$ efflux (22).

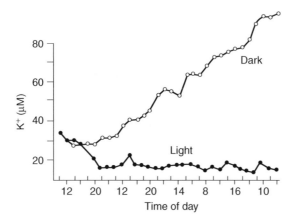

FIGURE 4.2 Potassium concentration changes in the nutrient solution with young intact maize plants exposed to light or dark over 4 days. (Adapted from K. Mengel, in *Frontiers in Potassium Nutrition: New Perspectives on the Effects of Potassium on Physiology of Plants.* Norcross, GA: Potash and Phosphate Institute, 1999, pp. 1–11.)

4.2.1.3.2 Potassium Transport within Tissues

Opening and closure of K^+ channels are of particular relevance for guard cells (23), and the mechanism of this action is controlled by the reception of red light, which induces stomatal opening (24). Diurnal rhythms of K^+ uptake were also found by Le Bot and Kirkby (25) and by MacDuff and Dhanoa (26), with highest uptake rates at noon and lowest at midnight. Energy supply is not the controlling mechanism, which still needs elucidation (26). Owing to the low-affinity channels, K^+ can be quickly transported within a tissue, and also from one tissue to another. This feature of K^+ does not apply for other plant nutrients. The low-affinity channel transport requires a relatively high K^+ concentration in the range of >0.1 mM (17). This action is mainly the case in leaf apoplasts, where the xylem sap has K^+ concentrations > 1 mM (27). At the root surface, the K^+ concentrations may be lower than 0.1 mM, and here high-affinity K^+ channels are required, as well as low-affinity channels, for K^+ uptake.

The principle of high-affinity transport is a *symport* or a *cotransport*, where K^+ is transported together with another cationic species such as H^+ or even Na^+. The K^+–H^+ or K^+–Na^+ complex behaves like a bivalent cation and has therefore a much stronger driving force along the electrochemical gradient. Hence, K^+ present near the root surface in micromolar concentrations is taken up.

Because of these selective K^+ transport systems, K^+ is taken up from the soil solution at high rates and is quickly distributed in plant tissues and cell organelles (28). Potassium ion distribution in the cell follows a particular strategy, with a tendency to maintain a high K^+ concentration in the cytosol, the so-called *cytoplasmic potassium homeostasis*, and the vacuole functions as a storage organelle for K^+ (29). Besides the H^+-ATPase, a pyrophosphatase (V-PPase) is also located in the tonoplast, for which the substrate is pyrophosphate. The enzyme not only pumps H^+ but also K^+ into the vacuole, and thus functions in the cytoplasmic homeostasis (Figure 4.3). This mechanism is an uphill transport because the vacuole liquid is less negatively charged than the cytosol. In Table 4.3, the typical pattern of K^+ concentration in relation to K^+ supply is shown (30). The cytosolic K^+ concentration remains at a high level almost independently of the K^+ concentration in the nutrient solution, whereas the vacuolar K^+ concentration reflects that of the nutrient solution.

4.2.1.3.3 Osmotic Function

The high cytosolic K^+ concentration required for polypeptide synthesis is particularly important in growing tissues; the K^+ in the vacuole not only represents K^+ storage but also functions as an indispensable osmoticum. In most cells, the volume of the vacuole is relatively large, and its turgor is essential for the tissue turgor. The osmotic function is not a specific one as there are numerous

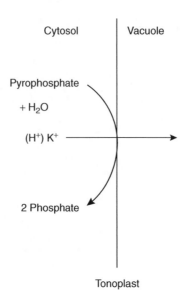

FIGURE 4.3 Pyrophosphatase located in the tonoplast and pumping H^+ or K^+ from the cytosol into the vacuole.

TABLE 4.3
K^+ Concentrations in the Cytosol and Vacuole as Related to the K^+ Concentration in the Outer Solution

	K^+ Concentration (mM)	
Outer Solution	**Vacuole**	**Cytosol**
1.2	85	144
0.1	61	140
0.01	21	131

Source: From M. Fernando et al., *Plant Physiol.* 100:1269–1276, 1992.

organic and inorganic osmotica in plants. There is a question, however, as to whether these can be provided quickly to fast-growing tissues, and in most cases it is the K^+ that is delivered at sufficient rates. In natrophilic species, Na^+ may substitute for K^+ in this osmotic function. The high vacuolar turgor in expanding cells produces the pressure potential required for growth. This pressure may be insufficient ($p < 0.6$ MPa) in plants suffering from K^+ deficiency (31). In Figure 4.4, pressure potentials and the related cell size in leaves of common bean (*Phaseolus vulgaris* L.) are shown. Pressure potentials (turgor) were significantly higher in the treatment with sufficient K^+ compared with insufficient K^+ supply. This higher turgor (ψ_p) promoted cell expansion, as shown in the lower part of Figure 4.4. From numerous observations, it is well known that plants insufficiently supplied with K^+ soon lose their turgor when exposed to water stress. In recent experiments it was found that K^+ increased the turgor and promoted growth in cambial tissue (32). The number of expanding cells derived from cambium was reduced with insufficient K^+ nutrition.

4.2.1.4 Photosynthesis and Respiration

Potassium ion transport across chloroplast and mitochondrial membranes is related closely to the energy status of plants. In earlier work, it was shown that K^+ had a favorable influence on photoreduction and photophosphorylation (33). More recently, it was found that an ATPase located in the

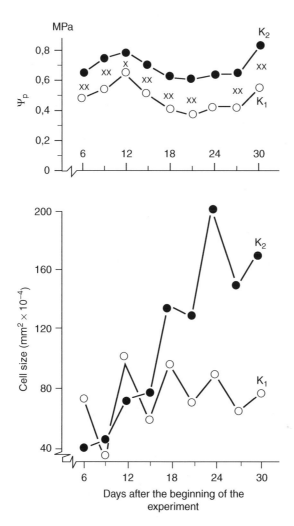

FIGURE 4.4 Pressure potential (ϕ_p) and cell size in leaves of common bean (*Phaseolus vulgaris* L.) insufficiently (K_1) and sufficiently (K_2) supplied with K^+. (Adapted from K. Mengel and W.W. Arneke, *Physiol. Plant* 54:402–408, 1982.)

inner membrane of chloroplasts pumps H^+ out of the stroma and thus induces a K^+ influx into the stroma via selective channels (34). The K^+ is essential for H^+ pumping by the envelope-located ATPase (35). Were it not for a system to pump H^+ from the illuminated chloroplast, the increase in stromal pH induced by the electron flow in the photosynthetic electron-transport chain would quickly dissipate (34). This high pH is a prerequisite for an efficient transfer of light energy into chemical energy, as was shown by a faster rate of O_2 production by photolysis in plants treated with higher K^+ concentration (36). The favorable effect of K^+ on CO_2 assimilation is well documented (37,38). An increase in leaf K^+ concentration was paralleled by an increase in CO_2 assimilation and by a decrease in mitochondrial respiration (38). Obviously, photosynthetic ATP supply substituted for mitochondrial ATP in the leaves with the high K^+ concentration. Thus, K^+ had a beneficial impact on the energy status of the plant.

4.2.1.5 Long-Distance Transport

Long-distance transport of K^+ occurs in the xylem and phloem vessels. Loading of the xylem occurs mainly in the root central cylinder, where protoxylem and xylem vessels are located adjacent to xylem parenchyma cells. The K^+ accumulates in the parenchyma cells (Figure 4.5) and is transported from there across the plasmalemma and the primary cell wall and through pits of the secondary cell wall into the xylem vessels (39). There is evidence that the outward-rectifying channels allow a K^+ flux (facilitated diffusion) from the parenchyma cells into the xylem vessel (40,41). The release of K^+ into the xylem sap decreases its water potential and thus favors the uptake of water (42). The direction of xylem sap transport goes along the transpiration stream and hence from root to leaves. The direction of the phloem sap transport depends on the physiological conditions and goes toward the strongest sinks. These may be young growing leaves, storage cells of roots, or fleshy fruits like tomato.

Phloem sap is rich in K^+, with a concentration range of 60 to 100 mM (43). Potassium ions are important for phloem loading and thus phloem transport. It was shown that K^+ particularly promotes the uptake of sucrose and glutamine into the sieve cells at high apoplastic pH (44). These metabolites presumably are taken up into the sieve vessels via a K^+ cotransport (Figure 4.5). This process is important, since in cases in which insufficient H^+ are provided by the plasmalemma H^+ pump, and thus the apoplastic pH is too high for a H^+ cotransport of metabolites, K^+ can substitute for H^+ and the most important metabolites required for growth and storage, sucrose and amino compounds, can be transported along the phloem. Hence the apoplastic K^+ concentration contributes much to phloem loading (Figure 4.5). This occurrence is in line with the observation that the phloem flow rate in castor bean (*Ricinus communis* L.) was higher in plants well supplied with K^+ than in plants with a low K^+ status (43). The favorable effect of K^+ on the transport of assimilates to growing plant organs has been shown by various authors (45).

Potassium ions cycle via xylem from roots to upper plant parts and via phloem from leaves to roots. The direction depends on the physiological demand. During the vegetative stage, the primary meristem is the strongest sink. Here, K^+ is needed for stimulating the plasmalemma ATPase that produces the necessary conditions for the uptake of metabolites, such as sucrose and amino acids. High K^+ concentrations are required in the cytosol for protein synthesis and in the vacuole for cell expansion (Figure 4.4). During the generative or reproductive phase, the K^+ demand depends on whether or not fruits rich in water are produced, such as apples or vine berries. These fruits need K^+ mainly for osmotic balance. Organs with a low water content, such as cereal grains, seeds, nuts, and cotton bolls, do not require K^+ to a great extent. Provided that cereals are well supplied with K^+ during the vegetative stage, K^+ supply during the generative stage has no major impact on grain formation (46).

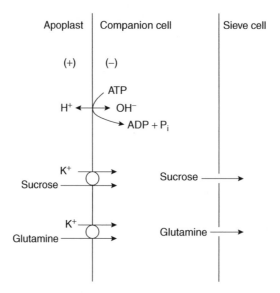

FIGURE 4.5 Cotransport of K^+/sucrose and K^+/glutamine from the apoplast into the companion cell, and from there into the sieve cell, driven by the plasmalemma ATPase.

However, for optimum grain filling, a high K^+ concentration in the leaves is required for the translocation of assimilates to the grains and for protein synthesis in these grains (47).

The generative phase of cereal growth requires hardly any K^+, but still appreciable amounts of N. In such cases, nitrate uptake of the plants is high and K^+ uptake low. The K^+ is recycled via the phloem from the leaves to the roots, where K^+ may enter the xylem again and balance the negative charge of the NO_3^- (48). Both the ionic species, K^+ and nitrate, are efficient osmotica and are thus of importance for the uptake of water into the xylem (49). In the phloem sap, K^+ balances the negative charge of organic and inorganic anions.

In storage roots and tubers, K^+ is required not only for osmotic reasons, but it may also have a more specific function. From work with sugar beet (*Beta vulgaris* L.) roots, a K^+-sucrose cotransport across the tonoplast into the vacuole, driven by an H^+/K^+ antiport cycling the K^+ back into the cytosol, was postulated (50).

4.3 DIAGNOSIS OF POTASSIUM STATUS IN PLANTS

4.3.1 SYMPTOMS OF DEFICIENCY

The beginning of K^+ deficiency in plants is growth retardation, which is a rather nonspecific symptom and is thus not easily recognized as K^+ deficiency. The growth rate of internodes is affected (51), and some dicotyledonous species may form rosettes (52). With the advance of K^+ deficiency, old leaves show the first symptoms as under such conditions K^+ is translocated from older to younger leaves and growing tips via the phloem. In most plant species, the older leaves show chlorotic and necrotic symptoms as small stripes along the leaf margins, beginning at the tips and enlarging along leaf margins in the basal direction. This type of symptom is particularly typical for monocotyledonous species. The leaf margins are especially low in K^+, and for this reason, they lose turgor, and the leaves appear flaccid. This symptom is particularly obvious in cases of a critical water supply. In some plant species, e.g., white clover (*Trifolium repens* L.), white and necrotic spots appear in the intercostal areas of mature leaves, and frequently, these areas are curved in an upward direction. Such symptoms result from a shrinkage and death of cells (53) because of an insufficient turgor. Growth and differentiation of xylem and phloem tissue is hampered more than the growth of the cortex. Thus, the stability and elasticity of stems is reduced so that plants are more prone to lodging (54). In tomato (*Lycopersicon esculentum* Mill.) fruits insufficiently supplied with K^+, maturation is disturbed, and the tissue around the fruit stem remains hard and green (55). The symptom is called *greenback* and it has a severe negative impact on the quality of tomato.

At an advanced stage of K^+ deficiency, chloroplasts (56) and mitochondria collapse (57). Potassium-deficient plants have a low-energy status (58) because, as shown above, K^+ is essential for efficient energy transfer in chloroplasts and mitochondria. This deficiency has an impact on numerous synthetic processes, such as synthesis of sugar and starch, lipids, and ascorbate (59) and also on the formation of leaf cuticles. The latter are poorly developed under K^+ deficiency (15). Cuticles protect plants against water loss and infection by fungi. This poor development of cuticles is one reason why plants suffering from insufficient K^+ have a high water demand and a poor *water use efficiency* (WUE, grams of fresh beet root matter per grams of water consumed). Sugar beet grown with insufficient K^+, and therefore showing typical K^+ deficiency, had a WUE of 5.5. Beet plants with a better, but not yet optimum, K^+ supply, and showing no visible K^+ deficiency symptoms, had a WUE of 13.1, and beet plants sufficiently supplied with K^+ had a WUE of 15.4 (60). Analogous results were found for wheat (*Triticum aestivum* L.) grown in solution culture (61). The beneficial effect of K^+ on reducing fungal infection has been observed by various authors (54,61,62). The water-economizing effect of K^+ and its protective efficiency against fungal infection are of great ecological relevance.

Severe K^+ deficiency leads to the synthesis of toxic amines such as putrescine and agmatine; in the reaction sequence arginine is the precursor (63). The synthetic pathway is induced by a low cytosolic pH, which presumably results from insufficient pumping of H^+ out of the cell by the plasmalemma H^+-ATPase, which requires K^+ for full activity. The reaction sequence is as follows:

- Arginine is decarboxylated to agmatine
- Agmatine is deaminated to carbamylputrescine
- Carbamylputrescine is hydrolyzed into putrescine and carbamic acid

4.3.2 SYMPTOMS OF EXCESS

Excess K^+ in plants is rare as K^+ uptake is regulated strictly (64). The oversupply of K^+ is not characterized by specific symptoms, but it may depress plant growth and yield (65). Excess K^+ supply has an impact on the uptake of other cationic species and may thus affect crop yield and crop quality. With an increase of K^+ availability in the soil, the uptake of Mg^{2+} and Ca^{2+} by oats (*Avena sativa* L.) was reduced (66). This action may have a negative impact for forage, where higher Mg^{2+} concentrations may be desirable. The relationship between K^+ availability and the Mg^{2+} concentrations in the aerial plant parts of oats at ear emergence is shown in Figure 4.6 (66). From the graph, it is clear that the plants took up high amounts of Mg^{2+} only if the K^+ supply was not sufficient for optimum growth. High K^+ uptake may also hamper the uptake of Ca^{2+} and thus contribute to the appearance of bitter pit in apple (*Malus pumila* Mill.) fruits (67) and of blossom-end rot in tomato fruits, with strong adverse effects on fruit quality (55).

The phenomenon that one ion species can hamper the uptake of another has been known for decades and is called *ion antagonism* or *cation competition*. In this competition, K^+ is a very strong competitor. If it is present in a relatively high concentration, it particularly affects the uptake of Na^+, Mg^{2+}, and Ca^{2+}. If K^+ is not present in the nutrient solution, the other cationic species are taken up at high rates. This effect is shown in Table 4.4 for young barley (*Hordeum vulgare* L.) plants grown in solution culture (68). In one treatment with the barley, the K^+ supply was interrupted for 8 days, having a tremendous impact on the Na^+, Mg^{2+}, and Ca^{2+} concentrations in roots and shoots as compared with the control plants with a constant supply of K^+. The sum of cationic equivalents in roots and shoots remained virtually the same. This finding is explained by the highly efficient uptake systems for K^+ as compared with uptake of the other cationic species. Uptake of K^+ leads to a partial depolarization of the plasmalemma (the cytosol becomes less negative due to the influx of K^+). This depolarization reduces the driving force for the uptake of the other cationic species, which are

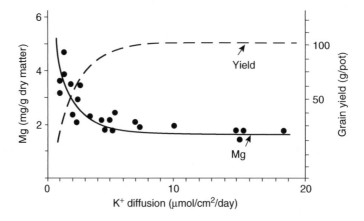

FIGURE 4.6   Effect of K^+ availability expressed as K^+ diffusion rate in soils on the Mg concentration in the aerial plant parts of oats at ear emergence and on grain yield (Adapted from H. Grimme et al., *Büntehof Abs.* 4:7–8, 1974/75.)

TABLE 4.4
Effect of Interrupting the K⁺ Supply for 8 Days on the Cationic Elemental Concentrations in Roots and Shoots of Barley Plants

	Elemental Concentration (me/kg dry weight)			
	Roots		Shoots	
Element	Control	Interruption	Control	Interruption
K	1570	280	1700	1520
Ca	90	120	240	660
Mg	360	740	540	210
Na	30	780	trace	120
Total	22,050	1920	2480	2510

Source: From H. Forster and K. Mengel, Z. *Acker-Pflanzenbau* 130:203–213, 1969.

otherwise taken up by facilitated diffusion. In the roots, the absence of K⁺ in the nutrient solution promoted especially the accumulation of Na⁺, and the shoots showed remarkably elevated Ca^{2+} and Mg^{2+} concentrations. Owing to the increased concentrations of cations except K⁺, the plants were able to maintain the cation–anion balance but not the growth rate. The interruption of K⁺ supply for only 8 days during the 2-to-3-leaf stage of barley significantly depressed growth and yield; the grain yield in the control treatment was 108 g/pot, and in the K⁺-interrupted treatment was 86 g/pot. This result shows the essentiality of K⁺ and demonstrates that its function cannot be replaced by other cationic species.

In this context, the question to what degree Na⁺ may substitute for K⁺ is of interest. The osmotic function of K⁺ is unspecific and can be partially replaced by Na⁺, as was shown for ryegrass (*Lolium* spp.) (69) and for rice (70). The Na⁺ effect is particularly evident when supply with K⁺ is not optimum (71). A major effect of Na⁺ can be expected only if plants take up Na⁺ at high rates. In this respect, plant species differ considerably (72). Beet species (*Beta vulgaris* L.) and spinach (*Spinacia oleracea* L.) have a high Na⁺ uptake potential, and in these species Na⁺ may substitute for K⁺ to a major extent. Cotton (*Gossypium hirsutum* L.), lupins (*Lupinus* spp. L.), cabbage (*Brassica oleracea capitata* L.), oats, potato (*Solanum tuberosum* L.), rubber (*Hevea brasiliensis* Willd. ex A. Juss.), and turnips (*Brassica rapa* L.) have a medium Na⁺ uptake potential; barley, flax (*Linum usitatissimum* L.), millet (*Pennisetum glaucum* R. Br.), rape (*Brassica napus* L.), and wheat have a low Na⁺ potential and buckwheat (*Fagopyrum esculentum* Moench), corn, rye (*Secale cereale* L.), and soybean (*Glycine max* Merr.) a very low Na⁺ uptake potential. However, there are also remarkable differences in the Na⁺ uptake potential between cultivars of the same species, as was shown for perennial ryegrass (*Lolium perenne* L.) (73). The Na⁺ concentration in the grass decreased with K⁺ supply and was remarkably elevated by the application of a sodium fertilizer. In sugar beet, Na⁺ can partially substitute for K⁺ in leaf growth but not in root growth (74). This effect is of interest since root growth requires phloem transport and thus phloem loading, which is promoted by K⁺ specifically (see above). The same applies for the import of sucrose into the storage vacuoles of sugar beet (50). Also, Na⁺ is an essential nutrient for some C4 species, where it is thought to maintain the integrity of chloroplasts (75). The Na⁺ concentrations required are low and in the range of micronutrients.

4.4 CONCENTRATIONS OF POTASSIUM IN PLANTS

Potassium in plant tissues is almost exclusively present in the ionic form. Only a very small portion of total K⁺ is bound by organic ligands via the e⁻ pair of O atoms. Potassium ions are dissolved in the liquids of cell walls, cytosol, and organelles such as chloroplasts and mitochondria and especially in vacuoles. From this distribution, it follows that the higher the K^+ content of a tissue the more water it contains. These tissues have a large portion of vacuole and a low portion of cell wall material. Plant organs rich in such tissues are young leaves, young roots, and fleshy fruits. Highest K^+ concentrations are in the cytosol, and they are in a range of 130 to 150 mM K^+ (76). Vacuolar K^+ concentrations range from about 20 to 100 mM and reflect the K^+ supply (30). The high cytosolic K^+ concentration is typical for all eukaryotic cells (29), and the mechanism that maintains the high K^+ level required for protein synthesis is described above.

If the K^+ concentration of plant tissues, plant organs, or total plants is expressed on a fresh weight basis, differences in the K^+ concentration may not be very dramatic. For practical considerations, however, the K^+ concentration is frequently related to dry matter. In such cases, tissues rich in water show high K^+ concentrations, since during drying the water is removed and the K^+ remains with the dry matter. This relationship is clearly shown in Figures 4.7a to 4.7c (77). In Figure 4.7a, the K^+ concentration in the tissue water of field-grown barley is presented for treatments with or without nitrogen supply. Throughout the growing period the K^+ concentration remained at a level of about 200 mM. In the last phase of maturation, the K^+ concentration increased steeply because of water loss during the maturation process. The K^+ concentrations in the tissue water were somewhat higher than cytosolic K^+ concentrations. This difference is presumably due to the fact that in experiments the water is not removed completely by tissue pressing. In Figure 4.7b, the K^+ concentration is based on the dry matter. Here, in the first phase of the growing period the K^+ concentration increased, reaching a peak at 100 days after sowing. It then declined steadily until maturation, when the concentration increased again because of a loss of tissue water. In the treatment with nitrogen supply, the K^+ concentrations were elevated because the plant matter was richer in water than in the plants not fertilized with nitrogen. Figure 4.7c shows the K^+ concentrations in the tissue water during the growing period for a treatment fertilized with K^+ and a treatment without K^+ supply. The difference in the tissue water K^+ concentration between both treatments was high and remained fairly constant throughout the growing period, with the exception of the maturation phase.

From these findings, it is evident that the K^+ concentration in the tissue water is a reliable indicator of the K^+ nutritional status of plants, and it is also evident that this K^+ concentration is independent of the age of the plant for a long period. This fact is an enormous advantage for analysis of plants for K^+ nutritional status compared with measuring the K^+ concentrations related to plant dry matter. Here, the age of the plant matter has a substantial impact on the K^+ concentration, and the optimum concentration depends much on the age of the plant.

Until now, almost all plant tests for K^+ have been related to the dry matter because dry plant matter can be stored easily. The evaluation of the K^+ concentration in dry plant matter meets with difficulties since plant age and also other factors such as nitrogen supply influence it (77). It is for this reason that concentration ranges rather than exact K^+ concentrations are denoted as optimum if the concentration is expressed per dry weight (see Table 4.6). Measuring K^+ concentration in the plant sap would be a more precise method for testing the K^+ nutritional status of plants.

Figure 4.7c shows the K^+ concentration in tissue water during the growing period for treatments with or without K fertilizer. There is an enormous difference in tissue water K^+ concentration since the treatment without K has not received K fertilizer since 1852 (Rothamsted field experiments). Hence, potassium deficiency is clearly indicated by the tissue water K^+ concentration. The increase in K^+ concentration in the late stage is due to water loss.

If the K^+ supply is in the range of deficiency, then the K^+ concentration in plant tissue is a reliable indicator of the K^+ nutritional status. The closer the K^+ supply approaches to the optimum, the smaller become the differences in tissue K^+ concentration between plants grown with

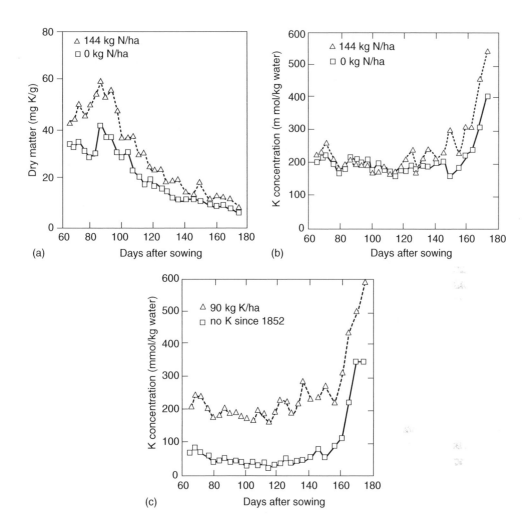

FIGURE 4.7 Potassium concentration in aboveground barley throughout the growing season of treatments with and without N supply (a) in the dry matter, (b) in the tissue water, and (c) in the tissue water with or without fertilizer K. (Adapted from A.E. Johnston and K.W. Goulding, in *Development of K Fertilizer Recommendations*. Bern: International Potash Institute, 1990, pp. 177–201.)

suboptimum and optimum supply. Such an example is shown in Table 4.5 (65). Maximum fruit yield was obtained in the K2 treatment at K^+ concentrations in the range of 25 to 35 mg K/g dry matter (DM). In the K^+ concentration range of 33 to 42 mg K/g DM, the optimum was surpassed.

The optimum K^+ concentration range for just fully developed leaves of 25 to 35 mg K/g DM, as noted for tomatoes, is also noted for fully developed leaves of other crop species, as shown in Table 4.6 (52). For cereals at the tillering stage, the optimum range is 35 to 45 mg K/g DM. From Table 4.5, it is evident that stems and fleshy fruits have somewhat lower K^+ concentrations than other organs. Also, roots reflect the K^+ nutritional status of plants, and those insufficiently supplied with K^+ have extremely low K^+ concentrations. Young roots well supplied with K^+ have even higher K^+ concentrations in the dry matter than well-supplied leaves (see Table 4.5). The K^+ concentrations for mature kernels of cereals including maize ranges from 4 to 5.5 mg/g, for rape seed from 7 to 9 mg/g, for sugar beet roots from 1.6 to 9 mg/g, and for potato tubers from 5 to 6 mg/g.

TABLE 4.5
Potassium Concentrations in Tomato Plants Throughout the Growing Season Cultivated with Insufficient K (K1), Sufficient K (K2), or Excess K (K3)

Plant Part		May 7	June 30	July 14	July 28	Aug 11	Aug 28
		Harvest Date					
		Potassium Concentration (mg K/g dry weight)					
Leaves	K1		10	13	15	10	11
	K2	29	25	34	31	30	35
	K3		33	41	40	39	41
Fruits	K1		22	22	23	18	18
	K2		28	30	28	26	26
	K3		27	27	33	29	28
Stems	K1		14	13	12	8	7
	K2	28	26	26	28	24	24
	K3		26	31	34	32	32
Roots	K1		8	12	6	4	5
	K2	17	47	44	22	27	43
	K3		43	52	44	37	39

Source: M. Viro, *Büntehof Abs.* 4:34–36, 1974/75.

TABLE 4.6
Range of Sufficient K Concentrations in Upper Plant Parts

Plant Species	Concentration Range (mg K/g DM)
Cereals, young shoots 5–8 cm above soil surface	
Wheat (*Triticum aestivum*)	35–55
Barley (*Hordeum vulgare*)	35–55
Rye (*Secale cereale*)	28–45
Oats (*Avena sativa*)	45–58
Maize (*Zea mays*)[a] at anthesis near cob position	20–35
Rice (*Oryza sativa*)[a] before anthesis	20–30
Dicotyledonous field crops	
Forage and sugar beets (*Beta vulgaris*)[a]	35–60
Potatoes (*Solanum tuberosum*)[a] at flowering	50–66
Cotton (*Gossypium*), anthesis to fruit setting	17–35
Flax (*Linum usitatissimum*), 1/3 of upper shoot at anthesis	25–35
Rape (*Brassica napus*)[a]	28–50
Sunflower (*Helianthus annuus*)[a] at anthesis	30–45
Faba beans (*Vicia faba*)[a] at anthesis	21–28
Phaseolus beans (*Phaseolus vulgaris*)	20–30
Peas (*Pisum sativum*)[a] at anthesis	22–35
Soya bean (*Glycine max*)	25–37
Red clover (*Trifolium pratense*)[a] at anthesis	18–30
White clover (*Trifolium repens*) total upper plant part at anthesis	17–25
Alfalfa (*Medicago sativa*) shoot at 15 cm	25–38
Forage grasses	
Total shoot at flowering 5 cm above soil surface, *Dactylis glomerata*, *Poa pratensis, Phleum pratense, Lolium perenne, Festuca pratensis*	25–35

TABLE 4.6 *(Continued)*

Plant Species	Concentration Range (mg K/g DM)
Vegetables	
Brassica species[a] *Brassica oleracea botrytis, B. oleracea capita,*	
B. oleracea gemmifera, B. oleracea gongylodes	30–42
Lettuce (*Lactuca sativa*)[a]	42–60
Cucumber (*Cucumis sativus*)[a] at anthesis	25–54
Carrot (*Daucus carota sativus*)[a]	27–40
Pepper (*Capsicum annuum*)[a]	40–54
Asparagus (*Asparagus officinalis*) fully developed shoot	15–24
Celery (*Apium graveolens*)[a]	35–60
Spinach (*Spinacia oleracea*)[a]	35–53
Tomatoes (*Lycopersicon esculentum*)[a] at first fruit setting	30–40
Watermelon (*Citrullus vulgaris*)[a]	25–35
Onions (*Allium cepa*) at mid vegetation stage	25–30
Fruit trees	
Apples (*Malus sylvestris*) mid-positioned leaves of youngest shoot	11–16
Pears (*Pyrus domestica*) mid-positioned leaves of youngest shoot	12–20
Prunus species[a], mid-positioned leaves of youngest shoots in summer	
P. armeniaca, P. persica, P. domestica, P. cerasus, P. avium	20–30
Citrus species[a], in spring shoots of 4–7 months, *C. paradisi, C. reticulata,*	
C. sinensis, C. limon	12–20
Berry fruits[a]	
From anthesis until fruit maturation *Fragaria ananassa, Rubus idaeus,*	
Ribes rubrum, Ribes nigrum, Ribes grossularia	18–25
Miscellaneous crops	
Vine (*Vitis vinifera*), leaves opposite of inflorescence at anthesis	15–25
Tobacco (*Nicotiana tabacum*)[a] at the mid of the vegetation season	25–45
Hop (*Humulus lupulus*)[a] at the mid of the vegetation season	28–35
Tea (*Camellia sinensis*)[a] at the mid of the vegetation season	16–23
Forest trees	
Coniferous trees, needles from the upper part of 1- or 2-year-old shoots,	
Picea excelsa, Pinus sylvestris, Larix decidua, Abies alba	6–10
Broad-leaved trees[a] of new shoots, species of *Acer, Betula, Fagus,*	
Quercus, Fraxinus, Tilia, Populus	12–15

[a]Youngest fully developed leaf.

Source: W. Bergmann, Ernährungsstörungen bei Kulturpflanzen, 3rd ed. Jena: Gustav Fischer Verlag, 1993, pp. 384–394.

4.5 ASSESSMENT OF POTASSIUM STATUS IN SOILS

4.5.1 POTASSIUM-BEARING MINERALS

The average potassium concentration of the earth's crust is 23 g/kg. Total potassium concentrations in the upper soil layer are shown for world soils and several representative soil groups in Table 4.7 (78). The most important potassium-bearing minerals in soils are alkali feldspars (30 to 20 g K/kg), muscovite (K mica, 60 to 90 g K/kg), biotite (Mg mica, 36 to 80 g K/kg), and illite (32 to 56 g K/kg). These are the main natural potassium sources from which K^+ is released by weathering and which feed plants. The basic structural element of feldspars is a tetrahedron forming a Si—Al–O framework in which the K^+ is located in the interstices. It is tightly held by covalent bonds (79). The weathering of the mineral begins at the surface and is associated with the release of K^+. This process is promoted by very low K^+ concentrations in the soil solution in contact with the mineral surface, and these low concentrations are

TABLE 4.7
Total K Concentrations in Some Soil Orders

Soil Order	Concentration of K (mg/g soil)
Entisols	26.3 ± 0.6
Spodosols	24.4 ± 0.5
Alfisol	11.7 ± 0.6
Mollisol	17.2 ± 0.5

Source: P.A. Helmke, in M.E. Sumner ed., *Handbook of Soil Science*, London: CRC Press, 2000, pp. B3-B24.

produced by K^+ uptake by plants and microorganisms and by K^+ leaching. The micas are phyllosilicates (80) and consist of two Si-Al-O tetrahedral sheets between which an M-O-OH octahedral sheet is located. M stands for Al^{3+}, Fe^{2+}, Fe^{3+}, or Mg^{2+} (81). Because of this 2:1 layer structure, they are also called 2:1 minerals. These three sheets form a unit layer, and numerous unit layers piled upon each other form a mineral. These unit layers of mica and illite are bound together by K^+ (Figure 4.8). K^+ is located in hexagonal spaces formed by O atoms, of which the outer electron shell attracts the positively charged K^+. During this attraction process, the K^+ is stripped of its hydration water. This dehydration is a selective process due to the low hydration energy of K^+. This action is in contrast to Na^+, which has a higher hydration energy than K^+; the hydrated water molecules are bound more strongly and hence are not stripped off, and the hydrated Na^+ does not fit into the interlayer. The same holds for divalent cations and cationic aluminum species. This selective K^+ bond is the main reason why K^+ in most soils is not leached easily, in contrast to Na^+. Ammonium has a similar low hydration energy as K^+ and can, for this reason, compete with K^+ for interlayer binding sites (82,83). This interlayer K^+ is of utmost importance for the release and for the storage of K^+. Equilibrium conditions exist between the K^+ concentration in the adjacent soil solution and the interlayer K^+. The equilibrium level differs much between biotite and muscovite, the former having an equilibrium at about 1 mM and the latter at about 0.1 mM K^+ in the soil solution (84). For this reason, the K^+ of the biotite is much more easily released than the K^+ from muscovite, and hence the weathering rate associated with the K^+ release of biotite is much higher than that of muscovite. The K^+ release is induced primarily by a decrease of the K^+ concentration in the adjacent solution caused by K^+ uptake of plant roots, or by K^+ leaching, or by both processes. The release of K^+ begins at the edge positions and proceeds into the inner part of the interlayer. This release is associated with an opening of the interlayer because the bridging K^+ is lacking. The free negative charges of the interlayer are then occupied by hydrated cationic species (Ca^{2+}, Mg^{2+}, Na^+, cationic Al species). From this process, it follows that the interlayer K^+ is exchangeable. The older literature distinguishes between p (planar), e (edge), and i (inner) positions of adsorbed (exchangeable) K^+ according to the sites where K^+ is adsorbed, at the outer surface of the mineral, at the edge of the interlayer, or in the interlayer. It is more precise, however, to distinguish between hydrated and nonhydrated adsorbed K^+ (79), the latter being much more strongly bound than the former. With the exception of the cationic aluminum species, hydrated cationic species may be replaced quickly by K^+ originating from the decomposition of organic matter or inorganic and organic (slurry, farm yard manure) K fertilizer. The dehydrated K^+ is adsorbed and contracts the interlayers and is thus 'fixed.' The process is called K^+ *fixation*. Fixation depends much on soil moisture and is restricted by dry (and promoted by moist) soils.

It is generally believed that H^+ released by roots also contributes much to the release of K^+ from K-bearing minerals. This process, however, is hardly feasible since in mineral soils the concentration of free protons is extremely low and is not reflected by the pH because of the very efficient H^+ buffer systems in mineral soils (85). It is the decrease of the K^+ concentration in the adjacent solution that mainly drives the K^+ release (86,87). Only high H^+ concentrations (pH < 3) induce a remarkable release of K^+, associated with the decomposition of the mineral (88). A complete removal of the

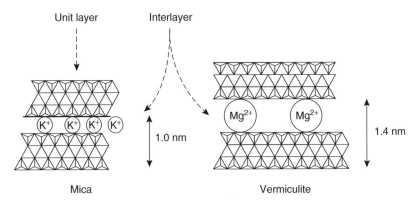

FIGURE 4.8 Scheme of a K^+-contracted interlayer of mica or illite and of vermiculite interlayer expanded by Mg^{2+}. (Adapted from K. Mengel and E.A. Kirkby, *Principles of Plant Nutrition*. 5th ed. Dordrecht: Kluwer Academic Publishers, 2001.)

interlayer K^+ by hydrated cations, including cationic aluminum species, leads to the formation of a new secondary mineral as shown in Figure 4.8 for the formation of vermiculite from mica (15). In acid mineral soils characterized by a relatively high concentration of cationic aluminum species, the aluminum ions may irreversibly occupy the interlayer sites of 2:1 minerals, thus forming a new secondary mineral called chlorite. By this process, the soil loses its specific binding sites for K^+ and hence the capacity of storing K^+ in a bioavailable form.

Under humid conditions in geological times, most of the primary minerals of the clay fraction were converted into secondary minerals because of K^+ leaching. The process is particularly relevant for small minerals because of their large specific surface. For this reason, in such soils the clay fraction contains mainly smectites and vermiculite, which are expanded 2:1 clay minerals. In soils derived from loess (Luvisol), which are relatively young soils, the most important secondary mineral in the clay fraction is the illite, which is presumably derived from muscovite. Its crystalline structure is not complete, it contains water, and its K^+ concentration is lower than that of mica (89). Mica and alkali feldspars present in the silt and sand fraction may considerably contribute to the K^+ supply of plants (90,91). Although the specific surface of these primary minerals in the coarser fractions is low, the percentage proportion of the silt and sand fraction in most soils is high and, hence, also the quantity of potassium-bearing minerals.

Cropping soils without replacing the K^+ removed from the soil in neutral and alkaline soils leads to the formation of smectites and in acid soils to the decomposition of 2:1 potassium-bearing minerals (92). Smectites have a high distance between the unit layers, meaning that there is a broad interlayer zone occupied mainly by bivalent hydrated cationic species and by adsorbed water molecules. For this reason, K^+ is not adsorbed selectively in the interlayers of smectites. The decomposition of K^+-selective 2:1 minerals results also from K^+ leaching. In addition, under humid conditions, soils become acidic, which promotes the formation of chlorite from K^+-selective 2:1 minerals. Thus, soils developed under humid conditions have a poor K^+-selective binding capacity and are low in potassium, for example, highly weathered tropical soils (Oxisols).

Organic soil matter has no specific binding sites for K^+, and therefore its K^+ is prone to leaching. Soils are generally lower in potassium, and their proportion of organic matter is higher. Soils with a high content of potassium are young soils, such as many volcanic soils, but also include soils derived from loess under semiarid conditions.

4.5.2 POTASSIUM FRACTIONS IN SOILS

Fractions of potassium in soil are (a) total potassium, (b) nonexchangeable (but plant-available) potassium, (c) exchangeable potassium, and (d) water-soluble potassium. The total potassium comprises the mineral potassium and potassium in the soil solution and in organic matter. Soil solution potassium plus organic matter potassium represent only a small portion of the total in mineral soils. The total potassium depends much on the proportion of clay minerals and on the type of clay minerals. Kaolinitic clay minerals, having virtually no specific binding sites for K^+, have low potassium concentrations in contrast to soils rich in 2:1 clay minerals. Mean total K^+ concentrations, exchangeable K^+ concentrations, and water-soluble K^+ are shown Table 4.8 (93). Soils with mainly kaolinitic clay minerals have the lowest, and those with smectitic minerals, which include also the 2:1 clay minerals with interlayer K^+, have the highest potassium concentration. The K^+ concentration of the group of mixed clay minerals, kaolinitic and 2:1 clay minerals, is intermediate. Water-soluble K^+ depends on the clay concentration in soils and on the type of clay minerals. As can be seen from Figure 4.9, the index of soluble K^+ decreases linearly with an increase in the clay concentration in soils and the level of soluble K^+ in the kaolinitic soil group is much higher than that of the mixed soil group and of the smectitic soil group (94).

The determination of total soil potassium requires a dissolution of potassium-bearing soil minerals. The digestion is carried out in platinum crucibles with a mixture of hydrofluoric acid, sulfuric acid, perchloric acid, hydrochloric acid, and nitric acid (95). Of particular importance in the available soil potassium is the exchangeable K^+, which is obtained by extracting the soil sample with a 1 M NH_4Cl or a 1 M NH_4 acetate solution (96). With this extraction, the adsorbed hydrated K^+ and some of the nonhydrated K^+ (K^+ at edge positions) is obtained. In arable soils, the exchangeable K^+ ranges between 40 to 400 mg K/kg. Soil extraction with $CaCl_2$ solutions (125 mM) extracts somewhat lower quantities of K^+ as the Ca^{2+} cannot exchange the nonhydrated K^+, in contrast to NH_4^+ of the NH_4^+-containing extraction solutions. For the determination of the nonexchangeable K^+, not obtained by the exchange with NH_4^+ and consisting of mainly interlayer K^+ and structural K^+ of the potassium feldspars, diluted acids such as 10 mM HCl (97) or 10 mM HNO_3 are used (98). These extractions have the disadvantage in that they extract a K^+ quantity and do not assess a release rate, the latter being of higher importance for the availability of K^+ to plants.

The release of K^+ from the interlayers is a first-order reaction (83) and is described by the following equations (99):

- Elovich function: $y = a + b \ln t$
- Exponential function: $\ln y = \ln a + b \ln t$
- Parabolic diffusion function: $y = b\, t^{1/2}$

where y is the quantity of extracted K^+, a the intercept on the Y-axis, and b the slope of the curve.

In this investigation, soils were extracted repeatedly by Ca^{2+}-saturated ion exchangers for long periods (maximum time 7000 h). Analogous results are obtained with electro-ultra-filtration (EUF), in which K^+ is extracted from a soil suspension in an electrical field (100). There are two successive extractions; the first with 200 V and at 20°C (first fraction) and a following extraction (second fraction)

TABLE 4.8

Representative K Concentrations in Soil Fractions Related to Dominating Clay Minerals

	K Concentration in Clay Types (mg K/kg soil)		
K Fraction	**Kaolinitic (26 Soils)**	**Mixture (53 Soils)**	**2:1 Clay Minerals (23 Soils)**
Total	3340	8920	15,780
Exchangeable	45	224	183
Water-soluble	2	5	4

Source: From N.C. Brady, and R.R. Weil, *The Nature and Properties of Soils*. 12th ed. Englewood Cliffs, NJ: Prentice-Hall, 1999.

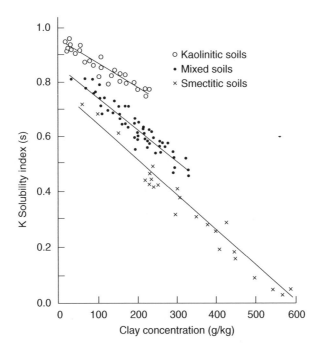

FIGURE 4.9 Potassium solubility of various soils related to their type of clay minerals (Adapted from A.N. Sharpley, *Soil Sci.* 149:44–51, 1990.)

with 400 V at 80°C. The first fraction contains the nonhydrated adsorbed K^+ plus the K^+ in the soil solution, whereas the second fraction contains the interlayer K^+. The extraction curves are shown for four different soils in Figure 4.10, from which it is clear that the K^+ release of the second fraction is a first-order reaction (101). The curves fit the first-order equation, the Elovich function, the parabolic diffusion function, and the power function, with the Elovich function having the best fit with $R^2 > 0.99$.

4.5.3   PLANT-AVAILABLE POTASSIUM

Several decades ago it was assumed that the 'activity ratio' between the K^+ activity and the Ca^{2+} plus Mg^{2+} activities in the soil solution would describe the K^+ availability in soils according to the equation (102)

$$AR = K^+/\sqrt{(Ca^{2+} \, Mg^{2+})}$$

In diluted solutions such as the soil solution, the K^+ activity is approximately the K^+ concentration. It was found that this activity ratio does not reflect the K^+ availability for plants (103). Of utmost importance for the K^+ availability is the K^+ concentration in the soil solution. The formula of the AR gives only the ratio and not the K^+ activity or the K^+ concentration. The K^+ flux in soils depends on the diffusibility in the medium, which means it is strongly dependent on soil moisture and on the K^+ concentration in the soil solution, as shown in the following formula (104):

$$J = D_1 \, (dc_1/dx) + D_2(dc_2/dx) + c_3 v;$$

where J is the K^+ flux toward root surface, D_1 the diffusion coefficient in the soil solution, c_1 the K^+ concentration in the soil solution, D_2 the diffusion coefficient at interlayer surfaces, c_2 the K^+ concentration at the interlayer surface, x the distance, dc/dx the concentration gradient, c_3 the K^+ concentration in the mass flow water, and v the volume of the mass flow water.

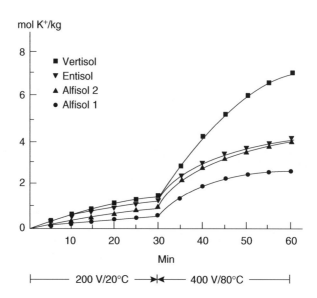

FIGURE 4.10 Cumulative K^+ extracted from four different soils by electro-ultra-filtration (EUF). First fraction extracted at 200 V and 20°C and the second fraction at 400 V and 80°C. (Adapted from K. Mengel and K. Uhlenbecker, *Soil Sci. Soc. Am. J.* 57:761–766, 1993.)

The hydrated K^+ adsorbed to the surfaces of the clay minerals can be desorbed quickly according to the equilibrium conditions, in contrast to the nonhydrated K^+ of the interlayer, which has to diffuse to the edges of the interlayer. The diffusion coefficient of K^+ in the interlayer is in the range of 10^{-13} m²/s, whereas the diffusion coefficient of K^+ in the soil solution is about 10^{-9} m²/s (105). The distances in the interlayers, however, are relatively short, and the K^+ concentrations are high. Therefore, appreciable amounts of K^+ can be released by the interlayers. The K^+ that is directly available is that of the soil solution, which may diffuse or be moved by mass flow to the root surface according to the equation shown above.

Growing roots represent a strong sink for K^+ because of K^+ uptake. Generally the K^+ uptake rate is higher than the K^+ diffusion, and thus a K^+ depletion profile is produced with lowest K^+ concentration at the root surface (106), as shown in Figure 4.11. This K^+ concentration may be as low as 0.10 μM, whereas in the equilibrated soil solution K^+, concentrations in the range of 500 μM prevail. Figure 4.11 shows such a depletion profile for exchangeable K^+. From this figure it is also clear that higher the value of dc/dx the higher the level of exchangeable K^+ (106). The K^+ concentration at the root surface is decisive for the rate of K^+ uptake according to the following equation (107):

$$Q = 2\pi a \alpha c t$$

where Q is the quantity of K^+ absorbed per cm root length, a the root radius in cm, α the K^+-absorbing power of the root, c the K^+ concentration at the root surface, and t the time of nutrient absorption.

The K^+-absorbing power of roots depends on the K^+ nutritional status of roots; plants well supplied with K^+ have a low absorbing power and vice versa. In addition, absorbing power depends also on the energy status of the root, and a low-energy status may even lead to K^+ release by roots (19). The K^+ concentration at the root surface also depends on the K^+ buffer power of soils, which basically means the amount of adsorbed K^+ that is in an equilibrated condition with the K^+ in solution. The K^+ buffer power is reflected by the plot of adsorbed K^+ on the K^+ concentration of the equilibrated soil solution, as shown in Figure 4.12. This relationship is known as the Quantity/Intensity relationship.

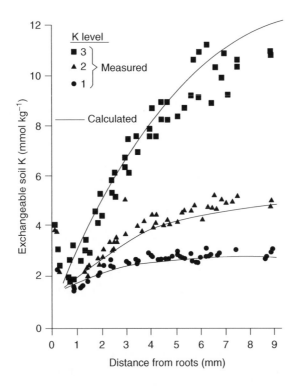

FIGURE 4.11 Potassium depletion profile produced by young rape roots in a Luvisol with three K^+ levels. (Adapted from A.O. Jungk, in *Plant Roots, the Hidden Half.* New York: Marcel Dekker, 2002, pp. 587–616.)

(Q/I relationship) in which the quantity represents the adsorbed K^+ (hydrated + nonhydrated K^+), and the intensity represents the K^+ concentration in the equilibrated soil solution. As can be seen from Figure 4.12, the quantity per unit intensity is much higher for one soil than the other, and the 'high' soil has a higher potential to maintain the K^+ concentration at the root surface at a high level than the 'medium' soil.

4.5.4 Soil Tests for Potassium Fertilizer Recommendations

The most common test for available K^+ is the exchangeable K^+ obtained by extraction with 1 M NH_4Cl or NH_4 acetate. This fraction contains mainly soil solution K^+ plus K^+ of the hydrated K^+ fraction and only a small part of the interlayer K^+. Exchangeable K^+ ranges between 40 and about 400 mg/kg soil and even more. Concentrations of <100 mg K/kg are frequently in the deficiency range; concentrations between 100 and 250 mg K/kg soil are in the range of sufficiently to well-supplied soils. Since one cannot distinguish between interlayer K^+ and K^+ from the hydrated fraction, this test gives no information about the contribution of interlayer K^+. The interpretation of the exchangeable soil test data therefore requires some information about further soil parameters, such as clay concentration and type of clay minerals. But even if these are known, it is not clear to what degree the interlayer K^+ is exhausted and to what degree mica of the silt fraction contributes substantially to the crop supply (90). Available K^+ is determined also by extraction with 1 mM HCl, by which the exchangeable K^+ and some of the interlayer K^+ are removed. Furthermore, with this technique the contribution of the interlayer K^+ also is not determined. The same is true for soil extraction with a mixture of 0.25 mM Ca lactate and HCl at a pH of 3.6 (108). Quantities of K^+ extracted with this technique are generally somewhat lower than the quantities of the exchangeable K^+

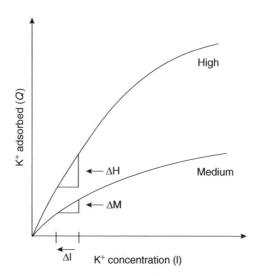

FIGURE 4.12   Potassium buffer power of a soil with a high or a medium buffer power [quantity–intensity (Q/I) ratio].

fraction. With the EUF technique, a differentiation between the nonhydrated exchangeable K^+ and the interlayer K^+ is possible, as shown in Figure 4.10. In the EUF, routine analysis extraction of the adsorbed hydrated K^+ lasts 30 minutes (200 V, 20°C); for the second fraction (400 V, 80°C), the soil suspension is extracted for only 5 minutes. The K^+ extracted during this 5-minute period is a reliable indicator of the availability of interlayer K^+ and is taken into consideration for the recommendation of the potassium fertilization rates. This EUF technique is nowadays used on a broad scale in Germany and Austria with much success for the recommendation of K fertilizer rates, particularly to crops such as sugar beet (109). With the EUF extraction procedure, not only are values for available K^+ obtained but the availability of other plant nutrients such as inorganic and organic nitrogen, phosphorus, magnesium, calcium, and micronutrients are also determined in one soil sample.

4.6   POTASSIUM FERTILIZERS

4.6.1   KINDS OF FERTILIZERS

The most important potassium fertilizers are shown in Table 4.9 (15). Two major groups may be distinguished, the chlorides and the sulfates. The latter are more expensive than the chlorides. For this reason, the chlorides are preferred, provided that the crop is not chlorophobic. Most field crops are not sensitive to chloride and should therefore be fertilized with potassium chloride (muriate of potash). Oil palm (*Elaeis guineensis* Jacq.) and coconut (*Cocos nucifera* L.) have a specific chloride requirement, with Cl^- functioning as a kind of plant nutrient because of its osmotic effect (110). Potassium nitrate is used almost exclusively as foliar spray. Potassium metaphosphate and potassium silicate have a low solubility and are used preferentially in artificial substrates with a low K^+-binding potential to avoid too high K^+ concentrations in the vicinity of the roots. Potassium silicates produced from ash and dolomite have a low solubility, but solubility is still high enough in flooded soils to feed a rice crop (111). The silicate has an additional positive effect on rice culm stability. Sulfate-containing potassium fertilizers should be applied in cases where the sulfur supply is insufficient; magnesium-containing potassium fertilizers are used on soils low in available magnesium. Such soils are mainly sandy soils with a low cation exchange capacity.

TABLE 4.9
Important Potassium Fertilizers

Fertilizer	Formula	Plant Nutrient Concentration (%)					
		K	K_2O^a	Mg	N	S	P
Muriate of potash	KCl	50	60	–	–	–	–
Sulfate of potash	K_2SO_4	43	52	–	–	18	–
Sulfate of potash magnesia	K_2SO_4 $MgSO_4$	18	22	11	–	21	–
Kainit	$MgSO_4+KCl+NaCl$	10	12	3.6	–	4.8	–
Potassium nitrate	KNO_3	37	44	–	13	–	–
Potassium metaphosphate	KPO_3	33	40	–	–	–	27

^{a}Expressed as K_2O, as in fertilizer grades.

Source: From K. Mengel and E.A. Kirkby, *Principles of Plant Nutrition*. 5th ed. Dordrecht: Kluwer Academic Publishers, 2001.

4.6.2 APPLICATION OF POTASSIUM FERTILIZERS

Chlorophobic crop species should not be fertilized with potassium chloride. Such species are tobacco (*Nicotiana tabacum* L.), grape (*Vitis vinifera* L.), fruit trees, cotton, sugarcane (*Saccharum officinarum* L.), potato, tomato, strawberry (*Fragaria x ananassa* Duchesne), cucumber (*Cucumis sativus* L.), and onion (*Allium cepa* L.). These crops should be fertilized with potassium sulfate. If potassium chloride is applied, it should be applied in autumn on soils that contain sufficiently high concentrations of K^+-selective binding sites in the rooting zone. In such a case, the chloride may be leached by winter rainfall, whereas the K^+ is adsorbed to 2:1 minerals and hence is available to the crop in the following season. On soils with a medium to high *cation exchange capacity* (CEC > 120 mgmol/kg) and with 2:1 selective K^+-binding minerals, potassium fertilizers can be applied in all seasons around the year since there is no danger of K^+ leaching out of the rooting profile (Alfisols, Inceptisols, Vertisols, and Mollisols, in contrast to Ultisols, Oxisols, Spodosols, and Histosols). In the latter soils, high K^+ leaching occurs during winter or monsoon rainfall. Histosols may have a high CEC on a weight basis but not on a volume basis because of their high organic matter content. In addition, Histosols contain few K^+-selective binding sites. Under tropical conditions on highly weathered soils (Oxisols, Ultisols), potassium fertilizer may be applied in several small doses during vegetative growth in order to avoid major K^+ leaching.

The quantities of fertilizer potassium required depend on the status of available K^+ in the soil and on the crop species, including its yield level. Provided that the status of available K^+ in the soil is sufficient, the potassium fertilizer rate should be at least as high as the quantity of potassium present in the crop parts removed from the field, which in many case are grains, seeds, tubers, roots or fruits. In Table 4.10 (15), the approximate concentrations of potassium in plant parts are shown. It is evident that the potassium concentrations in cereal grains are low compared with leguminous seeds, sunflower (*Helianthus annuus* L.) and rape seed. Potassium removal by fruit trees is shown in Table 4.11. The concept of assessing fertilizer rates derived from potassium removal is correct provided that no major leaching losses occur during rainy seasons. In such cases, the K^+ originating from leaves and straw remaining on the field may be leached into the subsoil at high rates. Such losses by leaching are the case for Spodosols, Oxisols, and Ultisols. Here, besides the K^+ removed from the soil by crop plants, the leached K^+ must also be taken into consideration. On the other hand, if a soil has a high status of available K^+, one or even several potassium fertilizer applications per crop species in the rotation may be omitted. As a first approach for calculating the amount of available K^+ in the soil, 1 mg/kg soil of exchangeable K^+ equals approximately 5 kg K/ha. In this calculation, interlayer K^+ is not taken into consideration. If the soil is low in available K^+, for most soils higher fertilization rates are required than 5 kg K/ha per mg exchangeable K^+, since with the

TABLE 4.10
Quantities of Potassium Removed from the Field by Crops

Crop and Product	Removal[a]	Crop and Product	Removal[a]
Barley, grain	4.5	Soybeans, grain	18
Barley, straw	12.0	Sunflower, seeds	19
Wheat, grain	5.2	Sunflower, straw	36
Wheat, straw	8.7	Flax, seeds	8
Oats, grain	4.8	Flax, straw	12
Oats, straw	15.0	Sugarcane, aboveground matter	3.3
Maize, grain	3.9	Tobacco, leaves	50
Maize, straw	13.5	Cotton, seed + lint	8.2
Sugar beet, root	2.5	Potato, tubers	5.2
Sugar beet, leaves	4.0	Tomatoes, fruits	3.0
Rape, seeds	11	Cabbage, aboveground matter	2.4
Rape, straw	40	Oil palm, bunches for 1000 kg oil	87
Faba beans, seeds	11	Coconuts	40
Faba beans, straw	21	Bananas, fruits	4.9
Peas, seeds	11	Rubber, dry	3.8
Peas, straw	21	Tea	23

[a] kg K/1000 kg (tonne) plant matter.

Source: From K. Mengel and E.A. Kirkby, *Principles of Plant Nutrition*. 5th ed. Dordrecht: Kluwer Academic Publishers, 2001.

TABLE 4.11
Potassium Removal by Fruits of Fruit Trees with Medium Yield

Fruit	K Removed (kg/ha/year)
Pome fruits	60
Stone fruits	65
Grapes	110
Oranges	120
Lemons	115

Source: From K. Mengel and E.A. Kirkby, *Principles of Plant Nutrition*. 5th ed. Dordrecht: Kluwer Academic Publishers, 2001.

exception of Histosols and Spodosols, sites of interlayer positions must be filled up by K^+ before the exchangeable K^+ will be raised. This problem is particularly acute on K^+-fixing soils. Here, high K fertilizer rates are required, as shown in Table 4.12 (112). From the discussion, it is clear that with normal potassium fertilizer rates, the yield and the potassium concentration in leaves were hardly raised and optimum yield and leaf potassium concentrations were attained with application of 1580 kg K/ha. As soon as the K^+-fixing binding sites are saturated by K^+, fertilizer should be applied at a rate in the range of the K^+ accumulation by the crop.

　　Plant species differ in their capability for exploiting soil K^+. There is a major difference between monocotyledonous and dicotyledonous species, the latter being less capable of exploiting

TABLE 4.12
Effect of Potassium Fertilizer Rates on Grain Yield of Maize, Potassium Concentrations in Leaves, and Lodging for Crops Grown on a K^+-Fixing Soil

Fertilizer Applied (kg K/ha)	Leaf K (mg K/g dry weight)	Grain Yield (1000 kg/ha)	Water in Grain (%)	Lodging (%)
125	6.4	1.75	31.5	42
275	7.8	2.57	28.7	21
460	8.6	4.66	28.6	18
650	10.3	6.95	29.2	20
835	14.3	7.76	29.7	5
1580	17.1	8.98	29.7	2
2200	18.6	8.88	29.3	2
LSD < 0.05	1.0	0.65	1.5	

Source: From V. Kovacevic and V. Vukadinovic, *South Afr. Plant Soil* 9: 10–13, 1992.

soil K^+, mainly interlayer K^+, than the former. In a 20-year field trial on an arable soil derived from loess (Alfisol), the treatment without potassium fertilizer produced cereal yields that were not much lower than those in the fertilized treatment, in contrast to the yields of potatoes, faba beans (*Vicia faba* L.), and a clover-grass mixture. With these crops, the relative yields were 73, 52, and 84, respectively, with a yield of 100 in the potassium-fertilized treatment (113). This different behavior is particularly true for grasses and leguminous species. Root investigations under field conditions with perennial ryegrass and red clover (*Trifolium pratense* L.) cultivated on an Alfisol showed considerable differences in root morphology, including root hairs and root length, which were much longer for the grass (114). Hence the root–soil contact is much greater for the grass than for the clover. The grass will therefore still feed sufficiently from the low soil solution K^+ concentration originating from interlayer K^+, a concentration that is insufficient for the clover. From this result, it follows that leguminous species in a mixed crop stand, including swards of meadow and pasture, will withstand the competition with grasses only if the soil is well supplied with available K^+.

This difference between monocots and dicots in exploiting soil K^+ implies that grasses can be grown satisfactorily on a lower level of exchangeable soil K^+ than dicots. It should be taken into consideration, however, that a major depletion of interlayer K^+ leads to a loss of selective K^+-binding sites because of the conversion or destruction of soil minerals (92), giving an irreversible loss of an essential soil fertility component.

Table 4.12 shows that the optimum K^+ supply considerably decreases the percentage of crop lodging. This action is an additional positive effect of K^+, which is also true with other cereal crops. As already considered above, K^+ favors the energy status of plants and thus the synthesis of various biochemical compounds such as cellulose, lignin, vitamins, and lipids. In this respect, the synthesis of leaf cuticles is of particular interest (15). Poorly developed cuticles and also thin cell walls favor penetration and infection by fungi and lower the resistance to diseases (115).

Heavy potassium fertilizer rates also may depress the negative effect of salinity since the excessive uptake of Na^+ into the plant cell is depressed by K^+. Table 4.13 presents such an example for mandarin oranges (*Citrus reticulata* Blanco) (116), showing that the depressive effect of salinity on leaf area was counterbalanced by higher potassium fertilizer rates. The higher the relative K^+ effect, the higher is the salinity level.

TABLE 4.13
Effect of Potassium Fertilizer on the Leaf Area of Satsuma Mandarins Grown at Different Salinity Levels Induced by NaCl

Salinity (dS/m)	Potassium Applied (g/tree)		
	0	70	150
	Leaf Area (cm²/tree)		
0.65	23.2	26.4	31.1
2.00	19.8	23.7	28.2
3.50	16.9	22.2	25.0
5.00	13.2	19.4	23.1
6.50	9.7	16.2	21.2

LSD ($P \le 0.05$) for the K effect = 0.5.

Source: From D. Anac et al., in *Food Security in the WANA Region, the Essential Need for Balanced Fertilization*. Basel: International Potash Institute, 1997, pp. 370–377.

REFERENCES

1. K. Scharrer. Die Bedeutung der Mineralstoffe für der Pflanzenbau; Historisches. In: G. Michael, ed. *Encyclopedia of Plant Physiology*. Berlin: Springer, 1958, pp. 851–866.
2. J. Boulaine. Lavoisier, his freschines domain and agronomy. *C. R. Agric. France* 80: 67–74, 1994.
3. W. Ostwald. In: G. Bugge, ed. *Das Buch der grossen* Chemiker: Weinheim Verlag Chemie, 1984, pp. 405–416.
4. E.J. Hewitt, T.A. Smith. *Plant Mineral Nutrition*. London: The English University Press, 1987.
5. R.R. van der Ploeg, W. Böhm, M.B. Kirkham. On the origin of the theory of mineral nutrition of plants and the Law of the Minimum. *Soil Sci. Am. J.* 63:1055–1062, 1999.
6. J. Liebig. *Die organische Chemie in ihrer Anwendung auf Agrikultur und Physiologie* Braunschweig: Viehweg-Verlag, 1840.
7. R.E. Nitsos, H.J. Evans. Effects of univalent cations on the activity of particulate starch synthetase. *Plant Physiol.* 44:1260–1266, 1969.
8. H. Kosegarten, F. Grolig, J. Wienecke, G. Wilson, B. Hoffmann. Differential ammonia-elicited changes of cytosolic pH in root hair cells of rice and maize as monitored by 2(,7(-bis-(2-carboxyethyl)-5 (and 6-)-carboxyfluorescein-fluorescence ratio. *Plant Physiol.* 113:451–461, 1997.
9. B. Demmig, H. Gimmler. Properties of the isolated intact chloroplast at cytoplasmic K^+ concentrations. I. Light-induced cation uptake into intact chloroplasts is driven by an electrical potential difference. *Plant Physiol.* 73:169–173, 1983.
10. C. Miller. Potassium selectivity in proteins–oxygen cage or in the F-ace. *Science* 261:1692–1693, 1993.
11. K.H. Mühling, S. Schubert, K. Mengel. Role of plasmalemma H ATPase in sugar retention by roots of intact maize and field bean plants. *Z. Pflanzenernähr Bodenk* 156:155–161, 1993.
12. R.G. Wyn Jones, A. Pollard. Proteins, enzymes and inorganic ions. In: A. Läuchli, R.L. Bieleski, eds. *Inorganic Plant Nutrition*. New York: Springer 1983, pp. 528– 562.
13. K. Koch, K. Mengel. The influence of the level of potassium supply to young tobacco plants (*Nicotiana tabacum* L.) on short term uptake and utilisation of nitrate nitrogen (15N). *J. Sci. Food Agric.* 25:465–471, 1974.
14. H.M. Helal, K. Mengel. Nitrogen metabolism of young barley plants as affected by NaCl salinity and potassium. *Plant Soil* 51:457–462, 1979.
15. K. Mengel, E.A. Kirkby. *Principles of Plant Nutrition*. 5th ed. Dordrecht: Kluwer Academic Publishers, 2001.
16. F.J.M. Maathuis. In: D. Oosterhuis, G. Berkowitz, eds. *Frontiers in Potassium Nutrition: New Perspectives on the Effects of Potassium on Physiology of Plants*. Norcross, Georgia, USA: Potash and Phosphate Institute, 1999, pp. 33–41.

17. F.J.M. Maathuis, D. Sanders. Regulation of K^+ absorption in plant root cells by external K^+ interplay of different plasma membrane K^+ transporters. *J. Exp. Bot.* 48:451–458, 1997.

18. K. Mengel, S. Schubert. Active extrusion of protons into deionized water by roots of intact maize plants. *Plant Physiol.* 79:344–348, 1985.

19. K. Mengel. Integration of functions and involvement of potassium metabolism at the whole plant level. In: D.M. Oosterhuis, G.A. Berkowitz, eds. *Frontiers in Potassium Nutrition: New Perspectives on the Effects of Potassium on Physiology of Plants.* Norcross, GA: Potash and Phosphate Institute, 1999, pp. 1–11.

20. F. Bouteau, U. Bousquet, A.M. Penarum, M. Convert, O. Dellis, D. Cornel, J.P. Rona. Time dependent K^+ currents through plasmalemma of laticifer protoplasts from *Hevea brasiliensis*. *Physiol. Plant* 98:97–104, 1996.

21. F.G. Viets. Calcium and other polyvalent cations as accelerators of ion accumulation by excised barley roots. *Plant Physiol.* 19:466–480, 1944.

22. K. Mengel, H.M.H. Helal. The influence of the exchangeable Ca^{2+} of young barley roots on the fluxes of K and phosphate–an interpretation of the Viets effect. *Z. Pflanzenphysiol.* 57:223–234, 1967.

23. L.T. Talbott, E. Zeiger. Central roles for potassium and sucrose in guard-cell osmoregulation. *Plant Physiol.* 111:1051–1057, 1996.

24. N. Roth-Bejerano, A. Nejidat. Phytochrome effects on K fluxes in guard cells of Commelina communis. *Physiol. Plant* 71:345–351, 1987.

25. J. Le Bot, E.A. Kirkby. Diurnal uptake of nitrate and potassium during the vegetative growth of tomato plants. *J. Plant Nutr.* 15:247–264, 1992.

26. J.H. MacDuff, M.S. Dhanoa. Diurnal and ultradian rhythms in K^+ uptake by *Trifolium repens* under natural light patterns: evidence for segmentation at different root temperatures. *Physiol. Plant* 98:298–308, 1996.

27. U. Schurr, E.D. Schulze. The concentration of xylem sap constituents in root exudates, and in sap from intact, transpiring castor bean plants (*Ricinus communis* L.). *Plant Cell Environ.* 118:409–420, 1995.

28. A.D.M. Glass. Regulation of ion transport. *Annu. Rev. Plant Physiol.* 34:311–326, 1983.

29. R.G. Wyn Jones. Cytoplasmic potassium homeostasis; review of the evidence and its implications. In: D.M. Oosterhuis, G.A. Berkowitz, eds. *Frontiers in Potassium Nutrition: New Perspectives on the Effect of Potassium on Physiology of Plants.* Norcross, GA: Potash and Phosphate Institute, 1999, pp. 13–22.

30. M. Fernando, J. Mehroke, A.D.M. Glass. De novo synthesis of plasma membrane and tonoplast polypeptides of barley roots during short-term K deprivation. In search of the high-affinity K transport system. *Plant Physiol.* 100:1269–1276, 1992.

31. K. Mengel, W.W. Arneke. Effect of potassium on the water potential, the pressure potential, the osmotic potential and cell elongation in leaves of *Phaseolus vulgaris*. *Physiol. Plant* 54:402–408, 1982.

32. W. Wind, M. Arend, J. Fromm. Potassium-dependent cambial growth in poplar. *Plant Biology* 6:30–37, 2004.

33. R. Pflüger, K. Mengel. Photochemical activity of chloroplasts from different plants fed with potassium. *Plant Soil* 36:417–425, 1972.

34. G.A. Berkowitz, J.S. Peters. Chloroplasts [spinach] inner-envelope Atase acts as a primary proton pump. *Plant Physiol.* 102:261–267, 1993.

35. R. Shingles, R.E. McCarty. Direct measurement of ATP-dependent proton concentration changes and characterization of a K^+-stimulated ATPase in pea chloroplast inner envelope vesicles. *Plant Physiol.* 106:731–737, 1994.

36. W. Wu, G.A. Berkowitz. Stromal pH and photosynthesis are affected by electroneutral K and H exchange through chloroplast envelope ion channels. *Plant Physiol.* 98:666–672, 1992.

37. P.A. Pier, G.A. Berkowitz. Modulation of water stress effects on photosynthesis by altered leaf K. *Plant Physiol.* 85:655–661, 1987.

38. T.R. Peoples, D.W. Koch. Role of potassium in carbon dioxide assimilation in *Medicago sativa* L. *Plant Physiol.* 63:878–881, 1979.

39. A. Läuchli, D. Kramer, M.G. Pitman, U. Lüttge. Ultrastructure of xylem parenchyma cells of barley roots in relation to ion transport to the xylem. *Planta* 119:85–99, 1974.

40. A.H. de Boer. Potassium translocation into the root xylem. *Plant Biol.* 1:36–45, 1999.

41. B. Köhler, K. Raschke. The delivery of salts to the xylem. Three types of anion conductance in the plasmalemma of the xylem parenchyma of roots of barley. *Plant Physiol.* 122:243–254, 2000.

42. D.A. Baker, P.E. Weatherley. Water and solute transport by exuding root systems of Ricinus communis. *J. Exp. Bot.* 20:485–496, 1969.

43. K. Mengel, H.E. Haeder. Effect of potassium supply on the rate of phloem sap exudation and the composition of phloem sap of *Ricinus communis. Plant Physiol.* 59:282–284, 1977.

44. A.J.E. van Bel, A.J. van Erven. A model for proton and potassium co-transport during the uptake of glutamine and sucrose by tomato internode disks. *Planta* 145:77–82, 1979.

45. K. Mengel, E.A. Kirkby. Potassium in crop production. *Adv. Agron.* 33:59–110, 1980.

46. K. Mengel, H. Forster. The effect of seasonal variation and intermittent potassium nutrition on crop and visual quality of barley. *Z. Acker- u Pflanzenbau* 127:317–326, 1968.

47. K. Mengel, M. Secer, K. Koch. Potassium effect on protein formation and amino acid turnover in developing wheat grain. *Agron. J.* 73:74–78, 1981.

48. H. Marschner, E.A. Kirkby, C. Engels. Importance of cycling and recycling of mineral nutrients within plants for growth and development. *Bot. Acta* 110:265–273, 1997.

49. G.I. McIntyre. The role of nitrate in the osmotic and nutritional control of plant development. *Aus. J. Plant Physiol.* 24:103–118, 1997.

50. D.P. Briskin, W.R. Thornley, R.E. Wyse. Membrane transport in isolated vesicles from sugar beet taproot. II. Evidence for a sucrose/H-antiport. *Plant Physiol.* 78:871–875, 1985.

51. M.D. de la Gardia, M. Benlloch. Effects of potassium and giberellic acid on stem growth of whole sunflower plants. *Physiol. Plant* 49:443–448, 1980.

52. W. Bergmann. *Ernährungsstörungen bei Kulturpflanzen–Entstehung und Diagnose* Jena: VEB Gustav Fischer, 3rd ed. 1993, pp. 384–394.

53. W. Bussler. Mangelerscheinungen an höheren Pflanzen. II Mangel an Hauptnährstoffen. *Z. Pflanzenkrankheiten Pflanzenschutz* 86:43–62,1979.

54. W. Krüger. The influence of fertilizer on the fungal diseases of maize. In: *Fertilizer Use and Plant Health*. Bern: International Potash Institute, 1976, pp. 145–156.

55. H. Forster, F. Ventner. The influence of potassium nutrition on greenback of tomato fruits. *Gartenbauwiss* 40:75–78, 1975.

56. H.P. Pissarek. The effect of intensity and duration of Mg-deficiency on the grain yields of oats. *Z. Acker-Pflanzenb* 148:62–71, 1979.

57. A.L. Kursanov, E. Vyskrebentzewa. The role of potassium in plant metabolism and the biosynthesis of compounds important for the quality of agricultural products. In: *Potassium and the Quality of Agricultural Products*. Bern: International Potash Institute, 1966, pp. 401–420.

58. H.E. Haeder. The effect of potassium on energy transformation. In: *Potassium Research and Agricultural Production*. Bern: International Potash Institute, 1974, pp. 153–160.

59. D. Anac, H. Colakoglu. Current situation of K fertilization in Turkey. In: K. Mengel, K. Krauss, eds. *K Availability of Soils in West Asia and North Africa–Status and Perspectives*. Basel: International Potash Institute, 1995 pp. 235–247.

60. K. Mengel, H. Forster. The effect of the potassium concentration of the soil solution on the yield, the water consumption and the potassium uptake rates of sugar beet (*Beta vulgaris* subsp. *esculenta* var. *altissima*). *Z. Pflanzenernähr Bodenk* 134:148–156, 1973.

61. I. Kovanci, H. Colakoglu. The effect of varying K level on yield components and susceptibility of young wheat plants to attack by Puccinia striiformis West. In: *Fertilizer Use and Plant Health*. Bern: International Potash Institute, 1976, pp. 177–181.

62. M. Olagnier, J.L. Renard. The influence of potassium on the resistance of oil palms to Fusarium. In: *Fertilizer Use and Plant Health*. Bern: International Potash Institute, 1976, pp. 157–166.

63. T.A. Smith, C. Sinclair. The effect of acid feeding on amine formation in barley. *Ann. Bot.* 31:103–111,1967.

64. A.D.M. Glass, M.Y. Siddiqi. The control of nutrient uptake in relation to the inorganic composition of plants. In: B.P. Tinker, A. Läuchli, eds. *Advances in Plant Nutrition*. New York: Praeger, 1984, pp. 103–147.

65. M. Viro. Influence of the K status on the mineral nutrient balance and the distribution of assimilates in tomato plants. *Büntehof Abs.* 4: 34–36, 1974/75.

66. H. Grimme, L.C. von Braunschweig, K. Nemeth. K, Mg and Ca interactions as related to cation uptake and yield. *Büntehof Abs.* 4:7–8, 1974/75.

67. B.G. Wilkinson. Mineral composition of apples. IX. Uptake of calcium by the fruit. *J. Sci. Food Agric.* 19:446–447, 1968.

68. H. Forster, K. Mengel. The effect of a short term interruption in the K supply during the early stage on yield formation, mineral content and soluble amino acid content. *Z. Acker-Pflanzenbau* 130:203–213, 1969.

69. P.B. Barraclough, R.A. Leigh. Critical plant K-concentrations for growth and problems in the diagnosis of nutrient deficiencies by plant analysis. *Plant Soil* 155/156: 219–222, 1993.

70. S. Yoshida, L. Castaneda. Partial replacement of potassium by sodium in the rice plant under weakly saline conditions. *Soil Sci. Plant Nutr.* 15:183–186, 1969.

71. J.V. Amin, H.E. Joham. The cations of the cotton plant in sodium substituted potassium deficiency. *Soil Sci.* 105:248–254, 1968.

72. H. Marschner. Why can sodium replace potassium in plants? In: *Potassium in Biochemistry and Physiology*. Bern: International Potash Institute, 1971, pp. 50–63.

73. K. Mengel, R. Pflüger. Sodium and magnesium contents of different ryegrass varieties in relation to temperature and air humidity. *Z. Acker-Pflanzenbau* 136:272–279, 1973.

74. M.G. Lindhauer. The role of K^+ in cell extension, growth and storage of assimilates. In: *Methods of K Research in Plants*. Bern: International Potash Institute, 1989, pp. 161–187.

75. P.F. Brownell, L.M. Bielig. The role of sodium in the conversion of pyruvate to phosphoenolpyruvate in mesophyll chloroplasts of C4 plants. *Aus. J. Plant Physiol.* 23: 171–177, 1996.

76. R.A. Leigh, D.J. Walker, W. Fricke, A. Deri Thomas, A.J. Miller. Patterns of potassium compartmentation in plant cells as revealed by microelectrodes and microsampling. In: D.J. Oosterhuis, G.A. Berkowitz, eds. *Frontiers in Potassium Nutrition: New Perspectives on the Effect of Potassium on Physiology of Plants*. Norcross, GA, 1999, pp. 63–70.

77. A.E. Johnston, K.W.T. Goulding. The use of plant and soil analysis to predict the potassium supplying capacity of soil. In: *Development of K Fertilizer Recommendations*. Bern: International Potash Institute, 1990, pp. 177–201.

78. P.A. Helmke. The chemical composition of soils. In: M.E. Sumner, ed. *Handbook of Soil Science*. London: CRC Press, 2000, pp. B3–B24.

79. P.M. Jardine, D.L. Sparks. Potassium-calcium exchange in a multireactive soil system. I. Kinetics. *Soil Sci. Soc. Am. J.* 48:39–45, 1984.

80. D.L. Sparks. *Environmental Soil Chemistry*. 2nd ed. London: Academic Press, 2003, pp. 52–53.

81. D.S. Fanning, V.Z. Keramidas, M.A. El Desoky. Micas. In: J.B. Dixon, S.B. Weed, eds. *Minerals in Soil Environments*. Madison: Soil Science Soc. America, 1989, pp. 551–634.

82. H.W. Scherer. Dynamics and availability of the non-exchangeable NH_4-N–a review. *Eur. J. Agron.* 2: 149–160, 1993.

83. D. Steffens, D.L. Sparks. Kinetics of nonexchangeable ammonium release from soils. *Soil Sci. Soc. Am. J.* 61:455–462, 1997.

84. H.W. Martin, D.L. Sparks. On the behaviour of nonexchangeable potassium in soils. *Commun. Soil Sci. Plant Anal.* 16:133–162, 1985.

85. U. Schwertmann, P. Süsser, L. Nätscher. Proton buffer compounds in soils. *Z. Pflanzenernähr Bodenk* 150:174–178, 1987.

86. P. Hinsinger, J.E. Dufey, B. Jaillard. Biological weathering of micas in the rhizosphere as related to potassium absorption by plant roots. In: B.L. McMichael, H. Persson, eds. *Plant Roots and Their Environment*. Amsterdam: Elsevier, 1991, pp. 98–105.

87. Rahmatullah, K. Mengel. Potassium release from mineral structures by H^+ ion resin. *Geoderma* 96: 291–305, 2000.

88. S. Feigenbaum, R.E. Edelstein, J. Shainberg. Release rate of potassium and structural cations from micas to ion exchangers in dilute solutions. *Soil Sci. Soc. Am. J.* 45:501–506, 1981.

89. M.S. Aktar. Potassium availability as affected by soil mineralogy. In: K. Mengel, A. Krauss, eds. *Potassium in Agriculture*. Basel: International Potash Institute, 1994, pp. 139–155.

90. K. Mengel, Rahmatullah, H. Dou. Release of potassium from the silt and sand fraction of loess-derived soils. *Soil Sci.* 163:805–813, 1998.

91. F. Wulff, V. Schulz, A. Jungk, N. Claassen. Potassium fertilization on sandy soils in relation to soil test, crop yield and K-leaching. *Z. Pflanzenernähr Bodenk* 161:591–599, 1998.

92. H. Tributh, E. von. Boguslawski, A. von Liers, S. Steffens, K. Mengel. Effect of potassium removal by crops on transformation of illitic clay materials. *Soil Sci.* 143:404–409, 1987.

93. N.C. Brady, R.R. Weil. *The Nature and Properties of Soils*. 12th ed. Englewood Cliffs, NJ: Prentice-Hall, 1999.

94. A.N. Sharpley. Reaction of fertilizer potassium in soils of differing mineralogy. *Soil Sci.* 149: 44–51, 1990.

95. L.R. Hossner. Dissolution for total elemental analysis. In: J.M. Bigham, ed. *Methods of Soil Analysis. Part 3, Chemical Methods*. Madison: Soil Sci. Soc. America, 1996, pp. 49–64.

96. P.A. Helmke, D.L. Sparks. Lithium, sodium, potassium, rubidium and cesium. In: J.M. Bigham, ed. *Methods of Soil Analysis. Part 3, Chemical Methods*. Madison: Soil Sci. Soc. America, 1996, pp. 551–574.

97. P. Schachtschabel. Fixation and release of potassium and ammonium ions. Assessment and determination of the potassium availability in soils. *Landw. Forsch.* 15:29–47, 1961.

98. P.F. Pratt. Potassium. In: C.A. Black, ed. *Methods in Soil Analysis*, Part 2. Madison: Am. Soc. Agron. 1965, pp. 1023–1031.

99. J.L. Havlin, D.G. Westfall, S.R. Olsen. Mathematical models for potassium release kinetics in calcareous soils. Soil Sci. Soc. Am. J. 49:371–376, 1985.

100. K. Nemeth. The availability of nutrients in the soil as determined by electro-ultrafiltration (EUF). *Adv. Agron.* 31:155–188, 1979.

101. K. Mengel, K. Uhlenbecker. Determination of available interlayer potassium and its uptake by ryegrass. *Soil Sci. Soc. Am. J.* 57:761–766, 1993.

102. P.H.T. Beckett. Studies on soil potassium. II. The 'immediate' Q/I relations of labile potassium in the soil. *J. Soil Sci.* 15:9–23, 1964.

103. A. Wild, D.L. Rowell, M.A. Ogunfuwora. Activity ratio as a measure of intensity factor in potassium supply to plants. *Soil Sci.* 108:432–439, 1969.

104. S.A. Barber. A diffusion and mass flow concept of soil nutrient availability. *Soil Sci.* 93:39–49, 1962.

105. P.H. Nye. Diffusion of ions and uncharged solutes in soils and soil clays. *Adv. Agron.* 31:225–272, 1979.

106. A.O. Jungk. Dynamics of nutrient movement at the soil-root interface. In: Y. Waisel, A. Eshel, U. Kafkafi, eds. *Plant Roots, the Hidden Half*. New York: Marcel Dekker, 2002, pp. 587–616.

107. M.C. Drew, P.H. Nye, L.V. Vaidyanathan. The supply of nutrient ions by diffusion to plant roots in soil. I. Absorption of potassium by cylindrical roots of onion and leek. *Plant Soil* 30:252–270, 1969.

108. G. Hoffmann. *Die Untersuchung von Böden*. Darmstadt: VDLUFA-Verlag, 1991.

109. K. Nemeth, H. Irion, J. Maier. Influence of EUF-K, EUK-Na and EUF-Ca fraction on the K uptake and the yield of sugar beet. *Kali-Briefe* 18:777–790, 1987.

110. H.R. von Uexküll. Response of coconuts to potassium in the Philippines. *Oleagineux* 27:31–91, 1972.

111. Y. Tokunaga. Potassium silicate–a slow-release fertilizer. *Fert. Res.* 30:55–59, 1991.

112. V. Kovacevic, V. Vukadinovic. The potassium requirements of maize and soybeans on a high K-fixing soil. *South Afr. Plant Soil* 9:10–13, 1992.

113. M. Schön, E.A. Niederbudde, A. Mahkorn. Results of a 20-year trial with mineral and organic fertilizer applications in the loess region of Landsberg (Lech). *Z. Acker-Pflanzenbau* 143:27–37, 1976.

114. K. Mengel, D. Steffens. Potassium uptake of ryegrass (*Lolium perenne*) and red clover (*Trifolium repens*) as related to root parameters. *Biol. Fert. Soils* 1:53–58, 1985.

115. R.L. Goss. The effects of potassium in disease resistance. In: V.J. Kilmer, S.E. Younts, N.C. Brady, eds. *The Role of Potassium in Agriculture*. Madison: American Potash Institute, 1968, pp. 221–241.

116. D. Anac, B. Okur, C. Cilic, U. Aksoy, Z. Hepaksoy, S. Anac, M.A. Ul, F. Dorsan. Potassium fertilization to control salinization effects. In: A.E. Johnston, ed. *Food Security in the WANA Region, the Essential Need for Balanced Fertilization*. Basel: International Potash Institute, 1997, pp. 370–377.

5 Calcium

David J. Pilbeam
University of Leeds, Leeds, United Kingdom

Philip S. Morley
Wight Salads Ltd., Arreton, United Kingdom

CONTENTS

5.1 Historical Information 121
    5.1.1 Determination of Essentiality 121
5.2 Functions in Plants 122
    5.2.1 Effects on Membranes..122
    5.2.2 Role in Cell Walls 122
    5.2.3 Effects on Enzymes..124
    5.2.4 Interactions with Phytohormones 125
    5.2.5 Other Effects 125
5.3 Diagnosis of Calcium Status in Plants 125
    5.3.1 Symptoms of Deficiency and Excess 125
    5.3.2 Concentrations of Calcium in Plants..128
        5.3.2.1 Forms of Calcium Compounds 128
        5.3.2.2 Distribution of Calcium in Plants 128
        5.3.2.3 Calcicole and Calcifuge Species 132
        5.3.2.4 Critical Concentrations of Calcium 133
        5.3.2.5 Tabulated Data of Concentrations by Crops 133
5.4 Assessment of Calcium Status in Soils..135
    5.4.1 Forms of Calcium in Soil 135
    5.4.2 Soil Tests 137
    5.4.3 Tabulated Data on Calcium Contents in Soils 137
5.5 Fertilizers for Calcium 137
    5.5.1 Kinds of Fertilizer 137
    5.5.2 Application of Calcium Fertilizers 139
Acknowledgment...140
References 140

5.1 HISTORICAL INFORMATION

5.1.1 DETERMINATION OF ESSENTIALITY

The rare earth element calcium is one of the most abundant elements in the lithosphere; it is readily available in most soils; and it is a macronutrient for plants, yet it is actively excluded from plant cytoplasm.

In 1804, de Saussure showed that a component of plant tissues comes from the soil, not the air, but it was considerably later that the main plant nutrients were identified. Liebig was the first person to be associated strongly with the idea that there are essential elements taken up from the soil (in 1840), although Sprengel was the first person to identify calcium as a macronutrient in 1828 (1). Calcium was one of the 20 essential elements that Sprengel identified.

Salm-Horstmar grew oats (*Avena sativa* L.) in inert media with different elements supplied as solutions in 1849 and 1851 and showed that omitting calcium had an adverse effect on growth (2). However, it was the discovery that plants could be grown in hydroponic culture by Sachs (and almost simultaneously Knop) in 1860 that made investigation of what elements are essential for plant growth much easier (2). Sachs' first usable nutrient solution contained $CaSO_4$ and $CaHPO_4$.

It has been well known since the early part of the twentieth century that there is a very distinct flora in areas of calcareous soils, comprised of so-called calcicole species. There are equally distinctive groups of plant species that are not found on calcareous soils, the calcifuge species (see Section 5.3.2.3).

5.2 FUNCTIONS IN PLANTS

Calcium has several distinct functions within higher plants. Bangerth (3) suggested that these functions can be divided into four main areas: (a) effects on membranes, (b) effects on enzymes, (c) effects on cell walls, and (d) interactions of calcium with phytohormones, although the effects on enzymes and the interactions with phytohormones may be the same activity. As a divalent ion, calcium is not only able to form intramolecular complexes, but it is also able to link molecules in intermolecular complexes (4), which seems to be crucial to its function.

5.2.1 EFFECTS ON MEMBRANES

Epstein established that membranes become leaky when plants are grown in the absence of calcium (5) and that ion selectivity is lost. Calcium ions (Ca^{2+}) bridge phosphate and carboxylate groups of phospholipids and proteins at membrane surfaces (6), helping to maintain membrane structure. Also, some effect occurs in the middle of the membrane, possibly through interaction of the calcium and proteins that are an integral part of membranes (6,7). Possibly, calcium may link adjacent phosphatidyl-serine head groups, binding the phospholipids together in certain areas that are then more rigid than the surrounding areas (8).

5.2.2 ROLE IN CELL WALLS

Calcium is a key element in the structure of primary cell walls. In the primary cell wall, cellulose microfibrils are linked together by cross-linking glycans, usually xyloglucan (XG) polymers but also glucoarabinoxylans in Poaceae (Gramineae) and other monocots (9). These interlocked microfibrils are embedded in a matrix, in which pectin is the most abundant class of macromolecule. Pectin is also abundant in the middle lamellae between cells.

Pectin consists of rhamnogalacturonan (RG) and homogalacturonan (HG) domains. The HG domains are a linear polymer of $(1{\rightarrow}4)$-α'-linked D-galacturonic acid, 100 to 200 residues long, and are deposited in the cell wall with 70 to 80% of the galacturonic acid residues methyl-esterified at the C6 position (9). The methyl-ester groups are removed by pectin methylesterases, allowing calcium ions to bind to the negative charges thus exposed and to form inter-polymer bridges that hold the backbones together (9). The whole structure can be thought of as resembling an eggbox (Figure 5.1).

Pectin is a highly hydrated gel containing pores; the smaller the size of these pores, the higher the Ca^{2+} concentration in the matrix and more cross-linking of chains occurs (11). This gel holds the XG molecules in position relative to each other, and these molecules in turn hold the cellulose microfibrils together (Figure 5.2). The presence of the calcium, therefore, gives

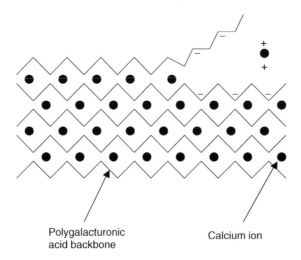

FIGURE 5.1   The 'eggbox' model of calcium distribution in pectin. (Based on E.R. Morris et al., *J. Mol. Biol.* 155: 507–516, 1982.)

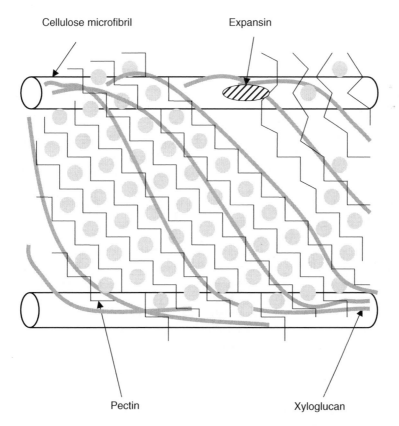

FIGURE 5.2   Diagrammatic representation of the primary cell wall of dicotyledonous plants. (Based on E.R. Morris et al., *J. Mol. Biol.* 155:507–516, 1982; F.P.C. Blamey, *Soil Sci. Plant Nutr.* 49:775–783, 2003; N.C. Carpita and D.M. Gibeaut, *Plant J.* 3:1–30, 1993.) To the right of the figure, Ca^{2+} ions have been displaced from the HG domains by H^+ ions, so that the pectin is no longer such an adhesive gel and slippage of the bonds between adjacent XG chains occurs and expansin is able to work on them. This loosens the structure and allows the cellulose microfibrils to be pushed further apart by cell turgor.

some load-bearing strength to the cell wall (13). It is suggested that when a primary cell wall is expanding, localized accumulation of H^+ ions may displace Ca^{2+} from the HG domains, thereby lowering the extent to which the pectin holds the XG strands together (11). In a root-tip cell, where the cellulose microfibrils are oriented transversely, slippage of the XG chains allows the cellulose microfibrils to move further apart from each other, giving cell expansion in a longitudinal direction.

Cell-to-cell adhesion may also be given by Ca^{2+} cross-linking between HG domains in the cell walls of adjacent cells, but this action is less certain as experimental removal of Ca^{2+} leads to cell separation in a only few cases (9). In the ripening of fruits, a loosening of the cells could possibly occur with loss of calcium. It has been postulated that decrease in apoplastic pH in ripening pome fruits may cause the release of Ca^{2+} ions from the pectin, allowing for its solubilization (14). However, in an experiment on tomato (*Lycopersicon esculentum* Mill.), the decline in apoplastic pH that occurred was not matched by a noticeable decrease in apoplastic Ca^{2+} concentration, and the concentration of the ion remained high enough to limit the solubilization of the pectin (15). It certainly seems that calcium inhibits the degradation of the pectates in the cell wall by inhibiting the formation of polygalacturonases (16), so the element has roles in possibly holding the pectic components together and in inhibiting the enzymes of their degradation. In a study on a ripening and a nonripening cultivar of tomato (Rutgers and *rin*, respectively), there was an increase in calcium concentration after anthesis in the *rin* cultivar, whereas in the Rutgers cultivar there was a noticeable fall in the concentration of bound calcium and an increase in polygalacturonase activity (17). In a study on calcium deficiency in potato (*Solanum tuberosum* L.), deficient plants had more than double the activity of polygalacturonase compared with normal plants (18).

5.2.3 EFFECTS ON ENZYMES

Unlike K^+ and Mg^{2+}, Ca^{2+} does not activate many enzymes (19), and its concentration in the cytoplasm is kept low. This calcium homeostasis is achieved by the action of membrane-bound, calcium-dependent ATPases that actively pump Ca^{2+} ions from the cytoplasm and into the vacuoles, the endoplasmic reticulum (ER), and the mitochondria (20). This process prevents the ion from competing with Mg^{2+}, thereby lowering activity of some enzymes; the action prevents Ca^{2+} from inhibiting cytoplasmic or chloroplastic enzymes such as phosphoenol pyruvate (PEP) carboxylase (21) and prevents Ca^{2+} from precipitating inorganic phosphate (22).

Calcium can be released from storage, particularly in the vacuole, into the cytoplasm. Such flux is fast (23) as it occurs by means of channels from millimolar concentrations in the vacuole to nanomolar concentrations in the cytoplasm of resting cells (24). The calcium could inhibit cytoplasmic enzymes directly, or by competition with Mg^{2+}. Calcium can also react with the calcium-binding protein calmodulin (CaM). Up to four Ca^{2+} ions may reversibly bind to each molecule of calmodulin, and this binding exposes two hydrophobic areas on the protein that enables it to bind to hydrophobic regions on a large number of key enzymes and to activate them (25). The Ca^{2+}–calmodulin complex also may stimulate the activity of the calcium-dependent ATPases (26), thus removing the calcium from the cytoplasm again and priming the whole system for further stimulation if calcium concentrations in the cytoplasm rise again.

Other sensors of calcium concentration are in the cytoplasm, for example, Ca^{2+}-dependent (CaM-independent) protein kinases (25). The rapid increases in cytoplasmic Ca^{2+} concentration that occur when the channels open and let calcium out of the vacuolar store and the magnitude, duration, and precise location of these increases give a series of calcium signatures that are part of the responses of a plant to a range of environmental signals. These responses enable the plant to respond to drought, salinity, cold shock, mechanical stress, ozone and blue light, ultraviolet radiation, and other stresses (24).

5.2.4 Interactions with Phytohormones

An involvement of calcium in the actions of phytohormones seems likely as root growth ceases within only a few hours of the removal of calcium from a nutrient solution (22). The element appears to be involved in cell division and in cell elongation (27) and is linked to the action of auxins. The loosening of cellulose microfibrils in the cell wall is controlled by auxins, giving rise to excretion of protons into the cell wall. Calcium is involved in this process, as discussed earlier. Furthermore, auxin is involved in calcium transport in plants, and treatment of plants with the indoleacetic acid (IAA) transport inhibitor, 2,3,5-triiodobenzoic acid (TIBA), results in restricted calcium transport into the treated tissue (28). As the relationship is a two-way process, it cannot be confirmed easily if calcium is required for the action of IAA or if the action of IAA gives rise to cell growth, and consequent cell wall development, with the extra pectic material in the cell wall then acting as a sink for calcium. It is also possible that IAA influences the development of xylem in the treated tissue (29).

Increase in shoot concentrations of abscisic acid (ABA) following imposition of water-deficit stress leads to increased cytoplasmic concentration of Ca^{2+} in guard cells, an increase that precedes stomatal closure (24). Further evidence for an involvement of calcium with phytohormones has come from the observation that senescence in maize (*Zea mays* L.) leaves can be slowed by supplying either Ca^{2+} or cytokinin, with the effects being additive (30). There is also a relationship between membrane permeability, which is strongly affected by calcium content and ethylene biosynthesis in fruit ripening (31).

5.2.5 Other Effects

It has been known for a long time that calcium is essential for the growth of pollen tubes. A gradient of cytoplasmic calcium concentration occurs along the pollen tube, with the highest concentrations being found in the tip. The fastest rate of influx of calcium occurs at the tip, up to 20 pmol cm^{-2} s^{-1}, but there are oscillations in the rate of pollen tube growth and calcium influx that are approximately in step (32). It seems probable that the calcium exerts an influence on the growth of the pollen tube mediated by calmodulin and calmodulin-like domain protein kinases (25), but the growth and the influx of calcium are not directly linked as the peaks in oscillation of growth precede the peaks in uptake of calcium by 4 s (32). Root hairs have a high concentration of Ca^{2+}, and root hair growth has a similar calcium signature to pollen tube growth (24). Slight increases in cytoplasmic Ca^{2+} concentration can close the plasmodesmata in seconds, with the calcium itself and calmodulin being implicated (33). Many sinks, such as root apices, require symplastic phloem unloading through sink plasmodesmata, so this action implies that calcium has a role as a messenger in the growth of many organs.

It seems that calcium can be replaced by strontium in maize to a certain extent (34), but despite the similarities in the properties of the two elements, this substitution does not appear to be common to many plant species. In general, the presence of abundant calcium in the soil prevents much uptake of strontium, and in a study on 10 pasture species, the concentration of strontium in the shoot was correlated negatively with the concentration of calcium in the soil (35).

5.3 DIAGNOSIS OF CALCIUM STATUS IN PLANTS

5.3.1 Symptoms of Deficiency and Excess

Plants deficient in calcium typically have upper parts of the shoot that are yellow-green and lower parts that are dark green (36) (Figure 5.3). Given the abundance of calcium in soil, such a condition is unusual, although it can arise from incorrect formulation of fertilizers or nutrient solutions.

FIGURE 5.3 Calcium-deficient maize (*Zea mays* L.). The younger leaves which are still furled are yellow, but the lamina of the older, emerged leaf behind is green. (Photograph by Allen V. Barker.) (For a color presentation of this figure, see the accompanying compact disc.)

However, despite the abundance of calcium, plants suffer from a range of calcium-deficiency disorders that affect tissues or organs that are naturally low in calcium. These include blossom-end rot (BER) of tomato (Figure 5.4 and Figure 5.5), pepper (*Capsicum annuum* L.), and water melon (*Cucumis melo* L.) fruits, bitter pit of apple (*Malus pumila* Mill.), black heart of celery (*Apium graveolens* L.), internal rust spot in potato tubers and carrot (*Daucus carota* L.) roots, internal browning of Brussels sprouts (*Brassica oleracea* L.), internal browning of pineapple (*Ananas comosus* Merr.), and tip burn of lettuce (*Lactuca sativa* L.) and strawberries (*Fragaria x ananassa* Duch.) (22,37,38). Recently, it has been suggested that the disorder 'crease' in navel and Valencia oranges (*Citrus aurantium* L.) may be caused by calcium deficiency in the albedo tissue of the rind (39).

In these disorders, the shortage of calcium in the tissues causes a general collapse of membrane and cell wall structure, allowing leakage of phenolic precursors into the cytoplasm. Oxidation of polyphenols within the affected tissues gives rise to melanin compounds and necrosis (40). With the general breakdown of cell walls and membranes, microbial infection is frequently a secondary effect. In the case of crease, calcium deficiency may give less adhesion between the cells of the rind, as the middle lamella of these cells is composed largely of calcium salts of pectic acid (39).

Local excess of calcium in the fruit gives rise to goldspot in tomatoes, a disorder that mostly occurs late in the season and that is pronounced with high temperature (41). The disorder 'peteca'

FIGURE 5.4 Fruit of tomato (*Lycopersicon esculentum* Mill. cv Jack Hawkins) (Beefsteak type) showing blossom-end rot (BER). (Photograph by Philip S. Morley.) (For a color presentation of this figure, see the accompanying compact disc.)

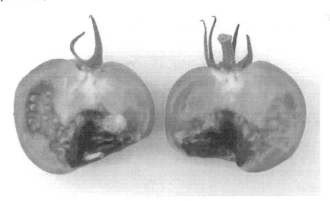

FIGURE 5.5 Cross section of fruit of tomato (*Lycopersicon esculentum* Mill. cv Jack Hawkin) showing advanced symptoms of BER. (Photograph by Philip S. Morley.) (For a color presentation of this figure, see the accompanying compact disc.)

that gives rise to brown spots on the rind of lemons (*Citrus limon* Burm. f.) is associated with localized high concentrations of calcium (as calcium oxalate crystals) and depressed concentrations of boron, although this phenomenon has not yet been shown to be the cause of the disorder (42).

Given the suggestion that calcium may be involved in cell-to-cell adhesion and in the ripening of fruit, it is hardly surprising that in pome fruits, firmness of the fruit is correlated positively with the concentration of calcium present (43). However, this relationship is by no means straightforward; in a study of Cox's Orange Pippin apples grown in two orchards in the United Kingdom, there were lower concentrations of cell wall calcium in the fruit from the orchard that regularly produced firmer fruits than in fruits from other orchards (44). The fruits from this orchard contained higher concentrations of cell wall nitrogen.

Other studies have shown no relationship between calcium concentration in apples at harvest and their firmness after storage, but it is definitely the case that fruit with low Ca^{2+} concentrations are more at risk of developing bitter pit while in storage (45).

5.3.2 CONCENTRATIONS OF CALCIUM IN PLANTS

5.3.2.1 Forms of Calcium Compounds

Within plants, calcium is present as Ca^{2+} ions attached to carboxyl groups on cell walls by cation-exchange reactions. As approximately one third of the macromolecules in the primary cell wall are pectin (9), it can be seen that a large proportion occurs as calcium pectate. Pectin may also join with anions, such as vanadate, and serve to detoxify these ions. The Ca^{2+} cation will also join with the organic anions formed during the assimilation of nitrate in leaves; these anions carry the negative charge that is released as nitrate is converted into ammonium (46). Thus, there will be formation of calcium malate and calcium oxalacetate and, also very commonly, calcium oxalate in cells.

Calcium oxalate can occur within cells and as extracellular deposits. In a study of 46 conifer species, all contained calcium oxalate crystals (47). All of the species in the Pinaceae family accumulated the compound in crystalliferous parenchyma cells, but the species not in the Pinaceae family had the compound present in extracellular crystals.

This accumulation of calcium oxalate is common in plants in most families. Up to 90% of total calcium in individual plants is in this form (48,49). Formation of calcium oxalate crystals occurs in specialized cells, crystal idioblasts, and as the calcium oxalate in these cells is osmotically inactive their formation serves to lower the concentration of calcium in the apoplast of surrounding cells without affecting the osmotic balance of the tissue (48). A variety of different forms of the crystals occur (49), and they can be composed of calcium oxalate monohydrate or calcium oxalate dihydrate (50).

5.3.2.2 Distribution of Calcium in Plants

Calcium moves toward roots by diffusion and mass flow (51,52) in the soil. A number of calcium-specific ion channels occur in the membranes of root cells, through which influx occurs, but these channels appear to be more involved in enabling rapid fluxes of calcium into the cytoplasm and organelles as part of signalling mechanisms (53). This calcium is then moved into vacuoles, endoplasmic reticulum, or other organelles, with movement occurring by means of calcium-specific transporters (20).

The bulk entry of calcium into roots occurs initially into the cell walls and in the intercellular spaces of the roots, giving a continuum between calcium in the soil and calcium in the root (54). For calcium to move from the roots to the rest of the plant, it has to enter the xylem, but the Casparian band of the endodermis is an effective barrier to its movement into the xylem apoplastically. However, when endodermis is first formed, the Casparian band is a cellulosic strip that passes round the radial cell wall (state I endodermis), so calcium is able to pass into the xylem if it passes into the endodermal cells from the cortex and then out again into the pericycle, through the plasmalemma abutting the wall (55). This transport seems to occur, with the calcium moving into the endodermal cells (and hence into the symplasm) through ion channels and from the endodermis into the pericycle (and ultimately into the much higher concentration of calcium already present in the xylem) by transporters (56,57). Highly developed endodermis has suberin lamellae laid down inside the cell wall around the entire cell (state II endodermis), and in the oldest parts of the root, there is a further layer of cellulose inside this (state III) (55). Although some ions such as K^+ can pass through state II endodermal cells, Ca^{2+} cannot. There are plasmodesmata between endodermis and pericycle cells, even where the Casparian band is well developed, but although phosphate and K^+ ions can pass, the plasmodesmata are impermeable to Ca^{2+} ions.

This restriction in effect limits the movement of calcium into the stele to the youngest part of the root, where the endodermis is in state I. Some movement occurs into the xylem in older parts of the root, and this transport can occur by two means. It is suggested that movement of calcium through state III endodermis might occur where it is penetrated by developing lateral roots, but the Casparian band rapidly develops here to form a complete network around the endodermal cells of the main and lateral roots (55). The second site of movement of calcium into the stele is through passage cells (55). During the development of state II and state III endodermis some cells remain in state I. These are passage cells. They tend to be adjacent to the poles of protoxylem in the stele, and they are the site of calcium movement from cortex to pericycle.

In some herbaceous plants (e.g., wheat, barley, oats), the epidermis and cortex are lost from the roots, especially in drought, so the passage cells are the only position where the symplast is in contact with the rhizosphere (55). Most angiosperms form an exodermis immediately inside the epidermis, and the cells of this tissue also develop Casparian bands and suberin lamellae, with passage cells in some places (55). These passage cells are similarly the only place where the symplasm comes in contact with the rhizosphere.

Because of this restricted entry into roots, calcium enters mainly just behind the tips, and it is mostly here that it is loaded into the xylem (Figure 5.6). Absorption of calcium into the roots may be passive and dependent on root cation-exchange capacity (CEC) (58). Transfer of calcium into roots is hardly affected by respiratory uncouplers, although its transfer into the xylem is affected (54,59).

Once in the xylem the calcium moves in the transpiration stream, and movement around the plant is restricted almost entirely to the xylem (60,61) as it is present in the phloem only at similarly low concentrations to those that occur in the cytoplasm.

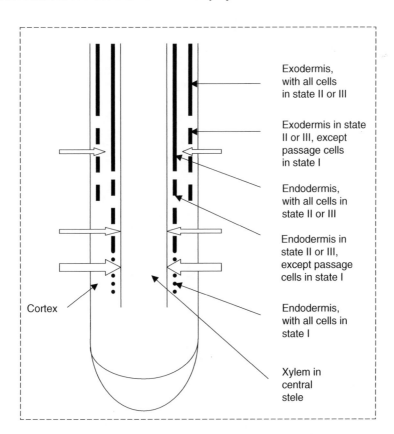

FIGURE 5.6 Diagrammatic representation of longitudinal section of root, showing development of endodermis and exodermis, and points of entry of calcium. (Based on C.A. Peterson and D.E. Enstone, *Physiol. Plant* 97: 592–598, 1996.)

As calcium is not mobile in the phloem, it cannot be retranslocated from old shoot tissues to young tissues, and its xylem transport into organs that do not have a high transpiration rate (such as fruits) is low (22). Its flux into leaves also declines after maturity, even though the rate of transpiration by the leaf remains constant (62), and this response could be related to a decline in nitrate reductase activity as new leaves in the plant take over a more significant assimilatory role (22,63). When a general deficiency of calcium occurs in plants, because of the low mobility of calcium in phloem, it is the new leaves that are affected, not the old leaves, as calcium in a plant remains predominantly in the old tissues (Figure 5.7).

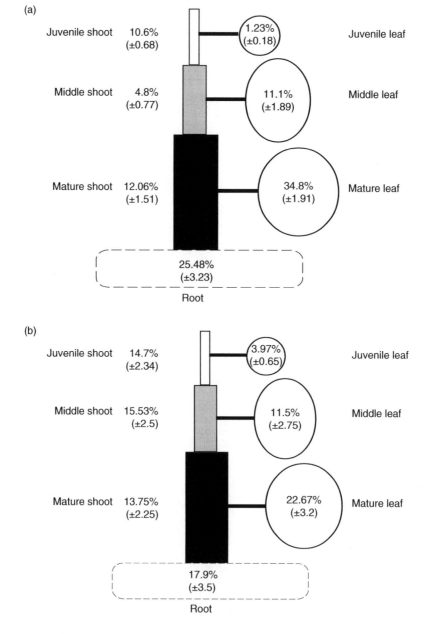

FIGURE 5.7 Distribution of calcium (a) and distribution of dry mass (b) in *Capsicum annuum* cv Bendigo plants grown for 63 days in nutrient solution (values are means of values for nine plants ± standard error).

It was long thought that a direct connection occurs between the amount of transpiration that a plant carries out and the amount of Ca^{2+} that it accumulates. For example, in a study of five tomato cultivars grown at two levels of electrical conductivity (EC) there was a linear, positive relationship between water uptake and calcium accumulation over 83 days (64). However, with the movement of Ca^{2+} in the symplasm of the endodermis apparently being required for xylem loading, it became accepted that Ca^{2+} is taken up in direct proportion to plant growth, as new cation-exchange sites are made available in new tissue. The link with transpiration could therefore be incidental, because bigger plants transpire more. Thus the plant acts as a giant cation exchanger, taking up calcium in proportion to its rate of growth.

Supplying calcium to decapitated plants at increased ion activity (concentration) leads to increased uptake of the ion, a process that appears to contradict this concept. However, in intact plants, the rate of uptake is independent of external ion activity, as long as the ratios of activities of other cations are constant relative to the activity of Ca^{2+} (65,66).

The theory that calcium travels across the root in the apoplastic pathway, until it reaches the Casparian band of the endodermis and at which its passage to the xylem becomes symplastic, is not entirely without problems. White (56,67) calculated that for sufficient calcium loading into xylem, there must be two calcium-specific ion channels per μm^2 of plasmalemma on the cortex side of the endodermis. This possibility is plausible. However, for the flux of calcium to continue from the endodermis into the pericycle there must be $0.8\,ng$ Ca^{2+}-ATPase protein per cell, equivalent to $1.3\,mg$ per gram of root fresh weight. This concentration is greater than the average total root plasmalemma protein concentration in plants. Furthermore, there is no competition between Ca^{2+}, Ba^{2+}, and Sr^{2+} for transport to mouse-ear cress (*Arabidopsis thaliana* Heynh.) shoots, as would be expected if there was protein-mediated transport in the symplast. Some apoplastic transport to the xylem cannot be ruled out.

The walls of xylem vessels have cation-exchange sites on them; in addition to the whole plant having a CEC, the xylem represents a long cation-exchange column with the Ca^{2+} ions moving along in a series of jumps (54). The distance between each site where cation exchange occurs depends on the velocity of the xylem sap and the concentration of Ca^{2+} ions in it (54). Thus, for transpiring organs such as mature leaves, the calcium moves into them quickly, but for growing tissues such as the areas close to meristems, the supply of calcium is dependent on the deposition of cell walls and the formation of new cation-exchange sites (54). It has been suggested that transpiring organs receive their calcium in the transpiration stream during the day, and growing tissues receive their calcium as a result of root pressure during the night (54).

The restriction in movement of calcium to the xylem gives rise to most of the calcium-deficiency disorders in plants. For example, BER (Figure 5.4 and Figure 5.5) in tomatoes occurs because the developing fruits are supplied solutes better by phloem than by xylem as the fruits do not transpire. Xylem fluid goes preferentially to actively transpiring leaves, giving a lower input of calcium into developing fruits (68). A period of hot, sunny weather not only gives rise to so much transpiration that calcium is actively pulled into leaves, but gives rates of photosynthesis that are enhanced to the extent that fruits expand very rapidly. Under these conditions, it is likely that localized deficiencies of calcium will occur in the distal end of the fruits, furthest from where the xylem enters them (the 'blossom' end) (Figure 5.4 and Figure 5.5). Typically, tomatoes grown for harvest in trusses are more susceptible to BER than 'single-pick' types, presumably because the calcium has to be distributed to several developing sinks at the same time. Conditions that promote leaf transpiration, such as low humidity, lower the import of calcium into developing fruits and increase the risk of BER.

It has also been thought in the past that salinity, which increases water potential in the root medium, would likewise restrict calcium import into the fruit, accounting for increased incidence of BER that is known to occur under saline conditions. This effect of salinity could be important in some natural soils, but is also important in glasshouse production of tomatoes as high-electroconductivity (EC) nutrient solutions are sometimes used because they increase dry matter production in fruits and improve flavor. However, it has been observed that if the ion activity ratios $a_K/\sqrt{(a_{Ca} + a_{Mg})}$ and a_{Mg}/a_{Ca} are kept below critical values, the risks of BER developing in high-EC nutrient solutions are lowered (69). It seems as if one of the causes of increased BER with salinity is normally due to increased uptake of K^+ and Mg^{2+}, which restricts the uptake and distribution of Ca^{2+} ions.

Cultivars differ in susceptibility to BER, with beefsteak and plum types of tomato being particularly susceptible. Susceptibility is related partly to fruit yield, and two susceptible cultivars of tomato (Calypso and Spectra) were shown to have a higher rate of fruit set than a nonsusceptible cultivar (Counter) (70). The so-called calcium-efficient strains of tomato do not have lower incidence of BER, since although they accumulate more dry matter than Ca-inefficient strains, this accumulation is predominantly in the leaves (64). Cultivars with relatively small fruits, such as Counter (70), and with xylem development in the fruit that is still strong under saline conditions (71), are able to accumulate comparatively high proportions of their calcium in the distal end of the fruits under such conditions and are less susceptible to BER (64). However, cultivars with low yields of fruits per plant may show even lower incidence of BER than those with high yields (64).

Losses of tomatoes to BER in commercial horticulture can reach 5% in some crops, representing a substantial loss of potential income. The main approaches to prevent BER are to use less-susceptible cultivars and to cover the south-facing side of the glasshouse (in the northern hemisphere) with white plastic or whitewash to limit the amount of solar radiation of the nearest plants and prevent their fruits from developing too quickly in relation to their abilities to accumulate calcium.

5.3.2.3 Calcicole and Calcifuge Species

In general, calcicole species contain high concentrations of intracellular calcium, and calcifuge species contain low concentrations of intracellular calcium. The different geographic distributions of these plants seem to be largely determined by a range of soil conditions other than just calcium concentration in the soil *per se*. In the calcareous soils favored by calcicoles, in addition to high concentration of Ca^{2+}, pH is high, giving low solubility of heavy metal ions and high concentrations of nutrient and bicarbonate ions. In contrast, the acid soils favored by calcifuges have low pH, high solubility of heavy metal ions, and low availability of nutrients (5).

The growth of calcicole species is related strongly to the concentration of calcium in the soil, but the inability of calcicole species to grow in acid soils is linked strongly to an inability to tolerate the high concentrations of ions of heavy metals, in particular Al^{3+}, Mn^{2+}, and Fe^{3+} (5,72). For calcifuge species, the difficulty in growing in a calcareous soil stems from an inability to absorb iron, although in some calcareous soils low availability of phosphate may also be a critical factor.

In an experiment with tropical soils in which the sorption of phosphate from $Ca(H_2PO_4)_2$ solution (and its subsequent desorption) were measured, pretreating the soil with calcium sulfate solution increased the sorption of phosphate (73). In the most acid of the soils tested, sorption of phosphate was increased by 93%. Because the extracts of the soil became more acid following calcium sulfate treatment, it appears that the calcium was attracted to the sites previously occupied by H^+ ions, and when present, itself offered more sites for sorption of phosphate ions. Where the supply of phosphorus to plants is limited because it is sorbed to soil inorganic fractions, it seems as if sorption to calcium is more difficult to break than sorption to other components. In an experiment in which wheat (*Triticum aestivum* L.) and sugar beet (*Beta vulgaris* L.) were grown in a fossil Oxisol, with mainly Fe/Al-bound P, and in a Luvisol, a subsoil from loess with free $CaCO_3$ and mainly Ca-bound P, both species (but particularly the sugar beet) were able to mobilize the Fe/Al-bound P more than the Ca-bound P (74).

Some plants are much more efficient than others at taking up phosphate from calcium-bound pools in the soil. One efficient species is buckwheat (*Fagopyrum esculentum* Moench). In a comparison of this species and wheat, the buckwheat took up 20.1 mg P per pot compared with 2.1 mg P per pot for wheat if nitrogen was supplied as nitrate (75). Changing the nitrogen supply to ammonium nitrate increased phosphorus accumulation by the wheat largely, with very little effect on the buckwheat, indicating that it is the capacity of buckwheat to acidify the rhizosphere even when the nitrogen supply is nitrate that makes buckwheat able to utilize this firmly bound source of phosphorus.

For calcifuge species growing on calcareous soils, it seems as if the availability of iron is the most significant factor affecting plant growth, with chlorosis occurring due to iron deficiency. However, this deficiency is caused largely by immobilization of iron within the leaves, not necessarily a restricted absorption of iron (76,77). Calcicole species seem to make iron and phosphate available in calcareous soils by exudation of oxalic and citric acids from their roots (78). The high concentrations of bicarbonate ions in calcareous soils seem to be important in inhibition of root elongation of some calcifuge species (79).

5.3.2.4 Critical Concentrations of Calcium

The concentrations of calcium in plants are similar to the concentrations of potassium, in the range 1 to 50 mg Ca g^{-1} dry matter (Mengel, this volume). Most of the calcium is located in the apoplast, and where it is present in the symplast, it tends to be stored in organelles or vacuoles or is bound to proteins. The concentration of free Ca^{2+} in a root cortical cell is of the order of 0.1 to 1.0 mmol m^{-3} (54).

In general, monocotyledons contain much less calcium than dicotyledons. In an experiment comparing the growth of ryegrass (*Lolium perenne* L.) and tomato, the ryegrass reached its maximum growth rate when the concentration of calcium supplied gave a tissue concentration of 0.7 mg g^{-1} dry mass, whereas tomato reached its maximum growth rate only when tissue concentration was 12.9 mg g^{-1} (80,81). This difference between monocotyledons and dicotyledons is dictated by the CEC of the two groups of plants. In algal species, where the cell wall is absent and CEC is consequently low, calcium is required only as a micronutrient (82).

Tissue concentrations of calcium can vary considerably according to the rate of calcium supply. In a study by Loneragan and Snowball (81), internal Ca^{2+} concentrations were reasonably constant for 0.3, 0.8, and 2.5 μM calcium in the flowing nutrient solutions for each plant species tested, but with 10, 100, or 1000 μM Ca^{2+} supply, internal Ca^{2+} concentrations were noticeably higher. In a recent study of chickpea (*Cicer arietinum* L.), nine different Kabuli (large-seeded) accessions had a mean concentration of Ca^{2+} in nodes 4 to 7 of the shoot of 17.4 mg g^{-1} dry mass after 33 days of growth, and 10 different Desi (small-seeded) accessions had a mean Ca^{2+} concentration of 17.1 mg g^{-1} dry mass (83). In the Kabuli accessions, the range was between 13.5 and 20.6 mg g^{-1}, compared with between 13.1 and 19.0 mg g^{-1} in the Desi accessions, so different genotypes of the same species grown under the same conditions seem to contain very similar shoot calcium concentrations.

There are considerable amounts of data regarding what the critical concentrations of calcium are in different plants and different species. For data on these concentrations in a large number of species, the reader is referred to some special publications (84,85).

In a study of three cultivars of bell pepper, mean tissue concentrations ranged only from 1.5 to 1.8 mg g^{-1} dry mass in the proximal parts and from 0.95 to 1.3 mg g^{-1} dry mass in the distal part of healthy fruits. concentrations in fruits suffering BER were between 0.6 and 1.0 mg g^{-1} (86). Concentrations of calcium in fruits of cucumber (*Cucumis sativus* L.), a plant that is not susceptible to BER, are typically three to seven times these values (87).

There is one important exception to the finding that internal calcium concentrations are relatively constant regardless of how plants are grown. Plants supplied with nitrogen as ammonium tend to have much lower concentrations of cations, including calcium, than plants supplied with nitrate (22). Thus, tomato plants supplied with ammonium-N are more prone to BER than plants grown on nitrate.

5.3.2.5 Tabulated Data of Concentrations by Crops

Concentrations of Ca^{2+} in shoots and fruits of some crop species are reported in Table 5.1 and Table 5.2.

TABLE 5.1
Deficient and Adequate Concentrations of Calcium in Leaves and Shoots of Various Plant Species

Plant Species	Plant Part	Type of Culture	Concentration in Dry Matter (mg kg^{-1})		Reference	Comments
			Deficient	Adequate		
Avena sativa L. (oat)	Tops	Pot culture, soil	1100–1400	2600	88	Plants at flowering
	Straw	Sand culture	1000–1400	3600–6400	88	At harvest
Bromus rigidus Roth	Shoot	Flowing nutrient solution	900	1010	81	Plants grown in 0.3 and 1000 mmol m^{-3} Ca^{2+}, respectively
Capsicum annuum L. (pepper)	Leaves	Nutrient solution		Up to 30000 5000	89	Mature leaves Juvenile leaves
Citrus aurantium L. (orange)	Leaves Shoots	Sand culture	1400–2000 2300–2800	14800 11700	88	Measurements taken in September
Ficus carica L. (fig)	Leaves	Orchard		30000 30000 29000 35000	90	Values for May, July, September and October. 10 trees surveyed in 9 areas of 2 orchards, for 3 years
Fragaria x ananassa Duchesne (strawberry)	Leaves	Sand culture	2300/9000	15000	91	'Adequate' plants had 1% of leaves with tipburn. 'Deficient' plants had 33.2% of leaves with tipburn (plants supplied 1/40th control Ca and 3x K) or 9% of leaves with tipburn (plants supplied control Ca and 3x K)
Hordeum vulgare L. (barley)	Shoots	Flowing nutrient solution	1100	7300	81	Plants grown in 0.3 and 1000 mmol m^{-3} Ca^{2+}, respectively
Linum usitatissimum L. (flax)	Tops	Field	2000–4500	3700–5200	88	
Lolium perenne L. (perennial ryegrass)	Shoots	Flowing nutrient solution	600	10800	81	Plants grown in 0.3 and 1000 mmol m^{-3} Ca^{2+}, respectively
Lupinus angustifolius L.	Shoots	Flowing nutrient solution	1400	13900	81	Plants grown in 0.3 and 1000 mmol m^{-3} Ca^{2+}, respectively
Lycopersicon esculentum Mill. (tomato)	Leaf blade	Sand culture	1700	16100	36	Upper leaves (yellow in deficient plants)
	Leaf blade		11000	38400		Lower leaves (still green in deficient plants)
	Petioles		1100	10800		Upper petioles
	Petioles		2600	22300		Lower petioles
	Stem		Trace	6700		Upper stems

TABLE 5.1 *(Continued)*

Plant Species	Plant Part	Type of Culture	Concentration in Dry Matter (mg kg^{-1})		Reference	Comments
			Deficient	Adequate		
	Stem		5300	9900		Lower stems
	Shoots	Flowing nutrient solution	2700	24900	81	Plants grown in 0.3 and 1000 mmol m^{-3} Ca^{2+}, respectively
Malus pumila Mill. [*M. domestica* Borkh.] (apple)	Leaves		7200		88	Leaves of terminal shoot, stated value below which deficiency symptoms occur
Medicago sativa L. (alfalfa)	Shoots	Flowing nutrient solution	1100	15000	81	One cultivar, in 0.3 and 1000 mmol m^{-3} Ca^{2+}, respectively
Nicotiana tabacum L. (tobacco)	Leaves	Field trial	9400–13000	13300–24300	88	
Phaseolus lunatus L. (lima bean)	Stem		6000	9000	88	Poor seed set below first value, good seed set above second value
Prunus persica (L.) Batsch (peach)	Leaves	Orchard		14500 17000 18200	92	Soil pH 5.6 Soil pH 5.9 Soil pH 6.2
Prunus insititia L. *Prunus domestica* L. *Prunus salicina* (Lindl.) × *Prunus cerasifera* (Ehrh.) (plum)	Leaves	Nutrient solution		5300/8200 6600/10300 6300/10100	93	Values for days 45 and 96
Secale cereale L. (rye)	Shoots	Flowing nutrient solution	900	8300	81	Plants grown in 0.3 and 1000 mmol m^{-3} Ca^{2+}, respectively
Solanum tuberosum L. (potato)	Young leaves	Nutrient solution	Below 900	Above 4500	18	21-day-old plants
Trifolium subterraneum L. (subterranean clover)	Shoots	Flowing nutrient solution	1400	19100	81	One cultivar, in 0.3 and 1000 mmol m^{-3} Ca^{2+}, respectively
Triticum aestivum L. (wheat)	Shoots	Flowing nutrient solution	800	4700	81	One cultivar, in 0.3 and 1000 mmol m^{-3} Ca^{2+}, respectively
Zea mays L. (corn)	Shoots	Flowing nutrient solution	300	9200	81	Plants grown in 0.3 and 1000 mmol m^{-3} Ca^{2+}, respectively

Note: Values in dry matter.

5.4 ASSESSMENT OF CALCIUM STATUS IN SOILS

5.4.1 FORMS OF CALCIUM IN SOIL

Calcium is the main exchangeable base of clay minerals and, as such, is a major component of soils. One of the most important natural sources of calcium is underlying limestone or chalk, where it occurs as calcium carbonate (calcite). Calcium in rocks also occurs as a mixture of calcium and magnesium carbonates (dolomite). Soils over such rocks often contain large amounts of calcium carbonate, although not invariably so. The soils may not have been derived from the rock, but have

TABLE 5.2
Deficient and Adequate Concentrations of Calcium in Fruits of Various Plant Species

Plant Species	Plant Part	Type of Culture	Concentration in Fresh Matter (mg kg⁻¹)		Reference	Comments
			Deficient	Adequate		
Capsicum annuum L. (pepper)	Fruits	Nutrient solution		1500–1800 (dry wt)	86	Proximal pericarp tissue
				1000–1200 (dry wt)		Distal pericarp tissue (healthy)
			600 (dry wt)			Distal pericarp tissue (BER-affected)
Cucumis sativus L. (cucumber)	Fruits	Rockwool and nutrient solution		3000–6000 (dry wt)	87	Range of values according to salinity treatment and size of fruit
Fragaria x ananassa Duchesne (strawberry)	Fruits	Sand culture		65/120/201 (559/1192/2060) (dry wt)	91	Values from left to right for plants that had 33.2% of leaves with tipburn (plants supplied 1/40th control Ca and 3x K), 9% of leaves with tipburn (plants supplied control Ca and 3x K) 1% of leaves with tipburn (control)
Lycopersicon esculentum Mill. (tomato)			210/240 (dry wt)	280 (dry wt)	94	For 'deficient' values, first value is for an experiment in which 44.5% of fruit had BER, second value for an experiment in which 18.9% of fruit had BER. For 'adequate' value 0.9% of fruit had BER
Malus pumila Mill. [*M. domestica* Borkh.] (apple) cv Jonagold	Fruitlets in July	34 different orchards	105	190	95	Fruitlets with 'deficient' concentration showed much higher incidence of physiological disorders in storage
cv Cox's Orange Pippin	Fruit at harvest	Orchard grown	33 36 38	64 64 62	45	Range found in fruit harvested in 3 consecutive years. Fruit with the lower values had higher incidence of bitter pit
cv Cox's Orange Pippin			45		96	Minimum level for recommending fruit for controlled atmosphere storage. Below this level bitter pit is common
Pyrus communis (pear)	Fruit	4 Orchards	60	76	97	Values of 60 and 67 mg kg⁻¹ fresh weight in fruit from different orchards linked with high incidence of internal breakdown and cork spot

Note: Values in fresh matter, unless shown to contrary.

come from elsewhere and been deposited by glaciers, and furthermore, although calcium carbonate is sparingly water soluble, it can be removed by leaching so that the overlying soil may be depleted of calcium carbonate and be acidic.

Some soils contain calcium sulfate (gypsum), but mostly only in arid regions. A further source of calcium in soils is apatite $[Ca(OH_2).3Ca(PO_4)_2]$ or fluorapatite $[Ca_5(PO_4)_3F]$. Chlorapatite $[Ca_5(PO_4)_3Cl]$ and hydroxyapatite $[Ca_5(PO_4)_3OH]$ also exist in soils (98). Calcium is also present in the primary minerals augite $[Ca(Mg,Fe,Al)(Al,Si)_2O_6]$, hornblende $[NaCa_2(Mg,Fe,Al)_5(Si,Al)_8O_{22}(OH)_2]$, and the feldspar plagioclase (any intermediate between $CaAl_2Si_2O_8$ and $NaAlSi_3O_8$) (98).

Within the fraction of soils where particles are as small as clay particles, calcium occurs in gypsum, calcite, hornblende, and plagioclase. Sherman and Jackson (99) arranged the minerals in the clay fractions of the A horizons of soils in a series according to the time taken for them to weather away to a different mineral. These calcium-containing minerals are all early in this sequence, meaning that calcium is lost from the minerals (and becomes available to plants) early in the weathering process, but has been entirely lost as a structural component in more mature soils (98). Any calcium present in these more mature soils will be present attached to cation-exchange sites, where it usually constitutes a high proportion of total exchangeable cations, so the amounts present depend on the CEC of the soil.

Concentrations of Ca^{2+} in soils may be affected by ecological disturbance. Acid depositions are known to decrease Ca^{2+} concentrations in soils, which while not necessarily affecting plant yields directly may have a big impact on ecosystem dynamics. Acid deposition on the coniferous forests of the Netherlands has been shown to give rise to fewer snails, and the birds that feed on the snails have fewer surviving offspring due to defects in their eggs (100). This effect seems to be related largely to the abundance of snails being depressed by low calcium concentrations in the plant litter. In terms of how serious this problem might prove to be, it should be noted that changes in soil Ca^{2+} concentration caused by acid rain are less than $1\,g\,Ca^{2+}\,m^{-2}\,year^{-1}$. This change is small compared with a transfer of 3.3 to $4.7\,g\,Ca^{2+}\,m^{-2}\,year^{-1}$ from mineral soil to young forest stands (101).

Experiments on the Hubbard Brook Experimental Forest in New Hampshire, USA, have shown that calcium is lost from ecosystems following deforestation. This loss is true for other cations and also for nitrate. In the Hubbard Brook experiment, during the 4 years following deforestation, the watershed lost $74.9\,kg\,Ca^{2+}\,ha^{-1}\,year^{-1}$ as dissolved substances in the streams, compared with $9.7\,kg\,Ca^{2+}\,ha^{-1}\,year^{-1}$ in a watershed where the vegetation had not been cut down (102). This increased loss was attributed partly to increased water flows due to decreased water loss by transpiration, but more importantly through the breakdown of the plant material enhancing the turnover of the nitrogen cycle and the consequent generation of H^+ ions, thereby releasing cations from the cation-exchange sites of the soil (102). Recent studies have shown that calcium loss continues for at least 30 years, with the longer-term loss possibly occurring because of the breakdown of calcium oxalate in the forest soil after removal of the trees (103).

5.4.2 SOIL TESTS

The main test for soil calcium is to calculate the amount of the limestone required for a particular crop on a particular soil (see 5.5.2 below).

5.4.3 TABULATED DATA ON CALCIUM CONTENTS IN SOILS

Concentrations of Ca^{2+} in soils typical of a range of soil orders are shown in Table 5.3.

5.5 FERTILIZERS FOR CALCIUM

5.5.1 KINDS OF FERTILIZER

The most common application of calcium to soils is as calcium carbonate in chalk or lime. This practice occurred in Britain and Gaul before the Romans (Pliny, quoted in Ref. (105)). It does not

TABLE 5.3
Calcium Concentration, Cation Exchange Capacity and pH of Top Layers of Some Representative Soils

Soil	Soil Order	Ca^{2+} Concentration (mmol kg^{-1})	CEC (cmol$_c$ kg^{-1})	pH
Typic Cryoboralf, Colorado, 0–18 cm depth	Alfisol	30.5	13.3	5.9
Typic Gypsiorthid, Texas, 5–13 cm depth	Aridisol	100.0	21.6	7.9
Typic Ustipsamment, Kansas, 0–13 cm depth	Entisol	9.5	52.0	6.6
Typic Dystrochrept, West Virginia, 5–18 cm depth	Inceptisol	5.0	11.4	4.9
Typic Argiustoll, Kansas, 0–15 cm depth	Mollisol	73.5	23.8	6.6
Typic Acrustox, Brazil, 0–10 cm depth (low CEC below 65 cm)	Oxisol	2.1	20.5	5.0
Typic Haplorthod, New Hampshire, 0–20 cm depth	Spodosol	14.5	25.7	4.9
Typic Umbraquult, North Carolina, 0–15 cm depth	Ultisol	2.0	26.2	3.9
Typic Chromoxerert, California, 0–10 cm depth	Vertisol	84.0	24.6	7.8

Source: Data from USDA, *Soil Taxonomy: A Basic System of Soil Classification for Making and Interpreting Soil Surveys*. Agricultural Handbook Number 436. Washington, DC: USDA, 1975.

come strictly under the definition of fertilizer, as the main functions of the calcium carbonate are to make clay particles aggregate into crumbs, thereby improving drainage, and to lower soil acidity.

Despite the observation that addition of gypsum to tropical soils may increase the sorption of phosphate (73), it seems as if this effect is not universal, and it is the change in pH brought about by limestone or dolomite that is more important in aiding phosphate sorption than the provision of Ca^{2+} ions. In an experiment on addition of calcium carbonate, dolomite, gypsum, and calcium chloride to the Ap horizon of a Spodosol, all additions increased the retention of phosphorus in the soil except the calcium chloride (106). The order of this increase was calcium carbonate > dolomite > gypsum, which followed the order of increase in pH. Gypsum is not expected to increase pH of soil, but it is likely that this pH change, and the consequent effect on phosphorus sorption, was due to impurities, likely lime, in the gypsum used.

Following an addition of lime, Ca^{2+} from the calcium carbonate ($CaCO_3$) exchanges for $Al(OH)_2^+$ and H^+ ions on the cation-exchange sites. The $Al(OH)_2^+$ ions give rise to insoluble $Al(OH)_3$ that precipitates; the H^+ ions react with bicarbonate (HCO_3)$^-$ that arises during the dissolution of calcium carbonate in the soil water. This reaction leads to the formation of carbon dioxide, lost from the soil as a gas, and water, both of which are neutral products (107).

In very acid soils, there is a shortage of available calcium, and application of calcium carbonate will help rectify this problem. One of the outcomes of adding calcium would be to displace Al^{3+} and H^+ ions from the root plasmalemma, where they would otherwise be displacing Ca^{2+} ions (108). Experiments with alfalfa (*Medicago sativa* L.) grown on acid soils showed that while application of lime increased calcium concentrations in the shoots, it also decreased concentrations of aluminum, manganese, and iron. As those cultivars that were the least sensitive to the acid soil had lower concentrations of these three elements anyway, it seems as if the beneficial effect of the lime was in modifying soil pH rather than supplying additional Ca (109).

The more neutral or alkaline pH brought about by liming gives a more favorable environment for the microorganisms of the nitrogen cycle, enhancing the cycling of nitrogen from organic matter. It also increases the availability of molybdenum, and it restricts the uptake of heavy metals (107).

Another action of lime is to decrease the concentration of fluoride in tea (*Camellia sinensis* L.) plants. This crop accumulates high concentrations of fluoride from soils of normal fluoride concentration. The action of liming in limiting fluoride concentrations in tea plants is surprising given that the uptake of fluoride is higher from more neutral soil than from acid soil and given that liming may increase the water-soluble fluoride content of the soil (110). In this case, it appears that the Ca^{2+} in the lime either affects cell wall and plasmalemma permeability or changes the speciation of the fluoride in the soil.

In some instances calcium sulfate (gypsum) may be applied as a fertilizer, but this application is more for a source of sulfur than calcium or to improve soil structure. Apatite (applied as rock phosphate) and superphosphate contain twice as much calcium by weight as the phosphorus that they are used primarily to supply, and triple superphosphate contains two thirds as much calcium as phosphorus (98). One situation where gypsum is particularly useful is in the reclamation of sodic soils, where the calcium ions replace the sodium on the cation-exchange sites and the sodium sulfate that results is leached out of the soil (107).

Calcium nitrate and calcium chloride are regularly used as sprays on developing apple fruits to prevent bitter pit (111). Of the two calcium forms, nitrate is less likely to cause leaf scorch, but some varieties of apple are susceptible to fruit spotting with nitrate. Dipping the fruit in $CaCl_2$ immediately after harvest supplements the regular sprays (111). Spraying apple trees with calcium nitrate during the cell expansion phase of fruit growth increases the nitrogen and the calcium concentrations in the fruit at harvest and gives firmer fruit at harvest and after storage (112).

Application of calcium salts to sweet cherry (*Prunus avium* L.) fruits just before harvest may also decrease the incidence of skin cracking that follows any heavy rainfall at this time (43). Multiple applications throughout the summer give better protection, and $CaCl_2$ is better than $Ca(OH)_2$, as the latter can cause fruit to shrivel in hot seasons (113). Recent research has shown that spraying $CaCl_2$ and boron with a suitable surfactant on strawberry plants at 5-day intervals from the time of petal fall gives fruits that are firmer and more resistant to botrytis rot at harvest, or after 3 days storage, than untreated fruits; after the 3 days, they have a higher concentration of soluble solids and more titratable acidity (114). Treating pineapples with lime during their growth seems to lower the incidence of internal browning that arises in the fruit in cold storage, and increases their ascorbic acid content (38). The fruit of tomato cultivars particularly susceptible to BER (e.g., the beefsteak cultivar Jack Hawkins) may be sprayed with calcium salts, although the efficacy of this treatment is doubtful.

There are also calcium treatments for improving shelf life and fruit quality that are used after harvest. For example, dipping cherry tomatoes in 25 mM $CaCl_2$ after harvest increases apoplastic calcium concentrations and decreases incidence of skin cracking (115). Vacuum infiltration of Ca^{2+} increases the time of ripening of peaches, so that they can be stored for longer periods before sale, and such use of calcium salts is common for tomatoes, mangoes (*Mangifera indica* L.), and avocadoes (*Persea americana* L.) (116). The firmness of plums (*Prunus domestica* L.) is increased by pressure infiltration of 1 mM $CaCl_2$ (117).

There is some evidence that supply of supplementary calcium nitrate partially alleviates the effects of NaCl salinity in strawberry in hydroponic culture (118) and in cucumber and melon (*Cucumis* spp. L.) in irrigated fields (119).

5.5.2 APPLICATION OF CALCIUM FERTILIZERS

Liming is carried out by application of $CaCO_3$ in limestone, a process that is described in some detail in Troeh and Thompson (98). The neutralizing capacity of the limestone used is measured by comparing it to calcite, which is $CaCO_3$, with a *calcium carbonate equivalent* (CCE) of 100%. The fineness of the lime affects its efficiency for liming, and the CCE and fineness and hardness of the lime together give the *effective calcium carbonate equivalent* or *reactivity*. Application should occur when the soil is dry or frozen, to avoid damage to the soil by the vehicles carrying the lime. Although soil testing will determine if an application is required, it is often the practice to apply lime a year ahead of a crop in a rotation that has a strong lime requirement (often a legume). An application once every 4 to 8 years is usually effective. Limestone, burned lime (CaO), or slaked lime [$Ca(OH)_2$] can also be used. Burned lime has a CCE of 179% and slaked lime a CCE of 133%.

The amount of lime required is determined from soil analysis, either by a pH base saturation method or a buffer solution method (98,120). The soil requirement for lime, defined, for example, as the number of tonnes of calcium carbonate required to raise the pH of a hectare of soil 200 mm deep to pH 6.5 (120), will depend on the initial pH and also on CEC of the soil. Most soils have a much greater proportion of their cations attached to cation-exchange sites than in solution, meaning that a high proportion of the H^+ ions present are not measured in a simple pH test. Adding lime to the soil neutralizes the acidity in the soil solution, but the Ca^{2+} ions displace H^+ ions from the exchange sites, with the potential to make the pH of the soil acidic once more, and this acidity is neutralized by reaction of the H^+ with the lime. The H^+ in soil solution is called the *active acidity*, and the H^+ held to the exchange sites on soil colloids is called the *reserve acidity* The greater the CEC, the greater the reserve acidity and the greater the lime requirement (98).

In the pH-base saturation method, the percent base saturation of the soil, the CEC of the soil and the initial pH all have to be measured. To calculate how much lime should be added the percent base saturation at the initial and at the target pH value are read off a graph, and the amount of $CaCO_3$ to be added is calculated from the difference in percent base saturation at the two pH values multiplied by the CEC (98).

In the buffer solution method, a sample of the soil is mixed with a buffer, and the amount of lime required is read off a table from the value of decrease in buffer pH on adding the soil (120).

ACKNOWLEDGMENT

We thank Dr. Paul Knox for the invaluable discussion on the structure of cell walls.

REFERENCES

1. R.R. van der Ploeg, W. Böhm, M.B. Kirkham. On the origin of the theory of mineral nutrition of plants and the Law of the Minimum. *Soil Sci. Am. J.* 63:1055–1062, 1999.
2. E.J. Hewitt. *Sand and Water Culture Methods Used in the Study of Plant Nutrition*. Farnham Royal, UK: CAB, 1966, pp. 5, 189–190.
3. F. Bangerth. Calcium related physiological disorders of plants. *Annu. Rev. Phytopath.* 17: 97–122, 1979.
4. D.T. Clarkson. Movement of ions across roots. In: D.A. Baker, J.L. Hall, eds. *Solute Transport in Plant Cells and Tissues. Monographs and Surveys in Biosciences*. New York: Wiley, 1988, pp. 251–304.
5. E. Epstein. *Mineral Nutrition of Plants: Principles and Perspectives*. New York: Wiley, 1972.
6. R.L. Legge, J.E. Thompson, J.E. Baker, M. Lieberman. The effect of calcium on the fluidity of phase properties of microsomal membranes isolated from postclimacteric golden delicious apples. *Plant Cell Physiol.* 23:161–169, 1982.
7. N. Duzgunes, D. Papahadjopoulos. Ionotrophic effects on phospholipid membranes: calcium/magnesium specificity in binding, fluidity, and fusion. In: R.C. Aloia, ed. *Membrane Fluidity in Biology*. New York: Academic Press, 1983, pp. 187–212.
8. C W Grant. Lateral phase separation and the cell membrane. In: R.C. Aloia, ed. *Membrane Fluidity in Biology*. New York: Academic Press, 1983, pp. 131–150.
9. W.G.T. Willats, L. McCartney, L. Mackie, J.P. Knox. Pectin: cell biology and prospects for functional analysis. *Plant Mol. Biol.* 47:9–27, 2001.

10. E.R. Morris, D.A. Powell, M.J. Gidley, D.A. Rees. Conformations and interactions of pectins. *J. Mol. Biol.* 155:507–516, 1982.

11. F.P.C. Blamey. A role for pectin in the control of cell expansion. *Soil Sci. Plant Nutr.* 49:775–783, 2003.

12. N.C. Carpita, D.M. Gibeaut. Structural models of primary cell walls in flowering plants: consistency of molecular structure with the physical properties of the walls during growth. *Plant J.* 3:1–30, 1993.

13. S.S. Virk, R.E. Cleland. Calcium and mechanical properties of soybean hypocotyl cell walls: possible role of calcium and protons in cell wall loosening. *Planta* 176:60–67, 1988.

14. M. Knee. Fruit softening III. Requirement for oxygen and pH effects. *J. Exp. Bot.* 33:1263–1269, 1982.

15. D.P.F. Almeida, D.J. Huber. Apoplastic pH and inorganic ion levels in tomato fruit: a potential means for regulation of cell wall metabolism during ripening. *Physiol. Plant* 105:506–512, 1999.

16. W. Bussler. Die Entwicklung von Calcium-mangelsymptomen. *Z. Pflanzenernaehrung Bodenkunde* 100:53–58, 1963.

17. B.W. Poovaiah. Role of calcium in ripening and senescence. *Commun. Soil Sci. Plant Anal.* 10:83–88, 1979.

18. S. Seling, A.H. Wissemeier, P. Cambier, P. van Cutsem. Calcium deficiency in potato (*Solanum tuberosum* ssp. *tuberosum*) leaves and its effects on the pectic composition of apoplastic fluid. *Physiol. Plant* 109:44–50, 2000.

19. L. Rensing, G. Cornelius. Biological membranes as components of oscillating systems. *Biol. Rundschau* 18:197–209, 1980.

20. D.S. Bush. Calcium regulation in plant cells and its role in signalling. *Annu. Rev. Plant Physiol. Plant Mol. Biol.* 46:95–122, 1995.

21. N.A. Gavalas, Y. Manetas. Calcium inhibition of phosphoenolpyruvate carboxylase: possible physiological consequences for 4-carbon-photosynthesis. *Z. Pflanzenphysiol.* 100:179–184, 1980.

22. E.A. Kirkby, D.J. Pilbeam. Calcium as a plant nutrient. *Plant Cell Environ.* 7:397–405, 1984.

23. D.T. Britto, H.J. Kronzucker. Ion fluxes and cytosolic pool sizes: examining fundamental relationships in transmembrane flux regulation. *Planta* 217:490–497, 2003.

24. J.J. Rudd, V.E. Franklin-Tong. Unravelling response-specificity in Ca^{2+} signalling pathways in plant cells. *New Phytol.* 151:7–34, 2001.

25. W.A. Snedden, H. Fromm. Calmodulin as a versatile calcium signal transducer in plants. *New Phytol.* 151:35–66, 2001.

26. D. Marmé. Calcium transport and function. In: A. Läuchli, R.L. Bieleski, eds. *Inorganic Plant Nutrition.* Berlin: Springer, 1983, pp. 599–625.

27. H.G. Burström. Calcium and plant growth. *Biol. Rev.* 43:287–316, 1968.

28. H. Marschner, H. Ossenberg-Neuhaus. Wirkung von 2,3,5-trijodobenzoesäure (TIBA) auf den Calciumtransport und die Kationenauschkapazität in Sonnenblumen. *Z. Pflanzenphysiol.* 85:29–44, 1977.

29. L.C. Ho, P. Adams. Calcium deficiency–a matter of inadequate transport to rapidly growing organs. *Plants Today* 2:202–207, 1989.

30. B.W. Poovaiah, A.C. Leopold. Deferral of leaf senescence with calcium. *Plant Physiol.* 52:236–239, 1973.

31. A.K. Mattoo, M. Lieberman. Localization of the ethylene-synthesising system in apple tissue. *Plant Physiol.* 60:794–799, 1977.

32. T.L. Holdaway-Clarke, P.K. Heppler. Control of pollen tube growth: role of ion gradients and fluxes. *New Phytol.* 159:539–563, 2003.

33. F. Baluska, F. Cvrckova, J. Kendrick-Jones, D. Volkmann. Sink plasmodesmata as gateways for phloem unloading. Myosin VIII and calreticulin as molecular determinants of sink strength. *Plant Physiol.* 126:39–46, 2001.

34. W.H. Queen, H.W. Fleming, J.C. O'Kelley. Effects on *Zea mays* seedlings of a strontium replacement of calcium in nutrient media. *Plant Physiol.* 38:410–413, 1963.

35. D.S. Veresoglou, N. Barbayiannis, T. Matsi, C. Anagnostopoulos, G.C. Zalidis. Shoot Sr concentration in relation to shoot Ca concentrations and to soil properties. *Plant Soil* 178:95–100, 1996.

36. G.T. Nightingale, R.M. Addoms, W.R. Robbins, L.G. Shermerhorn. Effects of calcium deficiency on nitrate absorption and on metabolism in tomato. *Plant Physiol.* 6:605, 1931.

37. C.B. Shear. Calcium-related disorders of fruits and vegetables. *Hort. Sci.* 10:361–365, 1975.

38. H.M.I. Herath, D.C. Bandara, D.M.G.A. Banda. Effect of pre-harvest calcium fertilizer application on the control of internal browning development during the cold storage of pineapple 'Mauritius' (*Ananas comosus* (L.) Merr.). *J. Hort. Sci. Biotech.* 78:762–767, 2003.

39. R. Storey, M.T. Treeby, D.J. Milne. Crease: another Ca deficiency-related fruit disorder? *J. Hortic. Sci. Biotechnol.* 77:565–571, 2002.

40. M. Faust, C.B. Shear. Corking disorders of apples. A physiological and biochemical review. *Bot. Rev.* 34:441–469, 1968.

41. L.C. Ho, D.J. Hand, M. Fussell. Improvement of tomato fruit quality by calcium nutrition. *Acta Hortic.* 481:463–468, 1999.

42. R. Storey, M.T. Treeby. Cryo-SEM study of the early symptoms of peteca in 'Lisbon' lemons. *J. Hortic. Sci. Biotech.* 77:551–556, 2002.

43. G.M. Glenn, B.W. Poovaiah. Cuticular properties and postharvest calcium applications influence cracking of sweet cherries. *J. Am. Soc. Hortic. Sci.* 114:781–788, 1989.

44. I.M. Huxham, M.C. Jarvis, L. Shakespeare, C.J. Dover, D. Johnson, J.P. Knox, G.B. Seymour. Electron-energy-loss spectroscopic imaging of calcium and nitrogen in the cell walls of apple fruits. *Planta* 208:438–443, 1999.

45. D.S. Johnson, M.J. Marks, K. Pearson. Storage quality of Cox's Orange Pippin apples in relation to fruit mineral composition during development. *J. Hortic. Sci.* 62:17–25, 1987.

46. E.A. Kirkby, A.H. Knight. Influence of the level of nitrate nutrition and ion uptake and assimilation, organic acid accumulation and cation-anion balance in whole tomato plants. *Plant Physiol.* 60:349–353, 1977.

47. J.W. Hudgins, T. Krekling, V.R. Franceschi. Distribution of calcium oxalate crystals in the secondary phloem of conifers: a constitutive defense mechanism. *New Phytol.* 159:677–690, 2003.

48. T.A. Kostman, V.R. Franceschi, P.A. Nakata. Endoplasmic reticulum sub-compartments are involved in calcium sequestration within raphide crystal idioblasts of *Pistia stratiotes* L. *Plant Sci.* 165:205–212, 2003.

49. P.A. Nakata. Advances in our understanding of calcium oxalate crystal formation and function in plants. *Plant Sci.* 164:901–909, 2003.

50. S.V. Pennisi, D.B. McConnell, L.B. Gower, M.E. Kane, T. Lucansky. Intracellular calcium oxalate crystal structure in *Dracaena sanderiana*. *New Phytol.* 150:111–120, 2001.

51. S.A. Barber. The role of root interception, mass flow and diffusion in regulating the uptake of ions by plants from soil. *Technical Report Series–IAEA* 65:39–45, 1966.

52. D.M. Hegde. Irrigation and nitrogen requirement of bell pepper. *Indian J. Agric. Sci.* 58:668–672, 1988.

53. P.J. White. Calcium channels in higher plants. *BBA- Biomembranes*, 1465(1–2):171–189, 2000.

54. D.T. Clarkson. Calcium transport between tissues and its distribution in the plant. *Plant Cell Environ.* 7:449–456, 1984.

55. C.A. Peterson, D.E. Enstone. Functions of passage cells in the endodermis and exodermis of roots. *Physiol. Plant* 97:592–598, 1996.

56. P.J. White. Calcium channels in the plasma membrane of root cells. *Ann. Bot.* 81:173–183, 1998.

57. D.T. Clarkson. Roots and the delivery of solutes to the xylem. *Philos. Tran. Roy. Soc. B* 341:5–17, 1993.

58. M.J. Armstrong, E.A. Kirkby. The influence of humidity on the mineral composition of tomato plants with special reference to calcium distribution. *Plant Soil* 52:427–435, 1979.

59. B. Bengtsson. Uptake and translocation of calcium in cucumber. *Physiol. Plant* 54:107–111, 1982.

60. E.W. Simon. The symptoms of calcium deficiency in plants. *New Phytol.* 80:1–15, 1978.

61. O. Biddulph, F.S. Nakayama, R. Cory. Transpiration stream and ascension of calcium. *Plant Physiol.* 36:429–436, 1961.

62. H.V. Koontz, R.E. Foote. Transpiration and calcium deposition by unifoliate leaves of *Phaseolus vulgaris* differing in maturity. *Physiol. Plant* 14:313–321, 1966.

63. N. Bellaloui, D.J. Pilbeam. Reduction of nitrate in leaves of tomato during vegetative growth. *J. Plant Nutr.* 13:39–55, 1990.

64. L.C. Ho, P. Adams, X.Z. Li, H. Shen, J. Andrews, Z.H. Xu. Responses of Ca-efficient and Ca-inefficient tomato cultivars to salinity in plant growth, calcium accumulation and blossom-end rot. *J. Hortic. Sci.* 70:909–918, 1995.

65. N.E. Nielsen, C.B. Sørensen. Macronutrient cation uptake by plants. I. Rate determining steps in net inflow of cations into intact and decapitated sunflower plants and intensity factors of cations in soil solution. *Plant Soil* 77:337–346, 1984.

66. N.E. Nielsen, E.M. Hansen. Macronutrient cation uptake by plants. II. Effects of plant species, nitrogen concentration in the plant, cation concentration, activity and activity ratio in soil solution. *Plant Soil* 77:347–365, 1984.

67. P.J. White. The pathways of calcium movement to the xylem. *J. Exp. Bot.* 52:891–899, 2001.
68. P. Adams, L.C. Ho. Effects of environment on the uptake and distribution of calcium in tomato and on the incidence of blossom-end rot. *Plant Soil* 154:127–132, 1993.
69. J. Willumsen, K.K. Petersen, K. Kaack. Yield and blossom-end rot of tomato as affected by salinity and cation activity ratios in the root zone. *J. Hortic. Sci.* 71:81–91, 1996.
70. P. Adams, L.C. Ho. The susceptibility of modern tomato cultivars to blossom-end rot in relation to salinity. *J. Hortic. Sci.* 67:827–839, 1992.
71. R. Belda, L.C. Ho. Salinity effects on the network of vascular bundles during tomato fruit development. *J. Hortic. Sci.* 68:557–564, 1993.
72. J.A. Lee. The calcicole-calcifuge problem revisited. *Adv. Bot. Res.* 29:1–30, 1999.
73. L.J. Cajuste, R.J. Laird, B. Cuevas, J. Alvarado. Phosphorus retention by tropical soils as influenced by sulfate application. *Commun. Soil Sci. Plant Anal.* 29:1823–1831, 1998.
74. P.S. Bhadoria, B. Steingrobe, N. Claassen, H. Liebersbach. Phosphorus efficiency of wheat and sugar beet seedlings grown in soils with mainly calcium, or iron and aluminium phosphate. *Plant Soil* 246:41–52, 2002.
75. Y.-G. Zhu, Y.-Q. He, S.E. Smith, F.A. Smith. Buckwheat (*Fagopyrum esculentum* Moench) has high capacity to take up phosphorus (P) from a calcium (Ca)-bound source. *Plant Soil* 239:1–8, 2002.
76. A. Zohlen, G. Tyler. Immobilization of tissue iron on calcareous soil: differences between calcicole and calcifuge plants. *Oikos* 89:95–106, 2000.
77. A. Zohlen. Chlorosis in wild plants: is it a sign of iron deficiency? *J. Plant Nutr.* 25:2205–2228, 2002.
78. L. Strom. Root exudation of organic acids: importance to nutrient availability and the calcifuge and calcicole behaviour of plants. *Oikos* 80:459–466, 1997.
79. E. Peiter, Y. Feng, S. Schubert. Lime-induced growth depression in *Lupinus* species: are soil pH and bicarbonate involved? *J. Plant Nutr. Soil Sci.* 164:165–172, 2001.
80. J.F. Loneragan, J.S. Gladstones, W.J. Simmons. Mineral elements in temperate crops and pasture plants. II. Calcium. *Aust. J. Agric. Res.* 19:353–364, 1968.
81. J.F. Loneragan, K. Snowball. Calcium requirements of plants. *Aust. J. Agric. Res.* 20:465–478, 1969.
82. A.H. Knight, W.M. Crooke. Cation exchange capacity and chemical composition of the floral parts of *Antirrhinum* and *Lilium*. *Ann. Bot.* 37:155–157, 1973.
83. H. Ibrikci, S.J.B. Knewston, M.A. Grusak. Green vegetables for humans: evaluation of mineral composition. *J. Sci. Fd. Agric.* 83:945–950, 2003.
84. D.J. Reuter, F.W. Smith, J.B. Robinson, T.J. Piggott, G.H. Price. In: D.J. Reuter, J.B. Robinson. *Plant Analysis. A Plant Interpretative Manual*. Melbourne: Inkata Press, 1986.
85. W. Bergmann, ed. *Colour Atlas. Nutritional Disorders of Plants: Visual and Analytical Diagnosis*. Jena: Fischer, 1992.
86. P.S. Morley, M. Hardgrave, M. Bradley, D.J. Pilbeam. Susceptibility of sweet pepper (*Capsicum annuum* L.) cultivars to the calcium deficiency disorder 'Blossom end rot'. In: M.A.C. Fragoso, M.L. van Beusichem, eds. *Optimization of Plant Nutrition*. Dordrecht: Kluwer Academic Publishers, 1993, pp. 563–567.
87. L.C. Ho, P. Adams. The physiological basis for high fruit yield and susceptibility to calcium deficiency in tomato and cucumber. *J. Hortic. Sci.* 69:367–376, 1994.
88. D.W. Goodall, F.G. Gregory. *Chemical Composition of Plants as an Index of their Nutritional Status*. East Malling, UK: IAB, 1947.
89. P.S. Morley, D.J. Pilbeam. Unpublished results.
90. P.H. Brown. Seasonal variations in fig (*Ficus carica* L.) leaf nutrient concentrations. *HortScience* 29:871–873, 1994.
91. T.F. Chiu, C. Bould. Effects of calcium and potassium on ^{45}Ca mobility, growth and nutritional disorders of strawberry plants (*Fragaria* spp.). *J. Hortic. Sci.* 51:525–531, 1976.
92. H.S. Xie, G.A. Cummings. Effect of soil pH and nitrogen source on nutrient status in peach: I. Macronutrients. *J. Plant Nutr.* 18:541–551, 1995.
93. D.W. Reeves, J.H. Edwards, J.M. Thompson, B.D. Horton. Influence of Ca concentration on micronutrient imbalances in *in vitro* propagated *Prunus* rootstock. *J. Plant Nutr.* 8:289–302, 1985.
94. F. Bangerth. Investigations upon Ca related physiological disorders. *Phytopath. Z* 77:20–37, 1973.
95. R.D. Marcelle, W. Porreye, T. Deckers, P. Simon, G. Goffings, M. Herregods. Relationship between fruit mineral composition and storage life of apples, cv. Jonagold. *Acta Hortic.* 258:373–378, 1989.
96. Minerals: Still a Case for Analysis. *The Grower*, February 10, 1994, pp. 30–31.

97. R.L. Vaz, D.G. Richardson. Calcium effects on the postharvest quality of Anjou pears. *Proceedings of the 79th Annual Meeting of the Washington State Horticultural Society*, 1983, pp. 153–161.

98. F.R. Troeh, L.M. Thompson. *Soils and Soil Fertility*. 5th ed. New York: Oxford University Press, 1993.

99. M.L. Jackson, G.R. Sherman. Chemical weathering of minerals in soils. *Adv. Agron.* 5:221–318, 1953.

100. J. Graveland, R. van der Wal, J.H. van Balen, A.J. van Noordwijk. Poor reproduction in forest passerines from decline of snail abundance on acidified soils. *Nature* 368:446–448, 1994.

101. S.P. Hamburg, R.D. Yanai, M.A. Blum, T.G. Siccama. Biotic control of calcium cycling in northern hardwood forests: acid rain and aging forests. *Ecosystems* 6:399–406, 2003.

102. F.H. Bormann, G.E. Likens, T.G. Siccama, R.S. Pierce, J.S. Eaton. The export of nutrients and recovery of stable conditions following deforestation at Hubbard Brook. *Ecol. Monograph* 44:255–277, 1974.

103. S.W. Bailey, D.C. Buso, G.E. Likens. Implications of sodium mass balance for interpreting the calcium cycle of a forested ecosystem. *Ecology* 84:471–484, 2003.

104. USDA. *Soil Taxonomy: A Basic System of Soil Classification for Making and Interpreting Soil Surveys*. Agricultural Handbook Number 436. Washington, DC: USDA, 1975.

105. E.J. Russell. *The World of the Soil*. London: Collins, 1957, pp. 212–213.

106. L. Boruvka, J.E. Rechcigl. Phosphorus retention by the Ap horizon of a spodosol as influenced by calcium amendments. *Soil Sci.* 168:699–706, 2003.

107. R.W. Miller, R.L. Donahue. *Soils. An Introduction to Soils and Plant Growth*. 6th ed. Englewood Cliffs, NJ: Prentice-Hall, 1990.

108. T.B. Kinraide. Toxicity factors in acidic forest soils: attempts to evaluate separately the toxic effects of excessive Al^{3+} and H^+ and insufficient Ca^{2+} and Mg^{2+} upon root elongation. *Eur. J. Soil Sci.* 54: 323–333, 2003.

109. H.S. Grewal, R. Williams. Liming and cultivars affect root growth, nodulation, leaf to stem ratio, herbage yield, and elemental composition of alfalfa on an acid soil. *J. Plant Nutr.* 26:1683–1696, 2003.

110. J. Ruan, M. Lifeng, Y. Shi, W. Han. The impact of pH and calcium on the uptake of fluoride by tea plants (*Camellia sinensis* L.). *Ann. Bot.* 93:97–105, 2004.

111. P. Needham. The occurrence and treatment of mineral disorders in the field. In: C. Bould, E.J. Hewitt, P. Needham. *Diagnosis of Mineral Disorders in Plants. Volume 1. Principles*. London: Her Majesty's Stationery Office, 1983, p. 163.

112. D.S. Johnson, C.J. Dover, T.J. Samuelson, I.M. Huxham, M.C. Jarvis, L. Shakespeare, G.B. Seymour. Nitrogen, cell walls and texture of stored Cox's Orange Pippin apples. *Acta Hortic.* 564:105–112, 2001.

113. M. Meheriuk, G.H. Nielsen, D.-L. McKenzie. Incidence of rain splitting in sweet cherries treated with calcium or coating materials. *Can. J. Plant Sci.* 71:231–234, 1991.

114. P. Wojcik, M. Lewandowska. Effect of calcium and boron sprays on yield and quality of "Elsanta" strawberry. *J. Plant Nutr.* 26:671–682, 2003.

115. A. Lichter, O. Dvir, E. Fallik, S. Cohen, R. Golan, Z. Shemer, M. Sagi. Cracking of cherry tomatoes in solution. *Postharvest Biol. Tech.* 26:305–312, 2002.

116. R.B.H. Willis, M.S. Mahendra. Effect of postharvest application of calcium on ripening of peach. *Aust. J. Exp. Agric.* 29:751–753, 1989.

117. D. Valero, A. Pérez-Vicente, D. Martínez-Romero, S. Castillo, F. Guillén, M. Serrano. Plum storability improved after calcium and heat postharvest treatments: role of polyamines. *J. Food Sci.* 67:2571–2575, 2003.

118. C. Kaya, B.E. Ak, D. Higgs. Response of salt-stressed strawberry plants to supplementary calcium nitrate and/or potassium nitrate. *J. Plant Nutr.* 26:543–560, 2003.

119. C. Kaya, D. Higgs, H. Kirnak, I. Tas. Ameliorative effect of calcium nitrate on cucumber and melon plants drip irrigated with saline water. *J. Plant Nutr.* 26: 1665–1681, 2003.

120. MAFF. *The Analysis of Agricultural Materials: A Manual of the Analytical Methods Used by the Agricultural Development and Advisory Service*. London: Her Majesty's Stationery Office, 1986, pp. 98–101.

6 Magnesium

Donald J. Merhaut
University of California, Riverside, California

CONTENTS

6.1 Historical Information 146
    6.1.1 Determination of Essentiality 146
6.2 Function in Plants 146
    6.2.1 Metabolic Processes 146
    6.2.2 Growth 147
    6.2.3 Fruit Yield and Quality 147
6.3 Diagnosis of Magnesium Status in Plants 148
    6.3.1 Symptoms of Deficiency and Excess 148
        6.3.1.1 Symptoms of Deficiency 148
        6.3.1.2 Symptoms of Excess 149
    6.3.2 Environmental Causes of Deficiency Symptoms 149
    6.3.3 Nutrient Imbalances and Symptoms of Deficiency 150
        6.3.3.1 Potassium and Magnesium 150
        6.3.3.2 Calcium and Magnesium 151
        6.3.3.3 Nitrogen and Magnesium 151
        6.3.3.4 Sodium and Magnesium 152
        6.3.3.5 Iron and Magnesium 152
        6.3.3.6 Manganese and Magnesium 153
        6.3.3.7 Zinc and Magnesium 153
        6.3.3.8 Phosphorus and Magnesium 153
        6.3.3.9 Copper and Magnesium 154
        6.3.3.10 Chloride and Magnesium 154
        6.3.3.11 Aluminum and Magnesium 154
    6.3.4 Phenotypic Differences in Accumulation 155
    6.3.5 Genotypic Differences in Accumulation 155
6.4 Concentrations of Magnesium in Plants 156
    6.4.1 Magnesium Constituents 156
        6.4.1.1 Distribution in Plants 156
        6.4.1.2 Seasonal Variations 156
        6.4.1.3 Physiological Aspects of Magnesium Allocation 156
    6.4.2 Critical Concentrations 157
        6.4.2.1 Tissue Magnesium Concentration Associations with Crop Yields 157
        6.4.2.2 Tabulated Data of Concentrations by Crops 157
6.5 Assessment of Magnesium in Soils 165
    6.5.1 Forms of Magnesium in Soils 165
    6.5.2 Sodium Absorption Ratio 165

6.5.3   Soil Tests 170
6.5.4   Tabulated Data on Magnesium Contents in Soils...170
         6.5.4.1   Soil Types..170
6.6   Fertilizers for Magnesium..170
         6.6.1   Kinds of Fertilizers 170
         6.6.2   Effects of Fertilizers on Plant Growth 170
         6.6.3   Application of Fertilizers 172
References 172

6.1   HISTORICAL INFORMATION

6.1.1   DETERMINATION OF ESSENTIALITY

The word 'magnesium' is derived from 'magnesia' for the Magnesia district in Greece where talc (magnesium stone) was first mined (1,2). However, there are other cities that are also named after the magnesium deposits in local regions (3). In 1808, Sir Humphry Davy discovered magnesium, but named it magnium, because he considered magnesium to sound too much like manganese. However, in time, the word magnesium was adopted (3–6). Twenty years later, magnesium was purified by the French scientist, Bussy (7). The essentiality of magnesium in plants was established nearly 50 years later (around 1860) by scientists such as Knop, Mayer, Sachs, and Salm-Horstmar (4,8,9), and during the period 1904–1912, Willstatter identified magnesium as part of the chlorophyll molecule (3,6). For many years, magnesium was applied unknowingly to agricultural lands through manure applications or as an impurity with other processed fertilizers (10); therefore, incidences of magnesium deficiency were relatively uncommon. One of the first mentions of magnesium deficiency in plants was in 1923 on tobacco and was referred to as 'sand drown,' since the environmental conditions that were associated with magnesium deficiency occurred in excessively leached sandy soils (11). Over 100 years later, magnesium has become a global concern, as scientists suggest that magnesium deficiency may be one of the major factors causing forest decline in Europe and North America (12–17). This malady may be an indirect result of the acidification of soils by acid rain, which can cause leaching of magnesium as well as other alkali metals.

Magnesium is also an essential nutrient for animals. If forage crops, commonly grasses, are low in magnesium, grazing animals may develop hypomagnesia, sometimes called grass tetany. For this reason, many studies have been conducted on magnesium nutrition in forage crops, in an effort to prevent this disorder (18–24). Based on the review of fertilizer recommendations for field soils in the Netherlands by Henkens (25), the magnesium requirement for forage crops is closely associated with the concentration of potassium and crude protein in the crop. This relationship of magnesium with potassium and crude protein (nitrogen) for animal nutrition is not much different from the magnesium-potassium-nitrogen associations in plant nutrition.

6.2   FUNCTION IN PLANTS

6.2.1   METABOLIC PROCESSES

Magnesium has major physiological and molecular roles in plants, such as being a component of the chlorophyll molecule, a cofactor for many enzymatic processes associated with phosphorylation, dephosphorylation, and the hydrolysis of various compounds, and as a structural stabilizer for various nucleotides. Studies indicate that 15 to 30% of the total magnesium in plants is associated with the chlorophyll molecule (26,27). In citrus (*Citrus volkameriana* Ten. & Pasq.), magnesium deficiency was associated directly with lower total leaf chlorophyll (28); however, there were no effects on chlorophyll *a/b* ratios within the magnesium-deficient leaves.

The other 70 to 85% of the magnesium in plants is associated with the role of magnesium as a cofactor in various enzymatic processes (1,2,26,29), the regulation of membrane channels and receptor proteins (30,31), and the structural role in stabilizing proteins and the configurations of DNA and RNA strands (32,33). Since magnesium is an integral component of the chlorophyll molecule and the enzymatic processes associated with photosynthesis and respiration, the assimilation of carbon and energy transformations will be affected directly by inadequate magnesium. In nutrient film-grown potato (*Solanum tuberosum* L.), relatively low (0.05 mM) or high (4.0 mM) magnesium concentrations increased dark respiration rates and decreased photosynthetic rates relative to magnesium fertilization rates ranging from 0.25 to 1.0 mM (34). In hydroponically grown sunflower (*Helianthus annuus* L.), photosynthetic rates decreased in ammonium-fertilized, but not nitrate-fertilized plants when the magnesium concentration of nutrient solutions decreased below 2 mM (35). This effect was related to the decreased enzymatic activity as well as the decrease in photosynthetic capacity due to the loss in assimilating leaf area, occurring mainly as a consequence of leaf necrosis and defoliation (36).

Magnesium may also influence various physiological aspects related to leaf water relations (37,38). In hydroponically grown tomato (*Lycopersicon esculentum* Mill.), increasing magnesium fertilization from 0.5 to 10 mM resulted in an increase in leaf stomatal conductance (Gs) and turgor potential (Ψ_p) and a decrease in osmotic potential (Ψ_π) but had no effect on leaf water potential (Ψ_w) (37). In other studies (38) where low leaf water potentials were induced in sunflower (*Helianthus annuus* L.) leaves, the increased magnesium concentrations in the stroma, caused by decreased stroma volume due to dehydration, caused magnesium to bind to the chloroplast-coupling factor, thereby inhibiting the ATPase activity of the enzyme and inhibiting photophosphorylation. Other experiments (39–41) have indicated that even though up to 1.2 mM magnesium may be required in the ATPase complex of photophosphorylation, magnesium concentrations of 5 mM or higher result in conformational changes in the chloroplast-coupling factor, which causes inhibition of the ATPase enzyme.

As regards to the role of magnesium in molecular biology, magnesium is an integral component of RNA, stabilizing the conformational structure of the negatively charged functional groups and also concurrently neutralizing the RNA molecule (42–44). In many cases, the role of the magnesium ion in the configurations and stabilities of many polynucleotides is not replaceable with other cations, since the ligand configurations are of a specific geometry that are capable of housing only magnesium ions (45). In addition, magnesium serves as a cofactor for enzymes that catalyze the hydrolysis and formation of phosphodiester bonds associated with the transcription, translation, and replication of nucleic acids (1,2).

6.2.2 GROWTH

Magnesium deficiency may suppress the overall increase in plant mass or specifically suppress root or shoot growth. However, the extent of growth inhibition of roots and shoots will be influenced by the severity of the magnesium deficiency, plant type, stage of plant development, environmental conditions, and the general nutritional status of the crop. In tomato, suboptimal magnesium concentrations did not affect overall plant growth (37); however, an accumulation of assimilates occurred in the shoots, suggesting that assimilate transport from the shoots to the roots was impaired. For birch (*Betula pendula* Roth.) seedlings, decreased magnesium availability in the rhizosphere had no effect on root branching pattern but decreased root length, root diameter, and root dry weight (36). In addition, the fraction of dry matter allocated to the leaves increased even though overall leaf area decreased (36). In raspberry (*Rubus* spp. L.), enhanced shoot growth was correlated with increased magnesium in the leaves (46,47).

6.2.3 FRUIT YIELD AND QUALITY

Magnesium deficiencies and toxicities may decrease fruit yield and quality. In two cultivars of apple (*Malus pumila* Mill.), fruit magnesium concentrations were correlated negatively with fruit color, whereas fruit potassium concentrations were positively correlated with fruit color (48). The effects of magnesium on apple fruit quality may have been due to antagonistic effects on potassium uptake and accumulation. In tomato, even though increasing magnesium fertilization rates did not affect total shoot dry weight, overall fruit yield decreased with increased magnesium fertilization supply from 0.5 to 10 mM (37).

6.3 DIAGNOSIS OF MAGNESIUM STATUS IN PLANTS

6.3.1 SYMPTOMS OF DEFICIENCY AND EXCESS

6.3.1.1 Symptoms of Deficiency

In a physiological sense, magnesium deficiency symptoms are expressed first as an accumulation of starch in the leaves (49), which may be associated with early reductions in plant growth and decreased allocation of carbohydrates from leaves to developing sinks (50). This process is followed by the appearance of chlorosis in older leaves, patterns of which can be explained by the physiological processes associated with magnesium uptake, translocation, and metabolism in plants (3–5,49). Magnesium is physiologically mobile within the plant. Therefore, if insufficient magnesium is available from the rhizosphere, magnesium can be reallocated from other plant parts and transported through the phloem to the actively growing sinks. Because of this mobility within the plant, symptoms of deficiency will first be expressed in the oldest leaves (Figure 6.1). Early symptoms of magnesium deficiency may be noted by fading and yellowing of the tips of old leaves (49,51,52), which progresses interveinally toward the base and midrib of leaves, giving a mottled or herringbone appearance (52). In later stages of development, deficiency symptoms may be difficult to distinguish from those of potassium deficiency. Under mild deficiencies, a 'V'-patterned interveinal chlorosis develops in dicots as a result of magnesium dissociating from the chlorophyll, resulting in chlorophyll degradation. In conifers, minor magnesium deficiency symptoms are browning of older needle tips (0.10% magnesium concentration) and in more severe deficiencies, the enter needle turns brown and senesce (0.07% magnesium concentration) (49,53). In some plants, a reddening of the leaves may occur, rather than chlorosis, as is the case for cotton (*Gossypium* spp.) (52,54), since other plant pigments may not break down as quickly as chlorophyll. The loss of protein from magnesium-deficient leaves, however, usually results in the loss of plastic pigments from most plants (55). On an individual leaf, as well as on a whole plant basis, deficiency

FIGURE 6.1 Symptoms of magnesium deficiency on (left) pepper (*Capsicum annum* L.) and (right) cucumber (*Cucumis sativus* L.). (Photographs by Allen V. Barker.) (For a color presentation of this figure, see the accompanying compact disc.)

symptoms may begin to appear only on the portions of a leaf or the plant that are exposed to the sun, with the shaded portions of leaves remaining green (49,56). Under severe deficiency symptoms, all lower leaves become necrotic and senesce (28,36) with symptoms of interveinal yellowing progressing to younger leaves (36,56).

Magnesium has functions in protein synthesis that can affect the size, structure, and function of chloroplasts (26). The requirement of magnesium in protein synthesis is apparent in chloroplasts, where magnesium is essential for the synthesis and maintenance of proteins in the thylakoids of the chlorophyll molecule (57–59). Hence, the degradation of proteins in chloroplasts in magnesium-deficient plants may lead to loss of chlorophyll as much as the loss of magnesium for chlorophyll synthesis.

On a cellular level, magnesium deficiency causes the formation of granules of approximately 80 nm in diameter in the mitochondria and leads to the disruption of the mitochondrial membrane (60). In the chloroplasts, magnesium deficiency results in reduced and irregular grana and reduced or nonexistent compartmentation of grana (61). Palomäki (53) noted that chloroplasts were rounded and thylakoids were organized abnormally in magnesium-deficient Scots pine (*Pinus sylvestris* L.) seedlings. In the vascular system, magnesium deficiency may cause swelling of phloem cells and collapse of surrounding cells, collapse of sieve cells, and dilation of proximal cambia and parenchyma cells in conifers (53). These alterations at the cellular level occurred before visual changes were evident and before a detectable decrease in leaf magnesium occurred.

6.3.1.2 Symptoms of Excess

During the early 1800s, symptoms of 'magnesium' toxicity in plants were described; however, during this time, manganese was called magnesium and magnesium was referred to as magnium or magnesia (3–5). Because of the confusion in nomenclature, early reports regarding magnesium and manganese should be read carefully. At the present time, no specific symptoms are reported directly related to magnesium toxicity in plants. However, relatively high magnesium concentrations can elicit deficiency symptoms of other essential cations. Plant nutrients that are competitively inhibited for absorption by relatively high magnesium concentrations include calcium and potassium and occasionally iron (62). Therefore, symptoms of magnesium toxicity may be more closely associated with deficiency symptoms of calcium or potassium.

6.3.2 ENVIRONMENTAL CAUSES OF DEFICIENCY SYMPTOMS

Conditions of the soil and rhizosphere such as drought or irregular water availability (63,64), poor drainage or excessive leaching (11), low soil pH (65–67), or cold temperatures (68,69) will exaggerate magnesium deficiency symptoms, as magnesium is not physically available under these environmental conditions or physiologically, the plant roots are not capable of absorbing adequate magnesium to sustain normal plant growth.

Conditions of the soil and rhizosphere such as drought or irregular water availability will impact magnesium uptake. In sugar maple (*Acer saccharum* Marsh.), foliar analysis indicated that magnesium deficiency occurred during drought (64). Likewise, Huang (63) reported that drought-stressed tall fescue (*Festuca arundinacea* Schreb.) had lower leaf magnesium concentrations than well-watered fescue.

Low soil pH is also associated with a low supply or depletion of magnesium, possibly due to leaching; however, research suggests that impairment of root growth in acid soils (pH 4.3 to 4.7) also may hinder magnesium absorption (67). In one study (65), low soil pH (3.0) resulted in increased accumulation of magnesium in the shoots, but decreased accumulation in the roots. Contradicting Marler (65) and Tan et al. (67), Johnson et al. (70) found no clear correlation between low soil pH and magnesium accumulation.

Relatively high and low root-zone temperatures affect magnesium uptake, but the degree of impact may be influenced by plant type and stage of plant development. Huang et al. (71) and

Huang and Grunes (68) reported that increasing root-zone temperature (10, 15, 20°C) linearly increased magnesium accumulation by wheat seedlings that were less than 30 days old but suppressed accumulation by seedlings that were more than 30 days old. Similarly, magnesium uptake decreased when temperatures in the rhizosphere decreased from 20 to 10°C (69).

Although any environmental condition such as unfavorable soil temperature or pH may reduce root growth and thus reduce magnesium uptake, other characteristics such as mycorrhizal colonization can increase magnesium uptake. Likewise, it has been shown that plants that have colonization of roots by mycorrhiza show higher amounts of magnesium accumulation relative to nonmycorrhizal plants (72–75).

Shoots exposed to environmental parameters such as high humidity (76), high light intensity (77,78), or high or low air temperatures (79) will decrease the ability of plants to absorb and translocate magnesium, since transpiration is reduced and the translocation of magnesium is driven by transpiration rates (63,76,80–84).

Light intensity can affect the expression of symptoms of magnesium deficiency. Partial shading of magnesium-deficient leaves has been shown to prevent or delay the development of chlorosis (77). Others (49,56) have also determined that magnesium deficiency symptoms may begin to appear only on the portions of a leaf or plant that are exposed to the sun, with the shaded portions of leaves remaining green. Zhao and Oosterhuis (78) also reported that shading (63% light reduction) increased leaf-blade concentrations of magnesium in cotton plants by 16% relative to unshaded plants.

6.3.3 Nutrient Imbalances and Symptoms of Deficiency

Magnesium deficiency symptoms may be associated with an antagonistic relationship between magnesium ions (Mg^{2+}) and other cations such as hydrogen (H^+), ammonium (NH_4^+), calcium (Ca^{2+}), potassium (K^+), aluminum (Al^{3+}), or sodium (Na^+). The competition of magnesium with other cations for uptake ranges from highest to lowest as follows: $K > NH_4^+ > Ca > Na$ (85,86). These cations can compete with magnesium for binding sites on soil colloids, increasing the likelihood that magnesium will be leached from soils after it has been released from exchange sites. Within the plant, there are also antagonistic relationships between other cations and magnesium regarding the affinity for various binding sites within the cell membranes, the degree of which is influenced by the type of binding site (lipid, protein, chelate, etc.), and the hydration of the cation (87). These biochemical interactions result in competition of other cations with magnesium for absorption into the roots and translocation and assimilation in the plant (88–92).

6.3.3.1 Potassium and Magnesium

Increased potassium fertilization or availability, relative to magnesium, will inhibit magnesium absorption and accumulation and vice versa (34,35,90,93–99). The degree of this antagonistic effect varies with potassium and magnesium fertilization rates, as well as the ratio of the two nutrients to one another. This phenomenon has been documented in tomato (62,96), soybean (*Glycine max* Merr.), (93,100), apple (101), poplar (*Populus trichocarpa* Torr. & A. Gray) (102), Bermuda grass (*Cynodon dactylon* Pers.) (103–105), perennial ryegrass (*Lolium perenne* L.) (18), buckwheat (*Fagopyrum esculentum* Moench) (93), corn (*Zea mays* L.) (98), and oats (*Avena sativa* L.) (93). Potassium chloride fertilization increased cotton (*Gossypium hirsutum* L.) plant size and seed and lint weight and increased efficiency of nitrogen use, but had suppressive effects on magnesium accumulation in various plant parts (106). Fontes et al. (107) reported that magnesium concentrations of potato (*Solanum tuberosum* L.) petioles declined as potassium fertilization with potassium sulfate increased from 0.00 to 800 kg K ha^{-1}. Legget and Gilbert (100) noted that with excised roots of soybean, magnesium uptake was inhibited if calcium and potassium were both present but not if calcium or potassium was present alone. The opposite also holds true in that potassium and calcium contents of roots were depressed with increasing rates of magnesium fertilization (100). Similar results were obtained in potatoes (*Solanum tuberosum* L.) where increasing magnesium fertilization from 0.05 to 4.0 mM decreased the potassium concentration in shoots from 76.6 to 67.6 mg g^{-1} shoot dry weight (34).

6.3.3.2   Calcium and Magnesium

High rhizosphere concentrations of calcium, relative to magnesium, are inhibitory to the absorption of magnesium and vice versa (34,35,37,86,90,108–110). In the early 1900s, the importance of proper ratios of magnesium to calcium in soils was emphasized through studies conducted by Loew and May (4) on the relationships of lime and dolomite. High calcium concentrations in solution or in field soils sometimes limit magnesium accumulation and may elicit magnesium deficiency symptoms (111–113). In tomato, the magnesium concentration in shoots (62) and fruits (114) decreased as the calcium fertilization rate increased. Similarly, it was shown that increased calcium concentrations inhibited magnesium uptake in common bean (*Phaseolus vulgaris* L.) (86). On the other hand, decreased accumulation of calcium in birch was directly correlated with the decreased absorption and accumulation of calcium as magnesium fertilization rates increased (36). The absorption of calcium decreased from 1.5 to 0.3 mmol g^{-1} root mass as magnesium fertilization increased (36). Morard et al. (115) reported a strong antagonism between calcium and magnesium, suggesting that calcium influenced magnesium translocation to leaves. Optimum leaf Ca/Mg ratios are considered to be approximately 2:1; however, Ca/Mg ratios >1:1 and <5:1 can produce adequate growth without the expression of magnesium deficiency (36,85). In a study with tomato, the root, stem, and leaf calcium concentrations decreased as fertilization rates increased from 0.50 to 10.0 mM Mg in solution culture (37). Similarly, with woody ornamentals, high fertilization rates of calcium relative to magnesium inhibited the accumulation of magnesium and decreased root and shoot growth, and inversely, high magnesium decreased calcium accumulation and plant growth (35,109). Clark et al. (116) used flue-gas desulfurization by-products to fertilize corn in greenhouse experiments. They noted that the materials needed to be amended with magnesium at a ratio of 1 part magnesium to 20 parts of calcium to avoid magnesium deficiency in the corn. In containerized crop production, general recommendations indicate sufficient calcium and magnesium additions to produce an extractable Ca/Mg ratio of 2:5 (117). Navarro et al. (118) reported an antagonist effect of calcium on magnesium accumulation in melon (*Cucumis melo* L.), regardless of salinity levels imposed by sodium chloride. In other studies (119–121), it was shown that even with the use of dolomitic lime, magnesium deficiency might occur. This occurrence is due to the different solubilities of magnesium carbonate (MgCO$_3$) and calcium carbonate (CaCO$_3$) in the dolomite. Therefore, during the first 4 months, both magnesium and calcium solubilized from the dolomite. However, after 4 months, all of the magnesium had dissolved from the dolomite, leaving only Ca from the CaCO$_3$ available for dissolution and availability to the plant (119,120). Based on these studies, it appears that the use of solid calcium and magnesium fertilizers with similar solubility rates may be important so that both elements are available in similar and sufficient levels throughout the entire crop production cycle (119–121).

6.3.3.3   Nitrogen and Magnesium

Nitrogen may either inhibit or promote magnesium accumulation in plants, depending on the form of nitrogen: with ammonium, magnesium uptake is suppressed and with nitrate, magnesium uptake is increased (35,101,122–124). In field soils, the chances of ammonium competing with magnesium for plant uptake are more likely to occur in cool rather than warm soils because in warmer soils, most ammonium is converted into nitrate by nitrification processes. In forests, high inputs of ammoniacal nitrogen amplified latent magnesium deficiency (125). In conditions of sand culture, ammonium-nitrogen of Norway spruce (*Picea abies* Karst.) resulted in significantly lower magnesium and chlorophyll concentrations in current-year and year-old needles compared to fertilization with nitrate-nitrogen (126). Similarly, in herbaceous plants such as wheat (*Triticum aestivum* L.) (127) and bean (*Phaseolus vulgaris* L.) (128), ammoniacal nitrogen reduced shoot accumulation of magnesium (127). In cauliflower (*Brassica oleracea* var. *botrytis* L.), increasing nitrate-nitrogen fertilization from 90 to 270 kg ha^{-1} increased yield response to increased magnesium fertilization rates (22.5 to 90 kg ha^{-1}) (129). Similarly, in hydroponically grown poinsettia (*Euphorbia pulcherrima* Willd.), magnesium concentrations in leaves increased as the proportion of nitrate-nitrogen to ammonium-nitrogen increased, even though all treatments received the same amount of total nitrogen (130). In a similar way, magnesium fertilization increased the plant accumulation of nitrogen, which was applied as urea, in rice (*Oryza sativa* L.) (131). As with other nutrients, the degree of impact of nitrogen on magnesium nutrition is influenced by the concentrations of the nutrients, relative to each other. For example, Huang et al. (71) demonstrated with hydroponically grown wheat that nitrogen form had no significant effect on shoot magnesium levels when magnesium concentrations in solutions were relatively high (97 mg L^{-1}); however, at low magnesium concentrations (26 mg L^{-1}) in solutions, increasing the proportion of ammonium relative to nitrate significantly decreased shoot Mg concentrations. In another study, Huang and Grunes (68) also noted that even though magnesium uptake rates were significantly higher for plants supplied with nitrate rather than ammonium, increasing the proportion of the nitrogen supply as nitrate decreased net magnesium translocation to the shoots.

6.3.3.4 Sodium and Magnesium

High soil or nutrient-solution salinity levels (with NaCl), relative to magnesium supply, may inhibit magnesium accumulation in plants (132–135). However, results are variable since salinity often inhibits plant growth; therefore, there may be a reduction in the total uptake of a nutrient into a plant. However, since the plant is smaller, the magnesium level, expressed in terms of concentration, may be higher. Application of sodium-containing fertilizers (chloride or nitrate) lowered the concentration of magnesium in white clover (*Trifolium repens* L.) leaves but increased the magnesium in perennial ryegrass (*Lolium perenne* L.) (133). In hydroponically grown taro (*Colocasia esculenta* Schott.) (136) and wheat (137), sodium chloride treatments resulted in a suppression of leaf magnesium. Use of sodium chloride to suppress root and crown rot in asparagus (*Asparagus officinalis* L. var. *altilis* L.) also suppressed magnesium accumulation in the leaves (138). Even in a halophyte such as *Halopyrum mucronatum* Stapf., increasing sodium chloride concentrations in nutrient solutions from 0.0 to 5220 mg L^{-1} significantly decreased magnesium concentrations in the shoots and roots (134). However, in hydroponically grown bean (*Phaseolus vulgaris* L.), sodium chloride increased leaf concentrations of magnesium, perhaps as a result of growth suppression (139). Growth suppression of rice was associated with salinity, but the levels of magnesium in the leaves were unaffected (140). Other research (141) found that sodium chloride increased accumulation of magnesium in shoots but suppressed magnesium accumulation in roots of strawberry (*Fragaria chiloensis* Duchesne var. *ananassa* Bailey). In fact, some (142) have attributed the salt tolerance of some soybean cultivars to the ability to accumulate potassium, calcium, and magnesium, in spite of saline conditions.

6.3.3.5 Iron and Magnesium

Uptake and accumulation of iron may be inhibited or unaffected by increased magnesium fertilization. In addition, the translocation of magnesium from the roots to the shoots may decrease in iron-deficient plants relative to iron-sufficient plants (143). The antagonistic relationship of iron with magnesium has been demonstrated in tomato (62) and radish (*Raphanus sativus* L.) (144). Nenova and Stoyanov (143) noted that the uptake and translocation of magnesium was reduced in iron-deficient plants compared to iron-sufficient plants. However, Bavaresco (145) reported that under lime-induced chlorosis, chlorotic grape (*Vitis vinifera* L.) leaves did not differ from green leaves in nutrient composition, but the fruits of chlorotic plants were different in that they had higher magnesium than fruits from normal plants. Iron concentrations did not differ among any of the tissues.

6.3.3.6   Manganese and Magnesium

Manganese, as a divalent cation, can compete with magnesium for binding sites on soil particles as well as biological membranes within plants (146). However, manganese is required in such small quantities (micromolar concentrations in nutrient solutions resulting in Manganese, as a divalent cation, can compete with magnesium for binding sites on soil particles as well as biological membranes within plants (146). However, manganese is required in such small quantities (micromolar concentrations in nutrient solutions resulting in $\approx$ 20 to 500 ppm in most plant tissues) that manganese toxicity usually occurs before quantities are high enough to significantly inhibit magnesium uptake to physiologically deficient levels (62,85). However, some experiments (147,148) have demonstrated that manganese can inhibit magnesium uptake. However, Alam et al. (147) and Qauartin et al. (148) did not indicate if the inhibition of magnesium was substantial enough to induce magnesium deficiency symptoms. On the other hand, increased magnesium fertilization has been shown to decrease manganese uptake and accumulation (34,80), and in some cases, magnesium fertilization may mitigate manganese toxicity (149,150). In one study (151), the tolerance of certain cotton (*Gossypium hirsutum* L.) cultivars to manganese appeared to be related to the ability to accumulate more magnesium than by the manganese-sensitive cultivars.

6.3.3.7   Zinc and Magnesium

As with manganese, zinc is a divalent cation that is required in minuscule quantities for normal plant growth. Therefore, plants usually suffer from zinc toxicity before concentrations are high enough to inhibit magnesium uptake. However, some research has indicated that as zinc increases to toxic levels in plants, the accumulation of magnesium is suppressed, but not to the degree of inducing magnesium deficiency symptoms. In hydroponically grown tomato (62), increasing zinc concentrations from 0.0 to 1.58 mg L^{-1} did not affect magnesium concentrations in shoots. Similarly, nontoxic levels of zinc applications through zinc-containing fungicides or fertilization (soil or foliar applied) did not affect magnesium concentrations in potato leaves, although zinc concentrations increased in leaves (152). However, at higher zinc concentrations (30 vs. 0.5 mg L^{-1}), magnesium accumulation in tomato leaves and fruit was inhibited (153). Similarly, with blackgram (*Vigna mungo* L.) grown in soil, accumulation of zinc in plants led to a suppression of magnesium, calcium, and potassium in leaves (154). Bonnet et al. (155) also reported that zinc fertilization of ryegrass (*Lolium perenne* L.) lowered magnesium content of leaves, in addition to lowering the efficiency of photosynthetic energy conversion, and elevating the activities of ascorbate peroxidase and superoxide dismutase. Conversely, pecan (*Carya illinoinensis* K. Koch) grown under zinc-deficient conditions had higher leaf magnesium than trees grown under zinc-sufficient conditions (156). However, in nutrient film-grown potatoes (*Solanum tuberosum* L.), increased levels of magnesium fertilization (1.2 to 96.0 mg L^{-1}) did not affect zinc concentrations in tissues.

6.3.3.8   Phosphorus and Magnesium

Phosphate ions have a synergistic effect on accumulation of magnesium in plants, and vice versa. This phenomenon is associated with the ionic balance related to cation and anion uptake into plants as well as the increased root growth sometimes observed with increased phosphorus fertilization. For example, with hydroponically grown sunflower (*Helianthus annuus* L.), phosphorus accumulation increased in tissues from 9.0 to 13.0 mg g^{-1} plant dry weight as magnesium concentrations in nutrient solutions were increased from 0.0 to 240 mg L^{-1} (35). Likewise, increasing phosphorus fertilization increases magnesium accumulation, as demonstrated in field-grown alfalfa (*Medicago sativa* L.) (157). The effect of phosphorus fertilization increasing magnesium uptake has also been documented in rice (*Oryza sativa* L.), wheat (*Triticum aestivum* L.), bean (*Phaseolus vulgaris* L.), and corn (*Zea mays* L.) (158). Reinbott and Blevins (82,159) reported that phosphorus fertilization of field-grown wheat (*Triticum aestivum* L.) and tall fescue (*Festuca arundinacea* Shreb.) increased leaf calcium and magnesium accumulation and concluded that proper phosphorus nutrition may be more important than warm root temperatures in promoting magnesium and calcium accumulation, particularly if soils have suboptimal phosphorus concentrations. Reinbott and Blevins (160) also showed a positive correlation between calcium and magnesium accumulation in shoots with increased phosphorus fertilization of hydroponically grown squash (*Cucurbita pepo* L.).

6.3.3.9 Copper and Magnesium

Like other micronutrients, copper is a plant nutrient, which is required in such low concentrations relative to the requirements for magnesium that high copper fertilization is more likely to induce copper toxicity before causing magnesium deficiency symptoms. However, some studies have shown that copper may competitively inhibit magnesium accumulation in plants (161,162). In taro (*Colocasia esculenta* Schott), increasing the nutrient solution copper concentrations from 0.03 to 0.16 mg L^{-1}, significantly decreased the accumulation of magnesium in leaves from 5.5 to 4.4 mg g^{-1} dry weight (161). In a study (162) using young spinach (*Spinacia oleracea* L.), where copper concentrations in nutrient solutions were increased from 0.0 to 10.0 mg L^{-1}, which is two orders of magnitude greater than the copper concentrations used in the study conducted by Hill et al. (2000), copper toxicity symptoms did occur, and there was a significant suppression in magnesium accumulation in the leaves and roots from 322 and 372 mg kg^{-1} to 41 and 203 mg kg^{-1}, respectively (162). However, the magnesium concentration reported in this study (162) is an order of magnitude lower than what is found typically in most herbaceous plants (85). On the other hand, effects of magnesium fertilization on copper uptake are not documented, although one study (34) indicated that increasing rates of magnesium fertilization did not significantly reduce the uptake and accumulation of copper.

6.3.3.10 Chloride and Magnesium

The effects of chloride on magnesium accumulation in plants have been studied in relation to the effects of salinity on growth and nutrient accumulation. In many of these studies, it is difficult to separate the effects of chloride from those of sodium ions; hence, many of the results show a depression of magnesium accumulation with increases in sodium chloride concentration in the root zone (132–135). In grapes (*Vitis vinifera* L.), salinity from sodium chloride did not affect magnesium concentrations in leaves, trunk, or roots (163). With tomato, increased magnesium fertilization rates did not increase the accumulation of chlorine in the leaves, stems, or roots (37). With soybean, uptake of chloride by excised roots was low from magnesium chloride solutions but was enhanced by the addition of potassium chloride (100).

6.3.3.11 Aluminum and Magnesium

Free aluminum in the soil solution inhibits root growth, which in turn will reduce ability of plants to take up nutrients (164). Research with red spruce (*Picea rubens* Sarg.) indicated that magnesium concentrations in roots and needles of seedlings were suppressed by exposure to ≈ 400 μM aluminum in nutrient solutions (165,166). Increasing concentrations of free aluminum have also been shown to reduce magnesium accumulation in taro (167), maize (*Zea mays* L.) (168,169), and wheat (*Triticum aestivum* L.) (170). Aluminum-induced magnesium deficiency may be one mechanism of expression of aluminum toxicity in plants, and aluminum tolerance of plants may be related to the capacity of plants to accumulate magnesium and other nutrients in the presence of aluminum (67,95,168,170–172). Some studies (173) have shown that the toxic effects of aluminum were reduced when magnesium was introduced into the nutrient solution and subsequently increased the production and excretion of citrate from the root tips. The authors (173) hypothesized that the citrate binds with free aluminum, forming nontoxic aluminum–citrate complexes. Keltjens (168) also reported that aluminum chloride in solution culture restricted magnesium absorption by corn but that aluminum citrate or organic complexes did not inhibit magnesium absorption and were not phytotoxic.

Sensitivity to aluminum toxicity may or may not be cultivar-specific. In a study (170) with wheat, differences in magnesium accumulation occurred for different cultivars, with a significantly greater accumulation of magnesium in the leaves of the aluminum-tolerant 'Atlas 66' compared to the aluminum-sensitive 'Scout 66' and increasing the magnesium concentration in nutrient solutions relative to aluminum and potassium concentrations increased the aluminum tolerance of 'Scout 66' (170). However, in another study (174) with aluminum-tolerant and aluminum-sensitive corn cultivars, increasing concentrations of aluminum resulted in higher nutrient concentrations in the shoots of aluminum-sensitive than in the aluminum-tolerant cultivar, probably the result of a greater suppression of growth in the sensitive cultivar.

6.3.4 Phenotypic Differences in Accumulation

The uptake and accumulation of magnesium may change during different stages of physiological development. Knowledge of these changes is important in managing nutritional regimes for plant growth and for sampling of plants to assess their nutritional status. In poinsettias, magnesium accumulation was greatest from the period of flower induction to the visible bud stage, but then accumulation decreased during the growth phase of visible bud to anthesis (130). With cotton (*Gossypium hirsutum* L.), maximum daily influx of magnesium into roots occurred at peak bloom (175). Accumulation (net influx) of magnesium in annual ryegrass (*Lolium multiflorum* Lam.) decreased with increasing plant age (176,177). Similarly, magnesium uptake rates by tomato decreased from 68 to 17.5 μ eq g^{-1} fresh weight per day as the plants aged from 18 to 83 days (110). With anthurium (*Anthurium andraeanum* Lind.), changes in the allocation of magnesium to different organs with increased plant age were attributed to transport of nutrients from lower leaves to the flowers, resulting in a lowering of magnesium concentrations in the lower leaves (178). Tobacco (*Nicotiana tabacum* L.) showed decreasing concentrations of leaf magnesium from base to top of the plants over the growing season, and stem magnesium concentrations also fell with plant age (179). Sadiq and Hussain (180) attributed the decline in magnesium concentration in bean (*Phaseolus vulgaris* L.) plants to a dilution effect from plant growth. However, Jiménez et al. (181) reported no significant differences in shoot-tissue magnesium concentrations throughout the different growth stages of different soybean cultivars.

6.3.5 Genotypic Differences in Accumulation

Variation in magnesium accumulation might occur for different cultivars or plant selections within a species. In a 2-year study with field-grown tomato plants in an acid soil, magnesium concentration of leaves was significantly greater in cultivar 'Walter' (1.1%) than in 'Better Boy' (0.9%) in a dry, warm year, but no differences (average 0.6%) occurred between the cultivars in a wetter, cooler year that followed (182). Mullins and Burmester (183) noted that cotton cultivars differed in concentrations of magnesium in leaves and burs under nonirrigated conditions. Differences in magnesium concentrations in different cultivars of Bermuda grass (*Cynodon dactylon* Pers.) have been reported (184). Rosa et al. (185) suggested that variation in calcium, magnesium, and sulfur among broccoli (*Brassica oleracea* var. *italica* Plenck) varieties justifies selection of a particular cultivar to increase dietary intake of these elements. Likewise, in different wheat (*Triticum aestivum* L.) (170) and barley (*Hordeum vulgare* L.) (171) cultivars, aluminum tolerance was associated with the ability to take up and accumulate magnesium under conditions of relatively high aluminum concentrations (1.35 to 16.20 mg L^{-1}) in the rhizosphere. Similar studies (94) have been conducted to select clonal lines of tall fescue (*Festuca arundinacea* Schreb.), which display higher accumulation of magnesium, in an effort to prevent magnesium tetany in grazing animals.

6.4 CONCENTRATIONS OF MAGNESIUM IN PLANTS

6.4.1 MAGNESIUM CONSTITUENTS

Magnesium is present in the plant in several biochemical forms. In studies with forage grasses, magnesium was measured in water-soluble, acetone-soluble, and insoluble constituents (18). These forms are present in the phloem, xylem, cytoplasm (water-soluble fraction), chlorophyll (acetone-soluble fraction), and cell wall constituents (insoluble fraction).

6.4.1.1 Distribution in Plants

The quantity of magnesium accumulated will differ for various plant organs, with a tendency toward greater allocation of magnesium in transpiring organs such as leaves and flowers, rather than the roots (186–188); however, this translocation to different plant parts may be affected by the status of other elements in the plant (143,164,189). Similarly, the ability of magnesium to remobilize and translocate out of a particular plant organ may vary among plant organs (186,187). In cucumber, magnesium concentrations were seven times higher in the shoots (70 μmol g^{-1} fresh weight) than in the roots (10 μmol g^{-1} fresh weight) (190). In native stands of 13-year-old Hooker's Banksia (*Banksia hookeriana* Meissn.), magnesium was distributed to different plant organs as follows (mg g^{-1} dry weight): 0.99 in stems, 1.41 in leaves, and 0.73 in reproductive structures, which account for 54, 21, and 25% of the total magnesium content, respectively (191). In walnut (*Juglans regia* L.), magnesium remobilization from catkins was less than that from leaves (186,187). Additional studies (192) indicate that the magnesium concentration in the seeds of several halophytes ranged from 0.22 to 0.90% for forbs and 0.07 to 0.97% for grasses (192). In corn (*Zea mays* L.), less magnesium was translocated from the roots to the shoots for iron-deficient plants than with plants with sufficient iron (143). In a similar manner for hydroponically grown tomatoes, increasing potassium concentrations of nutrient solutions resulted in decreased magnesium concentration in leaves and roots, but increased magnesium concentrations in fruits and seeds (193).

Although magnesium accumulates to higher levels in aboveground organs than in belowground organs, there may also be spatial differences in magnesium accumulation within a particular organ (194). In corn leaves, magnesium concentration decreased from the leaf tip to the leaf base (194). The relative distribution of magnesium within plants may be altered by magnesium fertilization rates as well as the fertilization rates of other nutrients. Other environmental stresses, such as iron deficiency, have also been shown to modify the spatial gradient of magnesium concentrations along the leaf blade of corn (194).

6.4.1.2 Seasonal Variations

In perennial ryegrass (18) and walnut (186,187), magnesium concentration increased throughout the growing season. For field-grown soybeans, there was an indication that magnesium was remobilized from stems and leaves and translocated to developing pods later in the growing season (195), since stems and leaf tissue magnesium concentrations decreased from approximately 0.70% to less than 0.50% as pod magnesium concentrations increased from 0.48 to 0.51%, indicating a remobilization of magnesium from vegetative to reproductive tissue. However, the degrees of differences were affected by soil type and irrigation frequency (195).

6.4.1.3 Physiological Aspects of Magnesium Allocation

Physiologically, certain stages of plant development, such as flowering and fruiting, may make plants more susceptible to magnesium deficiencies. In camellia (*Camellia sasanqua* Thunb.

'Shishi Gashira'), magnesium deficiency may be expressed after flowering, as the first vegetative flush commences in the spring (56). This expression appears to be attributed to the large flowers of 'Shishi Gashira' acting as sinks for magnesium. After flowering, when magnesium reserves in the plants are low, plants may be markedly susceptible to magnesium deficiency and may develop typical magnesium deficiency symptoms if sufficient magnesium is not available in the soil for uptake. Similarly, in cucumber, magnesium concentration in leaves increased with leaf age, until flowering and fruiting, at which point concentrations increased in the younger leaves (190). In grapes (*Vitis vinifera* L.), the magnesium concentration (10.1 mg/cluster) of ripening berries of 'Pinot Blanc,' a cultivar that is susceptible to lime-induced chlorosis during ripening, was significantly higher than the magnesium concentration (7.1 mg/cluster) for berries of the lime-tolerant cultivar 'Sauvignon Blanc' (145). However, in blades and petioles, there were no differences in magnesium concentrations (145). In other grape cultivars ('Canadian Muscat' and 'Himrod') that are susceptible to berry drop and rachis necrosis, spray applications of magnesium were shown to increase berry yield through the alleviation of rachis necrosis and berry drop (196). A similar observation was noted on grapefruit (*Citrus paradisi* Macfady) trees by Fudge (197). As fruit and seed development occurred, a depletion of magnesium from leaves near to the fruits was apparent, as only the leaves in proximity to the fruits expressed magnesium deficiency symptoms.

6.4.2 CRITICAL CONCENTRATIONS

6.4.2.1 Tissue Magnesium Concentration Associations with Crop Yields

The magnesium concentration of tissues considered as deficient, sufficient, or toxic depends on what growth parameter is being measured in the crops. In many food crops, classification of nutrient sufficiency is based on harvestable yields and quality of the edible plant parts (198). In ornamental plants, sufficiency values are based on plant growth rate and visual quality of the vegetative and reproductive organs. In forestry, ratings are based on rate of growth and wood quantity and quality. For example, in birch (*Betula pendula* Roth.) seedlings, magnesium sufficiency levels in leaves were correlated with relative growth rate (36). Based on their studies, maximum growth rate was correlated with a mature healthy leaf magnesium concentration of 0.14%, a concentration that was considered deficient for rough lemon (*Citrus jambhiri* Lush.) production (28). Austin et al. (199) reported that magnesium concentrations in taro (*Colocasia esculenta* Schott) varied from 0.07 to 0.42% with hydroponically grown plants and noted that growth parameters (biomass, leaf area, nutrient concentrations) did not vary as the magnesium in solution varied from 1.20 to 19.2 mg L^{-1}. In corn, optimal leaf magnesium concentrations were determined to range between 0.13 and 0.18% for maximum corn yields (198). With peach (*Prunus persica* Batsch.), the critical concentration or marginal level of magnesium in leaves was determined to be about 0.2% of the dry mass based on the appearance of symptoms of deficiency but with no growth suppression at this concentration (200).

6.4.2.2 Tabulated Data of Concentrations by Crops

In most commercially grown crops, magnesium concentrations average between 0.1 and 0.5% on a dry weight basis (29). However, total magnesium concentration may vary considerably between different plant families. The legumes (Leguminosae or Fabaceae) can have nearly double the magnesium concentration as most cereal crops (201). Likewise, oil seed crops and root crops can also contain high concentrations of magnesium (201). A tabulated description of magnesium concentrations for different crops is presented in Table 6.1.

TABLE 6.1
Ranges of Magnesium Concentrations in Different Crops, Which Were Considered Deficient, Sufficient, or Excessive, Depending on the Crop and the Crop Yield Component Being Considered

Type of Crop		Diagnostic Range (%)		
Latin Name	**Common Name**	**Low**	**Sufficient**	**High**
Abelia R. Br.	Abelia		0.25–0.36	
Abeliophyllum Nakai.	White forsythia		0.20–0.24	
Abies Mill.	Fir		0.06–0.16	
Acalypha hispida Burm.f.	Chenille plant		0.60	
Acer L.	Maple		0.10–0.77	
Achillea L.	Yarrow		0.18–0.27	
Acorus gramineus Ait.	Japanese sweet flag		0.23–0.37	
Actinidia Lindl.	Kiwi-fruit		0.35–0.80	
Aeschynanthus radicans Jack	Lipstick plant		0.25–0.30	
Aesculus L.	Buckeye, horsechestnut		0.17–0.65	
Aglaonema Schott	Chinese evergreen		0.30–1.00	
Agrostis L.	Bent grass		0.25–0.30	
Ajuga L.	Bugleweed		0.23–0.53	
Allamanda L.	Allamanda		0.25–1.00	
Allium cepa L.	Onion		0.25–0.50	
Allium sativum L.	Garlic		0.15–2.5	
Alocasia cucullata (Lour.) G. Don.	Chinese taro		0.87	
Aloe L.	Aloe		0.62–1.32	
Alstroemeria L.	Alstroemeria		0.20–0.50	
Amelanchier Medic.	Serviceberry		0.22–0.30	
Amsonia Walt.	Blue star		0.17–0.27	
Anacardium L.	Cashew		0.02–0.15	
Ananas Mill.	Pineapple		0.30–0.60	
Annona L.	Custard apple, soursop		0.30–0.50	
Anthurium Schott.	Anthurium		0.34–1.00	
Antirrhinum L.	Snapdragon		0.50–1.05	
Aphelandra squarrosa Nees.	Zebra plant		0.50–1.00	
Apium L.	Celery		0.20–0.50	
Arachis hypogaea L.	Peanut or groundnut		0.30–0.80	
Aralia spinosa L.	Devil's walkingstick		0.14–0.55	
Araucaria Juss.	Bunya-bunya, monkey puzzle tree, Norfolk Island pine		0.20–0.50	
Armoracia rusticana P. Gaertn., B. Mey. & Scherb.	Horseradish		0.25–3.0	
Artemisia L.	Dusty miller, wormwood, tarragon		0.19–0.62	
Asarum L.	Ginger or snakeroot		0.50–0.72	
Asclepias L.	Milkweed		0.22–0.40	
Asparagus L.	Asparagus		0.10–0.40	
Aspidistra elatior Blume	Cast-iron plant		0.12–0.33	
Aster L.	Aster		0.18–0.35	
Astilbe Buch.-Ham. Ex D. Don	Lilac rose		0.12–0.28	
Aucuba japonica Thunb.	Japanese laurel		0.13–0.26	

TABLE 6.1 (*Continued*)

Type of Crop		Diagnostic Range (%)		
Latin Name	Common Name	Low	Sufficient	High
Avena sativa L.	Oats	0.07–0.39	0.13–0.52	
Beaucarnea recurvata Lem.	Pony-tail palm		0.20–0.50	
Begonia L.	Begonia		0.30–0.88	
Berberis L.	Barberry		0.13–0.26	
Beta vulgaris L.	Beet		0.25–1.70	
Betula L.	Birch	0.14–0.37	0.16–1.00	
Bougainvillea glabra Choisy.	Paper flower		0.25–0.75	
Bouvardia Salisb.	Bouvardia		0.49–0.73	
Brassica L.	Mustard, kale, cauliflower, broccoli, cabbage		0.17–1.08	
Bromelia L.	Bromeliad		0.40–0.80	
Bromus L.	Bromegrass		0.08–0.30	
Buddleia L.	Butterfly bush		0.17–0.50	
Buxus L.	Boxwood		0.18–0.60	
Caladium Venten.	Fancy-leaf caladium		0.20–0.40	
Calathea G. F. Mey.	Feather calathea		0.25–1.30	
Callicarpa L.	Beautyberry		0.25–0.42	
Callisia L.	Wandering jew		0.92–1.40	
Calycanthus L.	Sweetshrub or Carolina allspice		0.12–0.17	
Camellia L.	Tea		0.12–0.33	
Campsis Lour.	Trumpet creeper		0.14–0.19	
Capsicum L.	Pepper		0.30–2.80	
Carex L.	Sedge		0.15–0.28	
Carica L.	Papaya		0.40–1.20	
Carissa grandiflora (E. H. Mey.) A. DC.	Natal plum		0.25–1.00	
Carpinus L.	Hornbeam		0.18–0.40	
Carya Nutt.	Hickory, pecan	0.04–0.12	0.18–0.82	
Caryopteris Bunge.	Bluebeard		0.16–0.17	
Catalpa Scop.	Catalpa		0.34–0.36	
Catharanthus G Don	Madagascar or rosy periwinkle		0.32–0.78	
Cattleya Lindl.	Orchid, cattleya		0.27–0.70	
Ceanothus impressus Trel.	Santa Barbara ceanothus		0.16–0.19	
Cedrus Trew.	Cedar		0.09–0.35	
Celosia L.	Celosia		1.36–4.05	
Celtis L.	Hackberry		0.47–0.53	
Cercis L.	Redbud		0.12–0.39	
Chaenomeles Lindl.	Flowering quince		0.20–0.30	
Chamaecyparis Spach	Falsecypress		0.07–0.39	
Chimonanthus praecox (L.) Link	Fragrant wintersweet		0.23–0.37	
Chionanthus Lindl.	Fringetree		0.13–0.31	
Chlorophytum Ker-Gawl.	Spider plant		0.25–1.50	
Chrysanthemum L.	Chrysanthemum		0.29–0.97	
Chrysobalanus L.	Coco plum		0.25–1.00	
Cichorium endiva L.	Endive		0.36–2.50	
Citrullus lanatus (Thunb.) Matsum. & Nakai	Watermelon		0.30–3.50	

Continued

TABLE 6.1   (*Continued*)

Type of Crop		Diagnostic Range (%)		
Latin Name	**Common Name**	**Low**	**Sufficient**	**High**
Citrus L.	Lime, orange, grapefruit, etc.		0.17–1.00	
Cladrastis Raf.	Yellowwood		0.24–0.32	
Clematis L.	Clematis		0.10–0.18	
Clethra L.	Summer-sweet		0.59–0.93	
Cocculus DC.	Laurel-leaf moonseed		0.13–0.21	
Codiaeum A. Juss.	Croton		0.40–0.75	
Coffea L.	Coffee		0.25–0.50	
Coleus Lour.	Coleus		1.27–1.48	
Cordyline terminalis (L.) Kunth	Ti plant		0.23–0.49	
Coreopsis L.	Coreopsis		0.46–0.50	
Cornus L.	Dogwood		0.23–0.90	
Coronilla L.	Crownvetch		0.42–0.65	
Corylopsis sinensis Hemsl.	Chinese winterhazel		0.11–0.21	
Corylus L.	Hazelnut, Filbert		0.22–0.59	
Cotinus Mill.	Smoke tree		0.19–0.41	
Cotoneaster Medic.	Cotoneaster		0.17–0.45	
Crassula Thunb.	Jade plant		0.33–0.82	
Crataegus L.	Hawthorn		0.29–0.33	
Crossandra Salisb.	Crossandra or firecracker flower		0.40–0.60	
Cucumis L.	Cantaloupe, honeydew, cucumber		0.35–0.80	
Cucurbita L.	Pumpkin, squash		0.30–2.50	
Cymbidium Swartz	Orchid, cymbidium		0.19–1.00	
Cynodon L.	Bermuda grass		0.10–0.50	
Dactylis L.	Orchard grass		0.15–0.30	
Daphne odora Thunb.	Winter daphne		0.10–0.18	
Daucus L.	Wild carrot		0.25–0.60	
Desmodium Desv.	Tick trefoil		0.14–0.17	
Dianthus L.	Carnation		0.19–1.05	
Dicentra Bernh.	Dutchman's breeches, bleeding heart		0.19–0.35	
Dieffenbachia Schott.	Dumb cane		0.30–1.30	
Digitalis L.	Foxglove		0.24–0.40	
Diospyros L.	Persimmon		0.18–0.74	
Dizygotheca N. E. Br.	False aralia		0.20–0.40	
Draceana L.	Dracaena		0.20–1.00	
Dypsis Noronha ex Mart.	Areca palm		0.20–0.80	
Elaeagnus pungens Thunb.	Thorny elaeagnus		0.17–0.22	
Elaeis Jacq.	Oil palm	0.12–0.27	0.23–0.50	
Epipremnum Schott.	Devil's ivy		0.30–1.00	
Eriobotrya Lindl.	Loquat	0.05		
Eruca Mill.	Arugula		0.28–0.29	
Eucalyptus L'Hér.	Mindanao gum or bagras		0.13–0.42	
Euonymus L.	Spindle tree		0.10–0.47	
Euphorbia milii Desmoul.	Crown-of-thorns		0.25–1.00	
Euphorbia pulcherrima Willd. ex Klotzsch	Poinsettia		0.20–1.00	

TABLE 6.1 (*Continued*)

Type of Crop		Diagnostic Range (%)		
Latin Name	**Common Name**	**Low**	**Sufficient**	**High**
Fagus L.	Beech		0.13–0.36	
Feijoa sellowiana O. Berg.	Pineapple guava		0.15–0.22	
Festuca L.	Fescue		0.24–0.35	
Ficus L.	Fig		0.20–1.00	
Forsythia Vahl.	Golden-bells		0.12–0.26	
Fothergilla L.	Witchalder		0.20–0.42	
Fragaria L.	Strawberry		0.25–0.70	
Fraxinus L.	Ash		0.17–0.49	
Gardenia Ellis	Gardenia		0.25–1.00	
Gelsemium sempervirens (L.) Ait	Carolina jasmine		0.13–0.20	
Geranium L.	Cranesbill		0.24–0.37	
Gerbera L.	Transvaal daisy		0.24–0.63	
Ginkgo biloba L.	Ginkgo		0.25–0.41	
Gladiolus L.	Gladiolus		0.50–4.50	
Gleditsia L.	Honeylocust		0.22–0.35	
Glycine max (L.) Merrill	Soybean		0.25–1.00	
Gossypium L.	Cotton		0.30–0.90	
Gynura Cass.	Royal velvet plant		0.70–0.94	
Gypsophila L.	Baby's breath		0.40–1.30	
Halesia L.	Silverbell		0.14–0.37	
Hamamelis L.	Witchhazel		0.15–0.18	
Hedera L.	Ivy		0.15–0.70	
Helianthus annuus L.	Sunflower		0.25–1.00	
Heliconia L.	Parrot flower		0.33–0.74	
Heliotropium L.	Heliotrope		0.57–0.73	
Helleborus L.	Lenten rose		0.21–0.33	
Hemerocallis L.	Daylily		0.13–0.38	
Heuchera L.	Alumroot		0.20–0.30	
Hibiscus syriacus L.	Rose-of-Sharon		0.36–1.12	
Hordeum L.	Barley		0.15–0.40	
Hosta Tratt.	Hosta		0.11–0.51	
Hydrangea L.	Hydrangea		0.22–0.70	
Hypericum L.	St. Johnswort		0.18–0.35	
Iberis L.	Candytuft		0.36–0.53	
Ilex L.	Holly		0.16–1.00	
Illicium L.	Anise-tree		0.11–0.32	
Impatiens L.	Impatiens, New Guinea		0.30–3.64	
Ipomoea batatas L. Lam.	Sweet potato		0.35–1.00	
Iris L.	Iris		0.17–0.45	
Itea virginica L.	Sweetspire		0.13–0.20	
Ixora L.	Flame-of-the-woods or Indian jasmine		0.20–1.00	
Jasminum L.	Jasmine		0.25–1.00	
Juglans L.	Walnut		0.29–1.01	
Juniperus L.	Juniper		0.08–0.41	
Kalanchoe Adans.	Kalanchoe		0.24–1.50	
Kalmia L.	Laurel		0.11–0.98	
Kerria DC.	Japanese rose		0.35–0.41	

Continued

TABLE 6.1   *(Continued)*

Type of Crop		Diagnostic Range (%)		
Latin Name	**Common Name**	**Low**	**Sufficient**	**High**
Koelreuteria Laxm.	Goldenraintree		0.21–0.31	
Lactuca sativa L.	Lettuce		0.24–3.50	
Lagerstroemia L.	Crepe myrtle		0.23–0.72	
Larix Mill.	Larch		0.11–0.15	
Leea L.	West Indian holly		0.25–0.80	
Leucothoe D. Don	Fetterbush		0.23–0.32	
Liatris Gaertn. ex Schreb.	Gayfeather		0.41–0.45	
Ligustrum L.	Privet		0.13–0.32	
Lilium L.	Lily, Asiatic		0.19–0.70	
Limonium Mill.	Statice, sea lavender		0.50–2.13	
Lindera Thunb.	Spicebush		0.16–0.49	
Liquidambar L.	Sweetgum		0.19–0.53	
Liriope Lour.	Lily-turf		0.10–0.49	
Litchi Sonn.	Lychee fruit		0.20–0.40	
Lolium L.	Ryegrass		0.16–0.32	
Lonicera L.	Honeysuckle		0.20–0.48	
Loropetalum R. Br.	Fringeflower		0.13–0.20	
Lotus L.	Bird's-foot trefoil		0.40–0.60	
Lycopersicon lycopersicum (L.) Karst. ex Farw.	Tomato		0.30–2.50	
Lysimachia L.	Loosestrife		0.28–0.54	
Macadamia F. J. Muell.	Macadamia nut		0.08–0.30	
Magnolia L.	Magnolia		0.12–0.45	
Mahonia Nutt.	Oregon holly		0.11–0.25	
Malpighia glabra L.	Barbados cherry		0.25–0.80	
Malus Mill.	Apple	0.01–0.47	0.12–0.72	
Mandevilla Lindl.	Mandevilla		0.25–0.50	
Mangifera L.	Mango		0.20–0.50	
Manihot Mill.	Cassava		0.25–0.60	
Maranta L.	Prayer plant		0.25–1.00	
Medicago L.	Alfalfa or lucerne		0.30–1.00	
Metasequoia glyptostroboides H. H. Hu & Cheng.	Dawn redwood		0.24–0.31	
Monstera Adans.	Swiss-cheese plant or Mexican breadfruit		0.25–0.65	
Murraya paniculata (L.) Jack	Orange jasmine		0.25–0.40	
Musa L.	Banana	0.04–0.09	0.25–0.80	
Myrica cerifera L.	Wax myrtle		0.11–0.35	
Nandina Thunb.	Heavenly bamboo		0.11–0.24	
Nasturtium officinale R. Br.	Watercress		1.00–2.00	
Nephrolepis Schott.	Sword fern		0.20–1.20	
Nicotiana L.	Tobacco		0.20–0.86	
Nyssa L.	Tupelo		0.23–0.51	
Olea L.	Olive		0.20–0.60	
Ophiopogon Ker-Gawl.	Mondo grass		0.15–0.67	
Oryza sativa L.	Rice		0.15–0.30	
Osmanthus Lour.	Devilweed		0.08–0.29	
Ostyra Scop.	Hornbeam		0.11–0.54	
Oxydendrum DC.	Sourwood		0.24–0.29	
Pachysandra Michx.	Spurge		0.16–0.73	
Pandanus L.	Screwpine		0.22–0.35	

TABLE 6.1 (*Continued*)

Type of Crop		Diagnostic Range (%)		
Latin Name	**Common Name**	**Low**	**Sufficient**	**High**
Pandanus L.	Screwpine		0.22–0.35	
Panicum L.	Switchgrass		0.14–0.33	
Parrotia C.A. Mey.	Persian ironwood		0.09–0.17	
Parthenocissus Planch.	Woodbine		0.14–0.33	
Passiflora L.	Passionfruit		0.25–0.35	
Pelargonium zonale L.	Geranium, Zonal		0.19–0.51	
Pennisetum L.	Fountain grass		0.18–0.19	
Peperomia Ruiz & Pav.	Peperomia		0.24–1.50	
Persea Mill.	Avocado		0.25–0.80	
Petunia Juss.	Petunia		0.36–1.37	
Phalaenopsis Blume.	Orchids, moth		0.40–1.07	
Phalaris arundinacea L.	Ribbon grass		0.19–0.22	
Phaseolus L.	Bean		0.25–1.00	
Philodendron Schott.	Philodendron		0.25–1.80	
Phleum L.	Timothy		0.16–0.25	
Phlox L.	Phlox		0.16–0.57	
Photinia Lindl.	Photinia		0.17–0.30	
Picea A. Dietr.	Spruce		0.08–0.63	
Pieris D. Don	Lily-of-the-valley bush		0.14–0.23	
Pilea Lindl.	Aluminum plant		0.53–1.80	
Pinus L.	Pine		0.09–0.50	
Pistacia L.	Pistachio, Mastic		0.18–1.25	
Pisum L.	Pea		0.27–0.70	
Pittosporum Banks ex Gaertn.	Mock orange		0.18–0.75	
Platanus L.	Sycamore		0.15–0.30	
Platycodon A. DC.	Balloonflower		0.28–0.32	
Poa L.	Bluegrass		0.13–0.37	
Podocarpus L'Hér.	Yew-pine		0.25–0.80	
Polyscias J. R. Forst & G. Forst	Ming aralia		0.43–0.47	
Populus L.	Cottonwood		0.14–0.72	
Prunus L.	Apricot, cherry, plum, almond, peach, nectarine		0.25–1.20	
Psidium L.	Guava		0.25–0.50	
Pulmonaria L.	Lungwort		0.18–0.27	
Pyracantha M. J. Roem.	Firethorn		0.22–0.23	
Pyrus L.	Pear	0.05	0.21–0.80	
Quercus L.	Oak		0.09–0.42	
Rhapis L.f.	Lady palm		0.20–0.30	
Rhododendron L.	Azalea		0.14–1.00	
Rhus L.	Sumac		0.18–0.27	
Ribes L.	Currant, gooseberry,		0.20–0.50	
Rosa L.	Rose, hybrid tea		0.22–0.64	
Rosmarinus officinalis L.	Rosemary		0.17–0.40	
Rubus L.	Blackberry, raspberry		0.25–0.80	
Rudbeckia L.	Coneflower		0.51–0.69	
Ruscus aculeatus L.	Butcher's broom		0.16–0.17	
Saccharum officinarum L.	Sugarcane		0.10–0.20	
Saintpaulia H. Wendl.	African violet		0.35–0.85	
Salix L.	Willow		0.15–0.35	

Continued

TABLE 6.1 (*Continued*)

Type of Crop		Diagnostic Range (%)		
Latin Name	**Common Name**	**Low**	**Sufficient**	**High**
Salvia L.	Sage		0.25–0.86	
Sansevieria Thunb.	Mother-in-law tongue		0.30–1.40	
Sarcococca Lindl.	Sweetbox		0.24–0.55	
Saxifraga L.	Strawberry begonia		0.45–0.66	
Schefflera J. R. Forst & G. Forst	Umbrella or octopus tree		0.25–1.00	
Schlumbergera Lem.	Christmas cactus		0.40–2.00	
Secale cereale L.	Rye		0.35–0.56	
Sedum L.	Stonecrop		0.24–0.67	
Sinningia Nees	Gloxinia		0.35–0.70	
Solanum melongena L.	Eggplant		0.30–1.00	
Solanum tuberosum L.	Potato		0.50–2.50	
Solidago L.	Goldenrod		0.30–0.43	
Sophora L.	Pagoda tree, mescal		0.27–0.40	
Sorghum Moench.	Sorghum		0.10–0.50	
Spathiphyllum Schott.	Peace lily		0.20–1.00	
Spigelia marilandica L	Indian pink		0.57–1.43	
Spinacia oleracea L.	Spinach		0.60–1.80	
Spiraea L.	Bridal-wreath		0.11–0.38	
Stachys byzantina C. Koch	Lamb's ears		0.28–0.31	
Stenotaphrum secundatum (Walt.) O. Kuntze	St. Augustine grass		0.15–0.25	
Stewartia L.	Stewartia		0.26–0.34	
Strelitzia Ait.	Bird-of-paradise		0.18–0.75	
Stromanthe Sond.	Stromanthe		0.30–0.50	
Styrax L.	Snowbell		0.08–0.24	
Syringa L.	Lilac		0.20–0.40	
Tagetes L.	Marigold		1.33–1.56	
Taxodium L. Rich.	Baldcypress		0.19–0.27	
Taxus L.	Yew		0.16–0.30	
Ternstroemia Mutis ex L.f.	False cleyera		0.29–0.33	
Teucrium L.	Wall germander		0.05–0.14	
Thalictrum L.	Meadow-rue		0.26–0.31	
Theobroma cacao L.	Cocoa or chocolate		0.20–0.50	
Thuja L.	Arborvitae		0.09–0.39	
Thymus L.	Thyme		0.29–0.40	
Tilia L.	Basswood		0.18–0.81	
Torenia L.	Wishbone flower		0.90–0.93	
Trachelospermum Lem.	Star jasmine		0.18–0.28	
Tradescantia L.	Spiderwort		0.33–1.32	
Trifolium L.	Clover		0.20–0.60	
Tripogandra Raf.	Tahitian bridal-veil or fern–leaf inch plant		0.42–0.46	
Triticum L.	Wheat		0.15–1.00	
Tsuga Carrière.	Hemlock		0.16–0.26	
Ulmus L.	Elm		0.22–0.58	
Vaccinium L.	Blueberry, cranberry		0.12–0.40	
Verbena L.	Verbena		0.53–1.58	
Veronica L.	Speedwell		0.23–0.72	
Viburnum L.	Arrowwood		0.15–1.00	

TABLE 6.1 (*Continued*)

Type of Crop		Diagnostic Range (%)		
Latin Name	**Common Name**	**Low**	**Sufficient**	**High**
Vigna unguiculata ssp. *unguiculata* (L.) Walp.	Black-eyed pea		0.30–0.50	
Vinca L.	Periwinkle		0.17–0.47	
Viola L.	Pansy		0.36–0.49	
Vitex L.	Chaste tree		0.22–0.33	
Vitis L.	Grape		0.13–1.50	
Yucca L	Soft yucca		0.20–1.00	
Zamia L.	Coontie fern		0.22–0.26	
Zea L.	Corn or maize		0.13–1.00	
Zelkova Spach.	Saw-leaf		0.13–0.20	
Zingiber Boehmer.	Ginger		0.35–0.47	
Zoysi Willd.	Zoysiagrass		0.11–0.15	

6.5 ASSESSMENT OF MAGNESIUM IN SOILS

6.5.1 FORMS OF MAGNESIUM IN SOILS

Approximately 1.3, 4.7, and 4.3% of the earth's continental upper layer, lower layer, and the ocean crust is made up of magnesium, respectively (202). However, in surface soils, magnesium concentrations usually range from 0.03 to 0.84%, with sandy soils typically having the lowest magnesium concentrations ($\approx 0.05\%$), and clay soils containing the highest magnesium concentrations ($\approx 0.50\%$) (10,29). Like other metallic elements, the soil magnesium pool consists of three fractions: nonexchangeable, exchangeable, and water-soluble fractions. The nonexchangeable fraction consists of the magnesium present in the primary minerals and many of the secondary clay minerals (Table 6.2) (29). In many cases these compounds may be hydrated with one to several water molecules. The exchangeable fraction may make up approximately 5% of the total magnesium in the soil, accounting for 4 to 20% of the cation-exchange capacity of the soil (29). Magnesium concentrations in the soil solution typically range from 0.7 to 7.0 mM, but may be as high as 100 mM, with the soil solutions of acid soils generally having a lower magnesium concentration (about 2.0 mM) than soil solutions derived from neutral soils (about 5.0 mM) (29,203–207).

6.5.2 SODIUM ABSORPTION RATIO

Magnesium is also an important component in evaluating the sodium absorption ratio (SAR) of irrigation waters and soil extracts. The SAR is calculated as

$$SAR = (Na^+)/\sqrt{(Ca^{2+} + Mg^{2+})/2}$$

In this equation, the concentrations of sodium (Na^+), calcium (Ca^{2+}), and magnesium (Mg^{2+}) ions are expressed in meq L^{-1}. When concentrations of magnesium, calcium, or both elements are increased, relative to sodium, the SAR decreases. Many soils in arid climates are affected by SAR in that as the SAR increases, the permeability of the soil decreases since the sodium reacts with clay, causing soil particles to disperse resulting in reduced water penetration into the soil (208). In most cases, a soil is considered sodic when the SAR > 13 (209). However, at lower SAR values, some crops may still be susceptible to the adverse effects of sodium on nutrient uptake rather than to the physiochemical effects on soil permeability.

TABLE 6.2
Primary and Secondary Minerals, Nonminerals, and Gems Containing Magnesium

Name	Chemical Formula	Magnesium Concentration (%)
Actinolite	$Ca_2(Mg, Fe)_5Si_8O_{22}(OH)_2$	15
Adelite	$CaMg(AsO_4)(OH)$	11
Admontite	$MgB_6O_{10} \cdot 7H_2O$	6
Amesite (Serpentine Group)	$Mg_2Al(SiAl)O_5(OH)_4$	–
Amianthus	See Parachrysotile	–
Ankerite	$Ca(Fe, Mg, Mn)(CO_3)_2$	13
Anthophyllite	$Mg_7Si_8O_{22}(OH)_2$	22
Antigorite	See Genthite	–
Arfvedsonite	$Na_3(Fe, Mg)_4FeSi_8O_{22}(OH)_2$	12
Artinite	$Mg_2(CO_3)(OH)_2 \cdot 3H_2O$	25
Asbestos	See Tremolite	–
Ascharite	See Camsellite	–
Astrakanite	$MgSO_4 \cdot Na_2SO_4 \cdot 4H_2O$	7
Augite	$(Ca, Na)(Mg, Fe, Al, Ti)(Si, Al)_2O_6$	12
Axinite	See Magnesio-axinite	–
Bayleyite	$Mg_2(UO_2)(CO_3)_3 \cdot 18H_2O$	6
Benstonite	$(Ba, Sr)_6(Ca, Mn)_6Mg(CO_3)_{13}$	2
Berthierine (Serpentine Group)	$(Fe, Fe, Mg)_2(Si, Al)_2O_5(OH)_4$	–
Bischofite	$MgCl_2 \cdot 6H_2O$	12
Biotite	$K(Mg, Fe)_3(Al, Fe)Si_3O_{10}(OH, F)_2$	17
Blodite	$Na_2Mg(SO_4)_2 \cdot 4H_2O$	7
Boracite	$Mg_3B_7O_{13}Cl$	19
Botryogen	$MgFe(SO_4)_2(OH) \cdot 7H_2O$	6
Boussingaultite	$(NH_4)_2Mg(SO_4)_2 \cdot 6H_2O$	7
Brandesite	See Seybertite	–
Brindleyite (Serpentine Group)	See Nimesite	–
Bronzite	See Hypersthene	–
Brucite	$Mg(OH)_2$	42
Calciotalc	See Seybertite	–
Camsellite	See Szaibelyite	–
Carnallite	$KMgCl_3 \cdot 6(H_2O)$	9
Caryopilite (Serpentine Group)	$(Mn, Mg)_3Si_2O_5(OH)_4$	4
Cebollite[a]	$Ca_2(Mg, Fe, Al)Si_2(O, OH)_7$	9
Chlorite[b]	$(Mg, Fe)_6(AlSi_3)O_{10}(OH)_8$	26
Chloritoid	$(Fe, Mg, Mn)_2Al_4Si_2O_{10}(OH)_4$	11
Chlorophoenicite	$(Mn, Mg)_3Zn_2(AsO_4)(OH, O)_6$	13
Chrysolite	See Olivine	–
Clinochlore	$(Mg, Fe)_5Al(Si_3Al)O_{10}(OH)_8$ (see Colerainite)	22
Clinochrysotile (Serpentine Group)	See Deweylite	–
Clinoenstatite	$Mg_2Si_2O_6$	24
Clintonite	See Xanthophyllite	–
Colerainite[c]	$4MgO \cdot Al_2O_3 \cdot 2SiO_2 \cdot 5H_2O$	21
Collinsite	$Ca_2(Mg, Fe)(PO_4)_2 \cdot 2H_2O$	7
Cordierite	$Mg_2Al_4Si_5O_{18}$	8
Corrensite	$(Ca, Na, K)(Mg, Fe, Al)_9$ $(Si, Al)_8O_{20}(OH)_{10} \cdot H_2O$	23
Crossite	$Na_2(Mg, Fe)_3(Al, Fe)_2Si_8O_{22}(OH)_2$	9
Cummingtonite	$(Mg, Fe)_7Si_8O_{22}(OH)_2$	22
Deweylite[d]	$Mg_3Si_2O_5(OH)_4$	4
Dickinsonite	$(K, Ba)(Na, Ca)_5(Mn, Fe, Mg)_{14}Al(PO_4)_{12}(OH, F)_2$	20

TABLE 6.2 (*Continued*)

Name	Chemical Formula	Magnesium Concentration (%)
Diopsode	$CaMgSi_2O_6$	11
Dolomite	$CaMg(CO_3)_2$	13
Dypingite	$Mg_5(CO_3)_4(OH)_2 \cdot 5H_2O$	25
Edenite	$NaCa_2(Mg, Fe)_5Si_7AlO_{22}(OH)_2$	15
Elbaite	$Na(Al, Fe, Li,$	7
	$Mg)_3B_3Al_3(Al_3Si_6O_{27})(O, OH, F)_4$	
Enstatite	$Mg_2Si_2O_6$	24
Epsomite	$MgSO_4 \cdot 7H_2O$	10
Falcondoite	See Genthite	–
Fayalite	See Hortonolite	–
Ferrierite	$(Na, K)_2Mg(Si, Al)_{18}O_{36}(OH) \cdot 9H_2O$	2
Fluoborite	$Mg_3(BO_3)(F, OH)_3$	40
Forsterite	Mg_2SiO_4	35
Gageite	$(Mn, Mg, Zn)_{42}Si_{16}O_{54}(OH)_{40}$	34
Galaxite	$(Mn, Fe, Mg)(Al, Fe)_2O_4$	17
Ganophyllite	$(K, Na)_2(Mn, Al, Mg)_8$	15
	$(Si, Al)_{12}O_{29}(OH)_7 \cdot 8\text{-}9H_2O$	
Garnierite[e]	$(Ni, Mg)_3Si_2O_5(OH)_4$	26
Genthite[f]	$2NiO \cdot 2MgO \cdot 3SiO_2 \cdot 6H_2O$	9
Glauconite	$(K, Na)(Fe, Al, Mg)_2(Si, Al)_4O_{10}(OH)_2$	13
Glaucophane	$Na_2(Mg, Fe)_3Al_2Si_8O_{22}(OH)_2$	9
Gordonite	$MgAl_2(PO_4)_2(OH)_2 \cdot 8H_2O$	5
Griffithite	$4(Mg, Fe, Ca)O \cdot (Al, Fe)_2O_3 \cdot 5SiO_2 \cdot 7H_2O$	14
Griphite	$Na_4Li_2Ca_6(Mn, Fe,$	13
	$Mg)_{19}Al_8[(F,OH)(PO_4)_3]_8$	
Grunerite	$(Fe, Mg)_7Si_8O_{22}(OH)_2$	24
Harkerite	$Ca_{24}Mg_8Al_2Si_8(O,OH)_{32}(BO_3)_8$	7
	$(CO_3)_8(H_2O, Cl)$	
Hastingsite	$NaCa_2(Fe, Mg)_4Fe(Si_6Al_2)O_{22}(OH)_2$	12
Hectorite	$Na_{0.3}(Mg, Li)_3Si_4O_{10}(F, OH)_2$	19
Hexahydrite	$MgSO_4 \cdot 6H_2O$	11
Högbomite	$(Mg, Fe)_2(Al, Ti)_5O_{10}$	14
Holdenite	$(Mn, Mg)_6Zn_3(AsO_4)_2(SiO_4)(OH)_8$	17
Hornblende	$Ca(Mg, Fe)_4AlSi_7AlO_{22}(OH)_2$	13
Hortonolite[g]	$(Fe, Mg, Mn)_2SiO_4$	35
Hulsite	$(Fe, Mg)_2(Fe, Sn)BO_5$	25
Huntite	$CaMg_3(CO_3)_4$	21
Hydroboracite	$CaMgB_6O_8(OH)_6 \cdot 3H_2O$	6
Hydromagnesite	$Mg_5(CO_3)_4(OH)_2 \cdot 4H_2O$	26
Hydrotalcite	$Mg_6Al_2(CO_3)(OH)_{16} \cdot 4H_2O$	24
Hypersthene[h]	$(Fe, Mg)SiO_3$	24
Iddingsite	$MgO \cdot Fe_2O_3 \cdot 3SiO_2 \cdot 4H_2O$	5
Jurupaite[i]	$(Ca, Mg)_6Si_6O_{17}(OH)_2$	24
Kainite	$MgSO_4 \cdot KCl \cdot 3H_2O$	10
Kammererite-Red	See Colerainite	–
Kerolite[j]	$(Mg, Ni)_3Si_4O_{10}(OH)_2 \cdot H_2O$	18
Kieserite	$MgSO_4 \cdot H_2O$	18
Kurchatovite	$Ca(Mg, Mn, Fe)B_2O_5$	15
Landesite	$(Mn, Mg)_9Fe_3(PO_4)_8(OH)_3 \cdot 9H_2O$	16
Langbeinite	$K_2Mg_2(SO_4)_3$	13
Lansfordite	$MgCO_3 \cdot 5H_2O$	14
Lazulite	$MgAl_2(PO_4)_2(OH)_2$	8
Leonite	$K_2Mg(SO_4)_2 \cdot 4H_2O$	7
Lizardite (Serpentine Group)	See Clinochrysotile	–
Löweite	$Na_{12}Mg_7(SO_4)_{13} \cdot 15H_2O$	9

Continued

TABLE 6.2   (*Continued*)

Name	Chemical Formula	Magnesium Concentration (%)
Ludwigite	Mg_2FeBO_5 (see Magnesioludwigite)	25
Magnesio-axinite[k]	$Ca_2MgAl_2BO_3Si_4O_{12}(OH)$	5
Magnesioludwigite[l]	$3MgO \cdot B_2O_3 \cdot MgO \cdot Fe_2O_3$	25
Magnesite	$MgCO_3$	30
Mcgovernite	$Mn_9Mg_4Zn_2As_2Si_2O_{17}(OH)_{14}$	7
Meerschaum[m]	$Mg_4Si_6O_{15}(OH)_2 \cdot 6H_2O$	15
Melilite	$(Ca, Na)_2(Al, Mg)(Si, Al)_2O_7$	10
Merwinite	$Ca_3Mg(SiO_4)_2$	7
Monticellite	$CaMgSiO_4$	16
Montmorillonite	$(Na, Ca)_{0 \cdot 33}(Al, Mg)_2Si_4O_{10}(OH)_2 \cdot nH_2O$	13
Mooreite	$(Mg, Zn, Mn)_{15}(SO_4)_2(OH)_{26} \cdot 8H_2O$	32
Népouite	See Garnierite	–
Nesquehonite	$Mg(HCO_3)(OH) \cdot 2H_2O$	18
Nimesite[n] (Serpentine Group)	$(Ni, Mg, Fe)_2Al(Si, Al)_5(OH)_4$	6
Norbergite	$Mg_3(SiO_4)(F,OH)_2$	37
Northupite	$Na_3Mg(CO_3)_2Cl$	10
Novacekite	$Mg(UO_2)_2(AsO_4)_2 \cdot 12H_2O$	2
Odinite (Serpentine Group)	$(Fe, Mg, Al, Fe, Ti, Mn)_{2.4}$ $(Si, Al)_2O_5(OH)_4$	22
Olivine[o]	$(Mg, Fe)_2SiO_4$	35
Orthoantigorite (Serpentine Group)	See Lizardite	–
Orthochrysotile[p] (Serpentine Group)	$Mg_3Si_2O_5(OH)_4$	26
Parachrysotile[p] (Serpentine Group)	$Mg_3Si_2O_5(OH)_4$	–
Pargasite	$NaCa_2(Mg, Fe)_4Al(Si_6Al_2)O_{22}(OH)_2$	12
Penninite[q]	$(Fe, Mg)_5Al(Si_3Al)O_{10}(OH)_8$	22
Periclase	MgO	60
Peridot	See Olivine	–
Phlogopite	$KMg_3(Si_3Al)O_{10}(F,OH)_2$	17
Pickeringite	$MgAl_2(SO_4)_4 \cdot 22H_2O$	3
Picromerite	See Schoenite	–
Pimelite	See Kerolite	–
Polyhalite	$K_2Ca_2Mg(SO_4)_4 \cdot 2H_2O$	4
Prochlorite	See Penninite	–
Pyrope	$Mg_3Al_2(SiO_4)_3$	18
Rabbittite	$Ca_3Mg_3(UO_2)_2(CO_3)_6(OH)_4 \cdot 18H_2O$	5
Ralstonite	$(Na)x(Mg)x(Al)(2-x)(F,OH)_6 \cdot H_2O$	13
Redingtonite	$(Fe, Mg, Ni)(Cr, Al)_2(SO_4)_4 \cdot 22H_2O$	3
Rhodonite	$(Mn, Fe, Mg, Ca)SiO_3$	24
Riebeckite	$Na_2(Fe, Mg)_3Fe_2Si_8O_{22}(OH)_2$	9
Ripidolite	See Penninite	–
Roscoelite	$K(V, A, Mg)_2AlSi_3O_{10}(OH)_2$	12
Saleeite	$Mg(UO_2)_2(PO_4)_2 \cdot 10H_2O$	3
Saponite	$Ca_{0 \cdot 25}(Mg, Fe)_3(Si, Al)_4O_{10}(OH)_2 \cdot nH_2O$	18
Sapphirine	$(Mg, Al)_8(Al, Si)_6O_{20}$	29
Sarcopside	$(Fe, Mn, Mg)_3(PO_4)_2$	28
Schoenite[r]	$K_2Mg(SO_4)_2 \cdot 6H_2O$	4
Sepiolite	See Meerschaum	15
Serpentine	$(Mg, Fe)_3Si_2O_5(OH)_4$	26
Seybertite[s]	$Ca(Mg, Al)_3(Al_3Si)O_{10}(OH)_2$	18
Sheridanite	See Penninite	–
Sklodowskite	$Mg(UO_2)_2(SiO_3OH)_2 \cdot 5H_2O$	3

TABLE 6.2 (*Continued*)

Name	Chemical Formula	Magnesium Concentration (%)
Spadaite	$MgSiO_2(OH)_2 \cdot H_2O$	18
Spinel	$MgAl_2O_4$	17
Staurolite	$(Fe, Mg, Zn)_2Al_9(Si, Al)_4O_{22}(OH)_2$	6
Stevensite	$(Ca, Na)_xMg_3Si_4O_{10}(OH)_2$	18
Stichtite	$Mg_6Cr_2(CO_3)(OH)_{16} \cdot 4H_2O$	22 *
Stilpnomelane	$K(Fe, Mg)_8(Si, Al)_{12}(O, OH)_{27}$	20
Swartzite	$CaMg(UO_2)(CO_3)_3 \cdot 12H_2O$	3
Szaibelyite[t]	$MgBO_2(OH)$	29
Tachyhydrite	$CaMg_2Cl_6 \cdot 12H_2O$	9
Taeniolite	$KLiMg_2Si_4O_{10}F_2$	12
Talc	$Mg_3Si_4O_{10}(OH)_2$	19
Tilasite	$CaMg(AsO_4)F$	11
Tremolite[p]	$Ca_2(Mg, Fe)_5Si_8O_{22}(OH)_2$	15
Triplite	$(Mn, Fe, Mg, Ca)_2(PO_4)(F, OH)$	30
Tychite	$Na_6Mg_2(CO_3)_4(SO_4)$	9
Uvite	$(Ca, Na)(Mg, Fe)_3Al_5Mg(BO_3)_3Si_6O_{18}(OH, F)_4$	10
Vanthoffite	$Na_6Mg(SO_4)_4$	4
Vesuvianite	$Ca_{10}(Mg, Fe)_2Al_4Si_9O_{34}(OH)_4$	3
Vosenite	$3(Fe, Mg) \cdot B_2O_3 \cdot FeO \cdot Fe_2O_3$	19
Wagnerite	$(Mg, Fe)_2(PO_4)F$	30
Xanthophyllite[u]	$Ca(Mg, Al)_3(Al_3Si)O_{10}(OH)_2$	18
Xonotlite	See Jurupaite and Stevensite	–

Note: The concentration of magnesium in these products is calculated from the chemical formula. The magnesium concentration presented in the table is based on the highest amount of magnesium possible in the compound (when magnesium occupies all potential sites in the formula).

[a] Cebollite (synonym: Cebollit or Cebollita) may be referred to as $Ca_5Al_2(SiO_4)_3(OH)_4$ with no Mg.

[b] There are several different minerals apart from the Chlorite group of minerals.

[c] Colerainite may be referred to as a synonym for Clinochlore.

[d] Deweylite may be referred to as a synonym for Clinochrysotile and Lizardite.

[e] Népouite may be referred to as a synonym for Garnierite with the same chemical formula and it may also be referred to as Falcondoite as a synonym for Garnierite and Genthite with different chemical formulas.

[f] Antigorite may be referred to as a synonym for Genthite with the chemical formula $(Mg, Fe)_3Si_2O_5(OH)_4$.

[g] Fayalite may be referred to as a synonym for Hortonolite.

[h] Bronzite may be referred to as a synonym for Hypersthene.

[i] Mg-bearing Xonotlite may be referred to as a synonym for Jurupaite or Stevensite with different chemical formulas.

[j] Kerolite (Ni) may be referred to as a synonym for Pimelite.

[k] Axinite may be referred to as a synonym for Magnesio-axinite.

[l] Ludwigite may be referred to as a synonym for Magnesioludwigite but with chemical formula Mg_2FeBO_5.

[m] Sepiolite may be referred to as a synonym for Meerschaum.

[n] Nimesite may be referred to as a synonym for Brindley

[o] Olivine may be referred to as a synonym for Peridot or Chrysolite-light yellowish green.

[p] Tremolite, Orthochrysotile, and Parachrysotile are occasionally referred to as Asbestos.

[q] Colerainite, Kammererite–Red, Pennine, Prochlorite, Ripidolite, Sheridanite may all be referred to as synonyms for Penninite.

[r] Schoenite may be referred to as synonym for Picromerite.

[s] Brandesite, Calciotalc, Seybertite, and Xanthophyllite may be referred to as synonyms for Seybertite.

[t] Camsellite may be referred to as synonym for Szaibelyite and Ascharite.

[u] Clintonite, Brandesite, Calciotalc, and Seybertite may be referred to as synonyms for Xanthophyllite.

6.5.3 SOIL TESTS

Several methods have been developed to extract the exchangeable magnesium fraction from soils. When preparing soils for extractions, the drying temperatures of 40 to 105°C do not affect the extractability of magnesium (210). In most soils, magnesium can be extracted with a solution containing ammonium acetate (211–213), $CaCl_2$ (210) or with water (214). However, for soils with a low cation-exchange capacity, acidic extractions are recommended (215). For alkaline soils, a water extraction is utilized (214). Another extraction method (AB-DTPA, ammonium bicarbonate-diethyleneaminepentaacetate) is utilized for alkaline soils; however, this method is suitable only for the extraction of sodium and potassium, since magnesium as well as calcium will react and precipitate with the bicarbonate in the extraction reagent (216). In Sweden, soils are extracted with ammonium lactate at pH 3.75 (10), and in Turkey, chemical extractions methods include various concentration of hydrochloric acid in addition to the ammonium acetate procedure (212).

After proper extractions are performed, the magnesium concentration of solutions can be quantified by ion-selective electrodes, flame-plasma emission spectroscopy, or atomic absorption spectroscopy (217). The wavelength used in atomic absorption is 285.2 nm. In the United States, the Environmental Protection Agency (EPA) (218,219) guidelines indicate that magnesium concentrations of samples have to be determined by inductively coupled plasma (ICP) spectrophotometry according to methods described in EPA Method 200.7, by ICP-mass spectrometry in EPA Method 200.8 (218), or by atomic absorption method 7450 EPA 7-series (219).

6.5.4 TABULATED DATA ON MAGNESIUM CONTENTS IN SOILS

6.5.4.1 Soil Types

Considering surface soils, sandy soils typically have the lowest magnesium concentrations and clay soils typically have the highest magnesium concentrations (193,220). Common soil types high in magnesium include soils that are not leached heavily or soils in depressions where leached nutrients may accumulate. Leached soils such as lateritic soils and podzols tend to be low on magnesium (29). Soils derived from parent bedrock of dolomite or igneous rock tend to be high in magnesium (29,221).

6.6 FERTILIZERS FOR MAGNESIUM

6.6.1 KINDS OF FERTILIZERS

Magnesium-containing fertilizers are derived from the mining of natural mineral deposits or through synthetic processing. Organic magnesium sources include most manures (209). The magnesium availability to plants from different fertilizers will be dictated by the water solubility of the compounds, release rates from fertilizer coatings (where applicable), and particle size, with the finer particles solubilizing more quickly than the coarser-grade fertilizers. Magnesium concentrations and solubility characteristics for some common fertilizers are listed in Table 6.3.

6.6.2 EFFECTS OF FERTILIZERS ON PLANT GROWTH

Although the requirements for magnesium is low relative to other macronutrients such as nitrogen (222), the effect of magnesium fertilization on plant growth may vary with the form of magnesium used and the fertilizer texture (coarseness) (223). Therefore, the type of magnesium fertilizer to use will depend on variables such as the type of crop and the longevity of the production cycle. In studies with ryegrass (*Lolium perenne* L.), the highest magnesium uptake occurred from fertilizers as follows: magnesium sulfate > potassium magnesium sulfate ($K_2SO_4.2MgSO_4$) > ground dolomite > pelletized dolomite (224). Studies by Tayrien and Whitcomb (119,120,225) indicated that the use of calcium carbonate and magnesium oxide produced greater vegetative growth than

TABLE 6.3
Fertilizers Containing Magnesium and the Approximate Percentage of Magnesium

Fertilizer	Formula	% Mg	Solubility in Water (g L^{-1})	Reference
Epsom salts	$MgSO_4 \cdot 7H_2O$	10	1720 cold	227, 228
Kieserite	$MgSO_4 \cdot H_2O$	18	680 hot	227
Burned lime	$nCaO$ and $nMgO$	6.0–20.0	0.006 cold	227, 228
Sulphate of potash magnesia, Langbeinite	$K_2SO_4 \cdot 2MgSO_4$	12	Soluble	227
Magnesite	$MgCO_3$	29	0.11 cold	228
Dolomite	$CaCO_3 \cdot MgCO_3$	11.7–13.1	0.32	228, 229
Dolomitic limestone	$CaCO_3 \cdot MgCO_3$ mixtures	1.3–6.5	0.01	229
Hydrated lime	Mixture of $Ca(OH)_2$ and $Mg(OH)_2$	2.3–11	0.009 cold; 0.04 hot	228
Limestone, high Mg	$CaCO_3$ and $MgCO_3$	0.6–1.3	0.01	229
Limestone, high Ca	$CaCO_3$ and $MgCO_3$	0–0.6	0.01	229
Magnesium nitrate	$Mg(NO_3)_2 \cdot 6H_2O$	10	1250 cold	228
Magnesium ammonium phosphate	$MgNH_4PO_4 \cdot H_2O$	16	0.14	227, 230
Animal manures		0.8–2.9 kg/1000 kg		
Calcium magnesium phosphate	$(Ca, Mg)PO_4 \cdot nH_2O$	9.0 (typical)	Sparingly soluble	

Note: Cold water is 15 or 20°C; hot water is 100°C.

equivalent quantities of calcium and magnesium supplied with dolomitic limestone (calcium carbonate and magnesium carbonate intergrade). However, in studies with cotoneaster (*Cotoneaster dammeri* C.K. Schneid), the greatest vegetative growth of roots and shoots occurred with the use of dolomite rather than with combinations of other calcium and magnesium sources (109). In other experiments with containerized woody ornamentals, the use of calcium and magnesium sulfates resulted in equal or better quality plants than plants receiving the same amount of calcium and magnesium in the carbonate form, regardless of the grade of dolomite (223). The effects of powdered dolomite compared to pelletized dolomite on plant quality varied with the rate of dolomite application, plant type, and form of other nutrients used, but there tended to be a general trend of increased plant quality with powdered dolomite compared to pelletized dolomite at low fertilizer rates (2.97 kg dolomite m^{-3}), but higher plant quality with pelletized compared to powdered dolomite at higher fertilization rates (5.95 kg dolomite m^{-3}). The diversity of growth effects with different fertilizer types can be attributed to the different solubilities of magnesium compounds and the coarseness of the fertilizers. The more soluble and finer the particle size of the fertilizers are, the more quickly they will dissolve and be available for plant uptake, but also the more quickly magnesium will leach from the root zone. Therefore, although quickly soluble fertilizer forms are suitable for relatively short-term crops (a few weeks), they would not be suitable for long-term crops since fertilizer might not be available in the later stages of crop development.

6.6.3 APPLICATION OF FERTILIZERS

The primary goal is to have sufficient magnesium, relative to other nutrients, readily available for plant uptake throughout crop development. The type and rate of magnesium to apply depends upon the crop, soil type, and method of production (field, container, or hydroponics). If plants are grown hydroponically, a completely soluble form of magnesium would be required. For container-grown nursery crops, Whitcomb (119,120) suggested injecting dissolved Epsom salts (magnesium sulfate) into irrigation water at a rate to produce a calcium/magnesium ratio from 1:1 up to 5:1. In preliminary studies with juniper (*Juniperus* spp. L.), increased vegetative growth occurred when magnesium was supplied by applications of magnesium sulfate in the irrigation water versus equivalent magnesium applications through the incorporation of fine dolomitic lime into the planting media (119–121). Obatolu (226) reported that magnesium deficiency resulted in a loss of yield and quality of tea *(Camellia sinensis* O. Kuntze) in Nigerian plantations. A spray of 30% magnesium oxide corrected magnesium deficiency within 14 days and increased growth from 16 to 134%. Two applications of a 20% solution were required to correct deficiencies. A second application of the 30% solution was toxic to the tea plants.

REFERENCES

1. C.B. Black, J.A. Cowan. Magnesium-dependent enzymes in general metabolism. In: J.A. Cowan, ed. *The Biological Chemistry of Magnesium.* New York: VCH Publishers, Inc., 1995, pp. 159–183.
2. C.B. Black, J.A. Cowan. Magnesium-dependent enzymes in nucleic acid biochemistry. In: J.A. Cowan. ed. *The Biological Chemistry of Magnesium.* New York: VCH Publishers, Inc., 1995, pp. 137–158.
3. J.K. Aikawa. *The Role of Magnesium in Biological Processes.* Springfield, IL: Charles C. Thomas Publisher, 1975, pp. 3–14.
4. O. Loew, D.W. May. The relation of lime and magnesia to plant growth. *U.S. Dept. Agric. Bur. Plant Ind. Bull.* 1, 1901.
5. C.H. LaWall. *Four Thousand Years of Pharmacy.* Philadelphia, PA: Lippincott and Co., 1927.
6. M.E. Weeks. *Discovery of the Elements.* 5th ed. Easton, PA: J. Chem. Ed., 1945.
7. W.H. Gross. *The Story of Magnesium.* Cleveland, OH: American Society for Metals, 1949.
8. T.W. Embleton. Magnesium. In: H.D. Chapman, ed. *Diagnostic Criteria for Plants and Soils.* Berkeley: University of California, Division of Agricultural Sciences, 1966, pp. 225–263.
9. B.S. Meyer, D.B. Anderson. *Plant Physiology.* New York: D. Van Nostrand Co., Inc., 1939.
10. L.G. Nilsson. Magnesium in grassland production. *Dev. Plant Soil Sci.* 29:20–31, 1987.
11. W.W. Garner, J.E. McMurtrey, Jr., C.W. Bacon, E.G. Moss. Sand-drown, a chlorosis of tobacco due to magnesium deficiency, and the relation of sulphates and chlorides of potassium to the disease. *J. Agric. Res.* 23:27–40, 1923.
12. E.D. Schulze. Air pollution and forest decline in a spruce (*Picea abies*) forest. *Science* 244(4906):776–783, 1989.
13. T.M. Roberts, L.W. Blank, R.A. Skeffington. Causes of type 1 spruce decline in Europe. *Forestry* 62(3):179–222, 1989.
14. F. Baillon, S. Geiss, S. Dalschaert, X. Dalschaert. Spruce photosynthesis: possibility of early damage diagnosis due to exposure to magnesium or potassium deficiency. *Trees* 2(3):173–179, 1988.
15. R.E. Oren, D. Schulze, K.S. Werk, J. Mayer. Performance of two *Picea abies* (L.) Karst. Stands at different stages of decline: VII. Nutrient relations and growth. *Oecologia* 77:163–173, 1988.
16. R. Oren, K.S. Werk, N. Buchmann, R. Zimmermann. Chlorophyll nutrient relationships identify nutritionally caused decline in *Picea abies* stands. *Can. J. Forest Res.* 23(6):1187–1195, 1993.
17. F. Wikström, T. Ericsson. Allocation of mass in trees subject to nitrogen and magnesium limitation. *Tree Physiol.* 15:339–344, 1995.
18. S. McIntosh, K. Simpson, P. Crooks. Sources of magnesium for grassland. *J. Agric. Sci. Cambridge* 81(Dec):507–511, 1973.
19. K.F. Smith, M.W. Anderson, H.S. Easton, G.J. Rebetzke, H.A. Eagles. Genetic control of mineral concentration and yield in perennial ryegrass (*Lolium perenne* L.), with special emphasis on minerals related to grass tetany. *Aus. J. Agric. Res.* 50(1):79–86, 1999.

20. R.J. Crawford, H.F. Mayland, D.A. Sleper, M.D. Massie. Use of an experimental high-magnesium tall fescue to reduce grass tetany in cattle. *J. Prod. Agric.* 11(4):491–496, 1998.

21. K.H. Asay, H.F. Mayland, D.H. Clark. Response to selection for reduced grass tetany potential in crested wheatgrass. *Crop Sci.* 36(4):895–900, 1996.

22. S. Sabreen, S. Saiga, H. Saitoh, M. Tsuiki, H.F. Mayland. Performance of high-magnesium cultivars of three cool-season grasses grown in nutrient solution culture. *J. Plant Nutr.* 26(3):589–605, 2003.

23. D.C. Lewis, L.A. Sparrow. Implications of soil type, pasture composition and mineral content of pasture components for the incidence of grass tetany in the southeast of South Australia. *Aus. J. Exp. Agric.* 31(5):609–615, 1991.

24. J. Kubota, G.H. Oberly, E.A. Naphan. Magnesium in grasses of selected regions in the United States and its relation to grass tetany. *J. Agron.* 72(6):907–914, 1980.

25. C.H. Henkens. Fertilizer recommendations for magnesium, sodium and some trace elements in relation to soil analysis in the Netherlands. *Dev. Plant Soil Sci.* 29:239–255, 1987.

26. H. Marschner. *Mineral Nutrition of Higher Plants.* 2nd ed. New York: Academic Press, 1995.

27. T.F. Neales. Components of the total magnesium content within the leaves of white clover and perennial rye grass. *Nature* 177:388–389, 1956.

28. R. Lavon, R. Salomon, E.E. Goldschmidt. Effect of potassium, magnesium, and calcium deficiencies on nitrogen constituents and chloroplast components in *Citrus* leaves. *J. Am. Soc. Hort. Sci.* 124:158–162, 1999.

29. E.A. Kirkby, K. Mengel. The role of magnesium in plant nutrition. *Z. Pflanzenern Bodenk* 2:209–222, 1976.

30. H. Matsuda. Magnesium gating of the inwardly rectifying K^+ channel. *Annu. Rev. Physiol.* 53:289–298, 1991.

31. R.E. White, H.C. Hartzell. Magnesium ions in cardiac function: regulator of ion channels and second messengers. *Biochem. Pharmacol.* 38:859–867, 1989.

32. M. Horlitz, P. Klaff. Gene-specific *trans*-regulatory functions of magnesium for chloroplast mRNA stability in higher plants. *J. Biol. Chem.* 275:35638–35645, 2000.

33. E.I. Ochiai. Structural functions. In: E. Frieden, ed. *General Principles of Biochemistry of the Elements.* New York: Plenum Press, 1987, pp. 197–212.

34. W. Cao, T.W. Tibbitts. Growth, carbon dioxide exchange and mineral accumulation in potatoes grown at different magnesium concentrations. *J. Plant Nutr.* 15(9):1359–1371, 1992.

35. B. Lasa, S. Frechilla, M. Aleu, B. González-Moro, C. Lamsfus, P.M. Aparicio-Tejo. Effects of low and high levels of magnesium on the response of sunflower plants grown with ammonium and nitrate. *Plant Soil* 225:167–174, 2000.

36. T. Ericsson, M. Kähr. Growth and nutrition of birch seedlings at varied relative addition rates of magnesium. *Tree Physiol.* 15:85–93, 1995.

37. M. Carvajal, V. Martínez, A. Cerdá. Influence of magnesium and salinity on tomato plants grown in hydroponic culture. *J. Plant Nutr.* 22:177–190, 1999.

38. I.M. Rao, R.E. Sharp, J.S. Boyer. Leaf magnesium alters photosynthetic response to low water potentials in sunflower. *Plant Physiol.* 84:1214–1219, 1987.

39. G.E. Anthon, A.T. Jagendorf. Effect of methanol on spinach thylakoid ATPase. *Biochim. Biophys. Acta* 723:358–365, 1983.

40. J. Moyle, P. Mitchell. Active/inactive state transitions of mitochondrial ATPase molecules influenced by Mg^{2+}, anions and aurovertin. *FEBS Lett.* 56:55–61, 1975.

41. H.M. Younis, G. Weber, J.S. Boyer. Activity and conformational changes in chloroplast coupling factor induced by ion binding: formation of a magnesium-enzyme–phosphate complex. *Biochemistry* 22:2505–2512, 1983.

42. D. Labuda, D. Porschke. Magnesium ion inner sphere complex in the anticodon loop of phenylalanine transfer ribonucleic acid. *Biochemistry* 21:49–53, 1982.

43. P. Schimmel, A. Redfield. Transfer RNA in solution: selected topics. *Annu. Rev. Biophys. Bioeng.* 9:181–221, 1980.

44. M.M. Teeter, G.J. Quigley, A. Rich. In: T.G. Spiro, ed. *Nucleic Acid-Metal Ion Interactions.* New York: Wiley, 1980.

45. D. Porschke. Modes and dynamics of Mg^{2+}-polynucleotide interactions. In: J.A. Cowan, ed. *The Biological Chemistry of Magnesium.* New York: VCH Publishers, Inc., 1995, pp. 85–110.

46. J.M. Spiers. Calcium, magnesium, and sodium uptake in rabbiteye blueberries. *J. Plant Nutr.* 16(5):825–833, 1993.
47. J.M. Spiers. Nitrogen, calcium, and magnesium fertilizer affects growth and leaf elemental content of 'Dormanred' raspberry. *J. Plant Nutr.* 16(12):2333–2339, 1993.
48. E. Fallahi, B.R. Simons. Interrelations among leaf and fruit mineral nutrients and fruit quality in 'Delicious' apples. *J. Tree Fruit Prod.* 1(1):15–25, 1996.
49. B. Mehne-Jakobs. The influence of magnesium deficiency on carbohydrate concentrations in Norway spruce (*Picea abies*) needles. *Tree Physiol.* 15:577–584, 1995.
50. E.S. Fischer, E. Bremer. Influence of magnesium deficiency on rates of leaf expansion, starch and sucrose accumulation, and net assimilation in *Phaseolus vulgaris*. *Physiol. Plant* 89:271–276, 1993.
51. A.W. Meerow. *Betrock's Guide to Landscape Palms*. Hollywood, FL.: Betrock Information Systems, Inc., 2000.
52. C.B. Shear, M. Faust. Nutritional ranges in deciduous tree fruits and nuts. *Hort. Rev.* 2:142–164, 1980.
53. V. Palomäki. Effects of magnesium deficiency on needle ultrastructure and growth of Scots pine seedlings. *Can. J. Forest Res.* 25(11):1806–1814, 1995.
54. G. Hambidge. *Hunger Signs in Crops*. Washington, DC: The National Fertilizer Association, 1941.
55. T. Baszynski, M. Warcholowa, A. Krupa, A. Tukendorf, M. Krol, D. Wolinsak. The effect of magnesium deficiency on photochemical activities of rape and buckwheat chloroplasts. *Z. Phlanzenphysiol.* 99:295–303, 1980.
56. D.J. Merhaut. Effects of magnesium sulfate on plant growth and nutrient uptake and partitioning in *Camellia sasanqua* 'Shishi Gashira'. *J. Environ. Hort.* 22(3):161–164, 2004.
57. R.J. Deshaies, L.E. Fish, A.T. Jagendorf. Permeability of chloroplasts to Mg^{++}—effects on protein synthesis. *Plant Physiol.* 74:956–961, 1984.
58. J. Papenbrock, E. Pfündel, H.P. Mock, B. Grimm. Decreased and increased expression of the subunit CHL I diminishes Mg chelatase activity and reduces chlorophyll synthesis in transgenic tobacco plants. *Plant J.* 22(2):155–164, 2000.
59. R.E. Zielinski, C.A. Price. Relative requirements for magnesium of protein and chlorophyll synthesis in *Euglena gracilis*. *Physiol. Plant* 61(4):624–625, 1978.
60. N.G. Marinos. Studies on submicroscopic aspects of mineral deficiencies. II. Nitrogen, potassium, sulfur, phosphorus, and magnesium deficiencies in the shoot apex of barley. *Am. J. Bot.* 50:998–1005, 1963.
61. W.W. Thomson, T.E. Weier. The fine structure of chloroplasts from mineral-deficient leaves of *Phaseolus vulgaris*. *Am. J. Bot.* 49:1047–1055, 1962.
62. A. Gunes, M. Alpaslan, A. Inal. Critical nutrient concentrations and antagonistic and synergistic relationships among the nutrients of NFT-grown young tomato plants. *J. Plant Nutr.* 21:2035–3047, 1998.
63. B. Huang. Nutrient accumulation and associated root characteristics in response to drought stress in tall fescue cultivars. *HortScience* 36(1):148–152, 2001.
64. T.E. Kolb, L.H. McCormick. Etiology of sugar maple decline in four Pennsylvania stands. *Can. J. Forest Res.* 23(11):2395–2402, 1993.
65. T.E. Marler. Solution pH influences on growth and mineral element concentrations of 'Waimanalo' papaya seedlings. *J. Plant Nutr.* 21:2601–2612, 1998.
66. H. Melakeberhan, A.L. Jones, G.W. Bird. Effects of soil pH and *Pratylenchus penetrans* on the mortality of 'Mazzard' cherry seedlings and their susceptibility to *Pseudomonas syringae* pv. *syringae*. *Can. J. Plant Pathol.* 22:131–137, 2000.
67. K.Z. Tan, W.G. Keltjens, G.R. Findenegg. Acid soil damage in sorghum genotypes—role of magnesium-deficiency and root impairment. *Plant Soil* 139(2):149–155, 1992.
68. J.W. Huang, D.L. Grunes. Effects of root temperature and nitrogen form on magnesium uptake and translocation by wheat seedlings. *J. Plant Nutr.* 15(6/7):991–1005, 1992.
69. P. Jensén, J. Perby. Growth and accumulation of N, K^+, Ca^{2+}, and Mg^{2+} in barley exposed to various nutrient regimes and root/shoot temperatures. *Physiol. Plant* 67:159–165, 1986.
70. E.L. Johnson, T.A. Campbell, C.D. Foy. Effect of soil pH on mineral element concentrations of two *Erythroxylum* species. *J. Plant Nutr.* 20(11):1503–1515, 1997.
71. J.W. Huang, R.M. Welch, D.L. Grunes. Magnesium, nitrogen form, and root temperature effects on grass tetany potential of wheat forage. *Agron. J.* 82(3):581–587, 1990.
72. R.B. Clark, S.K. Zeto. Mineral acquisition by arbuscular mycorrhizal plants. *J. Plant Nutr.* 23(7):867–902, 2000.

73. D.H. Lambert, D.E. Baker, H. Cole, Jr. The role of mycorrhizae in the interactions of phosphorus with zinc, copper, and other elements. *Soil Sci. Soc. Am. J.* 43:976–980, 1979.

74. S.R. Saif. Growth responses of tropical forage plant species to vesicular-arbuscular mycorrhizae. *Plant Soil* 97:25–35, 1987.

75. J.O. Siqueira, W.F. Rocha. Jr., E. Oliveira, A. Colozzi-Filho. The relationship between vesicular-arbuscular mycorrhiza and lime: associated effects on the growth and nutrition of brachiaria grass (*Brachiaria decumbens*). *Biol. Fert. Soils* 10:65–71, 1990.

76. C. Sonneveld. Magnesium deficiency in rockwool-grown tomatoes as affected by climatic conditions and plant nutrition. *J. Plant Nutr.* 10:1591–1604, 1987.

77. I. Cakmak, H. Marschner. Magnesium-deficiency and high light intensity enhance activities of superoxide dismutase, ascorbate peroxidase, and glutathione reductase in bean leaves. *Plant Physiol.* 98(4):1222–1227, 1992.

78. D. Zhao, D.M. Oosterhuis. Influence of shade on mineral nutrient status of field-grown cotton. *J. Plant Nutr.* 21(8):1681–1695, 1998.

79. M.P. Harvey, M.H. Brand. Growth and macronutrient accumulation of *Chasmanthium latifolium* (Michx.) Yates and *Hakonechloa macra* Makino 'Aureola' in response to temperature. *HortScience* 37(5):765–767, 2002.

80. J.G. Davis. Soil pH and magnesium effects on manganese toxicity in peanuts. *J. Plant Nutr.* 19(3&4):535–550, 1996.

81. S.G. De Pascale, C. Barbieri, A. Maggio. Growth, water relations, and ion content of field-grown celery (*Apium graveolens* L. var. dulce (Mill.) Pers.) under saline irrigation. *J. Am. Soc. Hort. Sci.* 28(1):136–143, 2003.

82. T.M. Reinbott, D.G. Blevins. Phosphorus and temperature effects and magnesium, calcium, and potassium in wheat and tall fescue leaves. *Agron. J.* 86(3):523–529, 1994.

83. Y.S. Tsiang. Variation and inheritance of certain characters of brome grass, *Bromus inermis* Leyss. *J. Am. Soc. Agron.* 36(6):508–522, 1944.

84. M.I. Piha. Yield potential, fertility requirements, and drought tolerance of grain amaranth compared with maize under Zimbabwean conditions. *Tropical Agric.* 72(1):7–12, 1995.

85. H.A. Mills, J.B. Jones, Jr. *Plant Analysis Handbook II.* Athens, GA: MicroMacro Publishing, Inc., 1996.

86. J.M. Peñalosa, M.D. Cáceres, M.J. Sarro. Nutrition of bean plants in sand culture: influence of calcium/potassium ratio in the nutrient solution. *J. Plant Nutr.* 18:2023–2032, 1995.

87. H. Hauser, B.A. Levine, R.J.P. Williams. Interactions of ions with membranes. *Trends Biochem. Sci.* 1:278–281, 1976.

88. D.W. Johnson, W.T. Swank, J.M. Vose. Simulated effects of atmospheric sulfur deposition on nutrient cycling in a mixed deciduous forest. *Biogeochemistry* 23(3):169–196, 1993.

89. S. Labuda. A new index of cation saturation state in soil. *Commun. Soil Sci. Plant Anal.* 24(13-14):1603–1608, 1993.

90. R.E. Lucas, G.D. Scarseth. Potassium, calcium and magnesium balance and recriprocal (sic) relationship in plants. *J. Am. Soc. Agron.* 39:887–896, 1947.

91. P.H. Nygaard, G. Abrahamsen. Effects of long-term artificial acidification on the ground vegetation and soil in a 100-year-old stand of Scots pine (*Pinus sylvestris*). *Plant Soil* 131(2):151–160, 1991.

92. A. Thimonier, J.L. Dupouey, F. Le Tacon. Recent losses of base cations from soils of *Fagus sylvatica* L. stands in northeastern France. *AMBIO* 29(6):314–321, 2000.

93. C.A. Bower, W.H. Pierre. Potassium response of various crops on a high-lime soil in relation to their contents of potassium, calcium, magnesium, and sodium. *J. Am. Soc. Agron.* 36:608–614, 1944.

94. J.H. Edwards, J.F. Pedersen. Growth and magnesium uptake of tall fescue lines at high and low potassium levels. *J. Plant Nutr.* 9(12):1499–1518, 1986.

95. H. Grimme, L.C. von Braunschweig, K. Németh. Potassium, calcium and magnesium interactions as related to cation uptake and yield (German). *Landvirtschaftlich Forschung. Sonderheft* 30(II):93–100, 1974.

96. K.L. Kabu, E.W. Toop. Influence of potassium-magnesium antagonism on tomato plant growth. *Can. J. Plant Sci.* 50:711–715, 1970.

97. T. Ohno, D.L. Grunes. Potassium-magnesium interactions affecting nutrient uptake by wheat forage. *Soil Sci. Soc. Am. J.* 49:685–690, 1985.

98. M.W. Walker, T.R. Peck. Effect of potassium upon the magnesium status of the corn plant. *Commun. Soil Sci. Plant Anal.* 6(2):189–194, 1975.

99. T. Walsh, T.F. O'Donohue. Magnesium deficiency in some crop plants in relation to the level of potassium nutrition. *J. Agric. Sci.* 35:254–263, 1945.

100. J.E. Leggett, W.A. Gilbert. Magnesium uptake by soybeans. *Physiol. Plant* 44(8):1182–1186, 1969.

101. E.G. Mulder. Nitrogen-magnesium relationships in crop plants. *Plant Soil* 7:341–376, 1956.

102. B. Diem, D.L. Godbold. Potassium, calcium, and magnesium antagonism in clones of *Populus trichocarpa. Plant Soil* 156:411–414, 1993.

103. G.L. Miller. Potassium application reduces calcium and magnesium levels in bermudagrass leaf tissue and soil. *HortScience* 34:265–268, 1999.

104. D.P. Belesky, S.R. Wilkinson. Response of Tifton-44 and Coastal bermudagrass to soil-pH, K, and N source. *Agron. J.* 75(1):1–4, 1983.

105. J.E. Matocha. Influence of sulfur sources and magnesium on forage yields of Coastal bermudagrass (*Cynodon dactylon* (L) Pers). *J. Agron.* 63(3):493–496, 1971.

106. W.T. Pettigrew, W.R. Meredith, Jr. Dry matter production, nutrient uptake, and growth of cotton as affected by potassium fertilization. *J. Plant Nutr.* 20(4/5):531–548, 1997.

107. P.C.R. Fontes, R.A. Reis, Jr, P.R.G. Pereira. Critical potassium concentration and potassium/calcium plus magnesium ratio in potato petioles associated with maximum tuber yields. *J. Plant Nutr.* 19(3/4):657–667, 1996.

108. K.J. Appenroth, H. Gabrys, R.W. Scheuerlein. Ion antagonism in phytochrome-mediated calcium-dependent germination of turions of *Spirodela polyrhiza* (L.) Schleiden. *Planta* 208:583–587, 1999.

109. P.R. Hicklenton, K.G. Cairns. Calcium and magnesium nutrition of containerized *Cotoneaster dammeri* 'Coral Beauty'. *J. Environ. Hort.* 10(2):104–107, 1992.

110. S. Schwartz, B. Bar-Yosef. Magnesium uptake by tomato plants as affected by Mg and Ca concentration in solution culture and plant age. *J. Agron.* 75(2):267–272, 1983.

111. J.M. Spiers, J.H. Braswell. Response of 'Sterling' muscadine grape to calcium, magnesium, and nitrogen fertilization. *J. Plant Nutr.* 17(10):1739–1750, 1994.

112. J.L. Sims, W.S. Scholtzhauer, J.H. Grove. Soluble calcium fertilizer effects on early growth and nutrition of burley tobacco. *J. Plant Nutr.* 18(5):911–921, 1995.

113. W.W. Garner, J.E. McMurtrey, Jr., J.O. Bowling, Jr., E.G. Moss. Magnesium and calcium requirements of the tobacco crop. *J. Agric. Res.* 40:145–168, 1930.

114. E.A.S. Paiva, R.A. Sampaio, H.E.P. Martinez. Composition and quality of tomato fruit cultivated in nutrient solutions containing different calcium concentrations. *J. Plant Nutr.* 21:2653–2661, 1998.

115. P. Morard, A. Pujos, A. Bernadac, G. Bertoni. Effect of temporary calcium deficiency on tomato growth and mineral nutrition. *J. Plant Nutr.* 19:115–127, 1996.

116. R.B. Clark, V.C. Baligar, K.D. Ritchey, S.K. Zeto. Maize growth and mineral acquisition on acid soil amended with flue gas desulfurization by-products and magnesium. *Commun. Soil Sci. Plant Anal.* 28(15/16):1441–1459, 1997.

117. K.A. Handreck, N.D. Black. *Growing Media for Ornamental Plants and Turf.* Sydney, Australia: University of New South Wales Press, 1999.

118. J.M. Navarro, V. Matinez, A. Cerda, M.A. Botella. Effect of salinity x calcium interaction on cation balance in melon plants grown under two regimes of orthophosphate. *J. Plant Nutr.* 23(7):991–1006, 2000.

119. C.E. Whitcomb. Effects of dolomite levels and supplemental magnesium on growth of container nursery stock. *Okla. Agric. Exp. Sta. Res. Bull.* 855, 1984.

120. C.E. Whitcomb. Suggested calcium and magnesium levels for accelerating plant growth. *Okla. Agric. Exp. Sta. Res. Bull.* 855, 1984.

121. C.E. Whitcomb. Plant nutrition. In: C.E. Whitcomb, ed. *Plant Production in Containers.* Stillwater, Okla.: Lacebark Publications, 1988, pp. 240.

122. L. Puech, B. Mehne-Jakobs. Histology of magnesium-deficient Norway spruce needles influenced by nitrogen source. *Tree Physiol.* 17(5):301–310, 1997.

123. G.E. Wilcox, J.E. Hoff, C.M. Jones. Ammonium reduction of Ca and Mg content of tomato and sweet-corn leaf tissue and influence on incidence of blossom end rot of tomato fruit. *J. Am. Soc. Hort. Sci.* 98:–89, 1973.

124. W.O. Chance III, Z.C. Somda, H.A. Mills. Effect of nitrogen form during the flowering period on zucchini squash growth and nutrient element uptake. *J. Plant Nutr.* 22:597–607, 1999.

125. M. Gulpen, K.H. Feger. Magnesium and calcium nutrition of spruce on high altitude sites - yellowing status and effects of fertilizer application (German). *Z. Pflanzenernahrung Bodenkunde* 161(6):671–679, 1998.

126. B. Mehne-Jakobs, M. Gulpen. Influence of different nitrate to ammonium ratios on chlorosis, cation concentrations and the binding forms of Mg and Ca in needles of Mg-deficient Norway spruce. *Plant Soil* 188(2):267–277, 1997.

127. M.S. Irshad, S. Yamamoto, A.E. Eneji, T. Honna. Wheat response to nitrogen source under saline conditions. *J. Plant Nutr.* 25(12):2603–2612, 2002.

128. M.J. Sarro, J.M. Penalosa, J.M. Sãnchez. Influence of ammonium uptake on bean nutrition. *J. Plant Nutr.* 21(9):1913–1920, 1998.

129. K.M. Batal, D.M. Granberry, B.G. Mullinix. Nitrogen, magnesium, and boron applications affect cauliflower yield, curdmass and hollow stem disorder. *HortScience* 32(1):75–78, 1997.

130. H.L. Scoggins, H.A. Mills. Poinsettia growth, tissue nutrient concentration, and nutrient uptake as influenced by nitrogen form and stage of growth. *J. Plant Nutr.* 21(1):191–198, 1998.

131. T.M.A. Choudhury, Y.M. Khanif. Evaluation of effects of nitrogen and magnesium fertilization on rice yield and fertilizer nitrogen efficiency using ^{15}N tracer technique. *J. Plant Nutr.* 24(6):855–871, 2001.

132. M. Carvajal, V. Martínez, A. Cerdá. Modification of the response of saline stressed tomato plants by the correction of cation disorders. *Plant Growth Regul.* 30(1):37–47, 2000.

133. P.C. Chiy, C.J.C. Phillips. Effects of sodium fertilizer on the chemical composition of perennial ryegrass and white clover leaves of different physiological ages. *J. Sci. Food Agric.* 73(3):337–348, 1997.

134. M.A. Khan, I.A. Ungar, A.M. Showalter. Effects of salinity on growth, ion content, and osmotic relations in *Halopyrum mucronatum* (L.) stapf. *J. Plant Nutr.* 22:191–204, 1999.

135. X.J. Liu, Y.M. Yang, W.Q. Li, C.X. Li, D.Y. Duan, T. Tadano. Interactive effects of sodium chloride and nitrogen on growth and ion accumulation of a halophyte. *Commun. Soil Sci. Plant Anal.* 35(15–16):2111–2123, 2004.

136. S. Hill, R.S. Miyasaka, R. Abaidoo. Sodium chloride concentration affects early growth and nutrient accumulation in taro. *HortScience* 33(7):1153–1156. 1998.

137. C.M. Grieve, J.A. Poss. Wheat response to interactive effects of boron and salinity. *J. Plant Nutr.* 23(9):1217–1226, 2000.

138. D.D. Warncke, T.C. Reid, M.K. Hausbeck. Sodium chloride and lime effects on soil cations and elemental composition of asparagus fern. *Commun. Soil Sci. Plant Anal.* 33(15–18):3075–3084, 2002.

139. A.A. Carbonell-Barrachina, F. Burló-Carbonell, J. Mataix-Beneyto. Effect of sodium arsenite and sodium chloride on bean plant nutrition (macronutrients). *J. Plant Nutr.* 20:1617–1633, 1997.

140. M.C. Shannon, S.C. Scardaci, M.D. Spyres, J.D. Rhoades, J.H. Draper. Assessment of salt tolerance in rice cultivars in response to salinity problems in California. *Crop Sci.* 38(2):394–398, 1998.

141. E. Turhan, A. Eris. Effects of sodium chloride applications and different growth media on ionic composition in strawberry plant. *J. Plant Nutr.* 27(9):1653–1665, 2004.

142. T.A. Essa. Effect of salinity stress on growth and nutrient composition of three soybean (*Glycine max* L. Merrill) cultivars. *J. Agron. Crop Sci.* 188(2):86–93, 2002.

143. V. Nenova, I. Stoyanov. Physiological and biochemical changes in young maize plants under iron deficiency. 3. Concentration and distribution of some nutrient elements. *J. Plant Nutr.* 22:565–578, 1999.

144. S.C. Agarwala, S.C. Mehrotra. Iron-magnesium antagonism in growth and metabolism of radish. *Plant Soil* 80(3):355–361, 1984.

145. L. Bavaresco. Relationship between chlorosis occurrence and mineral composition of grapevine leaves and berries. *Commun. Soil Sci. Plant Anal.* 28(1/2):13–21, 1997.

146. J.P. White, C.R. Cantor. Role of magnesium in binding of tetracycline to *Escherichia coli* ribosomes. *J. Mol. Biol.* 58:397–400, 1971.

147. S. Alam, S. Kamei, S. Kawai. Amelioration of manganese toxicity in young rice seedlings with potassium. *J. Plant Nutr.* 26(6):1301–1314, 2003.

148. V.M.L. Quartin, M.L. Antunes, M.C. Muralha, M.M. Sousa, M.A. Nunes. Mineral imbalance due to manganese excess in triticales. *J. Plant Nutr.* 24(1):175–189, 2001.

149. M.P. Löhnis. Effect on magnesium and calcium supply on the uptake of manganese by various crop plants. *Plant Soil* 12:339–376, 1960.

150. E.V. Maas, D.P. Moore, B.J. Mason. Influence of calcium and magnesium on manganese absorption. *Physiol. Plant* 44:796–800, 1969.

151. C.D. Foy, R.R. Weil, C.A. Coradetti. Differential manganese tolerances of cotton genotypes in nutrient solutions. *J. Plant Nutr.* 18(4):685–706, 1995.

152. P.C.R. Fontes, M.A. Moreira, R.L.F. Fontes, A.A. Cardosa. Effects of zinc fungicides and different zinc fertilizer application methods on soluble and total zinc in potato shoots. *Commun. Soil Sci. Plant Anal.* 30(13–14):1847–1859, 1999.

153. C. Kaya, M.A.S. Burton, D.E.B. Higgs. Responses of tomato CVs growth to fruit-harvest stage under zinc stress in glasshouse conditions. *J. Plant Nutr.* 24(2):369–382, 2001.

154. S.B. Kalyanaraman, P. Sivagurunathan. Effect of zinc on some important macroelements and microelements in black gram leaves. *Commun. Soil Sci. Plant Anal.* 25(13–14):2247–2259, 1994.

155. M. Bonnet, O. Camares, P. Veisseire. Effects of zinc and influence of *Acremonium lolii* on growth parameters, chlorophyll *a* fluorescence and antioxidant enzyme activities of ryegrass (*Lolium perenne* L. cv Apollo). *J. Exp. Bot.* 51:945–953, 2000.

156. T. Kim, H.A. Mills, H.Y. Wetzstein. Studies on the effect of zinc supply on growth and nutrient uptake in pecan. *J. Plant Nutr.* 25(9):1987–2000, 2002.

157. D.W. James, C.J. Hurst, T.A. Tindall. Alfalfa cultivar response to phosphorus and potassium-deficiency—Elemental composition of the herbage. *J. Plant Nutr.* 18(11):2447–2464, 1995.

158. N.K. Fageria, F.J.P. Zimmermann, V.C. Baligar. Lime and phosphorus interactions on growth and nutrient uptake by upland rice, wheat, common bean, and corn in an oxisol. *J. Plant Nutr.* 18(11):2519–2532, 1995.

159. T.M. Reinbott, D.G. Blevins. Phosphorus and magnesium fertilization interaction with soil phosphorus leaves: Tall fescue yield and mineral element content. *J. Prod. Agric.* 10(2):260–265, 1997.

160. T.M. Reinbott, D.G. Blevins. Phosphorus nutritional effects on root hydraulic conductance, xylem water flow and flux of magnesium and calcium in squash plants. *Plant Soil* 209(2):263–273, 1999.

161. S.A. Hill, S.C. Miyasaka, R.S. Yost. Taro responses to excess copper in solution culture. *HortScience* 35(5):863–867, 2000.

162. G. Ouzounidou, I. Ilias, H. Tranopoulou, S. Karatglis. Amelioration of copper toxicity by iron on spinach physiology. *J. Plant Nutr.* 21:2089–2101, 1998.

163. I.N. Fisarakis, P. Nikolaou, I. Therios, D. Stavrakas. Effect of salinity and rootstock on concentration of potassium, calcium, magnesium, phosphorus, and nitrate-nitrogen in Thompson seedless grapevine. *J. Plant Nutr.* 27(12):2117–2134, 2004.

164. F. Adams, Z.F. Lund. Effect of chemical activity of soil solution aluminum on cotton root penetration of acid subsoils. *Soil Sci.* 101:193–198, 1966.

165. D.L. Godbold. Aluminum decreases root growth and calcium and magnesium uptake in *Picea abies* seedlings. *Dev. Plant Soil Sci.* 45:747–753, 1991.

166. P.G. Schaberg, J.R. Cumming, P.F. Murakami, C.H. Borer, D.H. DeHaves, G.J. Hawley, G.R. Strimbeck. Acid mist and soil Ca and Al alter the mineral nutrition and physiology of red spruce. *Tree Physiol.* 20(2):73–85, 2000.

167. S.C. Miyasaka, C.M. Webster, E.N. Okazaki. Differential response of 2 taro cultivars to aluminum. 2. Plant mineral concentrations. *Commun. Soil Sci. Plant Anal.* 24(11–12):1213–1229, 1993.

168. W.G. Keltjens. Magnesium uptake by Al-stressed maize plants with special emphasis on cation interactions at root exchange sites. *Plant Soil* 171:141–146, 1995.

169. F.C. Lidon, H.G. Azinheira, M.G. Barreiro. Aluminum toxicity in maize: Biomass production and nutrient uptake and translocation. *J. Plant Nutr.* 23(2):151–160, 2000.

170. J.W. Huang, D.L. Grunes. Potassium/magnesium ratio effects on aluminum tolerance and mineral composition of wheat forage. *Agron. J.* 84:643–650, 1992.

171. J. Mendoza, F. Borie. Effect of *Glomus etunicatum* inoculation on aluminum, phosphorus, calcium, and magnesium uptake of two barley genotypes with different aluminum tolerance. *Commun. Soil Sci. Plant Anal.* 29(5&6):681–695, 1998.

172. Z. Rengel, D.L. Robinson. Aluminum effects on growth and macronutrient uptake by annual ryegrass. *Agron. J.* 81:208–215, 1989.

173. I.R. Silva, T.J. Smyth, D.W. Israel, C.D. Raper, T.W. Rufty. Magnesium ameliorates aluminum rhizotoxicity in soybean by increasing citric acid production and exudation by roots. *Plant Cell Physiol.* 42:546–554, 2001.

174. J. Pintro, P. Fallavier, J. Barloy. Aluminum effects on the growth and mineral composition of corn plants cultivated in nutrient solution at low aluminum activity. *J. Plant Nutr.* 19(5):729–741, 1996.

175. G.J. Schwab, C.H. Burmester, G.L. Mullins. Growth and nutrient uptake by cotton roots under field conditions. *Commun. Soil Sci. Plant Anal.* 31(1/2):149–164, 2000.

176. Z. Rengel, D.L. Robinson. Modeling magnesium uptake from an acid soil. I. Nutrient relationships at the soil root interface. *Soil Sci. Soc. Am. J.* 54(3):785–791, 1990.

177. Z. Rengel, D.L. Robinson. Modeling magnesium uptake from an acid soil. II. Barber-Cushman model. *Soil Sci. Soc. Am. J.* 54(3):791–795, 1990.

178. H.A. Mills, H.L. Scoggins. Nutritional levels for anthurium: young versus mature leaves. *J. Plant Nutr.* 21:199–203, 1998.

179. J.B. Drossopoulos, J. Karides, S.N. Chorianopoulou, C. Kitsaki, D.L. Bouranis, S. Kintsios, G. Aivalakis. Effect of nitrogen fertilization on distribution profiles of selected macronutrients in oriental field-grown tobacco plants. *J. Plant Nutr.* 22(3):527–541, 1999.

180. M. Sadiq, G. Hussain. Effect of chelate fertilizers on dry matter and metallic composition of bean plants in a pot experiment. *J. Plant Nutr.* 7(9):1477–1488, 1994.

181. M.P. Jiménez, D. Effón, A.M. de la Horra, R. Defrieri. Foliar potassium, calcium, magnesium, zinc, and manganese content in soybean cultivars at different stages of development. *J. Plant Nutr.* 19(6):807–816, 1996.

182. C.A. Mullins, J.D. Wolt. Effects of calcium and magnesium line sources on yield, fruit quality, and elemental uptake of tomato. *J. Am. Soc. Hort. Sci.* 108:850–854, 1983.

183. G.L. Mullins, C.H. Burmester. Uptake of calcium and magnesium by cotton grown under dryland conditions. *J. Agron.* 84(4):564–569, 1992.

184. J.N. McCrimmon. Macronutrient and micronutrient concentrations of seeded bermudagrasses. *Commun. Soil Sci. Plant Anal.* 33(15/18):2739–2758, 2002.

185. E.A.S. Rosa, E. Schnug, S.H. Haneklaus. Mineral content of primary and secondary inflorescences of eleven broccoli cultivars grown in early and late seasons. *J. Plant Nutr.* 25(8):1741–1751, 2002.

186. B. Drossopoulos, G.G. Kouchaji, D.L. Bouranis. Seasonal dynamics of mineral nutrients by walnut tree reproductive organs. *J. Plant Nutr.* 19(2):421–434, 1996.

187. B. Drossopoulos, G.G. Kouchaji, D.L. Bouranis. Seasonal dynamics of mineral nutrients and carbohydrates by walnut tree leaves. *J. Plant Nutr.* 19(3&4):493–516. 1996.

188. S.M.E. Satti, R.A. Al-Yhyai, F. Al-Said. Fruit quality and partitioning of mineral elements in processing tomato in response to saline nutrients. *J. Plant Nutr.* 19:705–715, 1996.

189. C. Sonneveld, W. Voogt. Effects of Ca-stress on blossom-end rot and Mg-deficiency in rockwool grown tomato. *Acta Horticulturae* 294:81–88, 1991.

190. B. Bengtsson, P. Jensen. Uptake and distribution of calcium, magnesium and potassium in cucumber of different age. *Physiol. Plant* 57:428–434, 1983.

191. E.T.F. Witkowski, B.B. Lamont. Disproportionate allocation of mineral nutrients and carbon between vegetative and reproductive structures in *Banksia hookeriana*. *Oecologia* 105:38–42, 1996.

192. M.A. Khan, I.A. Ungar. Comparative study of chloride, calcium, magnesium, potassium, and sodium content of seeds in temperate and tropical halophytes. *J. Plant Nutr.* 19:517–525, 1996.

193. K. Mengel, E.A. Kirkby. *Principles of Plant Nutrition*. Bern: International Potash Institute, 1982.

194. A. Mozafar. Distribution of nutrient elements along the maize leaf: alteration by iron deficiency. *J. Plant Nutr.* 20:999–1005, 1997.

195. J.T. Batchelor, H.D. Scott, R.F. Sojka. Influence of irrigation and growth state on element concentrations of soybean plant parts. *Commun. Soil Sci. Plant Anal.* 15(9):1083–1109 1984.

196. R.A. Cline. Calcium and magnesium effects on rachis necrosis of interspecific hybrids of Euvitis grapes cv. Canada Muscat and cv. Himrod grapes. *J. Plant Nutr.* 10:1897–1905, 1987.

197. B.R. Fudge. Relation of magnesium deficiency in grapefruit leaves to yield and chemical composition of fruit. *Fla. Univ. Agric. Exp. Sta. Bull.* 331:1–36, 1939.

198. J.L. Walworth, S. Ceccotti. A re-examination of optimum foliar magnesium levels in corn. *Commun. Soil Sci. Plant Anal.* 21(13–16):1457–1473, 1990.

199. M.T. Austin, M. Constantinides, S.C. Miyasaka. Effect of magnesium on early taro growth. *Commun. Soil Sci. Plant Anal.* 25(11/12):2159–2169, 1994.

200. J.H. Edwards, B.D. Horton. Influence of magnesium concentrations in nutrient solution on growth, tissue concentration, and nutrient uptake of peach seedlings. *J. Am. Soc. Hort. Sci.* 106:401–405, 1981.

201. A. Jacob. *Magnesium: The Fifth Major Plant Nutrient.* Translated from the German by Dr. Norman Walker. London: Staples Press Ltd., 1958.

202. R.C. Selley, L.R.M. Cocks, I.R. Plimer. *Encyclopedia of Geology.* New York: Elsevier Academic Press, 2005, pp. 403–409.

203. J.S. Burd, J.C. Martin. Water displacement of soils and the soil solution. *J. Agric. Sci.* 13:265–295, 1923.

204. J.S. Burd, J.C. Martin. Secular and seasonal changes in the soil solution. *Soil Sci.* 18:151–167, 1924.

205. P.S. Burgess. The soil solution, extracted by Lipman's direct-pressure method, compared with 1:5 water extracts. *Soil Sci.* 14:191–215, 1922.

206. M. Fried, R.E. Shapiro. Soil–plant relationships in ion uptake. *Ann. Rev. Plant Physiol.* 12:91–112, 1961.

207. D.R. Hoagland, J.C. Martin, G.R. Stewart. Relation of the soil solution to the soil extract. *J. Agric. Res.* 20:381–404, 1920.

208. California Fertilizer Assn. *Western Fertilizer Handbook, Horticulture Edition.* Danville, IL: Interstate Publishers, Inc., 1990.

209. N.C. Brady. *The Nature and Property of Soils.* 10th ed. New York: Macmillan Publishing Company, 1990.

210. P.J. Van Erp, V.J.G. Houba, M.L. van Beusichem. Effect of drying temperature on amount of nutrient elements extracted with 0.01 M $CaCl_2$ soil extraction procedure. *Commun. Soil Sci. Plant Anal.* 32(1&2):33–48, 2001.

211. M.F. Morgan. Chemical soil diagnosis by the universal soil testing system. *Conn. Agric. Exp. Sta. Bull.* 450, 1941.

212. I. Ortas, N. Güzel, H. Ibrikci. Determination of potassium and magnesium status of soils using different soil extraction procedures in the upper part of Mesopotamia (in the Harran Plain). *Commun. Soil Sci. Plant Anal.* 30:2607–2625, 1999.

213. B. Wolf. An improved universal extracting solution and its use for diagnosing soil fertility. *Commun. Soil Sci. Plant Anal.* 13(12):1005–1033, 1982.

214. T.L. Yuan. A double buffer method for the determination of lime requirement of acid soils. *Soil Sci. Soc. Am. Proc.* 38:437–440, 1974.

215. A. Mehlich. New extractant for soil test evaluation of phosphorus, potassium, magnesium, calcium, sodium, manganese, and zinc. *Commun. Soil Sci. Plant Anal.* 13:1005–1033, 1978.

216. P.N. Soltanpour, A.P. Schwab. A new soil test for simultaneous extraction of macro-and micro-nutrients in alkaline soils. *Commun. Soil Sci. Plant Anal.* 8(3):195–207, 1977.

217. T. Drakenberg. Physical methods for studying the biological chemistry of magnesium. In: J.A. Cowan, ed. *The Biological Chemistry of Magnesium.* New York: VCH Publishers, Inc., 1995.

218. U.S. Environmental Protection Agency. Methods for the Determination of Metals in Environmental Samples—Supplement 1, EPA/600/R-94/111. USEPA, Washington, DC, 1994.

219. U.S. Environmental Protection Agency. SW-846, Test Methods for Evaluating Solid Waste, Physical/Chemical Methods. 955-001-00000-1. USEPA, Washington, DC, 2005.

220. F.E. Bear et al.. *Hunger Signs in Crops; A Symposium.* Am. Soc. Agron. National Fert. Assoc., Washington, DC, 1949.

221. S.S. Fernández, A. Merino. Plant heavy metal concentrations and soil biological properties in agricultural serpentine soils. *Commun. Soil Sci. Plant Anal.* 30(13&14):1867–1884, 1999.

222. K.D. Starr, R.D. Wright. Calcium and magnesium requirements of *Ilex crenata* 'Helleri'. *J. Am. Soc. Hort. Sci.* 109(6):857–860, 1984.

223. D.L. Fuller, W.A. Meadows. Effects of powdered vs. pelletized dolomite and two fertilizer regimes on pH of growth medium and quality of eight woody species in containers. *Southern Nurserymen's Assn.* 9(1):1–7, 1983.

224. R.P. White, D.C. Munro. Magnesium availability and plant uptake from different magnesium sources in a greenhouse experiment. *Can. J. Soil Sci.* 61:397–400, 1981.

225. R.C. Tayrien, C.E. Whitcomb. An evaluation of calcium and magnesium sources and water quality on container grown *Nandina*. *Okla. Agric. Exp. Sta. Res. Bull.* 855, 1984.

226. C.R. Obatolu. Correction of magnesium deficiency in tea plants through foliar applications. *Commun. Soil Sci. Plant Anal.* 30:1649–1655, 1999.

227. U.S. Jones. *Fertilizers & Soil Fertility.* Reston, VA: Reston Publishing Company, 1979.

228. R.C. Weast, ed. *Handbook of Chemistry and Physics*. Cleveland, OH: Chemical Rubber Company, 1970.
229. S.A. Barber. Liming materials and practices. In: F. Adams, ed. *Soil Acidity and Liming, Agronomy 12*, Am. Soc. Agron, Madison, WI: 1984, pp. 270–282.
230. S.I. Tisdale, H.G. Cunningham. Advances in manufacturing of secondary and micronutrient fertilizers. In: M. McVickar, G.L. Bridger, L.B. Nelson, eds. *Fertilizer Technology and Usage*. Madison, WI: Soil Science Society of America, 1963, pp. 269–286.

7 Sulfur

Silvia Haneklaus, Elke Bloem, and Ewald Schnug
Institute of Plant Nutrition and Soil Science, Braunschweig,
Germany

Luit J. de Kok and Ineke Stulen
University of Groningen, Haren, The Netherlands

CONTENTS

7.1 Introduction 183
7.2 Sulfur in Plant Physiology 184
    7.2.1 Uptake, Transport, and Assimilation of Sulfate 185
        7.2.1.1 Foliar Uptake and Metabolism of Sulfurous Gases 187
    7.2.2 Major Organic Sulfur Compounds 188
    7.2.3 Secondary Sulfur Compounds 192
    7.2.4 Interactions between Sulfur and Other Minerals 195
        7.2.4.1 Nitrogen–Sulfur Interactions 195
        7.2.4.2 Interactions between Sulfur and Micronutrients 197
7.3 Sulfur in Plant Nutrition 198
    7.3.1 Diagnosis of Sulfur Nutritional Status 198
        7.3.1.1 Symptomatology of Single Plants 198
        7.3.1.2 Symptomatology of Monocots 200
        7.3.1.3 Sulfur Deficiency Symptoms on a Field Scale 201
7.4 Soil Analysis 202
7.5 Plant Analysis 206
    7.5.1 Analytical Methods 206
    7.5.2 Assessment of Critical Nutrient Values 208
    7.5.3 Sulfur Status and Plant Health 217
7.6 Sulfur Fertilization 219
Acknowledgment 223
References 223

7.1 INTRODUCTION

Sulfur (S) is unique in having changed within just a few years, from being viewed as an undesired pollutant to being seen as a major nutrient limiting plant production in Western Europe. In East Asia, where, under current legislative restrictions, sulfur dioxide (SO_2) emissions are expected to increase further by 34% by 2030 (1), considerations of sulfur pollution are a major issue. Similarly in Europe, sulfur is still associated with its once detrimental effects on forests which peaked in the

1970s (2), and which gave this element the name 'yellow poison.' With Clean Air Acts coming into force at the start of the 1980s, atmospheric sulfur depositions were reduced drastically and rapidly in Western Europe, and declined further in the 1990s after the political transition of Eastern European countries. In arable production, sulfur deficiency can be retraced to the beginning of the 1980s (3). Since then, severe sulfur deficiency has become the main nutrient disorder of agricultural crops in Western Europe. It has been estimated that the worldwide sulfur fertilizer deficit will reach 11 million tons per year by 2012, with Asia (6 million tons) and the Americas (2.3 million tons) showing the highest shortage (4).

Severe sulfur deficiency not only reduces crop productivity and diminishes crop quality, but it also affects plant health and environmental quality (5). Yield and quality in relation to the sulfur nutritional status for numerous crops are well described in the literature. In comparison, research in the field of interactions between sulfur and pests and diseases is relatively new. Related studies indicate the significance of the sulfur nutritional status for both beneficial insects and pests.

Since the very early days of research on sulfur in the 1930s, significant advances have been made in the field of analysis of inorganic and organic sulfur compounds. By employing genetic approaches in life science research, significant advances in the field of sulfur nutrition, and in our understanding of the cross talk between metabolic pathways involving sulfur and interactions between sulfur nutrition and biotic and abiotic stresses, can be expected in the future.

This chapter summarizes the current status of sulfur research with special attention to physiological and agronomic aspects.

7.2 SULFUR IN PLANT PHYSIOLOGY

Sulfur is an essential element for growth and physiological functioning of plants. The total sulfur content in the vegetative parts of crops varies between 0.1 and 2% of the dry weight (0.03 to 0.6 mmol S g^{-1} dry weight). The uptake and assimilation of sulfur and nitrogen by plants are strongly interrelated and dependent upon each other, and at adequate levels of sulfur supply the organic N/S ratio is around 20:1 on a molar basis (6–9). In most plant species the major proportion of sulfur (up to 70% of the total S) is present in reduced form in the cysteine and methionine residues of proteins. Additionally, plants contain a large variety of other organic sulfur compounds such as thiols (glutathione; ~1 to 2% of the total S) and sulfolipids (~1 to 2% of the total S); some species contain the so-called secondary sulfur compounds such as alliins and glucosinolates (7,8,10,11). Sulfur compounds are of great significance in plant functioning, but are also of great importance for food quality and the production of phyto-pharmaceuticals (8,12).

In general, plants utilize sulfate (S^{6+}) taken up by the roots as a sulfur source for growth. Sulfate is actively taken up across the plasma membrane of the root cells, subsequently loaded into the xylem vessels and transported to the shoot by the transpiration stream (13–15). In the chloroplasts of the shoot cells, sulfate is reduced to sulfide (S^{2-}) prior to its assimilation into organic sulfur compounds (16,17). Plants are also able to utilize foliarly absorbed sulfur gases; hence chronic atmospheric sulfur dioxide and hydrogen sulfide levels of 0.05 μL L^{-1} and higher, which occur in polluted areas, contribute substantially to the plant's sulfur nutrition (see below; 18–21).

The sulfur requirement varies strongly between species and it may fluctuate during plant growth. The sulfur requirement can be defined as 'the minimum rate of sulfur uptake and utilization that is sufficient to obtain the maximum yield, quality, and fitness,' which for crop plants is equivalent to 'the minimum content of sulfur in the plant associated with maximum yield' and is regularly expressed as kg S ha^{-1} in the harvested crop. In physiological terms the sulfur requirement is equivalent to the rate of sulfur uptake, reduction, and metabolism needed per gram plant biomass produced over time and can be expressed as mol S g^{-1} plant day^{-1}. The sulfur requirement of a crop at various stages of development under specific growth conditions may be predicted by upscaling the sulfur requirement in μmol S g^{-1} plant day^{-1} to mol S ha^{-1} day^{-1} by estimating the crop biomass density per hectare (tons of plant biomass ha^{-1}). When a plant is in the vegetative growth period, the sulfur requirement ($S_{requirement}$, expressed as μmol S g^{-1} plant day^{-1}) can be calculated as follows (11):

$$S_{requirement} = S_{content} \times RGR$$

where $S_{content}$ represents the total sulfur concentration of the plant (μmol g^{-1} plant biomass) and RGR is the relative growth rate of the plant (g g^{-1} plant day^{-1}). The RGR can be calculated by using the following equation:

$$RGR = (\ln W_2 - \ln W_1)/(t_2 - t_1)$$

where W_1 and W_2 are the total plant weight (g) at time t_1 and t_2, respectively, and $t_2 - t_1$ the time interval (days) between harvests. In general, the sulfur requirement of different crop species grown at optimal nutrient supply and growth conditions ranges from 0.01 to 0.1 mmol g^{-1} plant dry weight day^{-1}. Generally, the major proportion of the sulfate taken up is reduced and metabolized into organic compounds, which are essential for structural growth. However, in some plant species, a large proportion of sulfur is present as sulfate and in these cases, for structural growth, the organic sulfur content may be a better parameter for the calculation of the sulfur requirement (see also Section 7.3.1.3).

7.2.1 UPTAKE, TRANSPORT, AND ASSIMILATION OF SULFATE

The uptake and transport of sulfate in plants is mediated by sulfate transporter proteins and is energy-dependent (driven by a proton gradient generated by ATPases) through a proton–sulfate (presumably 3H$^+$/SO$_4^{2-}$) co-transport (14). Several sulfate transporters have been isolated and their genes have been identified. Two classes of sulfate transporters have been identified: the so-called 'high- and low-affinity sulfate transporters,' which operate ideally at sulfate concentrations < 0.1 mM and ≥ 0.1 mM, respectively. According to their cellular and subcellular expression, and possible functioning, the sulfate transporter gene family has been classified into as many as five different groups (15,22–24). Some groups are expressed exclusively in the roots or shoots, or in both plant parts. Group 1 transporters are high-affinity sulfate transporters and are involved in the uptake of sulfate by the roots. Group 2 are vascular transporters and are low-affinity sulfate transporters. Group 3 is the so-called 'leaf group;' however, still little is known about the characteristics of this group. Group 4 transporters may be involved in the transport of sulfate into the plastids prior to its reduction, whereas the function of Group 5 sulfate transporters is not yet known. Regulation and expression of the majority of sulfate transporters are controlled by the sulfur nutritional status of the plants. A rapid decrease in root sulfate content upon sulfur deprivation is regularly accompanied by a strongly enhanced expression of most sulfate transporter genes (up to 100-fold), accompanied by a substantial enhanced sulfate uptake capacity. It is still questionable whether, and to what extent, sulfate itself or metabolic products of sulfur assimilation (viz O-acetylserine, cysteine, glutathione) act as signals in the regulation of sulfate uptake by the root and its transport to the shoot, and in the expression of the sulfate tranporters involved (15,22–24).

The major proportion of the sulfate taken up by the roots is reduced to sulfide and subsequently incorporated into cysteine, the precursor and the reduced sulfur donor for the synthesis of most other organic sulfur compounds in plants (16,17,25–27). Even though root plastids contain all sulfate reduction enzymes, reduction predominantly takes place in the chloroplasts of the shoot. The reduction of sulfate to sulfide occurs in three steps (Figure 7.1). First, sulfate is activated to adenosine 5′-phosphosulfate (APS) prior to its reduction, a reaction catalyzed by ATP sulfurylase. The affinity of this enzyme for sulfate is rather low (K_m ~1 mM) and the in situ sulfate concentration in the chloroplast may be rate-limiting for sulfur reduction (7). Second, the activated sulfate (APS) is reduced by APS reductase to sulfite, a reaction where glutathione (RSH; Figure 7.1) most likely functions as reductant (17,26). Third, sulfite is reduced to sulfide by sulfite reductase with reduced ferredoxin as reductant. Sulfide is

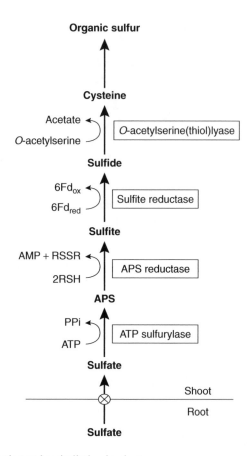

FIGURE 7.1 Sulfate reduction and assimilation in plants.

subsequently incorporated into cysteine, catalyzed by O-acetylserine(thiol)lyase, with O-acetylserine as substrate (Figure 7.1). The formation of O-acetylserine is catalyzed by serine acetyltransferase, and together with O-acetylserine(thiol)lyase it is associated as an enzyme complex named cysteine synthase (28,29). The synthesis of cysteine is a major reaction in the direct coupling between sulfur and nitrogen metabolism in the plant (6,9).

Sulfur reduction is highly regulated by the sulfur status of the plant. Adenosine phosphosulfate reductase is the primary regulation point in the sulfate reduction pathway, since its activity is generally the lowest of the enzymes of the assimilatory sulfate reduction pathway and this enzyme has a fast turnover rate (16,17,26,27). Regulation may occur both by allosteric inhibition and by metabolite activation or repression of expression of the genes encoding the APS reductase. Both the expression and activity of APS reductase change rapidly in response to sulfur starvation or exposure to reduced sulfur compounds. Sulfide, O-acetylserine, cysteine, or glutathione are likely regulators of APS reductase (9,16,17,26). The remaining sulfate in plant tissue is predominantly present in the vacuole, since the cytoplasmatic concentration of sulfate is kept rather constant. In general, the remobilization and redistribution of the vacuolar sulfate reserves is a rather slow process. Under temporary sulfur-limitation stress it may be even too low to keep pace with the growth of the plant, and therefore sulfur-deficient plants may still contain detectable levels of sulfate (13,15,22).

Cysteine is used as the reduced sulfur donor for the synthesis of methionine, the other major sulfur-containing amino acid present in plants, via the so-called trans-sulfurylation pathway (30,31). Cysteine is also the direct precursor for the synthesis of various other compounds such as glutathione, phytochelatins, and secondary sulfur compounds (12,32). The sulfide residue of the cysteine moiety in proteins is furthermore of great importance in substrate binding of enzymes, in metal–sulfur clusters in proteins (e.g., ferredoxins), and in regulatory proteins (e.g., thioredoxins).

7.2.1.1 Foliar Uptake and Metabolism of Sulfurous Gases

In rural areas the atmosphere generally contains only trace levels of sulfur gases. In areas with volcanic activity and in the vicinity of industry or bioindustry, high levels of sulfurous air pollutants may occur. Sulfur dioxide (SO_2) is, in quantity and abundance, by far the most predominant sulfurous air pollutant, but locally the atmosphere may also be polluted with high levels of hydrogen sulfide (18,19,21). Occasionally the air may also be polluted with enhanced levels of organic sulfur gases, viz carbonyl sulfide, methyl mercaptan, carbon disulfide, and dimethyl sulfide (DMS).

The impact of sulfurous air pollutants on crop plants appears to be ambiguous. Upon their foliar uptake, SO_2 and H_2S may be directly metabolized, and despite their potential toxicity used as a sulfur source for growth (18–21). However, there is no clear-cut transition in the level or rate of metabolism of the absorbed sulfur gases and their phytotoxicity, and the physiological basis for the wide variation in susceptibility between plants species and cultivars to atmospheric sulfur gases is still largely unclear (18–21). These paradoxical effects of atmospheric sulfur gases complicate the establishment of cause–effect relationships of these air pollutants and their acceptable atmospheric concentrations in agro-ecosystems.

The uptake of sulfurous gases predominantly proceeds via the stomata, since the cuticle is hardly permeable to these gases (33). The rate of uptake depends on the stomatal and the leaf interior (mesophyll) conductance toward these gases and their atmospheric concentration, and may be described by Fick's law for diffusion

$$J_{gas} \ (\text{pmol cm}^{-2} \ \text{s}^{-1}) = g_{gas} \ (\text{cm s}^{-1}) \times \Delta_{gas} \ (\text{pmol cm}^{-3})$$

where J_{gas} represents the gas uptake rate, g_{gas} the diffusive conductance of the foliage representing the resultant of the stomatal and mesophyll conductance to the gas, and Δ_{gas} the gas concentration gradient between the atmosphere and leaf interior (18,20,34). Over a wide range, there is a nearly linear relationship between the uptake of SO_2 and the atmospheric concentration. Stomatal conductance is generally the limiting factor for uptake of SO_2 by the foliage, whereas the mesophyll conductance toward SO_2 is very high (18,20,35). This high mesophyll conductance is mainly determined by chemical/physical factors, since the gas is highly soluble in the water of the mesophyll cells (in either apoplast or cytoplasm). Furthermore, the dissolved SO_2 is rapidly hydrated and dissociated, yielding bisulfite and sulfite ($SO_2 + H_2O \rightarrow H^+ + HSO_3^- \rightarrow 2H^+ + SO_3^{2-}$) (18,20). The latter compounds either directly enter the assimilatory sulfur reduction pathway (in the chloroplast) or are enzymatically or nonenzymatically oxidized to sulfate in either apoplast or cytoplasm (18,20). The sulfate formed may be reduced and subsequently assimilated or it is transferred to the vacuole. Even at relatively low atmospheric levels, SO_2 exposure may result in enhanced sulfur content of the foliage (18,20). The liberation of free H^+ ions upon hydration of SO_2 or the sulfate formed from its oxidation is the basis of a possible acidification of the water of the mesophyll cells, in case the buffering capacity is not sufficient. Definitely, the physical–biochemical background of the phytotoxicity of SO_2 can be ascribed to the negative consequences of acidification of tissue/cells upon the dissociation of the SO_2 in the aqueous phase of the mesophyll cells or the direct reaction of the (bi)sulfite formed with cellular constituents and metabolites (18,20).

The foliar uptake of H_2S even appears to be directly dependent on the rate of its metabolism into cysteine and subsequently into other sulfur compounds, a reaction catalyzed by O-acetylserine (thiol)lyase (19,21). The basis for the phytotoxicity of H_2S can be ascribed to a direct reaction of sulfide with cellular components; for instance, metallo-enzymes appear to be particularly susceptible to sulfide, in a reaction similar to that of cyanide (18,19,36).

The foliage of plants exposed to SO_2 and H_2S generally contains enhanced thiol levels, the accumulation of which depends on the atmospheric level, though it is generally higher upon exposure to H_2S than exposure to SO_2 at equal concentrations.

Changes in the size and composition of the thiol pool are likely the reflection of a slight overload of a reduced sulfur supply to the foliage. Apparently, the direct absorption of gaseous sulfur compounds bypasses the regulation of the uptake of sulfate by the root and its assimilation in the shoot so that the size and composition of the pool of thiol compounds is no longer strictly regulated.

7.2.2 MAJOR ORGANIC SULFUR COMPOUNDS

The sulfur-containing amino acids cysteine and methionine play a significant role in the structure, conformation, and function of proteins and enzymes in vegetative plant tissue, but high levels of these amino acids may also be present in seed storage proteins (37). Cysteine is the sole amino acid whose side-chain can form covalent bonds, and when incorporated into proteins, the thiol group of a cysteine residue can be oxidized, resulting in disulfide bridges with other cysteine side-chains (forming cystine) or linkage of polypeptides. Disulfide bridges make an important contribution to the structure of proteins. An impressive example for the relevance of disulfide bridges is the influence of the sulfur supply on the baking quality of bread-making wheat. Here, the elasticity and resistance to extensibility are related to the concentration of sulfur-containing amino acids and glutathione. First, it was shown in greenhouse studies that sulfur deficiency impairs the baking quality of wheat (38–41). Then, the analysis of wheat samples from variety trials in England and Germany revealed that decrease in the supply of sulfur affected the baking quality, before crop productivity was reduced (42,43). The sulfur content of the flour was directly related to the baking quality with each 0.1% of sulfur equalling 40 to 50 mL loaf volume. The data further revealed that a lack of either protein or sulfur could be partly compensated for by increased concentration of the other.

The crude protein of wheat can be separated into albumins and globulins, and gluten, which consist of gliadins and glutenins. The first, albumins and globulins, are concentrated under the bran and are thus present in higher concentrations in whole-grain flours. Their concentration is directly linked to the thousand grain weight. In the flour, gluten proteins are predominant and the gliadin/glutenin ratio influences the structure of the gluten, rheological features of the dough, and thus the baking volume (44). Gliadins are associated with the viscosity and extensibility, and glutenins with the elasticity and firmness of the dough (45). Here, the high-molecular-weight (HMW) glutenins give a higher proportion of the resistance of the gluten than low-molecular-weight (LMW) glutenins (46). Sulfur deficiency gives rise to distinctly firmer and less extensible doughs (Figure 7.2). Doughs from plants adequately supplied with sulfur show a significantly higher extensibility and lower resistance than do doughs made of flour with an insufficient sulfur supply (Figure 7.2). Sulfur-deficient wheat has a lower albumin content, but higher HMW-glutenin concentration and a higher HMW/LMW glutenin ratio (47).

Consequently the baking volume of sulfur-deficient wheat is reduced significantly. A comparison of British and German wheat varieties with similar characteristics for loaf volume and falling number is given in Table 7.1. In the German classification system, varieties C1 and C2 are used as feed or as a source for starch. Varieties B3, B4, and B5 are suitable for baking but are usually mixed with higher quality wheat. The highest bread-making qualities are in the A6–A9 varieties.

The results presented in Table 7.1 reveal that the quality of British and German varieties is similar. It is relevant in this context that the British varieties gave the same results in the baking experiment at lower protein concentrations than the German ones. The reason is that there was a higher sulfur concentration and thus a smaller N/S ratio in the British varieties. This means that higher sulfur concentrations can partially compensate for a lack of wheat protein and vice versa.

Sulfur supply has been recognized as a major factor influencing protein quality for a long time (48,49). Eppendorfer and Eggum (50,51), for instance, noted that the biological value of proteins in potatoes (*Solanum tuberosum* L.) was reduced from 94 to 55 by sulfur deficiency at high N supply, and from 65 to 40 and 70 to 61 in kale (*Brassica oleracea* var. *acephala* DC) and field beans

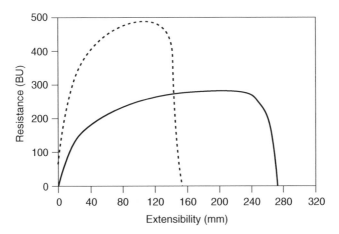

FIGURE 7.2 Extensographs for flour with average (continuous line) and low (broken line) sulfur content. +S flour: 0.146% S, 1.82% N, N:S = 12.5:1; −S flour: 0.089% S, 1.72% N, N:S = 19.3:1. (From Wrigley, C.W. et al., *J. Cereal Sci.*, 2, 15–24, 1984.)

TABLE 7.1
Comparison of Quality Parameters of German and British Wheat Varieties

Parameter	British D	German B4	British B	German A6/A7
Loaf volume (ml)	612	612	717	713
Falling number (s)	215	276	247	381
Protein content (%)	10.8	13.1	12.6	14.3
S content (mg g^{-1})	1.38	1.25	1.46	1.35
N:S ratio	12.6	16.6	14.0	17.8

Source: From Haneklaus, S. et al., *Sulphur Agric.*, 16, 31–35, 1992.

(*Vicia faba* L.), respectively. Whereas the essential amino acid concentrations declined due to sulfur deficiency, the content of amino acids of low nutritional value such as arginine, asparagine, and glutamic acid increased (50, 51). Figure 7.3 shows the relationship between sulfur supply to curly cabbage (*Brassica oleracea* var. *sabellica* L.), indicated by the total sulfur concentration in fully expanded younger leaves, and the cysteine and methionine concentration in leaf protein.

This example shows that a significant relationship between sulfur supply and sulfur-containing amino acids exists only under conditions of severe sulfur deficiency, where macroscopic symptoms are visible. The corresponding threshold is below leaf sulfur levels of 0.4% total sulfur in the dry matter of brassica species (52,53).

In comparison, sulfur fertilization of soybean significantly increased the cystine, cysteine, methionine, protein, and oil content of soybean grain (Table 7.2) (54).

The reason for these different responses of vegetative and generative plant tissue to an increased sulfur supply is that excess sulfur is accumulated in vegetative tissue as glutathione (see below) or as sulfate in vacuoles; the cysteine pool is maintained homeostatically because of its cytotoxicity (55). In comparison, the influence of sulfur supply on the seed protein content is related to the plant species. In oilseed rape, for instance, which produces small seeds, the total protein content is more or less not influenced by the sulfur supply (56). Species with larger seeds, which contain sulfur-rich proteins, such as soybean, respond accordingly to changes in the sulfur supply (5).

The most abundant plant sulfolipid, sulfoquinovosyl diacylglycerol, is predominantly present in leaves, where it comprises up to 3 to 6% of the total sulfur (10,57,58). This sulfolipid can occur in plastid membranes and is probably involved in chloroplast functioning. The route of biosynthesis

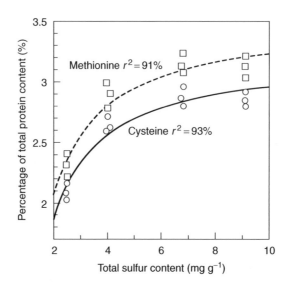

FIGURE 7.3 Relationship between the sulfur nutritional status of curly cabbage and the concentration of cysteine and methionine in the leaf protein. (From Schnug, E., in *Sulphur Metabolism in Higher Plants: Molecular, Ecophysiological and Nutritional Aspects*, Backhuys Publishers, Leiden, 1997, pp. 109–130.)

TABLE 7.2

Influence of Sulfur Fertilization on Sulfur-Containing Amino Acids, Total Protein, and Oil Content in Soybean Grains

	S-Containing Amino Acid (mg g^{-1})				
S Supply (mg kg^{-1})	Cystine	Cysteine	Methionine	Protein (%)	Oil (%)
0	1.9	1.2	7.6	40.3	19.6
40	2.4	1.6	10.5	41.0	21.0
80	2.9	1.9	13.9	41.6	20.6
120	2.9	2.0	16.4	42.2	20.8
LSD$_{5\%}$	0.14	0.10	1.13	0.99	0.19

Source: From Kumar, V. et al., *Plant Soil*, 59, 3–8, 1981.

of sulfoquinovosyl diacylglycerol is still under investigation; in particular, the sulfur precursor for the formation of the sulfoquinovose is not known, though from recent observations it is evident that sulfite is the likely candidate (58).

Cysteine is the precursor for the tripeptide glutathione (γGluCysGly; GSH), a thiol compound that is of great importance in plant functioning (32,59,60,61). Glutathione synthesis proceeds in a two-step reaction. First, γ-glutamylcysteine is synthesized from cysteine and glutamate in an ATP-dependent reaction catalyzed by γ-glutamylcysteine synthetase (Equation 7.1). Second, glutathione is formed in an ATP-dependent reaction from γ-glutamylcysteine and glycine (in glutathione homologs, β-alanine or serine) catalyzed by glutathione synthetase (Equation 7.2):

$$\text{Cys} + \text{Glu} + \text{ATP} \xrightarrow[\gamma\text{-glutamylcysteine synthetase}]{} \gamma\text{GluCys} + \text{ADP} + \text{Pi} \tag{7.1}$$

$$\gamma\text{GluCys} + \text{Gly} + \text{ATP} \xrightarrow[\text{glutathione synthetase}]{} \gamma\text{GluCysGly} + \text{ADP} + \text{Pi} \tag{7.2}$$

TABLE 7.3
Influence of Sulfur Fertilization on the Glutathione Content of the Vegetative Tissue of Different Crops

Crop Plant	Increase of Glutathione Concentration by S Supply	Reference
Asparagus spears	Field: 39–67 nmol g^{-1} (d.w.) per kg S[a] applied	62
Oilseed rape leaves	Field: 64 nmol g^{-1} (d.w.) per kg S[a] applied	63
	Pot: 3.9 nmol g^{-1} (d.w.) per mg S[b] applied	64
Spinach leaves	Pot: 656 nmol g^{-1} (f.w.) per µl l^{-1} H$_2$S[c]	65

[a]Maximum dose = 100 kg ha^{-1} S.
[b]Maximum dose = 250 mg pot^{-1} S.
[c]Maximum dose = 250 µl l^{-1} H$_2$S.

Glutathione and its homologs, for example, homoglutathione (γGluCysβAla) in Fabaceae and hydroxymethylglutathione (γGluCysβSer) in Poaceae, are widely distributed in plant tissues in concentrations ranging from 0.1 to 3 mM. The glutathione content is closely related to the sulfur nutritional status. In Table 7.3, the influence of the sulfur supply and sulfur status and the glutathione content is summarized for different crops. The possible significance of the glutathione content for plant health is discussed in Section 7.5.3.

Glutathione is maintained in the reduced form by an NADPH-dependent glutathione reductase, and the ratio of reduced glutathione (GSH) to oxidized glutathione (GSSG) generally exceeds a value of 7 (60–67). Glutathione fulfills various roles in plant functioning. In sulfur metabolism, glutathione functions as the reductant in the reduction of APS to sulfite (Figure 7.1). In crop plants, glutathione is the major transport form of reduced sulfur between shoot and roots, and in the remobilization of protein sulfur (e.g., during germination). Sulfate reduction occurs in the chloroplasts, and roots of crop plants mostly depend for their reduced sulfur supply on shoot–root transfer of glutathione via the phloem (59–61).

Selenium is present in most soils in various amounts, and its uptake, reduction, and assimilation strongly interact with that of sulfur in plants. Glutathione appears to be directly involved in the reduction and assimilation of selenite into selenocysteine (68). More detailed information about interactions between sulfur and other minerals is given in Section 7.2.4.

Glutathione provides plant protection against stress and a changing environment, viz air pollution, drought, heavy metals, herbicides, low temperature, and UV-B radiation, by depressing or scavenging the formation of toxic reactive oxygen species such as superoxide, hydrogen peroxide, and lipid hydroperoxides (61,69). The formation of free radicals is undoubtedly involved in the induction and consequences of the effects of oxidative and environmental stress on plants. The potential of glutathione to provide protection is related to the size of the glutathione pool, its oxidation–reduction state (GSH/GSSG ratio) and the activity of glutathione reductase.

Plants may suffer from an array of natural or synthetic substances (xenobiotics). In general, these have no direct nutritional value or significance in metabolism, but may, at too high levels, negatively affect plant functioning (70–72). These compounds may originate from either natural (fires, volcanic eruptions, soil or rock erosion, biodegradation) or anthropogenic (air and soil pollution, herbicides) sources. Depending on the source of pollution, namely air, water, or soil, plants have only limited possibilities to avoid their accumulation to diminish potential toxic effects. Xenobiotics (R-X) may be detoxified in conjugation reactions with glutathione (GSH) catalyzed by the enzyme glutathione S-transferase (70–72).

$$R\text{-}X + GSH \Rightarrow R\text{-}SG + X\text{-}H$$

The activity of glutathione S-transferase may be enhanced in the presence of various xenobiotics via induction of distinct isoforms of the enzyme. Glutathione S-transferases have great significance in herbicide detoxification and tolerance in agriculture. The induction of the enzyme by herbicide antidotes, the so-called safeners, is the decisive step for the induction of herbicide tolerance in many crop plants. Under normal natural conditions, glutathione S-transferases are assumed to be involved in the detoxification of lipid hydroperoxides, in the conjugation of endogenous metabolites, hormones, and DNA degradation products, and in the transport of flavonoids. However, oxidative stress, plant-pathogen infections, and other reactions, which may induce the formation of hydroperoxides, also may induce glutathione S-transferases. For instance, lipid hydroperoxides (R-OOH) may be degraded by glutathione S-transferases:

$$R\text{-}OOH + 2GSH \Rightarrow R\text{-}OH + GSSG + H_2O$$

Plants need minor quantities of essential heavy metals (zinc, copper, and nickel) for growth. However, plants may suffer from exposure to high toxic levels of these metals or other heavy metals, for example, cadmium, copper, lead, and mercury. Heavy metals elicit the formation of heavy-metal-binding ligands. Among the various classes of metal-binding ligands, the cysteine-rich metallothioneins and phytochelatins are best characterized; the latter are the most abundant ligands in plants (73–78). The metallothioneins are short gene-encoded polypeptides and may function in copper homeostasis and plant tolerance. Phytochelatins are synthesized enzymatically by a constitutive phytochelatin synthase enzyme and they may play a role in heavy metal homeostasis and detoxification by buffering the cytoplasmatic concentration of essential heavy metals, but direct evidence is lacking so far. Upon formation, the phytochelatins only sequester a few heavy metals, for instance cadmium. It is assumed that the cadmium–phytochelatin complex is transported into the vacuole to immobilize the potentially toxic cadmium (79). The enzymatic synthesis of phytochelatins involves a sequence of transpeptidation reactions with glutathione as the donor of γ-glutamyl-cysteine (γGluCys) residues according to the following equation:

$$(\gamma GluCys)_n Gly + (\gamma GluCys)_n Gly \Rightarrow (\gamma GluCys)_{n+1} Gly + (\gamma GluCys)_{n-1} Gly$$

The number of γ-glutamyl-cysteine residues $(\gamma GluCys)_n$ in phytochelatins ranges from 2 to 5, though it may be as high as 11. In species containing glutathione homologs (see above), the C-terminal amino acid glycine is replaced by β-alanine or serine (73–78). During phytochelatin synthesis, the sulfur demand is enhanced (80) so that it may be speculated that the sulfur supply is linked to heavy metal uptake, translocation of phytochelatins into root cell vacuoles, and finally transport to the shoot and expression of toxicity symptoms. The sulfur/metal ratio is obviously related to the length of the phytochelatin (81), which might offer a possibility to adapt to varying sulfur nutritional conditions. Hence, increasing cadmium stress (10 μmol Cd in the nutrient solution) yielded an enhanced sulfate uptake by maize roots of 100%, whereby this effect was associated with decreased sulfate and glutathione contents and increased phytochelatin concentrations (81). The studies of Raab et al. (82) revealed that 13% of arsenic was bound in phytochelatin complexes, whereas the rest occurred as nonbound inorganic compounds.

7.2.3 SECONDARY SULFUR COMPOUNDS

There are more than 100,000 known secondary plant compounds, and for only a limited number of them are the biochemical pathways, functions, and nutritional and medicinal significance known (84). Detailed overviews of the biochemical pathways involved in the synthesis of the sulfur-containing secondary metabolites, glucosinolates and alliins, are provided by Halkier (84) and Lancaster and Boland (85). Bioactive secondary plant compounds comprise various substances such as carotenoids, phytosterols, glucosinolates, flavonoids, phenolic acids, protease inhibitors, monoterpenes, phyto-estrogens, sulfides, chlorophylls, and roughages (87). Often, secondary metabolites are accumulated in plant tissues and concentrations of 1 to 3% dry weight have been determined (88). Secondary compounds in plants usually have a pharmacological effect on humans (87). Therefore, secondary metabolites contribute significantly to food quality, either as nutritives or antinutritives. Plants synthesize a great array of secondary metabolites as they are physically immobile (88), and the presence of secondary compounds may give either repellent or attractant properties.

The bioactive components in medicinal plants comprise the whole range of secondary metabolites and crop-specific cultivation strategies, which include fertilization, harvesting, and processing techniques, and which are required for producing a consistently high level of bioactive constituents. Ensuring a consistently high quality of the raw materials can be a problem, particularly if the active agent is unstable and decomposes after harvesting of the plant material, as is true for many secondary metabolites such as the sulfur-containing alliins and glucosinolates (89).

Glucosinolates are characteristic compounds of at least 15 dicotyledonous families. Of these, the Brassicaceae are the most important agricultural crops. Glucosinolates act as attractants, repellents, insecticides, fungicides, and antimicrobial protectors. The principal structure of a glucosinolate is given in Figure 7.4.

There are about 80 different glucosinolates, which consist of glucose, a sulfur-containing group with an aglucon rest, and a sulfate group (87). Alkenyl glucosinolates such as progoitrin and gluconapin have an aliphatic aglucon rest, whereas indole glucosinolates such as glucobrassicin and 4-hydroxyglucobrassicin in rape (*Brassica napus* L.) have an aromatic aglucon rest (Figure 7.4). Additional information about the characteristics of glucosinolate side-chains is given by Underhill (91), Larsen (92), and Bjerg et al. (93).

Glucosinolates are generally hydrolyzed by the enzyme myrosinase, which is present in all glucosinolate-containing plant parts. Bones and Rossiter (94) provided basic information about the biochemistry of the myrosinase–glucosinolate system. A proposed pathway for the recyclization of sulfur (and N) under conditions of severe sulfur deficiency is described by Schnug and Haneklaus (53).

The degradation of glucosinolates results in the so-called mustard oils, which are responsible for smell, taste, and biological effect. Glucosinolates are vacuolar defense compounds (95) of qualitative value (96) and are effective against generalist insects at low tissue concentrations (97). Isothiocyanates, the breakdown products after enzymatic cleavage of glucosinolates, may retard multiplication of spores but do not hamper growth of fungal mycelium (98), and fungi may overcome the glucosinolate–myrosinase system efficiently (99,100).

The influence of the sulfur nutritional status on the content of glucosinolates and other sulfur-containing secondary metabolites, which are related to nutritional and pharmaceutical quality, is shown in Table 7.4.

Generally, nitrogen fertilization reduces the glucosinolate content (104). However, under field conditions the effect of nitrogen fertilization on glucosinolate content varies substantially between seasons (105). Schnug (103) noted a distinct interaction between nitrogen and sulfur fertilization when nitrogen was supplied insufficiently, whereby the alkenyl, but not the indole, glucosinolate content in seeds of rape increased at higher nitrogen and sulfur rates. Kim et al. (106) also showed that nitrogen fertilization increased the alkenyl-glucosinolates, gluconapin, and glucobrassicanapin in particular, in rape.

More than 80% of the total sulfur in *Allium* species is present in secondary compounds. *Allium* species contain four S-alk(en)yl-L-cysteine sulfoxides, namely S-1-propenyl-, S-2-propenyl-,

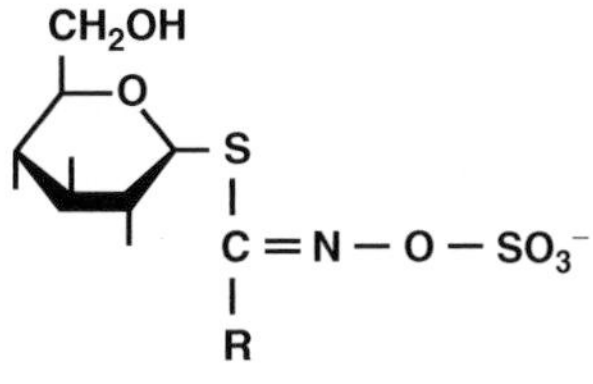

FIGURE 7.4 Basic structure of glucosinolates. (From Schnug, E., in *Sulfur Nutrition and Sulfur Assimilation in Higher Plants*, SPB Academic Publishing, The Hague, 1990, pp. 97–106.)

TABLE 7.4
Influence of Sulfur Fertilization on the Concentration of Sulfur-Containing Secondary Metabolites in Vegetative and Generative Tissues of Different Crops

Crop	Plant Part	S Metabolite	Influence of S Supply on Secondary Compound	Reference
Garlic	Leaves	Alliin	2.4 µmol g^{-1} (d.w.) per 10 mg S[a]	101
	Bulbs	Alliin	0.7 µmol g^{-1} (d.w.) per 10 mg S[a]	101
Mustard	Seeds	Glucosinolates	0.7 µmol g^{-1} per 10 kg S[b]	102
Nasturtium	Whole plant	Glucotropaeolin	3.4 µmol g^{-1} (d.w.) per 10 kg S[c]	89
	Leaves		4.3 µmol g^{-1} (d.w.) per 10 kg S[c]	89
	Stems		1.1 µmol g^{-1} (d.w.) per 10 kg S[c]	89
	Seeds		2.3 µmol g^{-1} per 10 kg S[c]	89
Oilseed rape	Leaves	Glucosinolates	0.04–1.5 µmol g^{-1} (d.w.) per 10 kg S[d]	63
	Seeds	Glucosinolates	0.3–0.6 µmol g^{-1} per 10 kg S[d]	63
			2.1 µmol g^{-1} per 10 kg S[e]	
			0.8 µmol g^{-1} per 10 kg S[f]	103
Onion	Leaves	(Iso)alliin	0.7 µmol g^{-1} (d.w.) per 10 mg S[a]	101
	Bulbs		0.4 µmol g^{-1} (d.w.) per 10 mg S[a]	101

[a]Maximum dose = 250 mg pot^{-1} S and 500 mg pot^{-1} N.
[b]Maximum dose = 185 kg ha^{-1} S.
[c]Maximum dose = 50 kg ha^{-1} S.
[d]Maximum dose = 100 and 150 kg ha^{-1} S.
[e]Severe S deficiency.
[f]Moderate S deficiency.

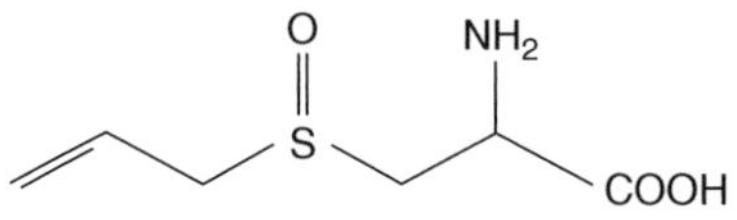

FIGURE 7.5 Chemical structure of alliin. (From Watzl, B., *Bioaktive Substanzen in Lebensmitteln*, Hippokrates Verlag, Stuttgart, Germany, 1999.)

S-methyl- and *S*-propyl-L-cysteine sulfoxides (107). Iso-alliin is the main form in onions, whereas alliin is the predominant form in garlic (108) (Figure 7.5). Alliins supposedly contribute to the defense of plants against pests and diseases. In vitro and in vivo experiments revealed a bactericidal effect against various plant pathogens (109).

The characteristic flavor of *Allium* species is caused after the enzyme alliinase hydrolyzes cysteine sulfoxides to form pyruvate, ammonia, and sulfur-containing volatiles. In the intact cell, alliin and related cysteine sulfoxides are located in the cytoplasm, whereas the C-S lyase enzyme alliinase is localized in the vacuole (110). Disruption of the cell releases the enzyme, which causes subsequent α,β-elimination of the sulfoxides, ultimately giving rise to volatile and odorous LMW organosulfur compounds (111). The cysteine sulfoxide content of *Allium* species is an important quality parameter with regard to sensory features, since it determines the taste and sharpness.

Alliin acts as an antioxidant by activating glutathione enzymes and is regarded as having an anticarcinogenic and antimicrobial effect (86). On average, 21% of sulfur, but only 0.9% of nitrogen, are present as (iso)alliin in onion bulbs at the start of bulb growth (101). The ratio between protein-S and sulfur in secondary metabolites of the *Allium* species is, at between 1:4 and 1:6, much wider than in members of the *Brassica* family (between 1:0.3 and 1:2). The reason for this difference is supposedly the fact that glucosinolates may be reutilized under conditions of sulfur deficiency whereas alliins are inert end products. Interactions between nitrogen and sulfur supply exist in such a way that nitrogen and sulfur fertilization has been shown to decrease total sulfur and nitrogen concentration, respectively, in onion (101).

7.2.4 INTERACTIONS BETWEEN SULFUR AND OTHER MINERALS

Interactions between sulfur and other minerals may significantly influence crop quality parameters (5,113,114). Sulfur and nitrogen show strong interactions in their nutritional effects on crop growth and quality due to their mutual occurrence in amino acids and proteins (see Section 7.2.3). Further examples of nitrogen–sulfur interactions that are not mentioned in previous sections of this chapter are shown below.

7.2.4.1 Nitrogen–Sulfur Interactions

Under conditions of sulfur starvation, sulfur deficiency symptoms are expressed moderately at low nitrogen levels but extremely with a high nitrogen supply. This effect explains the enhancement of sulfur deficiency symptoms in the field after nitrogen dressings (114). The question of why sulfur deficiency symptoms are more pronounced at high nitrogen levels is, however, still unanswered. For experimentation, these results are relevant as the adjustment of the nitrogen and sulfur nutritional status of plants is essential before any hypothesis on the effect of a nitrogen or sulfur treatment on plant parameters can be stated or proved.

The use of the nitrogen/sulfur ratio as a diagnostic criterion is problematic because the same ratio can be obtained at totally different concentration levels in the tissue. Surplus of one element may therefore be interpreted falsely as a deficiency of the other (see Section 7.3.1.3). Clear relationships between nitrogen/sulfur ratios and yield occur only in ranges of extreme ratios. Such ratios may be produced in pot trials but do not occur under field conditions. The effect of increasing nitrogen and sulfur supply on crop seed yield with increasing nitrogen supply is more pronounced with protein than with carbohydrate crops (Table 7.5).

TABLE 7.5
Seed Yield of Single (NIKLAS) and Double Low (TOPAS) Oilseed Rape Varieties in Relation to the Nitrogen and Sulfur Supply in a Glasshouse Experiment

	Seed Yield (g pot^{-1})			
	500 mg N		1000 mg N	
	NIKLAS	TOPAS	NIKLAS	TOPAS
Control	0 a	0 a	0 a	0 a
25 mg S	2.10 b	0.9 b	0 a	0 a
50 mg S	3.15 c	2.85 c	1.25 b	0.35 b
75 mg S	2.55 b	2.65 c	5.30 c	5.85 c
100 mg S	3.05 c	2.50 c	6.70 d	7.50 d

Note: Different characters after figures indicate statistically significant differences of means by Duncan's Multiple Range Test.

Source: From Schnug, E., Quantitative und Qualitative Aspekte der Diagnose und Therapie der Schwefelversorgung von Raps (*Brassica napus* L.) unter besonderer Berücksichtigung glucosinolatarmer Sorten. Habilitationsschrift, D.Sc. thesis, Kiel University, 1988.

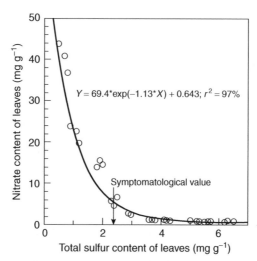

FIGURE 7.6 Nitrate concentrations in the dry matter of lettuce in relation to the sulfur nutritional status of the plants. (From Schnug, E., in *Sulphur Metabolism in Higher Plants: Molecular, Ecophysiological and Nutritional Aspects*, Backhuys Publishers, Leiden, 1997, pp. 109–130.)

Changes in the nitrogen supply affect the sulfur demand of plants and vice versa. Under conditions of sulfur deficiency, the utilization of nitrogen will be reduced and consequently nonprotein nitrogen compounds, including nitrate, accumulate in the plant tissue (Figure 7.6) (5,112).

The antagonistic relationship between sulfur supply and nitrate content exists in the range of severe sulfur deficiency, when macroscopic symptoms are visible. The higher the nitrogen level in the plants, the stronger the effect on the nitrate content will be. Thus, an adequate sulfur supply is vital for minimizing undesired enrichment with nitrate.

Photosynthesis and growth of pecan (*Carya illinoinensis* Koch) increased with N supply in relation to the nitrogen/sulfur ratio in pecan leaves (115). Both parameters were, however, reduced when combined leaf nitrogen and sulfur concentrations of <35 mg g^{-1} nitrogen and 3.7 mg g^{-1} sulfur were noted (115).

The initial supply of a crop with nitrogen and sulfur is decisive for its influence on the glucosinolate content, probably due to physiological or root-morphological reasons (103). Nitrogen fertilization to oilseed rape insufficiently supplied with nitrogen and sulfur will lead to decreasing glucosinolate concentrations because the demand of an increasing sink due to increasing numbers of seeds will not be met by the limited sulfur source. Only if the rooting depth or density is enhanced by the nitrogen supply, which increases the plant-available sulfur pool in the soil, does the glucosinolate content increase too. Higher glucosinolate concentrations in seeds can also be expected after nitrogen applications to crops with a demand for nitrogen but adequate sulfur supply due to the increased biosynthesis of sulfur-containing amino acids, which are precursors of glucosinolates. In the case of a crop already sufficiently supplied with nitrogen, there is no evidence for any specific nitrogen–sulfur interactions on the glucosinolate content (5,116).

In general, no significant influence of nitrogen fertilization on the alliin content has been found for onions (*Allium cepa* L.) and garlic (*Allium sativum* L.), but there is a tendency that a higher nitrogen supply results in a decreased alliin content (101). In comparison, an increasing sulfur supply has been related to an increasing alliin content in leaves and bulbs of both crops. There were also interactions between nitrogen and sulfur in such a way that the total sulfur content of onion leaves was correlated highly with nitrogen fertilization: the sulfur concentration of leaves decreased with increasing N fertilization, and the total nitrogen concentration of onion bulbs decreased with increasing sulfur fertilization. The same observations were made by Freeman and Mossadeghi (117) for garlic plants, where the nitrogen concentration decreased from 4.05 to 2.93% with sulfur fertilization, and by Randle et al. (118), who reported decreasing total bulb sulfur concentrations in response to increasing nitrogen fertilization.

7.2.4.2 Interactions between Sulfur and Micronutrients

Owing to antagonistic effects, sulfur fertilization reduces the uptake of boron and molybdenum. In soils with a marginal plant-available concentration of these two plant nutrients, sulfur fertilization may induce boron or molybdenum deficiency, particularly on coarse-textured sites where brassica crops are grown intensely in the crop rotation (119). In comparison, sulfur fertilization is an efficient tool to reduce the selenium, molybdenum, arsenic, bromine, and antimony uptake on contaminated sites. The influence of elemental sulfur applications on the concentration of trace elements of fully developed leaves of nasturtium (*Tropaeolum majus* L.) was tested on two sites in northern Germany (120). The results of this study reveal a significantly increased uptake of copper, manganese, cobalt, nickel, and cadmium, with increasing levels of sulfur. This increased uptake was caused by a higher availability of these elements due to the acidifying effect of elemental sulfur. At the same time, antagonistic effects were noted for arsenic, boron, selenium, and molybdenum in relation to the soil type.

The enzyme sulfite oxidase is a molybdo-enzyme, which converts sulfite into sulfate (121) and is thus important for sulfate reduction and assimilation in plants (see Figure 7.1). Stout and Meagher (122) have shown that the sulfate supply influences molybdenum uptake. Sulfate–molybdate antagonism can be observed at the soil–root interface and within the plant, as an increasing sulfur supply results in lower molybdenum concentrations in the tissues (123). The significance of sulfate–molybdate antagonism in agriculture is described comprehensively by Macleod et al. (124).

Selenium, like molybdenum, is chemically similar to sulfur. Comprehensive reviews about interactions between sulfate transporters and sulfur assimilation enzymes, and selenium–molybdenum uptake and metabolism, are given by Terry et al. (125) and Kaiser et al. (126). Accumulation of glutathione due to elevated levels of sulfate in the soil and SO_2/H_2S in the air was reduced drastically in spinach (*Spinacia oleracea* L.) leaf discs by selenate amendments (127). In those studies the uptake of sulfur was not influenced by the selenate treatment. Bosma et al. (128) suggested that selenate decreases sulfate reduction due to antagonistic effects during plant uptake, in combination with a rapid turnover of glutathione. An increasing sulfate supply gives higher sulfate concentrations in the plant tissue, so that the competition between sulfur and selenium for the enzymes of the sulfur assimilation pathway will finally result in less synthesis of selenoamino acids (129).

This antagonistic effect is of no practical significance on seleniferous soils, but it could be relevant on deficient and marginal sites (130). Field experiments with combined sulfur and selenium applications to grass-clover pastures, on selenium-deficient and high-selenium sites revealed that selenium concentrations in the different botanical species showed distinct differences in relation to the site (130).

On the high-selenium site, sulfur fertilization significantly decreased the selenium concentration in pasture. Spencer (130) attributed this action to a dilution effect, as the total selenium content remained constant. Studies on the pungency of onion bulbs in relation to the sulfur supply revealed that although sulfur content was increased at elevated selenium levels, the pungency was reduced (131). Kopsell and Randell (131) proposed that selenium had an impact on the biosynthetic pathway of flavor precursors.

A synergistic effect of sulfur and selenium on the shoot sulfur concentration was noted for hydroponically grown barley (*Hordeum vulgare* L.) and rice (*Oryza sativa* L.). With increasing selenium concentrations in the solution, a steep increase in the sulfur concentration of the shoots occurred even with a low sulfur supply (132).

Sulfur and phosphorus interactions in plants are closely related to plant species, because of the different root morphologies and nutrient demands of different species (133). A synergistic effect of sulfur and phosphorus on crop yield occurred for sorghum (*Sorghum vulgare* Pers.), maize (*Zea mays* L.), wheat (*Triticum aestivum* L.), and mustard (*Brassica* spp. L.) (134–137). A synergistic relationship between sulfur and potassium, which enhances crop productivity and quality, was determined in several studies (138–140).

7.3 SULFUR IN PLANT NUTRITION

7.3.1 DIAGNOSIS OF SULFUR NUTRITIONAL STATUS

7.3.1.1 Symptomatology of Single Plants

Visual diagnosis of sulfur deficiency in production fields requires adequate expertise and needs to involve soil or plant analysis (141). The literature describes symptoms of sulfur deficiency as being less specific and more difficult to identify than other nutrient deficiency symptoms (142–145). The symptomatology of sulfur deficiency is very complex and shows some very unique features. In this section, the basic differences in sulfur deficiency symptoms of species in the Gramineae representative of monocotyledonous, and species in the Cruciferae and Chenopodiaceae representative of dicotyledonous crops will be given for individual plants and on a field scale.

When grown side by side and under conditions of sulfur starvation, crops begin to develop sulfur deficiency symptoms in the order of oilseed rape (canola), followed by potato, sugar beet (*Beta vulgaris* L.), beans (*Phaseolus vulgaris* L.), peas (*Pisum sativum* L.), cereals, and finally maize. The total sulfur concentration in tissue corresponding to the first appearance of deficiency symptoms is highest in oilseed rape (3.5 mg g^{-1} S), and lowest in the Gramineae (1.2 mg g^{-1} S). Potato and sugar beet show symptoms at higher concentrations (2.1 to 1.7 mg g^{-1} S) than beans or peas (1 to 1.2 mg g^{-1} S).

Brassica species, such as oilseed rape, develop the most distinctive expression of symptoms of any crop deficient in sulfur. The symptoms are very specific and thus are a reliable guide to sulfur deficiency. There is no difference in the symptomatology of sulfur deficiency in high and low glucosinolate-containing varieties (103). The symptomatology of sulfur deficiency in brassica crops is characteristic during the whole vegetation period and is described below for specific growth stages according to the BBCH scale (146). Symptoms generally apply to dicotyledonous plants, except when specific variations are mentioned in the text. Colored guides of sulfur deficiency symptoms are provided by Bergmann (143) and Schnug and Haneklaus (53,114,147).

Even before winter, during the early growth of oilseed rape, leaves may start to develop visible symptoms of sulfur deficiency. As sulfur is fairly immobile within the plant (13), symptoms always show up in the youngest leaves. Though the plants are still small, symptoms can cover the entire plant. Sulfur fertilization before or at sowing will ensure a sufficient sulfur supply, particularly on light, sandy soils, and will promote the natural resistance of plants against fungal diseases (148).

Oilseed rape plants suffering from severe sulfur deficiency show a characteristic marbling of the leaves. Leaves begin to develop chlorosis (149–154), which starts from one edge of the leaves and spreads over intercostal areas; however, the zones along the veins always remain green (103,155). The reason for the green areas around the veins is most likely the reduced intercellular space in that part of the leaf tissue, resulting in shorter transport distances and a more effective transport of sulfate. Sulfur-deficient potato leaves show the same typical color pattern and veining as oilseed rape, whereas sugar beet, peas, and beans simply begin to develop chlorosis evenly spread over the leaf without any veining (156,157). A comparative evaluation of crop-specific, severe sulfur deficiency symptoms is given in Figure 7.7.

Chlorosis very rarely turns into necrosis (103,157) as it does with nitrogen and magnesium deficiencies, and is an important criterion for differential diagnosis. Even under conditions of extreme sulfur deficiency, an oilseed rape plant will not wither. The intensity of sulfur deficiency symptoms of leaves depends on the nitrogen supply of the plants (see Section 7.2.4.1). In general, a high nitrogen supply promotes the expression of sulfur deficiency symptoms and vice versa (158).

FIGURE 7.7 Macroscopic sulfur deficiency symptoms of oil seed rape (*Brassica napus* L.), cereals, and sugar beet (*Beta vulgaris* L.) at stem extension and row closing, respectively (from left to right). (For a color presentation of this figure, see the accompanying compact disc.)

FIGURE 7.8 Marbling, spoon-like leaf deformations and anthocyanin enrichments of sulfur-deficient oilseed rape plants (*Brassica napus* L.) (from left to right). (For a color presentation of this figure, see the accompanying compact disc.)

A characteristic secondary symptom of severe sulfur deficiency is a reddish-purple color due to the enrichment of anthocyanins in the chlorotic parts of brassica leaves (Figure 7.8). Under field conditions, the formation of anthocyanins starts 4 to 7 days after chlorosis. The phenomenon is initialized by the enrichment of carbohydrates in the cells after the inhibition of protein metabolism. Plants detoxify the accumulated carbohydrates as anthocyanates, which result from the reaction with cell-borne flavonols to avoid physiological disorders (159–165). Many other nutrient deficiencies are also accompanied by formation of anthocyanins, which therefore is a less specific indicator for sulfur deficiency.

In particular, leaves which are not fully expanded produce spoon-like deformations when struck by sulfur deficiency (Figure 7.8). The reason for this is a reduced cell growth rate in the chlorotic areas along the edge of the leaves, while normal cell growth continues in the green areas along the veins, so that sulfur-deficient leaves appear to be more succulent. The grade of the deformation is stronger the less expanded the leaf is when the plant is struck by sulfur deficiency. Marbling, deformations, and anthocyanin accumulation can be detected up to the most recently developed small leaves inserted in forks of branches (Figure 7.8).

FIGURE 7.9 White flowering (left) and morphological changes of petals (right) of sulfur-deficient oilseed rape (*Brassica napus* L.). (For a color presentation of this figure, see the accompanying compact disc.)

The higher succulence of sulfur-deficient plants (143,166) was suspected to be caused by enhanced chloride uptake due to an insufficient sulfate supply (159). However, with an increase of chloride concentrations by 0.4 mg Cl g^{-1} on account of a decrease of sulfur concentrations by 1 mg g^{-1} in leaves, this effect seems to be too small to justify the hypothesis (103). More likely, the above-explained mechanical effects of distortion, together with cell wall thickening, cause the appearance of increased succulence due to the accumulation of starch and hemicellulose (167).

During flowering of oilseed rape, sulfur deficiency causes one of the most impressive symptoms of nutrient deficiency: the 'white blooming' of oilseed rape (Figure 7.9). The white color presumably develops from an overload of carbohydrates in the cells of the petals caused by disorders in protein metabolism, which finally ends up in the formation of colorless leuco-anthocyanins (168). As with anthocyanins in leaves, the symptoms develop most strongly during periods of high photosynthetic activity. Beside the remarkable modification in color, size, and shape of oilseed rape, the petals change too (Figure 7.9). The petals of sulfur-deficient oilseed rape flowers are smaller and oval shaped, compared with the larger and rounder shape of plants without sulfur-deficiency symptoms (169). The degree of morphological changes, form, and color, are reinforced by the strength and duration of severe sulfur deficiency (53). The fertility of flowers of sulfur-deficient oilseed rape plants is not inhibited. However, the ability to attract honeybees may be diminished and can be of great importance for the yield of nonrestored hybrids, which need pollination by insect vectors (169).

The strongest yield component affected by sulfur deficiency in oilseed rape is the number of seeds per pod, which is significantly reduced (103). As described earlier for leaves, the branches and pods of S-deficient plants are often red or purple colored due to the accumulation of anthocyanins (Figure 7.10). Extremely low numbers of seeds per pod, in some cases even seedless 'rubber pods,' are characteristic symptoms of extreme sulfur deficiency (Figure 7.10).

7.3.1.2 Symptomatology of Monocots

The symptoms in gramineous crops such as cereals and corn are less specific than in cruciferous crops. In early growth stages, plants remain smaller and stunted and show a lighter color than plants without symptoms (170). The general chlorosis is often accompanied by light green stripes along the veins (Figure 7.11) (170–172). Leaves become narrower and shorter than normal (173).

There is no morphological deformation to observe, and usually no accumulation of anthocyanins either. Although the symptoms are very unspecific and are easily mistaken for symptoms of nitrogen deficiency, their specific pattern in fields provides good evidence for sulfur deficiency. Owing to an

FIGURE 7.10   Enrichment of anthocyanins during ripening of oilseed rape (*Brassica napus* L.) (left) and reduction of number of seeds per pod (right). (For a color presentation of this figure, see the accompanying compact disc.)

FIGURE 7.11   Macroscopic sulfur deficiency symptoms of winter wheat (*Triticum aestivum* L.) at stem extension. (For a color presentation of this figure, see the accompanying compact disc.)

early reduction of fertile flowers per head, sulfur-deficient cereals are characterized by a reduced number of kernels per head, which alone, however, is not conclusive evidence for sulfur deficiency (174).

7.3.1.3   Sulfur Deficiency Symptoms on a Field Scale

Some characteristic features in the appearance of fields can provide early evidence of sulfur deficiency. Sulfur deficiency develops first on the light-textured sections of a field. From above, these areas appear in an early oilseed rape crop as irregularly shaped plots with a lighter green color

FIGURE 7.12 Chlorotic patches in a field (left) and resultant effects on mature plants (right), indicating severe sulfur deficiency symptoms in relation to soil characteristics. (For a color presentation of this figure, see the accompanying compact disc.)

(wash outs). The irregular shape distinguishes the phenomenon from the regular shape of areas caused by nitrogen deficiency, which usually originates from inaccurate fertilizer application (Figure 7.12). Owing to frequent soil compaction and limited root growth, sulfur deficiency develops first along the headlands and tramlines or otherwise compacted areas of a field.

The appearance of sulfur-deficient oilseed rape fields is more obvious at the beginning of blooming; white flowers of oilseed rape are distinctively smaller and therefore much more of the green undercover of the crop shines through the canopy of the crop. Another very characteristic indicator of a sulfur-deficient site is the so-called second flowering of the oilseed rape crop. Even if a sulfur-deficient crop has finished flowering, it may come back to full bloom if sufficient sulfur is supplied. The typical situation for this action comes when a wet and rainy spring season up until the end of blooming is followed suddenly by warm and dry weather. During the wet period precipitation, water, which has only one-hundredth to one-tenth the sulfur concentrations of the entire soil solution, dilutes or leaches the sulfate from the rooting area of the plants, so that finally plants are under the condition of sulfur starvation. With the beginning of warmer weather, evaporation increases and sulfur-rich subsoil water becomes available to the plants and causes the second flowering of the crop. During maturity, sulfur deficiency in oilseed rape crops is revealed by a sparse, upright-standing crop.

Similarly, in cereals, sulfur deficiency develops first on light-textured parts of the field, yielding irregularly shaped 'wash-out' areas in images from above. Nitrogen fertilization promotes the expression of these irregularly distributed deficiency symptoms, such as uneven height and color. The irregular shape distinguishes these symptoms from areas caused by faulty nitrogen fertilizer application. In the field, these particular zones can be identified by a green yellowish glow in the backlight before sunset. Later, vegetation in these areas resembles a crop that is affected by drought. Owing to an inferior natural resistance (see also Section 7.5.2), the heads in sulfur-deficient areas can be infected more severely by fungal disease (e.g., *Septoria* species), which gives these areas a darker color as the crop matures.

7.4   SOIL ANALYSIS

A close relationship between the plant-available sulfur content of the soil and yield is a prerequisite for a reliable soil method. Such a significant correlation was verified in pot trials under controlled growth conditions (103,175–178). Several investigations have shown, however, that the relationship between inorganic soil sulfate and crop yield is only weak, or even nonexistent, under field conditions (103,179–181). Such missing or poor correlations are the major reason for the large number of different methods of soil testing, and they justify ongoing research for new methods (114,182–185). Soil analytical methods for plant-available sulfate differ in the preparation of the soil samples, concentration and type of extractant, duration of the extraction procedure, the soil-to-extractant ratio, the conditions of extraction, and the method that is used for the determination of sulfur or sulfate-S in the extract. A serious problem with regard to all laboratory methods is the treatment and preservation of soil samples prior to analysis. Increased temperature and aeration of the sample during storage increase the amount of extractable sulfur by oxidizing labile organic sulfur fractions, and occasionally mobilize reduced inorganic sulfur (186–188).

Besides water, potassium or calcium dihydrogenphosphate solutions are the most commonly used solvents to extract plant-available sulfate from soils (189,190). Soils with a high sulfate adsorption capacity are low in pH, so that phosphate-containing extractants extract more sulfate than other salt solutions because of ion-exchange processes. Sodium chloride is also used in countries where soils are frequently analyzed for available nitrate (183,191,192). Less frequently, magnesium chloride (193) or acetate solutions are employed (194,195). Other methodical approaches involve, for instance, anion-exchange resins (196,197) and perfusion systems (198).

In aerated agricultural soils, the organic matter is the soil-inherent storage and backup for buffering sulfate in the soil solution (199–201), and methods are described which focus on capturing organic sulfur fractions that might be mineralized during the vegetation period and thus contribute to the sulfate pool in soils (183,202–204). Such special treatments are, for example, the heating of the samples or employing alkaline conditions or incubation studies, which allow the measurement of either the easily mineralized organic sulfur pool or the rapidly mineralized organic sulfur. Most methods, however, extract easily soluble, plant-available sulfate.

The practical detection limit of sulfur determined by ICP-AES was $0.5\,mg\ S\ L^{-1}$, corresponding to $3.3\,mg\ S\ kg^{-1}$ (205) in the soil. On sulfur-deficient sites, however, sulfate-S concentrations of only $2\,mg\ S\ kg^{-1}$ were measured regularly in the topsoil by ion chromatography (206). Ion chromatography is much more sensitive, with a practical detection limit of $0.1\,mg\ SO_4\text{-}S\ L^{-1}$ (corresponding to $0.67\,mg\ S\ kg^{-1}$), allowing sulfate-S to be determined at low concentrations in soils. Additionally, this fact explains why soil sulfate-S measured by ICP-AES is usually below the detection limit. No matter which method is applied, and on which soils or crops the method is used, there is an astonishing agreement in the literature for approximately $10\,mg\ SO_4\text{-}S\ kg^{-1}$ as the critical value for available sulfur in soils (68,192,207). With the most common methods for the determination of sulfur (ICP and the formation of $BaSO_4$), values of $< 10\,mg\ S\ kg^{-1}$ will identify a sulfur-deficient soil with a high probability.

As expected, comparisons of different extractants and methods revealed that under the same conditions, all of these methods extract more or less the same amount of sulfate from the soil (178,182,183,185,198,203,207–209). Occasionally observed differences among methods were more likely to be caused by interferences due to the extractant itself (183) rather than by the method of sulfate-S determination (186,187).

As there is virtually no physicochemical interaction between the soil matrix and sulfate, the amount that is present and extractable from the soil is the main indicator commonly used to describe the sulfur nutritional status of a soil. Opinions in the literature on whether or not soil testing is a suitable tool for determining the sulfur status of soils vary from high acceptance (210–215) down to full denial (179,216–220).

Conclusions leading to high acceptance were always drawn from pot trials, which usually yield high correlation coefficients between soil analytical data, and give sulfur content or sulfur uptake of plants as the target value (103,178,183,185,192,194,198,212,221–223,225). Pot trials are always prone to deliver very high correlations between soil, and plant data or yield, as there is no uncontrolled nutrient influx and efflux. However, in the case of field surveys involving a greater range of sites and environmental factors, correlations are poor or fail to reach significance (103,180). For the relationship between available sulfur in soils and foliar sulfur, larger surveys employing a wide range of available sulfur in soils (5 to $250\,mg\ S\ kg^{-1}$), and plants (0.8 to $2.1\,g\ S\ kg^{-1}$), reported correlation coefficients for a total of 1701 wheat and 1870 corn samples of $r = 0.292$ ($P \le 0.001$) and $r = 0.398$ ($P \le 0.001$), respectively (195). Timmermann and coworkers (225) determined a correlation coefficient of $r = 0.396$ ($P < 0.05$) for 93 oilseed rape samples. In the field surveys conducted by Schnug (103), a significant relationship could not be verified for 489 oilseed rape samples ($r = 0.102$, $P > 0.05$) or for 398 cereal samples ($r = 0.098$, $P > 0.05$).

These results imply that a maximum of 16% of the variability of the sulfur concentrations in leaves can be explained by the variability of available sulfur in soils. However, Timmermann et al. (225) were able to improve the relationship between soil and plant data by using the ratio of available sulfur and nitrogen in soils (N_{min}/S_{min}) instead of just sulfur. This application gave a value of $r = -0.605$ ($P \leq 0.01$), which still explains less than one third of the variability.

The key problem of soil analysis for plant-available sulfur is that it is a static procedure that aims at reflecting the dynamic transfer of nutrient species among different chemical and biological pools in the soil. This concept is appropriate if the sample covers the total soil volume to which active plant roots have access and if no significant vertical and lateral nutrient fluxes occur to and from this specific volume. Sulfate, however, has an enormously high mobility in soils and can be delivered from sources such as subsoil or shallow groundwater, and sulfur has virtually no buffer fraction in the soil. Thus, the availability of sulfate is a question of the transfer among pools in terms of space and time rather than among biological or chemical reserves. Under field conditions sulfate moves easily in or out of the root zones so that close correlations with the plant sulfur status can hardly be expected. Attempts have been made to take subsoil sulfate into account by increasing the sampling depth (103,226–230), but the rapid vertical and lateral mobility of sulfate influences subsoils too. Thus, this procedure did not yield an improvement of the expressiveness of soil analytical data (103,225).

The soil sulfur cycle is driven by biological and physicochemical processes which affect flora and fauna. The variability of sulfate-S contents in the soil over short distances is caused by the high mobility of sulfate-S. Sulfate is an easily soluble anion, and it follows soil water movements. Significant amounts of adsorbed sulfate are found only in clay and sesquioxide-rich soil horizons with pH values < 5, which is far below the usual pH of northern European agricultural soils. Seasonal variations in mineralization, leaching, capillary rise, and plant uptake cause temporal variations in the sulfate-S content of the soil (205). The high spatiotemporal variation of sulfate in soils is the reason for the inadequacy of soil analysis in predicting the nutritional status of sulfur in soils. Thus, under humid conditions, the sulfur status of an agricultural site is difficult to assess (231). An overview of the factors of time and soil depth in relation to the variability of sulfate-S contents is given in Figure 7.13. The highest variability of sulfate-S could be observed on two sites in soil samples collected in April (Figure 7.13). On a sandy soil, the variability was distinctly higher at the second and third dates of sampling in comparison with a loamy soil, but time-dependent changes were significant only in the deeper soil layers. Though the range of sulfate-S contents measured was smaller on the loamy soil than on the sandy soil, the differences proved to be significant in all soil layers between the first and third and second and third dates of sampling respectively (Figure 7.13).

Sources and sinks commonly included in a sulfur balance are inputs by depositions from atmosphere, fertilizers, plant residues, and mineralization, and outputs by losses due to leaching. A frequent problem when establishing such simple sulfur balances is that the budget does not correspond to the actual sulfur supply. The reason is that under temperate conditions it is the spatiotemporal variation of hydrological soil properties that controls the plant-available sulfate-S content. A more promising way to give a prognosis of the sulfur supply is a site-specific sulfur budget, which includes information about geomorphology, texture, climatic data, and crop type and characteristics of the local soil water regime (Figure 7.14).

The results presented in Figure 7.14 reveal that plant sulfur status is distinctly higher on sites with access to groundwater than on sandy soils not influenced by groundwater. The significance of plant-available soil water as a source and storage for sulfur has been disregarded or underestimated so far. However, especially under humid growth conditions, plant-available soil water is the largest contributor to the sulfur balance (205). Leaching and import from subsoil or shallow groundwater sources (184,205) can change the amount of plant-available sulfate within a very short time. Groundwater is a large pool for sulfur, because sulfur concentrations of 5 to 100 mg S L^{-1} are common

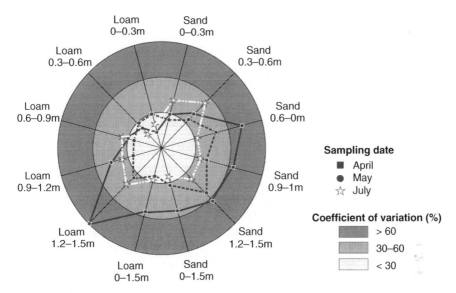

FIGURE 7.13 Spatiotemporal variability of the sulfate contents of different soil layers in two soil types. (From Bloem, E. et al., *Commun. Soil Sci. Plant Anal.*, 32, 1391–1403, 2001.)

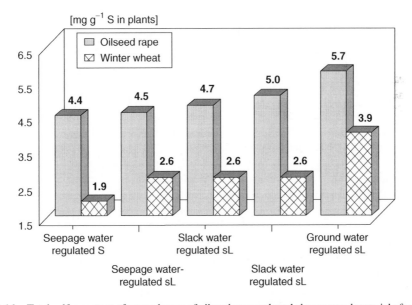

FIGURE 7.14 Total sulfur content of young leaves of oilseed rape and total aboveground material of winter wheat at stem extension in relation to soil hydrological parameters and soil texture (S=Sand; sL=sandy Loam) on the Isle of Ruegen. (From Bloem, E., Schwefel-Bilanz von Agraroekosystemen unter besonderer Beruecksichtigung hydrologischer und bodenphysikalischer Standorteigenschaften, Ph.D. thesis, TU-Braunschweig, Germany, 1998.)

in surfaces near groundwater (205,232). There are three ways in which groundwater contributes to the sulfur nutrition of plants. First, there is a direct sulfur input if the groundwater level is only 1 to 2 m below the surface, which is sufficient to cover the sulfur requirement of most crops as plants can utilize the sulfate in the groundwater directly by their root systems. Second, groundwater, which is used for irrigation, can supply up to 100 kg S ha^{-1} to the crop (205,233–235), but irrigation water will contribute significantly to the sulfur supply only if applied at the start of the main growth period of the crop. Third, the capillary rise of groundwater under conditions of a water-saturation deficit in the upper soil layers leads to a sulfur input. This process is closely related to climatic conditions. The sulfur supply of a crop increases with the amount of plant-available water or shallow groundwater. The higher the water storage capacity of a soil, the less likely are losses of water and sulfate-S by leaching and the greater is the pool of porous water and also the more likely is an enrichment of sulfate just by subsequent evaporation. Thus, heavy soils have a higher charging capacity for sulfate-S than light ones.

7.5 PLANT ANALYSIS

Plant families and species show great variabilities in sulfur concentrations. In general, gramineous species have lower sulfur levels than dicotyledonous crops (see Section 7.3.2). Within each genus, however, species producing S-containing secondary metabolites accumulate more sulfur than those without this capacity. The ratios of sulfur concentrations in photosynthetically active tissue of cereals, sugar beet, onion, and oilseed rape are approximately 1:1.5:2:3 (114,236). Thus plants with a higher tendency to accumulate sulfur, such as brassica species, are very suitable as monitor crops to evaluate differences between sites and environments, or for quick growing tests (176). Generative material is less suited for diagnostic purposes (237), because the sulfur concentration in seeds is determined much more by genetic factors (43,103,116). During plant growth, morphological changes occur and there is translocation of nutrients within the plant. Thus, changes in the nutrient concentration are not only related to fluctuations in its supply, but also to the plant part and plant age. These factors need to be taken into account when interpreting and comparing results of plant analysis (216,238–243). Basically, noting the time of sampling and analyzed plant part is simply a convention, but there are some practical reasons for it that should be considered: (a) photosynthetically active leaves show the highest sulfur concentrations of all plant organs, and as sulfur has a restricted mobility in plants sulfur concentrations in young tissues will respond first to changes in the sulfur supply; (b) sampling early in the vegetative state of a crop allows more time to correct sulfur deficiency by fertilization. It is relevant in this context that plant analysis is a reliable tool to evaluate the sulfur nutritional status, but usually it is not applicable as a diagnostic tool on production fields because of the shortcomings mentioned above.

In dicotyledonous crops, young, fully expanded leaves are the strongest sinks for sulfur, and they are available during vegetative growth. Therefore, they are preferable for tissue analysis (88,103,244). Oilseed rape, for instance, delivers suitable leaves for tissue analysis until 1 week after flowering, and sugar beet gives suitable leaves until the canopy covers the ground and the storage roots start to extend (103).

For the analysis of gramineous crops, either whole plants (1 cm above the ground) after the appearance of the first and before the appearance of the second node, or flag leaves are best suited for providing samples for analysis (142,143,245–249).

In all cases, care has to be taken to avoid contamination of tissue samples with sulfur from foliar fertilizers or sulfur-containing pesticides. Care is also needed when cleaning samples, because water used for washing may contain significant amounts of sulfate. Paper used for sample drying and storage contains distinct amounts of sulfate, originating from the manufacturing process. As sulfate bound in paper is more or less insoluble, the risk of contamination when washing plants is low, but adherent paper particles may significantly influence the results obtained.

7.5.1 Analytical Methods

Sulfur occurs in plants in different chemical forms (250), and nearly all of them have been tested as indicators for sulfur nutritional status. The parameters analyzed by laboratory methods for the purpose of diagnostics can be divided into three general classes: biological, chemical, and composed parameters.

Biological parameters are the sulfate and glutathione content. Many authors proposed the sulfate-S content as the most suitable diagnostic criterion for the sulfur supply of plants (241,242,251–255). They justify their opinion by referring to the role of sulfate as the major transport and storage form of sulfur in plants (256,257). Other authors, however, attribute this function also to glutathione (55,258,259). Based on this concept, Zhao et al. (260) investigated the glutathione content as a diagnostic parameter for sulfur deficiency.

Although indeed directly depending on the sulfur supply of the plant (64,103), neither of the compounds is a very reliable indicator for the sulfur status because their concentrations are governed by many other parameters, such as the actual physiological activity, the supply of other mineral nutrients, and the influence of biotic and abiotic factors (5,63,256,261). Biotic stress, for instance, increased the glutathione content by 24% (63). Amino acid synthesis is influenced by the deficiency of any nutrient and thus may indirectly cause an increase in sulfate or glutathione in the tissue. An example for this action is the increase in sulfate following nitrogen deficiency (103,262,263). Significant amounts of sulfate may also be physically immobilized in vacuoles (see Section 7.2.1).

In plant species synthesizing glucosinolates, sulfate concentrations can also be increased by the release of sulfate during the enzymatic cleavage of these compounds after sampling (103). As enzymatically released sulfate can amount to the total physiological level required, this type of post-sampling interference can be a significant source of error, yielding up to 10% higher sulfate concentrations (63,103). It is probably also the reason for some extraordinarily high critical values for sulfate concentrations reported for brassica species (220,264). The preference for sulfate analysis as a diagnostic criterion may also come from its easier analytical determination compared to any other sulfur compound or to the total sulfur concentration (265).

Hydrogen iodide (HI)-reducible S, acid-soluble sulfur, and total sulfur are chemical parameters used to describe the sulfur status of plants. None of them is related to a single physiological sulfur-containing compound. The HI-reducible sulfur or acid-soluble sulfur estimate approximately the same amount of the total sulfur in plant tissue (~50%). The acid-soluble sulfur is the sulfur extracted from plant tissue by a mixture of acetic, phosphoric, and hydrochloric acids according to Sinclair (167), who described this extractant originally for the determination of sulfate. Schnug (103) found in tissue samples from more than 500 field-grown oilseed rape and cereal plants that the acid-soluble sulfur content (y) is very closely correlated with the total sulfur content (x). The slope of the correlations is identical, but the intercept is specific for species with or without S-containing secondary metabolites:

$$\text{oilseed rape: } y = 0.58x - 1.25; \; r = 0.946 \qquad \text{cereals: } y = 0.58x - 0.39; \; r = 0.915$$

As the total sulfur content in Sinclair's (167) solution is easy to analyze by ICP, this extraction method seems to be a promising substitute for wet digestion with concentrated acids or using x-ray fluorescence spectroscopy for total sulfur determination (53,103,266–268).

The total sulfur content is most frequently used for the evaluation of the sulfur nutritional status (see Section 7.5.3). Precision and accuracy of the analytical method employed for the determination of the total sulfur content are crucial. In proficiency tests, X-ray fluorescence spectroscopy proved to be fast and precise (269,270). Critical values for total sulfur differ in relation to the growth stage (242,261), but this problem is also true for all the other parameters and can be overcome only by a strict dedication of critical values to defined plant organs and development stages (103). If this procedure is followed strictly, the total sulfur content of plants has the advantage of being less influenced by short-term physiological changes that easily affect fractions such as sulfate or glutathione.

Composed parameters are the nitrogen/sulfur (N:S) ratio, the percentage of sulfate-S from the total sulfur concentration, and the sulfate/malate ratio. The concept of the N/S ratio is based on the fact that plants require sulfur and nitrogen in proportional quantities for the biosynthesis of amino acids (271–273). Therefore, deviations from the typical N/S ratio were proposed as an indicator for sulfur deficiency (239,274–281). Calculated on the basis of the composition of amino acids in oilseed rape leaf protein, the optimum N/S ratio for this crop should theoretically be 12:1 (103,282), but empirically maximum yields were achieved at N/S ratios of 6:1 to 8:1 (216,242,253,283). Distinct relationships between N/S ratio and yield occur only in the range of extreme N/S ratios. Such N/S ratios may be produced in pot trials but do not occur under field conditions (see Figure 7.16).

There is no doubt that balanced nutrient ratios in plant tissues are essential for crop productivity, quality, and plant health, but the strongest argument against using the N/S ratio to assess the nutritional status is that it can result from totally different N and sulfur concentrations in the plant tissue. Surplus of one element may therefore falsely be interpreted as a deficiency of the other (284). The suitability of N/S ratios as a diagnostic criterion also implies a constancy (273,285–288), which is at least not true for species with a significant secondary metabolism of S-containing compounds such as *Brassica* and *Allium* species (289,290). Additionally, it requires the determination of two elements and thus is more laborious and costly.

The percentage of sulfate-S of the total sulfur content has been proposed as a diagnostic criterion (240–242,251–255). Except for laboratories operating x-ray fluorescence spectroscopy, which allows the simultaneous determination of sulfate-S and total sulfur (291,292), this determination doubles the analytical efforts without particular benefit. The sulfate/malate ratio is another example of a composed parameter (293). Though both parameters can be analyzed by ion chromatography in one run, the basic objection made with regard to sulfate (see above), namely its high variability, also applies to malate.

7.5.2 ASSESSMENT OF CRITICAL NUTRIENT VALUES

Critical values are indispensable for evaluating the nutritional status of a crop. Important threshold markers are: (a) the symptomatological value, which reflects the sulfur concentration below which deficiency symptoms become visible (see Section 7.3.1); (b) the critical nutrient value, which stands for the sulfur concentration above which the plant is sufficiently supplied with sulfur for achieving the maximum potential yield or yield reduced by 5, 10, or 20% (294); and (c) the toxicological value, which indicates the sulfur concentration above which toxicity symptoms can be observed. However, there is no one exclusive critical nutrient value for any crop, as it depends on the growth conditions, the developmental stage of the plant at sampling, the collected plant part, the determined sulfur species, the targeted yield, and the mathematical approach for calculating it. Smith and Loneragan (295) provided a comprehensive, general overview of the significance of relevant factors influencing the derivation of critical values. Numerous, differing critical sulfur values and ranges exist for each crop and have been compiled, for instance by Reuter and Robinson (294), for all essential plant nutrients and cultivated plants including forest plantations. In this section, an attempt was made to compile and categorize, from the literature, available individual data based on studies with varying experimental conditions of the variables, total sulfur and sulfate concentrations, and N/S ratios in relation to different groups of crops for facilitating an easy and appropriate evaluation of sulfur supply. Plant groups were assembled by morphogenetic and physiological features. Because of the wide heterogeneity of results for similar classes of sulfur supply and for a better comparability of results, concentrations were agglomerated into three major categories: deficient, adequate, and high, irrespective of the sampled plant part during vegetative growth (Table 7.6). A prior-made subdivision, which took these relevant criteria into consideration (see Section 7.3.1) next to additional characteristics of the sulfur supply (symptomatological and critical values of total S, sulfate, and N/S ratio), did not prove to be feasible as the variation of results was so high that no clear ranges, let alone threshold values, could be assigned for individual classes and crops, or crop groups. Smith and Loneragan (295) stressed that in addition to various biotic and abiotic factors, experimental conditions, plant age, and plant part, all influence the nutrient status; the procedure to derive a critical value itself has a significant impact, so that it is possible to define only ranges for different nutritional levels. This finding also implies that it is more or less impossible to compare results from different experiments. The integration of individual studies, which imply extreme values, are not suitable for a generalization of an affiliation to a certain class of sulfur supply and, more importantly, such interpretation may even yield an erroneous evaluation of the sulfur supply. In comparison, the compilation

TABLE 7.6
Mean Critical Values and Ranges of Sulfur Nutrition for Different Groups of Agricultural Crops

Deficient	Adequate	High	Parameter
	S Nutritional Status		

Poaceae: barley (*Hordeum vulgare*), corn (*Zea mays*), oats (*Avena sativa*), rice (*Oryza sativa*), sorghum (*Sorghum vulgare*), sugarcane (*Saccharum* ssp.), wheat (*Triticum aestivum; Triticum durum*)

Deficient	Adequate	High	Parameter
			S_{tot} (mg g^{-1})
0.94	1.7	4.7	Median
0.6	1.4	4.0	25% quartile
1.2	2.5	6.0	75% quartile
0.1–2.0	0.3–8.9	3.3–10.0	Range
41	145	18	(*n*)
			N/S ratio
24	16.0	—	Median
19.5	10.7	—	25% quartile
29.3	19.0	—	75% quartile
11.9–55	7–38	—	Range
15	45	—	(*n*)
			Sulfate (mg kg^{-1})
60	150	5400	Median
36.5	82.5	1500	25% quartile
235	1030	8300	75% quartile
23–400	30–6400	1200–11200	Range
4	20	5	(*n*)

Oil crops I: Mustard (*Brassica juncea*), oilseed rape, spring and winter varieties (*Brassica napus; Brassica campestris*)

Deficient	Adequate	High	Parameter
			S_{tot} (mg g^{-1})
1.6	4.8	—	Median
2.3	3.2	—	25% quartile
3.3	6.7	—	75% quartile
1.1–5.8	1.7–10.4	—	Range
8	54	—	(*n*)
			N:S ratio
—	6–7	—	Median
—	—	—	25% quartile
—	—	—	75% quartile
—	—	—	Range
—	1	—	(*n*)
			Sulfate (mg kg^{-1})
—	—	—	Median
—	—	—	25% quartile
—	—	—	75% quartile
—	—	—	Range
—	—	—	(*n*)

Oil crops II: Cotton (*Gossypium hirsutum*), linseed (*Linum usitatissimum*), peanut (*Arachis hypogaea*), soybean (*Glycine max*), sunflower (*Helianthus annuus*)

Deficient	Adequate	High	Parameter
			S_{tot} (mg g^{-1})
1.7	2.3	3	Median
0.9	2.0	—	25% quartile

Continued

TABLE 7.6 *(Continued)*

S Nutritional Status			
Deficient	**Adequate**	**High**	**Parameter**
2.0	3.1	—	75% quartile
0.8–2.9	1.1–9.9	—	Range
19	108	2	(*n*)
			N:S ratio
—	15.8	—	Median
—	13	—	25% quartile
—	20	—	75% quartile
—	12–25	—	Range
—	8	—	(*n*)
			Sulfate (mg kg^{-1})
10	360	—	Median
10	190	—	25% quartile
20	475	—	75% quartile
3–100	100–700	—	Range
6	5	—	(*n*)

Legumes: Chickpea (*Cicer arietinum*), Faba bean (*Vicia faba*), (field) pea (*Pisum sativum*), lentil (*Lens culinaris*), navy, bush, snap, green, dwarf, french beans (*Phaseolus vulgaris*), lupin (*Lupinus angustifolius, Lupinus albus, Lupinus cosentinii*), black gram (*Vigna mungo*), cowpea (*Vigna unguiculata*), pigeon pea (*Cajanus cajan*)

			S_{tot} (mg g^{-1})
1.1	2.7	—	Median
0.7	2.0	—	25% quartile
1.5	3.6	—	75% quartile
0.7–3.0	0.7–6.5	—	Range
7	62	—	(*n*)
			N:S ratio
—	15.5	—	Median
—	—	—	25% quartile
—	—	—	75% quartile
—	—	—	Range
—	2	—	(*n*)
			Sulfate (mg kg^{-1})
—	1600	11200	Median
—	500	—	25% quartile
—	3400	—	75% quartile
—	200–6400	—	Range
—	5	1	(*n*)

Root crops: Carrot *(Daucus* carota), cassava *(Manihot esculentum)*, potato (*Solanum tuberosum*), sugar beet, fodder beet, beetroot *(Beta vulgaris)*, sweet potato *(Ipomoea batatas)*

			S_{tot} (mg g^{-1})
1.4	3.0	3	Median
0.8	2.0	—	25% quartile
2.2	3.7	—	75% quartile
0.4–3.0	0.75–6.3	—	Range
8	45	1	(*n*)
			N:S ratio
—	11	—	Median
—	—	—	25% quartile
—	—	—	75% quartile

TABLE 7.6   (*Continued*)

Deficient	Adequate	High	Parameter
	S Nutritional Status		
—	—	—	Range
—	1	—	(*n*)
			Sulfate (mg kg^{-1})
150	400	2800	Median
50	250	—	25% quartile
200	3880	—	75% quartile
50–200	250–14000	—	Range
6	5	1	(*n*)

Fodder crops/pastures: Alfalfa (*Medicago sativa*), annual ryegrass (*Lolium rigidum*), Bahia grass (*Paspalum notatum*), Balansa cover (*Trifolium balansae*), barley grass (*Hordeum leporinum*), barrel medic (*Medicago truncatula*), Bermuda grass (*Cynodon dactylon*), Berseem clover (*Trifolium alexandrinum*), black medic (*Medicago lupulina*), Buffel grass (*Cechrus ciliaris*), burr/annual medic (*Medicago polymorpha*), Caribbean Stylo (*Stylosanthes hamata*), Centro (*Centrosema pubescens*), Cluster clover (*Trifolium glomeratum*), cocksfoot (*Dactylis glomerata*), dallis grass (*Paspalum dilatatum*), *Digitaria eriantha*, Dolichos lablab (*Lablab purpu*reus), glycine (*Neonotonia wightii*), *Glycine tabacina*, Great brome grass (*Bromus diandrus*), greenleaf desmodium (*Desmodium intortum*), Guinea grass (*Panicum maximum*), Kentucky bluegrass (*Poa pratensis*), Kenya white clover (*Trifolium semipilosum*), Kikuyu grass (*Pennisetum clandestinum*), Leucaena (*Leucaena leucocephala*), Lotonis (*Lotonis bainesii*), Murex medic (*Medicago murex*), Phalaris (*Phalaris aquatica*), perennial ryegrass (*Lolium perenne*), phasey bean (*Macroptilium lathroides*), purple bean (*Macroptilium atropurpureum*), Rhodes grass (*Chloris gayana*), Setaria (*Setaria sphacelata*), Shrubby Stylo (*Stylosanthes scabra*), silver leaf desmodium (*Desmodium uncinatum*), Sorghum-sudangrass (*Sorghum bicolor x S. sudanese*), Sticky Stylo (*Stylosanthes viscosa*), Stylo (*Stylosanthes guianensis*), subterranean clover (*Trifolium subterraneum*), Townsville Stylo (*Stylosanthes humilis*), white clover (*Trifolium repens*), wooly burr medic (*Medicago minima*)

Deficient	Adequate	High	Parameter
			S$_{tot}$ (mg g^{-1})
1.5	2.1	3.2	Median
1.1	1.7	3	25% quartile
3	2.7	5.6	75% quartile
0.6–3.1	0.7–6.5	2.3–7.5	Range
68	297	13	(*n*)
			N:S ratio
15	20	—	Median
—	16.3	—	25% quartile
—	20	—	75% quartile
—	10–29	—	Range
1	23	—	(*n*)
			Sulfate (mg kg^{-1})
109	500	10850	Median
98	209	—	25% quartile
146.5	1350	—	75% quartile
20–1300	20–3900	—	Range
16	64	2	(*n*)

Brassica vegetables: Broccoli (*Brassica oleracea* var. *italica*), brussels sprouts (*Brassica oleracea* var. *gemmifera*), cabbage (*Brassica oleracea*), cauliflower (*Brassica oleracea* var. *botrytis*), Chinese kale (*Brassica oleracea* var. *alboglabra*), Chinese cabbage (*Brassica rapa* var. *pekinensis*), kohlrabi (*Brassica oleracea* var. *gongylodes*), Pak-choi (*Brassica rapa* var. *chinensis*), spinach mustard (*Brassica pervirdis*), turnip (*Brassica rapa* var. *rapa*)

Deficient	Adequate	High	Parameter
			S$_{tot}$ (mg g^{-1})
—	7.5	6.5	Median
—	4	—	25% quartile

Continued

TABLE 7.6 (*Continued*)

Deficient	Adequate	High	Parameter
	S Nutritional Status		
—	12.8	—	75% quartile
—	2.5–19.2	—	Range
—	30	1	(*n*)
			N:S ratio
—	—	—	Median
—	—	—	25% quartile
—	—	—	75% quartile
—	—	—	Range
—	—	—	(*n*)
			Sulfate (mg kg^{-1})
—	—	—	Median
—	—	—	25% quartile
—	—	—	75% quartile
—	—	—	Range
—	—	—	(*n*)

Nonbrassica vegetables: Asparagus (*Asparagus officinalis*), Arugula salad (*Eruca sativa*), cantaloupe, honeydew (*Cucumis melo*), celery (*Apium graveolens*), cucumber (*Cucumis sativus*), endive (*Cichorium endiva*), fenugreek (*Trigonella foenum-graecum*), garden sorrel (*Rumex acetosa*), lettuce (*Lactuca sativa* spp.), onion (*Allium cepa*), spinach (*Spinacia oleracea*), tomato (*Lycopersicon esculentum*), wild radish (*Raphanus raphanastrum*), zucchini (*Cucurbita pepo*)

Deficient	Adequate	High	Parameter
			S_{tot} (mg g^{-1})
2.9	4.0	10	Median
1	3.0	7	25% quartile
3.9	7.0	10	75% quartile
0.6–4.9	1.6–14.0	7–10	Range
13	47	5	(*n*)
			N:S ratio
—	—	—	Median
—	—	—	25% quartile
—	—	—	75% quartile
—	—	—	Range
—	—	—	(*n*)
			Sulfate (mg kg^{-1})
1100	11750	—	Median
—	—	—	25% quartile
—	—	—	75% quartile
—	—	—	Range
1	2	—	(*n*)

Source: Compiled from references given in Schnug (103), Bergmann (143), Eaton (144), Reuter and Robinson (294), and Mills and Jones (296).

of the data in Table 7.6 indicates that the sampled plant part during the main vegetative development seems to be of minor relevance for generally addressing the sulfur nutritional status. However, for following up, for instance, nutritional or pathogen-related changes in sulfur metabolism, it might even be necessary to do so in defined parts of a plant organ or on a leaf cell level.

The results in Table 7.6 reveal that Poaceae and fodder crops have been studied intensely in relation to sulfur nutritional supply. For all crops, the total sulfur concentration was used most often to characterize the sulfur nutritional status. The range of variation was distinctly lower for total sulfur than for sulfate concentrations, independent of the crop type. It is also remarkable that the ranges in the three classes overlap regularly for all groups of crops and sulfur fractions. With the exception of the fodder crops, however, the 25 and 75% quartiles separate samples from the three nutritional levels efficiently if total sulfur concentrations were determined. For sulfate, such partition was feasible too, except in Poaceae. Generally, an insufficient sulfur supply is indicated by total sulfur concentrations of <1.7 mg g^{-1}. In the case of Poaceae and nonbrassica vegetables, this value may be lower at 0.94 mg S g^{-1} or higher at 2.9 mg S g^{-1} (Table 7.6; Section 7.3.1). Sulfate concentrations of <150 mg SO$_4$-S kg^{-1} indicate an insufficient sulfur supply. An adequate sulfur supply is reflected by total sulfur concentrations of 1.7 to 4 mg S g^{-1}; brassica crops show a higher optimum range with values of 4.8 (oil crops) to 7.5 (vegetables) mg S g^{-1} (Table 7.6). Values of 16 to 20 for N/S ratio, and 150 to 1600 for sulfate-S concentrations reflect a sufficient sulfur supply. In comparison, values of >2800 mg SO$_4$-S kg^{-1} denote an excessive sulfur supply (Table 7.6). Sulfate is usually not determined in brassica oil crops and vegetables as the degradation of glucosinolates might falsify the result (see Section 7.5). For fodder crops, total sulfur concentrations of even 3.2 mg S g^{-1} may be disproportionate, whereas the corresponding value for nonbrassica vegetables would equal 10 mg S g^{-1}.

The major criticism of critical values for the interpretation of tissue analysis is the small experimental basis, which often consists of not more than a single experiment (297). Besides the lack of data, the method of interpretation may also yield erroneous results. Methods based on regression analysis, like the 'broken stick method' by Hudson (298) and Spencer and Freney (241), or the 'vector analysis' by Timmer and Armstrong (299) investigate mathematical, but not necessarily causal, interactions between the nutrient content and yield, because the dictate of minimizing the sum of squared distances aims only to find a function that fits best across the data set. Like the method of Cate and Nelson (300,301), these methods have been designed primarily for the investigation of small data sets and plants grown under *ceteris paribus* conditions, where only the response to variations in the nutrient supply varied. Another quite significant disadvantage of critical values and critical ranges[*] (143,296,302), or 'no-effect values (NEV)'[†] (284) is that they ignore the nonlinearity of the Mitscherlich function describing the relationship between growth factors and yield (303). The ideal basis for critical values for the interpretation of tissue analysis are large sets of yield data and nutrient concentrations in defined plant organs that cover a wide range of growth factor combinations. The data may include samples from field surveys or field or pot experiments if the reference yield of 100% was obtained in all cases under optimum growth conditions. In Figure 7.15 and Figure 7.16, corresponding examples are given for the total sulfur concentration in shoots of cereals at stem extension and the N/S ratio in younger, fully developed leaves of oilseed rape at stem extension.

The data in Figure 7.15 reveal a characteristic bow-shaped bulk, which covers sulfur concentrations from 0.5 to 5.5 mg S g^{-1}. Sulfur deficiency can be expected at sulfur concentrations below 0.94 mg g^{-1} (Table 7.6). A symptomatological threshold for the expression of macroscopic symptoms of 1.2 mg S g^{-1} was determined for cereals by Schnug and Haneklaus (114). Total sulfur concentrations of 1.7 mg g^{-1} are considered as being adequate to satisfy the sulfur demand of cereal crops, whereas the data in Figure 7.15 show a further yield increase with higher sulfur concentrations. The reason is simply that the 100% yield margin corresponds to a grain yield of 10 t ha^{-1} (180), so that accordingly a total sulfur concentration of 1.7 mg S g^{-1} would be sufficient for 8.2 t ha^{-1}. A productivity level of 10 t ha^{-1} is extraordinarily high and restricted to areas of high fertility or inputs, whereas a level of 8 t ha^{-1} represents a high-yielding crop in many areas in the world. Thus, a total sulfur concentration of 4.7 mg g^{-1}, which is rated as reflecting a high sulfur supply, is marginal on high productivity sites.

Basic shortcomings of using, for instance, the N/S ratio for the evaluation of the sulfur nutritional status were discussed (Section 7.5) and are reflected in the data in Figure 7.16. Hence, there are no relationships between N/S ratio and yield in a way as was shown for total sulfur and cereals (Figure 7.15). Crop productivity seems to be fairly independent of variations in the N/S ratio within a range of 5:1 to 12:1 (Figure 7.16).

---

[*] Tissue concentration for 95% of maximum yield.

[†] Tissue concentration for maximum yield or the concentration above which no yield response occurs.

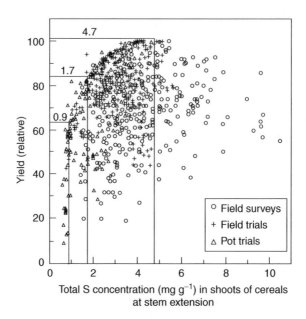

FIGURE 7.15 Scattergram of total sulfur in shoots and yield data for cereals in relation to experimental conditions (From Schnug, E. and Haneklaus, S., in *Sulphur in Agroecosystems*. Vol. 2, Part of the series 'Nutrients in Ecosystems', Kluwer Academic Publishers, Dordrecht, 1998, pp. 1–38.) and merged values thresholds for sulfur supply (see Table 7.7).

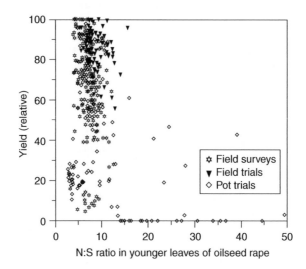

FIGURE 7.16 Relationship between N:S ratio in young leaves of oilseed rape at stem extension and relative seed yield. (From Schnug, E., Quantitative und Qualitative Aspekte der Diagnose und Therapie der Schwefelversorgung von Raps (*Brassica napus* L.) unter besonderer Berücksichtigung glucosinolatarmer Sorten. Habilitationsschrift, D.Sc. thesis, Kiel University, 1988.)

Comprehensive data sets like those presented in Figure 7.15 allow for the accurate calculations of so-called upper boundary line functions, which describe the highest yields observed over the range of nutrient values measured. Data points below this line relate to samples where some other factor limited the crop response to the nutrient. An overview of the scientific background and development of upper boundary lines is given by Schnug et al. (304).

The Boundary Line Development System (BOLIDES) was elaborated to determine the upper boundary line functions and to evaluate optimum nutrient values and ranges. The BOLIDES is based on a five-step algorithm (Figure 7.17) (304). For the identification of outliers, cell sizes are defined for nutrient and yield values together with an optional number of data points per cell (Figure 7.17a). The cell size can be chosen variably with proposed values for X (nutrient content) corresponding to the standard deviations and for Y (yield) with the coefficient of variation. If another variable, often a stable soil feature such as organic matter or clay content, has a significant effect on the response to the nutrient, its presence is indicated by two or more distinct concentrations of points, each with its own boundary line response to the nutrient (Figure 7.17b). The data can be classified on the basis of this third variable, and the boundary line can be determined separately for each class. Next, a boundary step function is calculated for each class, starting from the minimum nutrient content up to the point of maximum yield, as well as from the maximum nutrient content up to the maximum yield (Figure 7.17c). Then the boundary line, usually a first-order polynomial function, is fitted according to the least-squares method (Figure 7.17d). The first derivative of the fitted polynomial gives predicted yield response to fertilization in relation to the nutrient content (Figure 7.17). The last step is the classification of the nutrient supply to determine optimum nutrient levels or optimum nutrient ranges. The optimum nutrient value corresponds with the zero of the first derivative of the upper boundary line and the sign of the second derivative at this point. For the determination of the optimum ranges, that is, the range of nutrient concentration that gives 95% of the maximum yield, standard, numerical root-finding procedures are used for real polynomials of degree 4 with constant coefficients (Figure 7.17).

Thus boundary lines describe the 'pure effect of a nutrient' on crop yield under *ceteris paribus* conditions (246,247,305,306). The comparison of the boundary lines for total sulfur and yield

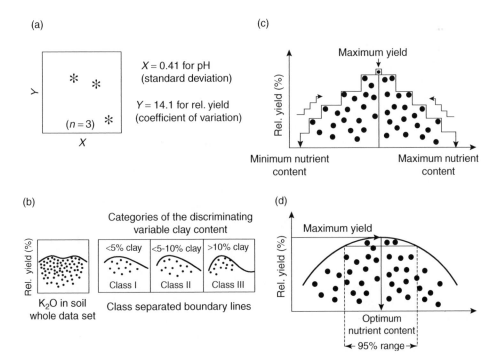

FIGURE 7.17 Structure of Boundary Line Development System (BOLIDES) for the determination of upper boundary line functions and optimum nutrient values and ranges in plants and soils: (a) identification of outliers; (b) discrimination against a third variable; (c) calculation of step functions; and (d) determination of the upper boundary line and calculation of optimum nutrient value and ranges. (From Haneklaus, S. and Schnug, E., *Aspects Appl. Biol.*, 52, 87–94, 1998.)

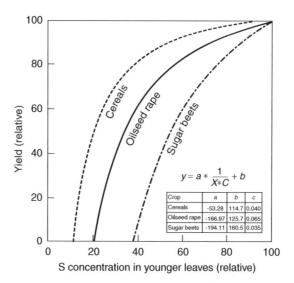

FIGURE 7.18 Comparison of boundary line functions for yield and total sulfur concentration in tissue of cereals, oilseed rape, and sugar beet. (From Schnug, E. and Haneklaus, S., in *Sulphur in Agroecosystems*, Vol. 2, Part of the series 'Nutrients in Ecosystems', Kluwer Academic Publishers, Dordrecht, 1998, pp. 1–38.)

(both relative) for oilseed rape, cereals, and sugar beet (Figure 7.18) reveals the physiological differences between these crops. The boundary lines for cereals and oilseed rape are for seed yields, and that for sugar beet for root yields. The optimum sulfur ranges proved to be the same for sugar beet root yield and sugar yield.

For all crops, the boundary lines show a steep increase at the beginning, which reflects the response of the photosynthetic system to sulfur deficiency. In cereals, the boundary line continues over a long range and asymptotically toward the value above which no further yield increase (NEV) is to be expected from increasing sulfur concentrations. This part of the boundary line most likely reflects the proportion of sulfur that is bound to the proteins of the cereal grain. In sugar beet, the boundary line reaches the NEV much faster after a steep increase, which is in line with the fact that sugar beet roots take up only small amounts of sulfur (205). Oilseed rape, with its internal storage system for S, which is based on the enzymatic recycling of glucosinolates (90,289), shows a steadier ascent of its boundary line. Therefore, within oilseed rape varieties, those with genetically low glucosinolate contents ('double low' or '00' varieties) show a steeper increase of their boundary lines than those with genetically high glucosinolate concentrations (103,116).

The nonlinearity of the boundary lines reveals once more the limited value of critical values. Above total sulfur concentrations of 6.5, 4.0, and 3.5 mg g^{-1} in foliar tissue of oilseed rape, cereals, and sugar beet, respectively, no further yield increases are to be expected by increasing tissue sulfur concentrations (NEVs). This result corresponds to the usually assigned 'critical values,' which are valid for 95% of the maximum yield, of 5.5, 3.2, and 3.0 mg S g^{-1} for rape, corn, and sugar beets, respectively. However, in this range of the response curve, there is still no linearity between tissue sulfur levels and yield.

The relationship between sulfur concentration in plant tissue and yield, which reflects the physiological patterns in the internal nutrient utilization, is specific for each plant species, and can be best established by boundary lines (Figure 7.17). In comparison, the relationship between fertilizer dose and sulfur concentration in plant tissues is much less dependent on physiological factors but is strongly influenced by factors affecting the physical mobility and losses of sulfur from soils. Therefore, this transfer function bears the largest part of insecurity for the effectiveness of sulfur fertilization. Thus, for the derivation of fertilizer recommendations, the common relationship between fertilizer dose and yield is best split into two partial relationships: (a) fertilizer dose versus nutrient uptake and (b) nutrient uptake versus yield (307). If tissue analysis is to be used for fertilizer recommendations, concentrations need to be calibrated against sulfur doses. This strategy was proved for nitrogen (308), and the setting up of sulfur response curves is recommended for sulfur too.

Professional Interpretation Program for Plant Analysis (PIPPA) software not only evaluates the status of individual plant nutrients but also appraises results from multiple elemental analyses (309). In PIPPA, boundary line and transfer functions are integrated for each element so that the yield-limiting effect is calculated for each specified nutrient, and finally fertilizer recommendations are given (309).

7.5.3 SULFUR STATUS AND PLANT HEALTH

Although the significance of individual nutrients for maintaining or promoting plant health saw some interest in the 1960s and 1970s (143), research in the field of nutrient-induced resistance mechanisms has been scarce because of its complexity, and because of its limited practical significance due to the availability of effective pesticides.

Since the beginning of the 1980s, atmospheric sulfur depositions have been declining drastically after Clean Air Acts came into force, and severe sulfur deficiency advanced to a major nutritional disorder in Western Europe (114,310,311). Increased infections of agricultural crops with fungal pathogens were observed, and diseases spread throughout the regions that were never infected before (312). Sulfur fertilization, applied to the soil as sulfate, proved to have a significant effect on the infection rate and infection severity of different crops by fungal diseases (148). Sulfur fertilization increased the resistance against various fungal diseases in different crops under greenhouse (313,314) and field conditions (315–317). Based on these findings, the concept of sulfur-induced resistance (SIR) was developed; research in this field has strengthened since then, and the advances made are discussed comprehensively by Bloem et al. (318) and Haneklaus et al. (148).

The term SIR stands for the reinforcement of the natural resistance of plants against fungal pathogens through triggering of the stimulation of metabolic processes involving sulfur by targeted fertilizer application strategies (148). A sufficient sulfur supply and an adequate availability of plant-available sulfate are presumably a prerequisite for inducing S-dependent resistance mechanisms in the plant so that the required sulfur rates and sulfur status may be higher than the physiological demand.

The mechanisms possibly involved in SIR may be related to processes of induced resistance (319), for example, via the formation of phytoalexins and glutathione, or the requirement of cysteine for the synthesis of salicylic acid by β-oxidation and the cysteine pool itself. Another option is the release of reduced sulfur gases, such as H_2S, which is described in the literature as being fungitoxic. The H_2S may be produced prior to or after cysteine formation (see Section 7.2 and (320)). Two enzymes that could be responsible for the H_2S release are L-cysteine desulfhydrase (LCD) and O-acetyl-L-serine(thiol)lyase (OAS-TL). The LCD catalyzes the decomposition of cysteine to pyruvate, ammonia, and H_2S. The OAS-TL is responsible for the incorporation of inorganic sulfur into the amino acid cysteine, which can be subsequently converted into other sulfur-containing compounds such as methionine or glutathione. The H_2S is evolved in a side reaction because of the nature of the pyridoxal 5′-phosphate cofactor and the specific reaction mechanism of the OAS-TL protein (321). There is wide variation with regard to specifications about the release of H_2S, ranging from 0.04 ng g^{-1} s^{-1} in whole soybean plants on a dry matter basis (322) to 100 pmol min^{-1} cm^{-1} in leaf discs of cucumber (323). Thus, H_2S emissions of cut plant parts may be 500 times higher than in intact plants (Table 7.7).

The release of H_2S by plants is supposedly regulated by interactions in the N and sulfur metabolic pathways. Lakkineni et al. (327) demonstrated a distinct increase in H_2S emissions when leaf discs of mustard, wheat, and groundnut (*Arachis hypogaea* L.) were fed with sulfate or cysteine (Table 7.8). Supply of additional nitrogen with the sulfate did not cause H_2S emissions to increase (Table 7.8). Lakkineni et al. (330) suggested a preferable synthesis of nitrogen- or sulfur-containing products at the level of substrate availability.

TABLE 7.7
Survey of Different Investigations of the Release of Hydrogen Sulfide from Terrestrial Plants

Measured H₂S Evolution	Plant/ Plant Part	Reference	Estimated H₂S Emission (nmol g⁻¹ d.w. h⁻¹)
0.04–0.08 ng g^{-1} d.w. s^{-1}	Soybeans (whole plant)	322	2.1–8.5
5.58–6.21 pmol kg^{-1} s^{-1}	Conifers (whole plant)	324	0.02
2.22 μg kg^{-1} h^{-1}	Spruce seedlings (*Picea abies* L. Karsten)	325	0.07
0.04–0.46 nmol min^{-1} leaves^{-1}	Attached leaves of different plants	326	8–92[a]
0.49–0.94 nmol g^{-1} f.w. h^{-1}	Leaf extract of *Brassica. napus*	327	3.3–6.3[b]
0.80–1.11 nmol g^{-1} f.w. h^{-1}	Leaf discs of mustard	327	5.3–7.4[b]
1.7–3.9 nmol min^{-1} leaves^{-1}	Detached leaves of different plants	326	340–780[a]
8 nmol g^{-1} f.w. min^{-1}	Maximum emission of detached leaves	326	3200[b]
2.4–3.9 nmol g^{-1} f.w. min^{-1}	Leaves of spinach and cucumber	65	960–1560[b]
40 pmol min^{-1} cm^{-2}	Leaf discs of different plants	323	800[c]
50–100 pmol min^{-1} cm^{-2}	Leaf discs of cucumber	328	1000–2000[c]
Total S emission from higher plants			Total S Emission (nmol S^{-1} d.w. h)
12–1062 ng S kg^{-1} d.w. min^{-1}	42 types of terrestrial plants	329	0.02–1.99

[a]Assuming a medium leaf weight of 2 g fresh weight and a leaf water content of 85%.
[b]Assuming a medium leaf water content of 85%.
[c]Assuming a dry weight of 3 mg cm^{-2}.
Source: From Bloem, E. et al., *J. Plant Nutr.*, 28, 763–784, 2005.

TABLE 7.8
Influence of Sulfate, Cysteine, and Nitrate on the Emission of H₂S from Leaf Discs of Mustard, Groundnut, and Wheat

Treatment	H₂S Emission (nmol g⁻¹ f.w. h⁻¹)		
	Mustard	Wheat	Groundnut
Control (H₂O)	0.80	1.27	0.25
Sulfate (5 mM)	1.15	1.85	—
Cysteine (5 mM)	1.11	2.19	0.80
Sulfate + nitrate (5 mM)	0.81	1.29	—
Cysteine + nitrate (5 mM)	0.72	2.63	—

Source: From Lakkineni, K.C. et al., in *Sulphur Nutrition and Sulphur Assimilation in Higher Plants; Fundamental, Environmental and Agricultural Aspects*, SPB Academic Publishing, The Hague, 1990, pp. 213–216.

H₂S and DMS emissions by plants are, however, supposedly not involved in SIR against fungal pathogens belonging to the class Basidiomycetes, as fumigation experiments with fungal mycelium of *Rhizoctonia solani* revealed that the pathogen metabolized both gases efficiently (331).

The amino acids cysteine and methionine are the major end products of sulfate assimilation in plants and bind up to 90% of the total sulfur (320). Conditions of sulfur deficiency will result in a decrease of sulfur-containing amino acids in proteins (5). As the amino acid composition is genetically determined, this effect is limited, and thereafter the total protein content will be reduced (5). Amino acid type and concentration in plant tissues are related to the susceptibility of plants to pathogens (332). Amino acids occur in the free state in plants, and the amino acids cysteine and methionine are enriched in resistant plant tissues. Soil-applied sulfur significantly increased the free cysteine content in the vegetative tissue from 0.5 to 1.2 μmol g^{-1} d.w. (63). Bosma et al. (333) reported a two- to five-fold increase in the content of water-soluble nonprotein sulfhydryl compounds in clover (*Trifolium* spp.) and spinach after fumigation with H$_2$S under field conditions, whereby the cysteine content increased 10-fold. De Kok (18) reported similar results for fumigation experiments with sulfur dioxide.

Glutathione is a major, free, low-molecular, nonprotein, thiol compound and is an important reservoir for nonprotein reduced sulfur in plants (66). A relationship between glutathione content and the extent of protection against fungal diseases exists (72). A low glutathione content in plants does not inevitably imply, however, a higher susceptibility of the plant, as a rapid accumulation of glutathione in response to pathogen attack was noted (334), and this observation proved to be decisive in pathogenesis (72). Sulfur-deficient plants have very low glutathione concentrations, and sulfur fertilization significantly increases the free thiol content (Table 7.3; Section 7.2.3). Basically, sulfur-deficient plants are expected to be more vulnerable to stress factors, which are usually compensated by the glutathione system so that sulfur fertilization should have a positive effect on resistance mechanisms.

Phytoalexins are important for plant defense (335). Phytoalexins are secondary plant metabolites which are synthesized de novo and accumulate in response to diverse forms of stress, including pathogenesis (336). The immunity is generally of short duration and is concentrated around the infected area. According to this definition, the formation of elemental sulfur, the stress-induced formation of pathogenesis-related (PR) proteins, and a novel class of LMW antibiotics, all come under the term phytoalexins. At the moment however, the influence of the sulfur nutritional status on phytoalexin synthesis can only be speculated from the dependency of their precursors on the sulfur supply. The influence of the sulfur nutritional status on the synthesis of PR-12, PR-13, and PR-14 proteins and elemental sulfur depositions in plant tissues remains obscure too (148).

7.6 SULFUR FERTILIZATION

The optimum timing, dose, and sulfur form used depends on the specific sulfur demand of a crop and application technique. Under humid conditions, the sulfur dose should be split in such a way that sulfur fertilization in autumn is applied to satisfy the sulfur demand on light, sandy soils before winter and to promote the natural resistance against diseases. At the start of the main vegetative growth, sulfur should be applied together with nitrogen. With farmyard manure, on an average 0.07 kg sulfur is applied with each kg of nitrogen. In mineral fertilizers and secondary raw materials, sulfur is available usually as sulfate, elemental sulfur, and sulfite. Sulfate is taken up directly by plant roots, whereas sulfite and elemental sulfur need prior oxidation to sulfate, whereby the speed of transformation depends on the particle size and dimension of the thiobacillus population in the soil (Figure 7.19) (337,338).

The main secondary-sulfur-containing raw materials from the flue gas desulfurization process are gypsum and spray dry absorption (SDA) products, which are a mixture of calcium sulfite and calcium sulfate in a mass ratio of about 8:1 (340). SDA products with fly ash contents < 8% may contain up to 68% calcium sulfite, whereas this percentage in products with fly ash contents between 20 and 85% will not exceed 47% (341). A phytotoxic effect of sulfite applied by SDA products was observed when it was used as a culture substrate and on soils with a pH < 4 (337). The time required for complete oxidation of sulfite is about 2 weeks (342). Sulfite oxidation proceeds faster with increasing oxygen content and soil pH, and decreasing soil moisture content (343,344). When sulfur was applied at rates of ≤80 kg ha^{-1} to exclusively satisfy the sulfur demand of agricultural crops, no negative impact on crop performance and subsequent crops in the rotation was detected (337,342,345,346).

In general, the efficiency of sulfur uptake by rape is highly dependent on the sulfur status of the shoots (Figure 7.20). There is a close relationship between the initial sulfur content and its increase

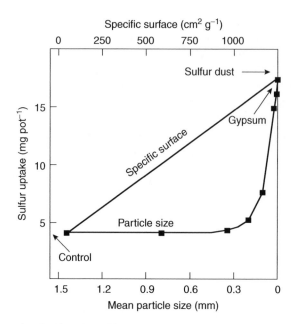

FIGURE 7.19 Sulfur uptake of maize plants 32 days after sowing, in relation to particle size and specific surface of elemental sulfur in a pot experiment. (From Fox, R.L. et al., *Soil. Sci. Soc. Am. Proc.*, 28, 406–408, 1964.)

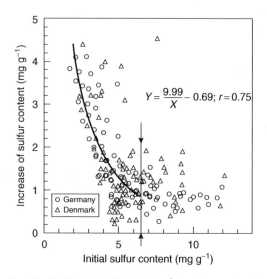

FIGURE 7.20 Influence of sulfur fertilization (20 kg S ha^{-1}) on the total sulfur concentration of oilseed rape leaves, in relation to the initial sulfur supply. (From Schnug, E. and Haneklaus, S., *Landbauforschung Völkenrode*, Sonderheft 144, 1994.)

by fertilization. Under sulfur-limiting growth conditions, root-expressed sulfur transporters are highly regulated and induced (see Section 7.2.1 and Section 7.2.2). Besides that, sulfur fertilization improved root growth and thus access to sulfate (53).

An insufficient sulfur supply will not only reduce crop productivity, diminish crop quality, and affect plant health, but it also will impair nitrogen-use efficiency (53,347). Under conditions of

TABLE 7.9
Influence of Sulfur Fertilization on the Nitrate Reductase Activity and N-Use Efficiency of Sugarcane

S Dose (kg ha^{-1})	Nitrate Reductase Activity (nmol NO$_2^-$ g^{-1} (f.w.) h^{-1})	Nitrogen-Use Efficiency (g (d.m.) g^{-1} (N) m^{-2})
0	1652	2.17
40	1775	2.23
80	1989	3.02
120	2020	2.54
160	1805	2.67

Source: From Shanmugam, K.S., *Fert. News*, 40, 23–26, 1995.

sulfur deficiency, nitrate and non-S-containing amino acids accumulate—actions which may reduce the nitrate reductase activity (see Section 7.2.4; 348). Sulfur fertilization promotes nitrate reduction and thus restricts nitrate accumulation in vegetative tissues. In Table 7.9, the influence of an increasing sulfur supply on the nitrate reductase activity and nitrogen-use efficiency is shown.

The highest nitrate reductase activity occurred at a sulfur dose of 120 kg S ha^{-1} and the highest N-use efficiency at 160 kg S ha^{-1} (Table 7.9) (349). This result corresponds to an increase of 18.2 and 18.7%, respectively, for the two doses. In comparison, the net nitrogen utilization of oilseed rape and cereals was significantly increased by sulfur fertilization by about 7 to 16%. A sulfur application rate of 100 kg S ha^{-1} yielded the best results for oilseed rape during three consecutive years of experimentation (347).

The sulfur demands of agricultural crops vary highly, as do the recommended sulfur doses (Table 7.10). Recommended sulfur rates vary between 30 and 100 kg S ha^{-1} for oilseed rape, and between 20 and 50 kg S ha^{-1} for cereals (103,337,348). For other crops such as sugar beet, grassland, rice, and soybean, the highest crop productivity occurred at sulfur rates of 25, 40, 45, and 60 kg S ha^{-1}, respectively (351–353).

Aulakh (364) gives a detailed overview of sulfur uptake and crop responses to sulfur fertilization in terms of yield and quality, with special attention being paid to crops grown in India. Sulfur fertilizer can be applied to the soil or given as foliar dressings. As the sulfur dose is limited when applied via the leaves, this form of fertilization can only be a complementary measure to correct severe sulfur deficiency. Usually, for foliar applications, either Epsom salts or elemental S are used. Calculated from changes in the sulfur uptake by seeds, only 0 to 3% of foliar-applied sulfate-S with Epsom salts was utilized, while 33 to 35% of sulfur applied as elemental sulfur product (Thiovit®) was utilized (338). Foliar-supplied sulfate moved into leaves much faster than elemental sulfur and was supposedly trapped in vacuoles so that it did not contribute to increased yield. The better results with elemental sulfur were explained by the fact that it needs to be oxidized before significant quantities can be absorbed by leaves. As oxidation is slow, sulfate supply from foliar-applied elemental sulfur fits better to the metabolic demand of the leaves and avoids excess sulfate concentrations in the cytosol and their deposition in vacuoles.

The problem of severe sulfur deficiency still exists on a large scale as the widespread regular appearance of macroscopic symptoms reveal, even more than 20 years after addressing this nutrient disorder (147). The reason is most likely the wide variation of official sulfur fertilizer recommendations in Europe (Table 7.11), recommendations, which only partly acknowledge site-specific features and productional peculiarities.

On-farm experimentation employing precision agriculture tools would be an ideal approach for setting up site-specific sulfur response curves (see Section 7.5.2 and (366)).

TABLE 7.10
Sulfur Demand (kg S t⁻¹) of Agricultural Crops

Crop	Based Plant Part	S Demand (kg S t⁻¹)	Reference
Poaceae			
Barley	Grain	1.2–1.9	354, 205
(winter varieties)	Straw	1.6–2.1[a]	354, 205
Barley	Grain	1.2–1.4	205
(summer varieties)	Straw	0.7–1.5[a]	205
Oats	Grain	1.7	354
Rice	Total	3.2	355
Sugarcane	Total	0.3	355
Wheat	Grain	1.6–2.2	354, 205
(winter varieties)	Straw	1.1–2.8[a]	205
Wheat	Grain and straw	4.3	355
Oil crops			
Mustard	Total	16.0–17.3	355, 356, 357, 358
Oilseed rape	Total	16	103
Groundnut	Pods	3.3–5.9	355, 357, 358,
		(20.9)	359, 360, 361
Soybean	Seeds	4.3–8.8	357, 358, 362
Sunflower	Seeds	7.1–12.7	356, 357, 358
Legumes			
Chickpea	Total	8.7	355
Pigeon pea	Total	7.5	355
Root crops			
Potato	Tuber	1.2–1.6	205
Sugar beet	Beet root	0.3–0.4	205
	Leaves	0.7–1.9[a]	205
Fodder crops			
Grass	Herbage	1.7	354
Red clover	1st cut	2.2–4.3	363
	2nd cut	2.0–4.0	363
	3rd cut	2.0–3.8	363
Vegetables			
Swedes	Roots[b]	3.0	354
	Tops[b]	1.4[a]	354
Turnip	Roots[b]	2.5	354
	Tops[b]	1.1[a]	354
Marrowstem kale	Whole plant[b]	4.0	354

[a]Yield of harvested product.
[b]Dry matter yield.

TABLE 7.11
Official Sulfur Fertilizer Recommendations and Optimum Fertilizer Doses Based on Scientific Experimentation for Various Crops in Europe

Crop	Range of Officially Recommended S Fertilizer Dose (kg ha^{-1})
Cabbage	30–50
Cereals	10–30
Grassland, cut	30–40
Grassland, grazed	0–30
Grass, silage	0–30
Oilseed rape	20–60
Peas	10–30
Potatoes	0–20
Sugar beet	0–40
Vegetables	20–40

Source: From Aulakh, M.S., in *Sulphur in Plants*, Kluwer Academic Publishers, Dordrecht, 2003, pp. 341–358.

ACKNOWLEDGMENT

The authors express their sincerest thanks to Mrs. Rose-Marie Rietz for the technical editing of this chapter.

REFERENCES

1. Ichikawa, Y.; Hayami, H.; Sugiyama, T.; Amann, M.; Schoepp, W. Forecast of sulfur deposition in Japan for various energy supply and emission control scenarios. *Water Air Soil Pollut.* 2001, 130, 301–306.
2. Ulrich, B. Die Waelder in Mitteleuropa: Messergebnisse ihrer Umweltbelastung, Theorie ihrer Gefaehrdung, Prognose ihrer Entwicklung. *Allgemeine Forstzeitschrift* 1980, 44 (special issue).
3. Schnug, E.; de la Sauce, L.; Pissarek, H.P. Untersuchungen zur Kennzeichnung der Schwefelversorgung von Raps. *Landwirtsch. Forsch.* 1984, 37, 662–673.
4. Messick, D.L.; Fan, M.X.; De Brey, C. Global sulfur requirement and sulfur fertilizers. *FAL–Agric. Res.* 2005, 283, 97–104.
5. Schnug, E. Significance of sulphur for the quality of domesticated plants. In *Sulphur Metabolism in Higher Plants: Molecular, Ecophysiological and Nutritional Aspects*; Cram, W.J., De Kok, L.J., Brunold, C., Rennenberg, H., Eds.; Backhuys Publishers: Leiden, 1997; pp. 109–130.
6. Brunold, C. Regulatory interactions between sulfate and nitrate assimilation. In *Sulfur Nutrition and Sulfur Assimilation in Higher Plants: Regulatory, Agricultural and Environmental Aspects*; De Kok, L.J., Stulen, I., Rennenberg, H., Brunold, C., Rauser, W., Eds.; SPB Academic Publishing: The Hague, 1993; pp. 125–138.
7. Stulen, I.; De Kok, L.J. Whole plant regulation of sulfur metabolism. In *Sulfur Nutrition and Sulfur Assimilation in Higher Plants: Regulatory, Agricultural and Environmental Aspects*; De Kok, L.J., Stulen, I., Rennenberg, H., Brunold, C., Rauser, W.E., Eds.; SPB Academic Publishing: The Hague, 1993; pp. 77–91.
8. Schnug, E. *Sulfur in Agroecosystems*. Kluwer Academic Publishers: Dordrecht, 1998.
9. Brunold, C.; Von Ballmoos, P.; Hesse, H.; Fell, D.; Kopriva, S. Interactions between sulfur, nitrogen and carbon metabolism. The plant sulfate transporter family. In *Specialized Functions and Integration*

with Whole Plant Nutrition; Davidian, J.-C., Grill, D., De Kok, L.J., Stulen, I., Hawkesford, M.J., Schnug, E., Rennenberg, H., Eds.; Backhuys Publishers: Leiden, 2003; pp. 45–46.

10. Heinz, E. Recent investigations on the biosynthesis of the plant sulfolipid. In: *Sulfur Nutrition and Sulfur Assimilation in Higher Plants: Regulatory, Agricultural and Environmental Aspects*; De Kok, L.J., Stulen, I., Rennenberg, H., Brunold, C., Rauser, W.E., Eds.; SPB Academic Publishing: The Hague, 1993; pp. 163–178.

11. De Kok, L.J.; Westerman, S.; Stuiver, C.E.E.; Stulen, I. Atmospheric H_2S as plant sulfur source: interaction with pedospheric sulfur nutrition–a case study with *Brassica oleracea* L. In *Sulfur Nutrition and Sulfur Assimilation in Higher Plants: Molecular, Biochemical and Physiological Aspects*; Brunold, C., Rennenberg, H., De Kok, L.J., Stulen, I., Davidian, J.-C., Eds.; Paul Haupt: Bern, 2000; pp. 41–56.

12. Haq, K.; Ali, M. Biologically active sulphur compounds of plant origin. In *Sulphur in Plants*; Abrol, Y.P., Ahmad, A., Eds.; Kluwer Academic Publishers: Dordrecht, 2003; pp. 375–386.

13. Cram, W.J. Uptake and transport of sulfate. In *Sulfur Nutrition and Sulfur Assimilation in Higher Plants: Fundamental, Environmental and Agricultural Aspects*; Rennenberg, H., Brunold C., De Kok L.J. and Stulen, I., Eds.; SPB Academic Publishing: The Hague, 1990; pp. 3–11.

14. Clarkson, D.T.; Hawkesford, M.J.; Davidian, J.C. Membrane and long-distance transport of sulfate. In *Sulfur Nutrition and Sulfur Assimilation in Higher Plants: Regulatory, Agricultural and Environmental Aspects*; De Kok, L.J., Stulen, I., Rennenberg, H., Brunold, C., Rauser, W., Eds.; SPB Academic Publishing: The Hague, 1993; pp. 3–19.

15. Davidian, J.-C.; Hatzfeld, Y.; Cathala, N.; Tagmount, A.; Vidmar, J.J. Sulfate uptake and transport in plants. In *Sulfur Nutrition and Sulfur Assimilation in Higher Plants: Molecular, Biochemical and Physiological Aspects*; Brunold, C., Rennenberg, H., De Kok, L.J., Stulen, I., Davidian, J.-C., Eds.; Paul Haupt: Bern, 2000; pp. 19–40.

16. Brunold, C. Reduction of sulfate to sulphide. In *Sulfur Nutrition and Sulfur Assimilation in Higher Plants. Fundamental, Environmental and Agricultural Aspects*; Rennenberg, H., Brunold, C., De Kok, L.J., Stulen, I., Eds.; SPB Academic Publishing: The Hague, 1990; pp. 13–31.

17. Kopriva, S.; Kopriviva, A. Sulphate assimilation: a pathway which likes to surprise. In *Sulphur in Plants*; Abrol, Y.P., Ahmad, A., Eds.; Kluwer Academic Publishers: Dordrecht, 2003; pp. 87–112.

18. De Kok, L.J. Sulfur metabolism in plants exposed to atmospheric sulfur. In *Sulfur Nutrition and Sulfur Assimilation in Higher Plants: Fundamental, Environmental and Agricultural Aspects*; Rennenberg, H., Brunold, C., De Kok, L.J., Stulen, I., Eds.; SPB Academic Publishing: The Hague, 1990; pp. 111–130.

19. De Kok, L.J.; Stuiver, C.E.E.; Stulen, I. Impact of atmospheric H_2S on plants. In *Responses of Plant Metabolism to Air Pollution and Global Change*. De Kok, L.J., Stulen, I., Eds.; Backhuys Publishers: Leiden, 1998; pp. 51–63.

20. De Kok, L.J.; Tausz, M. The role of glutathione in plant reaction and adaptation to air pollutants. In *Significance of Glutathione to Plant Adaptation to the Environment*; Grill, D., Tausz, M., De Kok, L.J., Eds.; Kluwer Academic Publishers: Dordrecht, 2001; pp. 185–201.

21. De Kok, L.J.; Stuiver, C.E.E.; Westerman, S; Stulen, I. Elevated levels of hydrogen sulfide in the plant environment: nutrient or toxin. In *Air Pollution and Biotechnology in Plants*; Omasa, K., Saji, H., Youssefian, S., Kondo, N., Eds.; Springer-Verlag: Tokyo, 2002; pp. 201–213.

22. Hawkesford, M.J. Plant responses to sulfur deficiency and the genetic manipulation of sulfate transporters to improve S-utilization efficiency. *J. Exp. Bot.* 2000, 51, 131–138.

23. Hawkesford, M.J.; Buchner, P.; Hopkins, L.; Howarth, J.R. The plant sulfate transporter family: Specialized functions and integration with whole plant nutrition. In *Sulfur Transport and Assimilation in Plants: Regulation, Interaction and Signalling*; Davidian, J.-C., Grill, D., De Kok, L.J., Stulen, I., Hawkesford, M.J., Schnug, E., Rennenberg, H, Eds.; Backhuys Publishers: Leiden, 2003; pp. 1–10.

24. Hawkesford, M.J.; Buchner, P.; Hopkins, L.; Howarth, J.R. Sulphate uptake and transport. In *Sulphur in Plants*; Abrol, Y.P., Ahmad, A., Eds.; Kluwer Academic Publishers: Dordrecht, 2003; pp. 71–86.

25. Hell, R. Molecular physiology of plant sulfur metabolism. *Planta* 1997, 202, 138–148.

26. Leustek, T.; Saito, K. Sulfate transport and assimilation in plants. *Plant Physiol.* 1999, 120, 637–643.

27. Saito, K. Molecular and metabolic regulation of sulfur assimilation: initial approach by the post-genomics strategy. In *Sulfur Transport and Assimilation in Plants: Regulation, Interaction and Signalling*; Davidian, J.-C., Grill, D., De Kok, L.J., Stulen, I., Hawkesford, M.J., Schnug, E., Rennenberg, H., Eds.; Backhuys Publishers: Leiden, 2003; pp. 11–20.

28. Droux, M.; Ruffet, M.L.; Douce, R.; Job, D. Interactions between serine acetyltransferase and O-acetylserine(thiol)lyase in higher plants: structural and kinetic properties of the free and bound enzymes. *Eur. J. Biochem.* 1998, 155, 235–245.

29. Hell, R. Metabolic regulation of cysteine synthesis and sulfur assimilation. In *Sulfur Transport and Assimilation in Plants: Regulation, Interaction and Signalling*; Davidian, J.-C., Grill, D., De Kok, L.J., Stulen, I., Hawkesford, M.J., Schnug, E., Rennenberg, H., Eds.; Backhuys Publishers: Leiden, 2003; pp. 21–31.

30. Giovanelli, J. Regulatory aspects of cysteine and methionine synthesis. In *Sulfur Nutrition and Sulfur Assimilation in Higher Plants. Fundamental, Environmental and Agricultural Aspects*; Rennenberg, H., Brunold, C., De Kok, L.J., Stulen, I., Eds.; SPB Academic Publishing: The Hague, 1990; pp. 33–48.

31. Noji, M.; Saito, K. Sulfur amino acids: biosynthesis of cysteine and methionine. In *Sulphur in Plants*; Abrol, Y.P., Ahmad, A. Eds.; Kluwer Academic Publishers: Dordrecht, 2003, pp. 135–144.

32. Grill, D.; Tausz, M.; De Kok, L.J. *Significance of Glutathione to Plant Adaptation to the Environment*. Kluwer Academic Publishers: Dordrecht, 2001.

33. Lendzian, K.L. Permeability of plant cuticles to gaseous air pollutants. In *Gaseous Air Pollutants and Plant Metabolism*; Koziol M.J., Whatley F.R.; Eds.; Butterworths: London, U.K., 1984, pp. 77–81.

34. Baldocchi, D.D. Deposition of gaseous sulfur compounds to vegetation. In *Sulfur Nutrition and Assimilation in Higher Plants; Regulatory, Agricultural and Environmental Aspects*; De Kok, L.J., Stulen, I., Rennenberg, H., Brunold C., Rauser, W.E., Eds.; SPB Academic Publishing: The Hague, 1993; pp. 271–293.

35. Yang, L.; Stulen, I.; De Kok, L.J. Sulfur dioxide: relevance of toxic and nutritional effects for Chinese cabbage. *Environ. Exp. Bot.* in press; 2006.

36. Stulen, I.; Posthumus, F.; Amâncio, S.; Masselink-Beltman, I.; Müller, M.; De Kok, L.J. Mechanism of H$_2$S phytotoxicity. In *Sulfur Nutrition and Sulfur Assimilation in Higher Plants: Molecular, Biochemical and Physiological Aspects*; Brunold, C., Rennenberg, H., De Kok, L.J., Stulen, I., Davidian, J.C., Eds.; Paul Haupt: Bern, 2000, pp. 381–383.

37. Tabatabai, M.A. *Sulfur in Agriculture*. American Society of Agronomy: Madison, Wisconsin, 1986.

38. Byers, M.; Franklin, J.; Smith, S.J. The nitrogen and sulphur nutrition of wheat and its effect on the composition and baking quality of the grain. *Aspects Appl. Biol.* 1987, 15, pp. 327–344.

39. Randall, P.J.; Spencer, K.; Freney, J.R. Sulphur and nitrogen fertilizer effects on wheat. I. Concentrations of sulphur and nitrogen and the nitrogen to sulphur ratio in grain in relation to the yield response. *Aus. J. Agric. Res.* 1981, 32, 203–212.

40. Yoshino, D.; McCalla, A.G. The effects of sulphur content on the properties of wheat gluten. *Can. J. Biochem.* 1996, 44, 339–346.

41. Zhao, F.J.; McGrath, S.P.; Crosland, A.R.; Salmon, S.E. Changes in the sulfur status of British wheat-grain in the last decade, and its geographical distribution. *J. Sci. Food Agric.* 1995, 68, 507–514.

42. Haneklaus, S.; Evans, E.; Schnug, E. Baking quality and sulphur content of wheat. I. Influence of grain sulphur and protein concentrations on loaf volume. *Sulphur Agric.* 1992, 16, 31–35.

43. Haneklaus, S.; Schnug, E. Baking quality and sulphur content of wheat. II. Evaluation of the relative importance of genetics and environment including sulphur fertilisation. *Sulphur Agric.* 1992, 16, 35–38.

44. Hagel, I. Sulfur and baking quality of bread making wheat. *FAL–Agric. Res.* 2005, 283, 23–36.

45. Wieser, H.; Seilmeier, W.; Belitz, H.-D. Use of RP-HPLC for a better understanding of the structure and functionality of wheat gluten proteins. In *High-Performance Liquid Chromatography of Cereal and Legume Proteins*; Kruger, J.E.; Bietz, J.A, Eds.; American Association of Cereal Chemists Inc.: St Paul, Minnesota, 1994.

46. Wieser, H.; Zimmermann, G. Importance of amounts and proportions of high molecular weight sub-units of glutenin for wheat quality. *Eur. Food Res. Technol.* 2000, 210, 324–330.

47. Wrigley, C.W.; Du Cros, D.L.; Fullington, J.G.; Kasarda, D.D. Changes in polypeptide composition and grain quality due to sulfur deficiency in wheat. *J. Cereal Sci.* 1984, 2, 15–24.

48. Saalbach, E.; Judel, G.K.; Kessen, G. Ueber den Einfluß des Sulfatgehaltes im Boden auf die Wirkung einer Schwefelduengung. *Z. Pflanzenernaehr. Duengung und Bodenkde* 1962, 99, 177–182.

49. Saalbach, E. *Sulphur Fertilization and Protein Quality*. The Sulphur Institute: Washington, 1966.

50. Eppendorfer, W.H.; Eggum, B.O. Dietary fibre, sugar, starch and amino acid content of kale, ryegrass and seed of rape and field beans as influenced by S- and N-fertilization. *Plant Foods Human Nutr.* 1992, 42, 359–371.

51. Eppendorfer, W.H.; Eggum, B.O. Dietary fibre, starch and amino acids and nutritive value of potatoes as affected by sulfur, nitrogen, phosphorus, potassium, calcium and water stress. *Acta Agric. Scand. Sect. B. Soil Plant Sci.* 1994, 44, 1–11.

52. Haneklaus, S.; Schnug, E. Evaluation of critical values of soil and plant nutrient concentrations of sugar beet by means of boundary lines applied to a large data set from production fields. *Aspects Appl. Biol.* 1998, 52, 87–94.

53. Schnug, E.; Haneklaus, S. Sulphur deficiency in *Brassica napus*—biochemistry, symptomatology, morphogenesis. *Landbauforschung Völkenrode*, 1994, Sonderheft 144.

54. Kumar, V.; Singh, M.; Singh, N. Effect of S, P and Mo on quality of soybean grain. *Plant Soil* 1981, 59, 3–8.

55. Rennenberg, H. The fate of excess sulfur in higher plants. *Annu. Rev. Plant Physiol.* 1984, 35, 121–153.

56. Schnug, E.; Haneklaus, S. The sulphur concentration as a standard for the total glucosinolate content of rapeseed and meal and its determiation by X-ray fluorescence spectroscopy (X-RF method). *J. Sci. Food Agric.* 1988, 45, 243–254.

57. Benning, C. Biosynthesis and function of the sulfolipid sulfoquinovosyl diacylglycerol. *Annu. Rev. Plant Physiol. Plant Mol. Biol.* 1998, 49, 53–75.

58. Harwood J.L.; Okanenko, A.A. Sulphoquinovosyl diacylglycerol (SQDG)—the sulpholipid of higher plants. In *Sulphur in Plants*; Abrol, Y.P., Ahmad, A., Eds.; Kluwer Academic Publishers: Dordrecht, 2003; pp. 189–219.

59. De Kok, L.J.; Stulen, I.; Rennenberg, H.; Brunold, C.; Rauser, W. *Sulfur Nutrition and Sulfur Assimilation in Higher Plants: Regulatory, Agricultural and Environmental Aspects.* SPB Academic Publishing: The Hague, 1993.

60. Rennenberg, H. Molecular approaches to glutathione biosynthesis. In *Sulfur Metabolism in Higher Plants: Molecular, Ecophysiological and Nutritional Aspects*; Cram, W.J., De Kok, L.J., Brunold, C., Rennenberg, H., Eds.; Backhuys Publishers: Leiden, 1997; pp. 59–70.

61. Noctor, G.; Arisi, A.C.M.; Jouanin, L.; Kunert, K.J.; Rennenberg, H.; Foyer, C. Glutathione: biosynthesis, metabolism and relationship to stress tolerance explored in transformed plants. *J. Exp. Bot.* 1998, 49, 623–647.

62. Shalaby, T. Genetical and nutritional influences on the spear quality of white asparagus (*Asparagus officinalis* L.). *FAL–Agric. Res.* 2004, 265 (special issue).

63. Salac, I. Influence of the sulphur and nitrogen supply on sulfur metabolites involved in the Sulphur Induced Resistance (SIR) of *Brassica napus* L.. *FAL - Agric. Res.* 2005, special issue No 277.

64. Schnug, E.; Haneklaus, S.; Borchers, A.; Polle, A. Relations between sulphur supply and glutathione, ascorbate and glucosinolate concentrations in *Brassica napus* varieties. *J. Plant Nutr. Soil Sci.* 1995, 158, 67–70.

65. De Kok L.J.; Maas F.M.; Godeke J.; Haaksma A.B.; Kuiper P.J.C. Glutathione, a tripeptide which may function as a temporary storage compound of excessive reduced sulphur in H_2S fumigated spinach plants. *Plant Soil* 1986, 91, 349–352.

66. Foyer, C.H.; Noctor, G. The molecular biology and metabolism of glutathione. In *Significance of Glutathione to Plant Adaptation to the Environment*; Grill, D., Tausz, M., De Kok, L.J., Eds.; Kluwer Academic Publishers: Dordrecht, 2001; pp. 27–56.

67. Tausz, M. The role of glutathione in plant response and adaptation to natural stress. In *Significance of Glutathione to Plant Adaptation to the Environment*; Grill, D., Tausz, M., De Kok, L.J., Eds.; Kluwer Academic Publishers: Dordrecht, 2001; pp. 101–122.

68. Anderson, J.W.; McMahon, P.J. The role of glutathione in the uptake and metabolism of sulfur and selenium. In *Significance of Glutathione to Plant Adaptation to the Environment*; Grill, D., Tausz, M., De Kok, L.J., Eds.; Kluwer Academic Publishers: Dordrecht, 2001, pp. 57–99.

69. Tausz, M.; Gullner, G.; Kömives, T.; Grill, D. The role of thiols in plant adaptation to environmental stress. In: *Sulphur in Plants*. Abrol, Y.P., Ahmad, A., Eds.; Kluwer Academic Publishers: Dordrecht, 2003; pp. 221–244.

70. Schröder, P. Halogenated air pollutants. In *Responses of Plant Metabolism to Air Pollution and Global Change*; De Kok, L.J., Stulen, I., Eds.; Backhuys Publishers: Leiden, 1998; pp. 131–145.

71. Schröder P. The role of glutathione *S*-transferases in plant reaction and adaptation to xenobiotics. In *Significance of Glutathione to Plant Adaptation to the Environment*; Grill, D., Tausz, M., De Kok, L.J., Eds.; Kluwer Academic Publishers: Dordrecht, 2001; pp. 155–183.

72. Gullner, G.; Kömives, T. The role of glutathione and glutathione-related enzymes in plant-pathogen interactions. In *Significance of Glutathione to Plant Adaptation to the Environment*; Grill, D., Tausz, M., De Kok, L.J., Eds.; Kluwer Academic Publishers: Dordrecht, 2001, pp. 207–239.

73. Rauser, W.E. Metal-binding peptides in plants. In *Sulfur Nutrition and Sulfur Assimilation in Higher Plants: Regulatory, Agricultural and Environmental Aspects*; De Kok, L.J., Stulen, I., Rennenberg, H., Brunold, C., Rauser, W.E., Eds.; SPB Academic Publishing: The Hague, 1993; pp. 239–251.

74. Rauser, W.E. The role of thiols in plants under metal stress. In *Sulfur Nutrition and Sulfur Assimilation in Higher Plants: Molecular, Biochemical and Physiological Aspects*; Brunold, C., Rennenberg, H., De Kok, L.J., Stulen, I., Davidian, J.-C., Eds.; Paul Haupt: Bern, 2000; pp. 169–183.

75. Rauser, W.E. The role of glutathione in plant reaction and adaptation to excess metals. In *Significance of Glutathione to Plant Adaptation to the Environment*; Grill, D., Tausz, M., De Kok, L.J., Eds.; Kluwer Academic Publishers: Dordrecht, 2001; pp. 123–154.

76. Cobett, C.S. Metallothioneins and phytochelatins: molecular aspects. In *Sulphur in Plants*; Abrol, Y.P., Ahmad, A., Eds.; Kluwer Academic Publishers: Dordrecht, 2003; pp. 177–188.

77. Cobett, C.S. Genetic and molecular analysis of phytochelatin biosynthesis, regulation and function. The plant sulfate transporter family. In *Specialized Functions and Integration with Whole Plant Nutrition*; Davidian, J.-C., Grill, D., De Kok, L.J., Stulen, I., Hawkesford, M.J., Schnug, E., Rennenberg, H., Eds.; Backhuys Publishers: Leiden, 2003; pp. 1–10.

78. Verkleij, J.A.C.; Sneller. F.E.C.; Schat, H. Metallothioneins and phytochelatins: ecophysiological aspects. In *Sulphur in Plants*; Abrol, Y.P., Ahmad, A., Eds.; Kluwer Academic Publishers: Dordrecht, 2003; pp. 163–176.

79. Carginale, V.; Sorbo, S.; Capasso, C.; Trinchella, F.; Cafiero, G.; Bsaile, A. Accumulation, localisation, and toxic effects of cadmium in the liverwort *Lunularia cruciata*. *Protoplasma* 2004, 223, 53–61.

80. Nocito, F.F.; Pirovano, L.; Cocucci, M.; Sacchi, G.A. Cadmium-induced sulfate uptake in maize roots. *Plant Physiol.* 2002, 129, 1872–1879.

81. Kneer, R.; Zenk, M.H. The formation of Cd-phytochelatin complexes in plant cell cultures. *Phytochemistry* 1997, 44, 69–74.

82. Raab, A.; Feldmann, J.; Meharg, A.A. The nature of arsenic-phytochelatin complexes in *Holcus lanatus* and *Pteris cretica*. *Plant Physiol.* 2004, 134, 1113–1122.

83. Wink, M.. Biochemistry of plant secondary metabolism; CRC Press: Boca Raton, USA, 1999; *Ann. Plant Rev.* Vol. 2.

84. Halkier, B.A. Glucosinolates. In *Naturally Occurring Glycosides*; Ikan, R., Ed.; Wiley: Chichester, UK, 1999; pp. 193–223.

85. Lancaster, J.E.; Boland M.J. Flavour biochemistry. In *Onions and Allied Crops Vol III Biochemistry, Food Science, and Minor Crops*; Brewster, J.L., Rabinowitch, H.D., Eds.; CRC Press: Boca Raton, USA, 1990, pp. 33–72.

86. Watzl, B.; Leitzmann, C. *Bioaktive Substanzen in Lebensmitteln*. Hippokrates Verlag, Stuttgart, Germany, 1999.

87. Wink, M. Introduction: biochemistry, role and biotechnology of secondary metabolites. In *Functions of Plant Secondary Metabolites and Their Exploitation in Biotechnology*, Wink, M., Ed.; CRC Press: Boca Raton, USA, 1999; *Ann. Plant Rev.* Vol. 3.

88. Bell, R.W.; Rerkasem, S.; Keerati-Kasikorn, P.; Phechawee, N.; Hiranburana, S.; Ratanarat, S.; Pongsakul, P.; Loneragan, J.F. Mineral nutrition of food legumes in Thailand with particular reference to micronutrients. *ACIAR Tech. Rep.* 1990, 16, 39–41.

89. Bloem, E.; Haneklaus, S.; Ahmed, S.S.; Schnug, E. Beitrag der Schwefelversorgung zur Qualitätssicherung beim Anbau von Arzneipflanzen in unterschiedlichen Klimaten. Fachtagung für Heil- und Gewürzpflanzen, 12.-15.11.2001, Bad Neuenahr-Ahrweiler, 2002, pp. 149–155.

90. Schnug, E. Glucosinolates—fundamental, environmental and agricultural aspects. In *Sulfur Nutrition and Sulfur Assimilation in Higher Plants*; Rennenberg, H., Brunold, C., De Kok, L.J., Stulen, I., Eds.; SPB Academic Publishing: The Hague, 1990; 97–106.

91. Underhill, E.W. Glucosinolates. In *Encyclopedia of Plant Physiology, Vol. 8, Secondary Plant Products*; Bell, E.A., Charlwood, B.V., Eds.; Springer Verlag: Berlin, 1980, pp. 493–511.

92. Larsen, P.O. Glucosinolates. In *The Biochemistry of Plants*; Conn, E., Ed.; Academic: Toronto, 1981, pp. 501–525.

93. Bjerg, B.; Kachlicki, P.W.; Larsen, L.M.; Sorensen, H. Metabolism of glucosinolates. *International Rapeseed Congress*, Poznan 1987, Vol. 1/2, 496–506.

94. Bones, A.M.; Rossiter, J.T. The myrosinase-glucosinolate system, its organisation and biochemistry. *Physiol. Plant* 1996, 97, 194–208.

95. Wink, M.; Schimmer, O. Modes of action of defensive secondary metabolites. In *Functions of Plant Secondary Metabolites and Their Exploitation in Biotechnology*; Wink, M., Ed.; CRC Press: Boca Raton, USA, *Ann. Plant Rev.* 1999; 3, 17–133.

96. Rosenthal, G.A.; Janzen, D.H. *Herbivores: Their Interaction with Secondary Plant Metabolites*. Academic Press, New York, 1979.

97. Larsen, L.M.; Nielsen, J.K.; Ploeger, A.; Sørensen, H. Responses of some beetle species to varieties of oilseed rape and to pure glucosinolates. In *Advances in the Production and Utilization of Cruciferous Crops*; Nijhoff, M., Jungk, E., Eds.; Kluwer Academic Publishing: Dordrecht, The Netherlands, 1985; pp. 230–244.

98. Drobnica, L.; Zemanová, M.; Nemec, P.; Antos, K.; Kristián, P.; Stullerová, S.; Knoppavá, V.; Nemec, P. Antifungal activity of isothiocyanates and related compounds. I. Naturally occurring isothiocyanates and their analogues. *Appl. Microbiol.* 1967, 15, 701–709.

99. Wu, X.-M.; Meijer, J. In vitro degradation of intact glucosinolates by phytopathogenic fungi of *Brassica*. *Proceedings of the 10th International Rapeseed Congress*, Sept. 26–29, 1999, Canberra, (CD-ROM).

100. Sexton, A.C.; Howlett, B.J. Characterization of a cyanide hydratase gene in the phytopathogenic fungus *Leptosphaeria maculans*. *Mol. Gen. Genet.* 2000, 263, 463–470.

101. Bloem, E.; Haneklaus, S.; Schnug, E. Influence of nitrogen and sulphur fertilisation on the alliin content of onions (*Allium cepa* L.) and garlic (*Allium sativum* L.). *J. Plant Nutr.* 2004, 27, 1827–1839.

102. Haneklaus, S.; Paulsen, H.M.; Gupta, A.K.; Bloem, E.; Schnug, E. Influence of sulfur fertilization on yield and quality of oilseed rape and mustard. *Proceedings of the 10th International Rapeseed Congress*, Sept. 26–29, 1999, Canberra, (CD-ROM).

103. Schnug, E. Quantitative und Qualitative Aspekte der Diagnose und Therapie der Schwefelversorgung von Raps (*Brassica napus* L.) unter besonderer Berücksichtigung glucosinolatarmer Sorten. Habilitationsschrift (D.Sc. thesis). Kiel University, 1988.

104. Rosa, E.A.S.; Heaney, R.K.; Fenwick, G.R.; Portas, C.A.M. Glucosinolates in crop plants. *Hortic. Rev.* 1997, 19, 99–215.

105. Asare, E.; Scarisbrick, D.H. Rate of nitrogen and sulphur fertilizers on yield, yield components and seed quality of oilseed rape (*Brassica napus* L.). *Field Crops Res.* 1995, 44, 41–46.

106. Kim, S.J.; Matsuo, T.; Watanabe, M.; Watanabe, Y. Effect of nitrogen and sulphur application on the glucosinolate content in vegetable turnip rape (*Brassica rapa* L.). *Soil Sci. Plant Nutr.* 2002, 48, 43–49.

107. Block, E. The organosulfur chemistry of the genus Allium. Implications for organic sulfur chemistry. *Angew. Chem.* 1992, 31, 1135–1178.

108. Kawakishi, S.; Morimitsu, Y. Sulfur chemistry of onions and inhibitory factors of the arachidonic- acid cascade. *American Chemical Society Symposium Series ACS* 1994, 546, 120–127.

109. Curtis, H.; Noll, U.; Stormann, J.; Slusarenko, A.J. Broad-spectrum activity of the volatile phytoanticipin allicin in extracts of garlic (*Allium sativum* L.) against plant pathogenic bacteria, fungi and oomycetres. *Physiol. Mol. Plant Pathol.* 2004, 65, 79–89.

110. Lancaster, J.E.; Collin, H.A. Presence of alliinase in isolated vacuoles and alkyl cysteine sulphoxides in the cytoplasm of bulbs in onion (*Allium cepa*). *Plant Sci. Lett.* 1981, 22, 169–176.

111. Block, E.; Calvey, E.M. Facts and artifacts in *Allium* chemistry. *American Chemical Society Symposium Series ACS* 1994, 564, 63–79.

112. Matula, J. The effect of chloride and sulphate application to soil on changes in nutrient content in barley shoot biomass at an early phase of growth. *Plant Soil Environ.* 2004, 50, 295–302.

113. Kowalenko, C.G. Variations in within-season nitrogen and sulfur interaction effects on forage grass response to combinations of nitrogen, sulfur, and boron applications. *Commun. Soil Sci. Plant Anal.* 2004, 35, 759–780.

114. Schnug, E.; Haneklaus, S. Diagnosis of sulphur nutrition. In *Sulphur in Agroecosystems*. Vol. 2, Part of the series 'Nutrients in Ecosystems'; Schnug, E., Ed.; Kluwer Academic Publishers: Dordrecht, 1998; pp. 1–38.

115. Hu, H.N.; Sparks, D. Nitrogen and sulfur interaction influences net photosynthesis and vegetative growth of pecan. *J. Am. Soc. Hortic. Sci.* 1992, 117, 59–64.

116. Schnug, E. Double low oilseed rape in West Germany: sulphur nutrition and levels. In production and protection of oilseed rape and other *Brassica* crops. *Aspects Appl. Biol.* 1989, 23, 67–82.

117. Freeman, G.G.; Mossadeghi, N. Influence of sulphate nutrition on the flavour components of garlic (*Allium sativum*) and wild onion (*A. vineale*). *J. Sci. Food Agric.* 1971, 22, 330–334.

118. Randle, W.M.; Kopsell, D.E.; Kopsell D.A. Sequentially reducing sulfate fertility during onion growth and development affects bulb flavor at harvest. *HortScience* 2002, 37, 118–121.

119. Schnug, E; Haneklaus, S. Molybdänversorgung im intensiven Rapsanbau. *Raps* 1990, 8, 188–191.

120. Haneklaus, S.; Bloem, E.; Hayfa, S.; Schnug, E. Influence of elemental sulphur and nitrogen fertilisation on the concentration of essential micronutrients and heavy metals in *Tropaeolum majus* L. *FAL – Agric. Res.* 2005, Special issue no. 286, 25–35.

121. Mendel, R.; Haensch, R. Molybdoenzymes and molybdenum cofactor in plants. *J. Exp. Bot.* 2002, 53, 1689–1698.

122. Stout, P.R.; Meagher, W.R. Studies of the molybdenum nutrition of plants with radioactive molybdenum. *Science* 1948, 108, 471–473.

123. Singh, M.; Kumar, V. Sulfur, phosphorus, and molybdenum interactions on the concentration and uptake of molybdenum in soybean plants. *Soil Sci.* 1979, 127, 307–312.

124. Macleod, J.A.; Gupta, U.C.; Stanfield, B. Molybdenum and sulfur relationships in plants. In *Molybdenum in Agriculture*; Gupta,U.C., Ed.; Cambridge University Press: UK, 1997.

125. Terry, N.; Zayed, A.M.; de Souza, M.P.; Tarun, A.S. Selenium in higher plants. *Ann. Rev. Plant Physiol. Plant Mol. Biol.* 2000, 51, 401–432.

126. Kaiser, B.N.; Grindley, K.L.; Tyerman, S.D.; Ngaire Bradey, J. Molybdenum nutrition in plants. *Ann. Bot.* in press; 2005.

127. De Kok, L.J.; Kuiper, P.J.C. Effect of short term dark incubation with sulfate, chloride and selenate on the glutathione content of spinach leaf discs. *Physiol. Plant* 1986, 68, 477–482.

128. Bosma, W.; Schupp, R.; De Kok, L.J.; Rennenberg, H. Effect of selenate on assimilatory sulfate reduction and thiol content of spruce needles. *Plant Physiol. Biochem.* 1991, 29, 131–138.

129. Zayed, A.M.; Terry, N. Selenium volatilization in broccoli as influenced by sulfate supply. *J. Plant Physiol.* 1992, 140, 646–652.

130. Spencer, K. Effect of sulfur application on selenium content of subterranean clover plants grown at different levels of selenium supply. *Aus. J. Exp. Agric. Anim. Husb.* 1982, 22, 420–427.

131. Kopsell, D.A.; Randle, W.M. Short-day onion cultivars differ in bulb selenium and sulfur accumulation which can affect bulb pungency. *Euphytica* 1997, 96, 385–390.

132. Mikkelsen, R.L.; Wan, H.F. The effect of selenium on sulfur uptake by barley and rice. *Plant Soil* 1990, 121, 151–153.

133. Abdin, M.Z.; Ahmad, A.; Khan, N.; Khan, I.; Jamal, A.; Iqbal, M. Sulphur interaction with other nutrients. In *Sulphur in Plants*; Abrol, Y.P., Ahmad, A. Eds.; Kluwer Academic Publishers: Dordrecht, 2003; pp. 177–188.

134. Naphada, G.D.; Mutalka, V.K. Effect of phosphorus fertilizer in Saurashtra soil. Effect of sulphur and phosphatic fertilizers on the growth of groundnut and maize. *Saurashtra J. Agric. Soil* 1984, 7, 5–10.

135. Noble, J.C.; Kleinig, C.R. Response by irrigated grain sorghum to broadcast gypsum and phosphorus on a heavy clay soil. *Aus.. J. Exp. Agric. Anim. Husb.* 1971, 11, 53–58.

136. Marok, A.S.; Dev, G. Phosphorus and sulphur interrelationship in wheat. *J. Indian Soc. Soil Sci.* 1980, 28, 184–188.

137. Joshi, D.C.; Seth, S.P.; Parekh, B.L. Studies on sulfur and P uptake by mustard. *J. Indian Soc. Soil Sci.* 1973, 21, 167–172.

138. Singh, V.; Rathore, S.S. Effect of applied potassium and sulphur on yield, oil content and their uptake by linseed. *J. Pot. Res.* 1994, 10, 407–410.

139. Prasad, R.; Prasad, U.S.M.; Sakal, R. Effects of potassium and sulphur on yield and quality of sugarcane grown in calcareous soils. *J. Pot. Res.* 1996, 12, 29–38.

140. Umar, S.; Debnath, G.; Bansal, S.K. Groundnut pod yield and leaf spot disease as affected by potassium and sulphur nutrition. *Indian J. Plant Physiol.* 1997, 2, 59–64.

141. Bennett, W.F. Plant nutrient utilization and diagnostic plant symptoms. In *Nutrient Deficiencies and Toxicities in Crop Plants*; Bennett, W.F., Ed.; APS Press: St. Paul, 1993; pp. 1–7.

142. Bergmann, W. *Nutritional Disorders of Plants—Visual and Analytical Diagnosis.* Gustav Fischer Verlag: Jena, 1992.

143. Bergmann, W. *Ernaehrungsstoerungen bei Kulturpflanzen. 3. Aufl.*, Gustav Fischer Verlag: Jena, 1993.

144. Eaton, F.M. Sulfur. In *Diagnostic Criteria for Plants and Soils*; Chapman, H.D., Ed.; University of California, Division of Agricultural Sciences: Riverside, CA, 1966.

145. Saalbach, E. Ueber die Bestimmung des Schwefelversorgungsgrades von Hafer. *Z. Pflanzenernaehr. Bodenkde* 1970, 127, 92–100.

146. Strauss, R.; Bleiholder, H.; Van der Boom, T.; Buhr, L.; Hack, H.; Hess, M.; Klose, R.; Meier, U.; Weber, E. *Einheitliche Codierung der phänologischen Entwicklungsstadien mono- und dikotyler Pflanzen*. Ciba Geigy AG, Basel, 1994.

147. Schnug, E.; Haneklaus, S. Sulphur deficiency symptoms in oilseed rape (*Brassica napus* L.) - the aesthetics of starvation. *Phyton* 2005, 45, 79–95.

148. Haneklaus, S.; Bloem, E.; Schnug, E. Sulfur and plant disease. In *Mineral Nutrition and Plant Diseases*; Datnoff, L., Elmer, W., Huber, D., Eds.; APS Press: St. Paul, MN, in press; 2005.

149. Burke, J.J.; Holloway, P.; Dalling, M.J. The effect of sulfur deficiency on the organisation and photosynthetic capability of wheat leaves. *J. Plant Physiol.* 1986, 125, 371–375.

150. Dietz, K.-J. Leaf and chloroplast development in relation to nutrient availability. *J. Plant Physiol.* 1989, 134, 544–550.

151. Dietz, K.-J. Recovery of spinach leaves from sulfate and phosphate deficiency. *J. Plant Physiol.* 1989, 134, 551–557.

152. Ergle, D.R.; Eaton, F.M. Sulphur nutrition of cotton. *Plant Physiol.* 1951, 26, 639–654.

153. Haq, I.U.; Carlson, R.M. Sulphur diagnostic criteria for French prune trees. *J. Plant Nutr.* 1993, 16, 911–931.

154. Stuiver, C.E.E.; De Kok, L.J.; Westermann, S. Sulfur deficiency in *Brassica oleracea* L.: development, biochemical characterization, and sulfur/nitrogen interactions. *Russian J. Plant Physiol.* 1997, 44, 505–513.

155. Lobb, W.R.; Reynolds, D.G. Further investigations in the use of sulphur in North Otago. *New Zealand J. Agric.* 1956, 92, 17–25.

156. Hall, R.; Schwartz, H.F. Common bean. In *Nutrient Deficiencies and Toxicities in Crop Plants*; Bennett, W.F., Ed.; APS Press: St. Paul, 1993; pp. 143–147.

157. Ulrich, A.; Moraghan, J.T.; Whitney, E.D. Sugar beet. In *Nutrient Deficiencies and Toxicities in Crop Plants*; Bennett, W.F., Ed.; APS Press: St. Paul, 1993; pp. 91–98.

158. Walker, K.C.; Booth, E.J. Sulphur deficiency in Scotland and the effects of sulphur supplementation on yield and quality of oilseed rape. *Norwegian J. Agric. Sci. Suppl.* 1994, 15, 97–104.

159. Deloch, H.W.; Bussler, W. Das Wachstum verschiedener Pflanzenarten in Abhaengigkeit von der Sulfatversorgung. *Z. Pflanzenernaehr. Bodenkde* 1964, 108, 232–244.

160. Eaton, S.V. Influence of sulphur deficiency on the metabolism of the soybean. *Bot. Gaz.* 1935, 97, 68–100.

161. Eaton, S.V. Influence of sulphur deficiency on the metabolism of the sunflower. *Bot. Gaz.* 1941, 102, 533–556.

162. Eaton, S.V. Effects of sulphur deficiency on growth and metabolism of tomato. *Bot. Gaz.* 1951, 112, 300–307.

163. Harborne, J.B. *Comparative Chemistry of the Flavonoid Compounds*. Academic Press: London, 1967.

164. Harborne, J.B. Flavonoids in the environment: structure-activity relationship. In *Flavonoids in Biology and Medicine*; Alan, R., Ed.; Liss Inc.: New York, 1968; 2, pp. 12–27.

165. Nightingale, G.T.; Schermerhorn, L.G.; Robbins, W.R. Effect of sulphur deficiency on metabolism in tomato. *Plant Physiol.* 1932, 7, 565–595.

166. Bugakova, A.N.; Beleva, V.I.; Tulunina, A.K.; Topcieva, V.T. Einfluß von Schwefel auf die morphologischen und anatomischen Bau sowie auf physiologische und biochemische Eigenschaften von Erbsen. *Agrochim.* 1969, 11, 128–130.

167. Sinclair, A.G. An auto-analyzer method for determination of extractable sulphate in plant material. *Plant Soil* 1974, 40, 693–697.

168. Schnug, E.; Haneklaus, S. Sulphur deficiency in oilseed rape flowers—symptomatology, biochemistry and ecological impacts. *Proceedings of the 9th International Rapeseed Congress*, Cambridge, UK, 1995, 1, 296–299.

169. Haneklaus, S.; Brauer, A.; Bloem, E.; Schnug, E. Relationship between sulfur deficiency in oilseed rape (*Brassica napus* L.) and its attractiveness for honeybees. *FAL–Agric. Res.* 2005, 283, 37–43.

170. Voss, R.D. Corn. In *Nutrient Deficiencies and Toxicities in Crop Plants*; Bennett, W.F., Ed.; APS Press: St. Paul, 1993; pp. 11–14.

171. Bloem, E.; Paulsen, H.-M.; Schnug, E. Schwefelmangel in Getreide. *DLG-Mitteilungen* 1995, 8, 17–18.

172. Knudsen, L.; Oestergaard, H.S. Goedskning og kalkning. In *Oversigt over Landsvforsoegene*; Pedersen C.A., Ed.; Brabrand Bogtryk ApS: Aarhus Denmark, 1992; pp. 86–88.

173. Gascho, G.J.; Anderson, D.L.; Bowen, J.E. Sugarcane. In *Nutrient Deficiencies and Toxicities in Crop Plants*; Bennett, W.F., Ed.; APS Press: St. Paul, 1993; pp. 37–42.

174. Haneklaus, S.; Murphy, D.P.; Nowak, G.; Schnug, E. Effects of the timing of sulphur on yield and yield components of wheat. *J. Plant Nutr. Soil Sci.* 1995, 158, 83–86.

175. Blair, G.J.; Chinoim, N.; Lefroy, R.D.B.; Anderson, G.C.; Crocker, G.J. A soil sulfur test for pastures and crops. *Aus. J. Soil Res.* 1991, 29, 619–626.

176. Sanford, J.O; Lancaster, J.D. Biological and chemical evaluation of the readily available sulfur status of Mississippi soils. *Soil Sci. Am. Proc.* 1962, 26, 63–65.

177. Westermann, D.T. Indexes of sulfur deficiency in alfalfa. I. Extractable soil SO_4-S. *Agron. J.* 1974, 66, 578–580.

178. Yli-Halla, M. Assessment of extraction and analytical methods in estimating the amount of plant available sulfur in soils. *Acta Agric. Scand.* 1987, 37, 419–425.

179. Freney, J.R.; Spencer, K. Diagnosis of sulphur deficiency in plants by soil and plant analysis. *J. Aus. Inst. Agric. Sci.* 1967, 33, 284–288.

180. Haneklaus, S.; Fleckenstein, J.; Schnug, E. Comparative studies of plant and soil analysis for the evaluation of the sulphur status of oilseed rape and wheat. *J. Plant Nutr. Soil Sci.* 1995, 158, 109–112.

181. Mitchell, C.C.; Mullins, G.L. Sources, rates and time of sulfur application to wheat. *Sulphur Agric.* 1990, 14, 20–24.

182. Alewell, C.; Matzner, E. Water, $NaHCO_3$-, NaH_2PO_4- and NaCl- extractable $SO_4{}^{2-}$ in acid forest soils. *Z. Pflanzenernähr. Bodenk.* 1996, 159, 235–240.

183. Anderson, G.; Lefroy, R.; Chinoim, N.; Blair, G. Soil sulphur testing. *Sulphur Agric.* 1992, 16, 6–14.

184. Eriksen, J.; Murphy, M.; Schnug, E. The soil sulphur cycle. In *Sulphur in Agroecosystems*. Part of the series 'Nutrients in Ecosystems'; Schnug, E., Ed.; Kluwer Academic Publishers: Dordrecht, 1998, 2, pp. 39–73.

185. Warman, P.R.; Sampson, H.G. Evaluation of soil sulfate extractants and methods of analysis for plant available sulfur. *Commun. Soil Sci. Plant Anal.* 1992, 23, 793–803.

186. Alewell, C. Effects of organic sulfur compounds on extraction and determination of inorganic sulfate. *Plant Soil* 1993, 149, 141–144.

187. Tan, Z.; McLaren, R.G.; Cameron, K.C. Forms of sulfur extracted from soils after different methods of sample preparation. *Aus. J. Soil Res.* 1994, 32, 823–834.

188. Duynisveld, W.H.M.; Strebel, O.; Boettcher, J. Prognose der Grundwasserqualitaet in einem Wassereinzugsgebiet mit Stofftransportmodellen. Forschungsbericht Nr. 10204371 UBA-FB 92–106. 1993, UBA Text 5/93, Berlin.

189. Cottenie, A. Soil and plant testing as a basis of fertilizer recommendations. *FAO Soils Bull.* 1980, 38/2.

190. Reisenauer, H.M.L.; Walsh, L.M.; Hoeft, R.G. Testing soils for sulfur, boron, molybdenum and chlorine. In *Soil Testing and Plant Analysis*; Walsh, L.M., Beaton, J. D., Eds.; Soil Sci. Soc. Am.: Madison, 1973; pp. 173–200.

191. Saalbach, E.; Aigner, H. Zum Diagnosewert der $NaCl + CaCl_2$ -extrahierbaren Sulfatmengen von Boeden. *Landwirtsch. Forschung* 1987, 40, 8–12.

192. Saalbach, E.; Kessen, G.; Judel, G.K. Untersuchungen ueber die Bestimmung des Gehaltes an pflanzenverfuegbarem Schwefel im Boden. *Landwirtsch. Forschung* 1962b, 15, 6–15.

193. Grunwaldt, H.-S. Untersuchungen zum Schwefelhaushalt schleswig-holsteinischer Boeden. Diss. Agrarwiss. Fak., 1969, Kiel.

194. Bansal, K.N.; Pal, A.R. Evaluation of a soil test method and plant analysis for determining the sulphur status of alluvial soils. *Plant Soil* 1987, 98, 331–336.

195. Jansson, H. Status of Sulphur in Soils and Plants of Thirty Countries. FAO World Soil Resources Reports 1995, p. 79.

196. Ribeiro, A.C.; Accioly, L.J.O.; Alvarez, V.V.H.; Braga, J.M.; Alves, V.M.C. Availiacao do enxofre disponivel pelo metodo da resina trocadora de anions. *Revista Braisleira de Ciencia do Solo* 1992, 15, 321–327.

197. Searle, P.L. The extraction of sulphate and mineralisable sulphur from soils with an anion exchange membrane. *Commun. Soil Sci. Plant Anal.* 1992, 23, 17–20.

198. Banerjee, M.R.; Chapman, S.J.; Sinclair, A.M.; Kilham, K. Evaluation of a perfusion system for investigation of the sulphur supplying capacity of soils. *Commun. Soil Sci. Plant Anal.* 1994, 25, 2613–2625.

199. Burns, G.R. Oxidation of Sulphur in Soils. The Sulphur Institute Technical Bulletin. 1967, p. 13.

200. Starkey, R.L. Oxidation and reduction of sulfur compounds in soils. *Soil Sci.* 1966, 101, 297–306.

201. Swift, R.S. Mineralization and immobilization of sulphur in soil. *Sulphur Agric.* 1985, 9, 20–25.
202. Blair, G. The development of the KCl-40 sulfur soil test. *Proceedings of the 15th World Congress Soil Science*, Acapulco, Mexico, 1994, 5a, pp. 351–363.
203. Santoso, D.; Lefroy, R.D.B.; Blair, G.J. A comparison of sulfur extractants for weathered acid soils. *Aus. J. Soil Res.* 1995, 33, 125–133.
204. Williams, C.H.; Steinbergs, A. Soil sulphur fractions as chemical indices of available sulphur in some Eastern Australian soils. *Aus. J. Agric. Res.* 1959, 10, 340–352.
205. Bloem, E. Schwefel-Bilanz von Agraroekosystemen unter besonderer Beruecksichtigung hydrologischer und bodenphysikalischer Standorteigenschaften. Ph.D. thesis, TU-Braunschweig, Germany, 1998.
206. Bloem, E.; Haneklaus, S.; Schnug, E. Optimization of a method for soil sulphur extraction. *Commun. Soil Sci. Plant Anal.* 2000, 33, 41–51.
207. Ajwa, H.A.; Tabatabai, M.A. Comparison of some methods for determination of sulphate in soils. *Commun. Soil Sci. Plant Anal.* 1993, 24, 1817–1832.
208. Shan, X.Q.; Chen, B.; Jin, L.Z.; Zheng, Y.; Hou, X.P.; Mou, S.F. Determination of sulfur fractions in soils by sequential extraction inductively couple plasma-optical emission spectroscopy and ion chromatography. *Chem. Speciation Bioavail.* 1992, 4, 97–103.
209. Sharp, G.S.; Hoque, S.; Kilham, K.; Sinclair, A.H.; Chapman, S.J. Comparison of methods to evaluate the sulphur status of soils. *Commun. Soil Sci. Plant Anal.* 1989, 20, 1821–1832.
210. Bolton, J. Effects of sulphur fertilizers and of copper on the yield and composition of spring wheat grown in a sandy soil prone to surface compaction. *J. Agric. Sci.* 1975, 84, 159–165.
211. Hamm, J.W.; Bettany, J.R.; Halstead, E.H. A soil test for sulphur and interpretive criteria for Saskatchewan. *Commun. Soil Sci. Plant Anal.* 1973, 4, 219–231.
212. Scott, N.M. Evaluation of sulphate status of soils by plant and soil tests. *J. Sci. Food Agric.* 1981, 32, 193–199.
213. Scott, N.M. Sulphur in agriculture—IV Miscellaneous and special topics. *Proceedings of the Conference*, London, UK, Nov. 14–17, More, A.I., Ed.; Vol. 1; 1982; preprint.
214. Scott, N.M.; Watson, M.E. Agricultural sulphur research and responses to sulphur in North Scotland. *Proceedings of the International Sulphur Conference*, London, 1982, Vol. 1, pp. 579–586.
215. Scott, N.M.; Watson, M.E.; Caldwell, K.S. Response of grassland to the applications of sulphur at two sites in north-east Scotland. *J. Sci. Food Agric.* 1983, 34, 357–361.
216. Bettany, J.R.; Janzen, H.H.; Stewart, J.W.B. Sulphur deficieny in the Prairie Provinces of Canada. *Proceedings of the International Sulphur Conference*, London 1982, Vol. 2, pp. 787–799.
217. Blair, G.J.; Mamaril, C.P.; Momuat, E. Sulfur Nutrition of Wetland Rice. IRPS 1978, 21.
218. Hoque, S.; Heath, S.B.; Killham, K. Evaluation of methods to assess adequacy of potential soil sulfur supply to crops. *Plant Soil* 1987, 101, 3–8.
219. Lee, R.; Speir, T.W. Sulphur uptake by ryegrass and its relationship to inorganic and organic sulphur levels and sulphatase activity in soil. *Plant Soil* 1979, 53, 407–425.
220. Saalbach, E. Zur Bestimmung des Schwefelversorgungsgrades von Boeden und landwirtschaftlichen Nutzpflanzen. *Landwirtschaftl. Forschung* 1964, Sonderh. 18, 84–90.
221. Skinner, R.J. Growth responses in grass to sulphur fertiliser. *Proceedings of the International Symposium on Elemental Sulphur in Agriculture*, Marseille, 1987, Vol. 2, pp. 525–535.
222. Spencer, K.; Jones, M.B.; Freney, J.R. Diagnostic indices for sulphur status of subterranean clover. *Aus. J. Agric. Res.* 1977, 28, 401–412.
223. Tsuji, T.; Goh, K.M. Evaluation of soil sulphur fractions as sources of plant-available sulphur using radioactive sulphur. *New Zealand J. Agric. Res.* 1979, 22, 595–602.
224. Zhao, F.; Mc Grath, S.P. Extractable sulphate and organic sulphur in soils and their availability to plants. *Plant Soil* 1994, 164, 243–250.
225. Timmermann, F.; Kluge, R.; Pfliehinger, A. Schwefel-Bedarfsermittlung anhand des N/S_{min} Verhaeltnisses im Boden. *VDLUFA-Schriftenreihe* 1995, 40, 303–306.
226. Bole, J.B.; Pittman, U.J. Availability of subsoil sulphates to barley and rapeseed. *Can. J. Soil Sci.* 1984, 64, 301–312.
227. Bullock, D.G.; Goodroad, L.L. Effect of sulfur rate application method and source on yield and mineral content of corn. *Commun. Soil Sci. Plant Anal.* 1989, 20, 1209–1218.
228. Holz, F. Bestimmung des Gehaltes an Nitrat und Sulfat in Boeden und Niederschlaegen durch simultane Durchflußanalyse. *Landwirtsch. Forschung* 1984, 41, 105–126.

229. Link, A.; Kuhlmann, H.; Lammel, J. S_{min}-Bodenuntersuchung zur Ermittlung des Schwefelduengebedarfs von Raps. *Raps* 1996, 13, 17–19.

230. Roberts, T.L.; Bettany, J.R. The influence of geomorphology on the nature and distribution of soil sulfur across a narrow environmental gradient. *Can. J. Soil Sci.* 1985, 65, 415–434.

231. Bloem, E.; Haneklaus, S.; Sparovek, G.; Schnug, E. Spatial and temporal variability of sulphate concentration in soils. *Commun. Soil Sci. Plant Anal.* 2001, 32, 1391–1403.

232. Isermann, K. Loeslicher N, Sulfat-S und (DO)C im (un-)gesaettigten Untergrund von Porengrundwasserleitern bei unterschiedlicher Landbewirtschaftung/Duengung. *Mitt- Dt. Bodenkd. Ges.* 1993, 71, 141–144.

233. Preuschoff, M. Untersuchungen zur Schwefelversorgung von Weißkohl an zwei Loeßstandorten. Ph.D., University of Hanover, Verlag Ulrich E. Grauer, Stuttgart, 1995.

234. Schlichting, M. Der Sulfatgehalt des Grundwassers in Abhaengigkeit von bodennutzungsspezifischen Stoffeintraegen und dessen Bedeutung fuer die Nutzung des Grundwassers zur Feldberegnung–Beispiel Fuhrberger Feld. Master thesis, University of Hanover, 1996.

235. Pedersen, C.A.; Knudsen, L.; Schnug, E. Sulphur fertilisation. In *Sulphur in Agroecosystems*; Schnug, E., Ed.;. Kluwer Academic Publishers: Dordrecht, 1998, pp. 115–134.

236. Sillanpää, M.; Jansson, H. Cadmium and sulphur contents of different plant species grown side by side. *Ann. Agric. Fenn.* 1991, 30, 407–413.

237. Gupta, U.C. Tissue sulfur levels and additional sulfur needs for various crops. *Can. J. Plant Sci.* 1976, 56, 651–657.

238. Andrew, C.S. The effect of sulphur on the growth, sulphur and nitrogen concentrations, and critical sulphur concentrations of some tropical and temperate pasture legumes. *Aus. J. Agric. Res.* 1977, 28, 807–820.

239. Freney, J.R.; Spencer, K.; Jones, M.B. Determining the sulphur status of wheat. *Sulphur Agric.* 1978, 2, 231.

240. Scaife, A.; Burns, I.G. The sulphate-S/ total sulfur ratio in plants as an index of their sulphur status. *Plant Soil* 1986, 91, 61–71.

241. Spencer, K.; Freney, J.R. Assessing the sulfur status of field-grown wheat by plant analysis. *Agron. J.* 1980, 72, 469–472.

242. Spencer, K.; Freney, J.R.; Jones, M.B. A preliminary testing of plant analysis procedures for the assesment of the sulfur status of oilseed rape. *Aus. J. Agric. Res.* 1984, 35, 163–175.

243. Widdowson, J.P.; Blakemore, L.C. The sulphur status of soils of the south-west Pacific area. *Proceedings of the International Sulphur Conference*, London 1982, Vol. 2, pp. 805–819.

244. Haneklaus, S.; Schnug, E. Nährstoffversorgung von Zuckerrüben in Schleswig-Holstein und Jütland. *Zuckerrübe* 1996, 45, 182–184.

245. Franck, E.v. Ermittlung von Zink-Ertragsgrenzwerten fuer Hafer und Weizen, Beurteilung der Zinkversorgung von Getreide in Schleswig-Holstein und Untersuchungen ueber Ursachen unzureichender Zinkversorgung auf Hochleistungsfeldern. Diss. Agrarwiss. Fak., 1978, Kiel.

246. Moeller-Nielsen, J. Kornplanters erbaerinstilstand vurderet og reguleret udfra planternes kemiske sammensaetning. Diss. Kgl. Veterinaer- og Landbohoiskole, Kopenhagen 1973.

247. Moeller-Nielsen, J.; Frijs-Nielsen. B. Evaluation and control of the nutritional status of cereals. II. Pure effect of a nutrient. *Plant Soil* 1976, 45, 339–351.

248. Thiel, H. Ermittlung von Grenzwerten optimaler Kupferversorgung fuer Hafer und Sommergerste in Gefaeßversuchen und unter Feldbedingungen Schleswig-Holsteins. Diss. Agrarwiss. Fak., 1972, Kiel.

249. Wichmann, W. Ermittlung von Grenzwerten der Pflanzenanalyse zur Kennzeichnung der Magnesium-Versorgung von Getreide in Schleswig-Holstein. Diss. Agrarwiss. Fak., 1976, Kiel.

250. Hell, R.; Rennenberg, H. The plant sulfur cycle. In *Sulphur in Agroecosystems*. Part of the series 'Nutrients in Ecosystems'; Vol. 2, Schnug, E., Ed.; Kluwer Academic Publishers: Dordrecht, 1997; pp. 135–174.

251. Cerdá, A.; Martinez, V.; Caro, M.; Fernández, F.G. Effect of sulfur deficiency and excess on yield and sulfur accumulation in tomato plants. *J. Plant Nutr.* 1984, 7, 1529–1543.

252. Freney, J.R.; Randall, P.J.; Spencer, K. Diagnosis of sulphur deficiency in plants. *Proceedings of the International Sulphur Conference*, London, 1982, Vol. 1, pp. 439–444.

253. Maynard, D.G.; Stewart, J.W.B.; Bettany, J.R. Use of plant analysis to predict sulfur deficiency in rapeseed (*Brassica napus* and *B. campestris*). *Can. J. Soil Sci.* 1983, 63, 387–396.

254. Scott, N.M. Sulphur responses in Scotland. *Sulphur Agric.* 1985, 9, 13–18.
255. Smith, F.W.; Dolby, G.R. Derivation of diagnostic indices for assessing the sulphur status of *Panicum maximum* var. *trichoglume*. *Commun. Soil Sci. Plant Anal.* 1977, 8, 221–240.
256. Saalbach, E. Sulphur requirements and sulphur removals of the most important crops. *Symposium Int. sur le Soufre en Agric.*, Versailles, 1970.
257. Syers, J.K.; Skinner, R.J.; Curtin, D. Soil and fertilizer sulphur in U.K. agriculture. *The Fert. Soc.* 1987, 264, 1–43.
258. Morris, R.J.; Tisdale, S.L.; Platou, J. The importance of sulphur in crop quality. *J. Fert. Issues* 1984, 1, 139–145.
259. Yoshida, S.; Chaudry, M.R. Sulfur nutrition of rice. *Soil Sci. Plant Nutr.* 1979, 25, 121–134.
260. Zhao, F.; Hawkesford, M.J.; Warrilow, A.G.S.; McGrath, S.P.; Clarkson, D.T. Diagnosis of sulphur deficiency in wheat. In *Sulfur Metabolism in Higher Plants: Fundamental Molecular, Ecological and Agricultural Aspects*; Cram et al., Eds.; SPB Academic Publishing: The Hague, 1997; pp. 349–351.
261. Freney, J.R. How much sulfur do plants require? *Fert. Solutions* 1966, 10, 14–15.
262. Janzen, H.H.; Bettany, J.R. The effect of temperature and water potential on sulfur oxidation in soils. *Soil Sci.* 1987, 144, 81–89.
263. Metson, A.J.; Collie, T.W. Iron pyrites as sulphur fertilisers III. Nitrogen-sulphur relationships in grass and clover seperates of pasture herbage in a field trial at Golden Bay, Nelson. *New Zealand J. Agric. Res.* 1972, 15, 585–604.
264. Marquard, R. Der Einfluß der Schwefelernaehrung auf den Senf- und Lauchoelgehalt bei einigen Pflanzen aus den Familien der Cruziferen, Tropaeolaceen und Liliaceen. Diss. Agrarwiss. Fak., 1967, Gießen.
265. Beaton, J.D.; Burns, G.R.; Platou, J. Determination of Sulphur in Soils and Plant Matertial. The Sulphur Institute Technical Bulletin, 1968, No 14.
266. Schnug, E.; Haneklaus, S. Diagnosis of the nutritional status and quality assessment of oilseed rape by x-ray spectroscopy. *Proceedings of the 10th International Rapeseed Congress*, Canberra, Australia, Sept. 26–29, 1999 (CD-ROM).
267. Reynolds, S.B.; Martin, A.D.E.; Bucknall, B.; Chambers, B.J. A simplified x-ray fluorescence (XRF) procedure for the determination of sulphur in graminaceous materials. *J. Sci. Food. Agric.* 1989, 47, 327–336.
268. Schnug, E.; Murray, F.; Haneklaus, S. Preparation techniques of small sample sizes for sulphur and indirect total glucosinolate analysis in *Brassica* seeds by x-ray fluorescence spectroscopy. *Fat Sci. Technol.* 1993, 95, 334–336.
269. Schnug, E.; Kallweit, P. Ergebnisse eines Ringversuches zur röntgenfluoreszenz-analytischen Bestimmung des Gesamtglucosinolatgehaltes von Rapssamen. *Fett-Wissenschaft Technologie* 1987, 89, 377–381.
270. Wagstaffe, P.; Boenke, J.; Schnug, E.; Lindsey, A.S. Certification of the sulphur content of three rape-seed reference materials. *Fresenius Z. Anal. Chem.* 1992, 344, 1–7.
271. Dijkshoorn, W.; Lampe, J.E.M.; Burg van, P.F.J. A method of diagnosting the sulphur nutrition status of herbage. *Plant Soil* 1960, 13, 227–241.
272. Dijkshoorn, W.; Wijk van, A.L. The sulphur requirements of plants as evidenced by the sulphur- nitrogen ratio in the organic matter. A review of published data. *Plant Soil* 1967, 26, 129—157.
273. Stewart, B.A. N:S Ratios. A guideline to sulphur needs. *The Sulphur Institute J.* 1969, 5, 12–15.
274. Brogan, J.C.; Murphy, M.D. Sulphur nutrition in Ireland. *Sulphur Agric.* 1980, 4, 2–6.
275. Cowling, D.W.; Jones, L.H.P. A deficiency in soil sulfur supplies for perennial ryegrass in England. *Soil Sci.* 1970, 110, 346–354.
276. Murphy, M.D. Essential micronutrients. In *III: Sulphur. Applied Soil Trace Elements.* Davis, B.E., Ed.; Wiley: Chichester, 1980; pp. 235–258.
277. Pumphrey, F.V.; Moore, D.P. Diagnosing sulfur deficiency of alfalfa (*Medicago sativa* L.) from plant analysis. *Agron. J.* 1965, 57, 364–366.
278. Salette, J.H. Sulphur content of grasses during primary growth. *Sulphur Agric.* 1978, 143–153.
279. Standford, G.; Jordan, H.V. Sulfur requirements of sugar, fiber, and oil crops. *Soil Sci.* 1966, 101, 258–266.
280. Stewart, B.A.; Porter, L.K. Nitrogen-sulfur relationships in wheat (*Triticum aestivum* L.), corn (*Zea mays*) and beans (*Phaseolus vulgaris*). *Agron. J.* 1969, 61, 267–271.

281. Whitehead, D.C.; Jones, L.H.P. Nitrogen/sulphur relationships in grass and legumes. *Sulphur in Forages* 1978, pp. 127–141.

282. Zhao, F.; Evans, E.J.; Bilsborrow, P.E.; Schnug, E.; Syers, K.J. Correction for the protein content in the determination of total glucosinolate content of rapeseed by the X-RF method. *J. Sci. Food Agric.* 1992, 58, 431–433.

283. Hester, B. Sulphur—The fourth major nutrient? *Fert. Solut.* 1979, 23, 44–50.

284. Finck, A. Die Pflanzenanalyse als Hilfsmittel zur Ermittlung des Duengerbedarfes. Sonderdruck aus Chemie und Landw. *Produktion* 1970, pp. 183–188.

285. Bansal, K.N.; Singh, D. Nitrogen-sulphur ratio for diagnosing sulphur status of alfalfa. *J. Indian Soc. Sci.* 1979, 27, 452–456.

286. Freney, J.R.; Spencer, K.; Jones, M.B. On the constancy of the ratio of nitrogen to sulphur in the protein of subterranean clover tops. *Commun. Soil Sci. Plant Anal.* 1977, 8, 241–249.

287. Koronowski, P. Schwefel. In *Handbuch der Pflanzenkrankheiten*. Bd. I; Sorauer, P. Ed.; Verlag Parey: Berlin, Hamburg, 1969; pp. 114–131.

288. Walker, D.R.; Bentley, C.F. Sulphur fractions of legumes as indicators of sulphur deficiency. *Can. J. Soil Sci.* 1961, 41, 164–168.

289. Schnug, E. Physiological functions and environmental relevance of sulphur-containing secondary metabolites. In *Sulfur Nutrition and Sulfur Assimilation in Higher Plants*; De Kok, L.J., Ed.; SPB Academic Publishing: The Hague, 1993; pp. 179–190.

290. Haneklaus, S.; Hoppe, L.; Bahadir, M.; Schnug, E. Sulphur nutrition and alliin concentrations in *Allium* species. In: *Sulfur Metabolism in Higher Plants: Fundamental Molecular, Ecological and Agricultural Aspects*. Cram et al., Eds.; SPB Academic Publishing: The Hague, 1997; pp. 331–334.

291. Hurley, R.G.; White, E.W. New soft X-ray method for determining the chemical forms of sulfur in coal. *Anal. Chem.* 1974, 46, 2234–2237.

292. Pinkerton, A.; Randall, P.J.; Norrish, K. Estimation of sulfate and amino acid sulfur in plant material by x-ray spectrometry. *Commun. Soil Sci. Plant Anal.* 1989, 20, 1557–1574.

293. Blake-Kalff, M.M.A.; Hawkesford, M.J.; Zhao, F.J.; McGrath, S.P. Diagnosing sulfur deficiency in field-grown oilseed rape (*Brassica napus* L.) and wheat (*Triticum aestivum* L.). *Plant Soil* 2000, 95–107.

294. Reuter, D.J.; Robinson, J.B. *Plant Analysis—An Interpretation Manual*. CSIRO Publishing: Collingwood, Australia, 1997.

295. Smith, F.W.; Loneragan, J.F. Interpretation of plant analysis: concepts and principles. In *Plant Analysis—An Interpretation Manual*; Reuter, D.J., Robinson, J.B., Eds.; CSIRO Publishing: Collingwood, Australia, 1997; pp. 3–26.

296. Mills, H.A.; Benton Jones, J. Jr. *Plant Analysis Handbook II*. Micro Macro Publ.: Athens, Georgia, 1997.

297. Vielemeyer, H.-P.; Neubert, P.; Hundt, I.; Vanselow, G.; Weissert, P. Ein neues Verfahren zur Ableitung von Pflanzenanalyse-Grenzwerten fuer die Einschaetzung des Ernaehrungszustandes land-wirtschaftlicher Kulturpflanzen. Arch. Acker- u. *Pflanzenbau u. Bodenkde* 1983, 27, 445–453.

298. Hudson, D.J. Fitting segmented curves whose join points have to be estimated. *J. Am. Stat. Assoc.* 1966, 61, 1097–1129.

299. Timmer, V.R.; Armstrong, G. Diagnosing nutritional status of containerized tree seedlings: comparative plant analyses. *Soil Sci. Soc. Am. J.* 1987, 51, 1082–1086.

300. Cate, B.R. Jr.; Nelson, L.A. A Rapid Method for Correlation of Soil Test Analyses with Plant Response Data. North Carolina State Agricultural Experimental Station Bulletin 1, Int. Soil Testing Series, 1965.

301. Cate, R.B.; Nelson, L.A. A simple statistical procedure for partitioning soil test correlation data into two classes. *Soil Sci. Soc. Am. Proc.* 1971, 35, 658–660.

302. Baier, J. Computer program for foliar fertilization. *Proceedings of the IAOPN Symposium*, Cairo, Egypt, 1995, pp. 23—28.

303. Wallace, A. Crop improvement through multidisciplinary approaches to different types of stresses-law of the maximum. *J. Plant Nutr.* 1990, 13, 313–325.

304. Schnug, E.; Heym, J.; Achwan, F. Establishing critical values for soil and plant analysis by means of the boundary line development system (BOLIDES). *Commun. Soil Sci. Plant Anal.* 1996, 27, 2739–2748.

305. Evanylo, G.K.; Sumner, M.E. Utilization of the boundary line approach in the development of soil nutrient norms for soybean production. *Commun. Soil Sci. Plant Anal.* 1987, 18, 1355–1377.

306. Walworth, J.L.; Letzsch, W.S.; Sumner, M.E. Use of boundary lines in establishing diagnostic norms. *Soil Sci. Soc. Am. J.* 1986, 50, 123–128.

307. Janssen, B.H.; Guiking, F.C.T.; Eijk, D.v.d.; Smaling, E.M.A.; Wolf, J.; Reuler, H.v. A system for quantitative evaluation of the fertility of tropical soils QUEFTS. *Geoderma* 1990, 46, 299–318.

308. Fotyma, E.; Fotyma, M. The agronomical and physiological efficiency of nitrogen applied for arable crops in Poland. In *Fertilizers and Environment*; Rodriguez-Barrueco, C., Ed.; Kluwer Academic Publishers: Dordrecht, 1996; pp. 27–30.

309. Schnug, E.; Haneklaus, S. PIPPA: un programme d'interprétation des analyses de plantes pour le colza et les céréales. *Supplément de Perspectives Agricoles* 1992, 171, 30–33.

310. Haneklaus, S.; Walker, K.C.; Schnug, E. A chronicle of sulfur research in agriculture. In *Physiological, Molecular Biochemical, Ecological, Environmental, Agricultural, Nutritional, Nutra-Pharmaceutical Aspects*; De Kok, L.J., Grill, D., Hawkesford, M.J., Rennenberg, H., Saito, K., Schnug, E., Stulen, I.; Eds.; Backhuys Publishers: Leiden, 2005; pp. 249–256.

311. Haneklaus, S.; Bloem, E.; Schnug, E. The global sulphur cycle and its links to plant environment. In *Sulphur in Plants*; Abrol, Y.P., Ahmad, A., Eds.; Kluwer Academic Publishers: Dordrecht, 2003; pp. 1–28.

312. Schnug, E.; Ceynowa, J. Crop protection problems for double low rape associated with decreased disease resistance and increased pest damage. *Proceedings of the Conference on Crop Protection in Northern Britain*, Dundee, 1990, pp. 275–282.

313. Luong, H.; Booth, E.J.; Walker, K.C. Utilisation of sulphur nutrition to induce the natural defence mechanisms of oilseed rape. Agriculture Group Symposium; Novel aspects of crop nutrition. *J. Sci. Food Agric.* 1993, 63, 119–120.

314. Wang, J.; Zhang, J.; Ma, Y.; Wang Li Yang, L.; Shi, S.; Liu, L.; Schnug, E. Crop resistance to diseases as influenced by sulphur application rates. *Proceedings of the 12th World Fertilizer Congress*, August 3–9, 2001, Beijing, China, 2003; pp. 1285–1296.

315. Bourbos, V.A.; Skoudridakis, M.T.; Barbopoulou, E.; Venetis, K. Ecological control of grape powdery mildew (*Uncinula necator*). 2000; http://www.landwirtschaft-mlr.baden-wuerttemberg.de/la/lvwo/kongress/SULFUR.html.

316. Klikocka, H.; Haneklaus, S.; Bloem, E.; Schnug, E. Influence of sulfur fertilization on infections of potato tubers (*Solanum tuberosum*) with *Rhizoctonia solani* and *Streptomyces scabies*. *J. Plant Nutr.* 2005, 28, 819–833.

317. Schnug, E.; Booth, E.; Haneklaus, S.; Walker, K.C. Sulphur supply and stress resistance in oilseed rape. *Proceedings of the 9th International Rapeseed Congress*, Cambridge UK, 1995, Vol. 1, pp. 229–231,

318. Bloem, E.; Haneklaus, S.; Schnug, E. Significance of sulfur compounds in the protection of plants against pests and diseases. *J. Plant Nutr.* 2005, 28, 763–784.

319. Agrawal, A.A.; Tuzun, S.; Bent, E. Editors' note on terminology. Page IX. In *Induced Plant Defenses Against Pathogens and Herbivores*; Agrawal A.A., Tuzun S., Bent, E. Eds.; APS Press: St. Paul, MN, 2000.

320. Giovanelli, J.; Mudd, S.H.; Datko, A.H. Sulfur amino acids in plants. In *The Biochemistry of Plants*; Vol. 5, Miflin, B.J., Ed.; Academic Press: New York, 1980, pp. 453–505.

321. Tai, C.H.; Cook, P.F. *O*-acetylserine sulfhydrylase. *Adv. Enzymol. Rel. Areas Mol. Biol.* 2000, 74, 185–234.

322. Winner, W.E.; Smith, C.L.; Koch, G.W.; Mooney, H.A.; Bewley, J.D.; Krouse, H.R. Rate of emission of H_2S from plants and patterns of stable sulphur isotope fractionation. *Nature* 1981, 289, 672–673.

323. Sekiya, J.; Schmidt, A.; Wilson, L.G.; Filner, P. Emission of hydrogen sulfide by leaf tissue in response to L-cysteine. *Plant Physiol.* 1982, 70, 430–436.

324. Kindermann, G.; Hüve, K., Slovik, S.; Lux, H.; Rennenberg, H. Emissions of hydrogen sulfide by twigs of coniferes—a comparison of Norway spruce (*Picea abies* (L.) Karst.), Scotch pine (*Pinus sylvestris* L.) and Blue spruce (*Picea pungens* Engelm.). *Plant Soil* 1995, 168–169, 421–423.

325. Spaleny, J. Sulphate transformation to hydrogen sulfide in spruce seedlings. *Plant Soil* 1977, 48, 557–563.

326. Wilson, L.G.; Bressan, R.A.; Filner, P. Light-dependent emission of hydrogen sulfide from plants. *Plant Physiol.* 1978, 61, 184–189.

327. Lakkineni, K.C.; Nair, T.V.R.; Abrol, Y.P. Sulfur and N interaction in relation to H$_2$S emission in some crop species. In *Sulphur Nutrition and Sulphur Assimilation in Higher Plants; Fundamental, Environmental and Agricultural Aspects*; Rennenberg, H., Brunold, C., De Kok, L.J., Stulen, I., Eds.; SPB Academic Publishing: The Hague, 1990; pp. 213–216.

328. Sekiya, J.; Schmidt, A.; Rennenberg, H.; Wilson, L.G.; Filner, P. Hydrogen sulfide emission by cucumber leaves in response to sulfate in light and dark. *Phytochemistry* 1982, 21, 2173–2178.

329. Kanda, K.; Tsuruta, H. Emissions of sulfur gases from various types of terrestrial higher plants. *Soil Sci. Plant Nutr.* 1995, 41, 321–328.

330. Lakkineni, K.C.; Ahmad, A.; Abrol, Y.P. Hydrogen sulphide: emission and utilization by plants. In *Sulphur in Plants*: Abrol, Y.P., Ahmad, A., Eds.; Kluwer Academic Publishers: Dordrecht, 2003; pp. 265–278.

331. Yang, Z.; Haneklaus, S.; De Kok, L.J.; Singh, B.R.; Schnug, E. Effect of H$_2$S and DMS on growth and enzymatic activities of *Rhizoctonia solani* and its implications for Sulfur-Induced Resistance (SIR) of agricultural crops. *phyton* 2006, (in press).

332. Vidhyasekaran, P. *Physiology of Disease Resistance in Plants*. Vol. II. CRC Press: Boca Raton, USA, 2000.

333. Bosma, W.; Kamminga, G.; De Kok, L.J. H$_2$S-induced accumulation of sulfhydryl-compounds in leaves of plants under field and laboratory exposure. In *Sulfur Nutrition and Sulfur Assimilation in Higher Plants*; Rennenberg, H., Brunold, C., De Kok, L.J., Stulen, I., Eds.; SPB Academic Publishing: The Hague, 1990; pp. 173–175.

334. Vanacker, H.; Carver, T.L.W.; Foyer, C. Early H$_2$O$_2$ accumlation in mesophyll cells leads to induction of glutathione during the hypersensitive response in the barley-powdery mildew interaction. *Plant Physiol.* 2000, 123, 1289–1300.

335. Sinha, A.K. Possible role of phytoalexin inducer chemicals in plant disease control. In *Handbook of Phytoalexin Metabolism and Action*; Daniel, M., Purkayastha, R.P., Eds.; Marcel Dekker Inc.: New York, 1995; pp. 555–592.

336. Kuc, J. Phytoalexins, stress metabolism and disease resistance in plants. *Ann. Rev. Phytopath.* 1995, 33, 275–297.

337. Paulsen, H.M. Produktionstechnische und oekologische Bewertung der landwirtschaftlichen Verwertung von Schwefel aus industriellen Prozessen. *FAL–Agric. Res.* 1998, 197 (special issue).

338. Schnug, E.; Paulsen, H.-M.; Untiedt, H.; Haneklaus, S. Fate and physiology of foliar applied sulphur compounds in *Brassica napus. Proceedings of the IAOPN Symposium*, Cairo, Egypt; 1995, pp. 91–100.

339. Fox, R.L.; Atesalp, H.M.L.; Kampbell, D.H.; Rhoades, H.F. Factors influencing the availlability of sulfur fertilizers to alfalfa and corn. *Soil. Sci. Soc. Am. Proc.* 1964, 28, 406–408.

340. Kolar, J. Verwertungsmoeglichkeiten für Reststoffe der Spruehabsorptionsverfahren. *VGB Kraftwerkstechnik* 1995, 75, Vol. 2, 167–173.

341. Anon VGB-TW 702, Verwertungskonzept für Reststoffe aus Kohlekraftwerken in der Bundesrepublik Deutschland. Teil 2 Rückstände aus der Verbrennung - Aschen - 1992.

342. Mortensen, J.; Nielsen, J.D. Use of a sulfite containing desulfurization product as sulfur fertilizer. *Z. Pflanzenernährung Bodenkunde* 1995, 158, 117–119.

343. Bertelsen, F.; Gissel-Nielsen, G. Toxicity of root applied sulphite in *Zea mays. Environ. Geochem. Health* 1987, 9, 12–16.

344. Ritchey, K.D.; Kinraide, T.B.; Wendell, R.R.; Clark, R.B.; Baligar, V.C. Strategies for overcoming temporary phytotoxic effects of calcium sulfite applied to agricultural soils. *Proceedings of the 11th Annual International Pittsburgh Coal Conference*, Pittsburgh, PA; Chiang, S.H., Ed.; 1994; pp. 457–462.

345. Anon Oversigt over Landsforsøgene. Forsog og undersøgelser i de landøkonomiske foreniger. Samlet og udarbejdet af Landsudvalget for Planteavl; Pedersen, C.A., Ed;. Arhus, 1993.

346. Anon, Oversigt over Landsforsøgene. Forsog og undersøgelser i de landøkonomiske foreniger. Samlet og udarbejdet af Landsudvalget for Planteavl; Pedersen, C.A., Ed.; Arhus, 1994.

347. Schnug, E.; Haneklaus, S.; Murphy, D. Impact of sulphur fertilisation on fertiliser nitrogen efficiency. *Sulphur Agric.* 1993, 17, 8–12.

348. Srivastava, H.S. Regulation of nitrate reductase activity in higher plants. *Phytochemistry* 1980, 19, 725–733.

349. Shanmugam, K.S. Sulphur nutrition of sugarcane. *Fert. News* 1995, 40, 23–26.

350. Zhao, F.J.; McGrath, S.P.; Blake-Kalff, M.A.; Link, A.; Tucker, M. Crop responses to sulphur fertilisation in Europe. *Proceedings of the International Fertilizer Society*, 2002, p. 504.

351. Murphy, M.D.; O'Donnell, T. Sulphur deficiency in herbage in Ireland. 2. Sulphur fertilisation and its effect on yield and quality of herbage. *Irish J. Agric. Res.* 1989, 28, 79–90.

352. Thomas, S.G.; Hocking, T.J.; Bilsborrow, P.E. Effects of sulphur fertilisation on the growth and metabolism of sugar beet grown on soils of differing sulphur status. *Field Crops Res.* 2002, 83, 223–235.

353. Li, S.; Lin, B.; Zhou, W. Crop response to sulfur fertilizers and soil sulfur status in some provinces of China. *FAL–Agric. Res.* 2005, 283, 81–84.

354. Singh, B.R. Sulphur requirement for crop production in Norway. *Norwegian J. Agric. Sci.* (Suppl.) 1994, 15, 35–44.

355. Katyal, J.C.; Sharma, K.L.; Srinivas, K. Sulphur in Indian agriculture. *Proceedings of the TSI/FAI/IFA Symposium on Sulphur in Balanced Fertilisation*, KS-2/1-KS-2/12, 1997.

356. Jain, G.L.; Sahu, M.P.; Somani, L.L. Balanced fertilization programme with special reference to secondary and micronutrients nutrition of crops under intensive cropping, *Proceedings of the FAI/NR Seminar*, Jaipur, 1984, pp. 147–174.

357. Aulakh, M.S.; Pasricha, N.S. Sulphur fertilization of oilseeds for yield and quality. *Sulphur in Indian Agriculture* 1988, SII/3-1-SII/3-14.

358. Aulakh, M.S.; Sidhu, B.S.; Arona, B.R.; Singh, B. Content and uptake of nutrients by pulses and oilseed crops. *Indian J. Ecol.* 1985, 12, 238–242.

359. Survase, D.N.; Dongale, J.H.; Kadrekar, S.B. Growth, yield, quality and composition of groundnut as influenced by F.Y.M., calcium, sulphur and boron in lateritic soil. *J. Maharashtra Agric. Univ.* 1986, 11, 49–51.

360. Naphade, P.S.; Wankhade, S.G. Effect of varying levels of sulphur and molybdenum on the content and uptake of nutrients and yield of mung *(Phaseolus* aureus L.). *PKV J. Res.* 1987, 11, 139–143.

361. Polaria, J.V.; Patel, M.S. Effect of principal and inadvertently applied nutrients through different fertilizer carriers on the yield and nutrient uptake by groundnut. *Gujarat Agric. Univ. Res. J.* 1991, 16, 10–15.

362. Nambiar, K.K.M.; Ghosh, A.B., Highlights of Research of a Long-Term Fertilizer Experiment in India (1971–82). Technical Bulletin No. 1, Longterm Fertilizer Experiment Project, 1984, IARI, p. 100

363. Saarela, I.; Hahtonen, M. Sulphur nutrition of field crops in Finland. *Norwegian J. Agric. Sci.* (Suppl.) 1994, 15, 119–126.

364. Aulakh, M.S. Crop responses to sulphur nutrition. In *Sulphur in Plants*; Abrol, Y.P., Ahmad, A., Eds.; Kluwer Academic Publishers: Dordrecht, 2003; pp. 341–358.

365. Walker, K.C.; Dawson, C. Sulphur fertiliser recommendations in Europe. *Proc. Int. Fert. Soc.* 2002, 506, 0–20.

366. Schroeder, D.; Schnug, E. Application of yield mapping to large scale field experimentation. *Aspects Appl. Biol.* 1995, 43, 117–124.

Section III

Essential Elements—Micronutrients

8 Boron

Umesh C. Gupta
Agriculture and Agri-Food Canada, Charlottetown,
Prince Edward Island, Canada

CONTENTS

8.1 Historical Information 242
    8.1.1 Determination of Essentiality 242
    8.1.2 Functions in Plants 242
        8.1.2.1 Root Elongation and Nucleic Acid Metabolism 243
        8.1.2.2 Protein, Amino Acid, and Nitrate Metabolism 243
        8.1.2.3 Sugar and Starch Metabolism 243
        8.1.2.4 Auxin and Phenol Metabolism...244
        8.1.2.5 Flower Formation and Seed Production..244
        8.1.2.6 Membrane Function 244
8.2 Forms and Sources of Boron in Soils 245
    8.2.1 Total Boron 245
    8.2.2 Available Boron..245
    8.2.3 Fractionation of Soil Boron 245
    8.2.4 Soil Solution Boron...245
    8.2.5 Tourmaline...246
    8.2.6 Hydrated Boron Minerals 246
8.3 Diagnosis of Boron Status in Plants 246
    8.3.1 Deficiency Symptoms 247
        8.3.1.1 Field and Horticultural Crops 247
        8.3.1.2 Other Crops 249
    8.3.2 Toxicity Symptoms 249
        8.3.2.1 Field and Horticultural Crops 249
        8.3.2.2 Other Crops 251
8.4 Boron Concentration in Crops 251
    8.4.1 Plant Part and Growth Stage 251
    8.4.2 Boron Requirement of Some Crops 252
8.5 Boron Levels in Plants 252
8.6 Soil Testing for Boron..257
    8.6.1 Sampling of Soils for Analysis 257
    8.6.2 Extraction of Available Boron...257
        8.6.2.1 Hot-Water-Extractable Boron...257
        8.6.2.2 Boron from Saturated Soil Extracts 258
        8.6.2.3 Other Soil Chemical Extractants 258
    8.6.3 Determination of Extracted Boron 259
        8.6.3.1 Colorimetric Methods 259
        8.6.3.2 Spectrometric Methods 259

8.7   Factors Affecting Plant Accumulation of Boron..260
         8.7.1   Soil Factors 260
                  8.7.1.1   Soil Acidity, Calcium, and Magnesium 260
                  8.7.1.2   Macronutrients, Sulfur, and Zinc 261
                  8.7.1.3   Soil Texture 263
                  8.7.1.4   Soil Organic Matter 263
                  8.7.1.5   Soil Adsorption...263
                  8.7.1.6   Soil Salinity 263
         8.7.2   Other Factors...264
                  8.7.2.1   Plant Genotypes 264
                  8.7.2.2   Environmental Factors 264
                  8.7.2.3   Method of Cultivation and Cropping...265
                  8.7.2.4   Irrigation Water 265
8.8   Fertilizers for Boron 266
         8.8.1   Types of Fertilizers 266
         8.8.2   Methods and Rates of Application 266
References 268

8.1   HISTORICAL INFORMATION

8.1.1   DETERMINATION OF ESSENTIALITY

Boron (B) is one of the eight essential micronutrients, also called trace elements, required for the normal growth of most plants. It is the only nonmetal among the plant micronutrients. Boron was first recognized as an essential element for plants early in the twentieth century. The essentiality of boron as it affected the growth of maize or corn (*Zea mays* L.) plants was first mentioned by Maze (1) in France. However, it was the work of Warington (2) in England that secured strong evidence of the essentiality of boron for the broad bean (*Vicia faba* L.), and later Brenchley and Warington (3) extended the study of boron to include several other plant species. The essentiality of boron to higher plants was decisively accepted after the experimental work of Sommer and Lipman (4), Sommer (5), and other investigators who followed them.

Since its discovery as an essential trace element, the importance of boron as an agricultural chemical has grown very rapidly. Its requirement differs markedly within the plant kingdom. It is essential for the normal growth of monocots, dicots, conifers, and ferns, but not for fungi and most algae. Some members of Gramineae, for example, wheat (*Triticum aestivum* L.) and oats (*Avena sativa* L.) have a much lower requirement for boron than do dicots and other monocots, for example, corn.

Of the known micronutrient deficiencies, boron deficiency in crops is most widespread. In the last 80 years, hundreds of reports have dealt with the essentiality of boron for a variety of agricultural crops in countries from every continent of the world.

8.1.2   FUNCTIONS IN PLANTS

Deficiency of boron can cause reductions in crop yields, impair crop quality, or have both effects. Some of the most severe disorders caused by a lack of boron include brown-heart (also called water core or raan) in rutabaga (*Brassica napobrassica* Mill.) and radish (*Raphanus sativus* L.) roots, cracked stems of celery (*Apium graveolens* L.), heart rot of beets (*Beta vulgaris* L.) brown-heart of cauliflower (*Brassica oleracea* var. *botrytis* L.), and internal brown spots of sweet potato (*Ipomoea batatas* Lam.). Some boron deficiency disorders appear to be physiological in nature and occur even when boron is in ample supply. These disorders are thought to be related to peculiarities in boron transport and distribution. The initial processes that control boron uptake in plants are located in the roots (6). Some of the main functions of boron are summarized below.

8.1.2.1 Root Elongation and Nucleic Acid Metabolism

Boron deficiency rapidly inhibits the elongation and growth of roots. For example, Bohnsack and Albert (7) showed that root elongation of squash (*Cucurbita pepo* L.) seedlings declined within 3 h after the boron supply was removed and stopped within 24 h. If boron was resupplied after 12 h, the rate of root elongation was restored to normal within 12 to 18 h. Josten and Kutschera (8) reported that the presence of boron resulted in the development of numerous roots in the lower part of the hypocotyl in sunflower (*Helianthus annuus* L.) cuttings. Consequently, the numerous adventitious roots entirely replaced the tap root system of the intact seedlings.

Root elongation is the result of cell elongation and cell division, and evidence suggests that boron is required for both processes (9). When boron is withheld for several days, nucleic acid content decreases. Krueger et al. (10) demonstrated that the decline and eventual cessation of root elongation in squash seedlings was correlated temporally with a decrease in DNA synthesis, but preceded changes in protein synthesis and respiration.

Lenoble et al. (11) concluded that boron additions may need to be increased under acid, high-aluminum soils, because applications of boron prevented aluminum inhibition of root growth on acid, aluminum-toxic soils.

8.1.2.2 Protein, Amino Acid, and Nitrate Metabolism

Protein and soluble nitrogenous compounds are decreased in boron-deficient plants (12). However, the influence of organ age, i.e., whether the organ was actively involved in the biosynthesis of amino acids and protein or remobilization of amino acids from protein reserves, has often been ignored (13). For example, Dave and Kannan (14) reported that 5 days of growth without boron increased the protein concentration of bean (*Phaseolus vulgaris* L.) cotyledons compared to control seedlings, suggesting that nitrogen remobilization is hindered due to boron deficiency. By contrast, protein concentrations in the actively growing regions could be reduced by lower rates of synthesis caused by boron deficiency (15,16).

Shelp (16) reported that the partitioning of nitrogen into soluble components (nitrate, ammonium, and amino acids) of broccoli (*Brassica oleracea* var. *botrytis* L.) was dependent on the plant organ and whether boron was supplied continuously at deficient or toxic levels. Boron deficiency did not substantially affect the relative amino acid composition (16) but did enhance the proportion of inorganic nitrogen, particularly nitrate, in plant tissues and translocation fluids (13). A number of researchers reported increases in nitrate concentration as well as corresponding decreases in nitrate reductase activity in sugar beet (*Beta vulgaris* L.), tomato (*Lycopersicon esculentum* Mill.), sunflower, and corn plants (17,18) due to boron deficiency. Boron deficiency in tobacco (*Nicotiana tabacum* L.) resulted in a decrease in leaf N concentration and reduced nitrate reductase activity (19). Boron-deficient soybeans (*Glycine max* Merr.) showed low acetylene reduction activities and damage to the root nodules (20).

8.1.2.3 Sugar and Starch Metabolism

Boron is thought to have a direct effect on sugar synthesis. In cowpeas (*Vigna unguiculata* Walp), acute boron deficiency conditions increased reducing and nonreducing sugar concentrations but decreased starch phosphorylase activity (21). Under boron deficiency, the pentose phosphate shunt comes into operation to produce phenolic substances (22). Boron-deficient sunflower seeds showed marked decrease in nonreducing sugars and starch concentrations, whereas the reducing sugars accumulated in the leaves (23). This finding indicates a specific role of boron in the production and deposition of reserves in sunflower seeds. High concentrations of nonreducing sugars were also found in boron-deficient mustard (*Brassica nigra* Koch) (24). Camacho and Gonzalas (19) also found higher starch concentration in boron-deficient tobacco plants. In low-boron sunflower leaves, starch decreased, but there was an increase in sugars and protein and nonprotein nitrogen fractions (25). In boron-deficient pea (*Pisum sativum* L.) leaves, the concentration of sugars and starch increased, but they decreased in the pea seeds and thus lowered the seed quality (26). Evidence on the impact of boron deficiency on starch concentration is conflicting. It is difficult to explain whether the differences are due to a variation in crop species.

8.1.2.4 Auxin and Phenol Metabolism

Boron regulates auxin supply in plants by protecting the indole acetic acid (IAA) oxidase system through complexation of *o*-diphenol inhibitors of IAA oxidase. Excessive auxin activity causes excessive proliferation of cambial cells, rapid and disproportionate enlargement of cells, and collapse of nearby cells (27). It has been established that adventitious roots develop on stem cuttings of bean only when boron is supplied (28,29). Auxin initiates the regeneration of roots, but boron must be supplied at relatively high concentrations 40 to 48 h after cuttings are taken, for primordial roots to develop and grow. It was initially proposed that boron acted by reducing auxin to concentrations that were not inhibitory to root growth (30,31), but more recently, Ali and Jarvis (28) reported that without boron, RNA synthesis decreases markedly within and outside the region from which roots ultimately develop.

There are many reports in the literature of phenol accumulation under long-term boron deficiency (32). Since boron complexes with phenolic compounds such as caffeic acid and hydroxyferulic acid, Lewis (33) proposed a role for boron in lignification. Absence of boron would therefore cause reactive intermediates of lignin biosynthesis and other phenolic compounds to affect changes in metabolism and membrane function, resulting in cell damage. However, the available evidence indicates that lignin synthesis may actually be enhanced by boron deficiency.

8.1.2.5 Flower Formation and Seed Production

The role of boron in seed production is so important that under moderate to severe boron deficiency, plants fail to produce functional flowers and may produce no seeds (34). Plants subjected to boron deficiency have been observed to result in sterility or low germination of pollen in alfalfa (*Medicago sativa* L.) (35), barley (*Hordeum vulgare* L.) (36), and corn (37). Even under moderate boron deficiency, plants may grow normally and the yield of the foliage may not be affected severely, but the seed yield may be suppressed drastically (38).

8.1.2.6 Membrane Function

Impairment of membrane function could affect the transport of all metabolites required for normal growth and development, as well as the activities of membrane-bound enzymes. Dugger (15) summarized early reports that illustrate changes in membrane structure and organization in response to boron deficiency. Boron may give stability to cellular membranes by reacting with hydroxyl-rich compounds. Consistent with this view is evidence suggesting that a major portion of the cellular boron is concentrated in protoplast membranes from mung bean (*Phaseolus aureus* Roxb.) (39).

The involvement of boron in inorganic ion flux by root tissue (40–42) and in the incorporation of phosphate into organic phosphate (43) was evident from earlier research. In general, the absorption of phosphate, rubidium, sulfate, and chloride was suppressed in boron-deficient root tissues, but it could be restored to normal or nearly normal rates by a concomitant addition of boron or pretreatment with boron for 1 h. This effect could be explained by a rapid reorganization of the carrier system, with boron functioning as an essential component of the membrane (15). The movement of monovalent cations is associated with membrane-bound ATPases. Boron-deficient corn roots had a limited ATPase activity, which could be restored by boron addition for only 1 h before enzyme extraction (40).

Recently, Tang and Dela Fuente (44,45) demonstrated that potassium leakage (as a measure of membrane integrity) from boron- or calcium-deficient sunflower hypocotyl segments was completely reversed by the addition of boron or calcium for 3 h. It was not possible to reverse the inhibited process by replacing one deficient element with the other. Seedlings deficient in both boron and calcium showed greater effects than seedlings deficient in one element only. Basipetal auxin transport was also inhibited by boron or calcium deficiency, but the addition of boron for 2 h did not restore the process reduced by boron deficiency. This reduction in auxin transport was not related to reduced growth rate, acropetal auxin transport, lack of respiratory substrates, or changes in calcium absorption, suggesting that boron had a direct effect on auxin transport.

8.2 FORMS AND SOURCES OF BORON IN SOILS

8.2.1 TOTAL BORON

The total boron content of most agricultural soils ranges from 1 to 467 mg kg^{-1}, with an average content of 9 to 85 mg kg^{-1}. Gupta (46) reported that total boron on Podzol soils from eastern Canada ranged from 45 to 124 mg kg^{-1}. Total boron in major soil orders, Inceptisol and Alfisol, in India ranged from 8 to 18 mg kg^{-1} (47). Such wide variations among soils in the total boron content are mainly ascribed to the parent rock types and soil types falling under divergent geographical and climatic zones. Boron is generally high in soils derived from marine sediments.

8.2.2 AVAILABLE BORON

Available boron, measured by various extraction methods (see Section 8.6.2), in agricultural soils varies from 0.5 to 5 mg kg^{-1}. Most of the available boron in soil is believed to be derived from sediments and plant material. Gupta (46) reported that available boron on Podzol soils from eastern Canada ranged from 0.38 to 4.67 mg kg^{-1}. Few studies have been conducted that attempt to identify solid-phase controls on boron solubility in soils. Most of the common boron minerals are much too soluble for such purposes (48).

8.2.3 FRACTIONATION OF SOIL BORON

Boron fractionation was studied in relation to its availability to corn in 14 soils (49). Up to 0.34% of the total boron was in a water-soluble form, 0 to 0.23% was nonspecifically adsorbed (exchangeable), and 0.05 to 0.30% was specifically adsorbed. Jin et al. (49) reported that most of the boron available to corn was in these three forms, and that boron in noncrystalline and crystalline aluminum and iron oxyhydroxides and in silicates was relatively unavailable for plant uptake. For the identification of different pools of boron in soils, Hou et al. (50) proposed a fractionation scheme, which indicated that readily soluble and specifically adsorbed boron accounted for <2% of the total boron. Various oxides–hydroxides, and organically bound forms constituted 2.3 and 8.6%, respectively. Most soil boron existed in residual or occluded form. Recent studies by Zerrari et al. (51) showed that the residual boron constituted the most important fraction at 78.75%.

8.2.4 SOIL SOLUTION BORON

In soil solution, boron mainly exists as undissociated acid H_3BO_3. Boric acid (also written as $B(OH)_3$) and $H_2BO_3^-$ are the most common geologic forms of boron, with boric acid being the predominant form in soils as reviewed by Evans and Sparks (52). They further reported that boric acid is the major form of boron in soils with $H_2BO_3^-$ being predominant only above pH 9.2. In their review, they stated that boron occurs in aqueous solution as boric acid $B(OH)_3$, which is a weak monobasic acid that acts as an electron acceptor or as a Lewis acid.

8.2.5 TOURMALINE

In most of the well-drained soils formed from acid rocks and metamorphic sediments, tourmaline is the most common boron-containing mineral identified (53). The name tourmaline represents a group of minerals that are compositionally complex borosilicates containing approximately 3% B. The tourmaline structure has rhombohedral symmetry and consists of linked sheets of island units. The boron atoms are found within BO_3 triangles, forming strong covalent B–O bonds (54). Tourmalines are highly resistant to weathering and virtually insoluble. Additions of finely ground tourmaline to soil failed to provide sufficient boron to alleviate boron deficiency of crop plants (55).

8.2.6 HYDRATED BORON MINERALS

Industrial deposits of boron are usually produced by chemical precipitation. Precipitation occurs following concentration on land, in brine waters in arid regions or as terrestrial evaporites and arid playa deposits (56). Precipitation also occurs as marine evaporites after concentration due to evaporation of seawater. Borates also form in salt domes and by further concentration of underground water in arid areas (56). The borate deposits of economic importance are restricted to arid areas because of the high solubility of these minerals.

Hydrated borates are formed originally as chemical deposits in saline lakes (57). The particular mineral suite formed is dependent on the chemical composition of the lake. Two kinds of borate deposits are formed in the arid western United States (57). Hydrated sodium borates form from lakes that have a high pH and that are high in sodium and low in calcium content. Hydrated sodium–calcium borates form from lakes of higher calcium content.

8.3 DIAGNOSIS OF BORON STATUS IN PLANTS

Boron deficiency in crops is more widespread than deficiency of any other micronutrient. This phenomenon is the chief reason why numerous reports are available on boron deficiency symptoms in plants. Because of its immobility in plants, boron deficiency symptoms generally appear first on the younger leaves at the top of the plants. This occurrence is also true of the other micronutrients except molybdenum, which is readily translocated.

Boron toxicity symptoms are similar for most plants. Generally, they consist of marginal and tip chlorosis, which is quickly followed by necrosis (58). As far as boron toxicity is concerned, it occurs chiefly under two conditions, owing to its presence in irrigation water or owing to accidental applications of too much boron in treating boron deficiency. Large additions of materials high in boron, for example, compost, can also result in boron toxicity in crops (59,60). Boron toxicity in arid and semiarid regions is frequently associated with saline soils, but most often it results from the use of high-boron irrigation waters. In the United States, the main areas of high-boron waters are along the west side of the San Joaquin and Sacramento valleys in California (61).

Boron does not accumulate uniformly in leaves, but typically concentrates in leaf tips of monocotyledons and leaf margins of dicotyledons, where boron toxicity symptoms first appear. In fact although leaf tips may represent only a small proportion of the shoot dry matter, they can contain sufficient boron to substantially influence total leaf and shoot boron concentrations. To overcome this problem, Nable et al. (62) recommended the use of grain in barley for monitoring toxic levels of boron accumulation. The main difficulty in using cereal grain for determining boron levels is the small differences in the grain boron concentration as obtained in response to boron fertilization (63). Low risk of boron toxicity to rice in an oilseed rape (*Brassica napus* L.)–rice (*Oryza sativa* L.) rotation was attributed to the relatively high boron removal in harvested seed, grain, and stubble, and the loss of fertilizer boron to leaching (64). Boron toxicity symptoms in zinc-deficient citrus (*Citrus aurantium* L.) could be mitigated with zinc applications. This finding is of practical importance as boron toxicity and zinc deficiencies are simultaneously encountered in some soils of semiarid zones.

8.3.1 Deficiency Symptoms

8.3.1.1 Field and Horticultural Crops

Alfalfa (*Medicago sativa* L.). Symptoms are more severe at the leaf tips, although the lower leaves remain a healthy green color. Flowers fail to form, and buds appear as white or light-brown tissue (65). Internodes are short; blossoms drop or do not form, and stems are short (66). Younger leaves turn red or yellow (67,68), and topyellowing of alfalfa occurs (69) (Figure 8.1).

Barley (*Hordeum vulgare* L.). No ears are formed (70). Flowers were opened by the swelling of ovaries caused by partial sterility due to B deficiency (36). Boron deficiency was also associated with the appearance of ergot.

Beet (*Beta vulgaris* L.). Boron deficiency results in a characteristic corky upper surface of the leaf petiole (69). Beet roots are rough, scabby (similar to potato scab) and off-color (71).

Broccoli (*Brassica oleracea* var. *botrytis* L.). Water-soaked areas occur inside the heads, and callus formation is slower on the cut end of the stems after the heads have been harvested (72). Symptoms of boron deficiency included leaf midrib cracking, stem corkiness, necrotic lesions, and hollowing in the stem pith (73).

Brussels sprouts (*Brassica oleracea* var. *gemmifera* Zenker). The first signs of boron deficiency are swellings on the stem and petioles, which later become suberised. The leaves are curled and rolled, and premature leaf fall of the older leaves may take place (58). The sprouts themselves are very loose instead of being hard and compact, and there is vertical cracking of the stem (74).

Carrot (*Daucus carota* L.). Boron deficiency results in longitudinal splitting of roots (75). Boron-deficient carrot roots are rough, small with a distinct white core in the center and plants show a browning of the tops (71).

Cauliflower (*Brassica oleracea* var. *botrytis* L.). The chief symptoms are the tardy production of small heads, which display brown, waterlogged patches, the vertical cracking of the stems, and rotting of the core (74) (Figure 8.2). When browning is severe, the outer and the inner portions of the head have a bitter flavor (76). Stems are stiff, with hollow cores, and curd formation is delayed (77). The roots are rough and dwarfed; lesions appear in the pith, and a loose curd is produced (69).

Clover (*Trifolium* spp.). Plants are weak, with thick stems that are swollen close to the growing point, and leaf margins often look burnt (78). Symptoms of boron deficiency in red and alsike clover may occur as a red coloration on the margins and tips of younger leaves; the coloration gradually spreads over the leaves, and the leaf tips may die (65).

FIGURE 8.1 Symptoms of boron deficiency in alfalfa (*Medicago sativa* L.) showing red and yellow color development on young leaves. (Photograph by Umesh C. Gupta.) (For a color presentation of this figure, see the accompanying compact disc.)

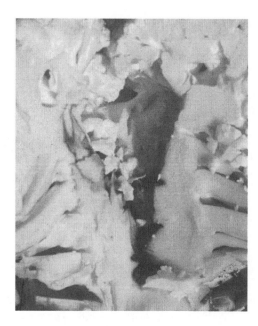

FIGURE 8.2 Symptoms of boron deficiency in cauliflower (*Brassica oleracea* var. *botrytis* L.) showing brown, waterlogged patches, and rotting of the core of the head. (Photograph by Umesh C. Gupta.) (For a color presentation of this figure, see the accompanying compact disc.)

Corn (*Zea mays* L.). Boron deficiency is seen on the youngest leaves as white, irregularly shaped spots scattered between the veins. With severe deficiency these spots may coalesce, forming white stripes 2.5 to 5.0 cm long. These stripes appear to be waxy and raised from the leaf surface (79). Interruption in the boron supply, from 1 week prior to tasselling until maturity, curtailed the normal development of the corn ear (80).

Oat (*Avena sativa* L.). Pollen grains are empty (70).

Peanuts or **groundnut** (*Arachis hypogaea* L.). Boron deficiency resulted in hollow-heart in peanut kernels at a few locations in Thailand (81).

Pea (*Pisum sativum* L.). Leaves develop yellow or white veins followed by some changes in interveinal areas; growing points die and blossoms shed (82). Unpublished data of Gupta and MacLeod (83) showed that boron deficiency in peas resulted in short internodes and small, shrivelled new leaves.

Potato (*Solanum tuberosum* L.). Deficiency results in the death of growing points, with short internodes giving the plant a bushy appearance. Leaves thicken and margins roll upward, a symptom similar to that of potato leaf roll virus (84). Boron deficiency resulted in rosetting of terminal buds and shoots, and the new leaves were malformed and chlorotic (85).

Radish (*Raphanus sativus* L.). Deficiency of boron in radish is also known as brown-heart, manifested first by dark spots on the roots, usually on the thickest parts (76). Roots upon cutting show brown coloration and have thick periderm (71).

Rutabaga (*Brassica napobrassica* Mill.). The boron deficiency disorder in rutabaga is generally referred to as brown-heart. Upon cutting, the roots show a soft, watery area (Figure 8.3). Under severe boron deficiency the surface of the roots is rough and netted, and often the roots are elongated (86). The roots are tough, fibrous, and bitter, and have a corky and somewhat leathery skin (58).

Snapbean (*Phaseolus vulgaris* L.). There is a yellowing of tops, slow flowering and pod formation (71).

Soybean (*Glycine max* Merr.). Boron deficiency results in necrosis of the apical growing point and young growth; the lamina is thick and brittle; and floral buds wither before opening (87). Boron

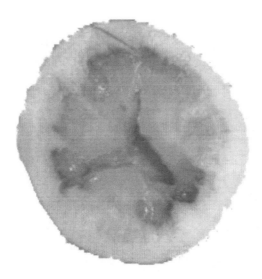

FIGURE 8.3 Symptoms of boron deficiency in rutabaga (*Brassica napobrassica* Mill.) showing a soft, watery area of a cut root. (Photograph by Umesh C. Gupta.) (For a color presentation of this figure, see the accompanying compact disc.)

deficiency induced a localized depression on the internal surface of one or both cotyledons of some seeds and resembled the symptoms of hollow-heart in groundnut seeds (88).

Sunflower (*Helianthus annuus* L.). There is basal fading and distortion of young leaves with soaked areas and tissue necrosis (25).

Tomato (*Lycopersicon esculentum* L.). The growing point is injured; flower injury occurs during the early stages of blossoming, and fruits are imperfectly filled (72). Failure to set fruit is common, and the fruit may be ridged, show corky patches, and ripen unevenly.

Wheat (*Triticum aestivum* L.). A normal ear forms but fails to flower (70). In the case of severe boron deficiency, the development of the inflorescence and setting of grains are restricted (87).

8.3.1.2   Other Crops

Cotton (*Gossypium hirsutum* L.). Boron deficiency causes retarded internodal growth (89). The terminal bud often dies, checking linear growth, and short internodes and enlarged nodes give a bushy appearance that is referred to as a rosette condition (90). Bolls are deformed and reduced in size. Root growth is severely inhibited, and secondary roots have a stunted appearance (91).

Sugar Beet (*Beta vulgaris* L.). Deficiency results in retarded growth, and young leaves curl and turn black (92). The old leaves show surface cracking, along with cupping and curling. When the growing point fails completely, it forms a heart rot (92).

Tobacco (*Nicotiana tabacum* L.). Boron deficiency results in interveinal chlorosis, dark and brittle newly emerging leaves, water-soaked areas in leaves, and delayed flowering, and formation of seedless pods (93). Tissues at the base of the leaf show signs of breakdown, and the stalk toward the top of the plant may show a distorted or twisted type of growth. The death of the terminal bud follows these stages (94).

8.3.2   Toxicity Symptoms

8.3.2.1   Field and Horticultural Crops

Alfalfa (*Medicago sativa* L.) and **red clover** (*Trifolium pratense* L.). Boron toxicity is marked by burnt edges on the older leaves (67,68) (Figure 8.4).

FIGURE 8.4 Symptoms of boron toxicity in alfalfa (*Medicago sativa* L.) showing scorch at margins of lower leaves. (Photograph by Umesh C. Gupta.) (For a color presentation of this figure, see the accompanying compact disc.)

Barley (*Hordeum vulgare* L.). Boron toxicity is characterized by elongated, dark-brown blotches at the tips of older leaves (79). Severe browning, spotting, and burning of older leaf tips occur, gradually extending to the middle portion of the leaf (59,63). There is a reduced shoot growth and increased leaf senescence (95).

Corn (*Zea mays* L.). Leaves show tip burn and marginal burning and yellowing between the veins (79,96). Burning of older leaf edges is more prominent (71).

Cowpea (*Vigna sinensis* Savi). Moderate boron toxicity results in marginal chlorosis and spotted necrosis, but under severe boron toxicity, trifoliate leaves show a slight marginal chlorosis (97).

Oat (*Avena sativa* L.). Boron toxicity in oats results in light-yellow bleached leaf tips (63).

Onion (*Allium cepa* L.). Boron toxicity results in burning of the tips of leaves, gradually increasing up to the base, and no development of bulb occurs (93).

Pea (*Pisum sativum* L.). Boron toxicity results in suppression of plant height and in the number of nodes (98). Unpublished data of Gupta and MacLeod (83) showed that boron toxicity results in burning of the edges of old leaves.

Potato (*Solanum tuberosum* L.). Boron toxicity symptoms include arching mid-rib and downward cupping of leaves and necrosis at leaf margins (85).

Rutabaga (*Brassica napobrassica* Mill.). The leaf margins are yellow in color and tend to curl and wrinkle. The symptoms on roots are similar to moderate boron deficiency symptoms—a water-soaked appearance of the tissues in the center of the root (99). Boron toxicity in turnip seedlings also results in marginal bleaching of the cotyledons and first leaves (100).

Bean (*Phaseolus vulgaris* L.). Boron toxicity results in marginal chlorosis of the older trifoliate leaves of snapbeans; unifoliate leaves are also chlorotic with intermittent marginal necrosis (97). Growth is suppressed, and old leaves have marginal burning (71). With faba beans (*Vicia faba* L.), stem growth was restricted, and the young leaves were wrinkled, thick, with a dark-blue color (101).

Strawberry (*Fragaria* x *ananassa* Duchesne). Slight boron toxicity was associated with marginal curling and interveinal bronzing and necrotic lesions. Under severe boron toxicity interveinal necrosis was severe, leaf margins became severely distorted and cracked, and overall plant growth was reduced (102).

Wheat (*Triticum aestivum* L.). Boron toxicity in wheat appears as light browning of older leaf tips converging into light greenish-blue spots (63). In durum wheat (*Triticum durum* Desf.), toxicity results in retarded growth, delayed heading, increase in aborted tillers, and suppressed grain yield per tiller (103).

8.3.2.2 Other Crops

Bajri (*Pennisetum typhoideum*). Boron toxicity results in the burning of leaf tips. On the basal leaves, small necrotic areas appear at the margins and proceed slowly toward the top of the plant (93).

Bean (*Phaseolus vulgaris* L.). Excess boron causes mottled and necrotic areas on the leaves, especially along the leaf margins (91). In faba bean (*Vicia faba* L.), symptoms first appeared as yellowing of the mature foliage, followed by a marginal necrosis and finally by the death of the whole plant (101).

Tobacco (*Nicotiana tabacum* L.). Boron toxicity results in brown circular spots on the periphery of the leaves, and stunted growth (93).

8.4 BORON CONCENTRATION IN CROPS

8.4.1 PLANT PART AND GROWTH STAGE

As extractants have not been developed fully to evaluate the availability of boron in soils, plant tissue testing continues to be the preferred means of delineating the boron deficiency and sufficiency levels in plants. It seems, therefore, desirable to sample the plant parts that contain the highest quantity of boron to characterize its status in crops. The use of plant parts containing the higher nutrient values should facilitate better differentiation between the deficiency and sufficiency levels.

The part of the leaf, its position in the plant, the plant age, and the plant part are some of the factors that affect the boron composition of plants. Studies by Vlamis and Ulrich (92) showed that young blades of sugar beets contained more boron than the mature and old blades of plants grown at low concentrations of boron in a nutrient solution. However, at higher boron concentrations in solution, no differences were found. The highest boron values in sugar beets occurred in the older leaves, but the lowest boron content occurred in the fibrous and storage roots (92). The boron concentration of corn leaves increased with age in seedling leaves (104). The uppermost corn leaves had higher concentrations than did leaves at positions below. Boron concentration in corn leaves and tassels of flowering corn plants increased with age, but boron in other plant parts remained low and relatively constant (105). Gorsline et al. (106) noted that boron concentration in the whole corn plant decreased during initial growth, remained unchanged during most of the vegetative period, and then decreased after silking.

Gupta and Cutcliffe (86) reported that boron level in leaf tissue of rutabaga was greater from early samplings than it was from late samplings. Older cucumber (*Cucumis sativus* L.) leaves contained more boron than the younger leaves; and within the leaf, boron accumulated in the marginal parts (107). Boron accumulation was greater in the marginal section of corn leaves than in the midrib section (108). Generally, boron in plants has a tendency to accumulate in the margin of leaves (109,110). Results of Miller and Smith (111) showed that alfalfa leaves had much higher boron content (75 to 98 mg kg^{-1}) than tips (47 mg kg^{-1}) or stems (22 to 27 mg kg^{-1}).

In a field study conducted in Prince Edward Island, Canada, the highest boron concentrations were in leaves and upper halves of plants of most species (Table 8.1). The boron concentrations were lowest in the stems. The lowest boron concentration was in alfalfa and the highest in Brussels sprouts and rutabaga. In a separate experiment, where the effect of not applying boron was studied against applied boron, the trend in boron accumulation in the various plant parts was similar. The boron content of pistils and stamens, although very high, was often lower than in leaves and sometimes of corollas (112).

Gupta (113) found that without added boron, the bottom third of the leaves of alfalfa and red clover contained significantly higher boron than did the upper leaves. In the case of stems the opposite was the case, i.e., the upper third of the stems contained more boron than the bottom third. This trend was similar for the unfertilized and boron-fertilized areas for leaves; however, in

TABLE 8.1

Variations in Boron Concentrations in Various Plant Parts of a Few Crop Species

		Plant Parts			
Crop	Leaves	Upper Stems	Lower Halves	Upper Halves	Means
		Boron Concentration (mg B kg^{-1})			
Alfalfa (*Medicago sativa* L.)	25	14	24	16	21
Broccoli (*Brassica oleracea* L.)	37	21	31	43	34
Brussels sprouts (*Brassica oleracea* var. *gemmifera* Zenker)	57	21	30	51	41
Cauliflower (*Brassica oleracea* var. *botrytis* L.)	36	19	25	39	30
Red Clover (*Trifolium pratense* L.)	23	16	21	18	20
Rutabaga (*Brassica napobrassica* Mill.)	52	24	37	48	41
Means	43	20	30	36	

Note: Standard error for plant parts = 4.0; for crops = 4; and for plant parts × crops = 10.0

Source: Adapted from Gupta U.C., *J. Plant Nutr.* 14:613–621, 1991.

the presence of added boron, differences in the boron content in the upper and lower stems were not significant.

The general theory is that boron translocates readily in the xylem, but once in the leaves, it becomes one of the least mobile of the micronutrients. Thus the boron immobility in leaves in terms of localized cyclic movement prevents escape and transport of this element over long distances (114). The results of Shelp (115) have also shown that younger leaves contain less boron than mature leaves; the authors assumed that the boron supply for mature leaves is delivered principally via the xylem.

The fact that boron deficiency exhibits in the younger leaves and not in the older leaves can be explained by the fact that the boron concentration is higher in the older leaves than in the younger leaves, as reported for alfalfa and red clover (113) and for broccoli (115). Since the boron concentration in the upper leaves was easily increased with boron fertilization (113), boron deficiency is controlled without much difficulty using boron applications.

It is suggested that leaves should be sampled to determine the boron status of the plants. Also, it is important to be consistent with the plant sampling technique in the field as well as the plant part sampled.

8.4.2 BORON REQUIREMENT OF SOME CROPS

Different crops have different requirements for boron; for example, rutabaga needs more boron than wheat. Boron requirement for crops varies considerably, and therefore boron recommendations must take these differences into account. A classification of a number of field and horticultural crops as having high, medium, or low boron requirement is given in Table 8.2.

8.5 BORON LEVELS IN PLANTS

Often when one talks about deficient, sufficient, and toxic levels of nutrients in crops, there is a range in values rather than one definite number that could be considered as critical. Therefore, the term *critical level* in crops is somewhat misleading. A nutrient content value considered critical by

---

TABLE 8.2
Boron Requirement of Some Field and Horticultural Crops

High	Medium	Low
Alfalfa	Asparagus	Barley
Apple	Carrot	Beans
Broccoli	Corn (sweet)	Blueberry
Brussels sprouts	Cotton	Cereals
Cabbage	Cherry	Citrus
Cauliflower	Lettuce	Corn
Celery	Onion	Cucumber
Clovers	Parsnip	Flax
Mustard	Peach	Grasses
Peanuts	Pear	Oat
Rape	Potato (sweet)	Peas
Red beet	Radish	Pepper
Rutabaga	Spinach	Potato (white)
Sugar beet	Tobacco	Raspberry
Sunflower	Tomato	Rye
Turnip		Sorghum
		Strawberry
		Wheat

Note: Based on rates of fertilizer application of boron recommended by state agricultural agencies in the United States, a high requirement is a recommended fertilization exceeding $2\,kg\,B\,ha^{-1}$; a medium requirement is fertilization with 1 to $2\,kg\,B\,ha^{-1}$; and a low requirement is fertilization with $<1\,kg\,B\,ha^{-1}$.

Source: Adapted from Mortvedt J.J. and Woodruff J.R., in *Boron and Its Role in Crop Production*. CRC Press, Boca Raton, FL, 1993, pp. 157–176.

---

workers in one area may not be considered critical in another area. Likewise, the term *optimum level* of a nutrient, as used in the literature by some researchers to express a relationship to maximum crop yield, is sometimes not clear. Theoretically, such a level for a given nutrient should be sufficient to produce the best possible growth of a crop. A range of values would be more appropriate to describe the nutrient status of the crop; therefore, the term sufficiency will be used, rather than critical or optimum.

The *critical level* of a nutrient has been defined as the concentration occurring in a specific plant part at 90% of the maximum yield (117). The concept is equally valid where crop quality is the main concern rather than yield (118). In this respect, rutabaga is an excellent example where deficiency of boron may not affect the mass of roots, but the quality of roots may be seriously impaired.

The ratio of toxic level to adequate level of boron is smaller than that for most other nutrient elements (119). Thus, excessive or deficient levels could be encountered in a crop during a single season. This occurrence emphasizes the fact that a critical value used to indicate the status of boron in crops would be unsuitable. In many cases the values referred to in this section overlap the deficiency and sufficiency ranges.

The deficient, sufficient, and toxic boron levels for specific crops as reported by various workers are given in Table 8.3. The deficient and toxic levels of boron as reported in this table are associated with plant disorders and suppressions of crop yields. For some crops, the deficiency and optimum levels seem to differ markedly. Differences in the techniques used and the locations of the various laboratories cannot be ruled out.

TABLE 8.3
Deficiency, Sufficiency, and Toxicity Levels of Boron in Field and Horticultural Crops

Crop	Plant Part Sampled	mg B kg^{-1} in Dry Matter			Reference
		Deficiency	Sufficiency	Toxicity	
Field Crops					
Alfalfa	Whole tops at early bloom	<15	20–40	200	120
(*Medicago*			15–20[a]		
sativa L.)	Top one third of plant shortly before flowering	<20	31–80	>100	121
	Upper stem cuttings in early flower stage		30[a]		122
	Whole tops in early bud		17–18[a]		123
	Whole tops	<15	15–20	200	124
	Whole tops at 10% bloom	8–12	39–52	>99	67
	Whole tops	<20			125
Barley	Boot-stage tissue	1.9–3.5	10	>20	63
(*Hordeum*	Boot-stage tissue			50–70[a]	95
vulgare L.)	Straw	7.1–8.6	21	>46	63
	Grain			>2–15	126
	Whole shoots at maturity			50–420	126
Corn	Whole plants when 25 cm tall		8–38	>98	71
(*Zea mays* L.)	Leaf at or opposite and below ear level at tassel stage		10[a]		122
	Total aboveground plant material at vegetative stage until ear formation	<9	15–90	>100	121
Oats	47-d-old plants			>105	127
(*Avena*	Boot-stage tissue		15–50	44–400	128
sativa L.)	Boot-stage tissue	<1	8–30	>30[b]	121
	Boot-stage tissue	1.1–3.5	37056	>35	63
	Straw	3.5–5.6	14–24	>50	63
Pasture grass (Gramineae family)	Aboveground part at first bloom at first cut		10–50	>800	121
Peanuts (*Arachis hypogaea* L.)	Shoot terminals		29		129
Peas	Young leaves	10.5	23	110	26
(*Pisum*	Seeds	7.6	10.5	51	26
sativum L.)					
Red Clover	Whole tops at bud stage	12–20	21–45	>59	67, 130
(*Trifolium*	Top one third of plant at bloom		20–60	>60[b]	121
pratense L.)	Whole tops at rapid growth		15–18[a]		123
Rice	Flag leaves	<7.3			131
(*Oryza sativa* L.)	Shoots	<3.6			131
Ryegrass (*Lolium perenne* L.)	Whole plants at rapid growth		9–38	>39–42	132

TABLE 8.3   (*Continued*)

Crop	Plant Part Sampled	mg B kg⁻¹ in Dry Matter			Reference
		Deficiency	Sufficiency	Toxicity	
Sorghum (*Sorghum bicolor* Moench.)	Whole shoots		17–18		133
	Recently matured leaves		25–31		133
Soybean (*Glycine max* Merr.)	Mature trifoliate leaves at early bloom	14–40		63	134
Spanish peanuts (*Arachis hypogaea* L.)	Young leaf tissue from 30-d-old plants		54–65 18–20[a]	>250	135
Sugar beets (*Beta vulgaris* L.)	Blades of recently matured leaves	12–40	35–200		136
	Middle fully developed leaf without stem taken at end of June or early July	<20	31–200	>800	121
Sunflower (*Helianthus annuus* L.)	Leaves	12.5	27	89	25
Timothy (*Phleum pratense* L.)	Whole plants at heading stage		3–93	>102	137
	Whole plants at rapid growth		11–46	47	132
Wheat (*Triticum aestivum* L.)	Boot-stage tissue	2.1–5.0	8	>16	63
	Straw	4.6–6.0	17	>34	63
	Leaves			>400	138
Winter wheat	Aboveground vegetative plant tissue when plants 40 cm high	<0.3	2.1–10.1	>10[b]	121
White clover (*Trifolium repens* L.)	Whole tops at rapid growth		13–16[a]		123
	Young plants		7.6[a]		139
	Whole plants at 6 weeks			53	140
White pea beans (*Phaseolus* spp.)	Aerial portion of plants 1 month after planting		36–94	144	141
Horticultural Crops					
Beans (*Phaseolus* spp.)	43-d-old plants		12	>160	127
Dwarf kidney beans (*Phaseolus* spp.)	Plants cut 50 mm above the soil Leaves and stems		44	132	60
Faba bean (*Vicia faba* L.)	Whole plants		25–100		101

Continued

TABLE 8.3   (*Continued*)

Crop	Plant Part Sampled	mg B kg⁻¹ in Dry Matter			Reference
		Deficiency	**Sufficiency**	**Toxicity**	
Snap beans	Pods		28	43	60
(*Phaseolus*	Recently matured leaves at			109	142
vulgaris L.)	prebloom				
	Plant tops at prebloom	<12	42	>125	71
Broccoli	Leaves		70		143
(*Brassica*	Leaf tissue when 5% heads	2–9	10–71		144, 145
oleracea var.	formed				
italica					
Plenck)					
Brussels	Leaf tissue when sprouts begin to	6–10	13–101		144, 145
sprouts	form				
(*Brassica*	Leaf tissue when sprouts begin to			161[b]	146
oleracea var.	form				
gemmifera					
Zenker)					
Cabbage	Mature leaf blade prior to head			132[b]	142
(*Brassica*	formation				
oleracea var.					
capitata L.)					
Carrots	Mature leaf lamina	<16	32–103	175–307	147
(*Daucus*	Leaves	18			75
carota L.)	Whole plants at swelling of roots	<28	54		148
Cauliflower	Whole tops before the appearance	3	12–23		130
(*Brassica*	of curd				
oleracea var.	Leaves	23	36		143
botrytis L.)	Leaf tissue when 5% heads formed	4–9	11–97		144, 145
Cucumber	Mature leaves from center of stem	<20	40–120	>300	121
(*Cucumis*	2 weeks after first picking				
sativus L.)					
Potatoes	32-d-old plants		12	>180	127
(*Solanum*	Fully developed first leaf at	<15	21–50	>50[b]	121
tuberosum L.)	75 days after planting				
	Shoots	<15	37–48	82–220	85
Radish	Whole plant when roots began to	<9	96–217		71
(*Raphanus*	swell				
sativus L.)					
Rutabaga	Leaf tissue at harvest	20–38	38–140	>250	99
(*Brassica*		<12 severely			99
napobrassica		deficient			
Mill.)	Leaf tissue when roots begin	32–40	40		86, 149
	to swell				
		moderately			86, 149
		deficient			
		<12 severely			86, 149
		deficient			

TABLE 8.3 (*Continued*)

| Crop | Plant Part Sampled | mg B kg^{-1} in Dry Matter | | | Reference |
		Deficiency	Sufficiency	Toxicity	
	Roots	<8 severely deficient	13		99
Strawberries (*Fragaria* x *ananassa* Duch.)	Old and young leaves at active growth stage			123	102
Tomatoes (*Lycopersicon* *esculentum* Mill.)	Mature young leaves from top of the plant	<10	30–75	>200	121
	63-d-old plants			>125	127
	Whole plants when 15 cm tall	<12	51–88	>172	71
	Whole plant			10–20	150

[a]Considered critical.
[b]Considered high.

8.6   SOIL TESTING FOR BORON

8.6.1   SAMPLING OF SOILS FOR ANALYSIS

Agricultural soils can be sampled by removing subsamples from uniform land areas to a depth of 15 to 20 cm. Uniform areas generally have similar soils and slopes, and do not include washed-out areas, bottomlands, or other dissimilar areas. Soil subsamples should be placed in a plastic container to avoid contamination and mixed together thoroughly. Generally, 25 to 50 subsamples per hectare are sufficient to obtain a representation of the soil.

8.6.2   EXTRACTION OF AVAILABLE BORON

Most procedures for extracting available boron from acid and alkaline soils are similar. The colorimetric and other methods of determining boron in the soil extract remain the same for testing on acid and alkaline soils. Methods have been extensively reviewed by Bingham (151). There are a number of methods for extracting available boron from soils (151). The most common extractant is hot water because soil solution boron is most important with regard to plant uptake. Hot water and other common extractants will be discussed in this section.

8.6.2.1   Hot-Water-Extractable Boron

The measurement of hot-water-soluble boron is a very popular method for determining available boron. Berger and Truog (152) established a hot-water method for determining available boron in soil that served as a reliable indicator of plant-available boron; however, the method was time-consuming. Additional modifications were made by Dible et al. (153), Baker (154), Wear (155), Jeffery and McCallum (156), and methods were summarized by Bingham (151).

Gupta (157) further modified the hot-water procedure by extracting soils with boiling water directly on a hot plate. Boron is then determined in the filtrates by a carmine colorimetric method (157) or by an azomethine-H procedure (158). However, Gupta found that a cooling period of more than 10 min before filtering the hot-water extracts resulted in slightly less recovery of boron. Yellow coloration that appears in some soil extracts interferes with the Azomethine-H procedure. The positive error due to yellow coloration can be reduced by refluxing soils in 10 mM $CaCl_2$. If the yellow color persists, the addition of not more than 0.16 g of charcoal per sample should be used. Too much charcoal tends to adsorb boron and reduce measured boron values (159,160). Gupta (158) reported that quantities of more than 0.8 g charcoal were necessary on soils containing more than 4.1% organic matter.

Extraction of hot-water-soluble boron is the most effective way to evaluate available boron to plants in most agricultural soils. Generally in the soil solution, less than 0.2 mg B L^{-1} is considered deficient for crops, whereas greater than 1 mg L^{-1} is considered toxic (161). On a soil mass basis, less than 1 mg B kg^{-1} is considered marginal for boron-sensitive crops whereas greater than 5 mg B kg^{-1} is considered toxic (119).

8.6.2.2 Boron from Saturated Soil Extracts

Saturation extracts of soils generally contain 0.1 to 10 mg B L^{-1}. The main advantage of a saturation extract is that it is easier to obtain than hot-water-soluble boron. Since the amount extracted by this method is less than that by hot-water extraction, this procedure has an advantage in determining the boron availability in toxic boron soils but would be less useful in soils containing low quantities of boron.

8.6.2.3 Other Soil Chemical Extractants

Li and Gupta (162) compared hot water, 0.05 M HCl, 1.5 M CH_3COOH, and hot 0.01 M $CaCl_2$ solution as boron extractants in relation to boron accumulation by soybean, red clover, alfalfa, and rutabaga. They concluded that 0.05 M HCl solution was the best extractant ($r = 0.82$) followed by 1.5 M CH_3COOH ($r = 0.78$), hot water ($r = 0.66$), and hot 0.01 M $CaCl_2$ solution ($r = 0.61$) for predicting the available boron status of acid soils. Aitken et al. (163) stated that hot water as well as hot 0.01 M $CaCl_2$ solution were far superior to mannitol and glycerol methods as a predictive test for plant boron requirement. They added that the levels of boron extracted with mannitol and glycerol were low compared to those displaced from the soil by the refluxing procedures. They suggested that mannitol would not be an effective extractant for boron in acid soils. Tsadilas et al. (164), working on high-boron soils, found that hot-water-soluble, 0.05 M mannitol in 0.1 M $CaCl_2$-extractable, 0.05 M HCl-soluble, and resin-extractable boron strongly correlated with each other. The coefficients of boron determination improved when the soil pH and clay content were included in the regression equation.

Mineral acid extraction of boron, especially with sulfuric acid, creates a number of problems for detection by complexing agents before the introduction of azomethine-H. Baker (165) found that phosphoric acid was a less suitable extractant than hot water for assessing the amount of soil boron available to sunflower during a short growing period. Gupta (166) found that sulfuric acid extraction of soils leads to high boron values due to interference with absorbance of the boron carmine complex. The HCl extracts were filtered easily, and no interference was encountered. Furthermore, the percentage recovery of added boron to soils was good and reproducible when extracted with 6 M HCl. No boron was lost when 6 M acid solutions were heated for 2 h at 100°C in a hot-water bath.

Another extractant, ammonium bicarbonate-diethylenetriaminepentaacetic acid (AB-DTPA), was suggested for determining boron in alkaline soils. The resultant filtrate is analyzed by inductively-coupled plasma spectroscopy (167). The AB-DTPA extractant has proven effective for determining boron and other nutrients on alkaline soils. It has been shown that this soil test alone was not as effective as the hot-water extractant in assessing boron availability to alfalfa (167). This soil test required the inclusion of percentage clay, organic matter, and soil pH to be effective. Gestring and Solanpour (168) further improved the AB-DTPA extractant on alkaline soils (pH 7.3 to 8.4) by the inclusion of ammonium acetate-extractable calcium into the regression equation of soil boron versus crop yield. This addition resulted in significantly increased correlation from $r^2 = 0.50$ to 0.77, suggesting a possible effect of calcium in boron toxicity. Studies conducted by Matsi et al. (169) showed that the AB-DTPA-extractable boron was significantly greater than the saturated extract and similar to the hot-water extract, and was correlated significantly with hot-water or with saturation extracts. They included cation-exchange capacity in the regression equation for boron determination.

Correlating an extractant for boron with plant growth is a key for determining the effectiveness of that extractant. The hot-water extraction method appears to be the most effective procedure for assessing B availability to plants on alkaline soils.

8.6.3 DETERMINATION OF EXTRACTED BORON

Several techniques are available to determine boron in soil extracts. Titrimetric, fluorometric, and bioassay methods were used earlier but are not commonly used now. In general, they are time-consuming, and some interferences are encountered. Colorimetric and spectrometric methods, which are more common, reliable, and accurate, will be discussed here.

8.6.3.1 Colorimetric Methods

Colorimetric methods for B determination are relatively inexpensive to perform and are somewhat free of interferences. The turmeric test (170,171) showed some promise earlier when it was discovered that dilute solutions of boric acid will change the color of turmeric paper from yellow to red. The procedure however, was long and required the precise control of temperature-regulated water baths. Berger and Truog (152) reported that the use of the turmeric paper test led to great difficulty because of its insensitivity due to its inability to differentiate between small amounts of boron.

The quinalizarine method is less laborious and more expeditious, whereas the curcumin method has the advantage of using easily prepared and easier to handle reagents (172). According to Berger (173), the mixing of 98% sulfuric acid–quinalizarin solution with the unknown solution generates a considerable amount of heat, and it was found that the higher the temperature, the redder is the color of the test solution. It was suggested that the solution be cooled to room temperature regardless of the temperature reached when the solutions were mixed. So it was possible and convenient to read unknown solutions in a colorimeter at a uniform temperature.

Porter et al. (174) saw the introduction of azomethine-H method as an answer to the handling difficulty involved in working with sulfuric acid for the carmine method. They added that the problem of having to concentrate boron in the solution of low boron concentration was also avoided. They concluded that an automated scheme improved the azomethine-H reagent method by overcoming the effect of sample color by dialysis.

Wolf (175) concluded that the results of boron determination using the azomethine-H method were in agreement with those of the curcumin method, and probably more reliable for soils high in nitrate. Also, the azomethine-H results (values) for plant boron agreed more closely with spectrographic analysis than the curcumin. Gestring and Soltanpour (176) found that the azomethine-H colorimetric method and inductively coupled plasma-atomic emission spectrophotometer (ICP) analysis were highly correlated. Both methods of analysis gave boron values comparable to National Bureau of Standards (NBS) values for dry-ashed plant samples; however, wet digestion using concentrated nitric acid resulted in interferences for the azomethine-H method but not for the ICP analysis.

8.6.3.2 Spectrometric Methods

The suitability (177) of the ICP spectrometer system for analysis of complex matrices was demonstrated by the high analytical precision and reproducibility of boron in alfalfa and in white bean (*Phaseolus coccineus* cv. Albus) (NBS samples). There was no interference from soluble organics observed in the complex soil solution matrices examined, although their presence would confound any colorimetric technique. It was possible to quantify boron in soil solutions to levels of 5 to 15 ng mL^{-1}, with extended integration periods utilizing the 249.773 nm emission line.

Parker and Gardner (178) employed ICP emission spectroscopic analysis of boron in distilled water and 0.02 M CaCl$_2$ solution, and indicated that the extractable boron level was not affected by the presence of CaCl$_2$. According to John et al. (179) the ICP method has the following advantages over the present colorimetric techniques: (a) carbon black is not needed since the color of the solution does not affect the analysis; (b) nitric acid digestion of samples can be utilized since ICP is not affected by the presence of nitrates; (c) other elements can be determined simultaneously; and (d) analysis by ICP is simple and rapid.

The use of Mehlich 3 extractant has been found to be simple, rapid, and practical in determining the availability of boron and a number of other nutrients in soils (180) with the ICP spectrophotometer. Using the ICP method, the Mehlich 3-extracted boron is well correlated with hot-water-soluble boron. The clear filtered extract (after shaking soil, Mehlich 3 reagent in 1:10 ratio for 5 min at 80 oscillations/min) is transferred into ICP tubes and analyzed by ICP at 249.678 nm (181). The ICP atomic emission spectrometry has also been used successfully in the determination of total soil B (182).

8.7 FACTORS AFFECTING PLANT ACCUMULATION OF BORON

8.7.1 SOIL FACTORS

8.7.1.1 Soil Acidity, Calcium, and Magnesium

Soil reaction or soil pH is an important factor affecting availability of boron in soils. Generally, boron becomes less available to plants with increasing soil pH. Several workers have observed negative correlations between plant boron accumulation and soil pH (67,183–185). In some studies in New Zealand, liming of the soil reduced boron concentration in the first cuts of alfalfa and red clover, particularly at higher rates of applied boron (123). Studies by Peterson and Newman (186) and Gupta and MacLeod (187) have shown that a negative relationship between soil pH and plant boron occurs when soil pH levels are greater than 6.3 to 6.5. The availability of boron to plants decreases sharply at higher pH levels, but the relationship between soil pH and plant boron at soil pH values below 6.5 does not show a definite trend.

Liming of soil decreased the plant boron accumulation when soil boron reserves were high (188). They attributed this effect to a high calcium content. Beauchamp and Hussain (189) in their studies on rutabagas, found that increased calcium concentration in tissue generally increased the incidence of brown-heart. Wolf (185) found that magnesium had a greater effect on boron reduction in plants than did calcium, sodium, or potassium, but the differences between calcium and magnesium effects were small. However, no distinction was made between the effects of soil pH and levels of calcium or magnesium on boron accumulation.

Experiments conducted to distinguish between the effects of soil pH and sources of calcium and magnesium showed that, in the absence of added boron, rutabaga roots and tops from calcium and magnesium carbonate treatments had more severe brown-heart condition than did roots from calcium and magnesium sulfate treatments (187). The leaf boron concentrations in rutabaga from treatments with no boron were lower at higher soil pH values where calcium or magnesium were applied as carbonates than they were at lower soil pH where sulfate was used as a source of calcium or magnesium (Table 8.4). In the presence of added boron, this trend was not clear, but the levels were well above the deficiency limit. The lower boron concentrations in the no-boron treatments with carbonates than in those with sulfates appear to be related to soil pH differences. These studies (187) showed no differences in boron accumulation whether the plants were fed with calcium or magnesium, as long as the corresponding anionic components were the same. Concentrations of calcium

TABLE 8.4
Effects of Calcium and Magnesium Sources and Boron Levels on Rutabaga (*Brassica napobrassica* Mill.) Leaf Tissue Boron Concentrations, and Soil pH.

Treatments			B (mg kg^{-1} tissue)[b]	Soil pH After Harvest
Cation[a]	Anion[a]	B (mg kg^{-1} soil)		
Control		0	33.5	5.6
Ca	CO$_3$	0	18.4	6.6
Mg	CO$_3$	0	17.4	6.3
Ca, Mg	CO$_3$	0	19.9	6.3
Ca	SO$_4$	0	31.6	4.8
Mg	SO$_4$	0	26.5	4.9
Ca, Mg	SO$_4$	0	29.9	4.9
Control		1	112	5.8
Ca	CO$_3$	1	118	6.5
Mg	CO$_3$	1	104	6.3
Ca, Mg	CO$_3$	1	108	6.6
Ca	SO$_4$	1	88	4.9
Mg	SO$_4$	1	92	5
Ca, Mg	SO$_4$	1	88	5
Means		0 boron	25.3b	
Means		1 boron	103a	

[a]Treatment consisted of 24 mol kg^{-1} soil either as a Ca or Mg salt or as a mixture in a 1:1 molar ratio of Ca and Mg. Control received 8 mmol each of CaCO$_3$ and MgCO$_3$ kg^{-1} soil.
[b]Values followed by a common letter do not differ significantly at $P \leq 0.05$ by Duncan's multiple range test.

Source: Adapted from Gupta U.C., in *Boron and Its Role in Crop Production*. CRC Press, Boca Raton, FL, 1993, pp. 87–104.

and magnesium, not shown in the table, were not related to the applications of boron. Table 8.4 shows that after the crop was harvested, lower quantities of hot-water-soluble boron were found in the soil that received calcium or magnesium sulfates than in soil that received calcium or magnesium carbonates.

Unpublished data (83) on podzol soils with a pH range of 5.4 to 7.8 showed that liming markedly decreased the boron content of pea plant tissue from 117 to 198 mg kg^{-1} at pH 5.4 to 5.6, to 36 to 43 mg kg^{-1} at pH 7.3 to 7.5. At pH values higher than 7.3 to 7.5, even tripling the amount of lime did not affect the boron content of plant tissue.

No clear relationship was found between the Ca/B ratio in the leaf blades and the incidence of brown-heart in rutabaga (189). However, it was noted that an application of sodium increased the calcium concentration in rutabaga tissue, thereby affecting the Ca/B ratio and possibly the incidence of brown-heart. It should be pointed out that use of the Ca/B ratio in assessing the boron status of plants should be viewed in relation to the sufficiency of other nutrients in the growing medium and in the plant.

8.7.1.2 Macronutrients, Sulfur, and Zinc

Among the macronutrients, nitrogen is of utmost importance in affecting boron accumulation by plants. Chapman and Vanselow (191) were among the pioneers in establishing that liberal nitrogen applications are sometimes beneficial in controlling excess boron in citrus. Under conditions of high boron, application of nitrogen depresses the level of boron in orange (*Citrus sinensis* Osbeck) leaves (192). Lysimeter experiments showed that tripled fertilization (NPK) rates and irrigation increased boron accumulation by plants on tested soils (193).

Boron concentrations in boot-stage tissue of barley and wheat increased significantly with increasing rates of compost additions (59). Such increases in boron were attributed to a high concentration of 14 mg B kg^{-1} in the compost. The authors reported that boron concentrations decreased with increasing rates of nitrogen. Additions of nitrogen decreased the severity of boron toxicity symptoms. The form of nitrogen can affect plant boron accumulation. Wojcik (194) reported that on boron-deficient, coarse-textured soils, nitrogen as calcium and ammonium nitrates increased the availability and uptake of boron by roots. This increase was attributed to the fact that nitrate inhibited boron sorption on iron and aluminum oxides, and increased boron in soil solution.

Increasing rates of nitrogen applied to initially nitrogen-deficient soils significantly decreased the boron concentration of boot-stage tissue in barley and wheat in a greenhouse study, but field experiments did not show any significant effect of nitrogen on boron concentration (195). The ineffectiveness of nitrogen in alleviating boron toxicity in cereals under field conditions is due to the fact that nitrogen failed to decrease the boron concentration in boot-stage tissue. Furthermore, nitrogen deficiency was more severe under greenhouse conditions than under field conditions. The decreases in boron concentrations were greater with the first level of added nitrogen than with the higher rates (195). This result may indicate that application of nitrogen is helpful in alleviating boron toxicity on soils low in available nitrogen.

Little difference in boron concentration of alfalfa was detected, and symptoms of boron deficiency progressed with increasing potassium concentration in the growth media (196). The authors suggested that the accentuating effect of high potassium on boron toxicity or deficiency symptoms might be due to the influence of potassium on cell permeability, which is presumably regulated by boron. Long-term experiments on cotton indicated positive yield responses to boron fertilization when associated with potassium applications (197). Yield increases were related to increased leaf potassium and boron concentrations.

The effects of phosphorus, potassium, and sulfur are less clear than those of nitrogen on the availability of boron to plants. Studies conducted in China (198) showed that rape (*Brassica napus* L.) plant boron concentration decreased with increasing potassium, and that lower potassium levels enhanced boron accumulation. The authors concluded that the optimum K/B ratio in rape plants was 1000:1.

Tanaka (199) showed that boron accumulation in radish increased with an increase in phosphorus supply. Malewar et al. (200) found that increasing the phosphorus fertilization rate resulted in higher phosphorus in cotton and groundnut. Experiments conducted on cotton also demonstrated that boron concentration in leaves was greatest with phosphorus fertilization (201). On the other hand, the presence of phosphorus can affect boron toxicity in calcareous soils. In studies on maize genotypes, boron was more toxic in the absence, rather than in the presence of, phosphorus, and thus boron toxicity in calcareous soils of the semiarid regions could be alleviated with applications of phosphorus (202).

Sulfate may have a slight effect on accumulation of boron in plant tissues (199). Field studies in Maharashtra, India, showed that boron applied with gypsum gave increased dry pod yield of groundnuts (203). The experimental results from a number of crops indicated that sulfur applications had no effect on boron concentration of peas, cauliflower, timothy (*Phleum pratense* L.), red clover, and wheat, but such applications significantly decreased the boron content of alfalfa and rutabaga (83). It is possible that various crops behave differently. For example, with soybean, applications of gypsum at 1000 kg ha^{-1} did not alleviate boron toxicity resulting from the application of 10 kg B ha^{-1} (204).

Recent studies showed that applied zinc played a role in partially alleviating boron toxicity symptoms by decreasing the plant boron accumulation (205). Zinc treatments partially depressed the inhibitory effect of boron on tomato growth (150).

8.7.1.3   Soil Texture

The texture of soil is an important factor affecting the availability of boron (206). A study on soils from eastern Canada showed that higher quantities of hot-water-soluble boron occurred in fine-textured soils than in coarse-textured soils (207). Studies in Poland showed that boron accumulation in potatoes and several cereals was less on sandy soils than on loamy soils (193).Page and Cooper (208) reported that leaching losses from acid, sandy soils after addition of 12.5 cm of water, account for as much as 85% of the applied boron. Movement is less rapid in heavy-textured soils because of increased fixation by the clay particles (119).

In Brazil, response to boron by cotton was significantly higher on Alic Cambisol, and the reverse was true for a dystrophic dark red latosol (209). It was suggested that high sand content (87%) and low clay (10%) and low organic matter (1.3%) in the latter soil could have resulted in toxic concentrations of boron in solution. The type of clay and the soil pH can significantly influence the amount of boron adsorbed. Hingston (210) reported that increasing pH resulted in an increase in the monolayer adsorption and a decrease in bonding energy for Kent sand kaolinite and Marchagee montmorillonite, and a slight increase in bonding energy for Willalooka illite up to pH 8.5. On a mass basis, illite adsorbed most boron over the range of pH values commonly occurring in soils; montmorillonite adsorbed appreciable amounts at higher pH, and kaolinite adsorbed the least.

Fine-textured soils generally require more boron than do the coarse-textured soils to produce similar boron concentrations in plants. Boron concentrations in solutions of 3.5 mg kg^{-1} in sandy loam and 4.5 mg kg^{-1} in clay loam resulted in similar boron concentrations in gram (*Cicer arietinum* L.) (211).

8.7.1.4   Soil Organic Matter

Organic matter is one of the chief sources of boron in acid soils, as relatively little boron adsorption on the mineral fraction occurs at low pH levels (212). The hot-water-soluble boron in soil has been positively related to the organic matter content of the soil (207). Addition of materials such as compost rich in organic matter resulted in large concentrations of boron in plant tissues and in phytotoxicity (60). Boron in organic matter is released in available form largely through the action of microbes (213). The complex formation of boron with dihydroxy compounds in soil organic matter is considered to be an important mechanism for boron retention (214). The influence of organic matter on the availability of boron in soils is amplified by increases in pH and clay content of the soil.

8.7.1.5   Soil Adsorption

When boron is released from soil minerals, mineralized from organic matter, or added to soils by means of irrigation or fertilization, part of the boron remains in solution, and part is adsorbed (fixed) by soil particles. An equilibrium exists between the solution and adsorbed boron (215). Usually more boron is adsorbed by soils than is present in solution at any one time (216), and fixation seems to increase with time (207).

Boron retention in soil depends upon many factors such as the boron concentration of the soil, soil pH, texture, organic matter, cation exchange capacity, exchangeable ion composition, and the type of clay and mineral coatings on clays (210,215,217,218). Of the clays, illite is the most reactive with boron, and kaolinite is the least reactive on a mass basis (210,219).

8.7.1.6   Soil Salinity

An antagonistic relationship existed between soil boron application levels and sodium adsorption ratio (SAR) of irrigation waters (220). Visible effects of boron toxicity developed in sugar beet plants at 0.5 SAR at high boron levels, and the symptoms intensified with plant age. However, effects of excess boron were markedly reduced at 20 and 40 SAR. Increasing soil salinity levels decreased the boron concentration in chickpea (gram) plants; such effects were accentuated at the higher boron levels (221).

8.7.2 OTHER FACTORS

8.7.2.1 Plant Genotypes

Data on the effect of plant genotypes on boron uptake are meager. Susceptibility to boron deficiency is controlled by a single recessive gene (222), as shown by the tomato cultivars T 3238 (B-inefficient) and Rutgers (B-efficient). Studies (222,223) have shown that T 3238 lacks the ability to transport boron to the top of the plants and confirms the differential response of T 3238 and Rutgers to a given supply of boron. Gorsline et al. (106) observed that corn hybrids exhibited genetic variability related to boron uptake and leaf concentration. One study conducted by E.G. Beauchamp, L.W. Kannenberg, and R.B. Hunter at the University of Guelph, Ontario (personal communication), indicated that the corn inbred CG 10, compared with several others, was the least efficient in boron accumulation as measured by the boron content of leaves sampled at the anthesis stage. These researchers, in a study of 11 hybrids, also found that decreased boron accumulation was associated with higher stover yield.

Some wheat cultivars in Asia, were tolerant of boron deficiency, whereas several sensitive genotypes failed to set grain in the absence of boron (224). Experiments conducted in China showed that roots of some wheat varieties secreted more organic acids, resulting in low pH and increased availability of boron, zinc, and phosphorus (225).

8.7.2.2 Environmental Factors

One of the chief environmental factors affecting the response of plants to the availability of nutrients is the intensity of light. The faster the plant grows, for example, under high light conditions, the faster it will develop boron deficiency symptoms in a particular growth period. Observations by Broyer (226) indicated that deficiencies as well as toxicities are revealed earliest or most intensely in the summer. Experiments conducted with duckweed (*Lemna paucicostata* Hegelm.) showed that reducing light intensity decreased the response to boron deficiency or toxicity (227). In the absence of boron, severe deficiencies were observed in cultures under continuous illumination from a daylight fluorescent lamp at 5500 lux, but not at 1000 lux. Over the range of 0.5 to 2.5 mg B L^{-1} in the culture solution, plant boron accumulation was reduced with decreasing light intensity. Studies conducted on young tomato plants grown in solution culture showed that in the absence of boron deficiency, symptoms developed more rapidly at high than at low light intensity (228). Plants supplied with boron did not exhibit symptoms.

An interaction appears to occur between temperature and lighting conditions. Rawson et al. (229) reported that low light alone reduced floret fertility in wheat by around 8%; however, in combination with a marginal boron supply, low light amplified the boron deficiency effect by some 60%. Furthermore, reduced light had the most deleterious effect at high temperature. Field studies in Bangladesh (230) demonstrated that some of the factors responsible for sterility in wheat are low temperatures over many days during flowering, and saturated or waterlogged soil. These factors affect transpiration, which in turn affects boron transport in the plant during the critical preflowering or flowering period.

Soil water appears to affect the availability of boron more than that of some other elements. Studies by Kluge (231) indicated that boron deficiency in plants during drought may be only partially associated with the level of hot-water-soluble boron in soil. Reduced soil solution in connection with reduced mass flow and reduced diffusion rate, as well as limited transpiration flow in the plants during drought periods, may be causative factors of boron deficiency in spite of an adequate supply of available boron in the soil. Boron deficiencies are generally found in dry soils where summer or winter drought is severe; when adequate moisture is maintained throughout the summer, deficiency symptoms may not be common (232). In an experiment on barley, soil water had a significant effect on plant boron accumulation after boron was applied to the soil (195). The boron concentration of barley, with added boron, ranged from 162 to 312 mg kg^{-1} under normal conditions, but only from 87 to 135 mg kg^{-1} when the area near the boron fertilizer band was kept dry. Mortvedt and Osborn (233) likewise reported that movement of boron from the fertilizer granules increased with concentration gradient and soil moisture content.

Boron concentration of some plants has been found to be a direct function of air temperature over the 8 to 37°C range. For example, Forno et al. (234) found that Cassava (*Manihot esculentum* Crantz) roots grew well when the solution temperature was maintained at 28 or 33°C, but developed severe boron deficiency symptoms at 18°C. Mild symptoms of boron deficiency were also obtained at a solution temperature of 23°C.

Relative humidity also affects boron accumulation, for example, an increase in percent relative humidity from 30 to 95 resulted in a decrease from 16.5 to 9.9 mg B per plant (235). Boron deficiency symptoms observed in birdsfoot trefoil (*Lotus corniculatus* L.) were caused by a temporary deficiency of available boron, induced by local drought conditions (236).

Generally, soils that have developed in humid regions have low amounts of plant-available boron because of leaching. Further, plant-available boron that is present in such soils is located in the top 15 cm and in the organic matter fraction (237,238). Thus, plants growing in regosols, sandy podzols, alluvial soils, organic soils, and low humic gleys tend to develop boron deficiencies because of low soil boron reserves.

At low temperatures in spring and fall in temperate regions, availability of boron is low, as evident in crops such as alfalfa and red clover. It has been suggested that during the cool season, plants may have an increased demand for B at a time when microbial activity in the soil is depressed (David Pilbeam, Personal communication, University of Leeds, England). The lower rate of root growth during the cool season may cause the rhizosphere to become depleted of boron, and falling temperatures may make cell membranes less fluid.

Sterility has become one of the most important wheat production constraints in Nepal (239). Among environmental factors, cold temperatures during the reproductive stages at higher altitudes coupled with low availability of boron are major factors causing sterility in wheat (239). Pot experiments conducted on spring wheat also showed that cold temperatures significantly reduced the response of plants to boron, and if a cold-susceptible cultivar was cold-stressed, it accumulated less boron (240).

8.7.2.3   Method of Cultivation and Cropping

The method of ploughing has been shown to affect plant boron accumulation. For example, Lal et al. (241) reported that boron concentration in corn leaf tissue was significantly higher with mouldboard plough and ridge till than with no-till and beds. Cropping systems influence the availability of boron in soil. In a continuous cropping study in China, available boron in soil was higher after three crops of soybeans than after three crops of wheat (242).

8.7.2.4   Irrigation Water

Gupta et al. (243) reported that only a few irrigation waters have enough boron to injure plants directly. The continued use of irrigation and concentration of boron in the soil due to evapotranspiration are the reasons for the eventual toxicity problems. In arid and semiarid regions, boron concentrations of irrigation waters, especially underground waters, are often elevated and in some cases may be as high as 5 mg L^{-1} (244). The majority of surface waters have boron concentrations of 0.1 to 0.3 mg L^{-1}, but well waters are more variable in boron content and often have excessive amounts (215). Some river waters used for irrigation may show high levels of boron at certain times of the year due to the contribution of spring drainage areas high in boron. Generally, ground waters emanating from light-textured soils are higher in boron than those from heavy-textured soils (245).

Boron movement in plants has been associated with transpiration. Therefore, any component of the environment that changes transpiration flux can affect boron availability. It has been proposed that decreased boron availability leading to sterility in wheat is due to water deficit as well as water-logging in the root zone (246).

8.8 FERTILIZERS FOR BORON

Modern crop production depends on addition of fertilizers to supplement natural soil fertility. Historically, crop production management has progressed to more intensive methods. Precise nutrient management has become essential for sustainable agricultural production systems. Addition of all plant nutrients must be considered for optimum crop production. With intensification of crop production, the need for micronutrient fertilization increases. Boron deficiency has been recognized as one of the most common micronutrient problems in agriculture.

8.8.1 Types of Fertilizers

Boron deposits of major economic importance are found only in arid regions of the world where volcanic action brought B and other volatile elements to the surface of the Earth during the Cenozoic era (56). Boron combined with alkali or alkaline earth elements to form rich deposits consisting chiefly of hydrous borates of calcium and sodium. The high water solubility of surface borate deposits precludes their existence in humid regions (56).

Concentrated borate deposits of commercial value were formed in continental enclosed basins by the evaporation of waters, which were boron-enriched by volcanic emanations. The locations of the major deposits are primarily in or near zones with histories of volcanic activity in arid regions. For example, a huge borate deposit, the Kramer deposit, was formed in a continental (nonmarine) basin in the Mojave Desert of California, associated with thermal spring activity during the Miocene epoch of the Cenozoic era. Similarly, significant boron deposits were formed in Argentina along the Andean mountain range near Salta. Studies have shown similarities between the hydrous borates of magnesium, calcium, and sodium formed in the Tincalayu deposit in the Province of Salta, Argentina, those in Kirka, Turkey, and the Kramer deposit in California (247).

Before the nineteenth century, Tibet was the world's source of borates. During the nineteenth century, commercially viable deposits were discovered in Italy, Turkey, South America, and the United States. The largest known borate deposits occur in the interior plateau of Turkey. The second largest occur in the Mojave Desert. Other countries having substantial borate deposits are the former Soviet Union, Argentina, Peru, Bolivia, Chile, Mexico, and China (248). Borax and solubor are the two most common boron fertilizers. Borax ($Na_2B_4O_7.10H_2O$) has been an important commercial mineral for centuries. A list of common fertilizers is shown in Table 8.5.

8.8.2 Methods and Rates of Application

The boron requirement of crops varies considerably, so recommendations must take these differences into account. Although plant species having high boron requirements are more likely to become boron deficient under boron-limiting conditions in the soil, their recommended boron rates may vary according to other conditions such as differences in root systems, effects of other soil parameters, and available soil calcium. Therefore, generalized boron recommendations must take all such factors into account.

Application of boron fertilizers at the recommended rate for a high-boron-requiring crop may provide excessive available boron for another crop. Tolerance to higher levels of available boron

TABLE 8.5
Boron Compounds Commonly Used as Fertilizers

B Source	Chemical Formula	Solubility in Water	% B
Borax	$Na_2B_4O_7.10H_2O$	Soluble	11.3
Fertilizer borate	$Na_2B_4O_7.5H_2O$	Soluble	14.3–14.9
Anhydrous borax	$Na_2B_4O_7$	Soluble	21.5
Solubor[a]	$Na_2B_8O_{13}.4H_2O$	Very soluble	20.5
Boric acid	H_3BO_3	Soluble	17.5
Colemanite	$Ca_2B_6O_{11}.5H_2O$	Slightly soluble	15.8
Ulexite	$NaCaB_5O_9.8H_2O$	Slightly soluble	13.3
Boron frits	Boric oxide glass	Very slightly soluble	2.0–11.0

[a]A registered trademark by U.S. Borax and Chemical Corporation.

Source: Adapted from Mortvedt J.J. and Woodruff J.R., in *Boron and Its Role in Crop Production*. CRC Press, Boca Raton, FL, 1993, pp. 157–176.

varies considerably, and species with high boron requirements do not necessarily have high tolerance and vice versa. For example, alfalfa and cabbage (*Brassica oleracea* var. *capitata* L.) have high boron requirements but are only semitolerant to high boron levels (249).

Recommended rates of boron application generally range from 0.25 to 3 kg ha^{-1}, depending on crop requirements and methods of application. Higher rates of boron generally are required for broadcast soil applications than for banded soil application or foliar sprays. Rates are usually similar for all boron sources, except for higher rates with slowly soluble sources such as colemanite or fritted products. Recommended boron rates and methods of application for some commonly fertilized crops are summarized by Mortvedt and Woodruff (116).

A primary consideration for soil application of boron is the soil surface texture and depth. In coarse-textured soils, under high rainfall, boron may move rapidly downward and from the root zone (250). In a loamy sand with the argillic horizon more than 40 cm deep, boron side-dressed is more effective than broadcast applications for corn (251). Fine-textured soils have the capacity to restrict boron leaching from the upper layers. Tap-rooted crops such as soybeans, may absorb nutrients from deeper layers, especially in dry weather, and benefit from boron in subsurface layers.

The two chief methods of boron fertilization are by adding it directly to the soil or by foliar spraying. Generally, soil and foliar applications of B are effective for crops. Soil applications are generally used for applying boron to field crops, but foliar sprays are more common on perennial crops such as fruit trees. Foliar application rates are usually about 50% lower than soil application rates. However, Murphy and Lancaster (252) obtained maximum yields of cotton with either 0.5 kg B ha^{-1} applied as a foliar spray (five times at 0.1 kg ha^{-1} each) or with >0.3 kg B ha^{-1} applied to the soil. For soybeans in a silt loam, foliar boron sprays were effective in increasing the number of pods per branch, but soil-applied boron had no effect (253).

Either broadcast or banded applications to soil are recommended, depending on the crop and soil conditions. Broadcast applications are used to establish and maintain alfalfa and other nonrow crops. Banded applications may result in greater efficiency of applied boron. Root growth may be depressed in soil near banded boron fertilizers. Mortvedt and Osborn (233) reported soluble boron concentrations as high as 75 mg kg^{-1} in soil near banded NP fertilizers with 1% B as $Na_2B_4O_7.5H_2O$. Root growth of alfalfa and oats was depressed in soil containing soluble boron concentrations >10 mg kg^{-1}. Soluble boron concentrations in soil would be much lower if the same boron rate was broadcast rather than banded to soil.

Applications of boron to the soil alone or with mixed fertilizers are common, and most data reported on plant boron accumulation have been obtained with boron-containing fertilizers applied broadcast or in bands. In field studies on rutabaga, band applications of boron resulted in greater boron concentrations in leaf tissue than did broadcast applications (254). In fact, boron applications of 1.12 kg ha⁻¹ applied in bands resulted in greater boron concentrations in leaf tissue than did 2.24 kg ha⁻¹ applied broadcast. Studies on rutabagas (254) and on corn (108,255) indicated that band- or foliar-applied boron resulted in greater boron accumulation by plants than did boron applied broadcast. Greater boron accumulation when it is applied in bands is likely due to the fact that a large quantity of the available nutrient is concentrated in the immediate root zone. Thus, boron applied in bands would be concentrated over a small area and would be taken up by the plants rapidly.

Applications of nutrients by foliar spray are effective in areas of California and Arizona where soil applications of micronutrients are ineffective because elements such as zinc, manganese, and copper are fixed in forms that are not readily available to certain crops (256). Foliar applications, besides resulting in higher boron accumulation in plants, could be used to advantage if a farmer omitted the addition of boron in the NPK bulk fertilizer or if boron deficiency was suspected. Foliar spray applications in the early growth stages resulted in greater absorption of boron than did those applied at later stages of growth (254). Mortvedt (257) stated that early-morning applications of foliar-applied nutrients may result in increased absorption, as the relative humidity is high and the stomata are open. It should however be pointed out that more than 98% of the boron applied as a foliar spray on white clover (*Trifolium repens* L.) remained at the point of application, and less than 2% was useful to the growth of the plant (258). This small but efficient portion of boron was quite mobile and was distributed to the different parts of the plants and then transferred from the oldest parts to the newly formed leaves. In a study on barley, soil applications of boron produced higher boron concentrations in the boot-stage tissue and grain, than similar amounts of boron applied as foliar spray (259). This result indicates that boron uptake, at least by barley, is more efficient through soil–root systems than through the leaves.

For some elements such as molybdenum, which plants require in extremely small amounts, seed treatment with a preparation containing molybdenum will prevent a deficiency. However, because of the comparatively higher requirement of boron than molybdenum, and because of the toxic effect of boron on seeds or seedlings, seed treatment for boron fertilization has not received attention.

REFERENCES

1. P. Maze. Influences respectives des element de la solution minerale sur le developpement du mais. *Ann. Inst. Pasteur* 28:21–68, 1914.
2. K. Warington. The influence of length of day on the response of plants to boron. *Ann. Bot.* 37:629–672, 1923.
3. W.E. Brenchley, K. Warington. The role of boron in the growth of plants. *Ann. Bot.* 41:167–187, 1927.
4. A.L. Sommer, C.B. Lipman. Evidence on the indispensable nature of zinc and boron for higher green plants. *Plant Physiol.* 1:231–249, 1926.
5. A.L. Sommer. The search for elements essential in only small amounts for plant growth. *Science* 66:482–484, 1927.
6. J.C. Brown, W.E. Jones. Differential transport of boron in tomato (*Lycopersicon esculentum* Mill.). *Physiol. Plant* 25:279–282, 1971.
7. C.W. Bohnsack, L.S. Albert. Early effects of boron deficiency on indoleacetic acid oxidase levels of squash root tips. *Plant Physiol.* 59:1047–1050, 1977.
8. P. Josten, U. Kutschera. The micronutrient boron causes the development of adventitious roots in sunflower cuttings. *Ann. Bot.* 84:337–342, 1999.
9. B.J. Shelp. Physiology and biochemistry of boron in plants. In: U.C. Gupta, ed. *Boron and Its Role in Crop Production.* Boca Raton, FL: CRC Press, 1993, pp. 53–85.
10. R.W. Krueger, C.J. Lovatt, L.S. Albert. Metabolic requirement of *Cucurbita pepo* for boron. *Plant Physiol.* 83:254–258, 1987.
11. M.E. Lenoble, D.G. Blevins, R.J. Miles. Prevention of aluminum toxicity with supplemental boron. II. Stimulation of root growth in acidic, high-aluminum subsoil. *Plant Cell Environ.* 19:1143–1148, 1996.

12. O.C. Artes, R.O.C. Ruiz. Influence of boron on amino acid contents in tomato plant. I. sap. *Agrochimica* 27:498–505, 1983.

13. B.J. Shelp. Boron mobility and nutrition in broccoli (*Brassica oleracea* var. *italica*). *Ann. Bot.* 61:83–91, 1988.

14. I.C. Dave, S. Kannan. Influence of boron deficiency on micronutrients absorption by *Phaseolus vulgaris* and protein contents in cotyledons. *Acta Physiol. Plant* 3:27–32, 1981.

15. W.M. Dugger. Boron in plant metabolism. In: A. Lauchli, R.I. Bieleski, eds. *Encyclopedia of Plant Physiology*, new series, New York: Springer, 1983, pp. 626–650.

16. B.J. Shelp. The influence of nutrition on nitrogen partitioning in broccoli plants. *Commun. Soil Sci. Plant Anal.* 21:49–60, 1990.

17. I. Bonilla, P. Mateo, A. Garate. Accion del boro sobre el metabolismo nitrogenado en Lycopersicon esculentum cv. Dombo, cultivado en hydroponica. *Agrochimica* 32:276–283, 1988.

18. R. Kastori, N. Petrovic. Effect of boron on nitrate reductase activity in young sunflower plants. *J. Plant Nutr.* 12:621–632, 1989.

19. C.J.J. Camacho, F.A. Gonzales. Boron deficiency causes a drastic decrease in nitrate content and nitrate reductase activity, and increases the content of carbohydrates in leaves from tobacco plants. *Planta* 209:528–536, 1999.

20. M. Yamagishi, Y. Yamamoto. Effects of boron on nodule development and symbiotic nitrogen fixation in soybean plants. *Soil Sci. Plant Nutr.* 40:265–274, 1994.

21. C. Chatterjee, P. Sinha, S.C. Agarwala. Boron nutrition of cowpeas. *Proc. Indian Acad. Plant Sci.* 100:311–318, 1990.

22. S.G. Lee, S. Arnoff. Boron in plants: a biochemical role. *Science* 158:798–799, 1967.

23. C. Chatterjee, N. Nautiyal. Developmental aberrations in seeds of boron deficient sunflower and recovery. *J. Plant Nutr.* 23:835–841, 2000.

24. P. Sinha, R. Jain, C. Chatterjee. Interactive effect of boron and zinc on growth and metabolism of mustard. *Commun. Soil Sci. Plant Anal.* 31:41–49, 2000.

25. B.K. Dube, P. Sinha, C. Chatterjee, P. Sinha. Boron stress affects metabolism and seed quality of sunflower. *Trop. Agric.* 77:89–92, 2000.

26. P. Sinha, C. Chatterjee, C.P. Sharma, P. Sinha. Changes in physiology and quality of pea by boron stress. *Ann. Agric. Res.* 20:304–307, 1999.

27. P.C. Srivastava, U.C. Gupta. Essential trace elements in crop production. In: P.C. Srivastava, U.C. Gupta, eds. *Trace Elements in Crop Production*. New Delhi, India: Oxford & IBH Publishing Cop. Pvt. Ltd., 1996, pp. 73–173.

28. A.H.N. Ali, B.C. Jarvis. Effects of auxin and boron on nucleic acid metabolism and cell division during adventitious root regeneration. *New Phytol.* 108:381–391, 1988.

29. B.C. Jarvis, S. Yasmin, M.T. Coleman. RNA and protein metabolism during adventitious root formation in stem cuttings of *Phaseolus aureus*. *Physiol. Plant* 64:53–59, 1985.

30. B.C. Jarvis, A.H.N. Ali, A.I. Shaheed. Auxin and boron in relation to the rooting response and ageing of mung bean cuttings. *New Phytol.* 95:509–518, 1983.

31. B.C. Jarvis, S. Yasmin, A.H.N. Ali, R. Hunt. The interaction between auxin and boron in adventitious root development. *New Phytol.* 97:197–204, 1984.

32. M. Ya Shkolnik. *Trace Elements in Plants*. Amsterdam: Elsevier, 1984, p. 463.

33. D.H. Lewis. Are there interrelations between the metabolic role of boron, synthesis of phenolic phytoalexins and the gemination of pollen? *New Phytol.* 84:261–270, 1980.

34. A. Mozafar. Role of boron in seed production. In: U.C. Gupta, ed. *Boron and its Role in Crop Production*. Boca Raton, FL: CRC Press, 1993, pp. 187–208.

35. S.M. Misra, B.D. Patil. Effect of boron on seed yield in lucern (*Medicago sativa* L.). *J. Agron. Crop Sci.* 158:34–37, 1987.

36. P. Simjoki. Boron deficiency in barley. *Ann. Agr Fenniae* 30:389–405, 1991.

37. S.C. Agarwala, P.N. Sharma, C. Chatterjee, C.P. Sharma. Development and enzymatic changes during pollen development in boron deficient maize plants. *J. Plant Nutr.* 3:329–336, 1981.

38. A. Hasler, A. Maurizio. Die Wirkung von Bor auf Samenansatz und Nektarsekretion bei Raps (*Brassica napus* L.). *Phytopathol. Z* 15:193–207, 1949.

39. T. Tanada. Localization of boron in membranes. *J. Plant Nutr.* 6:743–749, 1983.

40. A.S. Pollard, A.J. Parr, B.C. Loughman. Boron in relation to membrane function in higher plants. *J. Exp. Bot.* 28:831–841, 1977.

41. G.A. Robertson, B.C. Loughman. Rubidium uptake and boron deficiency. *J. Exp. Bot.* 24:1046–1052, 1973.

42. G.A. Robertson, B.C. Loughman. Reversible effects of boron and on the absorption and incorporation of phosphate in *Vicia faba* L. *New Phytol.* 73:291–298, 1974.

43. G.A. Robertson, B.C. Loughman. Response to boron deficiency: a comparison with responses produced by chemical methods of retarding root elongation. *New Phytol.* 73:821–832, 1974.

44. P.M. Tang, R.K. Dela Fuente. The transport of indole-3-acetic acid in boron-and calcium-deficient sunflower hypocotyl segments. *Plant Physiol.* 81:646–650, 1986.

45. P.M. Tang, R.K. Dela Fuente. Boron and calcium sites involved in indole-3-acetic acid transport in sunflower hypocotyl segments. *Plant Physiol.* 81:651–655, 1986.

46. U.C. Gupta. Relationship of total and hot-water soluble boron, and fixation of added boron, to properties of podzol soils. *Soil Sci. Soc. Am. Proc.* 32:45–48, 1968.

47. K. Borkakati, P.N. Takkar. Forms of boron in major orders of Assam under different land use. *J. Agric. Sci. Soc. North East India* 9:28–33, 1996.

48. W.L. Lindsay. Inorganic equilibria affecting micronutrients in soils. In: J.J. Mortvedt, F.R. Cox, L.M. Shuman, R.M. Welch, eds. *Micronutrients in Agriculture*. Soil Sci. Soc. Am. Book Series No. 4, 2nd ed. Madison, WI: Soil Science Society of America, 1991, pp. 89–112.

49. J. Jin, D.C. Martens, L.W. Zelazny. Distribution and plant availability of soil; boron fractions. *Soil Sci. Soc. Am. J.* 51:1228–1231, 1987.

50. J. Hou, L.J. Evans, G.A. Spiers. Boron fractionation in soils. *Commun. Soil Sci. Plant Anal.* 25:1841–1853, 1994.

51. N. Zerrari, D. Moustaoui, M. Verloo. The forms of boron in soils: importance, effect of soil characteristics and availability for the plants. *Agrochimica* 43:77–88, 1999.

52. C.M. Evans, D.L. Sparks. On the chemistry and mineralogy of boron in pure and in mixed systems: a review. *Commun. Soil Sci. Plant Anal.* 14:827–846, 1983.

53. R.R. Whetstone, W.O. Robinson, H.G. Byers. Boron Distribution in Soils and Related Data. U.S. Dep. Agric. Tech. Bull. No. 797, Washington, DC, 1942.

54. T. Tsang, S. Ghose. Nuclear magnetic resonance of ^{1}H, ^{7}Li, ^{11}B, ^{23}Na, ^{27}Al in tourmaline (elbaite). *Am. Mineralogist* 58:224–229, 1973.

55. G.A. Fleming. Essential micronutrients I: Boron and molybdenum. In: B.E. Davies, ed. *Applied Soil Trace Elements*. New York: Wiley, 1980, pp. 155–197.

56. T. Watanabe. Geochemical cycle and concentration of boron in the earth's crust. In: C.T. Walker, ed. *Geochemistry of Boron*. London: Halsted Press, 1975, pp. 388–399.

57. C.L. Christ, A.H. Truesdell, R.C. Erd. Borate mineral assemblages in the system Na_2O-CaO-MgO-B_2O_3-H_2O. *Geochim. Cosmochim. Acta* 31:313–337, 1967.

58. V.M. Shorrocks. *Boron Deficiency—Its Prevention and Cure*. Borax Consolidated Ltd. Borax House Carlisle Place London: The Soman-Wherry Press Ltd., Norwich 1974, pp. 3–53.

59. U.C. Gupta, J.D.E. Sterling, H.G. Nass. Influence of various rates of compost and nitrogen on the boron toxicity symptoms in barley and wheat. *Can. J. Plant Sci.* 53:451–456, 1973.

60. D. Purves, E.J. MacKenzie. Effects of applications of municipal compost on uptake of copper, zinc, and boron by garden vegetables. *Plant Soil* 39:361–373, 1973.

61. R.L. Branson. Soluble salts, exchangeable sodium, and boron in soils. In: H.M. Reisenauer, ed. *Soil and Plant Tissue Testing in California*. Univ. Calif. Div. Agric. Sci. Bull. 1879, 1976, pp. 42–45.

62. R.O. Nable, J.G. Paull, B. Cartwright. Problems associated with the use of foliar analysis for diagnosing boron toxicity in barley. *Plant Soil* 128:225–232, 1990.

63. U.C. Gupta. Boron and molybdenum nutrition of wheat, barley and oats grown in Prince Edward Island soils. *Can. J. Soil Sci.* 51:415–422, 1971.

64. K. Wang, Y. Yang, R.W. Bell, J.M. Xue, Z.O. Ye, Y.Z. Wei. Low risks of toxicity from boron fertiliser in oilseed rape-rice rotations in southeast China. *Nutr. Cycl. Agroecosyst.* 54:187–197, 1999.

65. W.L. Nelson, S.A. Barber. Nutrient deficiencies in legumes for grain and forage. In: H.B. Sprague, ed. *Hunger Signs in Crops*, 3rd ed. New York: David McKay Co., 1964, pp. 143–179.

66. K.C. Berger. Micronutrient deficiencies in the United States. *J. Agric. Food Chem.* 10:178–181, 1962.

67. U.C. Gupta. Effects of boron and lime on boron concentration and growth of forage legumes under greenhouse conditions. *Commun. Soil Sci. Plant Anal.* 3:355–365, 1972.

68. U.C. Gupta. Boron nutrition of alfalfa, red clover, and timothy grown on podzol soils of eastern Canada. *Soil Sci.* 137:16–22, 1984.

69. W. Bergmann. *Ernährungsstörungen bei Kulturpflanzen in Farbbildem*, VEB Gustav Fischer Verlag Jena, Jena, 1976, pp. 164–172.

70. M.P. Löhnis. Boriumbehoefte en boriumgehalte van cultuurplanten. *Chem. Weekblad* 33:59–61, 1936.

71. U.C. Gupta. Boron deficiency and toxicity symptoms for several crops as related to tissue boron levels. *J. Plant Nutr.* 6:387–395, 1983.

72. T. Inden. Minor Elements for Vegetables. Ext. Bull. 55, ASPAC Food Fert. Technol. Cent., 1975, pp. 1–20.

73. B.J. Shelp, R. Penner, Z. Zhu. Broccoli (*Brassica oleracea* var. *italica*) cultivar response to boron deficiency. *Can. J. Plant Sci.* 72:883–888, 1992.

74. F. Haworth. A note on boron deficiency of brassicae on St. Coombs. *J. Tea Res. Inst. Ceylon* 23:86, 1952.

75. K.W. Smilde, B. van Luit. Boron deficiency in roots. *Bedrijfsontwikkeling* 1:30–31, 1970.

76. E.R. Purvis, R.L. Carolus. Nutrient deficiencies in vegetable crops. In: H.B. Sprague, ed. *Hunger Signs in Crops*, 3rd ed. New York: David McKay Co., 1964, pp. 245–286.

77. O.N. Mehrotra, P.H. Misra. Micronutrient deficiencies in cauliflower (*Brassica oleracea* L. var. *botrytis* L.). *Prog. Hortic.* 5:33–39, 1974.

78. Di G. Janos. The significance of trace elements and their utilization in agriculture. *Magy Tud Akad Biol Orv Tud Oszt Kozl* 22:349–361, 1963.

79. B.A. Krantz, S.W. Melsted. Nutrient deficiencies in corn, sorghums, and small grains. In: H.B. Sprague, ed. *Hunger Signs in Crops*, 3rd ed. New York: David McKay Co., 1964, pp. 25–57.

80. A. Mozafar. Effect of boron on ear formation and yield components of two maize (*Zea mays* L.) hybrids. *J. Plant Nutr.* 10:319–332, 1987.

81. B. Rerkasem, R. Netsangtip, R.W. Bell, J.F. Loneragan, N. Hiranburana. Comparative species responses to boron on a Typic Tropaqualf in Northern Thailand. *Plant Soil* 106:15–21, 1988.

82. C.S. Piper. The symptoms and diagnosis of minor-element deficiencies in agricultural and horticultural crops, Pt. I. Diagnostic Methods. Boron. Manganese. *Emp. J. Exp. Agric.* 8:85–100, 1940.

83. U.C. Gupta, J.A. MacLeod. A Field Study on the Boron Nutrition of Peas, 1981, unpublished manuscript.

84. G.V.C. Houghland. Nutrient deficiencies in the potato. In: H.B. Sprague, ed. *Hunger Signs in Crops*, 3rd ed. New York: David McKay Co., 1964, pp. 219–244.

85. S. Roberts, J.K. Rhee. Boron utilization by potato in nutrient cultures and in field plantings. *Commun. Soil Sci. Plant Anal.* 21:921–932, 1990.

86. U.C. Gupta, J.A. Cutcliffe. Determination of optimum levels of boron in rutabaga leaf tissue and soil. *Soil Sci.* 111:382–385, 1971.

87. S.C. Agarwala, C.P. Sharma. *Recognizing Micronutrient Disorders of Crop Plants on the Basis of Visible Symptoms and Plant Analysis*, Lucknow University, India: Prem Printing Press, 1979, p. 72.

88. B. Rerkasem, R.W. Bell, S. Lodkaew, J.F. Loneragan. Boron deficiency in soybean (*Glycine max* (L.) Merr), peanut (*Arachis hypogaea* L.) and black gram (*Vigna mungo* (L.) Hepper): symptoms in seeds and differences among soybean cultivars in susceptibility to boron deficiency. *Plant Soil* 150:289–294, 1993.

89. K. Ohki. Manganese nutrition of cotton under two boron levels. I. Growth and development. *Agron. J.* 65:482–485, 1973.

90. L. Donald. Nutrient deficiencies in cotton. In: H.B. Sprague, ed. *Hunger Signs in Crops*. 3rd ed. New York: David McKay Co., 1964, pp. 59–98.

91. H.A. van de Venter, H.B. Currier. The effect of boron deficiency on callose formation and ^{14}C location in bean (*Phaseolus vulgaris* L.) and cotton (*Gossypium hirsutum* L.). *Am. J. Bot.* 64:861–865, 1977.

92. J. Vlamis, A. Ulrich. Boron nutrition in the growth and sugar content of sugarbeets. *J. Am. Soc. Sugarbeet Technol.* 16:428–439, 1971.

93. S.G. Gandhi, B.V. Mehta. Studies on boron deficiency and toxicity symptoms in some common crops of Gujarat. *Indian J. Agric. Sci.* 29:63–70, 1959.

94. J.E. McMurtrey Jr. Nutrient deficiencies in tobacco. In: H.B. Sprague, ed. *Hunger Signs in Crops*. 3rd ed. New York: David McKay Co., 1964, pp. 99–141.

95. M.M. Riley. Boron toxicity in barley. *J. Plant Nutr.* 10:2109–2115, 1987.

96. L.V. Wilcox. Boron Injury to Plants. Agric. Info. Bull. 211, U.S. Dept. Agric., Washington, DC, 1960, pp. 3–7.

97. L.E. Francois. Boron tolerance of snap bean and cowpea. *J. A. Soc. Hortic. Sci.* 114:615–619, 1989.

98. A. Bagheri, J.G. Paull, A.J. Rathjen, S.M. Ali, D.B. Moody. Genetic variation in the response of pea (*Pisum sativum* L.) to high soil concentrations of boron. *Plant Soil* 146:261–269, 1992.

99. U.C. Gupta, D.C. Munro. The boron content of tissues and roots of rutabagas and of soil as associated with brown-heart condition. *Soil Sci. Soc. Am. Proc.* 33:424–426, 1969.

100. F.B. Muller, G. McSweeney. Toxicity of borates to turnips. *N. Z. J. Exp. Agric.* 4:451–455, 1976.

101. D. Poulain, H. Al-Mohammad. Effects of boron deficiency and toxicity on faba bean (*Vicia faba* L.). *Eur. J. Agron.* 4:127–134, 1995.

102. G.F. Haydon. Boron toxicity of strawberry. *Commun. Soil Sci. Plant Anal.* 12:1085–1091, 1981.

103. S.K. Yau, M.C. Saxena. Variation in growth, development, and yield of durum wheat in response to high soil boron. *Aus. J. Agric. Res.* 48:945–949, 1997.

104. R.B. Clark. Mineral element concentrations in corn leaves by position on the plant and age. *Commun. Soil Sci. Plant Anal.* 6:439–450, 1975.

105. R.B. Clark. Mineral element concentrations of corn plant parts with age. *Commun. Soil Sci. Plant Anal.* 6:451–464, 1975.

106. G.W. Gorsline, W.I. Homas, D.E. Baker. Major Gene Inheritance of Sr, Ca, Mg, K, P, Zn, Cu, B, Al, Fe and Mn Concentration in Corn (*Zea mays* L.). Penn. State Univ. Agric. Exp. Stn. Bull. 746, 1968. p. 47.

107. D. Von Alt, W. Schwarz. Bor-Toxizität, bor-aufnahme und borverteilung bei Jungen gurkenflazen unter dem einflusz der N-form. *Plant Soil* 39:277–283, 1973.

108. J.T. Touchton, F.C. Boswell. Boron application for corn grown on selected southeastern soils. *Agron. J.* 67:197–200, 1975.

109. H.C. Kohl Jr, J.J. Oertli. Distribution of boron in leaves. *Plant Physiol.* 36:420–424, 1961.

110. J.B. Jones Jr. Distribution of fifteen elements in corn leaves. *Commun. Soil Sci. Plant Anal.* 1:27–33, 1970.

111. D.A. Miller, R.K. Smith. Influence of boron on other chemical elements in alfalfa. *Commun. Soil Sci. Plant Anal.* 8:465–478, 1977.

112. G. Lotti, A. Saviozzi, S. Balzini. The distribution of boron and major mineral elements in plant organs in relation to the content of their leaves. *Agrochimica* 33:129–142, 1989.

113. U.C. Gupta. Boron, molybdenum and selenium status in different plant parts in forage legumes and vegetable crops. *J. Plant Nutr.* 14:613–621, 1991.

114. J.J. Oertli, W.F. Richardson. The mechanism of boron immobility in plants. *Physiol. Plant* 23:108–116, 1970.

115. B.J. Shelp. Mineral nutrient distribution: significance in the nutrition of vegetable crops, *Highlights Agric. Res. Ontario* 12:21–24, 1989.

116. J.J. Mortvedt, J.R. Woodruff. Technology and application of boron fertilizers for crops. In: U.C. Gupta, ed. *Boron and Its Role in Crop Production*. Boca Raton, FL: CRC Press, 1993, pp. 157–176.

117. A. Ulrich, F.J. Hills. Plant analysis as an aid in fertilizing sugarbeet. In: R.L. Westerman, ed. *Soil Testing and Plant Analysis*. 3rd ed. SSSA Book Series 3, Madison, WI: Soil Sci. Soc. Am., Inc., 1990, pp. 429–447.

118. T.E. Bates. Factors affecting critical nutrient concentrations in plants and their evaluation: a Review. *Soil Sci.* 112:116–130, 1971.

119. H.M. Reisenauer, L.M. Walsh, R.G. Hoeft. Testing soils for sulphur, boron, molybdenum, and chlorine. In: L.M. Walsh, J.D. Beaton, eds. *Soil Testing and Plant Analysis*. Madison, WI: Soil Science Society of America, 1973, pp. 173–200.

120. R.D. Meyer, W.E. Martin. Plant Analysis as a Guide for Fertilization of Alfalfa. Univ. Calif. Div. Agric. Sci. Bull. 1879, 1976, pp. 1–32.

121. P. Neubert, W. Wrazidlo, H.P. Vielemeyer, I. Hundt, F. Gollmick, W. Bergmann. Tabellen zur pflanzenanalyse—Erste orientierende ubersicht—Inst. Fur Pflanzenernährung Jena, 69 Jena, Naumburger Strasse 98, 1970, pp. 1–40.

122. S.W. Melstead, H.L. Motto, T.R. Peck. Critical plant nutrient composition values useful in interpreting plant analysis data. *Agron. J.* 61:17–20, 1969.

123. C.G. Sherrell. Boron deficiency and response in white and red clovers and lucerne. *N. Z. J. Agric. Res.* 26:197–203, 1983.

124. W.E. Martin, V.V. Rendig, A.D. Haig, L.J. Berry. Fertilization of Irrigated Pasture and Forage Crops in California, Calif. Agric. Exp. Stn. Bull. 815, 1965, pp. 1–36.

125. S.A. Barber. Boron Deficiency in Indiana Soils. Lafayette, Indiana: Purdue Univ. Agric. Exp. Stn. Bull. 652, 1957.

126. M.M. Riley, A.D. Robson. Pattern of supply affects boron toxicity in barley. *J. Plant Nutr.* 17:1721–1738, 1994.

127. D.C. MacKay, W.M. Langille, E.W. Chipman. Boron deficiency and toxicity in crops grown on sphagnum peat soil. *Can. J. Soil Sci.* 42:302–310, 1962.

128. H.E. Jones, G.D. Scarseth. The calcium-boron balance in plants as related to boron needs. *Soil Sci.* 57:15–24, 1944.

129. A. Rashid, E. Rafique, N. Ali. Micronutrient deficiencies in rainfed calcareous soils of Pakistan. II. Boron Nutrition of the peanut plant. *Commun. Soil Sci. Plant Anal.* 28:149–159, 1997.

130. U.C. Gupta. Boron requirement of alfalfa, red clover, Brussels sprouts and cauliflower grown under greenhouse conditions. *Soil Sci.* 112:280–281, 1971.

131. X. Yu, P.F. Bell, X.O. Yu. Nutrient deficiency symptoms and boron uptake mechanisms of rice. *J. Plant Nutr.* 21:2077–2088, 1998.

132. C.G. Sherrell. Boron nutrition of perennial ryegrass, cocksfoot, and timothy. *N. Z. J. Agric. Res.* 26:205–208, 1983.

133. A. Rashid, E. Rafique, N. Bughio. Micronutrient deficiencies in rainfed calcareous soils of Pakistan. III. Boron nutrition of sorghum. *Commun. Soil Sci. Plant Anal.* 28:441–454, 1997.

134. J.R. Woodruff. Soil boron and soybean leaf boron in relation to soybean yield. *Commun. Soil Sci. Plant Anal.* 10:941–952, 1979.

135. L.G. Morrill, W.E. Hill, W.W. Chrudimsky, L.O. Ashlock, L.D. Tripp, B.B. Tucker, L. Weatherly. Boron Requirements of Spanish Peanuts in Oklahoma: Effects on Yield and Quality and Interaction with Other Nutrients. Agric. Exp. Stn. Oklahoma State University, MP-99, 1977, pp. 1–20.

136. F.J. Hills, A. Ulrich. Plant analysis as a guide for mineral nutrition of sugar beets. In: H.M. Reisenauer, ed. *Soil and Plant Tissue Testing.* Berkley, CA: Div. Agric. Sci. Univ. Calif. Bull. 1879. 1976, pp. 21–24.

137. U.C. Gupta, J.A. MacLeod. Boron nutrition and growth of timothy as affected by soil pH. *Commun. Soil Sci. Plant Anal.* 4:389–395, 1973.

138. C.M. Grieve, J.A. Poss. Wheat response to interactive effects of boron and salinity. *J. Plant Nutr.* 23:1217–1226, 2000.

139. R.N. Singh, H. Sinha. Evaluation of critical limits of available boron for soybean and maize in acid red loam soil. *J. Indian Soc. Soil Sci.* 35:456–459, 1987.

140. M. Prasad, E. Byrne. Boron source and lime effects on the yield of three crops grown in peat. *Agron. J.* 67:553–556, 1975.

141. L.S. Robertson, B.D. Knezek, J.O. Belo. A survey of Michigan soils as related to possible boron toxicities. *Commun. Soil Sci. Plant Anal.* 6:359–373, 1975.

142. U.C. Gupta, J.A. Cutcliffe. Effects of applied and residual boron on the nutrition of cabbage and field beans. *Can. J. Soil Sci.* 64:57–576, 1984.

143. T. Wallace. *The Diagnosis of Mineral Deficiencies in Plants.* London: HM Stationery Office, 1951, pp. 26–35,61–75.

144. U.C. Gupta, J.A. Cutcliffe. Boron nutrition of broccoli, Brussels sprouts, and cauliflower grown on Prince Edward Island soils. *Can. J. Soil Sci.* 53:275–279, 1973.

145. U.C. Gupta, J.A. Cutcliffe. Boron deficiency in cole crops under field and greenhouse conditions. *Commun. Soil Sci. Plant Anal.* 6:181–188, 1975.

146. U.C. Gupta, R. Cormier, J.A. Cutcliffe. Tolerance of Brussels sprouts to high boron levels. *Can. J. Soil Sci.* 67:205–207, 1987.

147. W.C. Kelly, G.F. Somers, G.H. Ellis. The effect of boron on the growth and carotene content of carrots. *Proc. Am. Soc. Hortic. Sci.* 59:352–360, 1952.

148. U.C. Gupta, J.A. Cutcliffe. Boron nutrition of carrots and table beets grown in a boron deficient soil. *Commun. Soil Sci. Plant Anal.* 16:509–516, 1985.

149. U.C. Gupta, J.A. Cutcliffe. Effects of lime and boron on brown-heart, leaf tissue calcium/boron ratios, and boron concentrations of rutabaga. *Soil Sci. Soc. Am. Proc.* 36:936–939, 1972.

150. A. Gunes, M. Alpaslan, Y. Cikili, H. Ozcan. The effect of zinc on alleviation of boron toxicity in tomato plants (*Lycopersicon esculenicon* l.). *Turkish J. Agr. Forestry* 24:505–509, 2000.

151. F.T. Bingham. Boron. In: A.L. Page, ed. *Methods of Soil Analysis.* Part II, 2nd ed. Agron. Monogr. 9, Madison, WI: Soil Science Society of America, 1982, pp. 501–538.

152. K.C. Berger, E. Truog. Boron determination in soils and plants using the quinalizarin reaction. *Ind. Eng. Chem. Anal. Ed.* 11:540–545, 1939.

153. W.T. Dible, E. Truog, K.C. Berger. Boron determination in soils and plants. Simplified curcumin procedure. *Anal. Chem.* 26:418–421, 1954.

154. A.S. Baker. Soil boron determination. Modifications in the curcumin procedure for the determination of boron in soil extracts. I. *Agric. Food Chem.* 12:367–370, 1964.

155. J.I. Wear. Boron. In: C.A. Black, ed. *Methods of Soil Analysis, Part 2, Chemical and Microbiological Properties.* Madison, WI: American Society of Agronomy, 1965, pp. 1059–1063.

156. A.J. Jeffrey, L.E. McCallum. Investigation of a hot 0.01 M $CaCl_2$ soil boron extraction procedure followed by ICP-AES. *Commun. Soil Sci. Plant Anal.* 19:663–673, 1988.

157. U.C. Gupta. A simplified method for determining hot water soluble B in podzol soils. *Soil Sci.* 103:424–427, 1967.

158. U.C. Gupta. Some factors affecting the determination of hot water soluble B from podzol soils using azomethine-H. *Can. J. Soil Sci.* 59:241–247, 1979.

159. S.L. McGeehan, K. Topper, D.V. Naylor. Sources of variation in hot water extraction and colorimetric determination of soil boron. *Commun. Soil Sci. Plant Anal.* 20:1777–1786, 1989.

160. B. Wolf. Improvement in the azomethine-H method for the determination of boron. *Commun. Soil Sci. Plant Anal.* 5:39–44, 1974.

161. D.C. Adriano. Trace elements in the terrestrial environment. In: D.C. Adriano, ed. *Boron*. New York: Springer-Verlag, 1986, pp. 73–105.

162. R. Li, U.C. Gupta. Extraction of soil B for predicting its availability to plants. *Commun. Soil Sci. Plant Anal.* 22:1003–1012, 1991.

163. R.L. Aitken, A.J. Jeffrey, B.L. Compton. Evaluation of selected extractants for B in some Queensland soils. *Aus. J. Soil Res.* 25:263–273, 1987.

164. C.D. Tsadilas, D. Dimoyianni, V. Samaras. Methods of assessing boron availability to kiwi fruit plants growing on high boron soils. *Commun. Soil Sci. Plant Anal.* 28:973–987, 1997.

165. A.S. Baker. Relation between available B and B extracted from soils by hot water or phosphoric acid. *Commun. Soil Sci. Plant Anal.* 2:311–320, 1971.

166. U.C. Gupta. A modified procedure for the determination of total B from soil fused with sodium carbonate. *Soil Sci. Soc. Am. Proc.* 30:655–656, 1966.

167. W.D. Gestring, P.N. Soltanpour. Evaluation of the ammonium bicarbonate-DTPA soil test for assessing B availability to alfalfa. *Soil Sci. Soc. Am. J.* 48:96–100, 1984.

168. W.D. Gestring, P.N. Soltanpour. Comparison of soil tests for assessing B toxicity to alfalfa. *Soil Sci. Soc. Am. J.* 51:1214–1218, 1987.

169. T. Matsi, V. Antoniadis, N. Barbayinnis. Evaluation of the NH4HCO3-DTPA soil test for assessing boron availability to wheat. *Commun. Soil Sci. Plant Anal.* 31:669–678, 2000.

170. J.A. Naftel. Colorimetric microdetermination of B. *Ind. Eng. Chem. Anal. Ed.* 11:407–409, 1939.

171. A.R.C. Haas. The turmeric determination of water soluble B in soils of citrus orchards in California. *Soil Sci.* 58:123–137, 1944.

172. K.C. Berger, E. Truog. Boron tests and determination for soils and plants. *Soil Sci.* 57:25–36, 1944.

173. K.C. Berger. Boron in soils and crops. *Adv. Agron.* 1:321–351, 1949.

174. S.R. Porter, S.C. Spindler, A.E. Widdowson. An improved automated colorimetric method for the determination of B in extracts of soils, soil-less peat-based compost, plant materials and hydroponic solutions with azomethine-H. *Commun. Soil Sci. Plant Anal.* 12:461–473, 1981.

175. B. Wolf. The determination of boron in soil extracts, plant materials, composts, manures, water, and nutrient solutions. *Commun. Soil Sci. Plant Anal.* 2:363–374, 1971.

176. W.D. Gestring, P.N. Soltanpour. Boron analysis in soil extracts and plant tissue by plasma emission spectroscopy. *Commun. Soil Sci. Plant Anal.* 12:733–742, 1981.

177. G.A. Spiers, L.J. Evans, S.W. McGeorge. Boron analysis of soil solutions and plant digests using a photodiode-array equipped ICP spectrometer. *Commun. Soil Sci. Plant Anal.* 12:1645–1661, 1990.

178. D.R. Parker, E.H. Gardner. The determination of hot water soluble B in some Oregon soils using a modified azomethine-H procedure. *Commun. Soil Sci. Plant Anal.* 12:1311–1322, 1981.

179. K.J. John, H.H. Chuah, J.H. Neufeld. Application of improved azomethine-H method to the determination of B in soils and plants. *Anal. Lett.* 8:559–568, 1975.

180. A. Mehlich. Mehlich 3 soil test extractant: a modification of Mehlich 2 extractant. *Commun. Soil Sci. Plant Anal.* 15:1409–1416, 1984.

181. W. Horwitz. Minerals in Animal Feed. Atomic Absorption Spectrophotometric Method (968.08) *AOAC Int*, Vol. 1, 2000, chap. 4, p. 40.

182. D. Sun, J.K. Waters, T.P. Mawhinney. Determination of total boron in soils by inductively coupled plasma atomic emission spectrometry using microwave-assisted digestion. *Commun. Soil Sci. Plant Anal.* 29:2493–2503, 1998.

183. O.L. Bennett, E.L. Mathias. Growth and chemical composition of crownvetch as affected by lime, boron, soil source, and temperature regime. *Agron. J.* 65:587–591, 1973.

184. R.J. Bartlett, C.J. Picarelli. Availability of boron and phosphorus as affected by liming an acid potato soil. *Soil Sci.* 116:77–83, 1973.

185. B. Wolf. Factors influencing availability of boron in soil and its distribution in plants. *Soil Sci.* 50:209–217, 1940.

186. L.A. Peterson, R.C. Newman. Influence of soil pH on the availability of added boron. *Soil Sci. Soc. Am. J.* 40:280–282, 1976.

187. U.C. Gupta, J.A. MacLeod. Influence of calcium and magnesium sources on boron uptake and yield of alfalfa and rutabagas as related to soil pH. *Soil Sci.* 124:279–284, 1977.

188. P. Eck, F.J. Campbell. Effect of high calcium application on boron tolerance of Carnation, *Dianthus caryophyllus. Am. Soc. Hort. Sci. Proc.* 81:510–517, 1962.

189. E.G. Beauchamp, I. Hussain. Brown heart in rutabagas grown on southern Ontario soils. *Can. J. Soil Sci.* 54:171–178, 1974.

190. U.C. Gupta. Factors affecting boron uptake by plants. In: U.C. Gupta, ed. *Boron and Its Role in Crop Production*. Boca Raton, FL: CRC Press, 1993, pp. 87–104.

191. H.D. Chapman, A.P. Vanselow. Boron deficiency and excess. *Calif. Citrogr.* 40:455–457, 1955.

192. W.W. Jones, T.W. Embleton, S.B. Boswell, M.L. Steinacker, B.W. Lee, E.L. Barnhart. Nitrogen control program for oranges and high sulfate and/or high boron. *Calif. Citrogr.* 48:128–129, 1963.

193. M. Ruszkowska, Z. Rebowska, M. Kusio, S. Sykut, A. Wojcikowska-Kapusta. Balance of micronutrients in a lysimeter experiment (1985–1989). II. Balance of boron and molybdenum. *Pamietnik Pulawski* 105:63–77, 1994.

194. P. Wojcik. Behavior of soil boron and boron uptake by M.26 apple rootstock as affected by application of different forms of nitrogen rates. *J. Plant Nutr.* 23:1227–1239, 2000.

195. U.C. Gupta, J.A. MacLeod, J.D.E. Sterling. Effects of boron and nitrogen on grain yield and boron and nitrogen concentrations of barley and wheat. *Soil Sci. Soc. Am. J.* 40:723–186, 1976.

196. A.F. El-Kholi, A.A. Hamdy. Boron potassium interrelationship in alfalfa plants. *Egypt J. Soil Sci.* 17:87–97, 1977.

197. N.M. Da Silva, L.H. Carvalho, O.C. Bataghia, C.A. De Abreu. Potassium and boron fertilization of cotton: long-term experiment. *Informacoes Agronomicas* 75:5–6, 1996.

198. M.D. Li, P.A. Dai, C.X. Luo, S.X. Zheng. Interactive effects of K and B on rape seed yield and the nutrient status in rape plants. *J. Soil Sci. China* 20:31–33, 1989.

199. H. Tanaka. Boron absorption by crop plants as affected by other nutrients of the medium. *Soil Sci. Plant Nutr.* 13:41–44, 1967.

200. G.U. Malewar, I.I. Syed, B.S. Indulkar. Phosphorus in cotton-groundnut cropping sequence. *Ann. Agric. Res.* 13:269–270, 1992.

201. C.S. Snyder, Q. Hornsby, J. Welch, L. Gordon, T. Franklin. Effect of Phosphate and Poultry Litter on Cotton Production on Recently Levelled Land in Lonoke County. Res. Ser., Arkansas Agric. Exp. Stn. No. 425, 1993, pp. 64–66.

202. A. Gunes, M. Alpaslan. Boron uptake and toxicity in maize genotypes in relation to boron and phosphorus supply. *J. Plant Nutr.* 23:541–550, 2000.

203. S.S. Ghulaxe. Effect of gypsum and boron on yield performance of groundnut. *J. Soils Crops* 5:157–159, 1995.

204. C.A. Pinyerd, J.W. Odom, F.J. Long, J.H. Dane. Boron movement in a Norfolk loamy sand. *Soil Sci.* 137:428–433, 1984.

205. A. Gunes, M. Alpaslan, Y. Cikili, H. Ozcan. Effect of zinc on the alleviation of boron toxicity in tomato. *J. Plant Nutr.* 22:1061–1068, 1999.

206. J.I. Wear, R.M. Patterson. Effect of soil pH and texture on the availability of water soluble B in the soil. *Soil Sci. Soc. Am. Proc.* 26:344–345, 1962.

207. U.C. Gupta. Relationship of total and hot-water soluble boron, and fixation of added boron, to properties of podzol soils. *Soil Sci. Soc. Am. Proc.* 32:45–48, 1968.

208. N.R. Page, H.P. Cooper. Less-soluble boron compounds for correcting boron nutritional deficiencies. *J. Agric. Food Chem.* 3:222–225, 1955.

209. E.B. Luchese, E. Lenzie, L.O.B. Favero, A. Lanziani. Response of cotton to application of boron to soils from the state of Parana. *Arquivos Biologia Technologia* 37:775–785, 1994.

210. F.J. Hingston. Reactions between boron and clays. *Aus. J. Soil Res.* 2:83–95, 1964.

211. D.V. Singh, R.P.S. Chauhan, R. Charan. Safe and toxic limits of boron for grain in sandy loam and clay loam soils. *Indian J. Agron.* 21:309–310, 1976.

212. E. Okazaki, T.T. Chao. Boron adsorption and desorption by some Hawaiian soils. *Soil Sci.* 105:255–259, 1968.

213. K.C. Berger, P.F. Pratt. Advances in secondary and micronutrient fertilization. In: M.H. McVickar, G.L. Bridger, L.B. Nelson, eds. *Fertilizer Technology and Usage.* Madison, WI: Soil Science Society of America, 1963, pp. 287–340.

214. W.L. Parks, J.L. White. Boron retention by clay and humus systems saturated with various cations. *Soil Sci. Soc. Am. Proc.* 16:298–300, 1952.

215. F.T. Bingham. Boron in cultivated soils and irrigation waters. In: E.L. Kothny, ed. *Trace Elements in the Environment.* Advances in Chemistry Series 123. Washington, DC: American Chemical Society, 1973, pp. 130–138.

216. Y.W. Jame, W. Nicholaichuk, A.J. Leyshon, C.A. Campbell. Boron concentration in the soil solution under irrigation. A theoretical analysis. *Can. J. Soil Sci.* 62:461–470, 1982.

217. E.L. Couch, R.E. Grim. Boron fixation by illites clays. *Clays Clay Miner.* 16:249–255, 1968.

218. F.T. Bingham, A. Elseewi, J.J. Oertli. Characteristics of boron absorption by excised barley roots. *Soil Sci. Soc. Am. Proc.* 34:613–617, 1970.

219. R. Keren, U. Mezuman. Boron adsorption by clay minerals using a phenomenalogical equation. *Clays Clay Miner.* 29:198–204, 1981.

220. N.K. Mehrotra, S.A. Khan, S.C. Agarwala. High SAR (sodium adsorption ratio) irrigation and boron phytotoxicity in sugar beet. *Ann. Arid Zone* 28:69–78, 1989.

221. H.D. Yadav, O.P. Yadav, O.P. Dhankar, M.C. Oswal. Effect of chloride salinity and boron on germination, growth and mineral composition of chickpea (*Cicer arietinum* L.). *Ann. Arid Zone* 28:63–67, 1989.

222. J.R. Wall, C.F. Andrus. The inheritance and physiology of boron response in tomato. *Am. J. Bot.* 49:758–762, 1962.

223. J.C. Brown, J.E. Ambler, R.L. Chaney, C.D. Foy. Differential responses of plant genotypes to micronutrients. In: J.J. Mortvedt, P.M. Giordano, W.L. Lindsay, eds. *Micronutrients in Agriculture.* Madison, WI: Soil Science Society of America, 1972, pp. 389–418.

224. B. Rerkasem, J.F. Loneragan. Boron deficiency in two genotypes in a warm, subtropical region. *Agron. J.* 86:887–890, 1994.

225. J.Y. Li, X.D. Liu, W. Zhou, J.H. Sun, Y.P. Tong, W.J. Liu, Z. Li, P.T. Wang, S.J. Yao. Technique of wheat breeding for efficiently utilizing soil nutrient elements. *Sci. China Ser. B, Chem. Life Sci. Earth Sci.* 38:1313–1320, 1995.

226. T.C. Broyer. Some factors affecting the intensity and locus of expression of mineral deficiency and toxicity symptoms by plants. *Commun. Soil Sci. Plant Anal.* 2:241–248, 1971.

227. H. Tanaka. Response of *Lemna pausicostata* to boron as affected by light intensity. *Plant Soil.* 25:425–434, 1966.

228. C.B. MacInnes, L.S. Albert. Effect of light intensity and plant size on rate of development of early boron deficiency symptoms in tomato root tips. *Plant Physiol.* 44:965–967, 1969.

229. H.M. Rawson, R.N. Noppakoonwong, K.D. Subedi. Effects of boron limitation in combination with changes in temperature, light and humidity on floret fertility in wheat. Sterility in wheat in subtropical Asia: extent, causes and solutions. In: H.M. Rawson, ed. *ACIAR Proceedings Workshop* Lumle, Pokhara, Nepal: Agric. Res. Centre, September 1995, No. 72, 1996, pp. 90–101.

230. M. Saifuzzaman, C.A. Meisner, K.D. Subedi. Wheat sterility in Bangladesh: an overview of the problem, research and possible solutions. Sterility in wheat in subtropical Asia: extent, causes and solutions: In: H.M. Rawson, ed. *ACIAR Proceedings Workshop* Lumle, Pokhara, Nepal: Agric. Res. Centre, September 1995, No. 72, 1996, pp. 104–108.

231. R. Kluge. Contribution to the problem of drought-induced B deficiency in agricultural crops. *Arch. Acker-Pflanzenbau Bodenkd* 15:749–754, 1971.

232. British Columbia Department of Agriculture Field Crops Branch. Boron. Soil Ser. No. 8, 1962, pp. 1–6.

233. J.J. Mortvedt, G. Osborn. Boron concentration adjacent to fertilizer granules in soil, and its effect on root growth. *Soil Sci. Soc. Am. Proc.* 29:187–191, 1965.

234. D.A. Forno, C.J. Asher, D.G. Edwards. Boron nutrition of cassava, and the boron × temperature interaction. *Field Crops Res.* 2:265–279, 1979.

235. J.E. Bowen. Effect of environmental factors on water utilization and boron accumulation and translocation in sugar cane. *Plant Cell Physiol.* 13:703–714, 1972.

236. J. MacQuarrie, W.E. Sackston, B.E. Coulman. Suspected boron deficiency in birdsfoot trefoil in field plots. *Can. Plant Dis. Survey* 63:23–24, 1983.

237. C.H.E. Werkhoven. Boron in some saline and non-saline soils in Southeastern Saskatchewan. *Soil Sci. Soc. Am. Proc.* 28:542–544, 1964.

238. J.S. Kanwar, S.S. Singh. Boron in normal and saline-alkali soils of the irrigated areas of the Punjab. *Soil Sci.* 92:207–211, 1961.

239. K.D. Joshi, B.R. Sthapit. Genetic variability and possible genetic advance for sterility tolerance in wheat (*Triticum aestivum* L.) through breeding. Seminar paper. Lumle Regional Agric. Res. Cent. No. 95-19, 1995.

240. K.D. Subedi, P.J. Gregory, R.J. Summerfield, M.J. Gooding. Cold temperatures and boron deficiency caused grain set failure in spring wheat (*Triticum aestivum* L.). *Field Crops Res.* 57:277–288, 1998.

241. R. Lal, N.R. Fausey, L.C. Brown. Drainage and tillage effects on leaf tissue nutrient contents of corn and soybeans on Crosby-Kokomo soils in Ohio. Drainage in the 21st century: food production and the environment. *7th International Drainage Symposium*, Orlando, FL, 1998, pp. 465–471.

242. X. Han, Y.L. Xu., XZHan, Y.L. Xu. A study of nutritional disorders in monoculture soyabeans and methods for their regulation. *Res. Agric. Modernization* 17:302–307, 1996.

243. U.C. Gupta, J.W. Jame, C.A. Campbell, A.J. Leyshon, W. Nicholaichuk. Boron toxicity and deficiency—a review. *Can. J. Soil Sci.* 65:381–409, 1985.

244. K.V. Paliwal, K.K. Mehta. Boron status of some soils irrigated with saline waters in Kota and Bhilwara regions of Rajasthan. *Indian J. Agric. Sci.* 43:766–772, 1973.

245. B.L. Jain, S.N. Saxena. Distribution of soluble salts and boron in soils in relation to irrigation water. *J. Indian Soc. Soil Sci.* 18:175–182, 1970.

246. H.M. Rawson, K.D. Subedi. Hypothesis for why sterility occurs in wheat in Asia. Sterility in wheat in subtropical Asia: extent, causes and solutions. In: H.M. Rawson, ed. *Proceedings Workshop*, Lumle, Pokhara, Nepal: Agric. Res. Centre, September 1995. No. 72, 1996, pp. 132–134.

247. P. Velasco, A.C. Gurmendi. The mineral industries of Southern South America, In: *Minerals Yearbook, Vol. III Area Reports*: International, U.S. Bureau of Mines, U.S. Department of Interior, Washington, DC: 1988, p. 1089.

248. J.C. Norman. Boron. *Mining Eng.* 50:6,28–30, 1998.

249. G.R. Bradford. Boron. In: H.D. Chapman, ed. *Diagnostic Criteria for Plants and Soils*. Riverside, CA: University of California, 1966, pp. 33–61.

250. C.M. Wilson, R.L. Lovvorn, W.W. Woodhouse Jr. Movement and accumulation of water-soluble boron within the soil profile. *Agron. J.* 43:363–367, 1951.

251. J.R. Woodruff, F.W. Moore, H.L. Musen. Potassium, boron, nitrogen, and lime effects on corn yield and ear leaf nutrient concentration. *Agron. J.* 79:520–524, 1987.

252. B.C. Murphy, J.D. Lancaster. Response of cotton to boron. *Agron. J.* 63:539–540, 1971.

253. M.K. Schon, D.G. Blevins. Foliar boron applications increase the final number of branches and pods on branches of field-grown soybeans. *Plant Physiol.* 92:602–607, 1990.

254. U.C. Gupta, J.A. Cutcliffe. Effects of method of boron application on leaf tissue concentration of boron and control of brown-heart of rutabaga. *Can. J. Plant Sci.* 58:63–68, 1978.

255. J.R. Peterson, J.M. MacGregor. Boron fertilization of corn in Minnesota. *Agron. J.* 58:141–142, 1966.

256. C.K. Labanauskas, W.W. Jones, T.W. Embleton. Low residue micronutrient sprays for citrus. *Proceedings of the 1st International Citrus Symposium* 3, 1969, pp. 1535–1542.

257. J.J. Mortvedt. Micronutrient sources and methods of application in the United States. *International Horticuture Congress 19th Warszawa*, 1974, pp. 497–505.

258. F. Martini, M. Thellier. Use of an (n, α) nuclear reaction to study the long-distance transport of boron in *Trifolium repens* after foliar application. *Planta* 150:197–205, 1980.

259. U.C. Gupta. Yield response to boron and factors affecting its uptake by crops. In: K. Singh, S. Mori, R. Welch, eds. *Perspectives on the Micronutrient Nutrition of Crops*. Jodhpur, India: Scientific Publishers, 2001, pp. 91–118.

9 Chlorine

Joseph R. Heckman
Rutgers University, New Brunswick, New Jersey

CONTENTS

9.1 Historical Information 279
    9.1.1 Determination of Essentiality 280
    9.1.2 Functions in Plants 280
9.2 Diagnosis of Chlorine Status in Plants 281
    9.2.1 Symptoms of Deficiency 281
    9.2.2 Symptoms of Excess 283
    9.2.3 Concentrations of Chlorine in Plants 283
        9.2.3.1 Chlorine Constituents...283
        9.2.3.2 Total Chlorine..283
        9.2.3.3 Distribution in Plants 284
        9.2.3.4 Critical Concentrations...285
        9.2.3.5 Chlorine Concentrations in Crops 285
9.3 Assessment of Chlorine Status in Soils 285
    9.3.1 Forms of Chlorine 285
    9.3.2 Soil Tests 286
    9.3.3 Chlorine Contents of Soil 286
9.4 Fertilizers for Chlorine 287
    9.4.1 Kinds 287
    9.4.2 Application 287
References 288

9.1 HISTORICAL INFORMATION

Chlorine is classified as a micronutrient, but it is often taken up by plants at levels comparable to a macronutrient. Supplies of chlorine in nature are often plentiful, and obvious symptoms of deficiency are seldom observed. In many crops it is necessary to remove chlorine from air, chemicals, and water to induce symptoms of chlorine deficiency. Using precautions to establish a relatively chlorine-free environment, Broyer et al. (1) was able to convincingly demonstrate that chlorine is an essential nutrient. Although crop responses to chlorine applications in the field were suspected as early as the mid-1800s, it was not until fairly recently that chlorine was considered a potentially limiting nutrient for crop production under field conditions. In the 1980s, the responsiveness of some crops to chlorine fertilization became recognized more widely (2). Even though chlorine has gained the attention of agronomists, much of the focus on chlorine in terms of crop production continues to be over the presence of excess levels of chloride salts in soils, water, and fertilizers (3,4). This chapter, however, is concerned primarily with chlorine as a plant nutrient.

9.1.1 Determination of Essentiality

Early observations of plant growth responses derived from the use of chlorine-containing fertilizers had suggested that chlorine was at least beneficial if not essential (5). Demonstrating the essentiality of chlorine is experimentally challenging because chlorine is present widely in the environment, and special precautions are necessary to remove chlorine from chemicals, water, and air to induce deficiency symptoms in most species (6). Solution culture experiments conducted in a relatively chlorine-free environment (1) provided the first recognition of chlorine as an essential microelement. These experiments further showed that chlorine deficiency symptoms were alleviated specifically by the addition of chloride. Using solution culture (7), acute chlorine deficiency or at least restricted growth was demonstrated in lettuce (*Lactuca sativa* L.), tomato (*Lycopersicum esculentum* Mill.), cabbage (*Brassica oleracea* var. *capitata* L.), carrot (*Daucus carota* L.), sugar beet (*Beta vulgaris* L.), barley (*Hordeum vulgare* L.), alfalfa (*Medicago sativa* L.), buckwheat (*Fagopyrum esculentum* Moench), corn (*Zea mays* L.), and beans (*Phaseolus vulgaris* L.). Under the same conditions however, squash (*Praecitrullus fistulosus* Pang.) plants failed to exhibit any signs of chlorine deficiency. Species not affected or least affected by low chlorine supply appear to accumulate more chlorine than provided by the culture solutions. It has been assumed that chlorine was absorbed from the atmosphere and that plants differed in this ability (6,7). More recently, low-chlorine solution studies have produced chlorine deficiency symptoms in red clover (*Trifolium pratense* L.) and in wheat (*Triticum aestivum* L.) (8–10). Thus, the essentiality of chlorine has been established by the observations of the deficiency in a wide range of species.

9.1.2 Functions in Plants

Chlorine is readily taken up by plants in the electrically charged form as chloride ion (Cl^-). Although chlorine occurs in plants as chlorinated organic compounds (11), chloride is the major form within plants, where it is bound only loosely to exchange sites or is a highly mobile free anion in the plant water. As an essential element, chlorine has several biochemical and physiological functions within plants.

Chloride appears to be required for optimal enzyme activity of asparagine synthethase (12), amylase (13), and ATPase (14). In photosynthesis, chloride is an essential cofactor for the activation of the oxygen-evolving enzyme associated with photosystem II (15,16). Chloride may bind (17) to the polypeptides associated with the water-splitting complex of photosystem II, and it may stabilize the oxidized state of manganese by acting as a bridging ligand (18–20). Chloride concentrations required for biochemical functions are relatively low in comparison to concentrations required for osmoregulation.

In rapidly expanding tissues such as elongating cells of roots and shoots, chloride accumulates in the tonoplast, to function as an osmotically active solute (21,22). This transport of chloride into the tonoplast occurs in association with the proton-pumping ATPase activity at the tonoplast, being specifically stimulated by chloride (14). This osmoregulatory function in specific tissues requires concentrations of chloride that are not typical of a micronutrient (23,24). The accumulation of chloride in plant cells increases tissue hydration (25) and turgor pressure (26). This osmotic function of chloride works closely with potassium to facilitate cell elongation and growth. The importance of this osmoregulatory role of chloride in plants depends on growing conditions and the presence of alternative anions, such as nitrate, which might function as substitutes for chloride.

Chloride along with potassium participates in stomatal opening by moving from epidermal cells to guard cells to act as an osmotic solute that results in water uptake into and a bowing apart of the guard cell pair (27). In many plant species, depending on the external supply of chloride, malate synthesis may occur in the guard cells and replace the need for chloride influx (28,29). Chloride, however, is essential for stomatal functioning in some plant species (30). In onion (*Allium cepa* L.), for example, where the guard cells are unable to synthesize malate, there is a requirement for an influx of chloride that is equivalent to potassium for stomatal opening to occur.

Relative differences in the uptake of cations (NH_4^+, Ca^{2+}, Mg^{2+}, K^+, Na^+) and anions (NO_3^-, Cl^-, SO_4^{2-}, $H_2PO_4^-$) by plants require the maintenance of electroneutrality in plant cells as well as in the external soil solution (31). As an anion, chloride serves to balance charges from cations. In plants well supplied with chloride, this inorganic anion may serve as an alternative to the formation of malate in its charge-balancing role (32). This role of chloride may be of greater importance when cation uptake exceeds anion uptake, as often occurs with plants provided with ammonium nutrition.

The functions of most of the over 130 chlorinated organic compounds (11) that have been identified in higher plants have not been determined. Some legume species contain chlorinated indole-3-acetic acid (IAA) in their seeds. The chlorinated form of IAA is more resistant to degradation, and this resistance may be responsible for increasing the rate of hypocotyl elongation over the rate of IAA production itself (4,33).

9.2 DIAGNOSIS OF CHLORINE STATUS IN PLANTS

9.2.1 SYMPTOMS OF DEFICIENCY

Visible deficiency symptoms for chlorine have been well characterized in several crops by growth of plants in chlorine-free nutrient solutions (1,7,8,10). The most commonly described symptom of chlorine deficiency is wilting of leaves, especially at the margins. As the deficiency becomes more severe, the leaves may exhibit curling, shriveling, and necrosis (Figure 9.1A). Roots of chlorine-deficient plants have been described as stubby with club tips. Deficiency symptoms of chlorine are not commonly exhibited visually in most crops growing in the field, but symptoms are sometimes observed in wheat and coconut palm (*Cocos nucifera* L.). In chlorine-deficient wheat, the symptoms are expressed as chlorotic or necrotic lesions on leaf tissue (Figure 9.1B). These symptoms that result from chlorine deficiency have been named 'Cl-deficient leaf spot syndrome' (9,10). It has also been shown that bromide (Figure 9.1C) does not substitute for chloride in the prevention of deficiency symptoms (10). In coconut palm, the symptoms are exhibited as wilting and premature senescence of leaves, frond fracture, and stem cracking and bleeding (34).

Chlorine deficiency is also indicated by yield increases that may occur with various crops in response to chloride fertilization. Wheat and barley often respond to chloride fertilization with increases in grain yield on soils with low chloride on the Great Plains of North America (2,35–41). Corn exhibited no response to chloride fertilization in some studies (2,42–44), but in a high-yield environment in New Jersey, fertilization of corn with 400 kg Cl ha^{-1} increased the 5-year average

FIGURE 9.1 (A) Wheat (*Triticum turgidum* L. Durum Group) grown with chloride added at 30 mmol in 15 liters of nutrient solution (0.002M KCl); (B) Wheat grown in the absence of halide; (C) Wheat grown in absence of chloride and with 1.5 mmol bromide in 15 liters of nutrient solution (0.0001M KBr). Photographs from Engel et al., (9). Reprinted with permission of the authors and Soil Science Society of America. (For a color presentation of this figure, see the accompanying compact disc.)

yield by 1000 kg ha^{-1} over the unfertilized control (45,46). Positive responses from chloride fertilization have also been observed with rice (*Oryza sativa* L.), sugarcane (*Saccharum edule* Hassk.), potato (*Solanum tuberosum* L.), kiwifruit (*Actinidia deliciosa* A. Chev.), coconut palm, sugar beet, and asparagus (*Asparagus officinalis* L.) (2,47). These responses indicate that chloride is sometimes a yield-limiting nutrient in field environments where chlorine inputs from rainfall and other natural sources are inadequate.

The beneficial effects of chloride fertilization are sometimes not the result of a plant response directly to enhanced chloride nutrition, but rather may result from suppression of plant diseases. Addition of chloride has been reported to reduce the severity of at least 15 different foliar and root diseases on 11 different crops (Table 9.1). Several possible mechanisms may explain the effects of chloride nutrition on disease suppression and host resistance.

In acid soils, chloride inhibits nitrification (48,49). Keeping nitrogen in the ammonium form can lower rhizosphere pH and influence microbial populations and nutrient availability in the rhizosphere (31,50). Competition between chloride and nitrate for uptake also tends to reduce nitrate concentrations in plant tissues (4,51). When plants take up more ammonium and less nitrate, it usually causes rhizosphere acidification, which in turn, may enhance manganese availability (52). Chloride can also enhance manganese availability by promoting manganese-reducing microorganisms in soil (53). Factors which increase manganese availability have been associated with improved host resistance to diseases such as take-all on grain crops (54). Higher concentrations of chloride in plant tissues can also enhance water retention and turgor when roots have been

TABLE 9.1
Diseases Suppressed by Chlorine Fertilization

Crop	Suppressed Disease	Reference
Asparagus (*Asparagus officinalis* L.)	Fusarium crown and root rot (*Fusarium oxysporum* and *Fusarium proliferatum*)	47, 53, 74, 75
Barley (*Hordeum vulgare* L.)	Common root rot (*Cochliobolus sativus* and *Fusarium* spp.)	55, 76, 77
	Fusarium crown and root rot (*Fusarium graminearum*)	70
	Spot blotch (*Bipolaris sorokiniana*)	77
Celery (*Apium graveolens* L.)	Fusarium yellows (*Fusarium oxysporum* f.sp. *apii*)	78
Coconut palm (*Cocos nucifera* L.)	Gray leaf spot (*Pestalotiopsis palmarum*; *Helminthosporium incurvatum*)	34
Corn (*Zea mays* L.)	Stalk rot (*Gibberella zeae*; *Colletotrichum graminicola*; *Diplodia maydis*)	46, 79
Durum (*Triticum durum* Desf.)	Common root rot (*Cochliobolus sativus* and *Fusarium* spp.)	70
Pearl millet (*Pennisetum glaucum* R. Br.)	Downy mildew (*Sclerospora graminicola*)	70
Spring wheat (*Triticum aestivum* L.)	Leaf rust (*Puccinia triticina*)	80
	Septoria (*Stagonospora nodorum*)	70
	Tanspot (*Pyrenophora triticirepentis*)	66
Table beets (*Beta vulgaris* L.)	Rhizoctonia crown and root rot (*Rhizoctonia solani*)	81
Winter wheat (*Triticum aestivum* L.)	Leafspot (*Pyrenophora triticirepentis*)	9, 10
	Leaf rust (*Puccinia triticina*)	82
	Stripe rust (*Puccinia striiformis*)	70
	Take-all root rot (*Gaeumannomyces graminis* var. *tritici*)	26, 83

attacked by pathogens (26). The amount of organic acids, such as malate, in plant tissues and exuded from roots, decreases with chloride supply; this action deprives pathogens of an organic substrate (55).

9.2.2 Symptoms of Excess

Chloride toxicity symptoms have been observed in many field, vegetable, and fruit crops (6,56). Curling of the leaf margins, marginal leaf scorch, leaf necrosis, and leaf drop are typical symptoms. Older leaves are usually the first to exhibit symptoms that may progress upward, affecting the entire foliage. Dieback of the terminal axis and small branches may occur in cases of severe toxicity. These symptoms of chloride toxicity occur in the absence of sodium, but they are also similar to symptoms of salt toxicity that occur when chloride is accompanied by sodium. Crops and cultivars within crops vary widely in tolerance to high levels of chloride, with corn being relatively tolerant to chloride (56) compared to soybean (*Glycine max* Merr.) (57).

9.2.3 Concentrations of Chlorine in Plants

9.2.3.1 Chlorine Constituents

Most of the chlorine in plants is present in the form of the anion, chloride. However, more than 130 natural chlorine-containing compounds have been isolated from plants (11). They may include polyacetylenes, thiophenes, iridoids, sesquiterpene lactones, pterosinoids, diterperenoids, steroids and gibberellins, maytansinoids, alkaloids, chlorinated chlorophyll, chloroindoles and amino acids, phenolics, and fatty acids. Although the functions of naturally occurring chlorine-containing compounds in plants have not received much attention in plant nutrition, the fact that these compounds often exhibit a strong biological activity suggests a need to investigate their potential importance. Some chlorine-containing compounds may behave as hormones in the plant, or they may have a function in protection against attack from other organisms.

9.2.3.2 Total Chlorine

The total chlorine accumulation by crops varies greatly, depending on chloride supply from soil. Many studies (45,56,58–62) of plant responses to applied chloride have shown that plant tissue chloride concentrations increase markedly with increasing application rates of chloride. A few studies have measured total chlorine uptake by crops, and these studies also indicate that chloride accumulation by crops increases with increasing amounts of chloride fertilization. A study (25) conducted in North Carolina with corn fertilized with 0, 50, 100, 150, and 200 kg Cl ha^{-1} in the form of KCl found that the aboveground biomass at 77 days after emergence accumulated 26, 50, 63, 79, and 81 kg Cl ha^{-1}, respectively. A Wisconsin study (62) found that alfalfa accumulated only 5 kg Cl ha^{-1} on unamended soil, but on soil fertilized with 1017 kg Cl ha^{-1} as KCl in the fall of the previous season, the herbage accumulated 86 kg Cl ha^{-1}. These accumulation values for chloride by corn or alfalfa indicate that the potential for total crop accumulation for this nutrient is potentially large on soils well supplied with chloride. Even though chlorine is classified as a micronutrient, total chlorine accumulation often exceeds the levels of crop accumulation of macronutrients such as phosphorus or sulfur.

The amount of chlorine accumulation required to prevent deficiency symptoms in most crops however, is much less than that which is typically accumulated (Table 9.2). A laboratory study (7) that determined the chlorine requirements of 11 different crop species estimated that plants require 1 lb of chlorine for each 10,000 lb of dry matter produced, or a concentration of about 0.1 g kg^{-1}. On a land area basis, large crops may need about 2.24 kg ha^{-1} or more of chlorine. This estimate for plant chlorine requirement is presumed to be for biochemical functions (2). The benefits that are

TABLE 9.2
Chloride Concentrations in Plants

Crop	Latin Name	Plant Part	Concentration Ranges of Tissue Cl (mg g⁻¹ DM)			Reference
			Deficient	Normal	Toxic[a]	
Alfalfa	*Medicago sativa* L.	Shoot	0.65	0.9–2.7	6.1	6, 72
Apple	*Malus domestica* Borkht.	Leaves	0.1		>2.1	6
Avocado	*Persea americana* Mill.	Leaves		~1.5–4.0	~7.0	84, 85
Barley	*Hordeum vulgare* L.	Heading shoot	1.2–4.0	>4.0		9, 86
Citrus	*Citrus* spp. L.	Leaves		~2.0	~4.0–7.0	84, 87
Coconut palm	*Cocos nucifera* L.	Leaves	2.5–4.5	>6.0–7.0		86
Corn	*Zea mays* L.	Ear leaves		>3.2		45
Corn	*Zea mays* L.	Ear leaves		1.1–10.0	>32.7	56
Corn	*Z. mays* L.	Shoots	0.05–0.11			7
Cotton	*Gossypium hirsutum* L.	Leaves		10.0–25.0	>25.0–33.1	88
Grapevine	*Vitis vinifera* L.	Petioles		0.7–8.0	10.0–11.0	6, 64
Kiwifruit	*Actinidia deliciosa* A. Chev.	Leaves	2.1	6.0–13.0	>15.0	60, 89
Lettuce	*Lactuca sativa* L.	Leaves	>0.14	2.8–19.8	>23.0	7, 90
Pear	*Pyrus communis* L.	Leaves		<0.50	>10.0	91
Peach	*Prunus persica* Batsch.	Leaves		0.9–3.9	10.0–16.0	6, 91
Peanut	*Arachis hypogaea* L.	Shoot		<3.9	>4.6	92
Potato	*Solanum tuberosum* L.	Mature shoot	<1.0	2.0–3.3	12.2	93
Potato	*Solanum tuberosum* L.	Petioles	0.71–1.42	18.0	44.8	58, 94
Red clover	*Trifolium pratense* L.	Shoot	0.15–0.21			8
Rice	*Oryza sativa* L.	Shoot	<3.0		>7.0–8.0	95
Rice	*O. sativa* L.	Mature straw		5.1–10.0	>13.6	73, 96
Soybean	*Glycine max* L. Merr.	Leaves		0.3–1.5	16.7–24.3	97, 98, 99
Spinach	*Spinacia oleracea* L.	Shoot	>0.13			100
Spring wheat	*Triticum aestivum* L.	Heading shoot	1.5	3.7–4.7	>7.0	66, 92
Strawberry	*Fragaria vesca*	Shoot		1.0–5.0	>5.3	91, 92
Subterranean clover	*Trifolium subterraneum* L.	Shoot	>1.0			101
Sugar beet	*Beta vulgaris* L.	Leaves	0.71–1.78			102, 103
Sugar beet	*B. vulgaris* L.	Petioles	<5.7	>7.1–7.2	>50.8	102, 104
Tobacco	*Nicotiana tabacum* L.	Leaves		1.2–10.0	>10.0	6, 105
Tomato	*Lycopersicon esculentum* Mill.	Shoot	0.25		~30.0	1, 106
Wheat	*Triticum aestivum* L.	Heading shoot	1.2–4.0	>4.0		9, 86

[a]The plant yields decline or the plant shows visible scorching symptoms in leaves.

sometimes observed from higher concentrations of chlorine are likely due to its osmoregulatory role in plants (36).

9.2.3.3 Distribution in Plants

Most of the chlorine in plants is not incorporated into organic molecules or dry matter, but remains in solution as chloride and is loosely bound to organic molecules. Chloride concentrations expressed on a tissue-water basis may typically range from 50 to 150 mmol L^{-1} (4). A study (25) that determined chloride in the tissue water and the dry matter of whole corn plants at 35 days after emergence found a concentration of 66 mmol Cl L^{-1} (1.83 g kg^{-1} dry matter basis) for corn grown on soil fertilized with 200 kg Cl ha^{-1} applied as KCl and only 10 mmol Cl L^{-1} (2.5 g kg^{-1} dry matter basis) for corn plants grown on unamended soil. In general, chloride concentrations are higher in tissues that have high water content. Chloride concentrations are presumably highest in the rapidly expanding zones of root and shoot tissue. Pulvini and guard cells also have higher concentrations of chloride than the bulk tissue (4).

Vegetative plant tissues usually accumulate increasing concentrations of chloride with increasing supply of chloride, but plants parts can also exclude chloride (4,25,63). Corn seed may have only 0.44 to 0.64 g Cl kg^{-1} on a dry weight basis, and chloride accumulation in the grain is not influenced by chloride supply (45). In many crops, chloride transport from roots to shoots is restricted by a mechanism that resides in the roots (4,64,65). Soybean cultivars that exclude chloride from the shoots are more salt-tolerant than cultivars that accumulate chloride (57).

9.2.3.4   Critical Concentrations

Reports on critical tissue concentrations of chloride for crops grown in the field are few in number (Table 9.2). Studies conducted in the Great Plains of the United States have examined the relationship between tissue chloride concentration and relative yield of wheat. In wheat plants at head emergence, a critical chloride concentration of 1.5 g kg^{-1} was given in a 1986 report (66). In a more recent and larger study (67) that was based on an assessment of 219 wheat cultivars, three zones of chloride status were identified: (i) a deficiency zone with a plant chloride concentration <1.0 g kg^{-1}, (ii) an adequate chloride status zone with concentrations ≥4.0 g kg^{-1}, (iii) and a transition, or critical range, between these two zones. A study (45) of corn grown in high-yield environments in New Jersey suggested a critical ear-leaf chloride concentration of 3.2 g kg^{-1}, derived from a comparatively small database.

9.2.3.5   Chlorine Concentrations in Crops

A review (4) of chlorine nutrition tabulated the concentrations of chloride in a wide variety of crops. The compilation of data in Table 9.2 shows that concentrations of chloride classified as deficient, normal, or toxic vary widely among plant species.

9.3   ASSESSMENT OF CHLORINE STATUS IN SOILS

9.3.1   Forms of Chlorine

Chlorine is present in the soil solution primarily in the anionic form as chloride. Chloride concentrations in soil extracts may range from <1 mg kg^{-1} to more than several thousand mg kg^{-1} (68). Chlorine may also be present in organic forms such as chlorinated hydrocarbon pesticide residues. Some of these chlorine-containing molecules are recalcitrant, whereas others can be metabolized or mineralized to release the chlorine.

Although plants can accumulate chlorine foliarly and from the atmosphere, the concentration of chlorine in plant tissue is often closely related to the supply or concentration of chloride in soil. Testing soils for chloride is routine in laboratories involved in salinity problems, but soil testing for chloride supply to predict crop response to fertilization is a fairly recent development. Soil test interpretations for chloride supply are currently conducted in the North American Great Plains and are limited to only a few crops (2).

In this large land-locked geographical region, little potassium fertilizer (KCl) is applied, and chloride input from rainfall is low. Soil test interpretations for chloride have not been developed outside this region because chloride inputs from various sources are often greater and because supplies of this nutrient are generally considered adequate for most crops.

9.3.2 SOIL TESTS

The solubility and mobility of chloride in soil is similar to nitrate, and soil sampling depths for chloride, like nitrate, are typically greater than for less mobile nutrients. Although the best soil sampling depth may vary depending on the rooting depth of the crop, a sampling to a depth of 60 cm has been found to be a good indicator of chloride availability to potato (58) and to spring wheat (2). Crops, such as sugar beet and winter wheat with deeper rooting depths, may need a deeper sampling depth (2,37).

Because chloride is highly soluble and only weakly adsorbed, it can be extracted from soil with water or any dilute electrolyte. The choice of extractant may depend on the analytical method employed to determine the concentration of chloride in the extract. Methods of analysis for quantifying extractable chloride may include colorimetric, potentiometric, or chromatographic procedures (69). Precautions should be taken to avoid potential sources of chloride contamination (e.g., perspiration, soil sample containers, dust, glassware, water) during soil sampling and laboratory analysis.

9.3.3 CHLORINE CONTENTS OF SOIL

In the Great Plains of the United States, soil tests are performed to assess the soil chloride level as a factor to be considered in decisions regarding application of chloride fertilizer. The relative responsiveness of the various wheat and barley cultivars to chloride is also considered. Some cultivars of spring wheat and barley frequently exhibit responses to chloride, while others seldom exhibit a response (41,66,70,71). Chloride response trials conducted at 36 locations found that a critical level of 43 kg Cl ha^{-1} in the top 60 cm layer of soil would generally separate responsive sites from nonresponsive sites (66,70). On the basis of this research, soils were classified as low ($\leq$34 kg Cl ha^{-1}), medium (35 to 67 kg Cl ha^{-1}), or high (>67 kg Cl ha^{-1}) in relation to the probability of observing a response to chloride addition. Chloride fertilization is recommended according to this equation: Cl$^-$ to apply (kg ha^{-1}) = 67 – Cl$^-$ (kg Cl ha^{-1} to 60 cm sampling depth). This recommendation is specific to wheat and barley crops grown in the region, and it should not be extrapolated to other areas under different climate, soil, and cultural conditions.

Soil test calibration data on chloride are unavailable for most crops and soils around the world. However, an observation of chloride deficiency in Australia provides some insight into concentrations of chloride in soil that may limit growth of some plants (72). In this instance, it was found that subterranean clover (*Trifolium subterraneum* L.) exhibited poor growth when the soil contained only 3 to 5 µeq of Cl per 100 g (1 to 2 mg kg^{-1}).

When other factors limit crop yield potential, the potential for a response to chloride fertilization is also limited. For example, corn grown in high-yield environments in New Jersey (18 miles from the Atlantic Ocean) exhibited yield increases from chloride addition on soils that held 20 kg Cl ha^{-1} in the top 60 cm layer of soil (45,46). In other studies with corn under less favorable conditions, yield increases due to chloride fertilization were either small or nil (2,42–44).

In many instances, chloride is frequently supplied to crops as a consequence of the widespread use of KCl-based fertilizers that are applied with the intention of providing potassium. Recommended application rates of potassium, when applied as KCl, will generally supply sufficient chloride to most crops. It is possible that the supply of chloride is sometimes limiting for crops grown on a wider range of soils but that the crop responses to chloride go unrecognized because they are attributed to potassium.

Chloride is widely distributed in soils. Concentrations normally range from 20 to 900 mg kg^{-1} with a mean concentration of 100 mg kg^{-1} (68). Because igneous rocks and parent materials in general contain only minor amounts of chloride, little of this nutrient arises from weathering. Most of the chloride present in soils arrive from rainfall, marine aerosols, volcanic emissions, irrigation waters, and fertilizers (4).

Chloride is not adsorbed by minerals at pH levels above 7.0 and is only weakly absorbed in kaolinitic and oxidic soils that have positive charges under acid conditions (68). Chloride accumulates primarily in soil under arid conditions where leaching is minimal and where chloride moves upward in the soil profile in response to evapotranspiration. Poorly drained soils and low spots receiving chloride from runoff, seepage, or irrigation water also may accumulate chloride (57). Near the ocean, soils have high levels of chloride, but with increasing distance from the ocean, chloride concentration in soils typically falls (2,4).

How a crop is harvested influences the amount of chloride in soil. When harvested only as seed, the amount of chloride removed is limited (<8 kg ha^{-1} for a corn yield of 11.3 Mg ha^{-1}), but when harvested as green biomass the amount of chloride removal may be substantial (81 kg ha^{-1} for corn as silage) (25). Because chloride leaches from aging leaves, harvest of mature biomass may remove only about half as much chloride as does harvest before the onset of senescence (59,61).

9.4 FERTILIZERS FOR CHLORINE

9.4.1 KINDS

Chlorine is added to soil from a wide variety of sources that include chloride from rainwater, irrigation waters, animal manures, plant residues, fertilizers, and some crop protection chemicals. The amount of chloride deposited annually from the atmosphere varies from 18 to 36 kg^{-1} ha^{-1} year^{-1} for continental areas to more than 100 kg^{-1} ha^{-1} year^{-1} for coastal areas (4). Most of the chloride applied as animal manures or plant residues is soluble and readily available for crop uptake. Because most of the chloride in animal manure is probably present in the liquid fraction, manure management and handling may influence the concentration of chloride.

Potassium chloride is the most widely applied chloride fertilizer. Although KCl is usually intended as a potassium fertilizer, it in effect supplies 0.9 kg of chloride for each kg of potassium. Other chloride fertilizers include NaCl, $CaCl_2$, $MgCl_2$, and NH_4Cl (Table 9.3). All these salts are soluble and readily available to supply chloride for plant uptake. Organic agriculture, which discourages the use of KCl and most salt-based fertilizers, obtains chloride primarily from manure and other natural sources.

9.4.2 APPLICATION

Chloride, like nitrate, is susceptible to loss from soil by leaching in areas of high rainfall (62,73). Management practices that minimize chloride leaching will enhance chloride accumulation by crops. When crops with high chloride requirements are grown, the application of chloride in the

TABLE 9.3
Sources Commonly Used as Chlorine Fertilizers

Source	Chlorine Concentrations (%)
Potassium chloride (KCl)	47
Sodium chloride (NaCl)	60
Ammonium chloride (NH_4Cl)	66
Calcium chloride ($CaCl_2$)	64
Magnesium chloride ($MgCl_2$)	74

spring or close to the time of plant growth should enhance chloride accumulation. Owing to the potential for salt injury, it is safer to broadcast chloride fertilizers than to apply them as a band.

REFERENCES

1. T.C. Broyer, A.B. Carlton, C.M. Johnson, P.R. Stout. Chlorine — a micronutrient element for higher plants. *Plant Physiol.* 29:526–532, 1954.
2. P.E. Fixen. Crop responses to chloride. *Adv. Agron.* 50:107–150, 1993.
3. Anonymous. The trends towards chloride-free specialty fertilizers: always fully justified? *New Ag. Int.* 2:36–49, 2001.
4. G. Xu, H. Magen, J. Tarchitzky, U. Kafkaf. Advances in chloride nutrition of plants. *Adv. Agron.* 28:97–150, 2002.
5. C.H. Lipman. Importance of silicon, aluminum, and chlorine for higher plants. *Soil Sci.* 45:189–198.
6. F.M. Eaton. Chlorine. In: H.D. Chapman, ed. *Diagnostic Criteria for Plants and Soils.* Riverside: University of California, 1966, pp. 98–135.
7. C.M. Johnson, P.R. Stout, T.C. Broyer, A.B. Carlton. Comparative chlorine requirements of different plant species. *Plant Soil* 8:337–353, 1957.
8. D.C. Whitehead. Chlorine deficiency in red clover grown in solution culture. *J. Plant Nutr.* 8:193–198, 1985.
9. R.E. Engel, P.L. Bruckner, D.E. Mathre, S.K.Z. Brumfield. A chloride-deficient leaf spot syndrome of wheat. *Soil Sci. Soc. Am. J.* 61:176–184, 1997.
10. R.E. Engel, L. Bruebaker, T.J. Emborg. A chloride deficient leaf spot of durum wheat. *Soil Sci. Soc. Am. J.* 65:1448–1454, 2001.
11. K.C. Engvild. Chlorine-containing natural compounds in higher plants. *Phytochemistry* 25:781–791, 1986.
12. S.E. Rognes. Anion regulation of lupin asparagine synthetase: chloride activation of the glutamine-utilizing reaction. *Phytochemistry* 19:2287–2293, 1980.
13. D.E. Metzler. *Biochemistry—The Chemical Reactions of Living Cells.* New York: Academic Press, 1979, pp. 357–370.
14. K.A. Churchill, H. Sze. Anion-sensitive, H^+-pumping ATPase of oat roots. Direct effects of Cl^-, NO_3^-, and a disulfonic stilbene. *Plant Physiol.* 76:490–497, 1984.
15. D.I. Arnon, F.R. Whatley. Is chloride a coenzyme of photosynthesis? *Science* 110:554–556, 1949.
16. S. Izawa, R.L. Heath, G. Hind. The role of chloride ion in photosynthesis. III. The effect of artificial electron donors upon electron transport. *Biochim. Biophys. Acta* 180:388–398, 1969.
17. I.C. Baianu, C. Critchley, Govindjee, H.S. Gutowsky. NMR study of chloride ion interactions with thylakoid membranes. *Proc. Natl. Acad. Sci. USA* 81:713–3717, 1984.
18. C. Critchley. The role of chloride in photosystem II. *Biochim. Biophys. Acta* 811:33–46, 1985.
19. W.J. Coleman, Govindjee, H.S. Gutowsky. The location of the chloride binding sites in the oxygen-evolving complex of spinach photosystem II. *Biochim. Biophys. Acta* 894:453–459, 1987.
20. P.H. Homann. Structural effects of chloride and other anions on the water oxidizing complex of chloroplast photosystem II. *Plant Physiol.* 88:194–199, 1988.
21. E.V. Maas. Physiological responses to chloride. In: T.L. Jackson, ed. *Special Bulletin on Chloride and Crop Production.* Atlanta, GA: Potash & Phosphate Institute, No. 2, 1986, pp. 4–20.
22. A. Hager, M. Helmle. Properties of an ATP-fueled, Cl-dependent proton pump localized in membranes of microsomal vesicles from maize coleoptiles. *Z. Naturforsch* 36C:997–1008, 1981.
23. D.F. Gerson, R.J. Poole. Chloride accumulation by mung bean root tips. A low affinity active transport system at the plasmalemma. *Plant Physiol.* 50:603–607, 1972.
24. T.J. Flowers. Chloride as a nutrient and as an osmoticum. In: B. Tinker, A. Lauchi, eds. *Advances in Plant Nutrition,* Vol. 3. New York: Praeger, 1988, pp. 55–78.
25. J.R. Heckman. Corn and Soybean Tissue Water Content, Nutrient Accumulation, Yield and Growth Pattern Responses to Potassium and Chloride Fertility Differences. Ph.D. Dissertation. North Carolina State University, Raleigh, NC, 1989.
26. N.W. Christensen, R.G. Taylor, T.L. Jackson, B.L. Mitchell. Chloride effects on water potentials and yield of winter wheat infected with take-all root rot. *Agron. J.* 73:1053–1058, 1981.

27. Y. Lee, S.M. Assmann. Diacylglycerols induce both ion pumping in patch-clamped guard-cell proto-plasts and opening of intact stomata. *Proc. Natl. Acad. Sci. USA* 88:2127–2131, 1991.

28. Z. Du, K. Aghoram, W.H. Outlaw Jr. In vivo phosphorylation of phosphoenolpyruvate carboxylase in guard cells of *Vicia faba* L. is enhanced by fusicoccin and suppressed by abscisic acid. *Arch. Biochem. Biophys.* 337:345–350, 1997.

29. K. Raschke, H. Schnabl. Availability of chloride affects the balance between potassium chloride and potassium malate in guard cells of *Vicia faba* L. *Plant Physiol.* 62:84–87, 1978.

30. H. Schnabl, K. Raschke. Potassium chloride as stomatal osmoticum in *Allium cepa* L. [onion], a species devoid of starch in guard cells. *Plant Physiol.* 65:88–93, 1980.

31. J.R. Heckman, J.E. Strick. Teaching plant-soil relationships with color images of rhizosphere pH. *J. Nat. Resour. Life Sci. Educ.* 25:13–17, 1996.

32. H. Beringer, K. Hoch, M.G. Lindauer. Source:sink relationships in potato (*Solanum tuberosum*) as influenced by potassium chloride or potassium sulphate nutrition. *Plant Soil* 124:287–290, 1990.

33. M. Hoffinger, M. Bottger. Identification by GC-MS of 4-chloroindolylacetic acid and its methyl ester in immature *Vicia faba* broad bean seeds. *Phytochemistry* 18:653–654, 1979.

34. H.R. von Uexküll, J.L. Sanders. Chloride in the nutrition of palm trees. In: T.L. Jackson, ed. *Special Bulletin on Chloride and Crop Production*. Atlanta, GA: Potash & Phosphate Institute, No. 2, 1986, pp. 84–99.

35. L.C. Bonczkowski. Response of Hard Red Winter Wheat to Chloride Application in Eastern Kansas. Ph.D. dissertation, Kansas State University, Manhattan, KS, 1989.

36. R.E. Engel, D.E. Mathre. Effect of fertilizer nitrogen source and chloride on take-all of irrigated hard red spring wheat. *Plant Dis.* 72:393–396, 1988.

37. R.E. Engel, W.E. Grey. Chloride fertilizer effects on winter wheat inoculated with *Fusarium culmorum*. *Agron. J.* 83:204–208, 1991.

38. P.E. Fixen. Chloride fertilization. *Crops Soil Mag.* 39:14–16, 1987.

39. P.E. Fixen, G.W. Buchenau, R.H. Gelderman, T.E. Schumacher, J.R. Gerwing, F.A. Cholik, B.G. Farber. Influence of soil and applied chloride on several wheat parameters. *Agron. J.* 78:736–740, 1986.

40. R.J. Goos, B.E. Johnson, R.W. Stack. Effect of potassium chloride, imazalil, and method imazalil application on barley infected with common root rot. *Can. J. Plant Sci.* 69:437–444, 1989.

41. R.M. Mohr. The Effect of Chloride Fertilization on Growth and Yield of Barley and Spring Wheat. M.S. thesis, University of Manitoba, Winnipeg, Manitoba, Canada, 1992.

42. W.K. Schumacher, P.E. Fixen. Residual effects of chloride application in a corn-wheat rotation. *Soil Sci. Soc. Am. J.* 53:1742–1747, 1989.

43. R.W. Teater, H.J. Mederski, G.W. Volk. Yield and mineral content of corn as affected by ammonium chloride fertilizer. *Agron. J.* 52:403–405, 1960.

44. S.E. Younts, R.B. Musgrave. Growth, maturity, and yield of corn as affected by chloride in potassium fertilizer. *Agron. J.* 50:423–462, 1958.

45. J.R. Heckman. Corn responses to chloride in maximum yield research. *Agron. J.* 87:415–419, 1995.

46. J.R. Heckman. Corn stalk rot suppression and grain yield response to chloride. *J. Plant Nutr.* 21:149–155, 1998.

47. W.H. Elmer, D.A. Johnson, G.I. Mink. Epidemiology and management of the diseases causal to asparagus decline. *Plant Dis.* 80:117–125, 1996.

48. B.E. Hahn, F.R. Olson, J.L. Roberts. Influence of potassium chloride on nitrification in Bedford silt loam. *Soil Sci.* 55:113–121, 1942.

49. R.J. Rosenberg, N.W. Christensen, T.L. Jackson. Chloride, soil solution osmotic potential, and soil pH effects on nitrification. *Soil Sci. Soc. Am. J.* 50:941–945, 1986.

50. N.W. Christensen, M. Brett. Chloride and liming effects on soil nitrogen form and take-all of wheat. *Agron. J.* 77:157–163, 1985.

51. D.W. James, D.C. Kidman, W.H. Weaver, R.L. Reeder. Factors affecting chloride uptake and implications of the chloride-nitrate antagonism in sugar beet mineral nutrition. *J. Am. Soc. Sugar Beet Technol.* 15:647–656, 1970.

52. D.C. Thompson, B.B. Clarke, J.R. Heckman. Nitrogen form and rate of nitrogen and chloride applications for the control of summer patch in Kentucky bluegrass. *Plant Dis.* 79:51–55, 1995.

53. W.H. Elmer. Association between Mn-reducing root bacteria and NaCl applications in suppression of Fusarium crown and root rot of asparagus. *Phytopathology* 85:1461–1467, 1995.

54. D.M. Huber. The role of nutrition in the take-all disease of wheat and other small grains. In: A. Englehard, ed. *Soilborne Plant Pathogens: Management of Diseases with Macro and Microelements.* St. Paul, MN: American Phytopathological Society, 1989, pp. 46–74.

55. R.J. Goos, B.E. Johnson, B.M. Holmes. Effect of potassium chloride fertilization on two barley cultivars differing in common root rot reaction. *Can. J. Plant Sci.* 67:395–401, 1987.

56. M.B. Parker, T.P. Gaines, G.J. Gascho. Chloride effects on corn. *Commun. Soil Sci. Plant Anal.* 16:1319–1333, 1985.

57. M.B. Parker, G.J. Gascho, T.P. Gaines. Chloride toxicity of soybeans grown on Atlantic coast flatwoods soils. *Agron. J.* 75:439–443, 1983.

58. D.W. James, W.H. Weaver, R.L. Reeder. Chloride uptake by potatoes and the effects of potassium chloride, nitrogen and phosphorus fertilization. *Soil Sci.* 109:48–53, 1970.

59. C. Metochis, P.I. Orphanos. Course of chloride concentration in tobacco leaves through the harvesting season. *J. Plant Nutr.* 13:485–493, 1990.

60. M. Prasad, G.K. Burge, T.M. Spiers, G. Fietje. Chloride induced leaf breakdown in kiwifruit. *J. Plant Nutr.* 16:999–1012, 1993.

61. W.K. Schumacher. Residual Effects of Chloride Fertilization on Selected Plant and Soil Parameters. M.S. thesis, South Dakota State University, Brookings, SD, 1988.

62. D. Smith, L.A. Peterson. Chlorine concentrations in alfalfa herbage and soil with KCl topdressing of a low fertility silt loam soil. *Commun. Soil Sci. Plant Anal.* 6:521–533, 1975.

63. G.H. Abel. Inheritance of the capacity for chloride inclusion and chloride exclusion by soybeans. *Crop Sci.* 9:697–698, 1969.

64. W.J.S. Downton. Growth and mineral composition of the Sultana grapevine as influenced by salinity and rootstock. *Aus. J. Agric. Res.* 36:425–434, 1985.

65. R.M. Stevens, G. Harvey. Effects of waterlogging, rootstock and salinity on Na, Cl and K concentrations of the leaf and root, and shoot growth of Sultana grapevines. *Aus. J. Agric. Res.* 46:541–551, 1995.

66. P.E. Fixen, R.H. Gelderman, J.R. Gerwing, F.A. Cholick. Response of spring wheat, barley, and oats to chloride in potassium chloride fertilizers. *Agron. J.* 78:664–668, 1986.

67. R.E. Engel, P.L. Bruckner, J. Eckhoff. Critical tissue concentration and chloride requirements for wheat. *Soil Sci. Soc. Am. J.* 62:401–405, 1998.

68. J.J. Mortvedt. Bioavailability of micronutrients. In: M.E. Sumner, ed. *Handbook of Soil Science.* New York: CRC Press, 2000, pp. D-71–D-88.

69. P.E. Fixen, R.H. Gelderman, J.L. Denning. Chloride tests. In: W.C. Dahnke, ed. *Recommended Chemical Soil Test Procedures for the North Central Region.* North Dakota Agric. Exp. Stn.: North Central Reg. Publ. 211 (revised), 1988, pp. 26–28.

70. P.E. Fixen, R.H. Gelderman, J.R. Gerwing, B.G. Farber. Calibration and implementation of a soil Cl test. *J. Fert. Issues* 4:91–97, 1987.

71. R. Gelderman, B. Farber, P. Fixen. Response of Barley Varieties to Chloride Fertilization. South Dakota State University, Brookings, SD, Soil PR 88-13, 1988.

72. P.G. Ozanne. Chlorine deficiency in soils. *Nature (London)* 182:1172–1173, 1958.

73. Y. Huang, Y.P. Rao, T.J. Liao. Migration of chloride in soil and plant. *J. Southwest Agric. Univ.* (in Chinese) 17:259–263, 1995.

74. W.H. Elmer. Suppression of Fusarium crown and root rot of asparagus with sodium chloride. *Phytopathology* 82:97–104, 1992.

75. T.C. Reid, M.K. Hausbeck, K. Kizilkaya. Effects of sodium chloride on commercial asparagus and of alternative forms of chloride salt on Fusarium crown and root rot. *Plant Dis.* 85:1271–1275, 2001.

76. P.A. Shefelbine, D.E. Mathre, G. Carlson. Effects of chloride fertilizers and systemic fungicide seed treatment on common root rot on barley. *Plant Dis.* 70:639–642, 1986.

77. C.A. Timm, R.J. Goos, B.E. Johnson, F.J. Sobolik, R.W. Stack. Effect of potassium fertilizers on malting barley infected with common root rot. *Agron. J.* 78:197–200, 1986.

78. R.W. Schneider. Suppression of Fusarium yellows of celery with potassium, chloride, and nitrate. *Phytopathology* 75:40–48, 1985.

79. S.E. Younts, R.B. Musgrave. Chemical composition, nutrient absorption, and stalk rot incidence of corn as affected by chloride in potassium fertilizer. *Agron. J.* 50:426–429, 1958.

80. G.E. Russell. Some effects of applied sodium and potassium on yellow rust in winter wheat. *Ann. Appl. Biol.* 90:163–168, 1978.

81. W.H. Elmer. Influence of chloride and nitrogen form on rhizoctonia root and crown rot of table beets. *Plant Dis.* 81:635–640, 1997.

82. D.W. Sweeney, G.V. Granade, M.G. Eversmeyer, D.A. Whitney. Phosphorus, potassium, chloride, and fungicide effects on wheat yield and leaf rust severity. *J. Plant Nutr.* 23:1267–1281, 2000.

83. R.G. Taylor, T.L. Jackson, R.L. Powelson, N.W. Christensen. Chloride, nitrogen form, lime, and planting date effects on take-all rot of winter wheat. *Plant Dis.* 67:1116–1120, 1983.

84. Y. Bar, A. Apelbaum, U. Kafkafi, R. Goren. Relationship between chloride and nitrate and its effect on growth and mineral composition of avocado and citrus plants. *J. Plant Nutr.* 20:715–731, 1997.

85. E. Lahav, R. Steinhardt, D. Kalmar, M.A.C. Fragoso, M.L.V. Beusichem. Effect of salinity on the nutritional level of the avocado: optimization of plant nutrition. *8th International Colloq. Optimum Plant Nutrition*, Lisbon, Portugal, 1992, pp. 593–596.

86. R.E. Engel, J. Eckhoff, R. Berg. Grain yield, kernel weight, and disease responses of winter wheat cultivars to chloride fertilization. *Agron. J.* 86:891–896, 1994.

87. P.F. Bell, J.A. Vaughn, W.J. Bourgeois. Leaf analysis finds high levels of chloride and low levels of zinc and manganese in Louisiana citrus. *J. Plant Nutr.* 20:733–743, 1997.

88. N.X. Tan, J.X. Shen. A study on the effect of Cl on the growth and development of cotton. *Soil Fert.* (in Chinese) 2:1–3, 1993.

89. G.S. Smith, C.J. Clark, P.T. Holland. Chloride requirement of kiwi-fruit (*Actinidia deliciosa*). *New Phytol.* 106:71–80, 1987.

90. S.Q. Wei, Z.F. Zhou, C. Liu. Effects of chloride on yield and quality of lettuce and its critical value of tolerance. *Chin. J. Soil Sci.* 30:262–264, 1989.

91. J.B. Robinson. Fruits, vines and nuts. In: D.J. Reuter, J.B. Robinson, eds. *Plant Analysis: An Interpretation Manual*. Sydney, Australia: Inkata Press, 1986, pp. 120–147.

92. D.Q. Wang, B.C. Guo, X.Y. Don. Toxicity effects of chloride on crops. *Chin. J. Soil Sci.* 30:258–261, 1989.

93. E.G. Corbett, H.W. Gausman. The interaction of chloride and sulfate in the nutrition of potato plants (*Solanum tuberosum*). *Agron. J.* 52:94–96, 1960.

94. L. Bernstein, A.D. Ayers, C.H. Wadleigh. The salt tolerance of white potatoes. *Proc. Am. Soc. Hortic. Sci.* 57:231–236, 1951.

95. M.J. Yin, J.J. Sun, C.S. Liu. Contents and distribution of chloride and effects of irrigation water of different chloride levels on crops. *Soil Fert.* (in Chinese) 1:3–7, 1989.

96. Q.S. Zhu, B.S. Yu. Critical tolerance of chloride of rice and wheat on three types of soils. *Wubei Agric. Sci.* (in Chinese) 5:22–26, 1991.

97. M.B. Parker, T.P. Gaines, G.J. Gascho. Sensitivity of Soybean Cultivars to Soil Chloride. University of Georgia, Res. Bull. No. 347, Tifton, GA, 1986.

98. M.B. Parker, T.P. Gaines, G.J. Gascho. The chloride toxicity problem in soybean in Georgia. In: T.L. Jackson, ed. *Special Bulletin on Chloride and Crop Production*. Atlanta, GA: Potash & Phosphate Institute, No. 2, pp. 100–108, 1986.

99. J. Yang, R.W. Blanchar. Differentiating chloride susceptibility in soybean cultivars. *Agron. J.* 85:880–885, 1993.

100. S.P. Robinson, J.S. Downton. Potassium, sodium, and chloride content of isolated intact chloroplasts in relation to ionic compartmentation in leaves. *Arch. Biochem. Biophys.* 228:197–206, 1984.

101. P.G. Ozanne, J.T. Woolley, T.C. Broyer. Chloride and bromine in the nutrition of higher plants. *Aus. J. Biol. Sci.* 10:66–79, 1957.

102. A. Ulrich, K. Ohki. Chlorine, bromine and sodium as nutrients for sugar beet plants. *Plant Physiol.* 31:171–181, 1956.

103. N. Terry. Photosynthesis, growth, and the role of chloride. *Plant Physiol.* 60:69–75, 1977.

104. B.K. Zhou, X.Y. Zhang. Effects of chloride on growth and development of sugarbeet. *Soil Fert.* (in Chinese) 3:41–43, 1992.

105. L.T. Li, D.H. Yuan, Z.J. Sun. Influence of a Cl-containing fertilizer on Cl concentration in tobacco leaves. *J. Southwest Agric. Univ.* (in Chinese) 16:415–418, 1994.

106. U. Kafkafi, N. Valoras, J. Letay. Chloride interaction with NO_3^- and phosphate nutrition in tomato. *J. Plant Nutr.* 5:1369–1385, 1982.

10 Copper

David E. Kopsell
University of Wisconsin-Platteville, Platteville, Wisconsin

Dean A. Kopsell
University of Tennessee, Knoxville, Tennessee

CONTENTS

10.1 The Element Copper 293
    10.1.1 Introduction 293
    10.1.2 Copper Chemistry 294
10.2 Copper in Plants 294
    10.2.1 Introduction 294
    10.2.2 Uptake and Metabolism 294
    10.2.3 Phytoremediation 313
10.3 Copper Deficiency in Plants 314
10.4 Copper Toxicity in Plants 315
10.5 Copper in the Soil 316
    10.5.1 Introduction 316
    10.5.2 Geological Distribution of Copper in Soils 317
    10.5.3 Copper Availability in Soils 317
10.6 Copper in Human and Animal Nutrition...321
    10.6.1 Introduction 321
    10.6.2 Dietary Sources of Copper 321
    10.6.3 Metabolism of Copper Forms 321
10.7 Copper and Human Health 322
    10.7.1 Introduction 322
    10.7.2 Copper Deficiency and Toxicity in Humans 322
References 323

10.1 THE ELEMENT COPPER

10.1.1 INTRODUCTION

Copper is one of the oldest known metals and is the 25th most abundant element in the Earth's crust. The words 'aes Cyprium' appeared in Roman writings describing copper, to denote that much of the metal at the time came from Cyprus. Refinement of copper metal dates back to 5000 BC. The metal by itself is soft, but when mixed with zinc produces brass and when mixed with tin produces bronze. Copper is malleable, ductile, and a good conductor of electricity. In its natural state, it is a reddish solid with a bright metallic luster.

10.1.2 COPPER CHEMISTRY

Copper has an atomic number 29 and atomic mass of 63.55. It belongs to Group I-B transition metals. The melting point of copper is 1084.6°C. Copper occurs naturally in the cuprous (I, Cu^+) and cupric (II, Cu^{2+}) valence states. There is a single electron in the outer 4s orbital. The $3d^{10}$ orbital does not effectively shield this outer electron from the positive nuclear charge, and therefore the $4s^1$ electron is difficult to remove from the Cu atom (1). The first ionization potential is 7.72 eV and the second is 20.29 eV. Because the second ionization potential is much higher than the first, a variety of stable Cu^+ species exist (2). The ionization state of copper depends on the physical environment, the solvent, and the concentration of ligands present. In solution, copper is present as Cu^{2+} or complexes of this ion. The cuprous ion Cu^{1+} is unstable in aqueous solutions at concentrations greater than 10^{-7} M (3). However, in wet soils, Cu^{1+} is moderately stable at typically expected conditions (10^{-6} to 10^{-7} M). Under such conditions, hydrated Cu^{1+} would be the dominant copper species (1). Copper can exist as two natural isotopes, ^{63}Cu and ^{65}Cu, with relative abundances of 69.09 and 30.91%, respectively (4). In the Earth's crust, copper is present as stable sulfides in minerals rather than silicates or oxides (3). The Cu^{1+} ion is present more commonly in minerals formed at considerable depth, whereas Cu^{2+} is present close to the Earth's surface (3).

The transition metals are noted for the variety of complexes they form with bases. In these complexes, Cu^{1+} and Cu^{2+} act as electron acceptors. Chelating bases are so named because they have two or more electron donor sites (often on O, S, or N atoms) that form a 'claw' around the copper ion (1). Such complexes are important in soil chemistry and in plant nutrition. The Cu^{1+} ion forms strong complexes with bases containing S, but Cu^{2+} does not. In the presence of these bases, Cu^{2+} acts as a strong oxidant (2).

10.2 COPPER IN PLANTS

10.2.1 INTRODUCTION

Copper was identified as a plant nutrient in the 1930s (5,6). Prior to this realization, one of the first uses of copper in agriculture was in chemical weed control (7). Despite its essentiality, copper is toxic to plants at high concentrations (8). Uptake of copper by plants is affected by many factors including the soil pH, the prevailing chemical species, and the concentration of copper present in the soil. Once inside the plant, copper is sparingly immobile. Accumulation and expression of toxic symptoms are often observed with root tissues. Extensive use of copper-containing fungicides in localized areas and contamination of soils adjacent to mining operations has created problems of toxicity in some agricultural regions. Because of this problem, remediation of copper and identification of tolerant plant species are receiving increased attention. Concentrations of copper in some plant species under different cultural conditions are reported in Table 10.1.

10.2.2 UPTAKE AND METABOLISM

The rate of copper uptake in plants is among the lowest of all the essential elements (9). Uptake of copper by plant roots is an active process, affected mainly by the copper species. Copper is most readily available at or below pH 6.0 (4). Most sources report copper availability in soils to decrease above pH 7.0. Increasing soil pH will cause copper to bind more strongly to soil components. Copper bioavailability is increased under slightly acidic conditions due to the increase of Cu^{2+} ions in the soil solution. On two soils in Spain, with similar pH values (8.0 and 8.1) but with different copper levels (0.64 and 1.92 mg Cu kg^{-1}, respectively), leaf content of willow leaf foxglove (*Digitalis obscura* L.) was equal, i.e., 7 mg kg^{-1} dry weight on both soils (10). Copper concentrations of tomato (*Lycopersicon esculentum* Mill.) and oilseed rape (canola, *Brassica napus* L.) roots and shoots were significantly higher in an acidic soil (pH 4.3) than in a calcareous soil (pH 8.7) (11). In contrast, however, if a mixture of Cd (II), Cu (II), Ni(II), and Zn(II) was applied to

TABLE 10.1

Copper Tissue Analysis Values of Various Plant and Crop Species

Plant						Copper Concentration in Dry Matter (mg kg⁻¹ Unless Otherwise Noted)[b]			
Common and Scientific Name	Variety	Type of Culture[a]	Type of Tissue Sampled	Age, Stage, Condition, or Date of Sample	Cu Treatment	Low	Medium	High	Reference
Alfalfa (*Medicago sativa* L.)	Mesa	Greenhouse soil	Shoot	15 days after planting	20 mg kg⁻¹, pH 4.5 20 mg kg⁻¹, pH 5.8 25 mg kg⁻¹, pH 7.1		~85 ~70 ~115		12
Artemesia, wormwood (*Artemisia absinthium* L.)		Native soil	Leaves Flowers Roots	Mature	1.03 ± 0.48 mg kg⁻¹	0.1 0.1 0.1	21.6 23.3 14.3	64.0 69.4 48.9	37
Artemesia, white sage (*Artemisia ludoviciana* Nutt.)		Native soil	Leaves Flowers Roots	Mature	1.68 ± 1.04 mg kg⁻¹	0.1 0.1 0.1	18.5 24.7 12.6	66.9 108.3 49.6	
Bean (*Phaseolus vulgaris* L.)	IAPAR 57	Greenhouse soil culture	Total plant	30 days old	0 mmol kg⁻¹ 0.1 mmol kg⁻¹ 0.2 mmol kg⁻¹ 0.5 mmol kg⁻¹ 1.0 mmol kg⁻¹ 2.0 mmol kg⁻¹ 0 mmol kg⁻¹, 1.0 chicken manure 0.1 mmol kg⁻¹, 1.0 chicken manure 0.2 mmol kg⁻¹, 1.0 chicken manure 0.5 mmol kg⁻¹, 1.0 chicken manure 1.0 mmol kg⁻¹, 1.0 chicken manure 2.0 mmol kg⁻¹, 1.0 chicken manure		7.5 7.5 7.5 10 21.5 38 7 9 9.5 10 13 17		77

Continued

TABLE 10.1 (*Continued*)

Plant Common and Scientific Name	Variety	Type of Culture[a]	Type of Tissue Sampled	Age, Stage, Condition, or Date of Sample	Cu Treatment	Low	Medium	High	Reference
						Copper Concentration in Dry Matter (mg kg⁻¹ Unless Otherwise Noted)[b]			
Dwarf bean modus	Native soil	Edible portion	Mature	18 ± 1 mg kg⁻¹, pH 6.1, 1.9% organic matter		6.6		38
					326 ± 15 mg kg⁻¹, pH 7.0, 3.4% organic matter		6.7		
					430 ± 20 mg kg⁻¹, pH 6.1, 2.3% organic matter		7.3		
Beet, Sugar (*Beta vulgaris* L.)		Native soil	Roots	Mature	90 mg kg⁻¹		5		29
					125 mg kg⁻¹		2.5		
					210 mg kg⁻¹		3.5		
Carrot (*Daucus carota* L.)	Rotin and sperlings	Native soil	Root	Mature	18 ± 1 mg kg⁻¹, pH 6.1, 1.9% organic matter		5.1		38
					326 ± 15 mg kg⁻¹, pH 7.0, 3.4% organic matter		8.1		
					430 ± 20 mg kg⁻¹, pH 6.1, 2.3% organic matter		7.2		
Celery (*Apium graveolens* var. *dulce* Pers.)		Native soil	Tuber	Mature	18 ± 1 mg kg⁻¹, pH 6.1, 1.9% organic matter		7.5		8
					326 ± 15 mg kg⁻¹, pH 7.0, 3.4% organic matter		12		

Plant	Cultivar	Growth medium	Plant part	Plant age	Treatment/conditions	Concentration	Reference
					430 ± 20 mg kg⁻¹, pH 6.1, 2.3% organic matter	13	33
Chickpea (*Cicer arietinum* L.)	Tyson	Soil pot culture	Shoots	62 days old, 5 plant per pot	0.06 mg kg⁻¹ DTPA-extractable + 0 µg pot⁻¹, pH 6.4	2.5 µg pot⁻¹	
					0.06 mg kg⁻¹ DTPA-extractable + 100 µg pot⁻¹, pH 6.4	3.8 µg pot⁻¹	
					0.06 mg kg⁻¹ DTPA-extractable + 200 µg pot⁻¹, pH 6.4	5.5 µg pot⁻¹	
					0.06 mg kg⁻¹ DTPA-extractable + 400 µg pot⁻¹, pH 6.4	8.0 µg pot⁻¹	
					0.06 mg kg⁻¹ DTPA-extractable + 800 µg pot⁻¹, pH 6.4	11.3 µg pot⁻¹	
Chinese cabbage (*Brassica pekinensis* Rupr.)	Nagaoka 50	Native soil	Leaves	35 days	16 mg kg⁻¹, 5 mg kg⁻¹ DTPA-extractable, calcareous soil, pH 8.6	21	119
				50 days		17	
				65 days		15	
				80 days		12	
				90 days		11	
	Xiayangbai	Nutrient solution culture	Shoots	15 days old	Full strength Hoagland solution + 0 mg Cu L⁻¹	18.4	22
					Full strength Hoagland solution + 0.5 mg Cu L⁻¹	40.1	
					Full strength Hoagland solution + 1 mg Cu L⁻¹	36.8	
					Full strength Hoagland solution + 4 mg Cu L⁻¹	200.0	
			Roots		Full strength Hoagland solution + 0 mg Cu L⁻¹	160.3	
					Full strength Hoagland solution + 0.5 mg Cu L⁻¹	278.8	

Continued

TABLE 10.1 (*Continued*)

Plant Common and Scientific Name	Variety	Type of Culture[a]	Type of Tissue Sampled	Age, Stage, Condition, or Date of Sample	Cu Treatment	Copper Concentration in Dry Matter (mg kg^{-1} Unless Otherwise Noted)[b] Low	Medium	High	Reference
Corn (*Zea mays* L.)		Native soil	Seeds	Mature	Full strength Hoagland solution + 1 mg Cu L^{-1}		349.7		29
					Full strength Hoagland solution + 4 mg Cu L^{-1}		2436.0		
		Native soil	Grain	Mature	90 mg kg^{-1}		2		15
					125 mg kg^{-1}		1.5		
					210 mg kg^{-1}		2.5		
					25.89 ± 2.78 mg kg^{-1}		4.13		
					37.19 ± 17.41 mg kg^{-1}		3.60		
					54.39 ± 8.70 mg kg^{-1}		4.53		
					181.68 ± 49.12 mg kg^{-1}		3.60		
			Stem	Mature	25.89 ± 2.78 mg kg^{-1}		5.40		
					37.19 ± 17.41 mg kg^{-1}		6.61		
					54.39 ± 8.70 mg kg^{-1}		10.14		
					181.68 ± 49.12 mg kg^{-1}		24.09		
			Roots	Mature	25.89 ± 2.78 mg kg^{-1}		16.74		
					37.19 ± 17.41 mg kg^{-1}		22.28		
					54.39 ± 8.70 mg kg^{-1}		25.37		
					181.68 ± 49.12 mg kg^{-1}		108.89		
Cucumber (*Cucumis sativus* L.)	Vert long mariacher	Sand/solution culture	Leaves	Expanding	0.5 µM CuCl$_2$·H$_2$O	11		14	34
				Mature	0.5 µM CuCl$_2$·H$_2$O	14		23	
				Expanding	10 µg g^{-1} substrate + 0.5 µM CuCl$_2$·H$_2$O	27		35	
				Mature	10 µg g^{-1} substrate + 0.5 µM CuCl$_2$·H$_2$O	23		25	
Bermudagrass (*Cynodon dactylon* Steud.)		Native soil	Shoot	Mature	2.55 ± 0.56 mg kg^{-1}, 1.10 ± 0.09 mg kg^{-1} DTPA-extractable, pH 5.32		14.81		59

Species	Cultivar/population	Culture	Plant part	Soil/treatment	Value		Ref.
			Roots	198 ± 22 mg kg⁻¹, 6.95 ± 2.15 mg kg⁻¹ DTPA-extractable, pH 6.13	22.26		
				2.55 ± 0.56 mg kg⁻¹, 1.10 ± 0.09 mg kg⁻¹ DTPA-extractable, pH 5.32	20.75		
				198 ± 22 mg kg⁻¹, 6.95 ± 2.15 mg kg⁻¹ DTPA-extractable, pH 6.13	45.56		
Willow-leaf foxglove (*Digitalis obscura* L.)	Wild population	Native soil	Leaves	0.87 mg kg⁻¹	10		10
				0.84 mg kg⁻¹	8		
				0.64 mg kg⁻¹	7		
				1.92 mg kg⁻¹	7		
Shiny elsholtzia (*Elsholtzia splendens* Nakai)		Nutrient solution culture	Shoots	500 μM	1133		55
				1000 μM	3417		
			Roots	0.12 μM	38		
				1000 μM	12,752		
			Leaves	0.12 μM	70		
				1000 μM	525		
Faba bean (*Vicia faba* L.)	Fiord	Soil pot culture	Shoots	0.06 mg kg⁻¹ DTPA-extractable + 0 μg pot⁻¹, pH 6.4	16 μg pot⁻¹	⋮	33
				0.06 mg kg⁻¹ DTPA-extractable + 100 μg pot⁻¹, pH 6.4	23 μg pot⁻¹	⋮	
				0.06 mg kg⁻¹ DTPA-extractable + 200 μg pot⁻¹, pH 6.4	34 μg pot⁻¹	⋮	
				0.06 mg kg⁻¹ DTPA-extractable + 400 μg pot⁻¹, pH 6.4	38 μg pot⁻¹	⋮	

62 days old, 5 plant per pot

Continued

TABLE 10.1 (*Continued*)

Plant Common and Scientific Name	Variety	Type of Culture[a]	Type of Tissue Sampled	Age, Stage, Condition, or Date of Sample	Cu Treatment	Copper Concentration in Dry Matter (mg kg^{-1} Unless Otherwise Noted)[b] Low	Medium	High	Reference
					0.06 mg kg^{-1} DTPA-extractable + 800 μg pot^{-1}, pH 6.4	…	50 μg pot^{-1}	…	
Grape (*Vitis vinifera* L.)	Merlot, 3309 Couderc root stock	Native soil	Leaves	Mature	75.1 mg kg^{-1}, DTPA-extractable		276		110
					61.8 mg kg^{-1}, DTPA-extractable		264		
					63.0 mg kg^{-1}, DTPA-extractable		279		
			Musts	Mature	75.1 mg kg^{-1}, DTPA-extractable		4.74 mg L^{-1}		
					61.8 mg kg^{-1}, DTPA-extractable		4.65 mg L^{-1}		
					63.0 mg kg^{-1}, DTPA-extractable		5.08 mg L^{-1}		
			Wine		75.1 mg kg^{-1}, DTPA-extractable		0.076 mg L^{-1}		
					61.8 mg kg^{-1}, DTPA-extractable		0.070 mg L^{-1}		
					63.0 mg kg^{-1}, DTPA-extractable		0.073 mg L^{-1}		
Kohlrabi (*Brassica oleracea* var. *gongylodes* L.)		Native soil	Edible portion	Mature	18 ± 1 mg kg^{-1}, pH 6.1, 1.9% organic matter		1.9		38
					326 ± 15 mg kg^{-1}, pH 7.0, 3.4% organic matter		2.8		

Plant	Cultivar	Culture	Plant part	Growth stage	Soil	Cu	Reference
					430 ± 20 mg kg⁻¹, pH 6.1, 2.3% organic matter	2.5	
Lentil (*Lens culinaris* Medik)	Digger	Soil pot culture	Shoots	62 days old, 5 plants per pot	0.06 mg kg⁻¹ DTPA-extractable + 0 µg pot⁻¹, pH 6.4	0.6 µg pot⁻¹	33
					0.06 mg kg⁻¹ DTPA-extractable + 100 µg pot⁻¹, pH 6.4	1.5 µg pot⁻¹	
					0.06 mg kg⁻¹ DTPA-extractable + 200 µg pot⁻¹, pH 6.4	2.0 µg pot⁻¹	
					0.06 mg kg⁻¹ DTPA-extractable + 400 µg pot⁻¹, pH 6.4	2.8 µg pot⁻¹	
					0.06 mg kg⁻¹ DTPA-extractable + 800 µg pot⁻¹, pH 6.4	3.5 µg pot⁻¹	
Lettuce (*Lactuca sativa* L.)	American gathering brown	Native soil	Leaves	Mature	18 ± 1 mg kg⁻¹, pH 6.1, 1.9% organic matter	11	38
					326 ± 15 mg kg⁻¹, pH 7.0, 3.4% organic matter	40	
					430 ± 20 mg kg⁻¹, pH 6.1, 2.3% organic matter	21	
Lucerne (Alfalfa, *Medicago sativa* L.)		Native soil	Leaves	Mature	90 mg kg⁻¹	12	29
					125 mg kg⁻¹	11.5	
					210 mg kg⁻¹	15	
Mangold (*Beta vulgaris* L. var. *macrorhiza*)		Native soil	Edible portion	Mature	18 ± 1 mg kg⁻¹, pH 6.1, 1.9% organic matter	11	38

Continued

TABLE 10.1 (*Continued*)

Plant Common and Scientific Name	Variety	Type of Culture[a]	Type of Tissue Sampled	Age, Stage, Condition, or Date of Sample	Cu Treatment	Copper Concentration in Dry Matter (mg kg⁻¹ Unless Otherwise Noted)[b]			Reference
						Low	Medium	High	
					326 ± 15 mg kg⁻¹, pH 7.0, 3.4% organic matter		18		
					430 ± 20 mg kg⁻¹, pH 6.1, 2.3% organic matter		23		
Indian mustard (*Brassica juncea* L.)		Native soil	Leaves	Mature	0 mg kg⁻¹, 0 g kg⁻¹ biosolid organic carbon		<10		106
					50 mg kg⁻¹, 0 g kg⁻¹ biosolid organic carbon		~40		
					100 mg kg⁻¹, 0 g kg⁻¹ biosolid organic carbon		~50		
					200 mg kg⁻¹, 0 g kg⁻¹ biosolid organic carbon		~85		
					400 mg kg⁻¹, 0 g kg⁻¹ biosolid organic carbon		~200		
Oat (*Avena sativa* L.)		Native soil	Stems	Mature	12.2 mg kg⁻¹		3.9		16
			Leaves		12.2 mg kg⁻¹		5.5		
			Flowers		12.2 mg kg⁻¹		7.9		
			Roots		12.2 mg kg⁻¹		11.5		
		Native soil	Tillers	Mature	3.1 mg kg⁻¹, DTPA-extractable		3.9		50
					3.5 mg kg⁻¹, DTPA-extractable		4.5		
					2.5 mg kg⁻¹, DTPA-extractable		6.1		
					3.3 mg kg⁻¹, DTPA-extractable		4.0		

Plant	Soil	Plant part	Growth stage	Soil condition	Cu		Reference
Onion (*Allium cepa* L.)	Native soil	Bulb	Mature	<400 mg kg⁻¹	7.1		36
		Stem	Mature	<400 mg kg⁻¹	6.4		
		Leaves	Mature	<400 mg kg⁻¹	6.6		
		Bulb	Mature	>400 mg kg⁻¹	8.2		
		Stem	Mature	>400 mg kg⁻¹	10.2		
		Leaves	Mature	>400 mg kg⁻¹	10.9		
Oregano (*Origanum vulgare* L. subsp. *hirtum* Soó)	Native soil	Upper leaves	Mature	12–26 μM g⁻¹	2.5 μmol g⁻¹	4.1 μmol g⁻¹	
		Lower leaves	Mature		3.5 μmol g⁻¹	5.5 μmol g⁻¹	
Knotgrass (*Paspalum disticum* L.)	Native soil	Shoots	Mature	2.55 ± 0.56 mg kg⁻¹, 1.10 ± 0.09 mg kg⁻¹ DTPA-extractable, pH 5.32	13.35		59
				99 ± 6.42 mg kg⁻¹, 10 ± 2.61 mg kg⁻¹ DTPA-extractable, pH 7.25	32.27		
				191 ± 33 mg kg⁻¹, 7.38 ± 3.2 mg kg⁻¹ DTPA-extractable, pH 7.38	8.79		
		Roots		2.55 ± 0.56 mg kg⁻¹, 1.10 ± 0.09 mg kg⁻¹ DTPA-extractable, pH 5.32	20.30		
				99 ± 6.42 mg kg⁻¹, 10 ± 2.61 mg kg⁻¹ DTPA-extractable, pH 7.25	21.48		
				191 ± 33 mg kg⁻¹, 7.38 ± 3.2 mg kg⁻¹ DTPA-extractable, pH 7.38	21.38		

Continued

TABLE 10.1 (*Continued*)

Plant Common and Scientific Name	Variety	Type of Culture[a]	Type of Tissue Sampled	Age, Stage, Condition, or Date of Sample	Cu Treatment	Copper Concentration in Dry Matter (mg kg^{-1} Unless Otherwise Noted)[b]			Reference
						Low	Medium	High	
Pea (*Pisum sativum* L.)	Fenomen	Solution culture	Roots	21 days old	6.42 µmol cumulative treatment		12 µg g^{-1} fresh weight		24
Radish (*Raphanus sativus* L.)	Rimbo	Solution culture	Above-ground part	28 days old	0.12 µM 5 µM 10 µM 15 µM		4 µg plant^{-1} 8 µg plant^{-1} 13 µg plant^{-1} 14 µg plant^{-1}		18
			Below-ground part		0.12 µM 5 µM 10 µM 15 µM		2.3 µg plant^{-1} 2.8 µg plant^{-1} 3.7 µg plant^{-1} 3.7 µg plant^{-1}		
		Native soil	Above-ground part	28 days old	591 ± 25 mg kg^{-1}, 200 ± 8 mg kg^{-1} EDTA-extractable, pH 6.3, 6.9% organic matter		7.0 µg plant^{-1}		
Red cover (*Trifolium pratense* L.)		Native soil	Stems Leaves Flowers Roots	Mature	12.2 mg kg^{-1}		8.6 16.1 20.2 24.2		16
Rhodegrass (*Chloris gayana* Kunth)	Kallide	Native soil	Tillers	Mature	3.1 mg kg^{-1}, DTPA-extractable 3.5 mg kg^{-1}, DTPA-extractable 2.5 mg kg^{-1}, DTPA-extractable 3.3 mg kg^{-1}, DTPA-extractable		8.5 7.2 10.1 10.0		50
Rice (*Oryza sativa* L.)		Native soil	Shoot	Mature	23 mg kg^{-1}, pH 6.2 90 mg kg^{-1}, pH 6.2 158 mg kg^{-1}, pH 7.0		12 28 37		117

Plant	Soil	Treatment	Part	Value	Ref
Rye (*Secale cereale* L.)	Native soil	12.2 mg kg⁻¹	Mature		
			Stems	6.9	16
			Leaves	5.2	
			Flowers	9.0	
			Roots	21.9	
Ryegrass (*Lolium multiflorum* Lam.)	Native soil	12.2 mg kg⁻¹	Mature		
			Stems	2.5	16
			Leaves	4.5	
			Flowers	7.5	
			Roots	10.9	
Willow (*Salix acmophylla* Boiss.)	Greenhouse soil pot culture	75 days old	Leaves		
		20.22 mg kg⁻¹		2.54	45
		20.22 mg kg⁻¹ + 500 mg kg⁻¹		10.8	
		20.22 mg kg⁻¹ + 1000 mg kg⁻¹		17.2	
		20.22 mg kg⁻¹ + 2000 mg kg⁻¹		49.4	
		20.22 mg kg⁻¹ + 5000 mg kg⁻¹		82.3	
		20.22 mg kg⁻¹ + 10,000 mg kg⁻¹		126.3	
		20.22 mg kg⁻¹	Stems	4.0	
		20.22 mg kg⁻¹ + 500 mg kg⁻¹		25.3	
		20.22 mg kg⁻¹ + 1000 mg kg⁻¹		73.3	
		20.22 mg kg⁻¹ + 2000 mg kg⁻¹		103.9	
		20.22 mg kg⁻¹ + 5000 mg kg⁻¹		179.2	
		20.22 mg kg⁻¹ + 10,000 mg kg⁻¹		203.7	
		20.22 mg kg⁻¹	Roots	6.85	
		20.22 mg kg⁻¹ + 500 mg kg⁻¹		24.8	

Continued

TABLE 10.1 (*Continued*)

Plant Common and Scientific Name	Variety	Type of Culture[a]	Type of Tissue Sampled	Age, Stage, Condition, or Date of Sample	Cu Treatment	Copper Concentration in Dry Matter (mg kg⁻¹ Unless Otherwise Noted)[b] Low	Medium	High	Reference
					20.22 mg kg⁻¹ + 1000 mg kg⁻¹		75.8		
					20.22 mg kg⁻¹ + 2000 mg kg⁻¹		177.5		
					20.22 mg kg⁻¹ + 5000 mg kg⁻¹		345.3		
					20.22 mg kg⁻¹ + 10,000 mg kg⁻¹		624.4		
Setaria, Forage (*Setaria sphacelata* Moss.)	Kazungula	Native soil	Tillers	Mature	3.1 mg kg⁻¹, DTPA-extractable		8.5		50
					3.5 mg kg⁻¹, DTPA-extractable		5.1		
					2.5 mg kg⁻¹, DTPA-extractable		10.4		
					3.3 mg kg⁻¹, DTPA-extractable		10.4		
Soybean (*Glycine max* Merr.)	Williams 82 and Pioneer 9391	Native soil	Main stem leaves	R5	N/A		9		35
			Branch stem leaves						
			Main stem seeds	Mature			11		
			Branch stem seeds				14		
							20		
Spinach (*Spinacia oleracea* L.)	Wonderful	Nutrient solution	Leaves	Mature	0.5 µM Cu		25		73
					160 µM Cu		729		
					160 µM Cu + 40 160 µM Fe		462		

		Roots	Mature	0.5 μM Cu	33			
				160 μM Cu	4727			
				160 μM Cu + 40 160 μM Fe	3800			128
Sunflower (*Helianthus annuus* L.)	Nutrient solution	Roots	6 days old	0.3 μM CuSO$_4$	42			
				10^{-5} M Cu^{2+}	108			
				10^{-4} M Cu^{2+}	138			
				10^{-3} M Cu^{2+}	1070			
		Hypocotyl		0.3 μM CuSO$_4$	20			
				10^{-5} M Cu^{2+}	52			
				10^{-4} M Cu^{2+}	49			
				10^{-3} M Cu^{2+}	165			
		Cotyledon		0.3 μM CuSO$_4$	24			
				10^{-5} M Cu^{2+}	47			
				10^{-4} M Cu^{2+}	66			
				10^{-3} M Cu^{2+}	95			
Tomato (*Lycopersicon esculentum* Mill.)	Greenhouse soil	Leaves	4th–5th fully expanded	7.79 mg kg^{-1} DTPA-extractable Cu	…	<5 mg kg^{-1}	1400 mg kg^{-1}	31
	Native soil	Fruit	Ripe	<400 mg kg^{-1}	14.7			36
		Stem	Mature	<400 mg kg^{-1}	19.5			
		Leaves	Mature	<400 mg kg^{-1}	35.7			
		Fruit	Ripe	>400 mg kg^{-1}	15.8			
		Stem	Mature	>400 mg kg^{-1}	26.2			
		Leaves	Mature	>400 mg kg^{-1}	64.4			
Wheat (*Triticum aestivum* L.)	Native soil	Stems	Mature	12.2 mg kg^{-1}	3.6			16
		Leaves			6.1			
		Flowers			7.9			
		Roots			7.5			
	Native soil	Shoot	Mature	14.5 g kg^{-1} organic Cu, 0.4 mg kg^{-1} DTPA extractable Cu	6.4	12.5		21
				37.6 g kg^{-1} organic Cu, 2.0 mg kg^{-1} DTPA extractable Cu	2.3	3.4		

Continued

TABLE 10.1 (*Continued*)

Plant Common and Scientific Name	Variety	Type of Culture[a]	Type of Tissue Sampled	Age, Stage, Condition, or Date of Sample	Cu Treatment	Copper Concentration in Dry Matter (mg kg⁻¹ Unless Otherwise Noted)[b]			Reference
						Low	Medium	High	
		Native soil	Grain	Mature	26.03 ± 2.56 mg kg⁻¹		3.47		15
					34.72 ± 16.38 mg kg⁻¹		6.84		
					87.40 ± 62.24 mg kg⁻¹		2.91		
					199.26 ± 66.54 mg kg⁻¹		6.84		
			Stem	Mature	26.03 ± 2.56 mg kg⁻¹		4.62		
					34.72 ± 16.38 mg kg⁻¹		4.54		
					87.40 ± 62.24 mg kg⁻¹		5.62		
					199.26 ± 66.54 mg kg⁻¹		6.12		
			Roots	Mature	26.03 ± 2.56 mg kg⁻¹		7.34		
					34.72 ± 16.38 mg kg⁻¹		8.68		
					87.40 ± 62.24 mg kg⁻¹		15.63		
					199.26 ± 66.54 mg kg⁻¹		40.10		
	Stretton	Soil pot culture	Shoots	62 days old, 5 plants per pot	0.06 mg kg⁻¹ DTPA-extractable + 0 µg pot⁻¹, pH 6.4		6 µg pot⁻¹		33
					0.06 mg kg⁻¹ DTPA-extractable + 100 µg pot⁻¹, pH 6.4		11 µg pot⁻¹		
					0.06 mg kg⁻¹ DTPA-extractable + 200 µg pot⁻¹, pH 6.4		18 µg pot⁻¹		
					0.06 mg kg⁻¹ DTPA-extractable + 400 µg pot⁻¹, pH 6.4		22 µg pot⁻¹		
					0.06 mg kg⁻¹ DTPA-extractable + 800 µg pot⁻¹, pH 6.4		27 µg pot⁻¹		
	Sunny	Nutrient solution culture	Roots	21 days old	18.02 µmol cumulative treatment		27 ± 1 µg g⁻¹ fresh weight		24

Plant	Cultivar	Growing medium	Stage	Part	Treatment	Concentration	
Wheat (*Triticum durum* Desf.)	Creso	Nutrient solution culture	15-day-old seedlings, 168 h after treatment	Shoots	Control—½ strength Hoagland's No. 2	15.1	92
					½ strength Hoagland's No. 2 + 150 µM	15.1	
				Roots	Control—½ strength Hoagland's No. 2	25.9	
					½ strength Hoagland's No. 2 + 150 µM	2900	
White clover (*Trifolium repens* L.)		Native soil	Mature	Stems	12.2 mg kg^{-1}	20.2	16
				Leaves		12.0	
				Flowers		38.0	
				Roots		28.4	

[a]'Native soil' denotes experiments or studies where crops were harvested from a field soil or natural environment and the copper level determined from a soil sample to estimate copper fertility.

[b]Information not available. When available in references, values have been expressed as an average concentration ± standard error.

a montmorillonite [$(Al,Mg)_2(OH)_2Si_4O_{10}$] soil at 50 mg kg^{-1} each, there were no differences in growth of alfalfa (*Medicago sativa* L.) between soil pH treatments of 4.5, 5.8, and 7.1, and plants grown at pH 7.1 accumulated the highest amount copper (12). However, if soil pH is above 7.5, plants should be monitored for copper deficiency.

Copper has limited transport in plants; therefore, the highest concentrations are often in root tissues (11,13,14,15). When corn (*Zea mays* L.) was grown in solution cultures at 10^{-5}, 10^{-4}, and 10^{-3} M Cu^{2+}, copper content of roots was 1.5, 8, and 10-fold greater respectively, than in treatments without copper additions, with little copper translocation to shoot tissues occurring (14). On a Savannah fine sandy loam pasture soil in Mississippi containing 12.3 mg Cu kg^{-1}, analysis of 16 different forage species revealed that root tissues accumulated the highest copper concentrations (28.8 mg kg^{-1}), followed by flowers (18.1 mg kg^{-1}), leaves (15.5 mg kg^{-1}), and stems (8.4 mg kg^{-1}) (16). Copper most likely enters roots in dissociated forms but is present in root tissues as a complex. Nielsen (17) observed that copper uptake followed Michaelis–Menten kinetics, with a $K_m = 0.11$ µmol L^{-1} and a mean $C_{min} = 0.045$ µmol L^{-1} over a copper concentration range of 0.08 to 3.59 µmol L^{-1}. Within roots, copper is associated principally with cell walls due to its affinity for carbonylic, carboxylic, phenolic, and sulfydryl groups as well as by coordination bonds with N, O, and S atoms (18). At high copper supply, significant percentages of copper can be bound to the cell wall fractions. Within green tissues, copper is bound in plastocyanin and protein fractions. As much as 50% or more of plant copper localized in chloroplasts is bound to plastocyanin (19). The highest concentrations of shoot copper usually occur during phases of intense growth and high copper supply (9).

Accumulation of copper can be influenced by many competing elements (Table 10.2). Copper uptake in lettuce (*Lactuca sativa* L.) in nutrient solution culture was affected by free copper ion activity, pH of the solution, and concentration of Ca^{2+} (20). Copper concentration of four Canadian wheat (*Triticum aestivum* L.) cultivars was affected by cultivar and applied nitrogen, but the variance due to applied nitrogen was fourfold greater than that due to cultivar (21). In Chinese cabbage (*Brassica pekinensis* Rupr.), iron and phosphorus deficiencies in nutrient solution stimulated copper uptake, but abundant phosphorus supply decreased copper accumulation (22). Fertilizing a calcareous soil (pH 8.7, 144 µg Cu g^{-1}) with an iron-deficient solution increased copper accumulation by roots and shoots in two wheat cultivars from 6 to 25 µg Cu g^{-1} (cv. Aroona) and 8 to 29 µg Cu g^{-1} (cv. Songlen) (13). In this same study, zinc deficiency did not significantly stimulate copper accumulation (13). Iron deficiency in nutrient solution culture increased copper and nitrogen leaf contents uniformly along corn leaf blades (23). Selenite (SeO_3^{-2}) and selenate (SeO_4^{-2}) depressed copper uptake, expressed as a percentage of total copper supplied, in pea (*Pisum sativum* L.), but not in wheat (*Triticum aestivum* L. cv. Sunny). However, copper uptake and tissue concentration were not affected by selenium (24).

Iron and copper metabolism appear to be associated in plants and in yeast (25,26). Ferric-chelate reductase is expressed on the root surface of plants and the plasma membrane of yeast under conditions of iron deficiency (25). Lesuisse and Labbe (27) reported that ferric reductase reduces Cu^{2+} in yeast and may be involved in copper uptake. Increases in manganese, magnesium, and potassium accumulation were associated with iron deficiency in pea, suggesting that plasma reductases may have a regulatory function in root ion-uptake processes via their influence on the oxidation–reduction status of the membrane (25,26). Evidence of this process was also supported by findings in a copper-sensitive mutant (*cup1-1*) of mouse-ear cress (*Arabidopsis thaliana* L. Heynh var. Columbia), suggesting that defects in iron metabolism may influence copper accumulation in plants (25).

The copper requirements among different plant species can vary greatly, and there can also be significant within-species variation of copper accumulation (28,29). The median copper concentration of forage plants in the United States was reported to be 8 mg kg^{-1} for legumes (range 1 to 28 mg kg^{-1}) and 4 mg kg^{-1} for grasses (range 1 to 16 mg kg^{-1}) (30). The copper content of native pasture plants in central southern Norway ranged from 0.9 to 27.2 mg kg^{-1} (28). Copper concentrations of tomato leaves from 105 greenhouses in Turkey ranged from 2.4 to 1490 mg kg^{-1} (31). Vegetables classified as having a low response to copper applications are asparagus (*Asparagus officinalis* L.), bean (*Phaseolus vulgaris* L.), pea, and potato (*Solanum tuberosum* L.). Vegetables classified as having a high response

TABLE 10.2
Descriptions of the Interaction of Copper in Plant Tissues with Various Elements

Element	Interaction with Cu in Plant Tissues[a]
Nitrogen (N)	Increasing levels of N fertilizers may increase requirement for Cu due to increased growth N fertilization linearly increases the Cu content of shoots High N levels may also inhibit translocation of Cu
Phosphorus (P)	Heavy use of P fertilizers can induce Cu deficiencies in citrus Excess P in solution culture decreased Cu accumulation in *Brassica*[b]
Potassium (K)	Foliar K sprays have reduced the copper content of pecan
Calcium (Ca)	Ca was shown to reduce Cu uptake in nutrient solution culture in lettuce[c] Increasing Ca in solution culture improved reduced growth due to Cu toxicity in mung bean[d]
Iron (Fe)	High levels of Fe have produced leaf chlorosis in citrus and lettuce Fe deficiency has stimulated copper uptake in solution culture in *Brassica*[y] and corn[e] Excess Fe in nutrient solution culture lessened the effects of Cu toxicity in spinach[f]
Zinc (Zn)	Cu significantly inhibits the uptake of Zn Zn will inhibit the uptake of Cu Zn is believed to interfere with the Cu absorption process
Manganese (Mn)	Cu has been shown to stimulate uptake of Mn in several plant species
Molybdenum (Mo)	Cu interferes with the role of Mo in the enzymatic reduction of nitrate A mutual antagonism has been found between Cu and Mo in several plant species
Aluminum (Al)	Al has been shown to adversely affect the uptake of Cu

[a]Reproduced from H.A. Mills, J.B. Jones, Jr., in *Plant Analysis Handbook II*, MicroMacro Publishing, Inc., Athens, GA, 1996, 422pp., unless otherwise noted. With permission.

[b]Adapted from Z. Xiong, Y. Li, B. Xu, *Ecotoxic Environ. Safety*, 53:200–205, 2002.

[c]Adapted from T. Cheng, H.E. Allen, *Environ. Toxic Chem.*, 20:2544–2511, 2001.

[d]Adapted from Z. Shen, F. Zhang, F. Zhang, *J. Plant Nutr.*, 21:1153–1162, 1998.

[e]Adapted from A. Mozafar, *J. Plant Nutr.*, 20:999–1005, 1997.

[f]Adapted from G. Ouzounidou, I. Illias, H. Tranopoulou, S. Karataglis, *J. Plant Nutr.*, 21:2089–2101, 1998.

to copper are beet (*Beta vulgaris* L. Crassa group), lettuce, onion (*Allium cepa* L.), and spinach (*Spinacia oleracea* L.) (32). In Australia, the critical copper concentration in young shoot tissue was 4.6 mg kg^{-1} for lentil (*Lens culinaris* Medik), 2.8 mg kg^{-1} for faba bean (*Vicia faba* L.), 2.6 mg kg^{-1} for chickpea (*Cicer arietinum* L.), and 1.5 mg kg^{-1} for wheat (*Triticum aestivum* L.) (33). Leaves of dwarf birch (*Betula nana* L.) had considerably lower copper levels than mountain birch (*Betula pubescens* Ehrh.) and willow (*Salix* spp.) in central southern Norway (28).

The response of many crops to copper addition depends on their growth stages (20,34). In soybean (*Glycine max* Merr.), the copper content of branch seeds was 20 µg g^{-1} whereas seeds from the main stems contained 14 µg g^{-1} (35). Addition of 10 µg $CuCl_2 \cdot 2H_2O$ g^{-1} to nutrient solution culture significantly suppressed leaf area in expanding cucumber (*Cucumis sativus* L.) leaves, whereas copper addition significantly limited photosynthesis in mature leaves (34). However, the suppression in photosynthesis was attributed to an altered source–sink relationship rather than the toxic effect of copper (34). Nitrogen and copper were the only elements that showed no gradation in concentration along the entire corn leaf blade (23).

The copper content of many edible plant parts is not correlated to the amount of soil copper (15,36,29,37,38). No correlations could be made between the level of applied copper and the amount of that metal in edible parts of corn grain, sugar beet (*Beta vulgaris* L.) roots, and alfalfa leaves (29). Despite differences of mean soil copper levels ranging from 160 to 750 mg kg^{-1}, copper concentrations of edible tomato fruit and onion bulbs were similar (36). Although soil copper levels ranged from 26 to 199 mg kg^{-1}, spring wheat (*Triticum aestivum* L.) grain accumulated only between 2.12 and 6.84 mg Cu kg^{-1} (15). Comparing a control soil containing 18 mg Cu kg^{-1} and a slag-contaminated soil containing 430 mg Cu kg^{-1}, the respective copper concentrations for bean (*Phaseolus vulgaris* L.) were 6.6 and 6.7 mg Cu kg^{-1} dry weight; for kohlrabi (*Brassica oleracea* var. *gongylodes* L.) were 1.9 and 2.8 mg Cu kg^{-1} dry weight; for mangold (*Beta vulgaris* L. cv. *macrorhiza*) were 11 and 18 mg Cu kg^{-1} dry weight; for lettuce were 11 and 40 mg Cu kg^{-1} dry weight; for carrot (*Daucus carota* L.) were 5.1 and 8.1 mg Cu kg^{-1} dry weight; and for celery (*Apium graveolens* var. *dulce* Pers.) were 7.5 and 12 mg Cu kg^{-1} dry weight (38).

Proportionally less accumulation of cadmium, lead, and copper occurred in *Artemisia* species in Manitoba, Canada, at high soil metal concentrations than in soils with low metal concentrations (37). Radish (*Raphanus sativus* L.) accumulated only 5 μg Cu plant^{-1} when grown on an agricultural soil (pH 6.3, 6.9% organic matter) contaminated with 591 mg Cu kg^{-1} (18). On the other hand, increasing copper treatments from 0.3 μM CuSO$_4$ to 10^{-5}, 10^{-4}, and 10^{-3} M Cu^{2+} increased root copper levels in sunflower (*Helianthus annuus* L.) from 42, 108, 138, and 1070 μg Cu g^{-1} dry weight, respectively, but not at the expense of growth (39). Contrary to results from many uptake and accumulation studies, the above ground portions of *H. annuus* in this study accumulated more copper than the roots (39).

Fertilizer sources of copper include copper chelate (Na$_2$CuEDTA [13% Cu]), copper sulfate (CuSO$_4$·5H$_2$O [25% Cu]), cupric oxide (CuO [75% Cu]), and cuprous oxide (Cu$_2$O [89% Cu]) (Table 10.3). The copper in micronutrient fertilizers is mainly as CuSO$_4$·5H$_2$O and CuO (40) with CuSO$_4$·5H$_2$O being the most common copper source because of its low cost and high water solubility (41). Copper can be broadcasted, banded, or applied as a foliar spray. Foliar application of chelated copper materials can be used to correct deficiency during the growing season (41).

TABLE 10.3
Copper Fertilizer Sources and Their Approximate Copper Content

Source	Chemical Formula	% Cu
Cuprous oxide	Cu$_2$O	89
Cupric oxide	CuO	80
Chalcocite	Cu$_2$S	80
Malachite, cupric carbonate	CuCO$_3$·Cu(OH)$_2$	57
Copper(II) sulfate-hydroxide	CuSO$_4$·3Cu(OH)$_2$	13–53
Copper chloride	CuCl$_2$	47
Copper frits	frits	40–50
Copper(II) oxalate	CuC$_2$O$_4$·2H$_2$O	40
Copper(II) sulfate monohydrate	CuSO$_4$·H$_2$O	35
Copper(II) sulfate pentahydrate	CuSO$_4$·5H$_2$O	25
Chalcopyrite	CuFeS$_2$	35
Copper(II) ammonium phosphate	Cu(NH$_4$)PO$_4$·H$_2$O	32
Copper(II) acetate	Cu(C$_2$H$_3$O$_2$)$_2$·H$_2$O	32
Cupric nitrate	Cu(NO$_3$)·nH$_2$O	31
Copper chelates	Na$_2$CuEDTA	13
	NaCuHEDTA	9
Organic forms	Animal manures	<0.5

Limitations may apply to the amount of copper to be applied to land during a growing season. For example, in Italy, additions of copper from fertilizers, including sewage sludge, cannot exceed $5 \, kg \, ha^{-1} \, year^{-1}$ (29). Cupric oxide was ineffective in correcting copper deficiency in the year of application but did show residual effects in subsequent years (42). Copper sulfate has been shown to increase the yield of plantlet regeneration from callus in tissue culture (43). In cereal crops, copper is required for anther and pollen development, and deficiencies can lead to pollen abortion and male sterility (44). When the concentration of copper sulfate was increased 100-fold over control treatments to $10 \, \mu M$, the rate of responding anthers in barley (*Hordeum vulgare* L.) increased from 57 to 72% and the number of regenerated plantlets per responding anther increased from 2.4 to 11% (44).

10.2.3 PHYTOREMEDIATION

Heavy metal contamination of agricultural soils, aquatic waters, and ground water can pose serious environmental and health concerns (45). Experimentation into the phyotoextraction of copper from soils is limited (46). However, approximately 24 copper-hyperaccumulating plant species have been reported, including members of Cyperaceae, Lamiaceae, Poaceae, and Scrophulariaceae families (46). Reportedly, the only true copper-accumulating plants are from the central African countries of Zaïre and Zambia (47,48). The political instability of these regions makes obtaining plant material for research experimentation difficult and has hindered the work in this area (47,48). Work by Morrison (49) with Zaïrian copper-tolerant plants showed mint species (*Aeollanthus biformifolius* De Wild) to accumulate $3920 \, \mu g \, Cu \, g^{-1}$ dry weight; figwort species, bluehearts, (*Buchnera metallorum* L.) to accumulate $3520 \, \mu g \, g^{-1}$ dry weight; gentian species (*Faroa chalcophila* P. Taylor) to accumulate $700 \, \mu g \, g^{-1}$ dry weight; and mint species (*Haumaniastrum robertii* (Robyns) Duvign. & Plancke) to accumulate $489 \, \mu g \, g^{-1}$ dry weight (47,48). Rhodegrass (*Chloris gayana* Kunth.), African bristlegrass or forage setaria (*Setaria sphacelata* Stapf. and C.E.Hubb), two indigenous grass species, and oat (*Avena sativa* L.) were evaluated for copper soil extraction in Ethiopian vegetable farms irrigated with wastewater from a textile factory, water from the Kebena and Akaki Rivers, and potable tap water. The maximum copper concentration of these plants was only $10.4 \, mg \, kg^{-1}$ dry weight. However, soil copper levels for the experiments ranged from 2.5 to 3.5 mg kg^{-1}, and these low values may indicate low copper delivery from these irrigation sources (50).

Phytochelatins are peptides $[(\gamma\text{-Glu-Cys})_n Gly]$ produced by plants in response to heavy metal ion exposure (51). These compounds function to complex and detoxify metal ions (52). A variety of metal ions such as Cu^{2+}, Cd^{2+}, Pb^{2+}, and Zn^{2+} induce phytochelatin synthesis (47,48). In addition, cations Hg^{2+}, Ag^+, Au^+, Bi^{3+}, Sb^{3+}, Sn^{2+}, and Ni^{2+}, and anions AsO_4^{3-} and SeO_3^{2-}, induce phytochelatin biosynthesis (52). Together with phytochelatin and metallothionein (cysteine-based proteins that transports metals) (53), internal coordination and vacuolar sequestration determine the tolerance of plant species and cultivars to heavy metals (18). No induction of phytochelatin synthesis was observed following exposure to Al^{3+}, Ca^{2+}, Co^{2+}, Cr^{2+}, Cs^+, K^+, Mg^{2+}, Mn^{2+}, MoO_4^{2-}, Na^+, or V^+ (52). Copper phytochelatins have been isolated from common monkeyflower (*Mimulus guttatus* Fisch. ex DC) (54). Exposure of serpentine roots (*Rauwolfia serpentina* Benth. ex Kurz) to $50 \, \mu M \, CuSO_4$ in hydroponic culture resulted in arrested plant growth for 10 h and rapid production of Cu^{2+}-binding phytochelatins. Two days after treatment, 80% of the copper in solution was depleted from the nutrient solution, and the intercellular phytochelatin concentration reached a constant level, and normal growth resumed (52).

Some plants have shown a strong potential for hyperaccumulation of copper in their tissues. A population of aromatic madder (*Elsholtzia splendens* Nakai) collected on a copper-contaminated site in the Zhejiang providence of China demonstrated phytoremediation potential after the species was noted to accumulate $12,752 \, \mu g \, Cu \, g^{-1}$ dry weight in roots and $3417 \, \mu g \, Cu \, g^{-1}$ dry weight in shoots when cultured in nutrient solutions containing $1000 \, \mu M \, Cu^{2+}$ (55). Alfalfa shoots accumulated as much as $12,000 \, mg \, Cu \, kg^{-1}$ (56). Roots of a willow species (*Salix acmophylla* Boiss.), an economically important tree which grows on the banks of water bodies, accumulated nearly 7 to

$624 \mu g$ Cu g^{-1} dry weight in response to increasing copper treatments in soil from 0 to 10,000 mg kg^{-1} (45). On three soils in Zambia, the roots of a grass species (*Stereochlanea cameronii* Clayton) accumulated 9 to 755 μg Cu g^{-1} dry weight in response to a range from 0.2 to 203 μg Cu g^{-1} in soil (57).

Evidence suggests quantitative genetic variation in the ability to hyperaccumulate heavy metals between- and within-plant populations (58). Populations of knotgrass (*Paspalum distichum* L.) and bermudagrass (*Cynodon dactylon* Pers.) located around mine tailings in China contained 99 to 198 mg Cu kg^{-1}. These native grass populations were more tolerant to increasing $CuSO_4$ concentrations in solution culture than similar genotypes collected from sites containing much lower levels of copper in soil (2.55 mg Cu kg^{-1}) (59). Legumes, *Lupinus bicolor* Lindl. and *Lotus purshianus* Clem. & Clem., growing on a copper mine site (abandoned in 1955) in northern California showed greater tolerance to 0.2 mg Cu L^{-1} in solution culture than genotypes growing in an adjacent meadow (60). Among ten Brassicaceae, only Indian mustard (*Brassica juncea* L.) and radish showed seed germination higher than 90% after 48 h exposure to copper concentrations ranging from 25 to 200 μM (18). As noted with other heavy metals, copper actually caused a slight increase in the degree of seed germination, possibly due to changes in osmotic potential that promote water flow into the seeds (18).

Copper toxicity limits have been established for grass species used to restore heavy metal-contaminated sites. Using sand culture, the lethal copper concentration for redtop (*Agrostis gigantea* Roth.) was 360 mg Cu L^{-1}, for slender wheatgrass (*Elymus trachycaulus* Gould ex Shiners) was 335 mg Cu L^{-1}, and for basin wildrye (*Leymus cinereus* A. Love) was 263 mg Cu L^{-1}, whereas tufted hairgrass (*Deschampsia caespitosa* Beauv.) and big bluegrass (*Poa secunda* J. Presl) displayed less than 50% mortality at the highest treatment level of 250 mg Cu L^{-1} (61).

Success has been shown with sodium-potassium polyacrylate polymers for copper remediation in solution and sand culture; however, the cost of application is often prohibitive. This polymer material at 0.07% dry mass in sand culture absorbed 47, 70, and 190 mg Cu g^{-1} dry weight at 0.5 μM, 1 μM, 0.01 M Cu (as $CuSO_4 \cdot 5H_2O$) in solution, respectively (62). In this experiment, the polyacrylate polymer increased the dry weight yield of the third and fourth cutting of perennial ryegrass (*Lolium perenne* L.) after 50 mg Cu kg^{-1} was applied.

10.3 COPPER DEFICIENCY IN PLANTS

Deficiencies of micronutrients have increased in some crop plants due to increases in nutritional demands from high yields, use of high analysis (N, P, K) fertilizers with low micronutrient quantities, and decreased use of animal manure applications (40). Copper deficiency symptoms appear to be species-specific and often depend on the stage of deficiency (7). Reuther and Labanauskas (7) give a comprehensive description of deficiency symptoms for 36 crops, and readers are encouraged to consult this reference. In general, the terminal growing points of most plants begin to show deficiency symptoms first, a result of immobility of copper in plants. Most plants will exhibit rosetting, necrotic spotting, leaf distortion, and terminal dieback (7,33). Many plants also will show a lack of turgor and discoloration of certain tissues (7,33). Copper deficiency symptoms in lentil, faba bean, chickpea, and wheat (*Triticum aestivum* L.) were chlorosis, stunted growth, twisted young leaves and withered leaf tips, and a general wilting despite adequate water supply (33).

Copper deficiency limits the activity of many plant enzymes, including ascorbate oxidase, phenolase, cytochrome oxidase, diamine oxidase, plastocyanin, and superoxide dismutase (63). Oxidation–reduction cycling between Cu(I) and Cu(II) oxidation states is required during single electron transfer reactions in copper-containing enzymes and proteins (64). Narrow-leaf lupins (*Lupinus angustifolius* L.) exhibited suppressed superoxide dismutase, manganese-superoxide dismutase, and copper/zinc-superoxide dismutase activity on a fresh weight basis under copper deficiency 24 days after sowing (65). Copper deficiency also depresses carbon dioxide fixation, electron transport, and thylakoid prenyl lipid synthesis relative to plants receiving full nutrition (66).

In brown, red, and green algae, the most severe damage in response to Cu^{2+} deficiency was a decrease in respiration, whereas oxygen production was much less affected (67).

Plants differ in their susceptibility to copper deficiency with wheat (*Triticum aestivum* L.), oats, sudangrass (*Sorghum sudanense* Stapf.), and alfalfa being highly sensitive; and barley, corn, and sugar beet being moderately sensitive. Copper tissues levels below $2\,mg\,kg^{-1}$ are generally inadequate for plants (9). A critical copper concentration for Canadian prairie soils for cereal crops production was reported as $0.4\,mg\,kg^{-1}$ (42).

10.4 COPPER TOXICITY IN PLANTS

Prior to the identification of copper as a micronutrient, it was regarded as a plant poison (7). Therefore, no discussion of copper toxicity can rightfully begin without mention of its use as a fungicide. In 1882, botanist Pierre-Marie-Alexis Millardet developed a copper-based formulation that saved the disease-ravaged French wine industry (68). Millardet's observation of the prophylactic effects against downy mildew of grapes by a copper sulfate–lime mixture led to the discovery and development of Bordeaux mixture $[CuSO_4 \cdot 5H_2O + Ca(OH)_2]$. Incidentally, this copper sulfate–lime mixture had been sprinkled on grapevines along the roadways for decades to prevent the stealing of grapes. The observation that Bordeaux sprays sometimes had stimulating effects on vigor and yield led to the experimentation that eventually proved the essentiality of copper as a plant micronutrient (7). It is likely that copper fungicides corrected many copper deficiencies before copper was identified as a required element (69).

The currently accepted theory behind the mode of action of copper as a fungicide is its nonspecific denaturation of sulfhydryl groups of proteins (70). The copper ion is toxic to all plant cells and must be used in discrete doses or relatively insoluble forms to prevent tissue damage (70). There are a multitude of copper-based fungicides and pesticides available to agricultural producers. Overuse or extended use of these fungicides in orchards and vineyards has produced localized soils with excessive copper levels (71).

The two general symptoms of copper toxicity are stunted root growth and leaf chlorosis. For ryegrass (*Lolium perenne* L.) seedlings in solution culture, the order of metal toxicity affecting root growth was $Cu + Ni + Mn + Pb + Cd + Zn + Al + Hg + Cr + Fe$ (72). This order is supported by earlier experiments with *Triticum* spp., white mustard (*Sinapis alba* L.), bent grass (*Agrostis* spp. L.), and corn (72). Stunted roots are characterized by poor development, reduced branching, thickening, and unusual dark coloration (7,14,72,73). Small roots and apices of large roots of spinach turned black in response to $160\,\mu M$ Cu in nutrient solution culture (73). Root growth was decreased progressively in corn when plants were exposed to 10^{-5}, 10^{-4}, 10^{-3} M Cu^{2+} in solution culture (14). However, due to the complexity of cell elongation in roots and influences of hormones, cell wall biosynthesis, and cell turgor, few research studies have defined the effect of copper on root growth (74).

Copper-induced chlorosis, oftentimes resembling iron deficiency, reportedly occurs due to Cu^{+} and Cu^{2+} ion blockage of photosynthetic electron transport (75). Chlorophyll content of spinach leaves was decreased by 45% by treatment of $160\,\mu M$ Cu in solution culture over control treatment (73). Increasing Cu^{2+} exposure to cucumber cotyledon and leaf tissue extracts decreased the amount of UV-light absorbing compounds (76). Chlorosis of bean (*Phaseolus vulgaris* L.) and barley was observed with copper toxicity (77,78). Energy capture efficiency and antenna size were decreased in spinach leaves exposed to toxic levels of copper (73). Copper toxicity symptoms of oregano (*Origanum vulgare* L.) leaves included thickening of the lamina and increases in number of stomata, glandular, and nonglandular hairs, as well as decreases in chloroplast number and disappearance of starch grains in chloroplasts of mesophyll cells (79). Copper ions also may be responsible for accelerating lipid peroxidation in chloroplast membranes (75).

In the photosynthetic apparatus, the donor and acceptor sites of Photosystem II (PSII) are sensitive to excess Cu^{2+} ions (80). The suggested sites of Cu^{2+} inhibition on the acceptor side of PSII are the primary quinone acceptor QA (81,82), the pheophytin–QA–Fe region (83), the non-heme Fe (82,84), and the secondary quinone acceptor QB (85). On the donor side of PSII, a reversible inhibition of oxidation of TyrZ (oxidation–reduction active tyrosine residue in a protein component of PSII) has been observed by Schröder et al. (86) and Jegerschöld et al. (81). However, Cu^{2+} ions in equal molar concentration to the number of PSII reaction centers stimulated oxygen evolution nearly twofold, suggesting that Cu^{2+} may be a required component of PSII (80). Substitution for magnesium in the chlorophyll heme by copper has been observed in brown and green alga under high or low irradiance during incubation at 10 to 30 μM $CuSO_4$ (67). High Cu^{2+} tissue concentrations inhibited oxygen evolution and quenched variable fluorescence (87). Brown and Rattigan (88) reported rapid and complete oxygen production in an aquatic macrophyte (*Elodea canadensis* Michx.) in response to copper toxicity. In fact, *E. canadensis* has been suggested to be a good biomonitor of copper levels in aquatic systems (89).

Excess heavy metals often alter membrane permeability by causing leakage of K^+ and other ions. Solution culture experiments noted that 0.15 μM $CuCl_2$ decreased hydrolytic activity of H^+-ATPase *in vivo* in cucumber roots, but stimulated H^+ transport in corn roots (90). During these experiments, Cu^{2+} also inhibited *in vitro* H^+ transport through the plasmalemma in cucumber roots but stimulated transport in corn roots (90). Copper toxicity also can produce oxidative stress in plants. Increased accumulation of the polyamine, putrescine, was detected in mung bean (*Phaseolus aureus* Roxb.) after copper was increased in solution culture (91). Fifteen-day-old wheat (*Triticum durum* Desf. cv. Cresco) roots exhibited a decrease in NADPH concentrations from 108 to 1.8 nmol g^{-1}, a 23% increase in glutathione reductase activity, and a 43-fold increase in ascorbate over control plants in response to 150 μM Cu in solution culture after a 168-h exposure (94).

In soil, copper toxicity was observed with upland rice (*Oryza sativa* L.) at an application of 51 mg Cu kg^{-1} to the soil, common bean at 37 mg kg^{-1}, corn at 48 mg kg^{-1}, soybean at 15 mg kg^{-1}, and wheat (*Triticum aestivum* L.) at 51 mg kg^{-1} (93). An adequate copper application rate was 3 mg kg^{-1} for upland rice, 2 mg kg^{-1} for common bean, 3 mg kg^{-1} for corn, and 12 mg kg^{-1} for wheat. In this study, an adequate soil test for copper was 2 mg kg^{-1} for upland rice, 1.5 mg kg^{-1} for common bean, 3 mg kg^{-1} for corn, 1 mg kg^{-1} for soybean, and 10 mg kg^{-1} for wheat, when Mehlich-1 extracting solution was used. The toxic level for the same extractor was 48 mg kg^{-1} for upland rice, 35 mg kg^{-1} for common bean, 45 mg kg^{-1} for corn, 10 mg kg^{-1} for soybean, and 52 mg kg^{-1} for wheat. Copper (Cu^{2+}) significantly inhibited growth of radish seedlings at 1 μM in solution culture (94). Addition of supplemental iron to nutrient solution culture lessened the effects of artificially induced copper toxicity in spinach (73). At 10 μM, Cu in the nutrient solution decreased epicotyl elongation and fresh weight of mung bean, but increasing the calcium concentration in the solution to 5 μM improved growth (91). Wheat net root elongation, in relation to the original length, was only 13% in solution culture in response to 1.75 μM Cu^{2+} as $Cu(NO_3)_2$, but additions of 240 μM malate with the $Cu(NO_3)_2$ increased root elongation to 27%; addition of 240 μM malonate increased root to 67%, and 240 μM citrate increased growth to 91%, indicating the potential of these organic ligands to complex Cu^{2+} and to lessen its toxicity (95).

10.5 COPPER IN THE SOIL

10.5.1 INTRODUCTION

Copper is regarded as one of the most versatile of all agriculturally important microelements in its ability to interact with soil mineral and organic components (96). Copper can occur as ionic and complexed copper in soil solution, as an exchangeable cation or as a specifically absorbed ion, complexed in organic matter, occluded in oxides, and in minerals (97). The type of soil copper extraction methodology greatly influences recovery (98). However, soil copper levels in soils correlate very poorly with plant accumulation and plant tissue levels.

10.5.2 Geological Distribution of Copper in Soils

Copper exists mainly as Cu (I) and Cu (II), but can occur in metallic form (Cu°) in some ores (40). Copper occurs in soils as sulfide minerals and less stable oxides, silicates, sulfates and carbonates (40). The most abundant copper-containing mineral is chalcopyrite ($CuFeS_2$) (3). Copper can also be substituted isomorphously for Mn, Fe, and Mg in various minerals (97).

Copper is most abundant in mafic (rich in Mg, Ca, Na, and Fe, commonly basalt and gabbro) rocks, with minimal concentration in carbonate rocks. Mafic rocks contain 60 to 120 mg Cu kg^{-1}; ultramafic rocks (deeper in the crust than mafic rocks) contain 10 to 40 mg kg^{-1}, and acid rocks (granites, gneisses, rhyolites, trachytes, and dacites) contain 2 to 30 mg kg^{-1}. Limestones and dolomites contain 2 to 10 mg Cu kg^{-1}; sandstones contain 5 to 30 mg kg^{-1}; shales contain about 40 mg kg^{-1}, and argillaceous sediments have about 40 to 60 mg kg^{-1} (9). Examples of copper-containing minerals include malachite ($Cu_2(OH)_2CO_3$), azurite ($Cu_3(OH)_2(CO_3)_2$), cuprite (Cu_2O), tenorite (CuO), chalcocite (Cu_2S), covellite (CuS), chalcopyrite ($CuFeS_2$), bornite (Cu_5FeS_4), and silicate chrysocolla ($CuSiO_3 \cdot 2H_2O$) (40). Chalcopyrite ($CuFeS_2$) is a brass-yellow ore that accounts for approximately 50% of the world copper deposits. These minerals easily release copper ions during weathering and under acidic conditions (9). The weathering of copper deposits produces blue and green minerals often sought by prospectors (3).

Because copper ions readily precipitate with sulfide, carbonate, and hydroxide ions, it is rather immobile in soils, showing little variation in soil profiles (9). Copper in soil is held strongly to organic matter, and it is common to find more copper in the topsoil horizons than in deeper zones. Four tropical agricultural soils (Bougouni, Kangaba, Baguinèda, and Gao) in Africa contained 3 to 5 mg Cu kg^{-1} despite differences in climatic zone and texture (99). Copper in these soils was associated mostly with the organic soil fraction. The minerals governing the solubility of Cu^{2+} in soils are not known (100).

The global concentration of total copper in soils ranges from 2 to 200 mg kg^{-1}, with a mean concentration of 30 mg kg^{-1} (40) (Table 10.4). Kabata-Pendias and Pendias (9) reported that worldwide copper concentrations in soils commonly range between 13 and 24 mg kg^{-1}. Reviews by Kubota (30), Adriano (4), and Kabata-Pendias and Pendias (9) present detailed descriptions of global copper distribution. The concentration of copper in soils of the United States ranges from 1 to 40 mg Cu kg^{-1}, with an average content of 9 mg kg^{-1} (30). Agricultural soils in central Italy ranged from 50 to 220 mg Cu kg^{-1} (29). Agricultural soils in central Chile were grouped into two categories: one cluster containing 162 mg Cu kg^{-1} and another cluster containing 751 mg kg^{-1} (36). However, much of this copper was associated with very sparingly soluble forms and was of low bioavailability to crop plants. Fifteen agricultural soils in China ranged from 5.8 to 66.1 mg Cu kg^{-1} (101). Eight soils classified as Alfisols, Inceptisols, or Vertisols in India ranged from 1.12 to 5.67 mg Cu kg^{-1} (102). On the other hand, alum shale and moraine soils from alum shale parent material in India contained 65 and 112 mg Cu kg^{-1}, respectively (103). Five grassland soils in the Xilin river watershed of Inner Mongolia ranged from 0.89 to 1.62 mg Cu kg^{-1} (101). Four calcareous soils from the Baiyin region, Gansu providence, China, ranged from 26 to 199 mg Cu kg^{-1}, the higher levels resulting from irrigation with wastewater from nonferrous metal mining and smelting operations in the 1950s (15). Similar copper soil concentrations were found in mine tailings (Pb–Zn) in Guangdong providence, China (59). The mean copper content of a Canadian soil at 3 to 6.3 km from a metal-processing smelter was 1400 to 3700 mg kg^{-1} (104).

10.5.3 Copper Availability in Soils

Parent material and formation processes govern initial copper status in soils. Atmospheric input of copper has been shown to partly replace or even exceed biomass removal from soils. Kastanozems, Chernozems, Ferrasols, and Fluvisols contain the highest levels of copper, whereas Podzols and Histosols contain the lowest levels.

TABLE 10.4
Copper Levels of Selected Soils from Around the World

Continent	Country	Location	Number of Soil Samples	Soil Copper Mean (mg kg^{-1})	Range	Reference[a]
North America	United States	Northeast	384	24	1–179	112
		North central	99	17	1–119	
		South central	119	19	8–191	
		Southeast	88	5	1–250	
		Pacific northwest	479	30	2–137	
		West	146	54	8–112	
	Canada	Alberta	4	1.1	0.3–2.0	21
		Manitoba	34	1.4	0.1–14.2	37
South America	Chile	Central region	150	256	26–1600	36
Europe	Italy	North central region	9		50–220	29
		Adige valley	1	194		110
	France			20		
		Roujan	2	164		11
	Spain	Granada	1	16		119
	Germany		1	18		38
	Great Britain				20	
Asia	Japan			93.1	26–151	
	India	Rayalaseema region	8		36–190	102
		Lucknow	1	20.2		45
	China	Inner Mongolia steppes	5		0.9–1.6	101
		Rural agricultural areas	15	25.2	5.8–66.1	124
		Gansu province	4		26–119	15
		Guangdong province	4		2–198	59
		Jiangsu province	3		14–98	117
	Turkey		210	7.8	0.8–88	31
	Russia				3–140	
		Eastern regions			5–55	
Africa		Western region	4	4	3–5	99
	Ethiopia		4	3.1	2.5–3.5	50
Australia	New Zealand	South end of North Island	1	11.0		106

[a]Adapted from J. Kubota, *Agron. J.*, 75:913–918, 1983 and D.C. Adriano, in *Trace Elements in the Terrestrial Environment*, Springer-Verlag, New York, 1986, 533pp., unless otherwise referenced.

Chelation and complexing govern copper behavior in most soils (9). For most agricultural soils, the bioavailability of Cu^{2+} is controlled by adsorption–desorption processes. Permanent-charge minerals such as montmorillonite carry a negative charge. Variable-charge minerals such as iron, manganese, and aluminum oxides can carry varying degrees of positive or negative charges depending on soil pH. Therefore, adsorption and desorption of Cu^{2+} is affected by the proportion of these minerals in soils (105). Adsorption of Cu^{2+} in variable charged soils is pH-dependent. Adsorption of Cu^{2+} in soils is often coupled with proton release, thereby lowering soil pH. Organic matter in soils has a strong affinity for Cu^{2+}, even at low Cu^{2+} concentrations. Copper adsorption capacity of a soil decreases in the order of concentration of organic matter + Fe, Al, and Mn oxides + clay minerals (105). In the Zhejiang providence of China, a Quaternary red earth soil (clayey, kaolinitic thermic plinthite Aquult, pH 5.39, 9.03 g organic C kg^{-1}) absorbed a higher percentage of Cu^{2+} added as Cu(NO$_3$)$_2$ than an arenaceous rock soil (clayey, mixed siliceous thermic typic Dystrochrept, pH 4.86, 6.65 g organic C kg^{-1}) (105).

The solubility of copper minerals follows this progression: CuCO$_3$ > Cu$_3$(OH)$_2$(CO$_3$) (azurite) > Cu(OH)$_2$ > Cu$_2$(OH)$_2$CO$_3$ (malachite) > CuO (tenorite) > Cu Fe$_2$O$_4$ cupric ferrite + soil-Cu. Increasing carbon dioxide concentrations decreases the solubility of the carbonate minerals. The solubility of cupric ferrite is influenced by Fe^{3+} and is not much greater than soil copper. Copper will form several sulfate and oxysulfate minerals; however, these minerals are too soluble in soils and will dissolve to form soil-Cu (100). Application of rare earth element fertilizers (23.95% lanthanum, 41.38% cerium, 4.32% praseodymium, and 13.58% neodymium oxides) increased the copper content of water-soluble, exchangeable, carbonate, organic, and sulfide-bound soil fractions, but not the Fe–Mn oxide-bound form (101).

Copper availability is affected substantially by soil pH, decreasing 99% for each unit increase in pH (40). In soil, Cu^{2+} dominates below pH 7.3, whereas CuOH$^+$ is most common at about pH 7.3 (40). The concentration of total soluble copper in the soil solution influences mobility, but the concentration of free Cu^{2+} determines the bioavailability of copper to plants and microorganisms (106). In an aquatic system, Cu^{2+} is the dominant form below pH 6.9, and Cu(OH)$_2$ dominates above that pH. Treatments of 87, 174, 348, and 676 mg CuSO$_4$ kg^{-1} to an alfisol soil (Oxic Tropudalf) in Nigeria significantly acidified the soil and reduced total bacterial counts, microbial respiration, nitrogen and phosphorus mineralization, short-term nitrification, and urease activity relative to untreated soils (107).

Copper ions are held very tightly to organic and inorganic soil exchange sites (9), and CuOH$^+$ is preferably sorbed over Cu^{2+}. The greatest amounts of adsorbed copper exist in iron and manganese oxides (hematite, goethite, birnessite), amorphous iron and aluminum hydroxides, and clays (montmorillonite, vermiculite, imogolite) (9). Microbial fixation is also important in copper binding to soil surfaces (9). Although Cu^{2+} can be reduced to Cu$^+$ ions, copper is not affected by oxidation–reduction reactions that occur in most soils (40). In neutral and alkaline soils, CuCO$_3$ is the major inorganic form, and its solubility is essentially unaffected by pH (108). The hydrolysis constant of copper is 10$^{-7.6}$ (109).

Copper forms stable complexes with phenolic and carboxyl groups of soil organic matter. Most organic soils can bind approximately 48 to 160 mg Cu g^{-1} of humic acid (9). These complexes are so strong that most copper deficiencies are associated with organic soils (40). Addition of composts (biosolids, farmyard manure, spent mushroom, pig manure, and poultry manure) increased the complexation of copper in a mineral soil in New Zealand, and addition of biosolids was effective in reducing the phytotoxicity of copper at high levels of copper addition (106). At the same level of total organic carbon addition, there were differences among these manure sources for copper adsorption (106). In this same study, a significant inverse relationship occurred between copper adsorption and dissolved organic carbon, indicating that copper forms soluble complexes with dissolved organic carbon. Addition of sewage sludge-bark and municipal solid waste compost at about 1000 kg ha^{-1} (containing 126 to 510 mg Cu kg^{-1} dry matter) to a vineyard soil in Italy did not affect total soil or ethylenediaminetetraacetic acid (EDTA)-extractable copper but did decrease diethylenetriaminepentaacetic acid (DTPA)-extractable copper (110). The copper content of grape (*Vitis vinifera* L.) leaves, musts, and wine were not affected by compost treatment over a six-year period but were affected by the nearly 15 to 20 kg Cu ha^{-1} applied through fungicidal treatments (110). Differences in copper accumulation by bean were observed in response to added poultry manure (1% by mass). After 2.0 mM Cu kg^{-1} as CuSO$_4$ was added to a Brazilian agricultural soil, bean plants accumulated 40.5 mg Cu kg^{-1} dry weight without manure additions, but plants grown on soil amended with poultry manure accumulated only 16.9 mg Cu kg^{-1} dry weight (77).

Kabata-Pendias and Pendias (9) report that copper is abundant in the soil solution of all types of soils, whereas Barber (97) notes that soil solution copper is rather low. According to Kabata-Pendias and Pendias (9), the concentration of copper in soil solutions range from 3 to 135 μg L^{-1}. Soils of similar texture do not have the same copper concentration (30). The most common forms of copper in the soil solution are organic chelates (9). Deficiencies are common on sandy soils that have been highly weathered, on mineral soils with high organic matter, and on calcareous mineral soils (111). Although Kubota and Allaway (69) generalized that crop yield responses to copper usually occur only on organic soils, Franzen and McMullen (112) reported that spring wheat yield significantly increased in response to 5 lb of 25% copper sulfate acre^{-1} (5 kg ha^{-1}) on a low organic matter, sandy loam in North Dakota and not on soils with more than 2.5% organic matter. Removal of copper from soils by plant growth is negligible compared to the total amount of copper in soils (9). An average cereal crop removes annually about 20 to 30 g ha^{-1}, and forest biomass annually removes about 40 g ha^{-1} (9).

Copper extraction from soils can differ by extraction method. Ethylenediaminetetraacetic acid has been shown to preferentially extract micronutrients associated with organic matter and bound to minerals (113). Copper extraction from soils in India was highest for 0.5 N ammonium acetate + 0.02 M EDTA, followed in order by 0.1 N HCl, a DTPA extraction mix (0.004 M DTPA, 0.1 M triethanolamine, and 0.01 M CaCl$_2$ at pH 7.3), and 0.05 N HCl + 0.025 N H$_2$SO$_4$ (102). In these Indian soils, soil solution fractions contained 0.38% of the total soil copper, exchangeable forms accounted for 1.00%; specifically absorbed, acid-soluble and Mn-occluded fraction accounted for 4.47%; and the amorphous Fe-occluded and crystal Fe-occluded fraction accounted for 9.94% (102). Increasing strengths of ammonium acetate (0.1, 0.3, 1 M) alone was a poor copper soil extractant; however, the addition of 1 M NH$_2$OH·HCl in acetic acid to the sequential extraction procedure removed 60 to 65% of the total soil copper and further extraction with 30% H$_2$O$_2$ in 1 M HNO$_3$ removed another 20%, which was likely associated with the organic soil fraction (103). The remaining soil copper is termed residual (the difference between extractable and total soil Cu) and is often approximately 50% of total soil copper (97). In contrast, Miyazawa et al. (77) report no differences in copper extraction from a sandy dystrophic dark red latosoil in Brazil by Mehlich-1 (0.05 N HCl + 0.025 N H$_2$SO$_4$), 0.005 M DTPA, pH 7.3 (15.0 g triethanolamine [TEA] + 2.0 g DTPA + 1.5 g CaCl$_2$·2H$_2$O), and 1 M NH$_4$OAc, pH 4.8. Atomic absorption spectrophotometry or colorimetry has been shown to work well in the analysis of ammonium acetate extraction methods (114).

Application of copper usually is not required every year, and residual effects of copper have been reported up to 12 years after soil application (115). Contamination of soils by excess copper occurs mainly by overapplication of fertilizers, sprays, and agricultural and municipal wastes containing copper and from industrial emissions (9). Copper hydroxide is the most widely used fungicide–bactericide for control of tomato diseases (116). Due to the intense use of foliar-applied, copper-containing chemicals, about 25% of tomato leaf samples from greenhouses in Turkey contained over the maximum accepted tolerance level of 200 mg Cu kg^{-1} (31). Due to overuse of copper-containing pesticides and fertilizers, 8.1% of 210 greenhouse soil samples in Turkey were shown to contain greater than 200 mg Cukg^{-1}, the critical soil toxicity level (31).

Localized excess soil copper levels occur in close proximity to industrial sites, but airborne fallout of copper is not substantial. Kabata-Pendias and Pendias (9) reported that atmospheric deposition of copper in Europe ranged from 9 to 224 g ha^{-1} year^{-1}. The average copper concentration of unpolluted river waterways was approximately 10 μg L^{-1}, whereas polluted water systems contained 30 to 60 μg L^{-1} (88). After soils were irrigated for one season with copper-enriched wastewater from a family-owned copper ingot factory in Jiangsu providence, China, copper levels increased sevenfold from 23 to 158 mg kg^{-1} compared to other soils in the region (116).

Runoff from tomato plots receiving 10 kg of 77% copper hydroxide solution ha^{-1} season^{-1} contained significantly more copper if polyethylene mulch was used between the rows instead of a vegetative mulch of vetch (*Vicia villosa* Roth.) (118). Incidentally, the particulate phase of the runoff contained 80% more copper than the dissolved phase. On a calcareous Fluvisol in Spain, Chinese cabbage (*Brassica pekinensis* Rupr.) accumulated 90% more copper under a perforated polyethylene, floating-row cover than plants in the bare-ground treatment. The floating-row cover increased the air temperature by 6.3°C and the root zone temperature by 5.2°C at a 5-cm depth and 4.3°C at a 15-cm depth (119).

10.6   COPPER IN HUMAN AND ANIMAL NUTRITION

10.6.1   INTRODUCTION

Copper was identified as an essential human dietary element approximately 65 years ago (120). Copper is a required catalytic cofactor of selective oxidoreductases and is important for ATP synthesis, normal brain development and neurological function, immune system integrity, cardio-vascular health, and bone density in elderly adults (120). Animals and humans exploit copper by cycling the element between the oxidized cupric ion and the reduced cuprous ion for single-electron transfer reactions (120). Because free or loosely bound copper has the potential to generate free radicals capable of causing tissue pathology, organisms have developed sophisticated mechanisms for its orderly acquisition, distribution, use, and excretion (120).

10.6.2   DIETARY SOURCES OF COPPER

Aside from a few select sources, most foods contain between 2 and 6 mg Cu kg^{-1} dry mass (120). Of the 218 core foods tested, 26 provided 65% of the required copper intake (121). This list included high copper-containing foods such as beef liver and oysters that are consumed infrequently and low copper-containing foods such as tea, potatoes, whole milk, and chicken, which are consumed frequently enough to be considered substantial dietary sources of copper (121). Whole fruits and vegetables contain 20 to 370 mg Cu kg^{-1}; dairy products, including whole milk, contain 3 to 220 mg Cu kg^{-1}; beef, lamb, pork, and veal contain 12 to 9310 mg Cu kg^{-1}; poultry contains 11 to 114 mg Cu kg^{-1}; and seafood and shellfish contain 11 to 79,300 mg Cu kg^{-1}, with cooked oysters having the maximum value (121). Although dietary copper varies regionally, geographically, and culturally, a balanced diet appears to provide an adequate intake of copper for most people. In some areas, additional daily intake of copper can be obtained from drinking water transmitted through copper pipes. In the United States, the current EPA limit for copper in drinking water is 1.3 mg L^{-1} (122). In developed and developing countries, adults, young children, and adolescents, who consume diets of grain, millet, tuber, or rice, along with legumes (beans), small amounts of fish or meat, some fruits and vegetables, and some vegetable oil, are likely to obtain enough copper if their total food consumption is adequate in calories. In developed countries where consumption of red meat is high, copper intake is also likely to be adequate (120).

Forage material containing 7 to 12 mg Cu kg^{-1} dry weight is considered a desirable range for most grazing ruminant animals (123). The copper content of Chinese leymus (*Leymus chinesis* Tzvelev), needlegrass (*Stipa grandis* P. Smirnow), and fringed sage (*Artemisia frigida* Willd.) on grasslands of Inner Mongolia ranged from 0.8 to 2.3 mg kg^{-1} dry matter, and this content was concluded to be severely deficient in copper for ruminant animals (124). The majority of mountain pasture plants examined in central southern Norway were unable to provide enough copper (28). Neonatal ataxia or 'swayback' is typical of copper deficiency in young lambs, and 'steely' or 'stringy' wool is a deficiency symptom in adult sheep (124).

10.6.3   METABOLISM OF COPPER FORMS

Copper is absorbed by the small intestinal epithelial cells by specific copper transporters or other nonspecific metal ion transporters on the brush-border surface (120). Once copper is absorbed, it is transferred to the liver. Copper is then re-secreted into the plasma bound to ceruloplasmin. Human patients who have abnormal ceruloplasmin production still exhibit normal copper metabolism. Therefore, ceruloplasmin is not thought to play a role in copper transportation into cells, and this process remains unknown (120). A well-supported theory is that copper is transported into cells by high-affinity transmembrane proteins. Once inside cells of animals, plants, yeast, and bacteria, copper is bound by protein receptor chaperones and delivered directly to target proteins in the cytoplasm and organelle membranes for incorporation into apocuproproteins (64,120). Liver, brain, and kidney tissues contain higher amounts of copper per unit weight than muscle or other bodily tissues. Copper is not usually stored in tissues and differences in amounts may be related more to concentrations of cuproenzymes. Aside from excretion of nonabsorbed copper, daily losses of copper are minimal in healthy individuals (120).

10.7 COPPER AND HUMAN HEALTH

10.7.1 INTRODUCTION

Copper has been used for medicinal purposes for thousands of years, dating back to the Egyptians and Chinese, who used copper salts therapeutically. Copper also has been used historically for the treatment of chest wounds and the purification of drinking water. Today, copper is used as an antibacterial, antiplaque agent in mouthwashes and toothpastes. The recommended dietary allowance (RDA) for copper was updated in 2001 to $900\,\mu g$ day^{-1}. Because copper is extremely important during fetal and infant development, during pregnancy and lactation, women are encouraged to consume 1000 to $1300\,\mu g$ Cu day^{-1}. The World Health Organization (WHO) and the Food and Agricultural Administration (FAA) suggest that the population mean intake of copper should not exceed $12\,mg$ day^{-1} for adult males and $10\,mg$ day^{-1} for adult females. The Tolerable Upper Intake Limit for copper intake is $10\,mg$ day^{-1}. The adult body can contain between 1.4 and $2.1\,mg$ Cu kg^{-1} of body weight (120).

Copper tends to be toxic to plants before their tissues can accumulate sufficient concentrations to affect animals or humans (125). Copper deficiency from foodstuffs derived from plants and animals exposed to low copper levels is more of a concern. The typical diet in the United States provides copper at just above the lower limits of current RDA levels. The richest food sources of copper include shellfish, nuts, seeds, organ meat, wheat bran cereals, whole-grain cereals, and naturally derived chocolate foods (120).

Roots, flowers, and leaves of the folk and naturopathic herb species, wormwood (*Artemisia absinthium* L.) and white sage (*A. ludoviciana* Nutt.) in Manitoba, Canada, accumulated considerable copper (14.3 to $24.7\,\mu g$ g^{-1} dry weight), indicating their potential importance for medicinal use (37).

10.7.2 COPPER DEFICIENCY AND TOXICITY IN HUMANS

Acquired copper deficiency in adults is quite rare (120), with most cases of deficiency appearing in premature and normal-term infants (126). This deficiency can lead to osteoporosis, osteoarthritis and rheumatoid arthritis, cardiovascular disease, chronic conditions involving bone, connective tissue, heart, and blood vessels, and possibly colon cancer. Other copper deficiency symptoms include anemia, neutropenia (a reduction in infection-fighting white blood cells), hypopigmentation (diminished pigmentation of the skin), and abnormalities in skeletal, cardiovascular, integumentary, and immune system functions (120). In infants and children, copper deficiency may result in anemia, bone abnormalities, impaired growth, weight gain, frequent infections (colds, flu, pneumonia), poor motor coordination, and low energy. Even a mild copper deficiency, which affects a much larger percentage of the population, can impair health in subtle ways. Symptoms of mild copper deficiency include lowered resistance to infections, reproductive problems, general fatigue, and impaired brain function (126).

Symptoms of copper toxicity, although quite rare, include metallic taste in the mouth and gastrointestinal distress in the form of stomach upset, nausea, and diarrhea. These symptoms usually stop when the high copper source is removed. Because copper household plumbing is a significant source of dietary copper, concern has developed for its contribution to elevated copper levels in drinking water (127). In most environments, copper concentrations in potable water delivered by copper-containing plumbing tubes are less than $1\,mg$ L^{-1}. Toxicity connected to copper-containing plumbing pipes is rare, but examples do exist. Toxicity symptoms were traced to contaminated drinking water in new copper plumbing pipes in an incident in Wisconsin (127). Water levels as high as 3.6 mg Cu L^{-1} from faucets connected to the new copper-containing pipes were detected. However, flushing the faucet for 1 min before each use decreased copper levels to <0.25 mg L^{-1}. After a few months, a protective layer of oxide and carbonate forms in copper tubing, and the amount of copper dissolved in the water is reduced. Given the population of the United States (almost 300 million people) and the widespread use of copper plumbing (85% of U.S. homes), the health-related cases from high levels of copper in drinking water are extraordinarily rare. In fact, the antimicrobial effects of copper can inhibit water-borne microorganisms in the drinking water that resides in the copper plumbing tubing (128).

REFERENCES

1. F.A. Cotton, G. Wilkinson, C.A. Murillo, M. Bochmann. *Advance Inorganic Chemistry*, 6th ed. Hoboken, NJ: Wiley, 1999, 1376pp.
2. A.J. Parker. Introduction: The chemistry of copper. In: J.F. Loneragan, A.D. Robson, R.D. Graham, eds. *Copper in Soils and Plants*. New York: Academic Press, 1981, pp. 1–22.
3. K.B. Krauskopf. Geochemistry of micronutrients. In: J.J. Mortvedt, P.M. Giordano, W.L. Lindsay, eds. *Micronutrients in Agriculture*. Madison, WI.: Soil Science Society of America, 1972, pp. 7–40.
4. D.C. Adriano. *Trace Elements in the Terrestrial Environment*. New York: Springer, 1986, 533pp.
5. A.L. Sommer. Copper as an essential for plant growth. *Plant Physiol.* 6:339–345, 1931.
6. C.B. Lipman, G. Mackinney. Proof of the essential nature of copper for higher green plants. *Plant Physiol.* 6:593–599, 1931.
7. W. Reuther, C.K. Labanauskas. Copper. In: H.D. Chapman, ed. *Diagnostic Criteria for Plants and Soils*. Berkeley, CA: University of California Division of Agricultural Sciences Press, 1966, pp. 394–404.
8. J. Delas. The toxicity of copper accumulated in soils. *Agrochemica* 7:258–288, 1963.
9. A. Kabata-Pendias, H. Pendias. *Trace Elements in Soils and Plants*, 2nd ed. Boca Raton, FL: CRC Press, 1992.
10. L. Roca- Pérez, P. Pérez-Bermúdez, R. Boluda. Soil characteristics, mineral nutrients, biomass, and cardenolide production in *Digitalis obscura* wild populations. *J. Plant Nutr.* 25:2015–2026, 2002.
11. V. Chaignon, F. Bedin, P. Hinsinger. Copper bioavailability and rhizosphere pH changes as affected by nitrogen supply for tomato and oilseed rape cropped on an acidic and calcareous soil. *Plant Soil.* 243:219–228, 2002.
12. J.R. Peralta-Videa, J.L. Gardea-Torresdey, E. Gomez, K.J. Tiemann, J.G. Parsons, G. Carrillo. Effect of mixed cadmium, copper, nickel and zinc at different pHs upon alfalfa growth and heavy metal uptake. *Environ. Pollut.* 119:291–301, 2002.
13. V. Chaignon, D. DiMalta, P. Hinsinger. Fe-deficiency increases Cu acquisition by wheat cropped in a Cu-contaminated vineyard soil. *New Phytol.* 154:121–130, 2002.
14. D. Liu, W. Jiang, W. Hou. Uptake and accumulation of copper by roots and shoots of maize. *J. Environ. Sci.* 13:228–232, 2001.
15. Z. Nan, G. Cheng. Copper and zinc uptake by spring wheat (*Triticum aestivum* L.) and corn (*Zea Mays* L.) grown in Baiyin region. *Bull. Environ. Contam. Toxicol.* 67:83–90, 2001.
16. G.A. Pederson, G.E. Brink, T.E. Fairbrother. Nutrient uptake in plant parts of sixteen forages fertilized with poultry litter: Nitrogen, phosphorus, potassium, copper, and zinc. *Agron. J.* 94:895–904, 2002.
17. N.E. Nielsen. A transport kinetic concept for ion uptake by plants. III. Test of a concept by results from water culture and pot experiments. *Plant Soil* 45:659–677, 1976.
18. M.F. Quartacci, E. Cosi, S. Meneguzzo, C. Sgherri, and F. Navari-Izzo. Uptake and translocation of copper in *Brassicaceae*. *J. Plant Nutr.* 26:1065–1083, 2003.
19. H. Marschner. *Mineral Nutrition of Higher Plant*, 2nd ed. San Diego, CA: Academic Press, 1995. 889pp.
20. T. Cheng, H.E. Allen. Prediction of uptake of copper from solution by lettuce (*Lactuca sativa* Romance). *Environ. Toxic Chem.* 20:2544–2551, 2001.

21. Y.K. Soon, G.W. Clayton, P.J. Clarke. Content and uptake of phosphorus and copper by spring wheat: Effect of environment, genotype, and management. *J. Plant Nutr.* 20:925–937, 1997.

22. Z. Xiong, Y. Li, B. Xu. Nutritional influence on copper accumulation by *Brassica pekinensis* Rupr. *Ecotoxic Environ. Safety* 53:200–205, 2002.

23. A. Mozafar. Distribution of nutrient elements along the maize leaf: Alteration by iron deficiency. *J. Plant Nutr.* 20:999–1005, 1997.

24. T. Landberg, M. Greger. Influence of selenium on uptake and toxicity of copper and cadmium in pea (*Pisum sativum*) and wheat (*Triticum aestivum*). *Physiol. Plant* 90:637–644, 1994.

25. C. van Vliet, C.R. Anderson, C.S. Cobbett. Copper-sensitive mutant of *Arabidopsis thaliana. Plant Physiol.* 109:871–878, 1995.

26. R.M. Welch, W.A. Norvell, S.C. Schaefer, J.E. Shaff, L.V. Kochian. Induction of iron(III) and copper(II) reduction in pea (*Pisum sativum* L.) roots by Fe and Cu status: Does the root cell plasmalemma Fe(III)-chelate reductase perform a general role in regulating cation uptake? *Planta* 190: 555–561, 1993.

27. E. Lesuisse, P. Labbe. Iron reduction and trans-plasma membrane electron transport in the yeast *Saccharomyces cerevisiae. Plant Physiol.* 100:769–777, 1992.

28. T.H. Garmo, A. Frøslie, R. Høie. Levels of copper, molybdenum, sulfur, zinc, selenium, iron and manganese in native pasture plants from a mountain area in southern Norway. *Acta Agric. Scand.* 36:147–161, 1986.

29. P. Mantovi, G. Bonazzi, E. Maestri, N. Marmiroli. Accumulation of copper and zinc from liquid manure in agricultural soils and crop plants. *Plant Soil* 250:249–257, 2003.

30. J. Kubota. Copper status of United States soils and forage plants. *Agron. J.* 75:913–918, 1983.

31. M. Kaplan. Accumulation of copper in soils and leaves of tomato plants in greenhouses in Turkey. *J. Plant Nutr.* 22:237–244, 1999.

32. J.M. Swiader, G.W. Ware. *Producing Vegetable Crops*, 5th ed. Danville, IL: Interstate Publishers, Inc., 2002, 658pp.

33. R.F. Brennan, M.D.A. Bolland. Comparing copper requirements for faba bean, chickpea, and lentil with spring wheat. *J. Plant Nutr.* 26:883–899, 2003.

34. F. Vinit-Dunand, D. Epron, B. Alaoui-Sossèm, P.-M. Badot. Effects of copper on growth and on photosynthesis of mature and expanding leaves in cucumber plants. *Plant Sci.* 163:53–58, 2002.

35. T.M. Reinbott, D.G. Blevins, M.K. Schon. Content of boron and other elements in main stems and branch leaves and seed of soybean. *J. Plant Nutr.* 20:831–842, 1997.

36. R. Babilla-Ohlbaum, R. Ginocchio, P.H. Rodrìguez, A Cèspedes, S. Gonzàlez, H.E. Allen, G.E. Lagos. Relationship between soil copper content and copper content of selected crop plants in central Chile. *Environ. Toxic Chem.* 20:2749–2757, 2001.

37. E. Pip, C. Mesa. Cadmium, copper, and lead in two species of *Artemisia* (Compositae) in southern Manitoba, Canada. *Bul.l Environ. Contam. Toxicol.* 69:644–648, 2002.

38. K. Bunzl, M. Trautmannsheimer, P. Schramel, W. Reifenhäuser. Availability of arsenic, copper, lead, thallium, and zinc to various vegetables grown in slag-contaminated soils. *J. Environ. Qual.* 30:934–939, 2001.

39. J. Lin, W. Jiang, D. Liu. Accumulation of copper by roots, hypocotyls, cotyledons and leaves of sunflower (*Helianthus annuus* L.). *Bioresource Tech.* 86:151–155, 2003.

40. J.J. Mortvedt. Bioavailability of micronutrients. In: ME Sumner, ed. *Handbook of Soil Science.* Boca Raton, Fla.: CRC Press, 2000, pp. D71–D88.

41. D. Mengel, G. Rehm. Fundamentals of fertilizer application. In: M.E. Sumner, ed. *Handbook of Soil Science.* Boca Raton, FL: CRC Press, 2000, pp. D155–D174.

42. R.E. Karamanos, G.A. Kruger, J.W.B. Stewart. Copper deficiency in cereal and oilseed crops in northern Canadian Prairie soils. *Agron. J.* 78:317–323, 1986.

43. L. Dahleen. Improved plant regeneration from barley callus cultures by increased copper levels. *Plant Cell Tissue Org. Cult.* 43:267–269, 1995.

44. G. Wojnarowiez, C. Jacquard, P. Devaux, R.S. Sangwan, C. Clément. Influence of copper sulfate on anther culture in barley (*Hordeum vulgare* L.). *Plant Sci.* 162:843–847, 2002.

45. M.B. Ali, P. Vajpayee, R.D. Tripathi, U.N. Rai, S.N. Singh, S.P. Singh. Phytoremediation of lead, nickel, and copper by *Salix acmophylla* Boiss: Role of antioxidant enzymes and antioxidant substances. *Bull. Environ. Contam. Toxicol.* 70:462–469, 2003.

46. S.P. McGrath. Phytoextraction for soil remediation. In: R.R. Brooks, ed. *Plants That Hyperaccumulate Heavy Metals.* Oxon, U.K.: CAB International, 1998, pp. 261–287.

47. R.R. Brooks. Phytochemistry of hyperaccumlators. In: R.R. Brooks, ed. *Plants That Hyperaccumulate Heavy Metals*. Oxon, U.K.: CAB International, 1998, pp.15–53.

48. R.R. Brooks. Geobotany and hyperaccumlators. In: R.R. Brooks, ed. *Plants That Hyperaccumulate Heavy Metals*. Oxon, U.K.: CAB International, 1998, pp. 55–94.

49. R.S. Morrison, R.R. Brooks, R.D. Reeves, F. Malaisse. Copper and cobalt uptake by metallophytes from Zaire. *Plant Soil* 53:535–539, 1979.

50. F. Itanna, B. Coulman. Phyto-extraction of copper, iron, manganese, and zinc from environmentally contaminated sites in Ethiopia, with three grass species. *Commun. Soil Sci. Plant Anal.* 34:111–124, 2003.

51. E. Grill, E.-L. Winnacker, M.H .Zenk. Phytochelatins: The principle heavy-metal complexing peptides of higher plants. *Science* 230:674–676, 1985.

52. E. Grill, E.-L. Winnacker, M.H. Zenk. Phytochelatins, a class of heavy-metal-binding peptides from plants, are functionally analogous to metallothioneins. *Proc. Natl. Acad. Sci.* 84:439–443, 1987.

53. W.E. Rauser. Structure and function of metal chelators produced by plants: The case for organic acids, amino acids, phytin, and metallothioneins. *Cell Biochem. Biophys.* 31:19–48, 1999.

54. D.E. Salt, D.A. Thurman, A.B. Tomsett, A.K. Sewell. Copper phytochelatins of *Mimulus guttatus*. *Proc. R. Soc. Lond. B* 236:79–89, 1989.

55. M.J. Yang, X.E. Yang, V. Römheld. Growth and nutrient composition of *Elsholtzia splendens* Nakai under copper toxicity. *J. Plant Nutr.* 25:1359–1375, 2002.

56. J.R. Peralta, J.L. Gardea-Torresdey, K.J. Tiemann, E. Gomez, S. Arteaga, E. Rascon, J.G. Parsons. Uptake and effects of five heavy metals on seed germination and plant growth in alfalfa (*Medicago sativa* L.). *Bull. Environ. Contam. Toxicol.* 66:727–734, 2001.

57. A. Reilly, C. Reilly. Zinc, lead, and copper tolerance in the grass *Stereochlaena cameronii* (Stapf) Clayton. *New Phytol.* 72:1041–1046, 1973.

58. A.J. Pollard, K.D. Powell, F.A. Harper, J.A.C. Smith. The genetic basis of metal hyperaccumulation in plants. *Critical Rev. Plant Sci.* 21:539–566, 2002.

59. W.S. Shu, Z.H. Ye, C.Y. Lan, Z.Q. Zhang, M.H. Wong. Lead, zinc, and copper accumulation and tolerance in populations of *Paspalum distichum* and *Cynodon dactylon*. *Environ. Pollut.* 120:445–453, 2002.

60. L. Wu, A.L. Kruckeberg. Copper tolerance in two legume species from a copper mine habitat. *New Phytol.* 99:565–570, 1985.

61. M.W. Paschke, E.F. Redente. Copper toxicity thresholds for important restoration grass species of the western United States. *Environ. Toxic Chem.* 21:2692–2697, 2002.

62. M.O. Torres, A. DeVarennes. Remediation of a sandy soil artificially contaminated with copper using a polyacrylate polymer. *Soil Use Mgt.* 14:106–110, 1998.

63. C.D. Walker, J. Webb. Copper in plants: Forms and behaviour. In: J.F. Loneragan, A.D. Robson, R.D. Graham, eds. *Copper in Soils and Plants*. New York: Academic Press, 1981, pp. 198–212.

64. M.D. Harrison, C.E. Jones, C.T. Dameron. Copper chaperones: Function, structure and copper-binding properties. *JBIC* 4:145–153, 1999.

65. Q. Yu, Z. Rengel. Micronutrient deficiency influences plant groth and activities of superoxide dismutases in narrow-leaf lupins. *Ann. Bot.* 83:175–182, 1999.

66. W. Bussler. Physiological functions and utilization of copper. In: J.F. Loneragan, A.D. Robson, R.D. Graham, eds. *Copper in Soils and Plants*. New York: Academic Press, 1981, pp. 213–234.

67. H. Küpper, I. Šetlík, M. Spiller, F.C. Küpper, O. Prášil. Heavy metal-induced inhibition of photosynthesis: Targets of *in vivo* heavy metal chlorophyll formation. *J. Phycol.* 38:429–441, 2002.

68. E.C. Large. *The Advance of the Fungi*. New York: Holt, 1940. 488pp.

69. J. Kubota, W.H. Allaway. Geographic distribution of trace element problems. In: J.J. Mortvedt, P.M. Giordano, W.L. Lindsay, eds. *Micronutrients in Agriculture*, Madison, WI: Soil Science Society of America, 1972, pp. 525–554.

70. G.W. Ware, D.M. Whitacre. *The Pesticide Book*, 6th ed. Willoughby, OH: MeisterPro Information Resources, 2004, 487pp.

71. E. Semu, B.R. Singh. Accumulation of heavy metals in soils and plants after long-term use of fertilizers and fungicides in Tanzania. *Fert. Res.* 44: 241—248, 1996.

72. M.H. Wong, A.D. Bradshaw. A comparison of the toxicity of heavy metals, using root elongation of rye grass, *Lolium perenne*. *New Phytol.* 91:255–261, 1982.

73. G. Ouzounidou, I. Illias, H. Tranopoulou, S. Karataglis. Amelioration of copper toxicity by iron on spinach physiology. *J. Plant Nutr.* 21:2089–2101, 1998.

74. H.W. Woolhouse, S. Walker. The physiological basis of copper toxicity and copper tolerance in higher plants. In: J.F. Loneragan, A.D. Robson, R.D. Graham, eds. *Copper in Soils and Plants*. New York: Academic Press, 1981, pp. 235–262.

75. G. Sandman, P. Boger. Copper-mediated lipid-peroxidation processes in photosynthetic membranes. *Plant Physiol.* 66:797–800, 1980.

76. C.R. Caldwell. Effect of elevated copper on the ultraviolet light-absorbing compounds of cucumber cotyledon and leaf tissues. *J. Plant Nutr.* 24:283–295, 2001.

77. M. Miyazawa, S.M.N. Giminez, M. Josefa, S. Yabe, E.L. Oliveira, M.Y. Kamogawa. Absorption and toxicity of copper and zinc in bean plants cultivated in soil treated with chicken manure. *Water Air Soil Pollut.* 138:211–222, 2002.

78. A. Vassilev, F.C. Lidon, M. do Céu Matos, J.C. Ramalho, I. Yordanov. Photosynthetic performance and content of some nutrients in cadmium- and copper-treated barley plants. *J. Plant Nutr.* 25:2343–2360, 2002.

79. H. Panou-Filotheou, A.M. Bosabalidis, S. Karataglis. Effects of copper toxicity on leaves of oregano (*Origanum vulgare* subsp. *hirtum*). *Ann. Bot.* 88:207–214, 2001.

80. K. Burda, J. Kruk, K. Strzalka, G.H. Schmid. Stimulation of oxygen evolution in photosystm II by copper(II) ions. *Z Naturforsch* 57:853–857, 2002.

81. C. Jegerschöld, J.B. Arellano, W.P. Schröder, P.J. van Kan, M. Barón, S. Styring. Copper(II) inhibition of electron transfer through photosystem II studied by EPR spectroscopy. *Biochemistry* 34:12747–12754, 1995.

82. C. Jegerschöld, F. McMillan, WLubitz, A.W. Rutherford. Effect of copper and zinc ions on photosystem II. Studies by EPR spectroscopy. *Biochemistry* 38:12439–12445, 1999.

83. I. Yruela, G. Gatzen, R. Picorel, A.R. Holzwarth. Cu(II)-inhibitory effect on photosystem II from higher plants. A picosecond time-resolved fluorescence study. *Biochemistry* 35:9469–9474, 1996.

84. D.P. Singh, S.P. Singh. Action of heavy metals on Hill activity and O_2 evolution. *Plant Physiol.* 83:12–14, 1987.

85. N. Mohanty, I. Vass, S. Demeter. Copper toxicity affects photosystem II electron transport at the secondary quinone accpetor (Q_B). *Plant Physiol.* 90:175–179, 1989.

86. W.P. Schröder, J.B. Arellano, T. Bittner, M. Barón, H.J. Eckert, G. Renger. Flash induced absorption spectroscopy studies of copper interaction with photosystem II in higher plants. *J. Biol. Chem.* 269:32856–32870, 1994.

87. J.B. Arellano, J.J. Lazaro, J. Lopez-Gorge, M. Baron. The donor side of photosystem II as the copper-inhibitory binding site. *Photosynth. Res.* 45:127–134, 1885.

88. B.T. Brown, B.M. Rattigan. Toxiciy of soluble copper and other metals to *Elodea canadensis*. *Environ. Pollut.* 20:303–314, 1979.

89. T.K. Mal, P. Adorjan, A.L. Corbett. Effect of copper on growth of an aquatic macrophyte, *Elodea canadensis*. *Environ Pollut.* 120:307–311, 2002.

90. M. Burzynski, E. Kolano. *In vivo* and *in vitro* effects of copper and cadmium on the plasma membrane H^+-ATPase from cucumber (*Cucumis sativus* L.) and maize (*Zea mays* L.) roots. *Acta Physiol. Plant* 25:39–45, 2003.

91. Z. Shen, F. Zhang, F. Zhang. Toxicity of copper and zinc in seedlings of mung bean and introducing accumulation of polyamine. *J. Plant Nutr.* 21:1153–1162, 1998.

92. C. Sgherri, M.F. Quartacci, R. Izzo, F. Navari-Izzo. Relation between lipoic acid and cell redox status in wheat grown in excess copper. *Plant Physiol. Biochem.* 40:591–597, 2002.

93. N.K. Fageria. Adequate and toxic levels of copper and manganese in upland rice, common bean, corn, soybean, and wheat grown on an oxisol. Commun. *Soil Sci. Plant Anal.* 32:1659–1676, 2001.

94. E.-L. Chen, Y.-A. Chen, L.-M. Chen, Z.-H. Liu. Effect of copper on peroxidase activity and lignin content if *Raphanus sativus*. *Plant Physiol. Biochem.* 40:429–444, 2002.

95. D.R. Parker, J.F. Pedler, Z.A.S. Ahnstrom, M. Resketo. Reevaluating the free-ion activity model of trace metal toxicity toward higher plants: Experimental evidence with copper and zinc. *Environ. Toxic Chem.* 20:899–906, 2001.

96. M.B. McBride. Forms and distribution of copper in solid and solution phases of soil. In: J.F. Loneragan, A.D. Robson, R.D. Graham, eds. *Copper in Soils and Plants*. New York: Academic Press, 1981, pp. 25–45.

97. S.A. Barber. *Soil Nutrient Bioavailability*, 2nd ed. New York: Wiley, 1995, 398pp.

98. D.E. Baker, M.C. Amacher. Nickel, copper, zinc, and cadmium. In: A.L. Page, ed. *Methods of Soil Analysis, Part 2, Chemical and Microbiological Properties*, 2nd ed., Agronomy 9. Madison, WI: American Society of Agronomy, 1982, pp. 323–336.

99. M. Soumarè, F.M.G. Tack, M.G. Verloo. Distribution and availability of iron, manganese, zinc, and copper in four tropical agricultural soils. *Comm. Soil Sci. Plant Anal.* 34:1023–1038, 2003.

100. W.L. Lindsay. *Chemical Equilibria in Soils*, 2nd ed. Caldwell, NJ: Blackburn Press, 2001, 449 pp.

101. Z. Wang, X. Shan, S. Zhang. Effects of exogenous rare earth elements of fractions of heavy metals in soils and bioaccumulation by plants. *Commun. Soil Sci. Plant Anal.* 34:1573–1588, 2003.

102. R. Rupa, L.M. Shukla. Comparison of four extractants and chemical fractions for assessing available zinc and copper in soils in India. Commun. *Soil Sci. Plant Anal.* 302579–2591, 1999.

103. R.P. Narwal, B.R. Singh, B. Salbu. Association of cadmium, zinc, copper, and nickel with components in naturally heavy metal-rich soils studied by parallel and sequential extractions. *Commun. Soil Sci. Plant Anal.* 30:1209–1230, 1999.

104. B. Freedman, T.C. Hutchinson. Pollutants inputs from the atmosphere and accumulation in soils and vegetation near a nickel-copper smelter at Sudbury, Ontario, Canada. *Can. J. Bot.* 58:108–132, 1980.

105. S. Yu, Z.L. He, C.Y. Huang, G.C. Chen, D.V. Calvert. Adsorption–desorption behavior of copper at contaminated levels in red soils from China. *J. Environ. Qual.* 31:1129–1136, 2002.

106. N. Bolan, D. Adriano, S. Mani, A. Khan. Adsorption, complexation, and phytoavailability of copper as influenced by organic manure. *Environ. Toxic Chem.* 22450–456, 2003.

107. A. Olayinka, G.O. Babalola. Effects of copper sulfate application on microbial numbers and respiration, nitritier and urease activities, and nitrogen and phosphorus mineralization in an alfisol. *Biol. Agric. Hort.* 19:1–8, 2001.

108. J.R. Sanders, C. Bloomfield. The influence of pH, ionic strength, and reactant concentration on copper complexing by humified organic matter. *J. Soil Sci.* 31:53–63, 1980.

109. B.G. Ellis, B.D. Knezek. Adsorption reactions of micronutrients in soils. In: J.J. Mortvedt, P.M. Giordano, W.L. Lindsay, eds. *Micronutrients in Agriculture*. Madison, WI: Soil Science Society of America, 1972, pp. 59–78.

110. F. Pinamonti, G. Nicolini, A. Dalpiaz, G. Stringari, G. Zorzi. Compost use in Viticulture: Effect on heavy metal levels in soil and plants. *Commun. Soil Sci. Plant Anal.* 30:1531–1549, 1999.

111. D.C. Martens, D.T. Westerman. Fertilizer applications for correcting micronutrient deficiencies. In: J.J. Mortvedt, F.R. Cox, L.M. Shurman, R.M. Welch, eds. *Micronutrients in Agriculture*, 2nd ed. Madison, WI.: Soil Science Society of America, 1991, pp. 549–592.

112. D.W. Franzen, M.V. McMullen. Spring Wheat Response to Copper Fertilization in North Dakota, North Dakota State University Ext Rep 50, 1999, 5pp.

113. R.J. Haynes, R.S. Swift. Amounts and forms of micronutrient cations in a group of loessial grassland soils of New Zealand. *Geoderma* 33:53–62, 1984.

114. H.A. Mills, J.B. Jones, Jr. *Plant Analysis Handbook II*. Athens, GA.: MicroMacro Publishing, Inc., 1996, 422pp.

115. J.W. Gartrell. Distribution and correction of copper deficiency in crops and pastures. In: J.F. Loneragan, A.D. Robson, R.D. Graham, eds. *Copper in Soils and Plants*. New York: Academic Press, 1981, pp. 313–349.

116. R.M. Davis, G. Hamilton, W.T. Lanini, T.H. Screen, C. Osteen. The Importance of Pesticides and Other Pest Management Practices in U.S. Tomato Production. Doc I-CA-98. US Dept Agric Nat Agric Pest Impact Asses Prog, Washington, DC, 1998.

117. Y. Luo, X. Jiang, L. Wu, J. Song, S. Wu, R. Lu, P. Christie. Accumulation and chemical fractionation of Cu in a paddy soil irrigated with Cu-rich wastewater. *Geoderma* 115:113–120, 2003.

118. P.J. Rice, L.L. McConnell, L.P. Heighton, A.M. Sadeghi, A.R. Isensee, J.R. Teasdale, A.A. Abdul-Baki, J.A. Harman-Fetcho, C.J. Hapeman. Comparison of copper levels in runoff from fresh-market vegetable production using polyethylene much or a vegetative mulch. *Environ. Toxic Chem.* 21:24–30, 2002.

119. D.A. Moreno, G. Villora, J. Hernández, N. Castilla, L. Romero. Accumulation of Zn, Cd, Cu, and Pb in Chinese cabbage as influenced by climatic conditions under protected cultivation. *J. Agric. Food Chem.* 50:1964–1969, 2002.

120. M.L. Failla, M.A. Johnson, J.R. Prohaska. Copper. In: B.A. Bowman, R.M. Russell, eds. *Present Knowledge in Nutrition*, 8th ed. Washington, DC: ILSI Press, 2001, pp. 373–383.

121. D.G. Lurie, J.M. Holden, A. Schubert, W.R. Wolf, N.J. Miller-Ihli. The copper content of foods based on a critical evaluation of published analytical data. *J. Food Comp. Anal.* 2:298–316, 1989.

122. United States Environmental Protection Agency. *The EPA Region III Risk-Based Concentration Table.* Philadelphia, PA: USEPA, 1999.

123. D.C. Blood, O.M. Radostits, J.A. Henderson. *Veterinary Medicine: A Textbook of the Diseases of Cattle, Sheep, Pigs, Goats and Horses.* 6th ed. London, U.K.: Bailliere Tindall, 1983, 1310pp.

124. S.P. Wang, Y.F. Wang, Z.Y. Hu, Z.Z. Chen, J. Fleckenstein, E. Schnug. Status of iron, manganese, copper, and zinc of soils and plants and their requirements for ruminants in Inner Mongolia steppes of China. *Commun. Soil Sci. Plant Anal.* 34:665–670, 2003.

125. D.M. Miller, W.P. Miller. Land application of wastes. In: M.E. Sumner, ed. *Handbook of Soil Science.* Boca Raton, FL.: CRC Press, 2000, pp. G217–G245.

126. C.A. Owen. *Copper Deficiency and Toxicity: Acquired and Inherited, in Plants, Animals, and Humans.* Park Ridge, NJ: Noyes Publications, 1981, 189pp.

127. L. Knobeloch, C. Shubert, J. Hayes, J. Clark, C. Fitzgerald, A.Fraundorff. Gastrointestinal upsets and new copper plumbing—is there a connection? *Wis. Med. J.* 97:49–53, 1998.

128. J. Linn, W. Jiang, D. Liu. Accumulation of copper by roots, hypocotyls, coteledons, and leaves of sunflower (*Helianthus annuus* L.). *Bioresource Technol.* 86:151–155, 2003.

11 Iron

Volker Römheld
University of Hohenheim, Stuttgart, Germany

Miroslav Nikolic
University of Belgrade, Belgrade, Serbia

CONTENTS

11.1 Historical Information 329
    11.1.1 Determination of Essentiality 329
11.2 Functions in Plants 330
11.3 Forms and Sources of Iron in Soils...330
11.4 Diagnosis of Iron Status in Plants...332
    11.4.1 Iron Deficiency 332
    11.4.2 Iron Toxicity 332
11.5 Iron Concentration in Crops 335
    11.5.1 Plant Part and Growth Stage 335
    11.5.2 Iron Requirement of Some Crops 335
    11.5.3 Iron Levels in Plants 336
        11.5.3.1 Iron Uptake 336
        11.5.3.2 Movement of Iron within Plants...338
11.6 Factors Affecting Plant Uptake 339
    11.6.1 Soil Factors 339
    11.6.2 Plant Factors 343
11.7 Soil Testing for Iron 344
11.8 Fertilizers for Iron 344
References 345

11.1 HISTORICAL INFORMATION

11.1.1 DETERMINATION OF ESSENTIALITY

Julius von Sachs, the founder of modern water culture experiments, included iron in his first nutrient cultures in 1860, and Eusèbe Gris, in 1844, showed that iron was essential for curing chlorosis in vines (1,2). Sachs had already shown that iron can be taken up by leaves, and within a few years L. Rissmüller had demonstrated that foliar iron is obviously translocated by phloem out of leaves before leaf fall in European beech (*Fagus sylvatica* L.). The early developments in the study of iron in plant nutrition were summarized by Molisch in 1892 (3).

It was another 100 years before the principal processes of the mobilization of iron in the rhizosphere started to be understood (4–8).

11.2   FUNCTIONS IN PLANTS

The ability of iron to undergo a valence change is important in its functions:

$$Fe^{2+} \rightleftharpoons Fe^{3+} + electron$$

It is also the case that iron easily forms complexes with various ligands, and by this modulates its redox potential. Iron is a component of two major groups of proteins. These are the heme proteins and the Fe-S proteins. In these macromolecules, the redox potential of the Fe(III)/Fe(II) couple, normally 770 mV, can vary across most of the range of redox potential in respiratory and photosynthetic electron transport (-340 to $+810$ mV). When iron is incorporated into these proteins it acquires its essential function (9).

The heme proteins contain a characteristic heme iron–porphyrin complex, and this acts as a prosthetic group of the cytochromes. These are electron acceptors–donors in respiratory reactions. Other heme proteins include catalase, peroxidase, and leghemoglobin.

Catalase catalyzes the conversion of hydrogen peroxide into water and O_2 (reaction A), whereas peroxidases catalyze the conversion of hydrogen peroxide to water (reaction B):

$$2H_2O_2 \rightarrow 2H_2O + O_2 \qquad\qquad\qquad (A)$$

$$H_2O_2 + AH_2 \rightarrow A + 2H_2O \qquad\qquad\qquad (B)$$

$$AH + AH + H_2O_2 \rightarrow A - A + 2H_2O$$

Catalase has a major role in the photorespiration reactions, as well as in the glycolate pathway, and is involved in the protection of chloroplasts from free radicals produced during the water-splitting reaction of photosynthesis. The reaction sequence of peroxidase shown above includes cell wall peroxidases, which catalyze the polymerization of phenols to form lignin. Peroxidase activity is noticeably depressed in roots of iron-deficient plants, and inhibited cell wall formation and lignification, and accumulation of phenolic compounds have been reported in iron-deficient roots.

As well as being a constituent of the heme group, iron is required at two other stages in its manufacture. It activates the enzymes aminolevulinic acid synthetase and coproporphorinogen oxidase. The protoporphyrin synthesized as a precursor of heme is also a precursor of chlorophyll, and although iron is not a constituent of chlorophyll this requirement, and the fact that it is also required for the conversion of Mg protoporphyrin to protochlorophyllide, means that it is essential for chlorophyll biosynthesis (10). However, the decreased chloroplast volume and protein content per chloroplast (11) indicate that chlorophyll might not be adequately stabilized as chromoprotein in chloroplasts under iron deficiency conditions, thus resulting in chlorosis.

Along with the iron requirement in some heme enzymes and its involvement in the manufacture of heme groups in general, iron has a function in Fe-S proteins, which have a strong involvement with the light-dependent reactions of photosynthesis. Ferredoxin, the end product of photosystem I, has a high negative redox potential that enables it to transfer electrons to a number of acceptors. As well as being the electron donor for the synthesis of NADPH in photosystem I, it can reduce nitrite in the reaction catalyzed by nitrite reductase and it is an electron donor for sulfite reductase.

11.3   FORMS AND SOURCES OF IRON IN SOILS

Iron occurs in concentrations of 7,000 to 500,000 mg kg^{-1} in soils (12), where it is present mainly in the insoluble Fe(III) (ferric, Fe^{3+}) form. Ferric ions hydrolyze readily to give $Fe(OH)_2^+$, $Fe(OH)_3$, and $Fe(OH)_4^-$, with the combination of these three forms and the Fe^{3+} ions being the total soluble inorganic iron, and the proportions of these forms being determined by the reaction (13):

$$Fe(OH)_3 \text{ (soil)} + 3H^+ \rightleftharpoons Fe^{3+} + 3H_2O$$

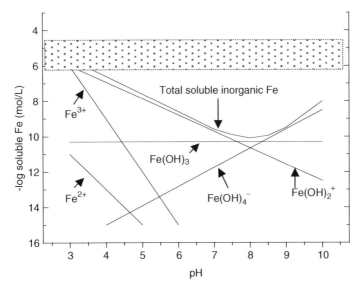

FIGURE 11.1   Solubility of inorganic Fe in equilibrium with Fe oxides in a well-aerated soil. The shaded zone represents the concentration range required by plants for adequate Fe nutrition. (Redrawn from Römheld, V., Marschner, H., in *Advances in Plant Nutrition*, Vol. 2, Praeger, New York, 1986, pp. 155–204 and Lindsay, W.L., Schwab, A.P., *J. Plant Nutr.*, 5:821–840, 1982.)

With an increase in soil pH from 4 to 8, the concentration of Fe^{3+} ions declines from 10^{-8} to 10^{-20} M. As can be seen from Figure 11.1, the minimum solubility of total inorganic iron occurs between pH 7.4 and 8.5 (14).

The various Fe(III) oxides are major components of a mineral soil, and they occur either as a gel coating soil particles or as fine amorphous particles in the clay fraction. Similar to the clay colloids, these oxides have colloidal properties, but no cation-exchange capacity. They can, however, bind some anions, such as phosphate, particularly at low pH, through anion adsorption. For this reason, the presence of these oxides interferes with phosphorus acquisition by plants, and in soils of pH above 6, more than 50% of the organically bound forms of phosphate may be present as humic-Fe(Al)-P complexes (15).

Although Fe(III) oxides are relatively insoluble in water, they can become mobile in the presence of various organic compounds. As water leaches through decomposing organic matter, it moves the Fe(III) oxide downwards, particularly at acidic pH, so that under such conditions podzols form. The iron is essentially leached from the top layers of soil as iron–fulvic acid complexes and forms an iron pan after precipitation lower down at higher pH. The upper layers are characteristically light in color, as it is the gel coating of Fe(III) oxide that, in conjunction with humus, gives soils their characteristic color. However, in soils in general, the intensity of the color is not an indication of iron content.

These organic complexes tend to make iron more available than the thermodynamic equilibrium would indicate (16), and in addition to iron-forming complexes with fulvic acid, it forms complexes with microbial siderophores (13), including siderophores released by ectomycorrhizal fungi (17). A water-soluble humic fraction extracted from peat has been shown to be able to form mobile complexes with iron, increasing its availability to plants (18).

In soils with a high organic matter content the concentration of iron chelates can reach 10^{-4} to 10^{-3} M (17,18). However, in well-aerated soils low in organic matter, the iron concentration in the soil solution is in the range of 10^{-8} to 10^{-7} M, lower than is required for adequate growth of most plants (13).

Under anaerobic conditions, ferric oxide is reduced to the Fe(II) (ferrous) state. If there are abundant sulfates in the soil, these also become oxygen sources for soil bacteria, and black Fe(II) sulfide is formed. Such reactions occur when a soil becomes waterlogged, but on subsequent drainage the Fe(II) iron is oxidized back to Fe(III) compounds. Alternate bouts of reduction and oxidation as the water table changes in depth give rise to rust-colored patches of soil characteristic of gleys. Ferrous iron, Fe^{2+}, and its hydrolysis species contribute toward total soluble iron in a soil only if the sum of the negative log of ion activity and pH together fall below 12 (equivalent to Eh of $+260\,mV$ and $+320\,mV$ at pH 7.5 and 6.5, respectively) (13,14). It is likely that the presence of microorganisms around growing roots causes the redox potential in the rhizosphere to drop because of the microbial oxygen demand, and this would serve to increase concentrations of Fe^{2+} ions for plant uptake (21).

Because the solubility of Fe^{3+} and Fe^{2+} ions decreases with increase in pH, growing plants on calcareous soils, and on soils that have been overlimed, gives rise to lime-induced chlorosis. The equilibrium concentration of Fe^{3+} in calcareous soil solution at pH 8.3 is $10^{-19}\,mM$ (22), which gives noticeable iron deficiency in plants not adapted to these conditions. It has been estimated that up to 30% of the world's arable land is too calcareous for optimum crop production (23,24).

Iron deficiency can also arise from excess of manganese and copper. Most elements can serve as oxidizing agents that convert Fe^{2+} ions into the less soluble Fe^{3+} ions (25), and excess manganese in acid soils can give rise to deficiencies of iron although it would otherwise be present in adequate amounts (26).

Corn (*Zea mays* L.) and sugarcane (*Saccharum officinarum* L.) may show iron deficiency symptoms when deficient in potassium. It seems that under these circumstances iron is immobilized in the stem nodes, a process that is accentuated by good phosphorus supply (27). Iron can bind a significant proportion of phosphate in well-weathered soil (as the mineral strengite), and as this substance is poorly soluble at pH values below 5, iron contributes to the poor availability of phosphorus in acid soils (25).

11.4   DIAGNOSIS OF IRON STATUS IN PLANTS

11.4.1   IRON DEFICIENCY

The typical symptoms of iron deficiency in plants are chlorotic leaves. Often the veins remain green whereas the laminae are yellow, and a fine reticulate pattern develops with the darker green veins contrasting markedly with a lighter green or yellow background (Figure 11.2, see also Figure 1.1 in Chapter 1). In cereals, this shows up as alternate yellow and green stripes (Figure 11.3). Iron deficiency causes marked changes in the ultrastructure of chloroplasts, with thylakoid grana being absent under extreme deficiency and the chloroplasts being smaller (27,28). As iron in older leaves, mainly located in chloroplasts, is not easily retranslocated as long as the leaves are not senescent, the younger leaves tend to be more affected than the older leaves (Figure 11.4). In extreme cases the leaves may become almost white. Plant species that can modify the rhizosphere to make iron more available can be classified as *iron-efficient* and those that cannot as *iron-inefficient*. It is among the iron-inefficient species that chlorosis is most commonly observed.

11.4.2   IRON TOXICITY

Iron toxicity is not a common problem in the field, except in rice crops in Asia (29). It can also occur in pot experiments, and in cases of oversupply of iron salts to ornamental plants such as azaleas. The symptoms in rice, known as 'Akagare I' or 'bronzing' in Asia, include small reddish-brown spots on the leaves, which gradually extend to the older leaves. The whole leaf may turn brown, and the older leaves may die prematurely (29). In other species, leaves may become darker in color and roots may turn brown (29). In rice, iron toxicity seems to occur above $500\,mg\ Fe\ kg^{-1}$ leaf dry weight (30) (Figure 11.5).

FIGURE 11.2 Iron-deficient cucumber (*Cucumis sativus* L.) plant. (Photograph by Allen V. Barker.) (For a color presentation of this figure, see the accompanying compact disc.)

FIGURE 11.3 Iron-deficient corn (*Zea mays* L.) plant. (Photograph by Allen V. Barker.) (For a color presentation of this figure, see the accompanying compact disc.)

FIGURE 11.4   Iron-deficient pepper (*Capsicum annuum* L.) plant. The young leaves are yellow, and the older leaves are more green. (Photograph by Allen V. Barker.) (For a color presentation of this figure, see the accompanying compact disc.)

FIGURE 11.5   Symptoms of iron toxicity in lowland rice (*Oryza sativa* L.) in Sri Lanka as a consequence of decreased redox potential under submergence. (Photograph by Volker Römheld.) (For a color presentation of this figure, see the accompanying compact disc.)

11.5 IRON CONCENTRATION IN CROPS

11.5.1 PLANT PART AND GROWTH STAGE

Most of the iron in plants is in the Fe(III) form (11). The Fe(II) form is normally below the detection level in plants (31). A high proportion of iron is localized within the chloroplasts of rapidly growing leaves (10). One of the forms in which iron occurs in plastids is as phytoferritin, a protein in which iron occurs as a hydrous Fe(III) oxide phosphate micelle (9), but phytoferritin is also found in the xylem and phloem (32). It also occurs in seeds, where it is an iron source that is degraded during germination (33). However, in general, concentrations of iron in seeds are lower than in the vegetative organs.

A large part of the iron in plants is in the apoplast, particularly the root apoplast. Most of this root apoplastic pool is in the basal roots and older parts of the root system (34). There is also a noticeable apoplastic pool of iron in the shoots.

In the iron hyperaccumulator Japanese blood grass (*Imperata cylindrica* Raeuschel), iron accumulates in rhizomes and leaves in mineral form, in the rhizomes in particular as jarosite, $KFe_3(OH)_6(SO_4)_2$, and in the leaves probably as phytoferritin (35). In the rhizome this accumulation is in the epidermis and the xylem, and in the leaves it is in the epidermis.

11.5.2 IRON REQUIREMENT OF SOME CROPS

Iron deficiency can be easily identified by visible symptoms, so this observation has made quantitative information on adequate concentrations of iron in plants more scarce (Table 11.1) (29).

TABLE 11.1
Fe Deficiency Chlorosis-Inducing Factors That Are Often Observed, and Synonyms for These Chlorosis Symptoms

Chlorosis-Inducing Factor	Synonym
Weather factors	
High precipitation	Bad-weather chlorosis
High soil water content	
Low soil temperature	
Soil factors	
High lime content	Lime-induced chlorosis
High bicarbonate concentration	Bicarbonate-induced chlorosis
Low O_2 concentration	
High ethylene concentration	Ethylene-induced chlorosis
High soil compaction	
High heavy metal content	
Management factors	
Soil compaction	'Tractor' chlorosis
High P fertilization	Phosphorus-induced chlorosis
High application of Cu-containing fungicides	Copper chlorosis
Inadequate assimilate delivery and late vintage (harvest)	Weakness chlorosis, stress chlorosis
Plant factors	
Low root growth	
High shoot:root dry matter ratio	
Low Fe efficiency	

Source: From Kirkby, E.A., Römheld, V. *Micronutrients in Plant Physiology: Functions, Uptake and Mobility*. Proceedings No. 543, International Fertiliser Society, Cambridge, U.K., December 9, 2004, pp. 1–54.

Furthermore, the so-called chlorosis paradox gives confusing results when critical levels are being determined. This confusion seems to be brought about by restricted leaf expansion due to shortage of iron, giving rise to similar concentrations of iron in the smaller, chlorotic leaves as in healthy green leaves (36). This paradox has been described in grapevine (*Vitis vinifera* L.) (37,38) and peach (*Prunus persica* Batsch) (39).

In general, the deficiency range is about 50 to 100 mg kg^{-1} depending on the plant species and cultivars (Table 11.2) (28). This range is somewhat complex to determine, as iron-efficient plant species are able to react to low availability of iron by employing mechanisms for its enhanced acquisition (see below), whereas iron-inefficient species are more dependent on adequate supplies of iron being readily available. In fact, it is apparent from simple calculations that plants must employ root-induced mobilization of iron to obtain enough element for normal growth (28). Calculations based on the iron concentration of crops at harvest compared with the concentration of iron in soil water indicate an apparent shortfall in availability of a factor of approximately 2000, and calculations based on the iron concentration of crops at harvest and their water requirements indicate a shortfall of a factor of approximately 36,000. Both are very crude calculations, but they clearly indicate that the presence of plants, at least iron-efficient plants, makes iron more available in the soil than would be expected. The data indicate a requirement of iron for an annual crop of 1 kg ha^{-1} year^{-1}, but even for tree species the requirement is considerable. It has been estimated that for a peach tree in northeastern Spain, the amount of iron in the prunings in particular, but also lost in the harvested fruit, in leaf and flower abscission and immobilized in the wood, is between 1 and 2 g per tree per year (40).

11.5.3 IRON LEVELS IN PLANTS

11.5.3.1 Iron Uptake

Transport of iron to plants roots is limited largely by diffusion in the soil solution (41,42), and thus the absorption is highly dependent on root activity and growth, and root length density.

The overall processes of iron acquisition by roots have been described in terms of different strategies to cope with iron deficiency (Figure 11.6) (10,43). Strategy 1 plants, such as dicots and other nongraminaceous species, reduce Fe(III) in chelates by a rhizodermis-bound Fe(III)-chelate reductase and take up released Fe^{2+} ions into the cytoplasm of root cells by a Fe^{2+} transporter. Strategy 2 plants, mostly grasses, release phytosiderophores that chelate Fe(III) ions and take up the phytosiderophore–Fe(III) complex by a transporter (44,45). A more recently postulated Strategy 3 may involve the uptake of microbial siderophores by higher plants (46), although this could be an indirect use of microbial siderophores through exchange chelation with phytosiderophores in Strategy 2 plants or through Fe^{III} chelate reductase in Strategy 1 plants (47,48).

In Strategy 1 plants, one of the major responses to iron deficiency is the acidification of the rhizosphere, brought about by differential cation–anion uptake (49), the release of dissociable reductants (8,50) and particularly by the action of an iron-deficiency-induced proton pump in the plasmalemma of rhizodermis cells of apical root zones (51). This acidification of the rhizosphere serves to make iron more available and to facilitate the required Fe(III)-chelate reductase activity (52). There is also an enhanced growth of root hairs (53) and the development of structures like transfer cells in the rhizodermis (10) as a response to iron deficiency.

In chickpea (*Cicer arietinum* L.) subjected to iron deficiency, anion and cation uptake were shown to be depressed, but anion uptake was depressed more than cation uptake (54). This effect gives rise to excess cation uptake, with consequent release of H$^+$ ions in a direct relationship to the extent of the cation–anion imbalance. The origin of the H$^+$ release in such circumstances could be through enhanced PEP carboxylase activity (55).

The release of reductants increases the reduction of Fe^{3+} to Fe^{2+} in the apoplast, and has been linked to compounds such as caffeic acid (56,57). These may reduce Fe^{3+} to Fe^{2+} ions, and also chelate the ions either for uptake or for reduction on the plasmalemma. Such reduction of Fe^{3+}

on the plasma membrane involves an iron-chelate reductase. It was thought at one time that there are two forms of such reductases, a constitutive form that works at a low capacity and is continuously present, and an inducible form that works with high capacity and is induced under iron deficiency (10). However, in tomato (*Lycopersicon esculentum* Mill.), iron deficiency gives rise to increased expression of constitutive Fe^{III}-chelate reductase isoforms in the root plasmalemma (58). Action of the Fe^{III}-chelate reductase is the rate-limiting step of iron acquisition of Strategy 1 plants under deficiency conditions (59–61). Genes encoding for proteins in Fe^{III}-chelate reductase and involved with the uptake of Fe^{2+} in Fe-deficient plants have been identified in the Strategy 1 plant *Arabidopsis thaliana*, and have been named *AtFRO2* and *AtIRT1*, respectively (62,63).

In Strategy 2 plants the phytosiderophores, nonprotein amino acids such as mugineic acid (64), are released in a diurnal rhythm following onset of iron deficiency (43,52). This release occurs particularly in the apical regions of the seminal and lateral roots (65). The phytosiderophores form stable complexes with Fe^{3+} ions, and these complexes are taken up by a constitutive transporter in the plasmalemma of root cells (66). Activity of this transporter also increases during iron deficiency. Mutants such as corn (*Zea mays* L.) *ys1/ys1* are very susceptible to iron chlorosis (44).

In the Strategy 1 species cucumber (*Cucumis sativus* L.), Fe^{3+} attached to the water-soluble humic fraction is apparently reduced by the plasmalemma reductase, allowing uptake to occur (67,68), whereas in Strategy 2 barley (*Hordeum vulgare* L.), there is an indirect method for uptake of this Fe^{3+} component that involves ligand exchange between the humic fraction and phytosiderophores released in response to iron deficiency (68). Uptake of iron then occurs as a Fe(III)–phytosiderophore complex. In Strategy II plants, iron deficiency also leads to a small increase in the capacity to take up Fe^{2+}, uptake previously thought only to occur in Strategy 1 plants (69).

It has been suggested in the past that the large root apoplastic pool of iron could be a source of iron for uptake into plants under iron deficiency. However, the apoplastic pool occurs largely in the older roots (34), yet the mobilization of rhizosphere iron and the uptake mechanisms that are induced under iron deficiency stress occur in the apical zones of the roots, so this seems unlikely (70). The Strategy 1 and Strategy 2 mechanisms are switched on by mild iron deficit stress, although under severe deficiency they become less effective. They are switched off within a day of resumption of iron supply to the plant.

The various iron transporters in plant cells have been well characterized. They include Nramp3 transporters on the tonoplast, and IRT1, IRT2 and Nramp transporters on the plasmalemma (71). Nramp (natural resistance associated macrophage proteins) transporters are involved in metal ion transport in many different organisms, and in *Arabidopsis* roots, three different Nramps are upregulated under iron deficiency. A model of iron transport in *Arabidopsis* has been shown elsewhere (72).

The transporter used by Strategy 1 plants is an AtIRT1 transporter, whereas Strategy 2 plants take up the phytosiderophore–Fe(III) complex by ZmYS1 transporters (44,45).

Uptake of zinc, and possibly manganese and copper also, may increase in Strategy 2 plants under iron deficiency, because although the iron-phytosiderophore transporter is specific to iron complexes, the presence of the phytosiderophores in the rhizosphere may increase the availability of these other ions both in the rhizosphere itself and in the apoplast (73).

As well as uptake through roots, iron is able to penetrate plant cuticles, at least at 100% humidity. Chelates of Fe^{3+} were shown to penetrate cuticular membranes from grey poplar (*Populus* x *Canescens* Moench.) leaves without stomata with a half-time of 20 to 30 h (74), although at 90% humidity Fe^{3+} chelated with lignosulfonic acid was the only chelate tested that still penetrated the membrane. Sachs himself showed that iron is taken up by plants after application to the foliage, and iron chelates have been applied to foliage to correct iron deficiencies because inorganic iron salts are unstable and phytotoxic (see (3)). Fe(III) citrate and iron-dimerum have been found to penetrate the leaves of chlorotic tobacco (*Nicotiana tabacum* L.) plants, and to be utilized by the cells (75), but it is the chelated forms of iron that enter most effectively.

Strategy 1: Dicotyledons and nongraminaceous plant species

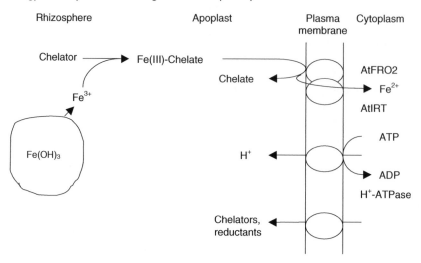

Strategy 2: Graminaceous plant species

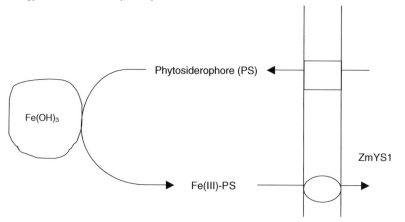

FIGURE 11.6 Strategies for acquisition of Fe in response to Fe deficiency in Strategy 1 and Strategy 2 plants. (Redrawn from Römheld, V., Schaaf, G., *Soil Sci. Plant Nutr.*, 50:1003–1012, 2004.)

11.5.3.2 Movement of Iron within Plants

Once taken up by root cells, iron moves within cells and between cells. The understanding of iron homeostasis at the subcellular level is incomplete, and the role of the vacuole is uncertain. A carrier called AtCCC1 may transport iron into vacuoles, and AtNRAMP3 and AtNRAMP4 are candidates for transporting it out (72). Of the cellular organelles, mitochondria and chloroplasts have a high requirement for iron, and the chloroplasts may be sites of storage of iron (76). Transport into chloroplasts is stimulated by light (77), and it occurs in the Fe(II) form (78).

Knowledge of the movement of iron between cells is also incomplete. Experiments in which ^{59}Fe-labelled iron-phytosiderophores were fed to roots of intact corn plants for periods of up to 2 h demonstrated intensive accumulation of iron in the rhizodermis and the endodermis (72,79). This accumulation was higher with iron deficiency stress, and probably reflected the role of increased number of root hairs and increased expression of the ZmYS1 iron-phytosiderophore transporter.

From the endodermis, the iron is loaded into the pericycle and from there into the xylem. Very little is known about these processes. Once in the shoots, much of the iron is present in the apoplast, from where it is loaded into the cytoplasm and into the organelles where it is required. It was thought at one time that high soil pH would raise shoot apoplastic pH and that this action would make iron unavailable for transport into leaf cells. However, this is not the case, as high root zone HCO_3^- has been shown not to increase apoplastic pH of leaves in both nutrient-solution-grown sunflower (*Helianthus annuus* L.) and field-grown grapevine (*Vitis vinifera* L.) (80), a result that is also in agreement with recent experiments of Kosegarten et al. (81,82). In experiments on grapevine, the presence of bicarbonate in the uptake medium was shown to inhibit uptake of iron and its translocation to the shoots, primarily by inhibiting the Fe(III) reduction capacity of the roots (83). Also, the recently discussed role of nitrate in iron inactivation in leaves and induction of chlorosis due to an assumed increased leaf apoplast pH (82) could not be confirmed (84). Probably, this nitrate-induced chlorosis in solution-cultured sunflower plants is a consequence of an impeded iron acquisition by roots as a consequence of a nitrate-induced pH increase at the uptake sites of the roots.

Movement of iron salts in phloem is obviously possible as Rissmüller observed retranslocation of iron from senescent leaves of beech trees long ago (3). However, it is usually thought that iron deficiency symptoms occur in young leaves rather than in old leaves because iron is not easily retranslocated in nonsenescent plants. However, such retranslocation is not confined to the senescent leaves of trees, as it has also been observed to occur out of young leaves of *Phaseolus vulgaris* subjected to iron deficiency (85,86).

Nicotianamine seems to be involved in phloem loading for retranslocation of iron and possibly in phloem unloading and uptake of iron into young leaves and reproductive organs. The maize ZmYS1 protein not only mediates transport of iron–phytosiderophore complexes (87,88), but experiments on this transporter in yeast and *Xenopus* have shown that it can also transport Fe(II)-nicotianamine and Fe(III)-nicotianamine (88). The AtYSL2 homolog of this protein has been implicated in lateral movement of iron in the vascular system of *Arabidopsis thaliana* (89,90), and its OsYSL2 homolog in rice has been suggested to be involved in transport of Fe(II)-nicotianamine in phloem loading and translocation of metals into the grain (91). Expression of a nicotianamine synthase gene from *Arabidopsis thaliana* in *Nicotiana tabacum* gave increased levels of nicotianamine, more iron in the leaves of adult plants, and improvement in the iron use efficiency of plants grown under iron deficiency stress (92).

11.6 FACTORS AFFECTING PLANT UPTAKE

11.6.1 SOIL FACTORS

The major factor affecting acquisition of iron by plants is soil pH, with high pH making iron less available and giving rise to chlorosis. Along with lime-induced chlorosis, there is a whole range of factors, including the weather, soil and crop management, and the plant genotypes themselves, that give rise to chlorosis by impeded uptake of iron (Table 11.1). In lime-induced chlorosis, it is the soil bicarbonate that is the key cause, largely due to the high pH in the rhizosphere and at the root uptake site, thereby affecting iron solubility and Fe(III)-chelate reductase activity (see Section 11.5.3.1).

One factor that may contribute to rhizosphere pH changes, other than the underlying substrate, is the nitrogen source. When plants take up nitrate as their predominant nitrogen source, they alkalinize the rhizosphere and this contributes to iron deficiency stress (84,93,94). It has been suggested that nitrate nutrition could actually raise the pH in the leaf apoplast, making iron less available for transport into leaf cells. However, this assumption was not experimentally confirmed (see Section 11.5.3.2).

Chlorosis in plant species with Strategy 1 is made worse by high soil moisture, particularly on calcareous soils, because of elevated concentrations of bicarbonates. A peach tree that was overirrigated in an orchard on a calcareous soil developed bicarbonate-induced chlorosis, whereas a tree that received proper irrigation showed no chlorosis (Figure 11.7). In addition, anaerobiosis may make root responses to iron deficiency stress more difficult (13). Organic matter content of the soil

TABLE 11.2
Deficient and Adequate Concentrations of Iron in Leaves and Shoots of Various Plant Species

Plant Species	Plant Part	Type of Culture	Concentration in Dry Matter (mg kg^{-1}) Deficient	Adequate	Reference	Comments
Allium sativum L. (onion)	Upper shoot	Sterile nutrient culture	24	224	117	
Avena sativa L. (oats)	Whole shoot	Solution culture	<50	50–80	118	
Brassica oleracea var. *italica* Plenck (broccoli)	Leaves	Farmers' fields		113	119	5% of heads formed
Brassica oleracea var. *gemmifera* Zenker (Brussels sprouts)	Leaves	Farmers' fields		105	119	Sprouts beginning to form
Brassica oleracea var. *botrytis* L. (cauliflower)	Leaves	Farmers' fields	117	119		5% of heads formed
Brassica napobrassica Mill. (rutabaga)	Leaves	Farmers' fields		159	119	Roots beginning to swell
Carya illinoinensis (pecan nut)	Leaf in July/ August	Field		62–92	120	40 named cultivars compared, values segregated into five ranges
Cicer arietinum L. (chickpea)	Shoot	Nutrient culture	60/70	130/170	54	Values for nitrate/ ammonium nutrition
	Root		210/180	1830/1570		
Daucus carota L. (carrot)	Whole shoot	Peat-grown	39–82		121	
Glycine max Merr. (soybean)	Seed	Field	42–45	70–77	116	Data for cultivars susceptible and resistant to Fe deficiency
Gossypium hirsutum L. (cotton)	Whole shoot	Soil-grown	<47		122	
Helianthus annuus L. (sunflower)	Leaves	Nutrient solution	34–50	78–100	84	Values for nitrate/ ammonium nutrition, buffered at pH 5.0 versus 7.5
Malus domestica Borkh (apple)	Leaf	Commercial orchards			123	
Cox's orange pippin				63		48–85 mg kg^{-1} range
Braeburn				66		53–91 mg kg^{-1} range
Medicago sativa L. (alfalfa)	Leaves	Farmers' fields	87	119		10% of plants in bloom

TABLE 11.2 (*Continued*)

Plant Species	Plant Part	Type of Culture	Concentration in Dry Matter (mg kg^{-1}) Deficient	Adequate	Reference	Comments
Prunus persica Batsch (peach)	Leaf	Field	44–58		66	124
Trifolium pratense L. (red clover)	Leaves	Farmers' fields		93	119	10% of plants in bloom
Vitis vinifera L. (grapevine) cv. Blauer Burgunder Faber Ruländer	Leaves	Field	40–60 80–140 50–90	65–100 90–160 90–120	37	Values for different cvs. and sites. No clear differentiation for Faber because of different extent of the chlorosis paradox.
Vitis vinifera cv. Syrah no inhibition severe inhibition (chlorosis paradox)	Young leaves	Field	65–100 140–170	100–140 90–100	38	Comparison of sites without and with severe leaf growth inhibition of chlorotic plants

Note: Values in dry matter. The concept of 'deficient' and 'adequate' concentrations is problematic because of the chlorosis paradox (see text).

FIGURE 11.7 Two peach (*Prunus persica* Batsch) trees in an orchard on a calcareous soil with drip irrigation. Left: over-irrigation by a defect dripper resulting in bicarbonate-induced chlorosis. Right: adequate irrigation, no chlorosis. (Photograph by Volker Römheld.) (For a color presentation of this figure, see the accompanying compact disc.)

can also be important, partly because of the increased tendency toward waterlogging in organic soils lowering iron availability, but also because of enhanced microbial activity and the presence of chelating agents in the organic matter making iron more available (13). Furthermore, soil organic matter, and also compaction of soil, could lower root growth and inhibit iron uptake because of generation of ethylene (95). Low temperature can make chlorosis more extreme because of the slower metabolic processes in the roots inhibiting the iron-deficiency responses; very high concentrations of soil phosphate can be deleterious through the adsorption of phosphates on to iron oxides; high soil solution osmotic strength appears to lower the effectiveness of iron chelation in Strategy 1 plants; and high concentrations of Cu, Zn, and Mn can induce iron chlorosis through replacement of iron in soil chelates and phytosiderophores and inhibition of the iron-deficiency responses (13). A summary of the interactions between environmental, edaphic and management conditions, and plant genotype, concerning the onset of chlorosis is shown in Figure 11.8.

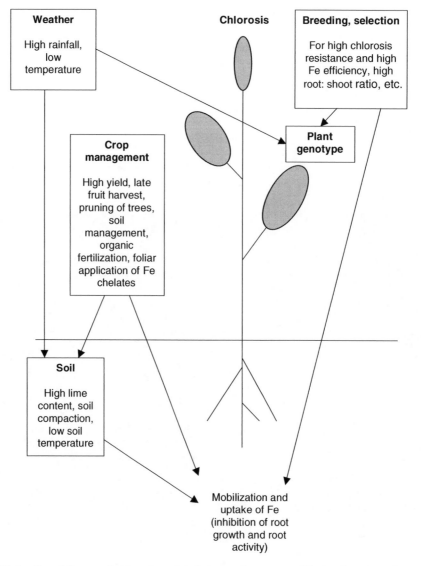

FIGURE 11.8 Causal factors of chlorosis and their interactions responsible for the onset of Fe-deficiency chlorosis in plants. (Redrawn from Kirkby, E.A., Römheld, V., *Micronutrients in Plant Physiology: Functions, Uptake and Mobility.* Proceedings No. 543, International Fertiliser Society, Cambridge, U.K., December 9, 2004, pp. 1–54.)

11.6.2 PLANT FACTORS

The two strategies for iron acquisition under iron deficiency stress are separated along taxonomic lines, with grasses (Gramineae, Poaceae) showing Strategy 2, and other plant families and orders, including some closely related to the grasses such as the Restionales, Eriocaules, Commelinales, and Juncales, showing Strategy 1 (13).

Iron deficiency does not occur in perennial woody plants such as grapevine or pear (*Pyrus communis* L.) grown on noncalcareous soils. For some plants such as sunflower, deficiency is uncommon even on calcareous soils. (In experiments in which sunflower has been used to examine the effects of iron deficiency, this effect has been achieved at conditions severely inhibiting iron acquisition, for example, by elevated bicarbonate concentrations.) In general, Strategy 1 plants show considerable sensitivity in their response to high bicarbonate and high soil pH, high soil moisture and poor aeration, high soil organic matter in calcareous soils, high concentrations of heavy metals, high ionic strength of the soil solution, and low soil temperature (13). In contrast, Strategy 2 plants have a lower sensitivity to these factors but a high sensitivity to high soil phosphate. Furthermore, high microbial activity in the rhizosphere can be deleterious due to a fast degradation of the released phytosiderophores (96,97).

The very term 'Fe-efficient' implies that the mechanisms of Strategy 1 and Strategy 2 for iron acquisition succeed in making sufficient iron available to plants for normal growth, and this result is indeed the case, particularly for Strategy 2 plants. For sunflower grown in calcareous soil, there is a rhythmic response to the low concentrations of available iron that is matched by a rhythmic uptake of iron (98). Calcicole plants growing in the wild are able to take up sufficient iron for normal growth, although it is probably adaptation to cope with the low availability of phosphorus that is more important in determining their ability to grow.

The whole concept of iron-efficient and iron-inefficient species raises the prospect of breeding for efficient acquisition of iron, and the level of knowledge about the genetics of the responses to onset of iron deficiency stress is making this improvement a distinct possibility. It has already been demonstrated that plants such as grapevines can be grown on iron-efficient rootstocks (Figure 11.9).

Resistance to chorosis may be brought about by engineering crops with increased iron acquisition capability in a number of ways. For example, transgenic rice with a genomic fragment containing *HvNAAT-A* and *HvNAAT-B* from barley exhibited enhanced release of phytosiderophores

FIGURE 11.9 Differences in chlorosis resistance of grapevines (*Vitis vinifera* L.) on different root stocks (left, 5BB; right, Fercal). (Photograph by Volker Römheld.) (For a color presentation of this figure, see the accompanying compact disc.)

and increased tolerance to low iron availability through the speeding up of a rate-limiting step of phytosiderophore biosynthesis (99). These plants had four times higher grain yield in alkaline soils than unmodified plants. The process of phytosiderophore release can also be crucial for iron acquisition (100), and this step could also be improved. In Strategy 1 *Arabidopsis thaliana*, increased iron acquisition has been achieved by overexpressing the FRO2 Fe(III) chelate reductase (61). Additionally, plants could be engineered to contain higher concentrations of nicotianamine (92).

In addition to increasing the efficiency of iron acquisition, it may be possible to increase the concentrations of iron in harvested crop plants for human nutrition. Much of the world suffers from iron deficiency in the diet, and breeding crops such as 'golden rice,' which has a higher iron concentration as well as more vitamin A precursors, would be of considerable benefit to human welfare (101,102). In wheat, it may be possible to breed from accessions of wild wheat ancestors, such as *Triticum turgidum* subsp. *dicoccoides*, which contain higher concentrations of iron in their seeds than *Triticum aestivum*, to improve the nutritional quality of human and livestock feedstuffs (103).

11.7 SOIL TESTING FOR IRON

Because of the major impact of soil pH and bicarbonate content on the availability of iron to plants, it is not common to test a soil for iron extractability. Tests of soil pH and lime content are much more valuable in assessing where lime chlorosis is likely to occur.

Where testing of iron content is desired, early methods were based on determining the exchangeable iron by extraction with ammonium acetate (104). Nowadays, soil iron is extracted by the use of a chelating agent, in some cases EDDHA but more commonly DTPA (diethylenetriaminepentacetic acid). This method, first proposed in 1967, is used for the analysis of zinc, iron, manganese, and copper in soils together, and involves adding DPTA to a soil solution buffered at pH 7.3 (105). The mixture contains $CaCl_2$ so that any $CaCO_3$ in the soil is not dissolved, with corresponding release of otherwise unavailable micronutrients.

The micronutrients in the extract are measured by atomic absorption spectrometry, inductively coupled plasma spectrometry, or neutron activation analysis.

11.8 FERTILIZERS FOR IRON

Formation of barely soluble iron hydroxides and oxides, particularly at high pH and in the presence of bicarbonate ions in the rooting medium, immobilizes iron supplied as inorganic salts. One way round this problem is to supply Fe(III) citrate, but this is photolabile. For these reasons the supply of iron in hydroponic culture is usually as a chelate (27). This can be as either FeEDTA (ethylenediaminetetraacetate) or FeEDDHA (ethylene diamine (di *o*-hydroxyphenyl) acetate). Both these chelates remain stable over a range of pH values, particularly FeEDDHA, although the iron is readily available to the plants. In fact, the whole chelate molecule can be taken up at high application rates, and as this absorption is by a passive mechanism it is probably at the root zone where the lateral roots develop (106). However, the main uptake of iron chelates in soils or nutrient solutions at realistic application rates takes place after exchange chelation in Strategy 2 plants (48) and after Fe(III) reduction and formation of Fe^{2+} in Strategy 1 plants (107). Interestingly, cucumber plants supplied with inorganic Fe seem to be more resistant to infection by mildew than plants supplied with FeEDDHA (106).

In terms of fertilizers for terrestrial plants, iron deficiency usually comes about because of alkaline pH in the soil, and supply of iron salts to the soil would have no effect. Foliar application of Fe(II) sulfate can be effective, typically as a 1% solution applied at regular intervals (25).

Where iron deficiency occurs in acid soils, supply of Fe(II) sulfate to the soil can be effective. Thus in ornamental horticulture, azaleas and other acid-loving plants benefit from application of this salt. However, in the field, supply to citrus trees on acid soils is not effective as other ions, particularly copper, interfere with the availability of iron (25). Application of iron can be made as FeEDTA or FeEDDHA, but the stability of FeEDTA at least is not high in calcareous soils (25). FeEDDHA and FeDTPA are the only commercially available iron chelates for soil application because of their stability at high pH. The synthetic iron phosphate vivianite $(Fe_3(PO_4)_2 \cdot 8H_2O)$ has been used on olive trees (108) and in kiwi orchards (109).

Therefore, the usual way in which lime-induced chlorosis is alleviated is by supply of iron chelates such as FeEDTA and FeHEDTA to the foliage. Usually more than one application is required (110). There is potential for supplying iron to the foliage of plants as iron-siderophores, as these microbial chelates are more biodegradable than the synthetic chelates, and so pose less environmental risk (111). FeEDTA may also damage the leaves of plants. It is also possible that these microbial siderophores could be used for root application, at least in hydroponics, as iron-rhizoferritin and Fe(III) monodihydroxamate and Fe(III) dihydroxamate siderophores have been shown to be taken up by a range of plant species by exchange chelation with phytosiderophores or via Fe(III) reduction in Strategy 2 and Strategy 1 plants, respectively (48,112,113).

Some of the effects of lime-induced chlorosis on the early stages of plant growth can be overcome by planting seeds that are high in iron. In the case of common bean (*Phaseolus vulgaris* L.), seeds from plants grown on acid soils are higher in iron than seeds from plants grown on calcareous soils, but the seed iron content can be increased by supply of iron to the soil at planting or after flowering (114). A preplanting application of FeEDDHA has a larger effect on seed yield of soybean (*Glycine max* L.) than an application at flowering, but the latter application has a more beneficial effect on iron concentration in the seeds of both common bean and soybean (115). There is other evidence that the iron concentration in soybean seeds is under very tight genetic control and is not influenced much by the supply of iron, but in that experiment the FeEDDHA was supplied at planting (116).

REFERENCES

1. E. Molz. *Untersuchungen über die Chlorose der Reben*. Jena: Gustav Fischer Verlag, 1907.
2. E.J. Hewitt, T.A. Smith. *Plant Mineral Nutrition*. London: The English Universities Press, 1975, p. 16.
3. H. Molisch. *Die Pflanze in ihren Beziehungen zum Eisen*. Jena: Gustav Fischer Verlag. 1892.
4. A. Wallace, D. Lunt. Iron chlorosis in horticultural plants, a review. *Am. Soc. Hortic. Sci.* 75:819–841, 1960.
5. R.L. Chaney, J.C. Brown, L.O. Tiffin. Obligatory reduction of ferric chelates in iron uptake by soybeans. *Plant Physiol.* 50:208–213, 1972.
6. S. Takagi. Naturally occurring iron-chelating compounds in oat- and rice-root washings. I. Activity measurement and preliminary characterization. *Soil Sci. Plant Nutr.* 22:423–433, 1976.
7. J.C. Brown. Mechanism of iron uptake by plants. *Plant Cell Environ.* 1:249–257, 1978.
8. H. Marschner, V. Römheld, M. Kissel. Different strategies in higher plants in mobilization and uptake of iron. *J. Plant Nutr.* 9:695–713, 1986.
9. H.F. Bienfait, F. van der Mark. Phytoferritin and its role in iron metabolism. In: D.A. Robb, W.S. Pierpoint, eds. *Metals and Micronutrients. Uptake and Utilization by Plants*. London: Academic Press: London, 1983, pp. 111–123.
10. H. Marschner. *Mineral Nutrition of Higher Plants*. London: Academic Press, 1995.
11. N. Terry, J. Abadia. Function of iron in chloroplasts. *J. Plant Nutr.* 9:609–646, 1986.
12. N.K. Fageria, V.C. Baligar, R.B. Clark. Micronutrients in crop production. In: D.L. Sparks, ed. *Advances in Agronomy*. San Diego: Academic Press, 2002, pp. 185–268.
13. V. Römheld, H. Marschner. Mobilization of iron in the rhizosphere of different plant species. In: B. Tinker, A. Läuchli, eds. *Advances in Plant Nutrition*. Vol. 2, New York: Praeger, 1986, pp. 155–204.

14. W.L. Lindsay, A.P. Schwab. The chemistry of iron in soils and its availability to plants. *J. Plant Nutr.* 5:821–840, 1982.

15. J. Gerke. Orthophosphate and organic phosphate in the soil solution of four sandy soils in relation to pH. Evidence for humic-Fe(Al)-phosphate complexes. *Commun. Soil Sci. Plant Anal.* 23:601–612, 1992.

16. M.O. Olomu, G.J. Racz, C.M. Cho. Effect of flooding on the Eh, pH and concentrations of Fe and Mn in several Manitoba soils. *Soil Sci. Soc. Am. Proc.* 37:220–224, 1973.

17. P.E. Powell, P.J. Staniszlo, G.R. Cline, C.P.P. Reid. Hydroxamate siderophores in the iron nutrition of plants. *J. Plant Nutr.* 5:653–673, 1982.

18. S. Cesco, V. Römheld, Z. Varanini, R. Pinton. Solubilization of iron by water-extractable humic substances. *J. Soil Sci. Plant Nutr.* 163:285–290, 2000.

19. A.S. Mashhady, D.L. Rowell. Soil alkalinity. II. The effects of Na_2CO_3 on iron and manganese supply to tomatoes. *J. Soil Sci.* 29:367–372, 1978.

20. G. Welp, U. Herms, G. Brümmer. Influence of soil reaction, redox conditions and organic matter on the phosphate content of soil solutions. *Z. Pflanzen Boden* 146:38–52, 1983.

21. G. Trolldenier. Secondary effects of potassium and nitrogen nutrition of rice: change in microbial activity and iron reduction in the rhizosphere. *Plant Soil* 38:267–279, 1973.

22. G. Julian, H.J. Cameron, R.A. Olsen. Role of chelation by ortho dihydroxy phenols in iron absorption by plant roots. *J. Plant Nutr.* 6:163–175, 1983.

23. H. Oki, K. SuYeon, H. Nakanishi, M. Takahashi, H. Yamaguchi, S. Mori, N.K. Nishizawa. Directed evolution of yeast ferric reductase to produce plants with tolerance to iron deficiency in alkaline soils. *Soil Sci. Plant Nutr.* 50:1159–1165, 2004.

24. M. Vasconcelos, V. Musetti, C.M. Li, S.K. Datta, M.A. Grusak. Functional analysis of transgenic rice (*Oryza sativa* L.) transformed with an *Arabidopsis thaliana* ferric reductase (AtFRO2). *Soil Sci. Plant Nutr.* 50:1151–1157, 2004.

25. F.R. Troeh, L.M. Thompson. *Soils and Soil Fertility.* 6th ed. Ames, Iowa: Blackwell, 2005, p. 293.

26. C.R. Lee. Interrelationships of aluminum and manganese on the potato plant. *Agron. J.* 64:546–549, 1972.

27. C. Bould, E.J. Hewitt, P. Needham. *Diagnosis of Mineral Disorders in Higher Plants. Volume 1. Principles.* London: Her Majesty's Stationery Office, 1983.

28. E.A. Kirkby, V. Römheld. Micronutrients in Plant Physiology: Functions, Uptake and Mobility. Proceedings No. 543, International Fertiliser Society, Cambridge UK, 9th December 2004, pp. 1–54.

29. W. Bergmann. *Nutritional Disorders of Plants. Visual and Analytical Diagnosis.* Jena: Gustav Fischer Verlag, 1992, p. 15.

30. M. Yamauchi. Rice bronzing in Nigeria caused by nutrient imbalances and its control by potassium sulfate application. *Plant Soil* 117:275–286, 1989.

31. B.A. Goodman, P.C. DeKock. Mössbauer studies of plant material. I. Duckweed, stocks, soybeans and pea. *J. Plant Nutr.* 5:345–353, 1982.

32. B.N. Smith. Iron in higher plants: storage and metabolic rate. *J. Plant Nutr.* 7:759–766, 1984.

33. S. Lobréaux, J.F. Briat. Ferritin accumulation and degradation in different organs of pea (*Pisum sativum*) during development. *Biochem. J.* 274:601–606, 1991.

34. O. Strasser, K. Köhl, V. Römheld. Overestimation of apoplastic Fe in roots of soil grown plants. *Plant Soil* 210:179–187, 1999.

35. N. Rodriguez, N. Menendez, J. Tornero, R. Amils, V. de la Fuente. Internal iron biomineralization in *Imperata cylindrica*, a perennial grass: chemical composition, speciation and plant localization. *New Phytol.* 165:781–789, 2005.

36. K. Venkat-Raju, H. Marschner. Inhibition of iron-stress reactions in sunflower by bicarbonate. *Z. Pflanzen Bodenk* 144:339–355, 1981.

37. M. Häussling, V. Römheld, H. Marschner. Beziehungen zwischen Chlorosegrad, Eisengehalten und Blattwachstum von Weinreben auf verschiedenen Standorten. *Vitis* 24:158–168, 1985.

38. V. Römheld. The chlorosis paradox: Fe inactivation as a secondary event in chlorotic leaves of grapevine. *J. Plant Nutr.* 23:1629–1643, 2000.

39. F. Morales, R. Grasa, A. Abadía, J. Abadía. Iron chlorosis paradox in fruit trees. *J. Plant Nutr.* 21:815–825, 1998.

40. J. Abadía, A. Álvarez-Fernández, A.D. Rombolà, M. Sanz, M. Tagliavini, A. Abadía. Technologies for the diagnosis and remediation of Fe deficiency. *Soil Sci. Plant Nutr.* 50:965–971, 2004.

41. G.A. O'Connor, W.L. Lindsay, S.R. Olsen. Diffusion of iron and iron chelates in soil. *Soil Sci. Soc. Am. Proc.* 35:407–410, 1971.

42. R.L. Chaney. Diagnostic practices to identify iron deficiency in higher plants. *J. Plant Nutr.* 1984, 7:46–67, 1984.

43. V. Römheld. Different strategies for iron acquisition in higher plants. *Plant Physiol.* 70:231–234, 1987.

44. N. von Wirén, S. Mori, H. Marschner, V. Römheld. Iron inefficiency in maize mutant *ysl* (*Zea mays* L. cv Yellow Stripe) is caused by a defect in uptake of iron phytosiderophores. *Plant Physiol.* 106:71–77, 1994.

45. N. von Wirén, H. Marschner, V. Römheld. Root of iron-efficient maize also absorb phytosiderophore-chelated zinc. *Plant Physiol.* 111:1119–1125, 1996.

46. H. Bienfait. Prevention of stress in iron metabolism of plants. *Acta Bot. Neerl* 38:105–129, 1989.

47. M. Shenker, R. Ghirlando, I. Oliver, M. Helman, Y. Hadar, Y. Chen. Chemical structure and biological activity of a siderophore produced by *Rhizopus arrhizus*. *Soil Sci. Soc. Am. J.* 59:837–843, 1995.

48. Z. Yehuda, M. Shenker, V. Römheld, H. Marschner, Y. Hadar, Y. Chen. The role of ligand exchange in the uptake of iron from microbial siderophores by graminaceous plants. *Plant Physiol.* 112:1273–1280, 1996.

49. K. Venkat-Raju, H. Marschner, V. Römheld. Effect of iron nutritional status on ion uptake, substrate pH and production and release of organic acids and riboflavin by sunflower plants. *Z. Pflanzen Boden* 132:177–190, 1972.

50. J.C. Brown, W.E. Jones. pH changes associated with iron-stress response. *Physiol. Plant* 30:148–152, 1974.

51. E.C. Landsberg. Organic acid synthesis and release of hydrogen ions in response to Fe deficiency stress of mono- and dicotyledonous plant species. *J. Plant Nutr.* 3:579–591, 1981.

52. V. Römheld. Existence of two different strategies for the acquisition of iron and other micronutrients in graminaceous species. In: G. Winkelmann, D. van der Helm, J.B. Neilands, eds. *Iron Transport in Microbes, Plants and Animals.* Weinheim: VCH, 1987, pp. 353–374.

53. W. Schmidt. From faith to fate: ethylene signalling in morphogenic responses to P and Fe deficiency. *J. Plant Nutr. Soil Sci.* 164:147–154, 2001.

54. G.A. Alloush, J. Le Bot, F.E. Sanders, E.A. Kirkby. Mineral nutrition of chickpea plants supplied with NO_3^- or NH_4-N. I. Ionic balance in relation to iron stress. *J. Plant Nutr.* 13:1575–1590, 1990.

55. C.R. Stocking. Iron deficiency in maize. *Plant Physiol.* 55:626–631, 1975.

56. R.A. Olsen, J.H. Bennett, D. Blune, J.C. Brown. Chemical aspects of the Fe stress response mechanism in tomatoes. *J. Plant Nutr.* 3:905–921, 1981.

57. N.H. Hether, N.R. Olsen, L.L. Jackson. Chemical identification of iron reductants exuded by plant roots. *J. Plant Nutr.* 7:667–676, 1984.

58. M.J. Holden, D.G. Luster, R.L. Chaney, T.J. Buckhout, C. Robinson. Fe^{3+}-chelate reductase activity of plasma membranes isolated from tomato (*Lycopersicon esculentum* Mill.) roots. *Plant Physiol.* 97:537–544, 1991.

59. M.A. Grusak, R.M. Welch, L.V. Kochian. Physiological characterization of a single-gene mutant of *Pisum sativum* exhibiting excess iron accumulation. 1. Root iron reduction and iron uptake. *Plant Physiol.* 93:976–981, 1990.

60. Y. Yi, M.L. Guerinot. Genetic evidence that induction of root Fe(III) chelate reductase activity is necessary for iron uptake under iron deficiency. *Plant J.* 10:835–844, 1996.

61. E.L. Connolly, N.H. Campbell, N. Grotz, C.L. Pritchard, M.L. Guerinot. Overexpression of the *FRO2* ferric chelate reductase confers tolerance to growth on low iron and uncovers posttranscriptional control. *Plant Physiol.* 133:1102–1110, 2003.

62. D. Eide, M. Broderius, J. Fett, M.L. Guerinot. A novel iron-regulated metal transporter from plants identified by functional expression in yeast. *Proc. Natl. Acad. Sci. USA* 93:5624–5628, 1996.

63. N.J. Robinson, C.M. Procter, E.L. Connolly, M.L. Guerinot. A ferric-chelate reductase for iron uptake from soils. *Nature* 397:694–697, 1999.

64. S. Takagi, K. Nomoto, T. Takemoto. Physiological aspect of mugineic acid, a possible phytosiderophore of graminaceous plants. *J. Plant Nutr.* 7:469–477, 1984.

65. H. Marschner, V. Römheld, M. Kissel. Localization of phytosiderophore release and of iron uptake along intact barley roots. *Physiol. Plant* 71:157–162, 1987.

66. V. Römheld, H. Marschner. Genotypical differences among graminaceous species in release of phytosiderophores and uptake of iron phytosiderophores. *Plant Soil* 123:147–153, 1990.

67. R. Pinton, S. Cesco, S. Santi, F. Agnolon, Z. Varanini. Water extractable humic substances enhance iron deficiency responses to Fe-deficient cucumber plants. *Plant Soil* 210:145–157, 1999.

68. S. Cesco, M. Nikolic, V. Römheld, Z. Varanini, R. Pinton. Uptake of ^{59}Fe from soluble ^{59}Fe-humate complexes by cucumber and barley plants. *Plant Soil* 241:121–128, 2002.

69. T. Zaharieva, V. Römheld. Specific Fe^{2+} uptake system in Strategy 1 plants inducible under Fe deficiency. *J. Plant Nutr.* 23:1733–1744, 2000.

70. S. Fiedler, O. Strasser, G. Neumann, V. Römheld. The influence of redox conditions in soils on extraplasmatic Fe-loading of plant roots. *Plant Soil* 264:159–169, 2004.

71. J.L. Hall, L.E. Williams. Transition metal transporters in plants. *J. Exp. Bot.* 54:2601–2613, 2003.

72. V. Römheld, G. Schaaf. Iron transport in plants: a future research in view of a plant nutritionist and a molecular biologist. *Soil Sci. Plant Nutr.* 50:1003–1012, 2004.

73. F.-S. Zhang, V. Römheld, H. Marschner. Diurnal rhythm of release of phytosiderophores and uptake rate of zinc in iron-deficient wheat. *Soil Sci. Plant Nutr.* 37:671–678, 1991.

74. J. Schönherr, V. Fernandez, L. Schreiber. Rates of cuticular penetration of chelated FeIII: role of humidity, concentration, adjuvants, temperature, and type of chelate. *J. Agric. Food Chem.* 53:4484–4492, 2005.

75. V. Fernandez, G. Winkelmann. Ebert G. Iron supply to tobacco plants through foliar application of iron citrate and ferric dimerum acid. *Physiol. Plant* 122:380–385, 2004.

76. J.F. Briat, S. Lobréaux. Iron transport and storage in plants. *Trends Plant Sci.* 2:187–193, 1997.

77. N. Bughio, M. Takahashi, E. Yoshimura, N.K. Nishizawa, S. Mori. Light-dependent iron transport into isolated barley chloroplasts. *Plant Cell Physiol.* 38:101–105, 1997.

78. R. Shingles, M. North, R.E. McCarty. Ferrous ion transport across chloroplast inner envelope membranes. *Plant Physiol.* 2002, 128:1022–1030, 2002.

79. G. Drecker. Lokalisation des spezifischen Aufnahemesystems für Fe(III)-Phytosiderophore in den Wurzeln von Gramineen. Masters thesis, Institute of Plant Nutrition, University of Hohenheim, Stuttgart, Germany, 1991.

80. M. Nikolic, V. Römheld. Does high bicarbonate supply to roots change availability of iron in the leaf apoplast? *Plant Soil* 241:67–74, 2002.

81. H. Kosegarten, B. Hoffmann, K. Mengel. Apoplastic pH and Fe^{3+} reduction in intact sunflower leaves. *Plant Physiol.* 121:1069–1079, 1999.

82. H. Kosegarten, B. Hoffmann, K. Mengel. The paramount influence of nitrate in increasing apoplastic pH of young sunflower leaf to induce Fe deficiency chlorosis, and the re-greening effect brought about by acidic foliar sprays. *J. Plant Nutr. Soil Sci.* 164:155–163, 2001.

83. M. Nikolic, V. Römheld, N. Merkt. Effect of bicarbonate on uptake and translocation of ^{59}Fe in grapevine rootstocks differing in their resistance to iron deficiency chlorosis. *Vitis* 39:145–149, 2000.

84. M. Nikolic, V. Römheld. Nitrate does not result in iron inactivation in the apoplast of sunflower leaves. *Plant Physiol.* 132:1303–1314, 2003.

85. C.D. Zhang, V. Römheld, H. Marschner. Retranslocation of iron from primary leaves of bean-plants grown under iron-deficiency. *J. Plant Physiol.* 146:268–272, 1995.

86. C.D. Zhang, V. Römheld, H. Marschner. Effect of primary leaves on ^{59}Fe uptake by roots and ^{59}Fe distribution in the shoot of iron sufficient and iron deficient bean (*Phaseolus vulgaris* L.) plants. *Plant Soil* 182:75–81, 1996.

87. C. Curie, Z. Panaviene, C. Loulergue, S.L. Dellaporta, J.F. Briat, E.L. Walker. Maize *yellow stripe1* encodes a membrane protein directly involved in Fe(III) uptake. *Nature* 409:346–349, 2001.

88. G. Schaaf, U. Ludewig, B.E. Erenoglu, S. Mori, T. Kitahara, N. von Wíren. ZmYS1 funtions as a proton-coupled symporter for phytosiderophore- and nicotianamide-chelated metals. *J. Biol. Chem.* 279:9091–9096, 2004.

89. R.J. DiDonato, L. Roberts, T. Sanderson, R.B. Eisley, E. Walker. Arabidopsis Yellow Stripe-Like2 (YSL2): a metal-regulated gene encoding a plasma membrane transporter of nicotianamine-metal complexes. *Plant J.* 39:403–414, 2004.

90. G. Schaaf, A. Schikora, J. Häberle, G. Vert, J.F. Briat, C. Curie, N. von Wíren. A putative function for the Arabidopsis Fe-phytosiderophore transporter homolog AtYSL2 in Fe and Zn homeostasis. *Plant Cell Physiol.* 46:762–774, 2005.

91. S. Koike, H. Inoue, D. Mizuno, M. Takahashi, H. Nakanishi, S. Mori, N.K. Nishizawa. OsYSL2 is a rice metal-nicotianamide transporter that is regulated by iron and expressed in the phloem. *Plant J.* 39:415–424, 2004.

92. D. Douchkov, C. Gryczka, U.W. Stephan, R. Hell, H. Baumlein. Ectopic expression of nicotianamine synthase genes results in improved iron accumulation and increased nickel tolerance in transgenic tobacco. *Plant Cell Environ.* 28:365–374, 2005.

93. K. Mengel, G. Guertzen. Iron chlorosis on calcareous soils: alkaline nutritional condition as the cause for the chlorosis. *J. Plant Nutr.* 9:161–173, 1986.

94. H. Kosegarten, G.H. Wilson, A. Esch. The effect of nitrate nutrition on chlorosis and leaf growth in sunflower (*Helianthus annuus* L.). *Eur. J. Agron.* 8:283–292, 1998.

95. P. Perret, W. Koblet. Soil compaction induced iron-chlorosis in grape vineyards: presumed involvement of exogenous soil ethylene. *J. Plant Nutr.* 7:533–539, 1984.

96. V. Römheld. The role of phytosiderophores in acquisition of iron and other micronutrients in graminaceous species: an ecological approach. *Plant Soil* 130:127–134, 1991.

97. N. von Wirén, V. Römheld, T. Shiviri, H. Marschner. Competition between micro-organisms and roots of barley and sorghum for iron accumulated in the apoplasm. *New Phytol.* 130:511–521, 1995.

98. V. Römheld, H. Marschner. Rhythmic iron stress reactions in sunflower at suboptimal iron supply. *Physiol. Plant* 53:347–353, 1981.

99. M. Takahashi, H. Nakanishi, S. Kawasaki, N.K. Nishizawa, S. Mori. Enhanced tolerance of rice to low iron-availability in alkaline soils using barley nicotianamine aminotransferase genes. *Nat. Biotechnol.* 19:466–469, 2001.

100. G. Neumann, V. Römheld. The release of root exudates as affected by the plant's physiological status. In: R. Pinton, Z. Varanini, P. Namiperi, eds. *The Rhizosphere. Biochemistry and Organic Substances at the Soil-Plant Interface.* New York: Marcel Dekker, 2000, pp. 41–93.

101. F. Goto, T. Yoshihara, N. Shigemoto, S. Toki, F. Takaiwa. Iron fortification of rice seed by the soybean ferritin gene. *Nat. Biotechnol.* 17:282–286, 1999.

102. P. Lucca, R. Hurrell, I. Potrykus. Fighting iron deficiency anemia with iron-rich rice. *J. Am. Coll. Nutr.* 21:184–190, 2002.

103. I. Cakmak, A. Torun, E. Millet, M. Feldman, T. Fahima, A. Korol, E. Nevo, H.J. Braun, H. Ozkan. *Triticum dicoccoides*: an important genetic resource for increasing zinc and iron concentration in modern cultivated wheat. *Soil Sci. Plant Nutr.* 50:1047–1054, 2004.

104. F.R. Cox, E.J. Kamprath. Micronutrient soil tests. In: K.K. Mortvedt, P.M. Giordano, W.L. Lindsay, eds. *Micronutrients in Agriculture.* Madison: Soil Sci. Soc. Am., 1972, pp. 289–317.

105. W.L. Lindsay, W.A. Norvell. Development of DTPA soil test for zinc, iron, manganese and copper. *Soil Sci. Soc. Am. J.* 42:421–428, 1978.

106. H.F. Bienfait, J. Garcia-Mina, A.M. Zamareño. Distribution and secondary effects of EDDHA in some vegetable species. *Soil Sci. Plant Nutr.* 50:1103–1110, 2004.

107. V. Römheld, H. Marschner. Mechanism of iron uptake by peanut plants. I. Fe^{III} reduction, chelate splitting, and release of phenolics. *Plant Physiol.* 71:949–954, 1983.

108. R. Rosado, M.C. del Campillo, M.A. Martinez, V. Barrón, J. Tarrent. Long-term effectiveness of vivianite in reducing iron chlorosis in olive trees. *Plant Soil* 241:139–144, 2002.

109. A.D. Rombolà, M. Toselli, J. Carpintero, T. Ammari, J. Quartieri, J. Torrent, B. Marangoni. Prevention of iron-deficiency induced chlorosis in kiwifruit (*Actinidia deliciosa*) through soil application of synthetic vivianite in a calcareous soil. *J. Plant Nutr.* 26:2031–2041, 2003.

110. K. Mengel, E.A. Kirkby. *Principles of Plant Nutrition.* 5th ed. Dordrecht: Kluwer, 2001, p. 569.

111. V. Fernandez, G. Ebert, G. Winkelmann. The use of microbial siderophores for foliar iron application studies. *Plant Soil* 272:245–252, 2005.

112. M. Shenker, I. Oliver, M. Helmann, Y. Hadar, Y. Chen. Utilization by tomatoes of iron mediated by a siderophore produced by *Rhizopus arrhizus*. *J. Plant Nutr.* 15:2173–2182, 1992.

113. W. Hördt, V. Römheld, G. Winkelmann. Fusarinines and dimerum acid, mono- and dihydrate siderophores from *Penicillium chrysogenum*, improve iron utilisation by strategy I and strategy II plants. *BioMetals* 13:37–46, 2000.

114. J.T. Moraghan, J. Padilla, J.D. Etchevers, K. Grafton, J.A. Acosta-Gallegos. Iron accumulation in seed of common bean. *Plant Soil* 246:175–183, 2002.

115. J.T. Moraghan. Accumulation and within-seed distribution of iron in common bean and soybean. *Plant Soil* 264:287–297, 2004.

116. J.V. Wiersma. High rates of Fe-EDDHA and seed iron concentration suggest partial solutions to iron deficiency in soybeans. *Agron. J.* 97:924–934, 2005.

117. J.A. Manthey, B. Tisserat, D.E. Crowley. Root response of sterile-grown onion plants to iron deficiency. *J. Plant Nutr.* 19:145–161, 1996.

118. J.C. Brown. Differential use of Fe^{3+} and Fe^{2+} by oats. *Agron. J.* 71:897–902, 1979.

119. U.C. Gupta. Levels of micronutrient cations in different plant parts of various crop species. *Commun. Soil Sci. Plant Anal.* 21:1767–1778, 1990.

120. R.E. Worley, B. Mullinix. Nutrient element concentration in leaves for 40 pecan cultivars. *Commun. Soil Sci. Plant Anal.* 24:2333–2341, 1993.

121. U.C. Gupta, E.W. Chipman. Influence of iron and pH on the yield and iron, manganese, zinc and sulphur concentrations of carrots grown on sphagnum peat soil. *Plant Soil* 44:559–566, 1976.

122. J.C. Brown, W.E. Jones. Fitting plants nutritionally to soil. II. Cotton. *Agron. J.* 69:405–409, 1977.

123. R.J. Haynes. Nutrient status of apple orchards in Canterbury, New Zealand. I. Levels of soil, leaves and fruit and the prevalence of storage disorders. *Commun. Soil Sci. Plant Anal.* 21:903–920, 1990.

124. A.T. Köseoglu. Investigation of relationships between iron status of peach leaves and soil properties. *J. Plant Nutr.* 18:1845–1859, 1995.

12 Manganese

Julia M. Humphries, James C.R. Stangoulis,
and Robin D. Graham
University of Adelaide, Adelaide, Australia

CONTENTS

12.1 Introduction 351
12.2 Forms of Manganese and Abundance in Soils 352
12.3 Importance to Plants and Animals 352
    12.3.1 Essentiality of Manganese to Higher Plants 352
    12.3.2 Function in Plants 352
    12.3.3 Importance to Animals 353
12.4 Absorption and Mobility 353
    12.4.1 Absorption Mechanisms 353
    12.4.2 Distribution and Mobility of Manganese in Plants 353
12.5 Manganese Deficiency 354
    12.5.1 Prevalence 354
    12.5.2 Indicator Plants 354
    12.5.3 Symptoms 354
    12.5.4 Tolerance 355
12.6 Toxicity 356
    12.6.1 Prevalence 356
    12.6.2 Indicator Plants 356
    12.6.3 Symptoms 356
    12.6.4 Tolerance 357
12.7 Manganese and Diseases 357
12.8 Conclusion 365
Acknowledgments 365
References 366

12.1 INTRODUCTION

The determination of manganese (Mn) essentiality in plant growth by McHargue (1914–1922) focused the attention of plant nutritionists on this nutrient, and led the way for further ground-breaking studies. Since then, research into the concentrations of manganese that confer deficiency or toxicity, and the variation between- and within-plant species in their tolerance or susceptibility to these afflictions has proliferated. The symptoms of toxicity and deficiency have also received much attention owing to their variation among species and their similarity to other nutrient anom-alies. The diversity of visual symptoms within a species that often confounds diagnosis has been

351

attributed to soil conditions. Soil pH is one of the most influential factors affecting the absorption of manganese by changing mobility from bulk soil to root surface. In addition to research on manganese diagnostics, workers have also focused on the role of manganese in resistance to pests and disease, revealing economically important interactions that further highlight the importance of this nutrient in optimal plant production.

This chapter reviews literature dealing with the identification of manganese deficiency and toxicity in various crops of economic importance, the physiology of manganese uptake and transport, and the interaction between manganese and diseases. In addition, a large table outlining deficient, adequate, and toxic concentrations for various crops is included.

12.2   FORMS OF MANGANESE AND ABUNDANCE IN SOILS

Manganese is the tenth-most abundant element on the surface of the earth. This metal does not occur naturally in isolation, but is found in combination with other elements to give many common minerals. The principal ore is pyrolusite (MnO_2), but lower oxides (Mn_2O_3, Mn_3O_4) and the carbonate are also known.

Manganese is most abundant in soils developed from rocks rich in iron owing to its association with this element [1]. It exists in soil solution as either the exchangeable ion Mn^{2+} or Mn^{3+}. Organic chelates derived from microbial activity, degradation of soil organic matter, plant residues, and root exudates can form metal complexes with micronutrient cations, and thereby increase manganese cation solubility and mobility [2]. Availability of manganese for plant uptake is affected by soil pH; it decreases as the pH increases. Divalent manganese is the form of manganese absorbed at the root surface cell membrane. As soil pH decreases, the proportion of exchangeable Mn^{2+} increases dramatically [3], and the proportions of manganese oxides and manganese bound to iron and manganese oxides decrease [4]. This action has been attributed to the increase in protons in the soil solution [5]. Acidification may also inhibit microbial oxidation that is responsible for immobilization of manganese. Manganese-oxidizing microbes are the most effective biological system oxidizing Mn^{2+} in neutral and slightly alkaline soils [6–8]. Relatively, as soil pH increases, chemical immobilization of Mn^{2+} increases [9], and chemical auto-oxidation predominates at pH above 8.5 to 9.0 [10,11].

12.3   IMPORTANCE TO PLANTS AND ANIMALS

12.3.1   Essentiality of Manganese to Higher Plants

The first reported investigations into the essentiality of manganese by Horstmar in 1851 [12] succeeded in identifying this nutrient as needed by oats, but only where iron was in excess. Further evidence for the essentiality of manganese was not made until some Japanese researchers reported that manganese stimulated the growth of several crops substantially [13,14]. These crops included rice (*Oryza sativa* L.), pea (*Pisum sativum* L.), and cabbage (*Brassica oleracea* var. *capitata* L.), and because of their economic importance, further interest was stimulated [15]. Supporting these field results were the physiological and biochemical studies of Bertrand [16–18]. His work reported manganese as having a catalytic role in plants, and that combinations with proteins were essential to higher plant life. This reported essentiality of manganese was supported by studies by Maze [19] in solution culture. Studies by McHargue [20,21], where the role of manganese in the promotion of rapid photosynthesis was determined, are regarded as having established that manganese is essential for higher plant growth.

12.3.2   Function in Plants

Manganese is involved in many biochemical functions, primarily acting as an activator of enzymes such as dehydrogenases, transferases, hydroxylases, and decarboxylases involved in respiration, amino acid and lignin synthesis, and hormone concentrations (22,23), but in some cases it may be replaced by other metal ions (e.g., Mg). Manganese is involved in oxidation–reduction (redox) reactions within the photosynthetic electron transport system in plants (24–26). Manganese is also involved in the photosynthetic evolution of O_2 in chloroplasts (Hill reaction). Owing to the key role in this essential process, inhibition of photosynthesis occurs even at moderate manganese deficiency; however, it does not affect chloroplast ultrastructure or cause chloroplast breakdown until severe deficiency is reached (27).

12.3.3 IMPORTANCE TO ANIMALS

In humans, manganese deficiency results in skeletal abnormalities (28,29). In the offspring of manganese-deficient rats, a shortening of the radius, ulna, tibia, and fibula is observed (30). Manganese deficiency during pregnancy results in offspring with irreversible incoordination of muscles, leading to irregular and uncontrolled movements by the animal, owing to malformation of the bones within the ear (30,31). Animals that are manganese-deficient are also prone to convulsions (32).

In contrast, manganese toxicity induces neurological disturbances that resemble Parkinson's disease, and the successful treatment of this disease with levodopa is associated with changes in manganese metabolism (33,34). In animals manganese is associated with several enzymes (35), including glycosyl transferase (36), superoxide dismutase (37,38), and pyruvate carboxylase (39).

Manganese requirement for humans is 0.035 to 0.07 mg kg^{-1}, with daily intake representing 2 to 5 mg day^{-1} in comparison to the body pool of 20 mg (30,40).

12.4 ABSORPTION AND MOBILITY

12.4.1 ABSORPTION MECHANISMS

As mentioned previously, manganese is preferentially absorbed by plants as the free Mn^{2+} ion from the soil solution (41–43). It readily complexes with plant and microbial organic ligands and with synthetic chelates. However, complexes formed with synthetic chelates are generally considered to be absorbed more slowly by roots than the free cation (44,45).

Manganese absorption by roots is characterized by a biphasic uptake. The initial and rapid phase of uptake is reversible and nonmetabolic, with other Mn^{2+} and Ca^{2+} being exchanged freely (46,47). In this initial phase, manganese appears to be adsorbed by the cell wall constituents of the root-cell apoplastic space. The second phase is slower; manganese is less readily exchanged (48), and its uptake is dependent on metabolism. Manganese is absorbed into the symplast during this slower phase (47,48). However, the exact dependence of manganese absorption on metabolism is not clear (46,49,50).

Uptake of manganese does not appear to be tightly controlled, unlike the major nutrient ions. Kinetic experiments have estimated manganese absorption to be at a rate of 100 to 1000 times greater than the need of plants (51). This may be due to the high capacity of ion carriers and channels in the transportation of manganese ions through the plasma membrane at a speed of several hundred to several million ions per second per protein molecule (52,53).

12.4.2 DISTRIBUTION AND MOBILITY OF MANGANESE IN PLANTS

The plant part on which symptoms of Mn deficiency is observed generally indicates the mobility of the nutrient within the plant. Manganese has been reported to be an immobile element, which is not re-translocated (54–59), and consequently symptoms do not occur on old leaves. In addition, symptoms of manganese deficiency regularly appear on fully expanded young leaves rather than on the newest leaf. This symptom may indicate an internal requirement in these leaves beyond that of the new leaves (60), or it may simply be a matter of supply and demand in what is the fastest growing tissue.

The location of manganese in plants is a significant factor in the expression of deficiency symptoms and is affected by its mobility in the xylem and phloem. Manganese moves easily from the root to the shoot in the xylem-sap transpirational stream (61). In contrast, re-translocation within the phloem is complex, with leaf manganese being immobile, but root and stem manganese being able to be re-mobilized (62). The net effect of the variable phloem mobility gives rise to a re-distribution of manganese in plant parts typical of a nutrient with low phloem mobility.

Studies into the mobility of manganese with wheat (*Triticum aestivum* L.) (63,64), lupins (*Lupinus* spp. L.) (55,65), and subterranean clover (*Trifolium subterraneum* L.) (56) have reported no re-mobilization from the old leaves to the younger ones. Further support for this lack of mobility was given in a study by Nable and Loneragan (57), in which plants provided with an early supply of ^{54}Mn failed to re-mobilize any of this radioactive element when their roots were placed in a solution with a low concentration of nonradioactive manganese. The apparent inconsistency with evidence that phloem is a major source of manganese from the roots and stems to developing seeds (59,66) can be explained by changes in carbon partitioning within the plant as Hannam and Ohki (67) reported a re-mobilization of manganese from the stem during the outset of the reproductive stages of plant development.

12.5   MANGANESE DEFICIENCY

12.5.1   PREVALENCE

Manganese deficiency is most prevalent in calcareous soils, the pH of which varies from 7.3 to 8.5, and the amounts of free calcium carbonate ($CaCO_3$) also vary (68). The pH of calcareous soils is well buffered by the neutralizing effect of calcium carbonate (69). Soils that have a high organic content, low bulk density, and a low concentration of readily reducible manganese in the soil are also susceptible to producing manganese deficiency. Climatically, cool and temperate conditions are most commonly associated with manganese deficiency, although there have been reports on the same from tropical to arid areas. Drier seasons have been reported to relieve (70) or to exacerbate (71) manganese deficiency.

12.5.2   INDICATOR PLANTS

Plants that have been reported to be sensitive to manganese deficiency are apple (*Malus domestica* Borkh.), cherry (*Prunus avium* L.), cirtus (*Citrus* spp. L.), oat (*Avena sativa* L.), pea, beans (*Phaseolus vulgaris* L.), soybeans (*Glycine max* Merr.), raspberry (*Rubus* spp. L.), and sugar beet (*Beta vulgaris* L.) (72–76).

Of the cereals, oats are generally regarded as the most sensitive to manganese deficiency, with rye (*Secale cereale* L.) being the least sensitive. However, there seems to be some discrepancy in the ranking of susceptibility to manganese deficiency of wheat and barley (*Hordeum vulgare* L.) (77–80). This occurrence might be attributed to a large within-species genetic variation that has been reported for several species, including wheat (77,81), oats (78,82), barley (70,78), peas (83), lupins (84), and soybeans (85).

Because of their sensitivity to manganese deficiency, several species previously considered susceptible to manganese deficiency have been the focus of breeding for more efficient varieties and may therefore not be considered susceptible species in more recent publications. It is generally agreed that grasses (Gramineae, Poaceae), clover (*Trifolium* spp. L.), and alfalfa (*Medicago sativa* L.) are not susceptible to manganese deficiency (76,86).

12.5.3   SYMPTOMS

Characteristic foliar symptoms of manganese deficiency become unmistakable only when the growth rate is restricted significantly (67) and include diffuse interveinal chlorosis on young expanded leaf blades (Figure 12.1) (60); in contrast to the network of green veins seen with iron

FIGURE 12.1   Manganese deficiency on crops: left, garden bean (*Phaseolus vulgaris* L.) and right, cucumber (*Cucumis sativus* L.). (For a color presentation of this figure, see the accompanying compact disc.)

deficiency (67). Severe necrotic spots or streaks may also form. Symptoms often occur first on the middle leaves, in contrast to the symptoms of magnesium deficiency, which appear on older leaves. With eucalyptus (*Eucalyptus* spp. L. Her.), the tip margins of juvenile and adult expanding leaves become pale green. Chlorosis extends between the lateral veins toward the midrib (60). With cereals, chlorosis develops first on the leaf base, while with dicotyledons the distal portions of the leaf blade are affected first (67).

With citrus, dark-green bands form along the midrib and main veins, with lighter green areas between the bands. In mild cases the symptoms appear on young leaves and disappear as the leaf matures. Young leaves often show a network of green veins in a lighter green background, closely resembling iron chlorosis (75). Manganese deficiency is confirmed by the presence of discoloration (marsh spot) on pea seed cotyledons (87), and split or malformed seed of lupins (84).

In contrast to iron deficiency chlorosis, chlorosis induced by manganese deficiency is not uniformly distributed over the entire leaf blade and tissue may become rapidly necrotic (88). The inability of manganese to be re-translocated from the old leaves to the younger ones designates the youngest leaves as the most useful for further chemical analysis to confirm manganese deficiency. Visual symptoms of manganese deficiency can easily be mistaken for those of other nutrients such as iron, magnesium, and sulfur (87), and vary between crops. However, they are a valuable basis for the determination of nutrient imbalance (87) and, combined with chemical analysis, can lead to a correct diagnosis.

12.5.4   TOLERANCE

Tolerance to manganese deficiency is usually conferred by an ability to extract more efficiently available manganese from soils that are considered deficient. Mechanisms that are involved in the improved extraction of manganese from the soil include the production of root exudates (89–91), differences in excess cation uptake thus affecting the pH of the rhizosphere (92,93), and changes in root density (94). The genotypic variation within species for manganese efficiency can be utilized by breeding programs to develop more efficient varieties (95,96).

Tolerance to manganese deficiency may be attributed to one or more of the following five adaptive mechanisms (96):

1. Superior internal utilization or lower functional requirement for manganese.
2. Improved internal re-distribution of manganese.
3. Faster specific rate of absorption from low manganese concentrations at the root–soil interface.
4. Superior root geometry.
5. Greater extrusion of substances from roots into the rhizosphere to mobilize insoluble manganese utilizing: (i) H^+; (ii) reductants; (iii) manganese-binding ligands; and (iv) microbial stimulants.

The importance of, and evidence for, each mechanism has been reviewed extensively by Graham (98), and so will not be re-analyzed here. It is concluded that mechanisms 1 and 2 are not important mechanisms of efficiency generally, mechanism 3 may be important in certain situations, while breeding for mechanism 4 is not thought to bring about rapid progress in improving tolerance. Mechanism 5 is thought to have some role, though this area requires further investigation.

12.6 TOXICITY

12.6.1 PREVALENCE

Manganese toxicity is a major problem worldwide and occurs mainly in poorly drained, acid soils owing to the interactions mentioned previously. However, not all poorly drained soils are sources of manganese toxicity as reported by Beckwith and co-workers (99), who noted that flooding often increased the pH, thus reducing the availability of manganese. Tropical, subtropical, and temperate soils have all been reported to be sources of manganese at concentrations high enough to produce visible symptoms of toxicity. In the tropics, toxicity has been reported in tropical grasses grown in the Catalina (basalt) and the Fajardo (moderately permeable) clayey soils of Puerto Rico (100), and in ryegrass (*Lolium* spp. L.) grown on red–brown clayey loam and granite–mica schists in Uganda, Africa (101). Among the subtropical regions, toxicity has been reported in subtropical United States in poorly drained soils and soils on limestone (102) and on ultisols. However, the impermeability of soils does not seem essential for manganese toxicity (103). In southeastern Australia, manganese toxicity has been reported in fruit trees grown in neutral-pH duplex soils (104), in French beans (*Phaseolus vulgaris* L.) grown in manganese-rich basaltic soil (105), and in pasture legumes (106). There is very little information available on manganese toxicity in temperate regions, though one report found toxicity on soils characterized by low pH and high concentrations of readily exchangeable manganese (107).

12.6.2 INDICATOR PLANTS

A number of crops are considered sensitive to manganese toxicity, and these include alfalfa, cabbage, cauliflower (*Brassica oleracea* var. *botrytis* L.), clover (*Trifolium* spp. L.), pineapple (*Ananas comosus* Merr.), potato (*Solanum tuberosum* L.), sugar beet, and tomato (*Lycopersicon esculentum* Mill.) (74,108). An excess of one nutrient can aggravate a deficiency of another, and so symptoms of manganese toxicity bear some features of deficiency of another nutrient. Additionally, toxicity of manganese is often confused with aluminum toxicity as both often occur in acid soils. However, in some species such as wheat (109) and rice (110), the tolerance to these two toxicities is opposite (111).

12.6.3 SYMPTOMS

The visual symptoms of manganese toxicity vary depending on the plant species and the level of tolerance to an excess of this nutrient. Localized as well as high overall concentrations of manganese are responsible for toxicity symptoms such as leaf speckling in barley (112), internal bark necrosis in apple (113), and leaf marginal chlorosis in mustard (*Brassica* spp. L.) (114).

The symptoms observed include yellowing beginning at the leaf edge of older leaves, sometimes leading to an upward cupping (crinkle leaf in cotton, (115)), and brown necrotic peppering on older leaves. Other symptoms include leaf puckering in soybeans and snap bean (116); marginal chlorosis and necrosis of leaves in alfalfa, rape (*Brassica napus* L.), kale (*Brassica oleracea* var. *acephala* DC.), and lettuce (*Lactuca sativa* L.) (116); necrotic spots on leaves in barley, lettuce, and soybeans (116); and necrosis in apple bark (i.e., bark measles) (60). Symptoms in soybeans include chlorotic specks and leaf crinkling as a result of raised interveinal areas (117,118); chlorotic leaf tips, necrotic areas, and leaf distortion (102) in tobacco (*Nicotiana tabacum* L.).

12.6.4 TOLERANCE

Reduction of manganese to the divalent and therefore more readily absorbed form is promoted in waterlogged soils, and tolerance to wet conditions has coincided with tolerance to excess manganese in the soil solution. Graven et al. (119) suggested that sensitivity to waterlogging in alfalfa may be partially due to manganese toxicity, and alfalfa has been shown to be more sensitive to manganese toxicity than other pasture species such as birdsfoot trefoil (*Lotus corniculatus* L.) (120). In support of this suggestion, several other pasture species have also been reported to have a relationship between waterlogging and manganese toxicity (121,122). For example, manganese-tolerant subterranean clover (*Trifolium subterraneum* cv. Geraldton) was reported to be more tolerant to waterlogging than the manganese-sensitive medic (*Medicago truncatula* Gaertner) (123). Increased tolerance to manganese toxicity by rice when compared with soybean is combined with increased oxidizing ability of its roots (124,125).

Tolerance to manganese toxicity has also been related to a reduction in the transport of manganese from the root to the shoot as shown by comparison between corn (tolerant) and peanut (*Arachis hypogaea* L.) (susceptible) (126,127). Furthermore, tolerance to manganese toxicity was observed in subterranean clover (compared with *Medicago truncatula*) and was associated with a lower rate of manganese absorption and greater retention in the roots (123). In an extensive study comparing eight tropical and four temperate pasture legume species, it was concluded that tolerance to manganese toxicity was partially attributable to the retention of excess manganese in the root system (128). This conclusion was also reached in comparing alfalfa clones that differed in manganese tolerance (129).

In rice, tolerance to high concentrations of manganese is a combination of the ability to withstand high internal concentrations of manganese with the ability to oxidize manganese, thus reducing uptake. This is in comparison with other grasses that are unable to survive the high concentrations found in rice leaves (130).

Tolerance is also affected by climatic conditions such as temperature and light intensity (131). For example, when comparing two soybean cultivars, Bragg (sensitive) and Lee (tolerant), an increase from 21 to 33°C day temperature and 18 to 28°C night temperature prevented the symptoms of manganese toxicity in both cultivars, despite the fact that manganese uptake was increased (132,133).

12.7 MANGANESE AND DISEASES

The manganese status of a plant can affect, and be affected by, disease infection, often leading to the misdiagnosis of disease infection as manganese deficiency or toxicity (134). The manganese concentration in diseased tissues has been observed to decrease as the disease progresses (135). This occurrence may be due to the pruning of the root system in the case of root pathogens, leading to a reduction in the absorptive surface with a resultant decrease in the plant concentration (136,137). Additionally, microbially induced changes in manganese status, such as that caused by the grey-speck disease (manganese deficiency) of oats have been reported to be due to the oxidizing bacteria in the rhizosphere causing the manganese to become unavailable (138,139). Manganese concentration at the site of infection also has been reported to increase, in direct contrast to the overall manganese plant concentration, which has decreased (140).

The most notable interaction between disease and manganese is that of the wheat disease take-all caused by the pathogen *Gaeumannomyces graminis* var. *tritici*, commonly referred to as *Ggt*. The importance of manganese in the defence against infection by *Ggt* was demonstrated by Graham (23). Manganese is the unifying factor in the susceptibility of varieties to *Ggt* under several soil conditions, including changing pH and nitrogen forms as shown in a table by Graham and Webb (141). The role of manganese fertilizer in the amelioration of *Ggt* has been reported in numerous papers (137,142,143). The effect of manganese fertilizer on infection by *Ggt* has been shown to impact before the onset of foliar symptoms (137,142).

TABLE 12.1
List of Critical Concentrations of Manganese in Various Agricultural Crops

Growth Stage	Plant Part	Type of Culture	Concentration of Manganese (mg kg^{-1})			Reference	Comments
			Deficient	Adequate	Toxic		
Barley (Hordeum vulgare L.)							
45 DAS	WS	Soil	13–21	24–50		149	Critical estimated at ~85% max. shoot yield
FS 5–6	WS	Literature review		30–100		150	Winter and summer barley
FS 7–8	WS	Literature review		25–100		150	Winter and summer barley
FS 10	WS	Soil		<140	>190	151	*H. distichon*
FS 10.1	WS	Literature review	<5	25–100		152	
Mid to late tillering	YMB	Field, survey	<12	25–300	700	153	
Veg.	YEB	Field, soil		12		154	Critical concentration
Black gram (Vigna mungo Hepper)							
25–33 DAT	WS	Solution culture			345–579	155	cv. Regur
Canola (Brassica napus L.)							
Veg.	ML	Literature review		40–100		150	*Brassica napus* var. *napobrassica*
Pre-anthesis	YML	Literature review		30–250	530–3650	153	*Brassica napus*, *B. campestris*
Early-anthesis	YML	Literature review		30–100		150	*Brassica napus* var. *oliefera*
Unknown	YML	Literature review	10	30		156	
Cassava (Manihot esculentum Crantz)							
30 DAS	WS	FSC		140–170		157	Toxic criteria at 90% max. yield
63 DAS	YMB	Solution culture	<14			158	Critical at 90% max. yield
Veg.	YMB	Field	<50	50–250	>1000	159	
3–4 months	YMB	Field	<45	50–120	>250	160	
Cereal rye (Secale cereale L.)							
Young plants	WS	Survey			200	161	Critical for acidic soils with pH values 4.1–4.4
22 DAS	WS	Soil		18–69		162	cv. did not respond to applied Mn, where other cereals did
Unknown	WS	Literature review		14–45		163	
FS 5–6	WS	Literature review		25–100		150	
FS 7–8	WS	Literature review		20–100		150	
Chickpea (Cicer arietinum L.)							
Veg.	YML	Literature review		60–300		153	
Cotton (Gossypium hirsutum L.)							
35 DAS	WS	Soil			494	164	
Before anthesis	YMB	Survey, diag.		50–350		165	

TABLE 12.1   (*Continued*)

Growth Stage	Plant Part	Type of Culture	Concentration of Manganese (mg kg⁻¹)			Reference	Comments
			Deficient	Adequate	Toxic		
36 DAS	YMB	RSC	2–8	11–247		166,167	Critical at 90% max. yield
Veg. to anthesis	YMB	Survey, Diag.	8	25–500	4000	153	
Anthesis to boll develop.	YML	Literature review		35–100		150	
33 DAS	3 YML	Soil		49–57	568–689	168	Data for 11 cotton genotypes
18 DAS	YL	Solution culture		55	962–3300	169	cv. 517
18 DAS	YL	Solution culture		45	1580–2660	169	cv. 307
21 DAT	3 young leaves (width <1 cm)	RSC		200–270	4030–10570	170	3 cultivars; peroxidase activity in leaves separated Mn toxic from adequate

Cowpea (Vigna unguiculata Walp.)

Growth Stage	Plant Part	Type of Culture	Deficient	Adequate	Toxic	Reference	Comments
25–33 DAT	WS	Solution culture		79–299		155	Data for 2 cv.
35 DAS	WS	Field		<1000	>2000	171	43 cv. examined; toxic at 50% max. yield
Pre-anthesis	YMB	Survey, diag.		70–300		153	
20 DAT	YMB	Solution culture		68		172	cv. TVu91, sensitive to Mn toxicity; symptoms in old leaves only
20 DAT	Old LB	Solution culture		183	310	172	cv. TVu91, sensitive to Mn toxicity; symptoms in old leaves only

Faba bean (Vicia faba L.)

Growth Stage	Plant Part	Type of Culture	Deficient	Adequate	Toxic	Reference	Comments
Unknown	YL	Literature review	3.3	55		173	Adequate plants no symptoms
Unknown	WS	Literature review		109	1083	173	
Onset of anthesis	YML	Literature review		40–100		150	
Early anthesis	YML	Literature review		50–300	1000–2020	153	

Field pea (Pisum sativum L.)

Growth Stage	Plant Part	Type of Culture	Deficient	Adequate	Toxic	Reference	Comments
Unknown	YL	Literature review	4.2	60–65		173	cv. Wirrega and Dinkum; adequate plants no symptoms
Unknown	WS	Literature review	85	1743–2988		173	cv. Wirrega and Dinkum; adequate plants no symptoms
Onset of anthesis	YML	Literature review		30–100		150	
Pre-anthesis	YML	Literature review		30–400	>1000	153	
First bloom	YML	Literature review	25–29	30–400		163	
Unknown	LB	Field	6–13	30–60		86	

Continued

TABLE 12.1   (*Continued*)

Growth Stage	Plant Part	Type of Culture	Concentration of Manganese (mg kg⁻¹)			Reference	Comments
			Deficient	Adequate	Toxic		
Ginger (*Zingiber officinale* Roscoe)							
2–3 months	Upper LB	Solution culture	20–23	125–250	950–990	174	
2–3 months	Lower LB	Solution culture	20–23	≤820	950–990	174	
Green gram (*Vigna radiata* R. Wilcz.)							
25–33 DAT	WS	Solution culture		247–259	784–901	155	cv. Berken
40 DAS	YML	Soil		20–38		175	cv. ML131; study on 14 soils
Guar (*Cyamopsis tetragonoloba* Taub.)							
25–33 DAT	WS	Solution culture		92–100		155	cv. Brooks
Hops (*Humulus lupulus* L.)							
Mid season	YML	Literature review		30–100		150	
Kenaf (*Hibiscus cannabinus* L.)							
Maturity	Stem	Literature review		14–23		163	
Linseed, Anthesisax (*Linum usitatissimum* L.)							
70 DAS	YL	Soil		56	1015	176	
Onset of anthesis	Upper third of shoots	Literature review		30–100		150	
49–70 DAS	WS	Soil		5–50	500–2000	176	
63 DAS	WS	Soil	14–18	108–145		176	
63 DAS	WS	Field		108–449		176	
70 DAS	WS	Soil		34	2295	176	
Lupin (*Lupinus angustifolius* L., *L. albus* L., *L. cosentinii* Guss.)							
40 DAS	WS	Literature review		277	>6164	177	
40 DAS	WS	Soil		245	>7724	177	*L. albus*
40 DAS	WS	Soil		277	>6164	177	
56 DAS	WS	Survey	31–55	318–1300		178	
Up to early anthesis	YFEL	Soil	<30			153,179	Diagnostic for shoot DW
Pre-anthesis	YML	Literature review		50–1200	1900–16000	153	Three *Lupinus* spp.
28 DAS	YOL	Literature review	5.6	245	>7724	177	*L. albus*
Anthesis	WS	Soil, field		>20		179	Predictive for absence of 'split seed' disorder. Buds and leaves poor predictors.
Maturity	Seed	Survey	4–9	7–53		178	
Maize; corn (*Zea mays* L.)							
30–45 DAE	WS	Unknown		50–160		180	
Six-leaf stage	WS	Field	8–9			181	
40–60 cm tall	YMB	Literature review		40–100		150	
Tassell—initial silk	Ear leaf	Field, diag.	<15	20–200	3000	153	Symptoms shown in toxic range
Initial silk	Ear leaf	Literature review	10–19	20–200		163	
Early silk	Ear leaf	Field	<11			182	
Early silk	Ear leaf	Field	<11			181	Critical at 90% max. grain yield

TABLE 12.1 *(Continued)*

Growth Stage	Plant Part	Type of Culture	Concentration of Manganese (mg kg^{-1})			Reference	Comments
			Deficient	Adequate	Toxic		
Silk	Ear leaf	Field	<15	20–150	>200	183	
40–60 cm tall	Leaf opposite ear	Literature review		35–100		150	
Before tassell	Leaf below whorl	Literature review	<15	15–300		163	
Before tassell	Leaf below whorl	Field, survey, diag.		20–300		165	
Navy bean (*Phaseolus vulgaris* L.)							
Veg.	YML	Literature review		20–100		184	
60 DAS	YMB	Survey			≥760	185	Plants with symptoms had highest levels of Fe and Mn.
Onset of anthesis	YML	Literature review		40–100		150	
Unknown	YML	Literature review	15–49	50–300		163	
Oats (*Avena sativa* L.)							
Young plants	WS	Survey			>300	161	Critical for acidic soils pH < 4.7
FS 5–6	WS	Literature review		40–100		150	
FS 7–8	WS	Literature review		35–100		150	
FS 6	WS	Field	<16			186	Critical at 90% max. grain yield
FS 10	WS	Survey	<15	>30		187	
FS 10.1	WS	Field, survey	<5	25–100		163,188	
Anthesis	WS	Survey	<14	14–150		189	
Mid to late tillering	YMB	Field, diag.	<12	25–300	700	153	Symptoms present in toxic range
Pre-head	Upper LB	Field, survey		25–100		165	
FS 10.5	Flag + next older LB	Survey	<12–15			190	
Peanut (*Arachis hypogaea* L.)							
25–33 DAT	WS	Solution culture		100–212		155	cv. Red Spanish
Pre-anthesis/ anthesis	YMB	Survey, diag.			600–800	165	
Unknown	YMB	Survey	<10			191	
Pre-anthesis to anthesis	YML	Survey, diag.		50–300	>700	153	
Anthesis	YML	Literature review		50–100		150	
Anthesis	YML	Literature review		20–350		192	
49 DAS	YML	Field	7–0	19–39		193	cv. Florunner; critical and deficient conc. Relate to plants grown at pH (water) = 6.8±0.1
63 DAS	YML	Field	7–12	26–64		193	cv. Florunner; critical and deficient conc. related to plants grown at pH (water) = 6.8±0.1

Continued

TABLE 12.1   (*Continued*)

Growth Stage	Plant Part	Type of Culture	Concentration of Manganese (mg kg⁻¹) Deficient	Adequate	Toxic	Reference	Comments
77 DAS	YML	Field	8–11	34–66		193	
91 DAS	YML	Field	9–11	37–100		193	
105 DAS	YML	Field	9–13	36–115		193	
119 DAS	YML	Field	9–12	33–118		193	
90 DAS approx.	YML	Field		83–170	244–687	194	Data from three sites; Mn toxic if Ca/Mn ratio <80

Pigeon pea (*Cajanus cajan* Huth.)

Veg.	WS	FSC		78–300	300	157	cv. Royes

Rice (*Oryza sativa* L.)

30 DAT	WS	RSC		57–130	770–7370	195	Adequate range for plants not affected by high Mn supply
Tillering	WS	Unknown			7000	196	
Various	WS	Solution culture	<20		>2500	197	
Panicle initiation	YB	Survey		252–792		188	
FS 3–5	YMB	Field, diag.		40–500	>5000	153	
Before anthesis	YMB	Literature review		40–100		150	

Safflower (*Carthamus tinctorius* L.)

70 DAS	YOL	Field		20–55		198	Predictive for seed yield
70 DAS	Upper S	Field		3.5–8		198	Predictive for seed yield
70 DAS	Upper S	Field		3.5–8		198	Predictive for seed yield
75 DAS	YOL	Field		20–75		198	Predictive for seed yield
75 DAS	Upper S	Field		3–4		198	Predictive for seed yield
Maturity	Seed	Field		6.5–8		198	Diagnostic for seed

Sorghum (*Sorghum bicolor* Moench.)

24 DAS	WS	Solution culture	24		217	199	
35 DAS	WS	Solution culture			>860	200	
GS 2	WS	Field		40–150		201	
GS 3	WS	Sand		40–70		201	Deficient, marginal, and adequate ranges <50%, 50–90%, and 90–100% max. yield, respectively
GS 3–5	YMB	Field		6–100		201	
Veg. and early anthesis	Third LB below head	Survey, diag.	<8	15–350		153	
63 DAS	Middle LB	Sand	12–15	20–30		202	
GS 6	3BBE	Field		8–190		201	
GS 7–8	3BBE	Field		8–40		201	
Anthesis	YML	Literature review		25–100		150	

Soybean (*Glycine max* Merr.) (Growth stages of soybean are as described by Fehr et al. (203))

37 DAE	WS	Soil		21–44	246–337	204	cv. Bragg
42 DAS	WS	Soil		13.349.2		205	cv. Bragg

TABLE 12.1 *(Continued)*

Growth Stage	Plant Part	Type of Culture	Concentration of Manganese (mg kg^{-1})			Reference	Comments
			Deficient	Adequate	Toxic		
Anthesis	YMB	Diag., Survey	<15	30–100	750–1000	153	
36–46 DAS	YMB	RSC	<11		>173	85	Seven cvv. compared
Late Anthesis	YMB	Literature review		30–100		150	
Early anthesis (R2)	YMB	Field	6–10	15–36		206	Critical conc. varies with soil
Pre-PS	YMB	Field, survey, diag.		21–100		207	
First pods	YMB	Field, survey, diag.	<20			208	
Early PF	YMB	Survey, diag.		30–200	>500	165	
21 DAT	YOL	Solution culture	10–13	43	402–648	133	cv. Bragg
21 DAT	YOL	Solution culture	8–13	38	541–686	133	cv. Lee
14 DAT	YML first trifoliate	Solution culture	9.5–18.5	33–69	865–1180	209	Data for four cvv.
38 D after tmt imposed	YL	Sand		103	1530	210	cv. Maple arrow; tmts imposed at 39 DAS
38 D after tmt imposed	Old leaves	Sand		144	2780	210	cv. Maple arrow; tmts imposed at 39 DAS
Unknown	Trifoliate leaf	Solution culture	9–13	44–69	479–945	211	cv. Williams
Maturity	Seed	Field	18.2–26.6			212	cv. Essex
Mature LB	Leaf	Field	10			213	cv. Bragg

Sugar beet (*Beta vulgaris* L.)

Growth Stage	Plant Part	Type of Culture	Deficient	Adequate	Toxic	Reference	Comments
Tenth leaf	WS	Soil	<35	30–62		214	Critical at 90% yield
Unknown	WS	Soil			>800	161	Linked with soil acidity
21 DAT	YMB	Soil, solution culture			>5000	167	Critical at 90% max. yield
Veg.	YMB	Literature review	4–20		>5500	215	
Unknown	YMB	Literature review	4–0	25–360		216	Plant growth less below critical; deficient = symptoms present; adequate = no symptoms
50–80 DAS	Leaf	Literature review	10–25	26–360		163	
50–60 DAS	ML	Literature review		35–100		150	

Sugar cane (*Saccharum spp.* L.)

Growth Stage	Plant Part	Type of Culture	Deficient	Adequate	Toxic	Reference	Comments
Rapid growth	TVD	Field, survey		12–100		217–219	
Rapid growth	TVD	Field, survey	<15	20–200		220	
Four months	Middle leaves (mid-portion less midrib)	Literature review		100–250		150	

Sunflower (*Helianthus annuus* L.) (Growth stages of sunflower, R1, R2, etc. are as described by Schneiter and Miller (221))

Growth Stage	Plant Part	Type of Culture	Deficient	Adequate	Toxic	Reference	Comments
R-2	YEL		<13	46–80		222	cv. Hysun 31
18–31 DAS	WS	FSC			5300	157	cv. Hysun 31
Florets about to emerge	Third fourth LB below flower bud	Diag.		41–850	>3000	223	

Continued

TABLE 12.1   (*Continued*)

Growth Stage	Plant Part	Type of Culture	Concentration of Manganese (mg kg^{-1}) Deficient	Adequate	Toxic	Reference	Comments
Tea (*Camellia sinensis* O. Kuntze)							
At plucking	Mature leaves	Field, survey	<50			224	
Tobacco (*Nicotiana tabacum* L.)							
Anthesis	YMB	Survey, diag.		30–250		165	
Anthesis	YMB	Field		33–156		225	
Veg (40–80 DAE)	YML	Survey, Diag.		35–350	1290–1420	153	
Various	Leaves	Various		160	933–11,000	75	
Veg.	Leaves (all)	Solution culture		33	797	226	cv. KY14
Veg.	Leaves (all)	Solution culture		41		226	cv. T.I.1112
42 DAT	Leaves	Sand			700–1200	227	D/N temp 22/18°C; cv. Coker 347
42 DAT	Leaves	Sand			2000–3500	227	D/N temp 26/22°C; cv. Coker 347
42 DAT	Leaves	Sand			5000–8000	227	D/N temp 30/26°C; cv. Coker 347
Mature	Cured leaves	Field		115		228	Yield ≥ 3.2 t/ha
Mature	Cured leaves	Sand			7000	229	
Triticale (*X Triticosecale*)							
22 DAS	WS	Soil	11–15			162	Concentration associated with reduced growth in two cvv.
25 DAS	WS	Solution culture			1100–3200	230	Toxic range associated with plant yield reduction in four cvv.
Wheat (*Triticum aestivum* L. and *Triticum durum* Desf.)							
18–31 DAS	WS	FSC			280	157	
22 DAS	WS	Soil	9–12			162	Conc. associated with plant symptoms and reduced growth in seven cvv.
25 DAS	WS	Soil	6	37–116		139	Three levels of Mn applied
Mid tillering	WS	Field	11	23		137	Two levels of Mn applied
FS 5–6	WS	Literature review		35–100		150	Winter and summer wheats
FS 7–8	WS	Literature review		30–100		150	Winter and summer wheats
FS 10.1	WS	Literature review	5–24	25–100		163	Spring wheat
Mid to late tillering	YMB	Field, survey	<12	25–300	700	223	Toxicity symptoms observed
Just before heading	Upper two leaves	Literature review		16–200		163	Winter wheat
Maturity	Grain	Field		18.2		231	
Maturity	Grain	Soil	<15.5		>24	232	Critical at max. grain yield

TABLE 12.1 (*Continued*)

Growth Stage	Plant Part	Type of Culture	Concentration of Manganese (mg kg⁻¹)			Reference	Comments
			Deficient	Adequate	Toxic		
Winged bean (*Psophocarpus tetragonolobus* DC.)							
25–33 DAT	WS	Solution culture		218–225		155	cv. UPS 31
42 DAS	WS	Sand		29–49		233	

Key

Growth stage

DAE, days after emergence; DAS, days after sowing; DAT, days after transplanting; FS, Feeke's scale of growth in cereals defined by Large 1954 (234); GS, growth stage; PF, pod fill/ grain fill; PS, pod set; Veg., vegetative.

Plant part

BBE, blade below ear; L, leaf; LB, leaf blade; ML, mature leaf; Trifol. L., trifoliate leaves; TVD, top visible dewlap (sugar cane); S, stem; WS, whole shoot; YEL, youngest expanded leaf; YFEL, youngest fully expanded leaf; YL, young leaves; YMB, youngest mature leaf blade; YML, youngest mature leaf; YOL, youngest open leaf; YOL +1, Next youngest open leaf.

Type of culture

Field, field experiment; sand, sand culture in glasshouse; RSC, solution culture where nutrients were replenished periodically; diag., diagnostic records from database; soil, soil culture in glasshouse; FSC, flowing solution culture; survey, survey from commercial crops; solution culture, solution culture in glasshouse.

Source: adapted from D.J. Reuter et al. *Plant Analysis: An Interpretation Manual.* Collingwood, Vic.: CSIRO Publishing, 1997, pp. 83–284.

Several mechanisms have been proposed for the interaction between manganese and disease resistance. These include lignification, with maximal levels reached at the same concentration of manganese as maximal biomass production (144); the concentration of soluble phenols, where manganese deficiency leads to a decrease in the their concentration (144); inhibition of aminopeptidase, which supplies essential amino acids for fungal growth, under manganese-deficient conditions (145); inhibition of pectin methylesterase, which is a fungal enzyme for degrading host cell walls, under manganese-deficient conditions (146); inhibition of photosynthesis leading to a decrease in root exudates and thus becoming more susceptible to invasion by root pathogens (142), though this mechanism has been shown not to be important in controlling *Ggt* by the lack of effect of foliar-applied manganese (137,147). A plant capable of mobilizing high concentrations of Mn^{2+} that are toxic to pathogens but not to plants in the rhizosphere may directly inhibit pathogenic attack (141).

12.8 CONCLUSION

This review has focused predominantly on the function of manganese in plants and its concentrations for maintaining optimal growth; the vast literature on diagnostics is heavily drawn on in Table 12.1. Developments in the last 10 years in manganese physiology and diagnostics have largely been refinements on the previous work rather than new radical developments. This may change with the emerging of new molecular technologies in the area of plant mineral nutrition.

ACKNOWLEDGMENTS

The authors thank Dr. Paul Lonergan for assistance in reviewing this chapter, and Margie Palotta for the photographs.

REFERENCES

1. R.J. Gilkes, R.M. McKenzie. Geochemistry and mineralology of manganese in soils. In: R.D. Graham, R.J. Hannam, N.C. Uren, eds. *Manganese in Soils and Plants.* Dordrecht: Kluwer Academic Publishers, 1988, pp. 23–35.

2. W. Shi, M. Chino, R.A. Youssef, S. Mori, S. Takagi. The occurrence of mugineic acid in the rhizosphere soil of barley plant. *Soil Sci. Plant Nutr.* 34:585–592, 1989.

3. S.M. Bromfield, R.W. Cumming, D.J. David, C.H. Williams. Change in soil pH, manganese and aluminium under subterranean clover pasture. *Aust. J. Exp. Agric. Anim. Husb.* 23:181–191, 1983.

4. J.T. Sims. Soil pH effects on the distribution and plant availability of manganese, copper and zinc. *Soil Sci. Soc. Amer. J.* 50:367–373, 1986.

5. W.L. Lindsay. Inorganic phase equilibria of micronutrients in soils. In: J.J. Mortvedt, P.M. Giordano, W.L. Lindsay, eds. *Micronutrients in Agriculture.* Madison, WI: Soil Science Society of America, 1972, pp. 41–57.

6. S.M. Bromfield. The effect of manganese-oxidising bacteria and pH on the availability of manganous ions and manganese oxides to oats in nutrient solutions. *Plant Soil* 49:23–39, 1978.

7. S.M. Bromfield, D.J. David. Sorption and oxidation of manganese ions and reduction of manganese oxide by cell suspension of a manganese-oxidising bacterium. *Soil Biol. Biochem.* 8:37–43, 1976.

8. N.C. Uren, G.W. Leeper. Microbial oxidation of divalent Mn. *Soil Biol. Biochem.* 10:85–87, 1978.

9. H.G. Dion, P.J.G. Mann. Three-valent manganese in soils. *J. Agric. Sci.* 36:239–245, 1946.

10. G.W. Leeper. *Six Trace Elements in Soils: Their Chemistry as Micronutrients.* Victoria, Australia: Melbourne University Press, 1970, pp. 20–22.

11. H.M. Reisenauer. Determination of plant-available soil manganese. In: R.D. Graham, R.J. Hannam, N.C. Uren, eds. *Manganese in Soil and Plants.* Dordrecht: Kluwer Academic Publishers, 1988, pp. 87–100.

12. P.S. Horstmar. Sur la nutrition de l'avoine, particulierement en ce qui concerne les matieres inorganiques qui sont necessaires a cette nutrition. *Ann. Chim. Phys.* 32:461–510, 1851.

13. K. Aso. The physiological influence of manganese compounds on plants. *Coll. Agric. Imp. Univ.Tokyo* 5:177–185, 1902.

14. O. Loew, S. Sawa. The action of manganese compounds on plants. *Bull. Coll. Agric. Imp. Univ. Tokyo* 5:161–172, 1902.

15. W.E. Brenchley. *Inorganic Plant Poisons and Stimulants.* Cambridge: The University Press, 1927, pp. 1–134.

16. G. Bertrand. Sur l'internention du manganese dans les oxydations provoquees par lalaccase. *Compt. Rend. Acad. Sci. Paris* 124:1032–1355, 1897.

17. G. Bertrand, M. Javillier. Action of manganese and the development of *Aspergillus niger. Bull. Soc. Chim. France* 4(IX):212–221, 1912.

18. G. Bertrand, M. Rosenblatt. Sur la présence générale du manganèse dans le règne végétal. *Compt. Rend.* 173:333–336, 1921.

19. P. Maze. Oxydation de l'ammoniaque ou nitrification par les vegetaux. *Compt. Rend. Soc. Biol. Paris* 77:98–102, 1914.

20. J.S. McHargue. The occurrence and significance of manganese in the seed coat of various seeds. *J. Amer. Chem. Soc.* 36:2532–2536, 1914.

21. J.S. McHargue. The role of manganese in plants. *J. Amer. Chem. Soc.* 44:1592–1598, 1922.

22. J.N. Burnell. The biochemistry of manganese in plants. In: R.D. Graham, R.J. Hannam, N.C. Uren, eds. *Manganese in Soils and Plants*: Dordrecht: Kluwer Academic Publishers, 1988, pp. 125–137.

23. R.D. Graham. Effect of nutrient stress on susceptibility of plants to disease with particular reference to the trace elements. *Adv. Bot. Res.* 10:221–276, 1983.

24. J. Amesz. The role of manganese in photosynthetic oxygen evolution. *Biochim. Biophys. Acta* 726:1–12, 1993.

25. G.T. Babcock. The photosynthetic oxygen-evolving process. In: J. Amesz, ed. *New Comparative Biochemistry: Photosynthesis.* Amsterdam: Elsevier Science Publishers, 1987, pp. 125–158.

26. R.C. Prince. Manganese at the active site of the chloroplast oxygen-evolving complex. *Trend Biochem. Sci.* 132:491–492, 1986.

27. P.E. Kriedemann, R.D. Graham, J.T. Wiskich. Photosynthetic dysfunction and *in vivo* changes in chlorophyll *a* fluorescence from manganese-deficient wheat leaves. *Aust. J. Agric. Res.* 36:157–169, 1985.

28. C.W. Asling, L.S. Hurley. The influence of trace elements on the skeleton. *Clin. Orthop.* 27:213, 1963.

29. R.M. Leach. Metabolism and function of manganese. In: A.S. Prasad, D. Oberleas, eds. *Trace Elements in Human Health.* New York: Academic Press, 1976, pp. 235–247.

30. L.S. Hurley. Manganese and other trace elements. In: B.A. Bowman, R.M. Russell, eds. *Nutrition Reviews' Present Knowledge in Nutrition.* Washington, DC: Nutrition Foundation, 1976, pp. 345–355.

31. L. Erway, L.S. Hurley, A. Fraser. Congenital ataxia and otolith defects due to manganese deficiency in mice. *J. Nutr.* 100:643, 1970.

32. P.S. Papavasiliou, S.T. Miller, G.C. Cotzias. Functional interaction between biogenic amines 3' 5' cAMP and manganese. *Nature* 220:74, 1968.

33. G.C. Cotzias, P.S. Papavasiliou, J. Ginas, A. Steck, S. Ruby. Metabolic modification of Parkinson's disease and of chronic manganese poisoning. *Ann. Rev. Med.* 22:305, 1971.

34. G.C. Cotzias, S.T. Miller, P.S. Papavasiliou, L.C. Tang. Interactions between manganese and brain dopamine. *Med. Clin. North Amer.* 60:729, 1976.

35. National Research Council (US). *Manganese.* Washington DC: National Academy of Sciences, Division of Medical Sciences Committee on Biological Effects of Atmospheric Pollutants, 1973, pp. 1–191.

36. R.M. Leach. Role of manganese in mucopolysaccharide metabolism. *Fed. Proc. Fed. Amer. Soc. Exp. Biol.* 30:991, 1971.

37. I. Fridovich. Superoxide dismutase. *Ann. Rev. Biochem.* 44:147–159, 1975.

38. J.M. McCord. Iron- and manganese-containing superoxide dismutase: structure, distribution and evolutionary relationships. *Adv. Exp. Med. Biol.* 74:540, 1976.

39. M.F. Utter. The biochemistry of manganese. *Med. Clin. North Amer.* 60:713, 1976.

40. H.A. Schroeder, J.J. Balassa, I.H. Tipton. Essential trace elements in man: manganese, a study in homeostasis. *J. Chronic Dis.* 19:545–571, 1966.

41. E.R. Page, E.K. Schofield-Palmer, A.J. McGregor. Studies in soil and plant manganese. I. Manganese in soil and its uptake by oats. *Plant Soil* 16:238–246, 1962.

42. H.R. Geering, J.R. Hodgson, C. Sdano. Micronutrient cation complexes in soil solution. IV. The chemical state of manganese in soil solution. *Soil Sci. Soc. Amer. Proc.* 33:81–85, 1969.

43. L.V. Kochian. Mechanism of micronutrient uptake and translocation in plants. In: J.J. Mortvedt, F.R. Cox, L.M. Shuman, R.M. Welch, eds. *Micronutrients in Agriculture.* Madison, WI: Soil Science Society America, 1991, pp. 229–296.

44. D.A. Barber, R.B. Lee. The effect of microorganisms on the absorption of manganese by plants. *New Phytol.* 73:97–106, 1974.

45. M.J. Webb, W.A. Norvell, R.M. Welch, R.D. Graham. Using a chelate-buffered nutrient solution to establish the critical solution activity of Mn^{2+} required by barley (*Hordeum vulgare* L.). *Plant Soil* 153:195–205, 1993.

46. E.R. Page, J. Dainty. Manganese uptake by excised oat roots. *J. Exp. Bot.* 15:428–443, 1964.

47. G.W. Garnham, G.A. Codd, G.M. Gadd. Kinetics of uptake and intercellular location of cobalt, manganese and zinc in the estuarine green alga *Chlorella salina. Appl. Microbiol. Biotechnol.* 37:270–276, 1972.

48. E.V. Maas, D.P. Moore. Manganese absorption by excised barley roots. *Plant Physiol.* 43:527–530, 1968.

49. E.V. Maas, D.P. Moore, B.J. Mason. Influence of calcium and magnesium on manganese absorption. *Plant Physiol.* 44:796–800, 1969.

50. S. Ratkovic, Z. Vucinic. The H^1 NMR relaxation method applied in studies of continual absorption of paramagnetic Mn^{2+} ions by roots of intact plants. *Plant Physiol. Biochem.* 28:617–622, 1990.

51. D.T. Clarkson. The uptake and translocation of manganese by plant roots. In: R.D. Graham, R.J. Hannam, N.C. Uren, eds. *Manganese in Soil and Plants.* Dordrecht: Kluwer Academic Publishers, 1988, pp. 101–112.

52. M. Tester. Plant ion channels: whole-cell and single-channel studies. *New Phytol.* 114:305–340, 1990.

53. S.D. Tyerman. Anion channels in plants. *Ann. Rev. Plant Physiol. Plant Mol. Biol.* 43:351–373, 1992.

54. J. Hill, A.D. Robson, J.F. Loneragan. The effect of copper supply on the senescence and retranslocation of nutrients of the oldest leaf of wheat. *Ann. Bot.* 44:279–287, 1979.

55. B.Radjagukguk. Manganese Nutrition in Lupins: Plant Response and the Relationship of Supply to Distribution. University of Western Australia, Perth. 1981.

56. R.O. Nable, J.F. Loneragan. Translocation of manganese in subterranean clover. I. Redistribution during vegetative growth. *Aust. J. Plant Physiol.* 11:101–111, 1984.

57. R.O. Nable, J.F. Loneragan. Translocation of manganese in subterranean clover. II. The effects of leaf senescence and of restricting supply of manganese to part of a split root system. *Aust. J. Plant Physiol.* 11:113–118, 1984.

58. F.K. El-Baz, P. Maier, A.H. Wissemeier, W.J. Horst. Uptake and distribution of manganese applied to leaves of *Vicia faba* (cv. Herzfreya) and *Zea mays* (cv. Regent) plants. *Zh. Pflansenernahr. Bodenk.* 153:279–282, 1970.

59. J.N. Pearson, Z. Rengel. Distribution and remobilisation of Zn and Mn during grain development in wheat. *J. Exp. Bot.* 45:1829–1835, 1994.

60. N.J. Grundon, A.D. Robson, M.J. Lambert, K. Snowball. Nutrient deficiency and toxicity symptoms. In: D.J. Reuter, J.B. Robinson, eds. *Plant Analysis: An Interpretation Manual*. Collingwood, Vic.: CSIRO Publishing, 1997, pp. 37–51.

61. S. Ramani, S. Kannan. Manganese absorption and transport in rice. *Physiol. Plant* 33:133–137, 1987.

62. J.F. Loneragan. Distribution and movement of manganese in plants. In: R.D.Graham, R.J. Hannam, N.C. Uren, eds. *Manganese in Soils and Plants*. Dordrecht: Kluwer Academic Publishers, 1988, pp. 113–121.

63. W.V. Single, I.F. Bird. The mobility of manganese in the wheat plant II. Redistribution in relation to foliar application. *Ann. Bot.* 22:479–488, 1958.

64. W.V. Single, I.F. Bird. The mobility of Mn in the wheat plant II. Redistribution in relation to manganese concentration and chemical state. *Ann. Bot.* 22:489–502, 1958.

65. R.J. Hannam, R.D. Graham, J.L. Riggs. Redistribution of manganese in maturing *Lupinus angustifolius* cv. Illyarrie in relation to levels of previous accumulation. *Ann. Bot.* 56:821–834, 1985.

66. J.N. Pearson, Z. Rengel. Uptake and distribution of ^{65}Zn and ^{54}Mn in wheat grown at sufficient and deficient leves of Zn and Mn. II during grain development. *J. Exp. Bot.* 46:841–845, 1995.

67. R.J. Hannam, K. Ohki. Detection of manganese deficiency and toxicity in plants. In R.D. Graham, R.J. Hannam, N.C. Uren, eds. *Manganese in Soils and Plants*. Dordrecht: Kluwer Academic Publishers, 1988, p. 244.

68. W.L. Lindsay. *Chemical Equilibria in Soils*. New York: Wiley, 1979, pp. 150–161.

69. M.A. Jauregui, H.M. Reisenauer. Calcium carbonate and manganese dioxide as regulators of available manganese and iron. *Soil Sci.* 134:105–110, 1982.

70. R.D. Graham, W.J. Davies, D.H.B. Sparrow, J.S. Ascher. Tolerance of barley and other cereals to mangenese-deficient soils of South Australia. In: M.R. Saric, B.C. Loughman, eds. *Genetic Aspects of Plant Nutrition*. The Hague: Marintus Nijhoff/Dr W Junk Publishers, 1983, pp. 339–345.

71. R.J. Hannam, W.J. Davies, R.D. Graham, J.L. Riggs. The effect of soil- and foliar-applied manganese in preventing the onset of manganese deficiency in *Lupinus angustifolius. Aust. J. Agric. Res.* 35:529–538, 1984.

72. H.D. Chapman, G.F.J. Liebig, E.R. Parker. Manganese studies on Californian soils and citrus leaf symptoms of deficiency. *Calif. Citrograph.* 24(12):427–454, 1939.

73. H.D. Chapman, G.F.J. Liebig, A.P. Vanselow. Some nutritional relationships as revealed by a study of mineral deficiency and excess symptoms on citrus. *Soil Sci. Soc. Amer. Proc.* 4:196–200, 1940.

74. T. Wallace. The Diagnosis of Mineral Deficiencies in Plants by Visual Symptoms: A Color Atlas and Guide. London: H.M. Stationery Office, 1961. p. 23.

75. C.K. Labanauskas. Manganese. In: H.D. Chapman, ed. *Diagnostic Criteria for Plants and Soils*. Riverside, CA: H.D. Chapman, 1966, pp. 264–285.

76. R.E. Lucas, B.D. Knezek. Climatic and soil conditions promoting micronutrients deficiencies in plants. In: J.J. Mortvedt, P.M. Giordano, W.L. Lindsay, eds. *Micronutrients in Agriculture*. Madison, WI: Soil Science Society of America, 1972, pp. 265–288.

77. P.H. Gallagher, T.Walsh. The susceptibility of cereal varieties to manganese deficiency. *J. Agri. Sci. Camb.* 33:197–203, 1943.

78. M. Nyborg. Sensitivity to manganese deficiency of different cultivars of wheat, oats and barley. *Can. Plant Sci.* 50:198–200, 1970.

79. O. Johansson, E. Ekman. Resultat ov de senasti arens svenska mikroelement-forsok II. Forsok med mangan. *St. Jordbr. Fosr. Medd.* 62:91–138, 1956.

80. D. Stenuit, R. Poit. Mangaangebrek en mangaanvergiftiging bij landbouwgewassen. *Agric. Heverlee.* 8:141–172, 1960.

81. R.D. Graham, W.J. Davies, J.S. Ascher. The critical concentration of manganese in field-grown wheat. *Aust. J. Agric. Res.* 36:145–155, 1985.

82. J.C. Brown, W.E. Jones. Differential response of oats to manganese stress. *Agron. J.* 66:624–626, 1974.

83. H.H. Glasscock, R.L. Wain. Distribution of manganese in the pea seed in relation to Marsh Spot. *J. Agric. Sci.* 30:132–140, 1940.

84. G.H. Walton. The effect of manganese on seed yield and the split seed disorder of sweet and bitter phenotypes of *Lupinus angustifolius* and *L. cosentinii. Aust. J. Agric. Res.* 29:1177–1189, 1978.

85. K. Ohki, D.O. Wilson, O.E. Anderson. Manganese deficiency and toxicity sensitivies of soybean cultivars. *Agron. J.* 72:713–716, 1980.

86. T. Batey. Manganese and boron deficiency. Minist. Agric. Fish Food, *Great Britain Tech. Bull.* 21:137–149, 1968.

87. A.D. Robson, K. Snowball. Nutrient deficiency and toxicity symptoms. In: D.J. Reuter and J.B. Robinson, eds. *Plant Analysis: An Interpretation Manual.* Melbourne: Inkata Press, 1986, pp. 13–19.

88. V. Romheld and H. Marschner. Function of micronutrients in plants. In: J.J. Mortvedt, F.R. Cox, L.M. Shuman, R.M. Welch, eds. *Micronutrients in Agriculture.* Madison, WI: Soil Science Society of America, 1991, pp. 297–328.

89. W.K. Gardner, D.G. Parberry, D.A. Barber. The acquisition of phosphorus by *Lupinus albus* L. I. Some characteristics of the soil/root interface. *Plant Soil* 69:19–32, 1982.

90. G.H. Godo, H.M. Reisenauer. Plant effects on soil manganese availability. *Soil Sci. Soc. Amer. J.* 44:993–995, 1980.

91. N.C. Uren. Chemical reduction of an insoluble higher oxide of manganese by plant roots. *J. Plant Nutr.* 4:65–71, 1981.

92. H. Marschner, V. Romheld. *In vivo* measurement of root-induced pH changes at the soil root interface: effect of plant species and nitrogen sources. *Zh. Pflanzenphysiol.* 111:241–251, 1983.

93. R.W. Smiley. Rhizosphere pH as influenced by plants, soils and nitrogen fertilizers. *Soil Sci. Soc. Amer. Proc.* 38:795–799, 1974.

94. C.D.J. Raper, S.A. Barber. Rooting systems of soybeans I. Differences in root morphology among varieties. *Agron. J.* 62:581–588, 1970.

95. R.D. Graham. Breeding for nutritional characteristics in cereals. *Adv. Plant Nutr.* 1:57–102, 1984.

96. R.D. Graham. Development of wheats with enhanced nutrient efficiency: progress and potential. *Wheat Production Constraints in Tropical Environments, Proceedings of the CIMMYT/INDP International Symposium* Chiang Mai, Thailand, January 19–23, 1988, pp. 305–320.

97. Reference deleted.

98. R.D. Graham. Genotypic differences in tolerance to manganese deficiency. In: In R.D. Graham, R.J. Hannam, N.C. Uren, eds. *Manganese in Soils and Plants.* Dordrecht: Kluwer Academic Publishers, 1988, pp. 261–276.

99. R.S. Beckwith, K.G. Tiller, E. Suwadji. The effects of flooding on the availability of trace metals to rice in soils of differing organic matter status. In: D.J.D. Nicholas, A.R. Egan, eds. *Trace Elements in Soil–Plant–Animal Systems.* New York: Academic Press, 1975, pp. 135–149.

100. F. Abruna, J. Vicente-Chandler, R.W. Pearson. Effects of liming on yields and composition of heavily fertilized grasses and on soil properties under humid tropical conditions. *Soil Sci. Soc. Amer. Proc.* 28:657–661, 1964.

101. P.H. Le Mare. Experiments on effects of phosphorus on the manganese nutrition of plants. I. Effects of monocalcium phosphate and its hydrolysis derivatives on ryegrass grown in two Buganda soils. *Plant Soil* 47:593–605, 1977.

102. C.E. Bortner. Toxicity of manganese to Turkish tobacco in acid Kentucky soils. *Soil Sci.* 39:15–33, 1935.

103. C.D. Foy, T.A. Campbell. Differential tolerances of *Amaranthus* strains to high levels of aluminium and manganese in acid soils. *J. Plant Nutr.* 7:1365–1388, 1984.

104. V.O. Grasmanis, G.W. Leeper. Toxic manganese in near-neutral soils. *Plant Soil* 25:41–48, 1966.

105. A. Siman, F.W. Cradock, P.J. Nicholls, H.C. Kirton. Effects of calcium carbonate and ammonium sulphate on manganese toxicity in an acid soil. *Aust. J. Agric. Res.* 22:201–214, 1971.

106. A. Siman, F.W. Cradock, A.W. Hudson. The development of manganese toxicity in pasture legumes under extreme climatic conditions. *Plant Soil* 41:129–140, 1974.

107. H.J. Snider. Manganese in some Illinois soils and crops. *Soil Sci.* 56:187–195, 1943.

108. G.D. Sherman. Manganese and soil fertility. In: US Department of Agriculture. *The Yearbook of Agriculture.* Washington, DC: The United States Government Printing Office, 1957, pp. 135–139.

109. C.D. Foy, A.L. Flemming, J.W. Schwartz. Opposite aluminium and manganese tolerances of two wheat varieties. *Agron. J.* 65:123–126, 1973.

110. L.E. Nelson. Tolerance of 20 rice cultivars to excess Al and Mn. *Agron. J.* 75:134–138, 1983.

111. C.D. Foy, B.J. Scott, J.A. Fisher. Genetic differences in plant tolerance to manganese toxicity. In: R.D. Graham, R.J. Hannam, N.C. Uren, eds. *Manganese in Soils and Plants.* Dordrecht: Kluwer Academic Publishers, 1988, p. 294.

112. D.E. Williams, J. Vlamis. The effect of silicon on yield and Mn^{54} uptake and distribution in the leaves of barley plants grown in culture solution. *Plant Physiol.* 32:404–409, 1957.

113. A.G. Fisher, G.W. Eaton, S.W. Porritt. Internal bark necrosis of Golden Delicious apple in relation to soil pH and leaf manganese. *Can. J. Plant Sci.* 57:297–299, 1977.

114. C.H. Williams, J. Vlamis, H. Hall, K.D. Gowans. Effects of urbanization, fertilization and manganese toxicity on mustard plants. *Calif. Agric.* 25(11):8–10, 1971.

115. D.C. Neal. Crinkle leaf, a new disease of cotton in Louisiana. *Phytopathology* 27:1171–1175, 1937.

116. C.D. Foy, R.L. Chanley, M.C. White. The physiology of metal toxicity in plants. *Ann. Rev. Plant Physiol.* 29:511–566, 1978.

117. K. Ohki. Manganese deficiency and toxicity levels for 'Bragg' soybeans. *Agron. J.* 68:861–864, 1976.

118. M.B. Parker, H.B. Harris, H.D. Morris, H.F. Perkins. Manganese toxicity of soybeans as related to soil and fertility treatments. *Agron. J.* 61:515–518, 1969.

119. E.H. Graven, O.J. Attoe, D. Smith. Effects of liming and flooding on manganese toxicity in alfalfa. *Soil Sci. Soc. Amer. Proc.* 29:702–706, 1965.

120. J.L. Dionne, A.R. Pesant. Effects of pH and water regime on the yield and Mn content of lucerne and birdsfoot trefoil grown in the glasshouse. *Can. J. Plant Sci.* 56:919–928, 1976.

121. I.J. Finn, S.H. Bourget, K.F. Neielson, B.K. Dow. Effects of different soil moisture tensions on grass and legume species. *Can. J. Plant Sci.* 41:16–23, 1961.

122. R.E. Mc Kenzie. Ability of forages to survive early spring flooding. *Sci. Agric.* 31:358–367, 1951.

123. A.D. Robson, J.F. Loneragan. Sensitivity of annual *Medicago* species to manganese toxicity as affected by calcium and pH. *Aust. J. Agric. Res.* 21:223–232, 1970.

124. Y. Doi. Studies on the oxidising power of roots of crop plants. 2. Interrelation between paddy rice and soybean. *Proc. Crop Sci. Soc. Jpn.* 12:14–15, 1952.

125. Y. Doi. Studies on the oxidising power of roots of crop plants. 1. The differences of crop plants and wild grasses. *Proc.Crop Sci. Soc. Jpn.* 21:12–13, 1952.

126. R. Benac. Effect of manganese concentration in the nutrient solution on groundnuts (*Arachis hypogaea* L.). *Oleagineau* 31:539–543, 1976.

127. R. Benac. Response of a sensitive (*Arachis hypogaea*) and a tolerant (*Zea mays*) species to different concentrations of manganese in the environment. *Can. ORSTOM Ser.* 11:43–51, 1976.

128. C.S. Andrew, M.P. Hegarty. Comparative responses to manganese excess of eight tropical and four temperate pasture legume species. *Aust. J. Agric. Res.* 20:687–696, 1969.

129. G.J. Ouellette, L. Dessureaux. Chemical composition of alfalfa as related to degree of tolerance to manganese and aluminium. *Can. J. Plant Sci.* 38:206–214, 1958.

130. J. Vlamis, D.E. Williams. Manganese and silicon interactions in the Gramineae. *Plant Soil* 28:31–40, 1967.

131. C.D. Foy. Manganese and plants. In: *Manganese.* Washington, DC: National Academies of Science, National Research Council, 1973, p. 191.

132. D.P. Heenan, O.G. Carter. Influence of temperature in the expression of manganese toxicities by two soybean varieties. *Plant Soil* 47:219–227, 1977.

133. D.P. Heenan, L.C. Campbell. The influence of temperature on the accumulation and distribution of manganese in two cultivars of soybean (*Glycine max* L. Merr.). *Aust. J. Agric. Res.* 41:835–843, 1990.

134. D.M. Huber. Disturbed mineral nutrition. In: J. Horsfall, E.B. Cowling, eds. *Plant Pathology — An Advanced Treatise.* New York: Academic Press, 1978, pp. 163–181.

135. D.M. Huber, N.S. Wilhelm. The role of manganese in resistance to plant diseases. In: R.D. Graham, R.J. Hannam, N.C. Uren, eds. *Manganese in Soils and Plants.* Dordrecht: Kluwer Academic Publishers, 1988, pp. 155–173.

136. N.S. Wilhelm, J.M. Fisher, R.D. Graham. The effect of manganese deficiency and cereal cyst nematode infection on the growth of barley. *Plant Soil* 85:23–32, 1985.

137. N.S. Wilhelm, R.D. Graham, A.D. Rovira. Application of different sources of manganese sulfate decreases take-all (*Greumannomyces graminis* var. *tritici*). *Aust. J. Agric. Res.* 39:1–10, 1988.

138. S.M. Bromfield. The oxidation of manganous ions under acid conditions by an acidophilous actino-mycete from acid soil. *Aust. J. Soil Res.* 16:91–100, 1978.

139. N.S. Wilhelm, R.D. Graham, A.D. Rovira. Control of Mn status and infection rate by genotypes of both host and pathogen in the wheat take-all interaction. *Plant Soil* 123:267–275, 1990.

140. H. Kunoh, H. Ishizaki, F. Kondo. Composition analysis of "halo" area of barley leaf epidermis incited by powdery mildew infection. *Ann. Phytopath. Soc. Jpn.* 41:33–39, 1975.

141. R.D. Graham, M.J. Webb. Micronutrients and disease resistance and tolerance in plants. In: J.J. Mortvedt, F.R. Cox, L.M. Shuman, R.M. Welch, eds. *Micronutrients in Agriculture.* Madison, WI: Soil Science Society of America, 1991, pp. 333–339.

142. R.D. Graham, A.D. Rovira. A role for manganese in the resistance of wheat plants to take-all. *Plant Soil* 78:441–444, 1984.

143. A.D. Rovira, R.D. Graham, J.S. Ascher. Reduction in infection of wheat roots by *Gaeumannomyces graminis* var. *tritici* with application of manganese to soil. In: C.A. Parker, ed. *Ecology and Management of Soil-Borne Plant Pathogens.* St. Paul, MN: American Phytopathology Society, 1985, pp. 212–214.

144. P.H. Brown, R.D. Graham, D.J.D. Nicholas. The effects of manganese and nitrate supply on the levels of phenolics and lignin in young wheat plants. *Plant Soil* 81:437–440, 1984.

145. D.M. Huber, R.R. Keeler. Alteration of wheat peptidase activity after infection with powdery mildew. *Proc. Amer. Phytopathol. Soc.* 4: 163, *69th Annual Meeting of the American Phytopathological Society,* East Lansing, MI, August 14–18, 1977.

146. T.S. Sadasivan. Effect of mineral nutrients on soil microorganisms and plant disease. In: C.A. Baker, W.C. Snyder, eds. *Ecology of Soil-Borne Pathogens. Proceedings of the International Symposium,* Berkeley: University of California Press, 1965, pp. 460–470.

147. E.M. Reis, R.J. Cook, B.L. McNeal. Effect of mineral nutrition on take-all of wheat. *Phytopathology* 72:224–229, 1982.

148. D.J. Reuter, D.G. Edwards, N.S. Wilhelm. Temperate and tropical crops. In: D.J. Reuter, J.B. Robinson, eds. *Plant Analysis: An Interpretation Manual.* Collingwood, Vic.: CSIRO Publishing, 1997, pp. 83–284.

149. R.L. Bansal, P.N. Takkar, V.K. Nayyar. Critical levels of Mn in coarse textured rice soils in India for predicting response of barley to Mn application. *Fert. Res.* 11:61–67, 1987.

150. W.E. Bergmann. *Nutritional Disorders of Plants: Development, Visual and Analytical Diagnosis.* Jena: Gustav Fischer, 1992, p. 26.

151. U.C. Gupta. Effect of manganese and lime on yield and on the concentrations of manganese, molybdenum, boron, copper and iron in the boot stage tissue of barley. *Soil Sci.* 114:131–136, 1972.

152. W.C. Dahnke. Barley. In: D.L. Pluncknett, H.B. Sprague, eds. *Detecting Mineral Nutrient Deficiencies in Tropical and Temperate Crops.* Boulder, CO: Westview Press, 1989, pp. 81–90.

153. R.G. Weir, G.C. Cresswell. *Plant Nutrient Disorders 4. Pastures and Field Crops.* Melbourne: Inkata Press, 1994, pp. 1–126.

154. R.J. Hannam, J.L. Riggs, R.D. Graham. The critical concentration of manganese in barley. *J. Plant Nutr.* 10:2039–2048, 1987.

155. R.W. Bell, D.G. Edwards, C.J. Asher. Growth and nodulation of tropical food legumes in dilute solution culture. *Plant Soil* 122:249–258, 1990.

156. S. Haneklaus, E. Schnug. Evaluation of the nutritional status of oilseed rape plants by leaf analysis. *Eighth International Rapeseed Congress,* Saskatoon, Canada, 1991, pp. 536–541.

157. D.G. Edwards, C.J. Asher. Tolerance of crop and pasture species to manganese toxicity. *Ninth International Colloquium on Plant Analysis and Fertility Problems,* Warwick, UK, 1982, pp. 145–150.

158. R.H. Howeler, D.G. Edwards, C.J. Asher. Micronutrient deficiencies and toxicities of cassava plants grown in nutrient solutions. I. Critical concentrations. *J. Plant Nutr.* 5:1059–1076, 1982.

159. R.H. Howeler. The mineral nutrition and fertilization of cassava. In: *Cassava Production Course.* Cali, Colombia: CIAT, 1978, pp 247–292.

160. R.H. Howeler. Potassium nutrition of cassava. In: R.D. Munson, ed. *Potassium in Agriculture.* Madison, WI: American Society of Agronomy, 1985, pp. 819–841.

161. W. Zorn, A. Paraube. Manganese content of cereals, maize and beet as indicator of soil acidity. *Zh Pflanzen Bodenk* 156:371–376, 1993.

162. N.E. Marcar, R.D. Graham. Tolerance of wheat, barley, triticale and rye to manganese deficiency during seedling growth. *Aust. J. Agric. Res.* 38:501–511, 1987.

163. J.B. Jones, Jr., B. Wolf, H.A. MIlls. *Plant Analysis Handbook: A Practical, Sampling, Preparation, Analysis and Interpretation Guide*. Athens, GA: Micro-Macro Press, 1991, pp. 1–213.

164. J.C. Brown, W.E. Jones. Fitting plants nutritionally to soils. II. Cotton. *Agron. J.* 69:405–409, 1977.

165. J.B. Jones, Jr. Plant analysis handbook for Georgia. *Univ. Georgia Coll. Agric. Bull.* 735, 1974.

166. K. Ohki. Manganese nutrition of cotton under two boron levels. *Agron. J.* 66:572–575, 1974.

167. K. Ohki, A. Ulrich. Manganese and zinc appraisal of selected crops by plant analysis. *Commun. Soil Sci. Plant Anal.* 8:297–312, 1977.

168. C.D. Foy, H.W. Webb, J.E. Jones. Adaptation of cotton genotypes to an acid, manganese toxic soil. *Agron. J.* 73:107–111, 1981.

169. C.D. Foy, R.R. Weil, C.A. Coradetti. Differential manganese tolerances of cotton genotypes in nutrient solution. *J. Plant Nutr.* 18:685–706, 1995.

170. C.W. Kennedy, J.E. Jones. Evaluating quantitative screening methods for manganese toxicity in cotton genotypes. *J. Plant Nutr.* 14:1331–1339, 1991.

171. B.T. Kang, R.L. Fox. A methodology for evaluating the manganese tolerance of cowpea (*Vigna unguiculata*) and some preliminary results of field trials. *Field Crops Res.* 3:199–210, 1980.

172. A.H. Wissemeier, W.J. Horst. Manganese oxidation capacity of homogenates of cowpea (*Vigna unguiculata* L. Walp.) leaves differing in manganese tolerance. *J. Plant Physiol.* 136:103–109, 1990.

173. K. Snowball, A.D. Robson. *Symptoms of Nutrient Deficiencies and Toxicities: Faba Beans and Field Peas*. Perth: University of Western Australia, School of Agriculture, Soil Science, and Plant Nutrition, 1991.

174. C.J. Asher, M.T. Lee. *Diagnosis and Correction of Nutritional Disorders in Ginger (Zingiber officinale)*. Brisbane: University of Queensland, Department of Agriculture, 1975.

175. R.L. Bansal, V.K. Nayyar. Critical level of Mn in Ustochrepts for predicting response of green gram (*Phaseolus surerus* L.) to manganese application. *Fert. Res.* 21:7–11, 1989.

176. P.J. Hocking, P.J. Randall, A. Pinkerton. Mineral nutrition of linseed and fibre flax. *Adv. Agron.* 41:221–296, 1987.

177. K. Snowball, A.D. Robson. *Symptoms of Nutrient Deficiencies: Lupins*. Perth: University of Western Australia, School of Agriculture, Soil Science, and Plant Nutrition, 1986.

178. M. Perry, J.W. Gartrell. Lupin "split-seed". A disorder of seed production in sweet, narrow-leafed lupins. *J. Agric. West Aust.* 17(4):20–25, 1976.

179. R.J. Hannam, R.D. Graham, J.L. Riggs. Diagnosis and prognosis of manganese deficiency in *Lupinus angustifolius* L. *Aust. Agric. Res.* 36:765–777, 1985.

180. R.B. Lockman. Relationship between corn yields and nutrient concentration in seedling whole-plant samples. *Agronomy Abstracts*. Madison, WI: American Society of Agronomy, 1969, p. 97.

181. E. Uribe, D.C. Martens, D.E. Brann. Response of corn (*Zea mays* L.) to manganese application on Atlantic costal plain soil. *Plant Soil* 112:83–88, 1988.

182. H.J. Mascagni, Jr., F.R. Cox. Diagnosis and correction of manganese deficiency in corn. *Comm. Soil Sci. Plant Anal.* 15:1323–1333, 1985.

183. J.B. Jones, Jr. Interpretation of plant analysis for several agronomic crops. In: G.W. Hardy, ed. *Soil Testing and Plant Analysis: Part 2*. Madison, WI: Soil Science Society of America, 1967, pp. 49–85.

184. R. Hall, H.F. Schwartz. Common bean. In: W.F. Bennett, ed. *Nutrient Deficiencies and Toxicities in Crop Plants*. St. Paul, MN: APS Press, 1993, pp. 91–98.

185. K.E. Giller, F. Amijee, S.J. Brodrick, S.P. McGrath, C. Mushi, O.T. Edje, J.B. Smithson. Toxic concentrations of iron and manganese in leaves of *Phaseolus vulgaris* L. growing on freely drained soils of pH 6.5 in northern Tanzania. *Commun. Soil Sci. Plant Anal.* 23:1663–1669, 1992.

186. R.E. Karamanos, G.A. Kruger, J.W.B. Stewart. Copper deficiency in cereal and oilseed crops in northern Canadian prairie soils. *Agron. J.* 78:317–323, 1986.

187. J.K. Hammes, K.C. Berger. Manganese deficiency in oats and correlation of plant manganese with various soil tests. *Soil Sci.* 90:239–244, 1958.

188. R.C. Ward, D.A. Whitney, D.G. Westfall. Plant analysis as an aid in fertilizing small grains. In: L.M. Walsh, J.D. Beaton, eds. *Soil Testing and Plant Analysis*. Madison, WI: Soil Science Society of America, 1973, pp. 329–348.

189. C.S. Piper. The symptoms and diagnosis of minor element deficiencies in agricultural and horticultural crops. I. Diagnostic methods, boron, manganese. *Emp. J. Exp. Agric.* 8:85–96, 1940.

190. J.W. Martens, R.I.H. McKenzie, V.M. Bendelow. Manganese levels in oats in western Canada. *Can J. Plant Sci.* 57:383–387, 1977.

191. C.I. Rich. Manganese content of peanut leaves as related to soil factors. *Soil Sci.* 82:353–363, 1956.

192. G.J. Gascho, J.G. Davis. Mineral nutrition. In: J. Smart, ed. *The Groundnut Crop. A Scientific Basis for Improvement.* London: Chapman & Hall, 1994, pp. 214–254.

193. M.B. Parker, M.E. Walker. Soil pH and manganese effects on manganese nutrition of peanut. *Agron. J.* 78:614–620, 1986.

194. A.W. Bekker, N.V. Hue, L.G.G. Yapa, R.G. Chase. Peanut growth as affected by liming, Ca–Mn interactions, and Cu plus Zn applications to oxidic Samoan soils. *Plant Soil* 164:203–211, 1994.

195. L.E. Nelson. The effect of temperature regime and substrate manganese on manganese concentration in rice. *J. Plant Nutr.* 5:1241–1257, 1982.

196. S. Yoshida. *Fundamentals of Rice Crop Science.* Los Banos, Laguna, Philippines: International Rice Research Institute, 1981.

197. A. Tanaka, S.A. Navasero. Manganese content of the rice plant under water culture conditions. *Soil Sci. Plant Nutr.* 12:21–26, 1966.

198. D.C. Lewis, J.D. McFarlane. Effect of foliar-applied manganese on the growth of safflower (*Carthamus tinctorius* L.) and the diagnosis of manganese deficiency by plant tissue and seed analysis. *Aust. J. Agric. Res.* 37:567–572, 1986.

199. R.B. Clark, P.A. Pier, D. Knudson, J.W. Maranville. Effect of trace element deficiencies and excesses on mineral nutrients in sorghum. *J. Plant Nutr.* 3:357–374, 1981.

200. S. Kuo, D.S. Mikkelsen. Effect of P and Mn on growth response and uptake of Fe, Mn and P by sorghum. *Plant Soil* 62:15–22, 1981.

201. R.B. Lockman. Mineral composition of grain sorghum plant samples. III. Suggested nutrient sufficiency limits at various stages of growth. *Commun. Soil Sci. Plant Anal.* 3:295–303, 1972.

202. P.N. Takkar, I.M. Chhibba, S.K. Mehta. *Twenty Years of Coordinated Research on Micronutrients in Soils and Plants: 1967–1987.* Bhopal, India: IISS.

203. W.R. Fehr, C.E. Caviness, D.T. Burmood, J.S. Pennington. Stage of development descriptions for soybeans, *Glycine max* L. Merrill. *Crop Sci.* 11:929–931, 1971.

204. J.T. Moraghan, T.P. Freeman, D. Whited. Influence of FeEDHA and soil temperature on the growth of two soybean varieties. *Plant Soil* 95:57–67, 1986.

205. R.L. Mikkelsen. Using hydrophilic polymers to improve uptake of manganese fertilizers by soybean. *Fert. Res.* 41:87–92, 1995.

206. H.J. Mascagni, Jr., F.R. Cox. Effective rates of fertilization for correcting manganese deficiency in soybeans. *Agron. J.* 77:363–366, 1985.

207. H.G. Small, A.J. Ohlrogge. Plant analysis as an aid in fertilizing soybeans and peanuts. In: L.M. Walsh, J.D. Beaton, eds. *Soil Testing and Plant Analysis.* Madison, WI: Soil Science Society of America, 1973, pp. 315–327.

208. S.W. Melsted, H.L. Motto, T.R. Peck. Critical plant nutrient composition values useful in interpreting plant analysis data. *Agron. J.* 61:17–20, 1969.

209. E.O. Leidi, M. Gomez, M.D. de la Guardia. Soybean genetic differences in response to Fe and Mn activity of metalloenzymes. *Plant Soil* 99:139–146, 1987.

210. S. Wu. Effect of manganese excess on the soybean plant cultivated under various growth conditions. *J. Plant Nutr.* 17:991–1003, 1994.

211. E.O. Leidi, M. Gomez, M.D. de la Guardia. Evaluation of catalase and peroxidase activity as indicators of Fe and Mn nutrition for soybean. *J. Plant Nutr.* 9:1239–1249, 1986.

212. S.W. Gettier, D.C. Martens, D.L. Hallock, M.J. Stewart. Residual Mn and associated soybean yield response from $MnSO_4$ application in a sandy loam soil. *Plant Soil.* 81:101–110, 1984.

213. K. Ohki. Manganese critical levels for soybean growth and physiological processes. *J. Plant Nutr.* 3(1–4):271–284, 1981.

214. R.F. Farley, A.P. Draycott. Manganese deficiency of sugar beet in organic soils. *Plant Soil.* 38(2):235–244, 1973.

215. A. Ulrich, J.T. Moraghan, E.D. Whitney. Sugar beet. In: W.F. Bennett, ed. *Nutrient Deficiencies and Toxicities in Crop Plants.* St. Paul, MN: APS Press, 1993, pp 91–98.

216. A. Ulrich, F.J. Hills. Sugarbeets. In: D.L. Plucknett, H.B. Sprauge, eds. *Detecting Mineral Nutrient Deficiencies in Tropical and Temperate Crops.* Boulder, CO: Westview Press, 1989, pp 225–240.

217. A.M.O. Elwali, G.J. Gascho. Soil testing, foliar analysis and DRIS as guides for sugarcane fertilization. *Agron. J.* 76:466–470, 1984.

218. G.J. Gascho, A.M.O. Elwali. Tissue testing of Florida sugarcane. *Sug. J.* 42:15–16, 1979.

219. D. Bassereau. Sugar cane. In: P. Martin-Prevel, J. Gagnard, P. Gautier, eds. *Guide to the Nutrient Requirements of Temperate and Tropical Crops.* New York, USA: Lavoisier Publishing Inc., 1987, pp. 513–525.

220. H. Evans. Elements other than nitrogen, potassium and phosphorus in the mineral nutrition of sugar-cane. *Proc. Int. Soc. Sugar Cane Technol.* 10:473–507, 1959.

221. A.A. Schneiter, J.F. Miller. Description of sunflower growth stages. *Crop Sci.* 21:901–903, 1981.

222. F.P.C. Blamey, R.K. Zollinger, A.A. Schneiter. Sunflower. In: *Sunflower Production and Culture.* Madison, WI: American Society of Agronomy, 1997.

223. R.G. Weir. Tissue Analysis for Pastures and Field Crops. Advisory note no. 11/83, NSW Department of Agriculture, 1983.

224. F.N. Fahmy. Soil and leaf analysis in relation to tree crop nutrition in Papua, New Guinea. In: Abstract in proceedings of conference on *Classification and Management of Tropical Soils.* Kuala Lumpur, 1977, pp. 309–318.

225. R.B. Lockman. Mineral composition of tobacco leaf samples. I. As affected by soil fertility, variety and leaf position. *Commun. Soil Sci. Plant Anal.* 1:95–108, 1970.

226. J. Wang, M.Y. Nielsen, B.P. Evangelou. A solution culture study of manganese-tolerant and sensitive tobacco genotypes. *J. Plant Nutr.* 17:1079–1093, 1994.

227. T.W. Rufty, G.S. Miner, C.D.J. Raper. Temperature effects on growth and manganese tolerance in tobacco. *Agron. J.* 71:638–644, 1979.

228. G.K. Evanylo, J.L. Sims, J.H. Grove. Nutrient norms for cured burly tobacco. *Agron. J.* 80:610–614, 1988.

229. A.D. Johnson, R.W. Knowlton. The effect of manganese on tobacco leaf quality and on the inorganic cation levels of tobacco leaves. *Aust. J. Exp. Agric. Anim. Husb.* 10:118–123, 1970.

230. L.M. Mugwira, M. Floyd, S.U. Pate. Tolerances of triticale lines to manganese in soil and nutrient solution. *Agron. J.* 73:319–322, 1981.

231. U.C. Gupta. Manganese nutrition of cereals and forages grown in Prince Edward Island. *Can. J. Soil Sci.* 66:59–65, 1986.

232. C. Ming, Y. Chungren. Effect of manganese and zinc fertilizer on nutrient balance and deficiency diagnosis of winter wheat crops in pot experiment. *International Symposium on the Role of Sulphur, Magnesium, and Micronutrients in Balanced Plant Nutrition*, China, 1992, pp. 369–378.

233. A.A. Csizinszky. Influence of total soluble salt concentration on growth and elemental concentration of winged bean seedlings, *Psophocarpus tetragonolobus* L. DC. *Commun. Soil Sci. Plant Anal.* 17:1009–1018, 1986.

234. E.C. Large. Growth stages in cereals. *Plant Path.* 3:128, 1954.

13 Molybdenum

Russell L. Hamlin
Coggins Farms and Produce, Lake Park, Georgia

CONTENTS

13.1 Historical Information 375
    13.1.1 Determination of Essentiality 375
    13.1.2 Function in Plants 376
        13.1.2.1 Nitrogenase 376
        13.1.2.2 Nitrate Reductase 377
        13.1.2.3 Xanthine Dehydrogenase 377
        13.1.2.4 Aldehyde Oxidase 378
        13.1.2.5 Sulfite Oxidase 378
13.2 Diagnosis of Molybdenum Status of Plants 378
    13.2.1 Deficiency 378
    13.2.2 Excess 379
    13.2.3 Molybdenum Concentration and Distribution in Plants 379
    13.2.4 Analytical Techniques for the Determination of Molybdenum in Plants 382
13.3 Assessment of Molybdenum Status of Soils 382
    13.3.1 Soil Molybdenum Content 382
    13.3.2 Forms of Molybdenum in Soils 384
    13.3.3 Interactions with Phosphorus and Sulfur 385
    13.3.4 Soil Analysis 386
        13.3.4.1 Determination of Total Molybdenum in Soil 386
        13.3.4.2 Determination of Available Molybdenum in Soil 386
13.4 Molybdenum Fertilizers 387
    13.4.1 Methods of Application 387
        13.4.1.1 Soil Applications 387
        13.4.1.2 Foliar Fertilization 388
        13.4.1.3 Seed Treatment 388
    13.4.2 Crop Response to Applied Molybdenum 388
References 389

13.1 HISTORICAL INFORMATION

13.1.1 DETERMINATION OF ESSENTIALITY

Molybdenum was discovered in 1778 by the Swedish chemist, Carl Wilhelm Scheele. However, its importance in biological systems was not established until 1930 when Bortels discovered that molybdenum was essential for the growth of *Azotobacter* bacteria in a nutrient medium (1). Subsequently in 1936, Steinberg determined that molybdenum was required for the growth of the fungus *Aspergillus niger* (2).

The essential nature of molybdenum for higher plants was first reported by Arnon and Stout in 1939 (3). In earlier experiments, Arnon observed that minute amounts of molybdenum improved the growth of plants in solution culture (4), and that a group of seven heavy metals, including molybdenum, increased the growth of lettuce (*Lactuca sativa* L.) and asparagus (*Asparagus officinalis* L.) (5). Prior to these studies (conducted in 1937 and 1938, respectively) only boron, copper, iron, manganese, and zinc were considered to be micronutrients. The observation that plant growth was improved by elements other than these led Arnon to believe that the list of essential elements was incomplete, and prompted him to test whether or not molybdenum was essential for the growth of higher plants (3).

In their studies, Arnon and Stout tested the molybdenum requirement of tomato (*Lycopersicon esculentum* Mill.) by their newly established criteria for essentiality (6). These criteria were (a) a deficiency of the essential element prevents plants from completing their life cycles; (b) the requirement is specific to the element, the deficiency of which cannot be prevented by any other element; and (c) the element is involved directly in the nutrition of plants. Plants grown in purified solution cultures developed deficiency symptoms in the absence of molybdenum, and symptoms were prevented by adding the equivalent of 0.01 mg Mo L^{-1} to the root medium (6). Normal growth was restored to deficient plants if molybdenum was applied to the foliage, thereby establishing that molybdenum exerted its effect directly on growth and not indirectly by affecting the root environment.

13.1.2　FUNCTION IN PLANTS

The transition element molybdenum is essential for most organisms and occurs in more than 60 enzymes catalyzing diverse oxidation–reduction reactions (7,8). Although the element is capable of existing in oxidation states from 0 to VI, only the higher oxidation states of IV, V, and VI are important in biological systems. The functions of molybdenum in plants and other organisms are related to the valence changes that it undergoes as a metallic component of enzymes (9).

With the exception of bacterial nitrogenase, molybdenum-containing enzymes in almost all organisms share a similar molybdopterin compound at their catalytic sites (7,8). This pterin is a molybdenum cofactor (Moco) that is responsible for the correct anchoring and positioning of the molybdenum center within the enzyme so that molybdenum can interact with other components of the electron-transport chain in which the enzyme participates (7). Molybdenum itself is thought to be biologically inactive until complexed with the cofactor, Moco.

Several molybdoenzymes including nitrogenase, nitrate reductase, xanthine dehydrogenase, aldehyde oxidase, and possibly sulfite oxidase are of significance to plants. Because of its involvement in the processes of N_2 fixation, nitrate reduction, and the transport of nitrogen compounds in plants, molybdenum plays a crucial role in nitrogen metabolism of plants (10).

13.1.2.1　Nitrogenase

The observation of Bortels (1) that molybdenum was necessary for the growth of *Azotobacter* was the first indication that molybdenum played a role in biological processes. It is now well established that molybdenum is required for biological N_2 fixation, an activity that is facilitated by the molybdenum-containing enzyme nitrogenase. Several types of asymbiotic bacteria, such as *Azotobacter*, *Rhodospirillum*, and *Klebsiella*, are able to fix atmospheric N_2, but of particular importance to agriculture is the symbiotic relationship between *Rhizobium* and leguminous crops (10). Nitrogenases from different organisms are similar in nature, and they catalyze the reduction of molecular nitrogen (N_2) to ammonia (NH_3) in the following reaction (11):

$$N_2+8H^++8e^-+16ATP \rightarrow 2NH_3+H_2+16ADP+16Pi$$

One of the great wonders in nature is how the process of N_2 fixation takes place biologically at normal temperatures and atmospheric pressure (12), when in the Haber–Bosch process, the same reaction performed chemically requires temperatures of 300 to 500°C and pressures of >300 atm (13).

According to Mishra et al. (11), nearly all nitrogenases contain the same two proteins, both of which are inactivated irreversibly in the presence of oxygen: an Mo–Fe protein (MW 200,000) and an Fe protein (MW 50,000 to 65,000). The Mo–Fe protein contains two atoms of molybdenum and has oxidation–reduction centers of two distinct types: two iron–molybdenum cofactors called FeMoco and four Fe-S (4Fe-4S) centers. The Fe–Mo cofactor (FeMoco) of nitrogenase constitutes the active site of the molybdenum-containing nitrogenase protein in N_2-fixing organisms (14).

The effect of biological N_2 fixation on the global nitrogen cycle is substantial, with terrestrial nitrogen inputs in the range of 139 to 170×10^6 tons of nitrogen per year (15). Despite the importance of molybdenum to N_2-fixing organisms and the nitrogen cycle, the essential nature of molybdenum for plants is not based on its role in N_2 fixation. The primary breach of the Arnon and Stout criteria of essentiality (6) is that many plants lack the ability to fix atmospheric N_2 and therefore do not require molybdenum for the activity of nitrogenase. In addition, the process of N_2 fixation is not essential for the growth of legumes if sufficient levels of nitrogen fertilizers are supplied (11,16).

13.1.2.2 Nitrate Reductase

The essential nature of molybdenum as a plant nutrient is based solely on its role in the NO_3^- reduction process via nitrate reductase. This enzyme occurs in most plant species as well as in fungi and bacteria (12), and is the principal molybdenum protein of vegetative plant tissues (17). However, the requirement of molybdenum for nitrogenase activity in root nodules is greater than the requirement of molybdenum for the activity of nitrate reductase in the vegetative tissues (18). Because nitrate is the major form of soil nitrogen absorbed by plant roots (19), the role of molybdenum as a functional component of nitrate reductase is of greater importance in plant nutrition than its role in N_2 fixation.

Like other molybdenum enzymes in plants, nitrate reductase is a homodimeric protein. Each identical subunit can function independently in nitrate reduction (9), and each consists of three functional domains: the N-terminal domain associated with a molybdenum cofactor (Moco), the central heme domain (cytochrome b_{557}), and the C-terminal FAD domain (7,20). This enzyme occurs in the cytoplasm and catalyzes the reduction of nitrate to nitrite (NO_2^-) in plants (19):

$$NO_3^- + 2H^+ + 2e_2^- \rightarrow NO_2^- + 2H_2O$$

Nitrate and molybdenum are both required for the induction of nitrate reductase in plants, and the enzyme is either absent (21), or its activity is reduced (22), if either nutrient is deficient. In deficient plants, the induction of nitrate reductase activity by nitrate is a slow process, whereas the induction of enzyme activity by molybdenum is much faster (10). It has been demonstrated that the molybdenum requirement of plants is higher if they are supplied nitrate rather than ammonium (NH_4^+) nutrition (23)—an effect that can be almost completely accounted for by the molybdenum in nitrate reductase (12).

13.1.2.3 Xanthine Dehydrogenase

In addition to the enzymes nitrogenase and nitrate reductase, molybdenum is also a functional component of xanthine dehydrogenase, which is involved in ureide synthesis and purine catabolism in plants (8). This enzyme is a homodimeric protein of identical subunits, each of which contains one molecule of FAD, four Fe-S groups, and a molybdenum complex that cycles between its Mo(VI) and Mo(IV) oxidation states (9,13). Xanthine dehydrogenase catalyzes the catabolism of purines to uric acid (7):

$$purines \rightarrow xanthine \rightarrow uric\ acid$$

In some legumes, the transport of symbiotically fixed N_2 from root to shoot occurs in the form of ureides, allantoin, and allantoic acid, which are synthesized from uric acid (10). Although xanthine dehydrogenase is apparently not essential for plants (10), it can play a key role in nitrogen metabolism for certain legumes for which ureides are the most prevalent nitrogen compounds formed in root nodules (9). The poor growth of molybdenum-deficient legumes can be attributed in part to poor upward transport of nitrogen because of disturbed xanthine catabolism (10).

13.1.2.4 Aldehyde Oxidase

Aldehyde oxidases in animals have been well characterized, but only recently has this molybdoenzyme been purified from plant tissue and described (24). In plants, aldehyde oxidase is considered to be located in the cytoplasm where it catalyzes the final step in the biosynthesis of the phytohormones indoleacetic acid (IAA) and abscisic acid (ABA) (8). These hormones control diverse processes and plant responses such as stomatal aperture, germination, seed development, apical dominance, and the regulation of phototropic and gravitropic behavior (25,26). Molybdenum may therefore play an important role in plant development and adaptation to environmental stresses through its effect on the activity of aldehyde oxidase, although other minor pathways exist for the formation of IAA and ABA in plants (7).

13.1.2.5 Sulfite Oxidase

Molybdenum may play a role in sulfur metabolism in plants. In biological systems the oxidation of sulfite (SO_3^{2-}) to sulfate (SO_4^{2-}) is mediated by the molybdoenzyme, sulfite oxidase (10). Although this enzyme has been well studied in animals (27), the existence of sulfite oxidase in plants is not well established. Marschner (9) explains that the oxidation of sulfite can be brought about by other enzymes such as peroxidases and cytochrome oxidase, as well as a number of metals and superoxide radicals. It is therefore not clear whether a specific sulfite oxidase is involved in the oxidation of sulfite in higher plants (28) and, consequently, also whether molybdenum is essential in higher plants for sulfite oxidation.

13.2 DIAGNOSIS OF MOLYBDENUM STATUS OF PLANTS

13.2.1 Deficiency

The discovery of molybdenum as a plant nutrient led to the diagnosis of the deficiency in a number of crop plants, with the first report of molybdenum deficiency in the field being made by Anderson (29) for subterranean clover (*Trifolium subterraneum* L.). The critical deficiency concentration in most crop plants is quite low, normally between 0.1 and 1.0 mg Mo kg^{-1} in the dry tissue (12). Symptoms of molybdenum deficiency are common among plants grown on acid mineral soils that have low concentrations of available molybdenum, but plants may occasionally become deficient in peat soils due to the retention of molybdenum on humic acids (19,30). Plants also may be prone to molybdenum deficiency under low temperatures and high nitrogen fertility (31).

Because molybdenum is highly mobile in the xylem and the phloem (32), its deficiency symptoms often appear on the entire plant. This appearance is unlike many of the other essential micronutrients where deficiency symptoms are manifest primarily in younger portions of the plant. Molybdenum deficiency is peculiar in that it often manifests itself as nitrogen deficiency, particularly in legumes. These symptoms are related to the function of molybdenum in nitrogen metabolism, such as its role in N_2 fixation and nitrate reduction. However, plants suffering from extreme deficiency often exhibit symptoms that are unique to molybdenum.

Legumes often require more molybdenum than other plants, particularly if they are dependent on N_2 as a source of nitrogen (9). Molybdenum-deficient legumes commonly become chlorotic, have stunted growth, and have a restriction in the weight or quantity of root nodules (33,34). In dicotyledonous species, a drastic reduction in leaf size and irregularities in leaf blade formation (whiptail) are the most typical visible symptoms, caused by local necrosis in the tissue and insufficient differentiation of vascular bundles at an early stage of leaf development (35). Marginal and interveinal leaf necrosis is a symptom of extreme molybdenum deficiency, and symptoms are often associated with high nitrate concentrations in the leaf, indicating that nitrate reductase activity is impaired (12).

The whiptail disorder is observed often in molybdenum-deficient cauliflower (*Brassica oleracea* var. *botrytis* L.), one of the most sensitive cruciferous crops to low molybdenum nutrition (36). In addition, molybdenum-deficient beans (*Phaseolus vulgaris* L.) often develop scald, where the leaves are pale with interveinal and marginal chlorosis, followed by burning of the leaf margin (36,37). In molybdenum-deficient tomatoes, lower leaves appear mottled and eventually cup upward and develop marginal necrosis (3). Molybdenum deficiency also decreases tasseling and inhibits anthesis and pollen formation in corn (*Zea mays* L.) (38). The inhibition of pollen formation with molybdenum deficiency may explain the lack of fruit formation in molybdenum-deficient watermelon (*Citrullus vulgaris* Schrad.) (9,39).

13.2.2 EXCESS

Most plants are not particularly sensitive to excessive molybdenum in the nutrient medium, and the critical toxicity concentration of molybdenum in plants varies widely. For instance, molybdenum is toxic to barley (*Hordeum vulgare* L.) if leaf tissue levels exceed 135 mg Mo kg^{-1} (40), but crops such as cauliflower and onion (*Allium cepa* L.) are able to accumulate upwards of 600 mg Mo kg^{-1} without exhibiting symptoms of toxicity (41). However, tissue concentrations >500 mg Mo kg^{-1} can lead to a toxic response in many plants (42), which is characterized by malformation of the leaves, a golden-yellow discoloration of the shoot tissues (9), and inhibition of root and shoot growth (43). These symptoms may, in part, be the result of inhibition of iron metabolism by molybdenum in the plant (12).

Toxicity symptoms in plants under field conditions are very rare, whereas toxicity to animals feeding on forages high in this element is well known (44). A narrow span exists between nutritional deficiency for plants and toxicity to ruminants (45). Molybdenum concentrations >10 mg Mo kg^{-1} (dry mass) in forage crops can cause a nutritional disorder called molybdenosis in grazing ruminants (9). This disorder is a molybdenum-induced copper deficiency that occurs when the consumed molybdate (MoO_4^{2-}) reacts in the rumen with sulfur to form thiomolybdate complexes, which inhibit copper metabolism (46).

Agricultural practices that can be used to decrease ruminant susceptibility to molybdenosis include field applications of copper and sulfur. The strong depressive effects of SO_4^{2-} on MoO_4^{2-} uptake can lower the molybdenum concentration in plants to levels that are nontoxic (47). Increasing the copper content of forages through fertilization may also help to reduce molybdenum-induced copper deficiency in animals (46).

13.2.3 MOLYBDENUM CONCENTRATION AND DISTRIBUTION IN PLANTS

The requirement of plants for molybdenum is lower than any other mineral nutrient except nickel (Ni) (9). Plants differ in their ability to absorb molybdenum from the root medium (48), and the sufficiency range for molybdenum in plants varies widely (Table 13.1). Most plants contain sufficient levels of molybdenum—in the range of 0.2 to 2.0 mg Mo kg^{-1}—in their dry tissue, but the difference between the critical deficiency and toxicity levels can vary up to a factor of 10^4 (e.g., 0.1 to 1000 mg Mo kg^{-1} dry mass) (9).

The source of nitrogen supplied to plants influences their requirement for molybdenum. Nitrate-fed plants generally have a high requirement for molybdenum (66), but there are conflicting reports as to whether plants supplied with reduced nitrogen have a molybdenum requirement. Cauliflower developed symptoms of molybdenum deficiency when grown with ammonium salts, urea, glutamate, or nitrate, in the absence of molybdenum (20). However, Hewitt (67) suggested that the molybdenum requirement, in the presence of reduced nitrogen, may result from the effects of traces of nitrate derived from bacterial nitrification. When cauliflower plants were supplied ammonium sulfate and no

TABLE 13.1
Deficient and Sufficient Concentrations of Molybdenum in Plants

Crop or Plant Type	Plant Part Sampled	Mo Concentration (mg kg⁻¹ dry mass)		Reference
		Deficient	Sufficient	
Agronomic Crops				
Alfalfa (*Medicago sativa* L.)	Upper portion of tops; prior to blossom	<0.4	0.5–5.0	49, 50
Barley (*Hordeum vulgare* L.)	Whole tops; boot stage		0.09–0.18	51
Canola (*Brassica napus* L.)	Mature leaves without petioles		0.25–0.60	52
Corn (*Zea mays* L.)	Stems	<0.12	1.4–7.0	53
	Ear leaves; silk stage	<1.1		54
Cotton (*Gossypium hirsutum* L.)	Fully mature leaves; after bloom		0.6–2.0	55
Oats (*Avena sativa* L.)	Whole tops		0.2–0.3	52
Peanuts (*Arachis hypogaea* L.)	Upper fully developed leaves	<1	0.5–1.0	55, 56
Red clover (*Trifolium pratense* L.)	Total aboveground plants; bloom	<0.15	0.3–1.59	50
	Whole plants; bud stage		0.46–1.08	41, 57
Rice (*Oryza sativa* L.)	Upper fully developed leaves; prior to flowering		0.4–1.0	55
Soybeans [*Glycine max* (L.) Merr.]	Whole plants	<0.2		58
	Upper fully developed leaves; end of blossom		0.5–1.0	55
Sugar beet (*Beta vulgaris* L. ssp. *vulgaris*)	Leaf blades	<0.16	0.2–20.0	59
	Fully developed leaf without stem	<0.15	0.2–20.0	50, 59
Sunflower (*Helianthus annuus* L.)	Mature leaves from new growth		0.25–0.75	52
Tobacco (*Nicotiana tabacum* L.)	Mature leaves from new growth		0.1–0.6	52
Wheat (*Triticum aestivum* L.)	Whole tops; boot stage		0.09–0.18	51
Vegetable Crops				
Beans (*Phaseolus vulgaris* L.)	Youngest fully expanded leaf; flowering	<0.2	0.2–5.0	36
Beets (*Beta vulgaris* L.)	Tops; 8 weeks old	<0.06		60
	Young mature leaves		0.15–0.6	36
Broccoli (*Brassica oleracea* L. convar. *botrytis*)	Tops; 8 weeks old	<0.05		60
	Mature leaves from new growth		0.30–0.50	52
Cabbage (*Brassica oleracea* L. var. *capitata*)	Wrapper leaves	<0.3	0.3–3.0	36, 52
Carrots (*Daucus carota* L.)	Mature leaves from new growth		0.5–1.5	52
Cauliflower (*Brassica oleracea* convar. *botrytis* var. *botrytis*)	Young leaves showing whiptail	0.07		58
	Aboveground portion of plants; appearance of curd	<0.26	0.68–1.49	61
Cucumber (*Cucumis sativus* L.)	Youngest fully mature leaves	<0.2	0.2–2.0	36
Lettuce (*Lactuca sativa* L.)	Leaves	<0.07	0.08–0.14	41, 62
Onion (*Allium cepa* L.)	Whole tops; maturity	<0.06	>0.1	63
Pea (*Pisum sativum* L.)	Recent fully developed leaves; onset of blossom		0.4–1.0	55
Potato (*Solanum tuberosum* L.)	Leaf blades	<0.16		64
	Fully developed leaves; early bloom		0.2–0.5	55

TABLE 13.1 (*Continued*)

Crop or Plant Type	Plant Part Sampled	Mo Concentration (mg kg^{-1} dry mass)		Reference
		Deficient	Sufficient	
Fruit Crops				
Apple (*Malus sylvestris* Mill.)	Mature leaves from new growth		0.10–2.00	52
Avocado (*Persea americana* Mill.)	Mature leaves from new flush		0.05–1.0	52
Orange (*Citrus sinensis* L.)	Mature leaves from nonfruiting		0.1–0.9	52
Pear (*Pyrus communis* L.)	Mid-shoot leaves from new growth		0.10–2.0	52
Peach (*Prunus persica* L. Batsch.)	Mid-shoot leaves		1.6–2.8	52
Strawberry (*Fragaria* x ananassa Duch.)	Mature leaves from new growth		0.25–0.50	52
Ornamental Plants				
New Guinea impatiens (*Impatiens* x hybrids)	Mature leaves from new growth		0.15–1.0	52
Poinsettia (*Euphorbia pulcherrima* Willd.)	Mature leaves from new growth	<0.5	0.12–0.5	52, 65
Rose, hybrid tea (*Rosa* x cultivars)	Upper leaflets from mature leaves		0.1–0.9	52
Salvia (*Salvia splendens*)	Mature leaves from new growth		0.2–1.08	52
Snapdragon (*Antirrhinum majus* L.)	Mature leaves from new growth		0.12–2.0	52
Verbena (*Verbena* x hybrids)	Mature leaves from new growth		0.14–0.8	52
Trees and Shrubs				
Common lilac (*Syringa vulgaris* L.)	Mature leaves from new growth		0.12–4.0	52
Douglass fir (*Pseudotsuga menziesii*)	Terminal cuttings		0.02–0.25	52
Loblolly pine (*Pinus taeda* L.)	Needles from terminal cuttings		0.12–0.56	52

Source: Adapted from U.C. Gupta, in *Molybdenum in Agriculture*, Cambridge University Press, New York, 1997, pp. 150–159. With permission from Cambridge University Press.

molybdenum under sterile conditions, Hewitt and Gundry (68) found that plants showed no abnormalities and apparently had no molybdenum requirement. On transfer to nonsterile conditions, whiptail symptoms appeared as a characteristic symptom of molybdenum deficiency. Hewitt (17) later stated that molybdenum is of very little importance for some plants if nitrate reduction is not necessary for nitrogen assimilation, but that it is impossible to say that an element is not required by plants given the limits of current analytical techniques.

Molybdenum is absorbed by plant roots in the form of the molybdate ion (MoO_4^{2-}), and its uptake is considered to be controlled metabolically (19). In long-distance transport in plants, molybdenum is readily mobile in the xylem and phloem (32). The form in which molybdenum is translocated is unknown, but its chemical properties indicate that it is most likely transported as MoO_4^{2-} rather than in a complexed form (9). The proportion of various molybdenum constituents in plants naturally depends on the quantity of molybdenum absorbed and accumulated in the tissue. Molybdenum-containing enzymes, such as nitrogenase and nitrate reductase, constitute a major pool for absorbed molybdenum, but under conditions of luxury consumption, excess molybdenum can also be stored in the vacuoles of peripheral cell layers of the plant (69).

The allocation of molybdenum to the various plant organs varies considerably among plant species, but generally the concentration of molybdenum is highest in seeds (12) and in the nodules of N_2-fixing plants (9). However, when molybdenum is limiting, preferential accumulation in root nodules may lead to considerably lower molybdenum content in the shoots and seeds of nodulated legumes (70). Molybdenum concentrations in leaves have been found to exceed concentrations in the stems of several crop species such as tomato, alfalfa (*Medicago sativa* L.), and soybeans (*Glycine max* Merr.) (12).

13.2.4 ANALYTICAL TECHNIQUES FOR THE DETERMINATION OF MOLYBDENUM IN PLANTS

The molybdenum status of crops is often overlooked by the farming community, probably because of the relatively low crop requirement for molybdenum and because of a lack of education on the necessity of molybdenum in fertility programs. In addition, many commercial soil and plant analysis laboratories fail to report this nutrient in routine tissue and soil analyses. This omission may be partially due to the difficulties in accurately determining the small quantities of molybdenum that are normally present in plant tissues. It is possible that many molybdenum deficiencies in crop plants are misdiagnosed as nitrogen deficiency because of the similarity in their deficiency symptoms.

The two most common methods of molybdenum extraction from plant tissues are dry ashing (71) and wet digestion (72), both of which give similar results (12). Dry ashing is often the preferred method of extraction due to the potential hazards involved with the use of perchloric acid ($HClO_4$) for wet digestion (72). Several analytical techniques have been proposed for the determination of molybdenum in the resulting extracts including the dithiol and thiocyanate colorimetric methods, determination by atomic absorption spectrometry (AAS), graphite furnace atomic absorption spectrometry (GF-AAS), and by inductively coupled plasma atomic emission spectrometry (ICP-AES). As the detection of molybdenum by ICP-AES is less sensitive than for other elements, this method should be used only for plant tissues suspected of having molybdenum concentrations >1.0 mg Mo kg^{-1} (dry mass) (73,74). The dithiol colorimetric method and the AAS method are probably the most commonly used techniques for determining molybdenum in soil and plant materials (12).

The dithiol method developed by Piper and Beckworth (75) and modified by Gupta and MacKay (76) is more sensitive and precise than other colorimetric methods used for the determination of molybdenum in plant tissues. This method is based on precipitation and extraction of a green-colored molybdenum dithiol complex after removal of interfering ions from the test solution (77). The molybdenum concentration is determined by comparing the absorbance of the sample with known standards on a light spectrophotometer. The detection limit of the dithiol method is about 20 ng Mo mL^{-1}, and the recovery of molybdenum added to the plant material has been greater than 90% (12). Although this method is relatively inexpensive, the procedure may be too tedious and time-consuming for use in many commercial analytical laboratories. For procedures of the dithiol method, readers are referred to Gupta (73).

Trace quantities of molybdenum in plant material have been determined by flame (78) or flameless AAS (79). These procedures provide adequate sensitivity for molybdenum and are relatively rapid, but are subject to matrix interferences (77). The GF-AAS method (80) improves the accuracy and precision of determining low concentrations of molybdenum, and the procedure is applicable to a range of different plant matrices (73). The detection limits for the determination of molybdenum by AAS using flame and graphite furnace are reported to be 10 and 2 ng mL^{-1}, respectively (78), and the recovery of molybdenum by these two methods is similar to that of the dithiol colorimetric method, ranging from 92 to 95% (12). For details of the flame and graphite furnace AAS methods, the reader is referred to Khan et al. (78) and Gupta (73).

13.3 ASSESSMENT OF MOLYBDENUM STATUS OF SOILS

13.3.1 SOIL MOLYBDENUM CONTENT

The amount of naturally occurring molybdenum in soils depends on the molybdenum concentrations in the parent materials. Igneous rock makes up some 95% of the Earth crust (81) and contains ~2 mg Mo kg^{-1}. Similar amounts of molybdenum are present in sedimentary rock (82). The total molybdenum content of soils differs by soil type and sometimes by geographical region (Table 13.2). Soils normally contain between 0.013 and 17.0 mg kg^{-1} total molybdenum (44), but molybdenum concentrations can exceed 300 mg Mo kg^{-1} in soils derived from organic-rich shale (83). Large quantities of molybdenum also occur in soils receiving applications of municipal sewage sludge (84) or in soils that are polluted by mining activities (46). Most agricultural soils contain a relatively low amount

TABLE 13.2
Molybdenum Content of Surface Soils of Different Countries

Soil	Country	Range (mg kg^{-1} dry weight)
Podzols and sandy soils	Australia	2.6–3.7
	Canada	0.40–2.46
	New Zealand	1–2[a]
	Poland	0.2–3.0
	Yugoslavia	0.17–0.51[b]
	Russia	0.3–2.9
Loess and silty soils	New Zealand	2.2–3.1[a]
	China	0.4–1.1
	Poland	0.6–3.0
	United States	0.75–6.40
	Russia	1.8–3.3
Loamy and clayey soils	Great Britain	0.7–4.5
	Canada	0.93–4.74
	Mali Republic	0.5–0.75
	New Zealand	2.1–4.2[a]
	Poland	0.1–6.0
	United States	1.2–7.2
	United States[c]	1.5–17.8
	Russia	0.6–4.0
Fluvisols	India	0.4–3.1[b]
	Czech Republic	2.8–3.5
	Mali Republic	0.44–0.65
	Yugoslavia	0.35–0.53[b]
	Russia	1.8–3.0
Gleysols	Australia	2.5–3.5
	India	1.1–1.8[b]
	Ivory Coast	0.18–0.60
	Yugoslavia	0.52–0.74
	Russia	0.6–2.0
Histosols and other organic soils	Canada	0.69–3.2
	Russia	0.3–1.9
Forest soils	Bulgaria	0.3–4.6
	Former Soviet Union	0.2–8.3
Various soils	Great Britain	1–5
	India	0.013–2.5
	Italy	0.4–2.2
	Japan	0.2–11.3
	United States	0.8–3.3
	Russia	0.8–3.6

[a]Soils derived from basalts and andesites.
[b]Data for whole soil profiles.
[c]Soils from areas of the western states of Mo toxicity to grazing animals.
Source: From A. Kabata-Pendias, H. Pendias, *Trace Elements in Soils and Plants*. 3rd ed., CRC Press, Boca Raton, FL. 2001, pp. 260–267. Copyright CRC Press.

of molybdenum by comparison, with an average of 2.0 mg kg^{-1} total molybdenum and 0.2 mg kg^{-1} available molybdenum (19).

Soils derived from granite, organic-rich shale, or limestone, and those high in organic matter are usually rich in molybdenum (85,86), and the available molybdenum content generally increases with alkalinity or fineness of the soil texture (85). In contrast, molybdenum is often deficient in well-drained coarse-textured soils or in soils that are highly weathered or acidic (83,87). The accumulation of molybdenum varies with depth in the soil, but molybdenum is normally highest in the A horizons of well-drained soils and is highest in the subsoil of poorly drained mineral soils (83). In soils, molybdenum can occur in four fractions: (a) dissolved molybdenum in the soil solution, (b) molybdenum occluded with oxides, (c) molybdenum as a mineral constituent, and (d) molybdenum associated with organic matter (85).

13.3.2 Forms of Molybdenum in Soils

The speciation and availability of molybdenum in the soil solution is a function of pH. At water pH >5.0, molybdenum exists primarily as MoO_4^{2-} (84), but at lower pH levels the $HMoO_4^-$ and $H_2MoO_4^0$ forms dominate (44). For each unit increase in soil pH above pH 5.0, the soluble molybdenum concentration increases 100-fold (88). Plants preferentially absorb MoO_4^{2-} and therefore the molybdenum nutrition of plants can be manipulated by altering soil acidity. Soil liming is commonly used to alleviate molybdenum deficiencies in plants by increasing the quantity of plant-available molybdenum in the soil solution (89), but the effect of liming on molybdenum nutrition varies by soil and plant type (Table 13.3). Excessive lime use may decrease the solubility of molybdenum through the formation of $CaMoO_4$ (44), but Lindsay (90) suggests that this complex is too soluble to persist in soils. Using lime to change the acidity of a clay loam from pH 5 to 6.5 resulted in greater molybdenum accumulation in cauliflower, alfalfa (*Medicago sativa* L.), and bromegrass (*Bromus inermis* Leyss.), but molybdenum accumulation was relatively unaffected if plants were grown in a sandy loam (Table 13.3) (87). For plants grown in sandy loam, lime and molybdenum were both required to significantly increase the molybdenum content of the plant tissue.

TABLE 13.3
Effects of Soil pH on Molybdenum Concentration in a Few Crops Grown on Two Soils

	Mo concentration (mg kg^{-1})					
	Cauliflower		Alfalfa		Bromegrass	
Soil pH[a]	No Mo	Mo (2.5 mg kg^{-1})	No Mo	Mo (2.5 mg kg^{-1})	No Mo	Mo (2.5 mg kg^{-1})
Silty clay loam						
5.0	Trace	0.02	Trace	0.43	0.11	0.95
5.5	Trace	0.21	0.51	4.40	0.30	1.80
6.0	0.11	1.62	0.91	4.63	0.27	1.67
6.5	0.56	6.43	1.48	4.93	0.62	2.30
Culloden sandy loam						
5.0	Trace	0.39	Trace	0.11	0.02	0.35
5.5	Trace	1.34	Trace	2.04	0.02	1.09
6.0	Trace	3.15	Trace	2.01	0.04	3.59
6.5	Trace	3.58	Trace	3.32	0.05	3.77

[a]Soil:water ratio 1:2.

Source: From U.C. Gupta, in *Molybdenum in Agriculture*, Cambridge University Press, New York, 1997, pp. 71–91. Reprinted with permission from Cambridge University Press.

Significant amounts of molybdenum can be bound, or fixed, in soils by iron and aluminum oxides, particularly under acidic conditions (19). These sesquioxides have a pH-dependent surface charge that becomes more electrically positive as soil pH decreases, and more negative as soil pH increases. Changes in the surface charge are due to the protonation and deprotonation of surface functional groups (91). Under acidic soil conditions, the molybdate anion is adsorbed strongly to the surface of iron and aluminum oxides by a ligand exchange mechanism (92), and adsorption is greatest at pH 4 (83). In acid soils the molybdenum concentration in the soil solution can be reduced greatly, but because molybdenum is adsorbed weakly to soils and hydrous oxides at alkaline pH, these soils have a relatively large proportion of molybdenum in the solution phase (93). Compared with adsorption on hydrous iron oxides, the strength of molybdenum adsorption to aluminum oxide is much weaker (94). Despite this difference, aluminum oxides play an important role in the sorption of molybdenum in soils. For instance, the adsorption capacity of montmorillonite increases in the presence of interlayered aluminum hydroxide polymers (85).

Molybdenum also exists in soils as a constituent of various molybdenum-containing minerals. The primary source of molybdenum in soils is molybdenite (MoS_2), but other minerals also contribute to the molybdenum content of soils, such as powellite ($CaMoO_4$), wulfenite ($PbMoO_4$), and ferrimolybdite ($Fe_2(MoO_4)_3 \cdot 8H_2O$) (95). Of these minerals, only molybdenite and ferrimolybdite are mined commercially (83). In water-saturated soils, the availability of molybdenum is influenced by its reaction with other redox-active elements such as sulfur. Under strongly reducing conditions molybdenum forms sparingly soluble thiomolybdate complexes, with MoS_2 being the most important mineral controlling molybdenum solubility (44). Other minerals whose ions are also affected by oxidation–reduction state, such as $MnMoO_4$ or $FeMoO_4$, are too soluble to precipitate in soils (92). Soil pH greatly influences the availability of molybdenum from these mineral sources; even $PbMoO_4$, the least soluble of the possible soil compounds, becomes more soluble as pH increases (87).

Soil organic matter has been found to complex or fix molybdenum in soils, but the mechanisms of sorption are not well understood. Molybdenum binds strongly to humic and fulvic acids (92). Owing to the great affinity of molybdenum to be fixed by organic matter, its concentration in forest litter can reach 50 mg Mo kg^{-1} (44). The accumulation of molybdenum in organic matter can be particularly high if soil drainage is impeded (95). Organic-matter-rich soils can supply adequate amounts of molybdenum for plant growth due to a slow release of molybdenum from the organic complex (44). However, there are conflicting reports concerning the effect of soil organic matter on the availability of molybdenum in the soil solution. Plant-available molybdenum has been reported to be low in soils having high quantities of organic matter (96), particularly on peat soils due to the strong fixation of molybdenum by humic acid (44). In contrast, Srivastiva and Gupta (85) suggested that soil organic matter increases the available molybdenum content of acid soils by inhibiting the fixation of MoO_4^{2-} by sesquioxides.

13.3.3 INTERACTIONS WITH PHOSPHORUS AND SULFUR

The molybdenum nutrition of plants can be affected by the interaction of molybdenum with other nutrients in the soil such as phosphorus and sulfur. It is well established that plant uptake of molybdenum is enhanced by the presence of soluble phosphorus and decreased by the presence of available sulfur (87). In comparison to MoO_4^{2-}, phosphate has a greater affinity for sorption sites in soils, such as on sesquioxides (92). Phosphorus fertilization often liberates soil-bound molybdenum into the soil solution and increases molybdenum accumulation by plants (85,97). Phosphorus may also stimulate molybdenum absorption through the formation of a phosphomolybdate complex in soils, which may be readily absorbed by plants (98). The effect of sulfur on molybdenum absorption by plants appears to be related to the direct competition between SO_4^{2-} and MoO_4^{2-} during root absorption. Stout and Meagher (99) showed that the addition of SO_4^{2-} to the culture medium reduced absorption of radioactive molybdenum by tomatoes, and decreased molybdenum absorption by tomatoes (*Lycopersicon esculentum* Mill.) and peas (*Pisum sativum* L.) in soil (100).

13.3.4 SOIL ANALYSIS

The use of soil testing to predict the soil's capacity to supply molybdenum for plant growth can be difficult because of the relatively small amounts of molybdenum in soil, the differences in plant requirement for molybdenum, and because of the importance of seed molybdenum reserves in supplying crop needs (74). In addition, the total molybdenum content of soils can differ considerably from the plant-available molybdenum fraction (77). The total molybdenum content in soils usually ranges between 0.013 and 17.0 mg Mo kg^{-1} (44) and is dependent on the molybdenum content of the parent material (101). However, the quantity of molybdenum available for plant uptake can be substantially less and is dependent on soil pH and other chemical and biological factors. For pollution monitoring, a method for determining the total molybdenum in soils is necessary. If the objective is to quantify the available molybdenum for plant uptake, then a method for determination of the mobile or readily extractable molybdenum is required (77).

Several excellent reviews on the determination of molybdenum in soils are provided by Sims (84), Eivazi and Sims (77), and Sims and Eivazi (74). The reader is referred to these references for detailed explanations of methods and procedures described here.

13.3.4.1 Determination of Total Molybdenum in Soil

Several extraction methods have been developed for the determination of molybdenum in soils. The most common method of soil extraction is by perchloric acid digestion (102). Dry ashing followed by acid extraction of the ash has also been used (103). Purvis and Peterson (104) proposed the sodium carbonate fusion method for extraction of total molybdenum.

The thiocyanate–stannous chloride spectrophotometric procedure revised by Johnson and Arkley (105) and modified by Sims (84), is used extensively for the determination of total molybdenum in soils. Details of the procedure are provided by Sims (84). Molybdenum in the soil extract reacts with thiocyanate and excess iron in the presence of stannous chloride to form the colored complex $Fe(MoO(SCN)_5)$. The complex is extracted from the aqueous phase with isoamyl alcohol that has been dissolved in carbon tetrachloride (CCl_4). The amount of molybdenum present is determined on a light spectrophotometer by comparison of the absorbance of the sample with appropriate standards. Difficulties associated with the thiocyanate method include interference from iron and the use of stannous chloride, which can vary in purity and consistency (77).

Graphite furnace atomic absorption spectrometry has also been used for the analysis of extract having a low concentration of molybdenum (<1.0 mg kg^{-1}) (106,107). For extracts high in molybdenum, AAS or ICP-AES have been used, but Sims (84) indicates that owing to low detection limits, interferences from other elements, or the enhancement of molybdenum readings, the usefulness of these methods is limited.

13.3.4.2 Determination of Available Molybdenum in Soil

According to Gupta and Lipsett (12), the first report on the available molybdenum in soils was given by Grigg (103) wherein soils were extracted with acid oxalate buffered at pH 3. Other extractants have been used with varying degrees of success for the determination of available molybdenum in soils including ammonium oxalate, hot water, anion-exchange resin, and ammonium bicarbonate-diethylenetriamine-pentaacetic acid (AB-DTPA) (84). The most common method for the determination of molybdenum in soil extracts is the thiocyanate method as described previously.

Although the ammonium oxalate procedure is the method most commonly used to determine available molybdenum in soils, the findings have not been consistent (77). Grigg (108) decided that the method was unreliable for diagnosis of molybdenum deficiencies, because oxalate extracts a portion of iron-bound molybdenum that is unavailable to plants. Water extraction has been shown to be well correlated with available molybdenum in some studies (109), but has failed to give positive results in others (110). Difficulties are encountered with water extraction because the quantities extracted are very low (12). Sims (84) indicates that anion-exchange resins have been used with success to extract molybdenum, but that the method has not been tested widely.

According to Sims and Eivazi (74), the AB-DTPA method was developed for the simultaneous soil extraction of macronutrients and micronutrients such as phosphorus, potassium, iron, manganese, copper, and zinc, and the method has been extended to include molybdenum. Molybdenum extracted with AB-DTPA increases with increasing soil pH (84), and the method has been used most often for soils or sediments high in molybdenum, such as calcareous or polluted soils (111,112). Because the extractant can be used in conjunction with ICP-AES, it offers the added potential for measuring molybdenum during routine analysis of multiple nutrients (74).

13.4 MOLYBDENUM FERTILIZERS

Several molybdenum sources can be used to prevent or alleviate molybdenum deficiency in crop plants (Table 13.4). These sources vary considerably in their solubility and in molybdenum content, and their effectiveness often depends primarily on the method of application, plant requirements, and on various soil factors (87). The relative solubilities of some molybdenum fertilizers are as follows: sodium molybdate > ammonium molybdate > molybdic acid > molybdenum trioxide > molybdenum sulfide (114). Molybdenum frits can also be used to supply Mo, but because of their limited solubility, they must be ground finely to be effective (89). Because of the low plant requirement for molybdenum and its mobility in plant tissues, several methods of molybdenum application are possible including soil application, foliar fertilization, and seed treatment with various molybdenum sources.

13.4.1 METHODS OF APPLICATION

13.4.1.1 Soil Applications

Molybdenum fertilizers can be incorporated into the soil by banding or by broadcast applications. Soluble sources of molybdenum such as sodium molybdate and ammonium molybdate may be sprayed onto the soil surface before tilling to obtain a more uniform coverage, but this practice is seldom used (89). Because the molybdenum requirement of plants is low, the quantities of molybdenum fertilizers needed for crop growth are less than for most other nutrients. Rates of 50 to 100 g Mo ha^{-1} are generally required for soil treatments of agronomic crops, but as much as 400 g Mo ha^{-1} may be needed for vegetable crops such as cauliflower (12). The uniform application of such small quantities of molybdenum is often achieved by combining molybdenum with phosphorus fertilizers or in mixed, complete (N-P-K) fertilizers, to increase the volume of applied material (89).

TABLE 13.4
Chemical Formulas of Various Molybdenum Sources and Percentage of Molybdenum in Them

Mo Source	Chemical Formula	Mo Concentration (%)
Molybdenum trioxide	MoO_3	66
Molybdenum sulfide	MoS_2	60
Ammonium molybdate	$(NH_4)_6Mo_7O_{24} \cdot 4H_2O$	54
Molybdic acid	$H_2MoO_4 \cdot H_2O$	53
Sodium molybdate	$Na_2MoO_4 \cdot 2H_2O$	39
Molybdenum frits	Fritted glass	20–30

Source: Adapted from U.C. Gupta, J. Lipsett, *Adv. Agron.*, 34:73–115, 1981 and D.C. Martens, D.T. Westermann, in *Micronutrients in Agriculture*. SSSA, Madison, WI, 1991, pp. 549–582.

13.4.1.2 Foliar Fertilization

Sodium molybdate and ammonium molybdate are the most commonly used molybdenum sources for foliar fertilization because of their high solubility in water. Foliar applications of molybdenum are most effective if applied at early stages of plant development, and generally a 0.025 to 0.1% solution of sodium or ammonium molybdate ($\sim$200 g Mo ha^{-1}), is recommended (85). Wetting agents may also be required in the spray solution to ensure adequate coverage on the foliage of crops such as onion and cauliflower (12). Foliar applications of molybdenum are often more effective than soil applications, particularly for acid soils (9) or under dry conditions (115).

13.4.1.3 Seed Treatment

Seed pelleting, or coating, is the most common method for supplying molybdenum to crops (89) and is an effective means of preventing deficiency in crops grown on soils having low concentration of available molybdenum (9). This method ensures a more uniform application in the field, and the amounts of molybdenum that can be coated onto seeds are sufficient to provide adequate molybdenum for plant growth (89). Sparingly soluble sources of molybdenum, such as molybdenum trioxide, are most often used to treat seeds of leguminous crops because soluble molybdenum sources can decrease the effectiveness of applied bacteria inoculum (85). Recommended rates for seed treatment are 7 to 100 g Mo ha^{-1} (9,85), and higher rates ($>$117 g Mo ha^{-1}) have been found to cause toxic effects in plants such as cauliflower (116).

13.4.2 CROP RESPONSE TO APPLIED MOLYBDENUM

The effect of molybdenum fertilization on increasing plant yield is often related to an increased ability of the plant to utilize nitrogen. The activities of nitrogenase and nitrate reductase are affected by the molybdenum status of plants, and their activities are often suppressed in plants suffering from molybdenum deficiency (22,117). Foliar application of molybdenum at 40 g ha^{-1} at 25 days after plant emergence greatly enhanced nitrogenase and nitrate reductase activities of common bean (*Phaseolus vulgaris* L.), resulting in an increase in total nitrogen accumulation in shoots (117). In addition, foliar fertilization of common bean with 40 g Mo ha^{-1} increased nodule size, but not the quantity of root nodules (118). Therefore, the main effect of molybdenum on nodulation was suggested to be the avoidance of nodule senescence, thus maintaining a longer period of effective N$_2$ fixation.

The application of molybdenum to soils with low amounts of available molybdenum can improve crop yield dramatically, particularly for legumes, which have a high molybdenum requirement (12). Large-seeded legumes often do not require molybdenum fertilization if their seeds contain enough molybdenum to meet the requirements of the plant (119). But for plants suffering from molybdenum deficiency, the response to molybdenum fertility often varies. The lack of response to molybdenum can be related to other nutritional problems, such as the toxic effects of aluminum and manganese in acid soils, which mask the effects of molybdenum nutrition (116). In addition, molybdenum can be rendered unavailable to plants in acid soils if molybdenum is fixed by iron and manganese oxides (120). Crop plants also vary in their requirement for molybdenum (Table 13.1) and thus require different levels of molybdenum fertilization to achieve maximum growth.

Soybean yields in southeastern United States have been shown to increase by 30 to 80% following molybdenum fertilization on acid soils (33,121). Similar results have been obtained for peanut (*Arachis hypogaea* L.) grown on acid soils in western Africa (122). However, Rhoades and Nangju (123) found that at soil pH 4.5, soybeans did not respond to molybdenum. Differences in the response of legumes to molybdenum may be related to the timing of fertilizer applications. During the lag phase between infection and active N$_2$ fixation (between 10 and 21 days) (9), the addition of molybdenum fertilizers may be ineffective because the growth response to added molybdenum is related primarily to the molybdenum requirements of the N$_2$-fixing bacteria (18). In other studies where molybdenum was seed-applied, cowpea (*Vigna sinensis* Endl.) yields increased by 25% (123), and oat (*Avena sativa* L.) yields increased by 48% (124). Molybdenum fertilization has also been shown to increase the production of melons (*Cucumis melo* L.), with treated test plots yielding 254 melons compared to 19 in the untreated plots (39).

The efficiency of molybdenum fertilizers can be affected by soil pH. In acid soils, the availability of applied molybdenum can be limited due to the fixation of MoO_4^{2-} by iron and aluminum oxides, but the quantity of molybdenum in the soil solution increases with increasing soil pH (120). Liming materials can be used in conjunction with molybdenum fertilization to increase molybdenum uptake by plants, but the effect on plant growth is limited to soil pH levels < 7.0 (48). Liming alone may liberate enough soil-bound molybdenum to sustain plant growth (89). However the effect of lime depends on the total molybdenum content of soils. On acid soils where aluminum toxicity can limit plant growth, adding both lime and molybdenum is often more beneficial than adding only one of them (125). Combined applications of lime and molybdenum to forage crops can lead to problems for grazing animals because the accumulation of molybdenum in plant tissues can be high enough to cause molybdenosis (126).

Other soil amendments such as phosphorus- or sulfur-containing fertilizers, may also influence the efficiency of molybdenum fertilizers by affecting the fixation of molybdenum in soils or its uptake by plant roots. The use of phosphate ($H_2PO_4^-$), which has a high affinity for iron oxides, can lead to the release of adsorbed molybdenum and to an increase in the water-soluble MoO_4^{2-} concentration of the soil (8). As a result, phosphorus fertilization often increases the molybdenum absorption by roots and its accumulation in plant tissues (12,87). In contrast, sulfate and MoO_4^{2-} are strongly competitive during root absorption, and sulfur fertilization has been shown to decrease the uptake of molybdenum by plants (127). Studies with peanut have shown that providing phosphorus in the form of triple superphosphate is superior to single superphosphate for plants grown in molybdenum-deficient soils (128). This difference was attributed to the sulfur component of single superphosphate and its effect on inhibiting molybdenum uptake and suppressing plant growth.

REFERENCES

1. H. Bortels. Molybdän als Katalysator bei der biologischen Stickstoffbindung. *Arch. Mikrobiol.* 1:333–342, 1930.
2. R.A. Steinberg. Relation of accessory growth substances to heavy metals, including molybdenum, in the nutrition of *Aspergillus niger. J. Agric. Res.* 52:439–448, 1936.
3. D.I. Arnon, P.R. Stout. Molybdenum as an essential element for higher plants. *Plant Physiol.* 14:599–602, 1939.
4. D.I. Arnon. Ammonium and nitrate nutrition of barley at different seasons in relation to hydrogen-ion concentration, manganese, copper, and oxygen supply. *Soil Sci.* 44:91–114, 1937.
5. D.I. Arnon. Microelements in culture-solution experiments with higher plants. *Am. J. Bot.* 25:322–325, 1938.
6. D.I. Arnon, P.R. Stout. The essentiality of certain elements in minute quantity for plants with special reference to copper. *Plant Physiol.* 14:371–375, 1939.
7. R.R. Mendel, G. Schwarz. Molybdoenzymes and molybdenum cofactor in plants. *Crit. Rev. Plant Sci.* 18:33–69, 1999.
8. W. Zimmer, R. Mendel. Molybdenum metabolism in plants. *Plant Biol.* 1:160–168, 1999.
9. H. Marschner. *Mineral Nutrition of Higher Plants.* 2nd ed. New York: Academic Press, 1995, pp. 369–379.
10. P.C. Srivastava. Biochemical significance of molybdenum in crop plants. In: U.C. Gupta, ed. *Molybdenum in Agriculture.* New York: Cambridge University Press, 1997, pp. 47–70.
11. S.N. Mishra, P.K. Jaiwal, R.P. Singh, H.S. Srivastiva. Rhizobium-legume association. In: H.S. Srivastava, R.P. Singh, eds. *Nitrogen Nutrition and Plant Growth.* Enfield, NH: Science Publishers, Inc., 1999, pp. 45–102.
12. U.C. Gupta, J. Lipsett. Molybdenum in soils, plants, and animals. *Adv. Agron.* 34:73–115, 1981.

13. D. Voet, J.G. Voet. *Biochemistry*. 2nd ed. New York: Wiley, Inc., 1995, pp. 820–822.

14. R.M. Allen, J.T. Roll, P. Rangaraj, V.K. Shah, G.P. Roberts, P.W. Ludden. Incorporation of molybdenum into the iron-molybdenum cofactor of nitrogenase. *J. Biol. Chem.* 274:15869–15874, 1999.

15. M.B. Peoples, E.T. Craswell. Biological nitrogen fixation: investments, expectations and actual contributions to agriculture. *Plant Soil* 141:13–39, 1992.

16. T. George, J.K. Ladha, R.J. Buresh, D.P. Garrity. Managing native and legume-fixed nitrogen in lowland rice-based cropping systems. *Plant Soil* 141:69–91, 1992.

17. E.J. Hewitt. A perspective of mineral nutrition: essential and functional metals in plants. In: D.A. Robb, W.S. Pierpoint, eds. *Metals and Micronutrients: Uptake and Utilization by Plants*. New York: Academic Press, 1983, pp. 277–326.

18. M.B. Parker, H.B. Harris. Yield and leaf nitrogen of nodulating and non-nodulating soybean as affected by nitrogen and molybdenum. *Agron. J.* 69:551–554, 1977.

19. K. Mengel, E.A. Kirkby. *Principles of Plant Nutrition*. 4th ed. Bern, Switzerland: International Potash Institute, 1987, pp. 551–558.

20. B.A. Notton. Micronutrients and nitrate reductase. In: D.A. Robb, W.S. Pierpoint, eds. *Metals and Micronutrients: Uptake and Utilization by Plants*. New York: Academic Press, 1983, pp. 219–240.

21. Z.Z. Li, P.M. Gresshoff. Developmental and biochemical regulation of 'constitutive' nitrate reductase activity in leaves of nodulating soybean. *J. Exp. Bot.* 41:1231–1238, 1990.

22. P.J. Randall. Changes in nitrate and nitrate reductase levels on restoration of molybdenum to molybdenum-deficient plants. *Aus. J. Agric. Res.* 20:635–642, 1969.

23. P.M. Giordano, H.V. Koontz, J.W. Rubins. C^{14} distribution in photosynthate of tomato as influenced by substrate copper and molybdenum level and nitrogen source. *Plant Soil* 24:437–446, 1966.

24. T. Koshiba, E. Saito, N. Ono, N. Yamamoto, M. Sato. Purification and properties of flavin- and molybdenum-containing aldehyde oxidase from coleoptiles of maize. *Plant Physiol.* 110:781–789, 1996.

25. J. Normanly, J.P. Slovin, J.D. Cohen. Rethinking auxin biosynthesis and metabolism. *Plant Physiol.* 107:323–329, 1995.

26. S. Merlot, J. Giraudat. Genetic analysis of abscisic acid signal transduction. *Plant Physiol.* 114:751–757, 1997.

27. C. Kisker, H. Schindelin, A. Pacheco, W.A. Wehbi, R.M. Garrett, K.V. Rajagopalan, J.H. Enemark, D.C. Rees. Molecular basis of sulfite oxidase deficiency from the structure of sulfite oxidase. *Cell* 91:973–983, 1997.

28. L.J. DeKok. Sulfur metabolism in plants exposed to atmospheric sulfur. In: H. Rennenburg, C. Brunold, L.J. Dekok, I. Stulen, eds. *Sulfur Nutrition and Sulfur Assimilation in Higher Plants*. Hauge, Netherlands: SPB Academic Publishing, 1990, pp. 111–130.

29. A.J. Anderson. Molybdenum deficiency on a South Australian ironstone soil. *J. Aus. Inst. Agric. Sci.* 8:73–75, 1942.

30. A. Szalay, M. Szilagyi. Laboratory experiments on the retention of micronutrients by peat humic acids. *Plant Soil* 29:219–224, 1968.

31. Z.Y. Wang, Y.L. Tang, F.S. Zhang. Effect of molybdenum on growth and nitrate reductase activity of winter wheat seedlings as influenced by temperature and nitrogen treatments. *J. Plant Nutr.* 22:387–395, 1999.

32. S. Kannan, S. Ramani. Studies on molybdenum absorption and transport in bean and rice. *Plant Physiol.* 62:179–181, 1978.

33. G.R. Hagstrom, K.C. Berger. Molybdenum status of three Wisconsin soils and its effect on four legume crops. *Agronomy J.* 55:399–401, 1965.

34. E.G. Mulder. Importance of molybdenum in the nitrogen metabolism of microorganisms and higher plants. *Plant Soil* 1:94–119, 1948.

35. W. Bussler. Die Entwicklung der Mo-Mangelsymptome an Blumenkohl. *Z. Pflanzenernähr Bodenk* 125:36–50, 1970.

36. R.G. Weir, G.C. Creswell. *Plant Nutrient Disorders 3—Vegetable Crops*. Melbourne: Inkata Press, 1993, pp. 85–99.

37. R.D. Wilson. Molybdenum in relation to the scald disease of beans. *Aus. J. Sci.* 11:209–211, 1949.

38. V. Romheld, H. Marschner. Function of micronutrients in plants. In: J.J. Mortvedt, F.R. Cox, L.M. Shuman, R.M. Welch, eds. *Micronutrients in Agriculture*. Madison, WI: SSSA, 1991, pp. 297–328.

39. W.D. Gubler, R.G. Gorgan, P.P. Osterli. Yellows of melons caused by molybdenum deficiency in acid soil. *Plant Dis.* 66:449–451, 1982.

40. R.D. Davis, P.H.T. Beckett, E. Wollan. Critical levels of twenty potentially toxic elements in young spring barley. *Plant Soil* 49:395–408, 1978.

41. U.C. Gupta, E.W. Chipman, D.C. MacKay. Effects of molybdenum and lime on the yield and molybdenum concentration of crops grown on acid sphagnum peat soil. *Can. J. Plant Sci.* 58:983–992, 1978.

42. U.C. Gupta. Deficient, sufficient, and toxic concentrations of molybdenum in crops. In: U.C. Gupta, ed. *Molybdenum in Agriculture*. New York: Cambridge University Press, 1997, pp. 150–159.

43. S. Kevresan, N. Petrovic, M. Popovic, J. Kandrac. Nitrogen and protein metabolism in young pea plants as affected by different concentrations of nickel, cadmium, lead, and molybdenum. *J. Plant Nutr.* 24:1633–1644, 2001.

44. A. Kabata-Pendias, H. Pendias. *Trace Elements in Soils and Plants*. 3rd ed. New York: CRC Press, 2001, pp. 260–267.

45. C. Neunhäuserer, M. Berreck, H. Insam. Remediation of soils contaminated with molybdenum using soil amendments and phytoremediation. *Water, Air, Soil Pollut.* 128:85–96, 2001.

46. J.M. Stark, E.F. Redente. Copper fertilization to prevent molybdenosis on retorted oil shale disposal piles. *J. Environ. Qual.* 19:502–504, 1990.

47. N.S. Pasricha, V.K. Nayyar, N.S. Randhawa, M.K. Sinha. Influence of sulphur fertilization on suppression of molybdenum uptake by berseem (*Trifolium alexandrium*) and oats (*Avena sativa*) grown on a molybdenum-toxic soil. *Plant Soil* 46:245–250, 1977.

48. J.J. Mortvedt. Nitrogen and molybdenum uptake and dry matter relationships of soybeans and forage legumes in response to applied molybdenum on acid soil. *J. Plant Nutr.* 3:245–256, 1981.

49. J.B. Jones Jr. Interpretation of plant analysis for several agronomic crops. In: G.W. Hardy, ed. *Soil Testing and Plant Analysis*. Madison, WI: SSSA, 1967, pp. 49–58.

50. P. Neubert, W. Wrazidlo, H.P. Vielemeyer, I. Hundt, F. Gollmick, W. Bergmann. *Tabellen zur pflanzenanalyse - Erste orientierende "Übersicht,"*. Jena: Institut für Pflanzenernährung, 1970, pp. 1–40.

51. U.C. Gupta. Boron and molybdenum nutrition of wheat, barley, and oats grown in Prince Edward Island soils. *Can. J. Soil Sci.* 51:415–422, 1971.

52. H.A. Mills, J.B. Jones Jr. *Plant Analysis Handbook II*. Athens, GA: MicroMacro Publishing, 1996, pp. 185–414.

53. R.V. Dios, T.V.I. Broyer. Deficiency symptoms and essentiality of molybdenum in crop hybrids. *Agrochimica* 9:273–284, 1965.

54. R.D. Vos. Corn. In: W.F. Bennett, ed. *Nutrient Deficiencies and Toxicities in Plants*. St. Paul, MN: APS Press, 1993, pp. 11–14.

55. W. Bergmann. *Nutritional Disorders of Plants. Visual and Analytical Diagnosis*. New York: Gustav Fischer, 1992, pp. 1–741.

56. D.H. Smith, M.A. Wells, D.M. Porter, F.R. Cox. Peanuts. In: W.F. Bennett, ed. *Nutrient Deficiencies and Toxicities in Crop Plants*. St. Paul, MN: APS Press, 1993, pp. 105–110.

57. U.C. Gupta. Molybdenum requirement of crops grown on a sandy clay loam soil in the greenhouse. *Soil Sci.* 110:280–282, 1970.

58. N.K. Peterson, E.R. Purvis. Development of molybdenum deficiency symptoms in certain crop plants. *Soil Sci. Soc. Am. Proc.* 25:111–117, 1961.

59. A. Ulrich, F.J. Hills. Plant analysis as an aid in fertilizing sugar crops. Part I. Sugarbeets. In: L.M. Walsh, J.D. Beaton, eds. *Soil testing and Plant Analysis*, 2nd ed. Madison, WI: SSSA, 1973, pp. 271–278.

60. C.M. Johnson, G.A. Pearson, P.R. Stout. Molybdenum nutrition of crop plants. II. Plant and soil factors concerned with molybdenum deficiencies in crop plants. *Plant Soil* 4:178–196, 1952.

61. E.W. Chipman, D.C. MacKay, U.C. Gupta, H.B. Cannon. Response of cauliflower cultivars to molybdenum deficiency. *Can. J. Plant Sci.* 50:163–167, 1970.

62. W. Plant. The Control of Molybdenum Deficiency in Lettuce Under Field Conditions. Long Ashton Res. Sta. Ann. Rep. 1951, pp. 113–115.

63. U.C. Gupta, P.V. LeBlanc. Effect of molybdenum application on plant molybdenum concentration and crop yields on sphagnum peat soils. *Can. J. Plant Sci.* 70:717–721, 1990.

64. A. Ulrich, J.T. Monaghan, E.D. Whitney. Sugarbeets. In: W.F. Bennett, ed. *Nutrient Deficiencies and Toxicities in Crop Plants*. St. Paul, MN: APS Press, 1993, pp. 91–98.

65. D.A. Cox. Foliar-applied molybdenum for preventing or correcting molybdenum deficiency of poinsettia. *Hortscience* 8:894–895, 1992.

66. S.C. Agarwala, E.J. Hewitt. Molybdenum as a plant nutrient. III. The interrelationships of molybdenum and nitrate supply in the growth and molybdenum content of cauliflower plants grown in sand culture. *J. Hort. Sci.* 29:278–290, 1954.

67. E.J. Hewitt. Essential nutrient elements for plants. In: F.C. Stewart, ed. *Plant Physiology. Vol. III. Inorganic Nutrition of Plants.* New York: Academic Press, 1963, pp. 137–360.

68. E.J. Hewitt, C.S. Gundry. The molybdenum requirement of plants in relation to nitrogen supply. *J. Hort. Sci.* 45:351–358, 1970.

69. K.L. Hale, S.P. McGrath, E. Lombi, S.M. Stack, N. Terry, I.J. Pickering, G.N. George, E.A.H. Pilon-Smits. Molybdenum sequestration in brassica species. A role for anthocyanins? *Plant Physiol.* 126:1391–1402, 2001.

70. J. Ishizuka. Characterization of molybdenum absorption and translocation in soybean plants. *Soil Sci. Plant Nutr.* 28:63–78, 1982.

71. R.O. Miller. High-temperature oxidation: dry ashing. In: Y.P. Kalra, ed. *Handbook of Reference Methods for Plant Analysis.* Boca Raton, FL: CRC Press, 1998, pp. 53–56.

72. R.O. Miller. Nitric-perchloric acid wet digestion in an open vessel. In: Y.P. Kalra, ed. *Handbook of Reference Methods for Plant Analysis.* Boca Raton, FL: CRC Press, 1998, pp. 57–62.

73. U.C. Gupta. Determination of boron, molybdenum, and selenium in plant tissue. In: Y.P. Kalra, ed. *Handbook of Reference Methods for Plant Analysis.* Boca Raton, FL: CRC Press, 1998, pp. 171–182.

74. J.L. Sims, F. Eivazi. Testing for molybdenum availability in soils. In: U.C. Gupta, ed. *Molybdenum in Agriculture.* New York: Cambridge University Press, 1997, pp. 111–130.

75. C.S. Piper, R.S. Beckworth. A new method for determination of molybdenum in plants. *J. Soc. Chem. Ind.* 67:374–378, 1948.

76. U.C. Gupta, D.C. MacKay. Determination of molybdenum in plant materials using 4-methyl-1,2-dimercaptobenzene (dithol). *Soil Sci.* 99:414–415, 1965.

77. F. Eivazi, J.L. Sims. Analytical techniques for molybdenum determination in plants and soils. In: U.C. Gupta, ed. *Molybdenum in Agriculture.* New York: Cambridge University Press, 1997, pp. 92–110.

78. S.U. Khan, R.O. Cloutier, M. Hidiroglou. Atomic absorption spectroscopic determination of molybdenum in plant tissue and blood plasma. *J. Assoc. Off. Anal. Chem.* 62:1062–1064, 1979.

79. W.M. Jarrell, M.D. Dawson. Sorption and availability of molybdenum in soils of Western Oregon. *Soil Sci. Soc. Am. J.* 42:412–415, 1978.

80. P.R. Curtis, J. Grusovin. Determination of molybdenum in plant tissue by graphite furnace atomic absorption spectrophotometry (GFAAS). *Commun. Soil Sci. Plant Anal.* 16:1279–1291, 1985.

81. R.L. Mitchell, Trace elements in soils. In: F.E. Bear, ed. *Chemistry of the Soil.* 2nd ed. New York: Reinhold Publishing Corp., 1964, pp. 320–368.

82. K. Norrish. The geochemistry and mineralogy of trace elements. In: D.J.D. Nicholas, A.R. Egan, eds. *Trace Elements in Soil-Plant-Animal Systems.* New York: Academic Press, Inc., 1975, pp. 55–81.

83. K.J. Reddy, L.C. Munn, L. Wang. Chemistry and mineralogy of molybdenum in soils. In: U.C. Gupta, ed. *Molybdenum in Agriculture.* New York: Cambridge University Press, 1997, pp. 4–22.

84. J.L. Sims. Molybdenum and cobalt. In: J.M. Bartels, J.M. Bingham, eds. *Methods of Soil Analysis. Part 3. Chemical Methods.* Madison, WI: SSSA, ASA, 1996, pp. 723–737.

85. P.C. Srivastiva, U.C. Gupta. *Trace Elements in Crop Production.* Labanon, NH: Science Publishers, Inc., 1996, pp. 101–104.

86. R.M. Welch, W.H. Allaway, W.A. House, J. Kubota. Geographic distribution of trace element problems. In: J.J. Mortvedt, F.R. Cox, L.M. Shuman, R.M. Welch, eds. *Micronutrients in Agriculture.* Madison, WI: SSSA, 1991, pp. 31–57.

87. U.C. Gupta. Soil and plant factors affecting molybdenum uptake by plants. In: U.C. Gupta, ed. *Molybdenum in Agriculture.* New York: Cambridge University Press, 1997, pp. 71–91.

88. W.L. Lindsay. Inorganic phase equilibria of micronutrients in soil. In: J.J. Mortvedt et al., eds. *Micronutrients in Agriculture.* Madison, WI: SSSA, 1972, pp. 41–57.

89. J.J. Mortvedt. Sources and methods for molybdenum fertilization in crops. In: U.C. Gupta, ed. *Molybdenum in Agriculture.* New York: Cambridge University Press, 1997, pp. 171–181.

90. W.L. Lindsay. *Chemical Equilibria in Soils.* New York: Wiley, 1979.

91. D.L. Sparks. *Environmental Soil Chemistry.* New York: Academic Press, 1995, pp. 99–140.

92. K.S. Smith, L.S. Balistrieri, S.M. Smith, R.C. Severson. Distribution and mobility of molybdenum in the terrestrial environment. In: U.C. Gupta, ed. *Molybdenum in Agriculture.* New York: Cambridge University Press, 1997, pp. 23–46.

93. H.M. Reisenauer, A.A. Tabikh, P.R. Stout. Molybdenum reactions with soils and hydrous oxides of iron, aluminum and titanium. *Proc. Soil Sci. Soc. Am.* 26:23–27, 1962.

94. L.H.P. Jones. Solubility of molybdenum in simplified systems and aqueous soil suspensions. *J. Soil Sci.* 8:313–327, 1957.

95. C.P. Sharma, C. Chatterjee. Molybdenum availability in alkaline soils. In: U.C. Gupta, ed. *Molybdenum in Agriculture.* New York: Cambridge University Press, 1997, pp. 131–149.

96. E.G. Mulder. Molybdenum in relation to growth of higher plants and microorganisms. *Plant Soil* 5:368–415, 1954.

97. R.J. Xie, A.F. Mackinzie. Molybdate sorption-desorption in soils treated with phosphate. *Geoderma* 48:321–333, 1991.

98. I. Barshad. Factors affecting the molybdenum content of pasture plants. II. Effect of soluble phosphates, available nitrogen and soluble sulfates. *Soil Sci.* 71:387–398, 1951.

99. P.R. Stout, W.R. Meagher. Studies of the molybdenum nutrition of plants with radioactive molybdenum. *Science* 108:471–473, 1948.

100. P.R. Stout, W.R. Meagher, G.A. Pearson, C.M. Johnson. Molybdenum nutrition of crop plants. I Influence of phosphate and sulfate on the absorption of molybdenum from soils and solution cultures. *Plant Soil* 3:51–87, 1951.

101. H.F. Massey, R.H. Lowe, H.H. Bailey. Relation of extractable molybdenum to soil series and parent rock in Kentucky. *Soil Sci. Soc. Am. Proc.* 31:200–202, 1967.

102. H.M. Reisenauer. Molybdenum. In: C.A. Black, ed. *Methods of Soil Analysis. Part 2. Chemical and Microbiological Properties.* Madison, WI: American Society of Agronomy, 1965, pp. 1050–1058.

103. J.L. Grigg. Determination of the available molybdenum in soils. *N.Z. J. Sci. Technol.* 34:405–414, 1953.

104. E.R. Purvis, N.K. Peterson. Methods of soil and plant analysis for molybdenum. *Soil Sci.* 81:223–228, 1956.

105. C.M. Johnson, T.H. Arkley. Determination of molybdenum in plant tissue. *Anal. Chem.* 26:572–574, 1954.

106. S. Henning, T.L. Jackson. Determination of molybdenum in plant tissue using flameless atomic absorption. *At. Absorpt. Newsl.* 12:100–101, 1973.

107. P.R. Curtis, J. Grusovin. Determination of molybdenum in plant tissue by graphite furnace atomic absorption spectrophotometry (GFAAS). *Commun. Soil Sci. Plant Anal.* 16:1279–1291, 1985.

108. J.L. Grigg. The distribution of molybdenum in the soils of New Zealand. I. Soils of the North Island. *N.Z. J. Agric. Res.* 3:69–86, 1960.

109. I. Barshad. Factors affecting the molybdenum content of pasture plants. I. Nature of soil molybdenum, growth of plants, and soil pH. *Soil Sci.* 71:297–313, 1951.

110. U.C. Gupta, D.C. MacKay. Extraction of water soluble copper and molybdenum in podzol soils. *Soil Sci. Soc. Am. Proc.* 29:323, 1965.

111. P.N. Soltanpour, A.P. Schwab. A new soil test for simultaneous extraction of macro and micro-nutrients in alkaline soils. *Commun. Soil Sci. Plant Anal.* 8:195–207, 1977.

112. L. Wang, K.J. Reddy, L.C. Munn. Comparison of ammonium bicarbonate-DTPA, ammonium carbonate, and ammonium oxalate to assess the availability of molybdenum in mine spoils and soils. *Commun. Soil Sci. Plant Anal.* 25:523–536, 1994.

113. D.C. Martens, D.T. Westermann. Fertilizer applications for correcting micronutrient deficiencies. In: J.J. Mortvedt, F.R. Cox, L.M. Shuman, R.M. Welch, eds. *Micronutrients in Agriculture.* Madison, WI: SSSA, 1991, pp. 549–582.

114. D.R. Lide. *Handbook of Chemistry and Physics.* 77th ed. Boca Raton, FL: CRC Press, 1996, pp. 4-35–4-98.

115. U.C. Gupta. Effect of methods of application and residual effect of molybdenum on the molybdenum concentration and yield of forages on podzol soils. *Can. J. Soil Sci.* 59:183–189, 1979.

116. J.F. Adams. Yield responses to molybdenum by field and horticultural crops. In: U.C. Gupta, ed. *Molybdenum in Agriculture.* New York: Cambridge University Press, 1997, pp. 182–201.

117. R.F. Vieira, C. Vieira, E.J.B.N. Cardoso, P.R. Mosquim. Foliar application of molybdenum in common bean. II. Nitrogenase and nitrate reductase activities in a soil of low fertility. *J. Plant Nutr.* 21:2141–2151, 1998.

118. R.F. Vieira, E.J.B.N. Cardoso, C. Vieira, S.T.A. Cassini. Foliar application of molybdenum in common bean. III. Effect on nodulation. *J. Plant Nutr.* 21:2153–2161, 1998.

119. E.J. Hewitt. Symptoms of molybdenum deficiency in plants. *Soil Sci.* 81:159–171, 1956.
120. U.C. Gupta, D.C. MacKay. Crop responses to applied molybdenum and copper on podzol soils. *Can. J. Soil Sci.* 48:235–242, 1968.
121. M.B. Parker, H.B. Harris. Soybean response to molybdenum and lime and the relationship between yield and chemical composition. *Agron. J.* 54:480–483, 1962.
122. H. Hafner, B.J. Ndunguru, A. Bationo, H. Marschner. Effect of nitrogen, phosphorus and molybdenum application on growth and symbiotic N_2-fixation of groundnut in an acid sandy soil in Niger. *Fert. Res.* 31:69–77, 1992.
123. E.R. Rhoades, D. Nangju. Effects of pelleting cowpea and soyabean seed with fertilizer dusts. *Exp. Agric.* 15:27–32, 1979.
124. J.N. Fitzgerald. Molybdenum on oats. *N.Z. J. Agric.* 89:619, 1954.
125. C.H. Burmester, J.F. Adams, J.W. Odom. Response of soybean to lime and molybdenum on Ultisols in northern Alabama. *Soil Sci. Soc. Am. J.* 52:1391–1394, 1988.
126. D.W. James, T.L. Jackson, H. Harward. Effect of molybdenum and lime on the growth and molybdenum content of alfalfa grown on acid soils. *Soil Sci.* 105:397–402, 1968.
127. J.A. MacLeod, U.C. Gupta, B. Stanfield. Molybdenum and sulfur relationships in plants. In: U.C. Gupta, ed. *Molybdenum in Agriculture*. New York: Cambridge University Press, 1997, pp. 229–244.
128. F.P. Rebafka, B.J. Ndunguru, H. Marschner. Single superphosphate depresses molybdenum uptake and limits yield response to phosphorus in groundnut (*Arachis hypogaea* L.) grown on an acid sandy soil in Niger, West Africa. *Fert. Res.* 34:233–242, 1993.

14 Nickel

Patrick H. Brown
University of California, Davis, California

CONTENTS

14.1 Introduction 395
14.2 Discovery of the Essentiality of Nickel 396
14.3 Physical and Chemical Properties of Nickel and Its Role in Animal and
     Bacterial Systems 397
     14.3.1 Nickel-Containing Enzymes and Proteins 397
     14.3.2 Essentiality and Function of Nickel in Plants 398
     14.3.3 Influence of Nickel on Crop Growth 400
14.4 Diagnosis of Nickel Status 401
     14.4.1 Symptoms of Deficiency and Toxicity 401
14.5 Concentration of Nickel in Plants 403
14.6 Uptake and Transport 404
14.7 Nickel in Soils 404
     14.7.1 Nickel Concentration in Soils 404
     14.7.2 Nickel Analysis in Soils 405
14.8 Nickel Fertilizers 405
14.9 Conclusion 406
References 406

14.1 INTRODUCTION

Nickel (Ni), the most recently discovered essential element (1), is unique among plant nutrients in that its metabolic function was determined well before it was determined that its deficiency could disrupt plant growth. Subsequent to the discovery of its essentiality in the laboratory, Ni deficiency has now been observed in field situations in several perennial species (2). The interest of plant scientists in the role of nickel was initiated following the discovery in 1975 (3) that it was a critical constituent of the plant enzyme, urease. The ultimate determination that nickel was essential for plant growth (1) depended heavily on the development of new techniques to purify growth media and to measure extremely low concentrations of nickel in plants. The establishment of nickel as an essential element, however, highlights the limitations of the current definition of essentiality of nutrients as applied to plants (4). It has been argued, for example, that even though nickel is clearly a normal and functional constituent of plants, it does not fulfill the definition of essentiality, since urease is not essential for plant growth and nickel deficiency apparently does not prevent the completion of the life cycle of all species, even though that criterion has not been explicitly satisfied for any element (5). Several authors (5,6) now suggest that the criteria for essentiality should be modified to include elements that are normal functional components of plants.

As our ability to determine the molecular structure, function, and regulation of biological systems improves, it is quite likely that additional elements will be shown to have irreplaceable functions in discrete biochemical processes that are important for plant life. This determination will be supplemented by advances in molecular and structural biology that will help predict the occurrence of similar processes across all organisms, allowing the relevance of discoveries made in bacterial systems to be immediately tested in plant and animal systems. The discovery of the essentiality of nickel is a good illustration of this principle and is likely to be repeated in the coming years. Nickel represents the first of several likely new essential elements that will be shown to be critical for certain metabolic processes normally active in plants, but not necessarily essential for the completion of the species' life cycle under all conditions.

The current definition of essentiality is clearly inadequate and its acceptance likely stifles the search for new essential elements. It is proposed, therefore, that the definition of essentiality be modified to more closely resemble that utilized in animal biology (7).

An element shall be considered essential for plant life if a reduction in tissue concentrations of the element below a certain limit results consistently and reproducibly in an impairment of physiologically important functions and if restitution of the substance under otherwise identical conditions prevents the impairment; and, the severity of the signs of deficiency increases in proportion to the reduction of exposure to the substance. (Nielson (7))

By this criterion, nickel is an essential element as are silicon and cobalt, which are essential elements for nitrogen-fixing plants.

14.2 DISCOVERY OF THE ESSENTIALITY OF NICKEL

The discovery in 1975 that nickel is a component of plant urease (3) prompted the first detailed studies on the essentiality of nickel for plant life. In 1977, Polacco (8) determined that tissue-cultured soybean (*Glycine max* Merr.) cells could not grow in the absence of nickel when provided with urea as the sole nitrogen source. Subsequently, many researchers demonstrated that plant growth is severely impacted by nickel deficiency when urea is the sole nitrogen source (9–14).

These results, though compelling, demonstrated a role for nickel only in certain species when grown with urea as the sole nitrogen source and as such did not satisfy the established criteria for essentiality, which state that an element is essential if without the element, the plant cannot complete its life cycle and the element is a constituent of an essential plant metabolite or molecule (4). Essentiality of nickel was subsequently established in 1987, when Brown et al. (1) demonstrated that barley (*Hordeum vulgare* L. cv. 'Onda') could not complete its life cycle in the absence of added nickel, even when plants were supplied with a nonurea source of nitrogen. In addition, it was shown that growth of oats (*Avena sativa* L. cv. 'Astro') and wheat (*Triticum aestivum* L. cv. 'Era') were significantly depressed under nickel-deficient conditions (15). The laboratory-based observations that Ni deficiency impacts a diversity of plant species has recently been verified in a diverse number of perennial species (*Carya, Betula, Pyracantha*) growing in the acidic low-nutrient soils of southeastern United States (2).

Nickel is now generally accepted as an essential ultra-micronutrient (16); however, the only defined role of nickel is in the metabolism of urea, a process that is not thought to be essential for plants supplied with a nitrogen source other than urea. The possibility that additional roles for nickel in plants exist was suggested by the results of Brown et al. (1,15), who demonstrated an effect of nickel deprivation in plants grown in the absence of urea and is implied in the work of Wood et al. (2), who demonstrated field responses to Ni supplementation in many ureide-transporting hydrophiles. A broader biological significance of nickel is also implied in the demonstration that nickel is essential for animal life and for a range of bacterial enzymes, including key enzymes in the nitrogen-fixing symbiont, *Bradyrhizobium japonicum* (17).

Our knowledge of the complete biological significance of nickel for plant productivity is still quite limited; however, with the demonstration of the essentiality of nickel in diverse species (1,2) and the increased use of urea as a nitrogen source, the importance of understanding the chemistry and biology of nickel and its potential impact on agricultural production has never been greater. Evidence that nickel plays an important function in animal and bacterial systems also suggests that nickel plays a larger role in plant productivity than is currently recognized. To obtain a full understanding of the potential role and management of nickel in agricultural systems, it is necessary to review the roles of nickel in other biological systems and to understand the plant and soil conditions under which nickel deficiency is likely to occur.

14.3 PHYSICAL AND CHEMICAL PROPERTIES OF NICKEL AND ITS ROLE IN ANIMAL AND BACTERIAL SYSTEMS

Nickel is a first-row transition metal with chemical and physical characteristics ideally suited to biological activity (18). Divalent nickel is the only oxidation state of nickel that is likely to be of any importance to higher plants. Nevertheless, Ni^{2+} forms a bewildering array of complexes with a variety of coordination numbers and geometries (19). Nickel readily binds, complexes, and chelates a number of substances of biological interest and is ubiquitous in all biological systems. Nickel is now known to be a functional constituent of seven enzymes, six of which occur in bacterial and animal systems, but not known to be active in plants, but the seventh enzyme, urease, is widely distributed in biology. The sensitivity of known biological nickel–complex equilibriums to temperature, concentration, and pH also make nickel an ideal element for the fine control of enzyme reactions (18).

14.3.1 NICKEL-CONTAINING ENZYMES AND PROTEINS

The field of nickel metallobiochemistry has seen tremendous growth over the preceding 10 years, and nickel is clearly a biologically important element in a diverse range of organisms. Indeed, it is highly likely that with the advent of molecular techniques to search for genetic and functional homology rapidly, the diversity of known functions of nickel in biology will increase substantially in the coming years. Advances in the field of bacterial and animal biology will rapidly flow to the plant sciences.

To date, seven nickel-dependent enzymes have been identified. Two of these enzymes have nonredox function (urease and glyoxylase), and the remaining five involve oxidation–reduction reactions (Ni-superoxide dismutase, methyl coenzyme M reductase, carbon monoxide dehydrogenase, acetyl coenzyme A synthase, and hydrogenase).

In all microorganisms that produce nickel-dependent metalloenzymes, there exist a number of proteins involved in nickel uptake, transport storage, and incorporation into the metalloenzyme. In bacteria, the transport of nickel into the cell involves two high-affinity transport systems, an ATP-dependent Nik family (Nik a–e) in *Escherichia coli* and a variety of nickel permeases (NixA, HoxN, etc.) in diverse species (17). Incorporation of nickel into the metalloenzyme involves a number of accessory proteins including metallo-chaperones (UreE, HypB, and CooJ) involved in nickel storage and in protein assembly (17).

Of the established nickel enzymes and proteins, urease is the sole nickel-specific enzyme known to function in plants; however, nickel-dependent hydrogenase also indirectly influences plant productivity through its role in nitrogen-fixing symbionts (20) and in leaf commensal bacteria (21). Currently, none of the bacterial proteins involved in nickel uptake and assimilation (NikA, NixA, UreE, etc.) is known to be present in plants. Interestingly, the hydrogenase and urease activities of leaf-surface symbionts are clearly inhibited when they colonize urease-deficient soybean mutants (21). The mechanism by which this inhibition occurs is unknown but may suggest that the urease-deficient mutants lack key nickel assimilatory proteins, thus preventing the transfer of nickel to the leaf-surface bacterial enzymes. This possibility would suggest that plants might contain nickel-dependent assimilatory proteins.

Nielsen reported the first description of a dietary deficiency of nickel in animals in 1970 for chickens and later for rats (*Rattus* spp.), goats (*Capra hircus*), sheep (*Ovis aries*), cows (*Bos taurus*), and mini pigs (*Sus scrofa*) (7). Nickel deficiency in these animals results in growth depression, physiological and anatomical disruption of liver function, and disruption of iron, copper, and zinc metabolism resulting in reduced levels of these enzymes in blood and various organs (22). Nickel deficiency also markedly reduces the activity of a number of hepatic enzymes, including several hydrogenases, urease, and glyoxylase, though a specific functional role for nickel in these enzymes in animals has not been determined.

One of the important and consistent findings from animal studies is that nickel deficiency induces iron deficiency, an observation that is also made in plants (15). In rats (22), and in sheep (23), nickel deprivation resulted in decreased iron uptake and reduced tissue-iron concentrations. Nielsen et al. (24) have suggested several possible roles for nickel in iron metabolism and oxidation–reduction (redox) shifts that draw upon the observation that nickel and iron are associated in a number of bacterial redox-based enzymes (17).

The suggestion that additional nickel-dependent enzymes and proteins are present in higher plants is supported by the observation that several of the known bacterial nickel-containing enzymes have analogs in plants and animals (including superoxide dismutase, glyoxylase, acetyl coenzyme A synthase, and hydrogenase). Our current failure to identify additional nickel-dependent enzymes in plants is likely a result of the relatively primitive state of plant enzymology, in contrast to bacterial enzymology, and the difficulty involved in research on complex organisms involving ultra-trace elements. The similarity between the effects of nickel deficiency in animals and plants also provides evidence of a common biological role for nickel in all organisms.

14.3.2 ESSENTIALITY AND FUNCTION OF NICKEL IN PLANTS

The first evidence of a response of a field crop to application of a nickel fertilizer was demonstrated in 1945 for potato (*Solanum tuberosum* L.), wheat (*Triticum aestivum* L.), and bean (*Phaseolus vulgaris* L.) crops (25). In these crops, the application of a dilute nickel spray resulted in a significant increase in yield. These experiments were conducted on the 'Romney Marshes' of England, a region that is well known for its trace mineral deficiencies, particularly of manganese and zinc. These experiments were conducted very carefully and excluded the possibility that the nickel applied was merely substituting for manganese, zinc, iron, copper, or boron, suggesting that the growth response was indeed due to the application of nickel. Interestingly, the soils of this region may be low in nickel since the conditions that limit manganese and zinc availability in these soils (acid sands of low mineral content) would also limit nickel availability to crops, and the concentrations of nickel provided were appropriate based on the current knowledge of nickel demand. These same soil types also dominate the region of southeast United States where Ni deficiency is now known to occur.

Mishra and Kar (26) and Welch (27) reviewed the evidence of the role of nickel in biological systems and cited many examples of yield increases in field-grown crops in response to the application of nickel to the crop or to the soil. The significance of these purported benefits of field applications of nickel is difficult to interpret since the majority of the reported experiments used very high nickel application rates. None of these reports considered the possibility that nickel influenced plant yield through its effect on disease suppression, nor was the nickel concentration in the crops determined. Indeed, prior to the availability of graphite-furnace atomic absorption spectrophotometers and inductively coupled plasma mass spectrometers (in the mid-1970s), it was exceedingly difficult to measure nickel at the concentrations (<0.1 mg Ni kg^{-1} dry weight) later shown to be critical for normal plant growth. In the absence of information on tissue-nickel concentrations, it is impossible to conclude that the observed yield increases were the result of a correction of a nickel deficiency in the plant.

Clear evidence that nickel application benefited the growth of nitrogen-fixing species of plant was demonstrated by Bertrand and DeWolf (28), who reported that soil-nickel application to field-grown soybean (*Glycine max* Merr.) resulted in a significant increase in nodule weight and seed yield. The authors suggested that the yield increase was the result of a nickel requirement of the nitrogen-fixing rhizobia. A specific role for nickel in nitrogen-fixing bacteria is now well established with the determination that a nickel-dependent hydrogenase is active in many rhizobial bacteria (20) and is thus essential for maximal nitrogen fixation (29). Nickel is also known to be essential for nitrogen fixation of the free-living cyanobacterium, *Nostoc muscorum* C.A. Adargh, though the specific mechanism has not been determined (30).

A role for nickel in plant disease resistance has long been observed and has been variously attributed to a direct phyto-sanitary effect of nickel on pathogens, or to a role of nickel on plant disease-resistance mechanisms. Mishra and Kar (26) concluded that nickel likely acted to reduce plant disease by direct toxicity to the pathogen. Nickel, however, is not particularly toxic when applied directly to microorganisms, and Graham et al. (31) demonstrated that nickel supplied to the roots of cowpea (*Vigna unguiculata* Walp.) that contained only 0.03 mg Ni kg^{-1} dry weight effectively reduced leaf-fungal infection by 50%. Whether this effect was directly due to a role of nickel in plant defense reactions (possibly involving superoxide dismutase-mediated processes) or a consequence of the alleviation of deficiency-induced changes in nitrogen metabolites (urea, amino acids, etc.) is uncertain. Regardless of the mechanism, a positive effect of nickel supplementation on disease tolerance was clearly documented.

The discovery that nickel is a component of the plant urease in 1975 (3) prompted a renewed interest in the role of nickel in plant life. In 1977, Polacco (32) determined that tissue-cultured soybean cells could not grow in the absence of nickel when provided with urea as the sole nitrogen source. Subsequently, an absolute nickel requirement was demonstrated for tissue-cultured rice (*Oryza sativa* L.) and tobacco (*Nicotiana tabacum* L.) (26,27). This finding was followed in 1981 by a review of nickel in biology that suggested that leguminous plants might have a unique requirement for nickel (28).

Using a novel chelation chromatography technique to remove nickel as a contaminant from the nutrient media, Eskew et al. (9,33,34) and Walker et al. (11) demonstrated that, under nickel-deficient conditions, urea accumulated to toxic levels in the leaves of soybean and cowpea. Leaflet tips of nickel-deficient plants contained concentrations of urea as high as 2.4% dry weight. The accumulation of urea occurred irrespective of the nitrogen source used and was assumed to have occurred as a result of urease-dependent disruption of the arginine-recycling pathway. Eskew et al. (9) concluded that nickel was an essential element for leguminous plants though they did not demonstrate a failure of nickel-deficient plants to complete their life cycles. Recently, Gerendas et al. (12–14), in a series of elegant studies demonstrated a profound effect of nickel deficiency on the growth of urea-fed tobacco, zucchini (*Cucurbita pepo* L.), rice, and canola (*Brassica napus* L.), but observed no growth inhibition when nitrogen sources other than urea were used.

Confirmation that nickel was essential for higher plants was provided by Brown et al. (1), who demonstrated that barley seeds from nickel-deprived plants were incapable of germination even when grown on a nitrogen source other than urea. Significant restrictions in shoot growth of barley, oats, and wheat (*Triticum aestivum* L.) were subsequently demonstrated under nickel-deficient conditions when the plants were supplied with mineral nitrogen sources (15). Brown et al. (15) also observed a marked suppression in tissue-iron concentrations in nickel-deficient plants, a response that is also observed in nickel-deficient animals (7). Reductions in tissue-malate concentrations have also been observed in nickel-deficient animals and plants (15,24,35). Confirmation of the essentiality of Ni under field conditions was provided in 2004 by Wood et al. (2), who observed a marked and specific positive response to application of Ni fertilizer to pecan (*Carya illinoinensis* K. Koch) and other species (2) that could not be corrected with any other known essential element.

The demonstration of a role for nickel in diverse plant species, the presence of nickel in a discrete metabolic process, and the failure of plants to complete their life cycles in the absence of nickel, satisfies the requirement for the establishment of essentiality (4).

Although nickel has been accepted generally as an essential element, there is reason to be cautious about this conclusion, and some authors suggest that nickel may not fully satisfy the most stringent interpretation of the laws of essentiality primarily since its role in a specific essential metabolic function has not been identified. Furthermore, even though nickel has a clear role in metabolism, it is now clear that urease is not, by itself, essential for plant life as evidenced by the observation that urease-null soybean mutants can complete their life cycles (37). There has also been no independent replication of the effect of nickel on barley grain viability though Horak (36) did observe a marked increase in seed viability with the addition of nickel to pea (*Pisum sativum* L.) seeds grown in nickel-deficient soils.

Regardless of these apparent contradictions, nickel is still clearly required for normal plant metabolism. As a component of urease, nickel is required for urea and arginine metabolism, and both of these metabolites are normal constituents of plants (5). Nickel is also an essential component of hydrogenases involved in nitrogen fixation and other associative bacterial processes, and nickel clearly influences plant response to disease. Nickel is clearly a normal constituent of plant life.

Many of the reported effects of nickel on plant growth cannot be attributed solely to the role of nickel in urease, and many symptoms of nickel deficiency (disrupted iron and malate metabolism) are also observed in animals (7). It is likely, therefore, that additional nickel-dependent enzymes and proteins await discovery and will help resolve the remaining questions on the function of nickel in plants.

14.3.3 INFLUENCE OF NICKEL ON CROP GROWTH

Many early reports of the role of nickel in agricultural productivity have been questioned since they did not adequately exclude the possibility that nickel was acting directly as a fungicidal element (27). Regardless of the many questionable reports, a compelling body of literature exists in which appropriate concentrations of nickel were applied or where the plant response is consistent with current knowledge of nickel functions including effects on nitrogen fixation, seed germination, and disease suppression (26,27,31,34,38,39).

The clearest agronomic responses to nickel have been observed when nitrogen is supplied as urea or by nitrogen fixation. The most illustrative example of the relationship between nickel and urea metabolism is provided from studies with foliar urea application and tissue-culture growth of plants. Plants without a supply of nickel have low urease activity in the leaves, and foliar application of urea leads to a large accumulation of urea and severe necrosis of the leaf tips (34). Nicoulaud and Bloom (40) observed that nickel, provided in the nutrient solution of tomato (*Lycopersicon esculentum* Mill.) seedlings growing with foliar urea as the only nitrogen source, significantly enhanced growth. The authors speculated that the effect of nickel was more consistent with its role in urea translocation than that on urease activity directly (40). This result is in agreement with the findings of Brown et al. (15), who suggest that nickel has a role in the transport of nitrogen to the seed thereby influencing plant senescence and seed viability.

The first demonstration of an agricultural Ni deficiency did not occur until 2004 (Wood et al., 2004), when it was observed in pecan (*Carya illinoinensis*). Nickel deficiency in pecan is associated with a physiological disorder 'mouse-ear' which occurs sporadically, but with increasing frequency, throughout the southeastern United States (portions of South Atlantic region) where it represents a substantial economic impact. In agreement with the results of Brown et al. (1), Ni deficiency in pecan results in a disruption of nitrogen metabolism and altered amino acid profiles (72).

The value of addition of nickel to Murashige and Skoog plant tissue-culture medium was shown by Witte et al. (41). These authors suggested that the lack of nickel and urease activity may represent a stress factor in tissue culture and recommended that the addition of 100 nM Ni be adopted as a standard practice. The benefits of adding nickel to solution cultures was also demonstrated by Khan et al. (42), who determined that a mixture of 0.05 mg Ni L^{-1} and 20% nitrogen as urea resulted in optimal growth of spinach (*Spinacia oleracea* L.) under hydroponic conditions.

14.4 DIAGNOSIS OF NICKEL STATUS

14.4.1 SYMPTOMS OF DEFICIENCY AND TOXICITY

In legumes and other dicotyledonous plants, nickel deficiency results in decreased activity of urease and subsequently in urea toxicity, exhibited as leaflet tip necrosis (9–11). With nitrogen-fixing plants or with plants grown on nitrate and ammonium, nickel deficiency results in a general suppression in plant growth with development of leaf tip necrosis on typically pale green leaves (9,10) (Figure 14.1 and Figure 14.2). These symptoms were attributed to the accumulation of toxic levels of urea in the leaf tissues.

In graminaceous species (Figure 14.3), deficiency symptoms include chlorosis similar to that induced by iron deficiency (1), including interveinal chlorosis and patchy necrosis in the youngest leaves. Nickel deficiency also results in a marked enhancement in plant senescence and a reduction in tissue-iron concentrations. In monocotyledons and in dicotyledons, the accumulation of urea in leaf tips is diagnostic of nickel deficiency. In early or incipient stages of nickel toxicity, no clearly visible symptoms develop, though shoot and root growth may be suppressed. Acute nickel toxicity results in symptoms that have variously been likened to iron deficiency (interveinal chlorosis in

FIGURE 14.1 Nitrogen-fixing cowpea seedlings (*Vigna unguiculata* Walp.) were grown for 40 days in nutrient solutions containing either 1 (left) or $0\,\mu g\,L^{-1}$ (right) and supplied with no inorganic nitrogen source. In the absence of nickel, plants developed pronounced leaf tip necrosis and marked yellowing and growth stunting. The observed symptoms closely resemble those of nitrogen deficiency. (Photograph by David Eskew.) (For a color presentation of this figure, see the accompanying compact disc.)

FIGURE 14.2 Leaf tip necrosis in soybean plants (*Glycine max* Merr.) grown in nutrient solution provided with equimolar concentrations of nitrate and ammonium. Solutions were made free from nickel by first passing solutions through a nickel-specific chelation resin. Leaf tip necrosis was observed coincident with the commencement of flowering. (Photograph by David Eskew.) (For a color presentation of this figure, see the accompanying compact disc.)

monocotyledons, mottling in dicotyledons) or zinc deficiency (chlorosis and restricted leaf expansion) (1,2,43). Severe toxicity results in complete foliar chlorosis with necrosis advancing in from the leaf margins, followed by plant death.

In pecan growing in the southeastern United States, the long-described but poorly understood symptoms of 'mouse-ear' or 'little-leaf disorder' (Figure 14.4) have recently been shown to be due

FIGURE 14.3   Nickel deficiency symptoms in barley (*Hordeum vulgare* L. cv. Onda) following 50 days growth in nutrient solution containing equimolar concentrations of nitrate and ammonium. Symptoms include leaf-tip chlorosis and necrosis, development of thin 'rat-tail' leaves, and interveinal chlorosis of young leaves. (Photograph by Patrick Brown.) (For a color presentation of this figure, see the accompanying compact disc.)

FIGURE 14.4   Branches of nickel-sufficient (left) and nickel-deficient (right) pecan (*Carya illinoinensis* K. Koch). Symptoms include delayed and decreased leaf expansion, poor bud break, leaf bronzing and chlorosis, rosetting, and leaf tip necrosis. (Photo courtesy of Bruce Wood.) (For a color presentation of this figure, see the accompanying compact disc.)

to nickel deficiency that can be cured by application of nickel (at 100 mg L^{-1}) (2). Nickel deficiency in pecan and in certain other woody perennial crops (e.g., plum, peach and pyracantha, and citrus) is characterized by

early-season leaf chlorosis, dwarfing of foliage, blunting of leaf or leaflet tips, necrosis of leaf or leaflet tips, curled leaf or leaflet margins, dwarfed internodes, distorted bud shape, brittle shoots, cold-injury-like death of over-wintering shoots, diminished root system with dead fibrous roots, failure of foliar lamina to develop, rosetting and loss of apical dominance, dwarfed trees, and tree death (Wood et al. (2))

Nickel deficiency was long unrecognized in this region because of its similarity to zinc deficiency and as a consequence of a complex set of factors that influences its occurrence. Nickel deficiency is induced by: (a) excessively high soil zinc, copper, manganese, iron, calcium, or magnesium; (b) root damage by root-knot nematodes; or (c) dry or cool soils at the time of bud break (2). The conditions under which Ni deficiency occurs also commonly result in a deficiency of zinc or copper, and this fact has resulted in the extensive use of copper and zinc fertilizers over many years further exacerbating the nickel deficiency. In many horticultural tree species, heavy application of fertilizers with zinc, copper, or both nutrients is common for their nutritional values and benefits for leaf removal and disease protection. In many orchard crops recalcitrant physiological disorders and poorly understood replant 'diseases' are frequent suggesting that induced nickel deficiency may be much more widespread than was previously recognized.

14.5 CONCENTRATION OF NICKEL IN PLANTS

The nickel concentration (Table 14.1) in leaves of plants grown on uncontaminated soil ranges from 0.05 to 5.0 mg Ni kg^{-1} dry weight (27,44,45). The adequate range for nickel appears to fall between 0.01 and 10 mg Ni kg^{-1} dry weight, which is an extremely wide range compared to that for the other elements (5). The critical nickel concentration required for seed germination in barley, shoot growth in oat, barley, and wheat, and shoot growth of urea-fed tomato, rice, and zucchini (*Cucumus pepo* var. *melopepo* Alef.) has been estimated independently by two groups to be approximately 100 mg Ni kg^{-1} (1,5), which is similar to the recently determined Ni requirement for pecan (2).

TABLE 14.1
Concentration Ranges of Nickel in Crop Species

Plant Species	Scientific Name	Concentrations of Nickel in Plants (mg Ni kg^{-1})				
		Deficient	Critical (deficiency)	Adequate	Critical (toxicity)	Reference
Barley	*Hordeum vulgare* L., *H. distichon* L.	—	0.1	—	—	1,15
Wheat	*Triticum aestivum* L., *T. durum* Desf	0.037		0.084	63–113	15,53
Cowpea	*Vigna unguiculata* Walp	<0.01–0.142		0.22–10.3		11
Beans	*Phaseolus vulgaris* L.				10–83	54
Oats	*Avena sativa* L.	0.017		0.10		15
Soybean	*Glycine max* Merr.		0.02–0.04			10
Italian ryegrass	*Lolium multiflorum* Lam.			0–8	>80	55
Pecan	*Carya illinoinensis* K. Koch		0.1			2

Nickel concentrations above the toxicity levels of $>10\,mg\;kg^{-1}$ dry weight in sensitive species, and $>50\,mg\;kg^{-1}$ dry weight in moderately tolerant ones (44,45,46) result in impaired root and shoot growth without any remarkable defining characteristics (47).

The nickel content of a plant is determined by the nickel availability in the soil, plant species, plant part, and season. Plants growing on serpentine soils (derived from ultramific rocks) or contaminated soils can accumulate high levels of nickel and other heavy metals (48,49). In naturally occurring high-nickel soils (serpentine soils) highly specialized plant species have evolved including several species that hyperaccumulate nickel, sometimes up to 1 to 5% of tissue dry weight (50,51). Species growing on the same soil can also vary dramatically in nickel content and within plant distribution. In general, nickel is transported preferentially to the grain, particularly under conditions of marginal nickel supply (52).

14.6 UPTAKE AND TRANSPORT

In bacterial systems, several families of nickel permeases and ATP-dependent nickel carriers have been characterized. No equivalent mechanism has yet been identified in animals or plants (17). In plant systems, most studies have been conducted at unrealistically high soil-nickel concentrations and as such may be relevant for nickel toxicity, but are not relevant for nickel uptake under normal conditions. Cataldo et al. (56) using ^{63}Ni indicated that a high-affinity Ni^{2+} carrier functioned at 0.075 or $0.25\,\mu M\;Ni^{2+}$ with a K_m of $0.5\,\mu M$ which approaches the nickel concentration in uncontaminated soils (48). Either Cu^{2+} or Zn^{2+} competitively inhibits Ni^{2+} uptake suggesting that all the three elements share a common uptake system (57). Uptake at higher nickel-supply levels (0.5 to $30\,\mu M$) was energy dependent and had a K_m of $12\,\mu M$ indicative of an active, low-affinity transport system.

No evidence suggests that associations with arbuscular mycorrhizal fungus increase nickel accumulation by plants (58,59).

Nickel, unlike many other divalent cations, is readily re-translocated within the plant likely as a complex with organic acids and amino acids (60). Nickel rapidly re-translocates from leaves to young tissues in the phloem, particularly during reproductive growth. Indeed, up to 70% of nickel in the shoots was transported to the seed of soybean (61). Nickel is associated primarily with organic acids and amino acids in the phloem. Above pH 6.5, histidine is the most significant chelator, whereas at pH <5, citrate is the most significant one (5).

14.7 NICKEL IN SOILS

14.7.1 Nickel Concentration in Soils

Nickel is abundant in the crust of the Earth, comprising about 3% of the composition of the earth. Nickel averages $50\,mg\;Ni\;kg^{-1}$ in soils and commonly varies from 5 to $500\,mg\;Ni\;kg^{-1}$ but ranges up to 24,000 to $53,000\,mg\;Ni\;kg^{-1}$ in soil near metal refineries or in dried sewage sludge, respectively. Agricultural soils typically contain 3 to $1000\,mg\;Ni\;kg^{-1}$, whereas soils derived from basic igneous rocks may contain from 2000 to $6000\,mg\;Ni\;kg^{-1}$ (62).

Total nickel content is, however, not a good measure of nickel availability. At pH>6.7, most of the nickel exists as sparingly soluble hydroxides, whereas at pH<6.5, most nickel compounds are relatively soluble (48). Depending on the soil type and pH, nickel may also be highly mobile in soil and is further mobilized by acid rain. The role of pH in nickel availability was illustrated by Van de Graaff et al. (63), who observed that long-term irrigation with sewage effluent increased heavy metal loading in soil, but that plant metal contents did not increase, apparently owing to the increased soil pH, iron complexation and coprecipitation, and precipitation of phosphorus–metal complexes.

Truly nickel-deficient soils have not been identified to date; however, Ni deficiency can occur as a result of excessive use of competing ions (Zn, Cu, and MgO and unfavorable growth conditions (2)).

Nickel is the 24th-most abundant element in the crust of the earth, and plant nickel requirement (<0.05 mg kg^{-1} dry weight) is the lowest of any essential element. Although a large number of analyses have been conducted for nickel in plant tissues, no recorded levels have been below 0.2 mg kg^{-1} dry weight in field-grown plants. Nickel can be supplied by atmospheric deposition, at rates that easily exceed the removal from the crops in the field (64). The ubiquitous nature of nickel is illustrated by the experiments that established the essentiality of nickel (1). In these experiments, the authors went to extraordinary lengths to purify or re-purify all chemical reagents, equipment, and water and to maintain contaminant-free growing conditions. Even under these conditions, it required three generations of crop growth to deplete the nickel carried over from the grain before the first evidence of nickel deficiency was observed.

The possibility that nickel-deficient soils exist, however, cannot be discounted particularly as purity of fertilizers is improved, the use of urea is increased, and atmospheric deposition of pollutant nickel is decreased. Plants grown under specialized conditions (greenhouses and tissue culture), particularly with urea as a nitrogen source, may be especially susceptible to nickel deficiency (40).

Nickel toxicity, which is usually associated with serpentine soils, sewage-sludge application, or industrial pollution, is a well-described constraint on crop production in many parts of the world. In serpentine soils (derived from basic igneous rocks), nickel concentrations may range from 1000 to 6000 mg kg^{-1} dry weight and are frequently associated with high concentrations of iron, zinc, and chromium and an unfavorable ratio of magnesium to calcium. Values for ammonium acetate-extractable nickel in these soils varies from 3 to 70 mg kg^{-1}; however, it is not always clear that poor plant growth can be ascribed to any single factor concerning nickel.

Similarly, in sewage-amended soils or in contaminated soils, it is often difficult to relate total nickel load with plant productivity as factors such as the chemical properties of the contaminant and base soil, pH, and oxidation–reduction state affect results (48,65). Indeed, the importance of considering soil pH is well illustrated by Kukier and Chaney (65 and references therein), who demonstrated that addition of limestone to raise soil pH is highly effective in immobilizing nickel *in situ* and in reducing phytotoxicity. Plant species also differ in their ability to obtain nickel from soils and hence any measurement of soil nickel must be interpreted with consideration of the plant species of interest.

14.7.2 NICKEL ANALYSIS IN SOILS

A large number of approaches, including diethyltriaminepentaacetic acid (DTPA), $BaCl_2$, $Sr(NO_3)_2$, and ammonium acetate among others (48,65) are used to extract metals from soils in an attempt to predict nickel availability to plants. The DTPA method, however, is probably the most commonly used (48,66,67) and has been shown to be quite effective for a variety of soils to define Ni excess. The DTPA method is improved significantly if factors such as soil pH and soil bulk density are incorporated into the resulting regression equation (65). Many authors (48,65), however, observe that plant species and soil environment (water, oxygen content, and temperature) can markedly affect the relationship between soil-extractable and plant-nickel concentrations (2). These results suggest that the condition under which the soil is collected and tested can significantly influence the interpretation of results. Nickel deficiency is also known to be exacerbated by environmental conditions that limit uptake (cold, wet weather) and by the oversupply of apparently competing elements such as Cu, Mn, Mg, Fe, Ca, and Zn (2). Nickel bioavailability can also be determined by the ion-exchange resin (IER) method, which has been used quite successfully in a limited number of soil types and facilitates the *in situ* assessment of exchangeable nickel (68).

14.8 NICKEL FERTILIZERS

Essentially under all normal field conditions, it is unlikely that application of nickel fertilizer will be required. Exceptions to this concept occur when urea is the primary source of nitrogen supply, in species in which ureides play an important physiological role (2), when excessive applications of

Zn, Cu, Mn, Fe, Ca, or Mg have been made over many years (2) and perhaps also in nitrogen-fixing crops grown on mineral-poor or highly nickel-fixing (high pH, high lime) soils. In experiments utilizing highly purified nutrient solutions or tissue-culture media, supplemental nickel may also be beneficial. In all of these cases, the nickel demand is quite low and can be satisfied easily with $NiSO_4$ or other soluble nickel sources including Ni–organic complexes (Bruce Wood, personal communication). In solution-grown plants and as a supplement to foliar urea applications, a nickel supply of 0.5 to 1 µM is sufficient.

Nickel is currently being applied to many fields in sewage sludge (48,69). In general, this usage does not represent a threat to human health, as its availability to crop plants is typically low. The total extractable nickel in these amended soils can also be controlled by selection of plant species and management of soil pH, moisture, and organic matter (65).

In recent years, a great deal of attention is being focused on nickel-accumulating plants that can tolerate otherwise nickel-toxic soils and accumulate substantial concentrations of nickel, up to 5% on a dry weight basis (70). Three nickel hyperaccumulators showed significantly increased shoot biomass with the addition of 500 mg Ni kg^{-1} to a nutrient-rich growth medium, suggesting that the nickel hyperaccumulators have a higher requirement for nickel than other plants (71). Considerable attention is also being focused on utilizing hyperaccumulating species for phytoremediation and phytomining, where they can be grown in a nickel-contaminated soil and then harvested and exported from the field. To date, however, this approach has not been successful owing to the small size and slow growth rate of many of the hyperaccumulating species. With a better understanding of the genetic basis of metal hyperaccumulation, it may be possible to transfer this trait into a fast-growing agronomic species and hence develop an effective phyoremediation strategy.

14.9   CONCLUSION

Nickel is the latest element to be classified as essential for plant growth in both laboratory and field conditions and an absolute requirement for nickel fertilizer under field conditions in perennial species growing in the southeast of the United States has now been established. Nickel clearly has a significant effect on the productivity of field-grown, nitrogen-fixing plants, those in which ureides are a significant form of nitrogen and those utilizing urea as a primary nitrogen source. The symptoms of nickel deficiency in barley, wheat, and oats observed by Brown et al. (1) and Wood et al. (2) are consistent with the observations made in nickel-deficient animals and are indicative of a role of nickel in nitrogen metabolism that cannot be easily explained through an exclusive role of nickel in urease. This finding in combination with the diverse known functions of nickel in bacteria suggests that nickel may indeed play a role in many, yet undiscovered processes in plants.

REFERENCES

1. P.H. Brown, R.M. Welch, E.E. Cary. Nickel: A micronutrient essential for higher plants. *Plant Physiol.* 85:801–803, 1987.
2. B.W. Wood, C.C. Reilly, A.P. Nyczepir. Mouse-ear of Pecan: A nickel deficiency. *HortScience* 39 (6):1238–1242, 2004.
3. N.E. Dixon, C. Gazzola, R.L. Blakeley, B. Zerner. Jack bean urease (EC 3.5.1.5). A metalloenzyme. A simple biological role for nickel. *J. Amer. Chem. Soc.* 97:4131–4133, 1975.
4. D.I. Arnon, P.R. Stout. The essentiality of certain elements in minute quantity for plants with special reference to copper. *Plant Physiol.* 14:371–375, 1939.
5. J. Gerendas, J.C. Polacco, S.K. Freyermuth, B. Sattelmacher. Significance of nickel for plant growth and metabolism. *J. Plant Nutr. Soil Sci.* 162:241–256, 1999.
6. E. Epstein, A.J. Bloom. Mineral Nutrition of Plants: Principles and Perspectives. 2nd edition. Sinauer Associates, Sunderland, MA, 2004, pp. 45.

7. F.H. Nielson. Nickel. In: E. Frieden, ed. *Biochemistry of the Essential Ultratrace Elements*. New York: Plenum Press, 1984, pp. 293–308.

8. J.C. Polacco. Nitrogen metabolism in soybean tissue culture: II Urea utilization and urease synthesis requires Ni^{2+}. *Plant Physiol.* 59:827–830, 1977.

9. D.L. Eskew, R.M. Welch, W.A. Norvell. Nickel: an essential micronutrient for legumes and possibly all higher plants. *Science* 222:621–623, 1983.

10. D.L. Eskew, R.M. Welch, W.A. Norvell. Nickel in higher plants. Further evidence for an essential role. *Plant Physiol.* 76:691–693, 1984.

11. C.D. Walker, R.D. Graham, J.T. Madison, E.E. Cary, R.M. Welch. Effects of nickel deficiency on some nitrogen metabolites in cow peas, *Vigna unguiculata*. *Plant Physiol.* 79:474–479, 1985.

12. J. Gerendas, S.B. Sattelmacher. Significance of Ni supply for growth, urease activity and the contents of urea, amino acids and mineral nutrients of urea-grown plants. *Plant Soil* 190:153–162, 1997.

13. J. Gerendas, S.B. Sattelmacher. Significance of N source (urea vs. NH_4NO_3) and Ni supply for growth, urease activity and nitrogen metabolism of zucchini (*Cucurbita pepo* convar. *giromontiina*). *Plant Soil* 196:217–222, 1997.

14. J. Gerendas, S.B. Sattelmacher. Influence of Ni supply on growth, urease activity and nitrogen metabolites of *Brassica napus* grown with NH_4NO_3 or urea as nitrogen source. *Ann. Bot.* 83:65–71, 1999.

15. P.H. Brown, R.M. Welch, E.E. Cary, R.T. Checkai. Beneficial effects of nickel on plant growth. *J. Plant Nutr.* 10:2125–2135, 1987.

16. H. Marschner. *Mineral Nutrition of Higher Plants,* 2nd edition. London: Academic Press, 1995, pp. 364–369.

17. M.J. Maroney. Structure/function relationships in nickel metallobiochemistry. *Curr. Opin. Chem. Biol.* 3:188–199, 1999.

18. D.A. Phipps. *Metals and Metabolism*. Oxford: Clarendon Press, 1976, pp. 28–56.

19. E. Frieden. *Biochemistry of the Essential Ultratrace Elements*. New York: Plenum Press, 1984, pp. 59–62.

20. R. Cammack. Splitting molecular hydrogen. *Nature* 373:556–557, 1995.

21. M.A. Holland, J.C. Pollacco. Urease-null and hydrogenase-null phenotypes of a phylloplane bacterium reveal altered nickel metabolism in two soybean mutants. *Plant Physiol.* 98:942–948, 1992.

22. Von A. Schnegg, M. Kirchgessner. Aktivitatsanderungen von enzymem der leber und niere im nickel- bzw. Eisen-Mangel. *Zh. Tierphysiol.*; *Tierernahrg u Futtermittelkde* 38:200–205, 1977.

23. M. Anke, P. Gropel, H. Kronemann, M. Gunn. Nickel—An essential element. In: F.W. Sunderman, Jr., A. Aito, eds. *Nickel in the Human Environment*. Lyon: IARC Scientific Publishers, 1984, pp. 339–365.

24. F.H. Nielson, T.J. Zimmerman, M.E. Collings, D.R. Myron. Nickel deprivation in rats: nickel-iron interactions. *J. Nutr.* 109:1623–1632, 1979.

25. W.A. Roach, C. Barclay. Nickel and multiple trace deficiencies in agricultural crops. *Nature* 157:696, 1946.

26. D. Mishra, M. Kar. Nickel in plant growth and metabolism. *Bot. Rev.* 40:395–452, 1974.

27. R.M. Welch. The biological significance of nickel. *J. Plant Nutr.* 31:345–356, 1981.

28. D. Bertrand, A. de Wolf. Nickel, a dynamic trace element for higher plants. *C R Academic Sci.* 265:1053–1055, 1967.

29. S.L. Albrecht, R.J. Maier, F.J. Hanus, S.A. Russel, D.W. Emerich, H.J. Evans. Hydrogenase in *Rhizobium japonicum* increases nitrogen fixation by nodulated soybeans. *Science* 203:1255–1257, 1979.

30. L.C. Rai, M. Raizada. Nickel-induced stimulation of growth heterocyst differentiation carbon-14 dioxide uptake and nitrogenase activity in *Nostoc muscorum*. *New Phytol.* 104:111–114, 1986.

31. R.D. Graham, R.M. Welch, C.D. Walker. A role of nickel in the resistance of plants to rust. *Proceedings of the Third Australian Agronomic Conference*, Hobart Tasmania, Australia, 1985.

32. J.C. Polacco. Is nickel a universal component of plant ureases? *Plant Sci. Lett.* 10:249–255, 1977.

33. D.L. Eskew, R.M. Welch, E.E. Cary. A simple plant nutrient solution purification method for effective removal of trace metals using controlled pore glass 8-hydroxyquinoline chelation column chromatography. *Plant Physiol.* 76:103–105, 1982.

34. D.L. Eskew, R.M. Welch. Nickel supplementation $1 \mu g \ L^{-1}$ prevents leaflet tip necrosis in soybeans grown in nutrient solutions purified using 8 hydroxy quinoline-controlled pore glass chromatography. *Plant Physiol.* (Supp.) 69:134, 1982.

35. M. Kirchgessner, Von A. Schnegg. Malate dehydrogenase and glucose-6-phosphate dehydrogenase activity in livers of Ni-deficient rats. *Bioinorg. Chem.* 6:155–161, 1976.

36. O. Horak. The importance of nickel in Fabaceae II. Uptake and requirement of nickel by *Pisum sativum*. *Phyton (Horn)* 25:301–307, 1985.

37. J.C. Polacco. A soybean seed urease-null produces urease in cell culture *Glycine max*. *Plant Physiol.* 66: 1233–1240, 1981.

38. O. Horak. The importance of nickel in Fabaceae I. Comparative studies on the content of nickel and certain other elements in vegetative parts and seeds. *Phyton (Horn)* 25:135–146, 1985.

39. P.K. Das, M. Kar, D. Mishra. Nickel nutrition of plants: I. Effect of nickel and some oxidase activities during rice (*Oryza sativa* L.) seed germination. *Zh. Pflanzenphysiol.* 90:225–233, 1978.

40. B.A.L. Nicoulaud, A.J. Bloom. Nickel supplements improve growth when foliar urea is the sole nitrogen source for tomato. *J. Amer. Soc. Hortic. Sci.* 123:556–559, 1998.

41. C.-P. Witte, S.A. Tiller, M.A. Taylor, H.V. Davies. Addition of nickel to Murashige and Skoog medium in plant tissue culture activates urease and may reduce metabolic stress. *Plant Cell Tissue Organ Culture* 68:103–104, 2002.

42. N.K. Khan, M. Wantanabe, Y. Wantanabe. Effect of different concentrations of urea with or without nickel on spinach (*Spinacia oleraceae* L.) under hydroponic culture. In: T. Ando, K. Fujita, T. Mae, S. Matsumoto, S. Mori, J. Sekiya, eds. *Plant Nutrition for Sustainable Food Production and Environment*. Dordrecht: Kluwer Academic Publishers, 1997, pp. 85–86.

43. H.D. Chapman. *Diagnostic Criteria for Plants and Soils*. Riverside: Division of Agricultural Science, University of California, 1966.

44. R.R. Brooks. Accumulation of nickel by terrestrial plants. In: J.O. Nriagu, ed. *Nickel in the Environment*. New York: Wiley, 1980, pp. 407–430.

45. E.G. Bollard. Involvement of unusual elements in plant growth and nutrition. In: A. Lauchli, R.L. Bielski, eds. *Encyclopedia of Plant Physiology*, New Series, Vol. 15B, 1983, pp. 695–755.

46. CJ Asher. Beneficial elements, functional nutrients, and possible new essential elements. In: JJ Mortvedt, ed. *Micronutrients in Agriculture*, 2nd edition. SSSA book series #4. Madison, WI: Soil Science Society of America, 1991, pp. 703–723.

47. R. Gabrielli, T. Pandolfini, O. Vergnano, M.R. Palandri. Comparison of two serpentine species with different nickel tolerance strategies. *Plant Soil* 122:271–277, 1990.

48. P.H. Brown, L. Dunemann, R. Schultz, H. Marschner. Influence of redox potential and plant species on the uptake of nickel and cadmium from soils. *Zh. Pflanzenernahr. Bodenkd.* 152:85–91, 1989.

49. D.E. Salt, M. Blaylock, N.P.B.A. Kumar, V. Dushenkov, B.D. Ensley, I. Chet, I. Raskin. Phytoremediation: A novel strategy for the removal of toxic metals from the environment using plants. *Biotech.* 13:468–474, 1995.

50. F.H. Nielson, H.T. Reno, L.O. Tiffin, R.M. Welch. Nickel. In: *Geochemistry and the Environmen,. Vol. II, The Relation of Other Trace Elements to Health and Disease*. Washington, DC: National Academy of Science, 1977, pp. 40–53.

51. R.D. Reeves, A.J.M. Baker, A. Borhidi, R. Berazain. Nickel hyperaccumulation in the serpentine flora of Cuba. *Ann. Bot.* 83:29–38, 1999.

52. M.M. Guha, R.L. Mitchell. Trace and major element composition of the leaves of some deciduous trees. II. Seasonal changes. *Plant Soil* 24:90–112, 1966.

53. B. Singh, Y.P. Dang, S.C. Mehta. Influence of nitrogen on the behavior of nickel in wheat. *Plant Soil* 127:213–218, 1990.

54. R.D. Macnicol, P.H.T. Beckett. Critical tissue concentrations of potentially toxic elements. *Plant Soil* 85:107–129, 1985.

55. A. Cottenie, A. Dhaese, R. Camerlynck. Plant quality response to uptake of polluting elements. *Qual. Plant.* 26:293–319, 1976.

56. D.A. Cataldo, T.R. Garland, R.E. Wildung. Nickel in plants. I. Uptake kinetics using intact soybean seedlings. *Plant Physiol.* 62:5636–5665, 1978.

57. L.V. Kochian. Mechanisms of micronutrient uptake and translocation in plants. In: J.J. Mortvedt, ed. *Micronutrients in Agriculture,* 2nd edition. SSSA book series #4. Madison, WI: Soil Science Society of America, 1991, pp. 229–296.

58. Y. Guo, E. George, H. Marschner. Contribution of an arbuscular mycorrhizal fungus to the uptake of cadmium and nickel in bean and maize plants. *Plant Soil* 184:195–205, 1996.

59. U. Ahonen-Jonnarth, R.D. Finlay. Effects of elevated nickel and cadmium concentrations on growth and nutrient uptake of mycorrhizal and nonmycorrhizal *Pinus sylvestris* seedlings. *Plant Soil* 236:129–138, 2001.

60. D.A. Cataldo, K.M. McFadden, T.R. Garland, R.E. Wildung. Organic constituents and complexation of nickel(II), iron(III), cadmium(II) and plutonium(IV) in soybean xylem exudates. *Plant Physiol.* 86:734–739, 1988.

61. L.O. Tiffin. Translocation of nickel in xylem exudates of plants. *Plant Physiol.* 48:273–277, 1971.

62. D.G. Barceloux. Nickel. *J. Toxicol.: Clin. Toxicol.* 37:239–242, 1999.

63. R.H.M. van de Graaff, H.C. Suter, S.J. Lawes. Long-term effects of municipal sewage on soils and pastures. *J. Environ. Sci. Health A* 37:745–757, 2002.

64. M.F. Hovmand. Cycling of Pb, Cd, Cu, Zn and Ni in Danish agriculture. In: S. Berglund, R.D. Davis, P.L. Hermite, eds. *Utilisation of Sewage Sludge on Land: Rates of Application and Long-Term Effects of Metals.* Dordrecht: D. Reidel, 1984, pp. 166–185.

65. U. Kukier, R.L. Chaney. Amelioration of nickel phytotoxicity in muck and mineral soils. *J. Environ. Qual.* 30:1949–1960, 2001.

66. D.R. Sauerbeck, A. Hein. The nickel uptake from different soils and its prediction by chemical extractions. *Water Air Soil Pollut.* 57–59:861–871, 1991.

67. A.U. Haq, M.H. Miller. Prediction of available soil Zn, Cu, Mn using chemical extractants. *Agron. J.* 64:779–782, 1972.

68. T. Becquer, F. Rigault, T. Jaffre. Nickel bioavailability assessed by ion exchange resin in the field. *Commun. Soil Sci. Plant Anal.* 33:439–450, 2002.

69. M. Khan, J. Scullion. Effects of metal (Cd, Cu, Ni, Pb or Zn) enrichment of sewage-sludge on soil microorganisms and their activities. *Appl. Soil Ecol.* 20:145–155, 2002.

70. R.S. Boyd, M.A. Davis, M.A. Wall, K. Balkwill. Nickel defends the South African hyperaccumulator *Senecio coronatus* (Asteraceae) against *Helix aspersa* (Mollusca: Pulmonidae). *Chemoecology* 12:91–97, 2002.

71. H. Kupper, E. Lombi, F.-J. Zhao, G. Wieshammer, S.P. McGrath. Cellular compartmentation of nickel in the hyperaccumulators *Alyssum lesbiacum, Alyssum bertolonii* and *Thlaspi goesingense. J. Exp. Bot.* 52:2291–2300, 2001.

72. C. Bai, C.C. Reilly, B.W. Wood. Nickel deficiency disrupts metabolism of ureides, amino acides, and organic acids of young pecan foliage. *Plant Physiol.* 140:433–443, 2006.

15 Zinc

J. Benton Storey
Texas A&M University, College Station, Texas

CONTENTS

15.1   Introduction 411
          15.1.1   Early Research on Zinc Nutrition of Crops 411
15.2   Absorption and Function of Zinc in Plants 412
15.3   Zinc Deficiency 412
15.4   Zinc Tolerance 415
15.5   Trunk Injection...422
15.6   Zinc in Soils 422
15.7   Phosphorus–Zinc Interactions 423
15.8   Tryptophan and Indole Acetic and Synthesis..423
15.9   Root Uptake 423
15.10 Foliar Absorption 424
          15.10.1   Influence of Humidity on Foliar Absorption 427
15.11 Role of Zinc in DNA and RNA Metabolism and Protein Synthesis 428
15.12 Zinc Transporters and Zinc Efficiency 428
15.13 Summary 429
References 430

15.1   INTRODUCTION

15.1.1   EARLY RESEARCH ON ZINC NUTRITION OF CROPS

Discovery of zinc as an essential element for higher plants was made by Sommer and Lipman (1) while working with barley (*Hordeum vulgare* L.) and sunflower (*Helianthus annuus* L.). However, Chandler et al. (2) stated that Raulin, as early as 1869, reported zinc to be essential in the culture media for some fungi, and speculated that zinc was probably essential in higher plants. Skinner and Demaree (3) reported on a typical Dougherty county pecan (*Carya illinoinensis* K. Koch) orchard in Georgia. Pecan trees that were placed in a study that started in 1918 increased in trunk diameter, but their tops had dieback each year, and their condition 'appeared hopeless' in 1922. Fertilizers (N, P, K), cover crops, and all known means were of no avail. Rosette, or related dieback, had been recognized since around 1900, but it was in 1932 before zinc was found to be the corrective element (4,5). The common assumption among pecan growers was that a deficiency of iron was responsible for rosette as pecans were brought into cultivation in the early 1900s. Alben used 0.8 to 1.0% solutions of $FeCl_2$ and $FeSO_4$ in his rosette treatments in 1931 and obtained conflicting results. The 1932 treatments included injections into dormant trees, soil applications while the trees were dormant and after the foliage was well developed, and foliar spraying and dipping. The only favorable results were obtained when Alben mixed the iron solutions in zinc-galvanized containers. Analysis proved that the solutions contained considerable quantities of zinc. These experiments led to the use of $ZnSO_4$ and $ZnCl_2$ solutions, which permitted normal development of new leaves. Satisfactory results were obtained with trees located on alkaline or acid soils. The most satisfactory results were obtained with a foliar spray of 0.18% $ZnSO_4$ and a 0.012% $ZnCl_2$ solution. Roberts and Dunegan (6) also observed a bactericidal effect when using a $ZnSO_4$-hydrated lime mixture that controlled bacterial leaf spot (*Xanthomonas pruni*), which later became a serious pest for susceptible peach (*Prunus persica* Batsch.) cultivars like 'Burbank July Elberta' in the 1940s, 'Sam Houston' in the 1960s, and 'O-Henry' in the 1990s (personal experience). Hydrated lime was necessary to prevent defoliation of peach trees by $ZnSO_4$ toxicity.

15.2   ABSORPTION AND FUNCTION OF ZINC IN PLANTS

Zinc is taken up predominantly as a divalent cation (Zn^{2+}), but at high pH it is probably absorbed as a monovalent cation ($ZnOH^+$) (7). Zinc is either bound to organic acids during long distance transport in the xylem or may move as free divalent cations. Zinc concentrations are fairly high in phloem sap where it is probably complexed to low-molecular-weight organic solutes (8). The metabolic functions of zinc are based on its strong tendency to form tetrahedral complexes with N-, O-, and particularly S-ligands, and thus it plays a catalytic and structural role in enzyme reactions (9).

Zinc is an integral component of enzyme structures and has the following three functions: catalytic, coactive, or structural (9,10). The zinc atom is coordinated to four ligands in enzymes with catalytic functions. Three of them are amino acids, with histidine being the most frequent, followed by glutamine and asparagine. A water molecule is the fourth ligand at all catalytical sites. The structural zinc atoms are coordinated to the S-groups of four cysteine residues forming a tertiary structure of high stability. These structural enzymes include alcohol dehydrogenase, and the proteins involved in DNA replication and gene expression (11). Alcohol dehyrogenase contains two zinc atoms per molecule, one with catalytic reduction of acetaldehyde to ethanol and the other with structural functions. Ethanol formation primarily occurs in meristematic tissues under aerobic conditions in higher plants. Alcohol dehyrdrogenase activity decreases in zinc-deficient plants, but the consequences are not known (7). Flooding stimulates the alcohol dehydrogenase twice as much in zinc-sufficient compared with zinc-deficient plants, which could reduce functions in submerged rice (12).

Carbonic anhydrase (CA) contains one zinc atom, which catalyzes the hydration of carbon dioxide (CO_2). The enzyme is located in the chloroplasts and the cytoplasm. Carbon dioxide is the substrate for photosynthesis in C_3 plants, but no direct relationship was reported between CA activity and photosynthetic CO_2 assimilation in C_3 plants (13). The CA activity is absent when zinc is extremely low, but when even a small amount of zinc is present, maximum net photosynthesis can occur. Photosynthesis by C_4 metabolism is considerably different (14,15) than that occurring in C_3 plants. For C_4 metabolism, a high CA activity is necessary to shift the equilibrium in favor of HCO_3^- for phosphoenolpyruvate carboxylase, which forms malate for the shuttle into the bundle sheath chloroplasts, where CO_2 is released and serves as substrate of ribulosebisphosphate carboxylase.

15.3   ZINC DEFICIENCY

Zinc deficiency is common in plants growing in highly weathered acid or calcareous soils (16). Roots of zinc-deficient trees often exude a gummy material. Major zinc-deficient sites are old barnyards or corral sites, where an extra heavy manure application accumulated over the years. Zinc ions become tied to organic matter to the extent that zinc is not available to the roots of peach trees (17,18). Zinc deficiency initially appears in all plants as intervenial chlorosis (mottling) in which lighter green to pale yellow color appears between the midrib and secondary veins (Figure 15.1 and Figure 15.2) Developing leaves are smaller than normal, and the internodes are short. Popular names describe these conditions as 'little leaf' and 'rosette' (19,20). Pecan trees in particular suffer

FIGURE 15.1   Zinc deficiency of peaches (*Prunus persica* Batsch) is expressed as developing leaves that are smaller than normal and the internodes are shorter causing leaves to be closer to each other and thence the popular names which describes the terminal branches as 'little leaf'. (Photograph by J.B. Storey.) (For a color presentation of this figure, see the accompanying compact disc.)

FIGURE 15.2   Zinc-deficient pecan (*Carya illinoinensis* K. Koch) leaves (left) can contain less than 30 mg Zn per kg compared to over 80 mg Zn per kg Zn in healthy leaves (right). The zinc-deficient leaves have small crinkled leaves that are mottled with yellow. Healthy zinc-sufficient leaves are dark green. Actual zinc concentration of each leaf is shown in the photograph. (Photograph by J.B. Storey.) (For a color presentation of this figure, see the accompanying compact disc.)

FIGURE 15.3   Zinc-deficient pecan (*Carya illinoinensis* K. Koch) trees have shorter internodes so that the leaves are closer together forming a rosette of poorly formed crinkled, chlorotic leaves. (Photograph by J.B. Storey.) (For a color presentation of this figure, see the accompanying compact disc.)

FIGURE 15.4   If the rosetted pecan (*Carya illinoinensis* K. Koch) trees are not treated, the terminals die followed by death of the entire tree. Dieback can occur on young or old trees. (Photograph by J.B. Storey.) (For a color presentation of this figure, see the accompanying compact disc.)

from shortened internodes (rosette) (Figure 15.3). Shoot apices die (shoot die-back) under severe zinc deficiency, as in a tree in Comanche county, Texas (Figure 15.4). Forest plantations in Australia have shown similar symptoms (21). Citrus often show diffusive symptoms (mottle leaf) (Figure 15.5). The ideal time to demonstrate citrus trace element deficiency symptoms is in winter months when the

FIGURE 15.5 Mottled leaf symptoms characterize zinc deficiency symptoms in citrus (*Citrus* spp. L.). (Photograph by J.B. Storey.) (For a color presentation of this figure, see the accompanying compact disc.)

soil is relatively cold. Treatment with zinc fertilizers is not necessary if the symptoms disappear when the soil temperature rises in the spring. Sorghum (*Sorghum bicolor* Moench) that is deficient in zinc forms chlorotic bands along the midrib and red spots on the leaves (22). Shoots are more inhibited by zinc deficiency than roots (23). For most plants, the critical leaf zinc deficiency levels range from 10 to 100 mg kg^{-1} depending on species (Table 15.1).

15.4 ZINC TOLERANCE

Zinc is the heavy metal most often in the highest concentrations in wastes arising in industrialized communities (21). Zinc exclusion from uptake, or binding in the cell walls, does not seem to contribute to zinc tolerance (24,25). Zinc exclusion might exist in Scots pine (*Pinus sylvestris* L.), where certain ectomycorrhizal fungi retain most of the zinc in their mycelia, resulting in the ability of the plant to tolerate zinc (26). Infections with ectomycorrhizal fungi are beneficial for the growth and development of pecan (27). These fungi are highly specialized parasites that do not cause root disease. They are symbiotic, thus gaining substance from the root and contributing to the health of the root.

Tolerance is achieved through sequestering zinc in the vacuoles, and zinc remains low in the cytoplasm of tolerant plants, whereas zinc is stored in the cytoplasm of non-tolerant clones (28). Positive correlation between organic acids such as citrate and malate with zinc in tolerant plants indicates a mechanism of zinc tolerance (29,30). Zinc tolerance in tufted hair grass (*Deschampsia caespitosa* Beauvois) was increased in plants supplied with ammonium as compared to nitrate nutrition. This effect apparently is caused by greater accumulation of asparagine in the cytoplasm of ammonium-fed plants, which form stable complexes with asparagines and zinc (31).

Foliar application of chelates is inefficient because of poor absorption of the large organic molecules through cuticles (32,33). Foliar ZnSO$_4$ treatments are toxic to peach leaves (34) and to many other species, probably because sulfur accumulates on leaves and results in salt burn. A zinc nitrate-ammonium nitrate-urea fertilizer (NZN^{TM}; 15% N, 5% Zn; Tessenderlo Kerley Group, Phoenix, AZ, U.S.A.) did not burn peach leaves. Apparently, NZN-treated peach leaves do not suffer from salt burn because the nitrate in NZN is readily absorbed in response to the need of leaves for nitrogen in protein synthesis thus not accumulating on the surface to cause leaf burn (34).

TABLE 15.1

Tissue Analysis Values Useful in Indicating Zinc Status

Plant	Conditions of Sampling			Concentration of Zinc in Dry Matter (mg kg^{-1})					Reference
	Type of Culture	Tissue Sampled	Age, Stage, Condition or Date of Sample	Showing Deficiency Symptoms	Low Range	Intermediate Range	High Range	Showing Toxicity Symptoms	
Asparagus (*Asparagus officinalis* L.)	Field		Spears at harvest time			52			99
Azalea (*Rhododendron indicum* Sweet)	Soil	Data bank	Flowering—youngest mature leaf	<15		15–60			100
Barley (*Hordeum vulgare* L.)	Soil	WS	Above ground portion at emergence of head at boot stage	<15		15–70	>70		101
Alfalfa (*Medicago sativa* L.)	Field	Tops	12 weeks old	13		39–48			102
Almond (*Prunus dulcis* D.A. Webb)	Field	Leaves (t)	Midshoots	<15		25–30			103
Apple (*Malus* spp.)	Field	Leaves		<20		35–50			104
Apricot (*Prunus armeniaca* L.)	Field	Leaves	Apical 6 to 8 in (September–October)	24–30		19–31			105
Avocado (*Persea americana* Mill.)	Field	Leaves	Mature	4–15		50			106
Clover, subterranean (*Trifolium subterraneum* L.)	Solution	Tops	12 weeks old	24–25		76–90			102
Beans (*Phaseolus vulgaris* L.)	Field	Mature leaf blade	Various ages	7–22		18–40			107
Beans	Field	Leaflets	Peak harvest			46			108
Beans	Field	Pods	Peak harvest			34			108
Beans	Field	Seed	Seed harvest			37			108
Beet (*Beta vulgaris* L.)	Field	Youngest mature leaf + petiole	Mature			15–30			109
Blueberry, High bush (*Vaccinium corymbosum* L.)	Field	Leaves	From 6th node from tip	<8		8–30	31–80	>80	110

Crop	Growth medium	Plant part	Growth stage						Ref.
Boston Fern (*Nephrolepis exaltata* Schott.)	Soil culture	Early sprout growth	Pinnae from whole fronds or 10–12 cm midsection			35–50			111
Brussels Sprouts (*Brassica oleracea* var. *gemmifera* Zenker)	Field	Upper leaves	Heart, 7 cm			26–35			112
Cabbage (*Brassica oleracea* var. *capitata* L.)	Field	Head (core sample)	Peak harvest			34			109
Carnation (*Dianthus caryophyllus* L.)			5th pair of leaves from apex of lateral before flowering	<15		25–75			100
Capsicum (*Capsicum annuum* L.) Bell Pepper	Soil and database	Youngest mature leaf + petiole	Early fruit	18–19		20–200			113
Carrot (*Daucus carota* var. *sativus* Hoffm.)	Peat	Above ground portion	Peak harvest			184–490			114
Cassava (*Manihot esculentum* Crantz)	Field	Leaves	63 days–youngest mature leaf	<35	35–50	40–100			115
Cassava	Field	Young leaf blade	43 days	<25	25–30	30–60	60–120	>120	116
Celery (*Apium graveolens* var. dulce Pers.)	Field	Petioles	Midgrowth			30–100			99
Cherry (*Prunus avium* L.)	Field	Midshoot leaves		<15	15–19	20–50	51–70	>70	117
Chrysanthemum (*Chrysanthemum morifolium* Ramat.)	Sand	Lower leaf on flower stem	Above ground portion 70 days after planting	<6.8	7	7.0–26.0	>100		118
Citrus (*Citrus* spp. L.)	Field	Midshoot leaves		<16	16–24	25–100	100–300	>300	119
Coffee (*Coffea arabica* L.)	Field	Leaves	Four pairs of leaves from tip of actively growing shoots	<10	10–15	15–30			120
Corn (*Zea mays* L.)	Field	Lower leaves	Tasseling	9–9.3		31.10–36.60			121
Corn	Field	Leaves	6th node from base At silking	15–24		25–100	101–150		113

Continued

TABLE 15.1 (*Continued*)

Plant	Conditions of Sampling			Concentration of Zinc in Dry Matter (mg kg⁻¹)					Reference
	Type of Culture	Tissue Sampled	Age, Stage, Condition or Date of Sample	Showing Deficiency Symptoms	Low Range	Intermediate Range	High Range	Showing Toxicity Symptoms	
Corn	Field	6th leaf above base	Full tasseling		15				122
Corn	Field	Ear leaf blade	Silking	<10		20–70	71–100	>100	123
Cotton (*Gossypium hirsutum* L.)	Soil culture	Youngest mature leaf blade	43 days		13–14	17–48		200	124
Cowpea (*Vigna unguiculata* Walp.)	Soil culture	Upper leaf blades	40 days	15–17	20	50–290			125
Cucumber (*Cucumis sativus* L.)	Field	Youngest mature leaf	Harvest			50–150			126
Dieffenbachia (*Dieffenbachia exotica*)	Database		Portion above ground			25–150			127
Fig (*Ficus carica* L.)	Field		Midsummer. 1st full size basal leaf		<15	>15			128
Flax (*Linum usitatissimum* L.)	Pots	Tops	71 days old	18		32–83			129
Geranium (*Pelargonium zonale* Ait.)	Flowering		All above ground portion	<6		8–40			100
Grape (*Vitis vinifera* L.)	Vineyard	1 petiole for each 100 vines	Petiole of basal leaf opposite bunch cluster	<15	15–26	>26			130, 131
Hazelnut (*Corylus avellana* L.)	Orchard		Midshoot leaves of current season's growth	<10		60–80	80–300	>300	128
Kiwi fruit (*Actinidia chinensis* Planch.)	Vineyard	Minimum of 10 leaves	1st leaf above fruit toward growing tip	<12		15–22	23–30	>30	132
Lettuce (*Lactuca sativa* L.)	Peat–vermiculite	Leaf	28 day old			39–71			133

Macadamia (*Macadamia integrifolia* Maiden and Betche and *M. tetraphylla* L.A.S. Johnson)	Mature leaves when hardened	4 pairs of leaves from 20 trees	Fruit set half developed	<10	10–15	15–50	>50		134
Mango (*Mangifera indica* L.)	Leaves after flowering	60 leaves in 2nd or 3rd position back of base of bloom		<15		20–150			135
Muskmelon (*Cucumis melo* L.)	Field	Youngest mature leaf	Harvest			30–80			109
Oat (*Avena sativa* L.)	Hydroponic	Plant tops		<15		15–70	>70		136
Olive (*Olea europea* L.)	Orchard	Fully expanded basal to midshoot leaves	Collect 4 leaves/tree from 25 trees			10–30			103
Onion (*Allium cepa* L.)	Field	First mature leaf	Midgrowth			30–100			99
Orange (*Citrus sinensis* Osbeck.)	Field	Leaves	4–7 months old	<15	16–24	25–100	110–200	300	137
Oil palm (*Elaeis guineensis* Jacq.)	Leaflets	6 upper and 6 lower leaflets from frond	Frond 17 mature or Frond 3 if young planting			15–20			138
Ground nuts (*Arachis hypogaea* L.)	Field	Young midleaf	Preflower to flower	18–20		25–80	>80		139
Pea (*Pisum sativum* L.)	Field	Above ground portion	Bud stage			34–36		236–665	140
Pea	Field	Pods	Early pod fill			24			108
Pea	Field	Seed	Seed harvest			61			108
Peach (*Prunus persica* Batsch.)	Orchard	4 leaves from 25 trees	Middle leaves from current season shoots	<15	15–19	20–50	51–70	>70	141
Pear (*Pyrus communis* L.)				15		15–30	>40		104
Pecan (*Carya illinoinensis* K. Koch)	Orchard	Leaflets	10 leaflets from midshoot of 10 trees	<30	30–49	50–100	>250		142

Continued

TABLE 15.1 (Continued)

Plant	Conditions of Sampling			Concentration of Zinc in Dry Matter (mg kg^{-1})					Reference
	Type of Culture	Tissue Sampled	Age, Stage, Condition or Date of Sample	Showing Deficiency Symptoms	Low Range	Intermediate Range	High Range	Showing Toxicity Symptoms	
Pecan	Orchard	100 leaflets from 50 midshoot leaves	Select leaves from mid shoot in midseason (July) at half tree height or 2 m.	<30	40–50	60–100	100–200		74
Pistachio (*Pistacia vera* L.)	Orchard	Single leaflets	6 subterminal leaflets near mid-non-bearing shoots 1 mo before harvest			7–14			143
Poinsettia (*Euphorbia pulcherrima* Willd.)			Upper most mature leaf just before flowering	<15		25–60			100
Potato (*Solanum tuberosum* L.)	Field, Sand and Database	Youngest mature leaf	Tubers half grown			20–40			109
Plum (*Prunus* spp. L.)	Orchard	Leaves from midcurrent season	Collect 4 leaves/tree in midseason	<15	15–19	20–50	51–70	>70	144
Raspberry, red (*Rubus idaeus* L.)	Leaves	5th to 12th leaves	Leaves taken 2–3 weeks after final pick	<13		34–80			145
Rice (*Oryza sativa* L.)	Soil	All top part of plant	Flowering		16	20–100		190	146
Rose, hybrid tea (*Rosa* spp. L.)		2nd and 3rd 5 leaflet leaves	1 day before flowering			24			147
Sorghum (*Sorghum bicolor* Moench)	Sand	All top part of plant	Stage 3	<11		40–50	>70		148
Soybean (*Glycine max* Merr.)		All top part of plant	Early flower			20–100			149
Spinach (*Spinacia oleracea* L.)	Field	Youngest mature leaf + petiole	30–50 days of age			50–75			109

Plant	Source	Plant part	Growth stage						Ref.
Strawberry (*Fragaria* spp. L.)	Field					58–73			104
Strawberry (*Fragaria* sp.)	Field	Blade + petiole	Select 30 or 40 leaves of 1 cultivar during growing season.	<20	20	30–50			150
Sugar beet (*Beta vulgaris* L.)	Solution culture	All top part of plant	83 days old	2–13	9	10–80			151
Sugar cane (*Saccharum* spp. L.)	Field	Sheaths 3–6	Rapid growth	<10	10	10–100			152
Sunflower (*Helianthus annuus* L.)	Soil and databank	3rd and 4th Leaves below flower bud	Florets about to emerge	20	30	190	240	>240	153
Sweet corn (*Zea mays rugosa* Bonaf.)	Field	Ear leaf	Postsilking			20–40			109
Tea (*Camellia sinensis* O. Kuntze)	Field	Mature leaves	At plucking	<3					154
Tobacco (*Nicotiana tabacum* L.)	Survey data	All top part of plant	Flowering			20–80			155
Tomato (*Lycopersicon esculentum* Mill.)	Field	All plant parts above ground	Mature fruit	17		24–60			156
Watermelon (*Citrullus lanatus* Matsum. and Nakai)	Field	Oldest mature leaf + petiole	Midgrowth		17	20–60			108
Walnut (*Juglans regia* L.)	Field	Leaves	Midgrowth			20–200			104
Wheat (*Triticum aestivum* L.)	Field and survey data	All top part of plant	Fleeks scale 10.1	<15		15–70	>70		136

15.5 TRUNK INJECTION

Experience with trunk injections of zinc has been disappointing in all cases despite rumors of success. It would seem logical that placement of any form of zinc in the secondary xylem of an actively transpiring tree would utilize the xylem vessels to rapidly transport the zinc to the actively growing meristems. However, many researchers including Millikan and Hanger (35,36) have proven that zinc transport is more complex than injecting zinc in any form into tree trunks. Millikan and Hanger (36) reported that ^{65}Zn moved from the injection point only when zinc was injected into the bark of 2-year-old apple trees. Supplying ethylenediaminetetraacetic acid (EDTA) enhanced ^{65}Zn movement in an acropetal (upward) direction only. The ^{65}Zn was distributed to spurs and laterals on the distal side of the injection point. Millikan and Hanger (36) also reported that ^{65}Zn accumulated at the nodes on lateral branches and in the petioles, midrib, and major veins of the leaves. Wadsworth (37) reported no significant effect of ZnEDTA applied via injection into the secondary xylem of mature 'Western' or 'Burkett' pecan tree leaves on nut quality or yield. He suggested that the volume of zinc was inadequate to influence such a large tree. The possibility of home owners using this means of applying zinc to their large pecan landscape trees, which would otherwise require large spray machines, was discounted by the danger of small children pulling them out of the trunks and inserting them in their mouths. The direct application of zinc chelates to the secondary xylem via injection was unsuccessful primarily because of the small volume of zinc injected (37).

15.6 ZINC IN SOILS

Zinc has a complete $3d^{10}4s^2$ outer electronic configuration and, unlike the other d block micronutrients such as such as manganese, molybdenum, copper, and iron, has only a single oxidation state and hence a single valence of II. The average concentration of zinc in the crust of the Earth, granitic, and basaltic igneous rock is approximately 70, 40, and 100 mg kg^{-1}, respectively (38), whereas sedimentary rocks like limestone, sandstone, and shale contain 20, 16, and 95 mg kg^{-1}, respectively (39). The total zinc content in soils varies from 3 to 770 mg kg^{-1} with the world average being 64 mg kg^{-1} (40).

There are five major pools of zinc in the soil: (a) zinc in soil solution; (b) surface adsorbed and exchangeable zinc; (c) zinc associated with organic matter; (d) zinc associated with oxides and carbonates; and (e) zinc in primary minerals and secondary alumino-silicate materials (41).

There is evidence that Zn^{2+} activities in the soil solution may be controlled by franklinite ($ZnFe_2O_4$), whose equilibrium solubility is similar to that of soil-held zinc over pH values of 6 to 9 (42,43). The mineral will precipitate whenever zinc concentration in the soil solution exceeds the equilibrium solubility of the mineral and will dissolve whenever the opposite is true. This process provides a zinc-buffering system.

Zinc may be associated with soil organic matter, which includes water-soluble and organic compounds. Zinc is bound via incorporation into organic molecules, exchange, chelation, or by specific and nonspecific adsorption (41).

Zinc is associated with hydrous oxides and carbonates via adsorption, surface complex formations, ion exchange, incorporation into the crystal lattice, and co-precipitation (41). Some of these reactions fix zinc rather strongly and are believed to be instrumental in controlling the amount of zinc in the soil solution (44). Zinc is complexed with $CaCO_3$ in alkaline (pH 8.2) soils in the western half of Texas where most of the pecans are grown in the state (45–47). Soil-incorporated $ZnSO_4$ at 91 kg per pecan tree did not bring the zinc content of the soils to an adequate level because the zinc was transferred from the sulfate form to sparingly soluble $ZnCO_3$ (48).

Five rates of $ZnSO_4$ and three rates of S were supplied to pecan trees in March 1966 in a single application to soil (deep Tivoli sand, pH 8.2; mixed thermic, Typic ustipamments) in Dawson county, Texas (south plains) (49). In the absence of applied sulfur, adding of $ZnSO_4$ in excess of 20 kg per tree was required to raise zinc concentrations in leaflets in June or September 1966 above the minimum optimum of 60 mg kg^{-1}. Additions of sulfur reduced the amount of ZnSO$_4$ required to reach 60 mg kg^{-1} to 18.8 kg per tree with 4.5 kg S per tree and to 16.2 kg per tree with 11.9 kg S per tree. Leaflets collected in September 1967 contained more than 60 mg Zn kg^{-1} if ZnSO$_4$ was applied in March 1966 at rates greater than 4.8 kg per tree. However, in 1967, at any given rate of ZnSO$_4$ (above 1.4 kg per tree), leaflet zinc concentration was reduced by the addition of sulfur, but the concentrations of zinc in the leaflets remained above the minimum optimum level. This study indicates that leaflet zinc of pecan trees in calcareous soils can be increased by soil applications of ZnSO$_4$, but that a larger increase will occur if S is applied with ZnSO$_4$. On the other hand, soil applications seemed impractical considering the fact that with a planting of 86 trees per ha, an application of 120 kg of ZnSO$_4$ ha^{-1} would be required. In acid soils of the southeastern United States, high rates of soil-applied zinc may be responsible for the elusive mouse-ear symptom in the acid soils of the southeastern United States (50). These results agree with Sommers and Lindsay (51), who reported that in soils with high concentrations of heavy metals, nickel will compete with zinc for chelation in acid soils and that cadmium and lead will do the same in alkaline soils.

15.7   PHOSPHORUS–ZINC INTERACTIONS

The higher phosphorus content in zinc-deficient plants supplied with high phosphorus can to some degree be attributed to a concentration effect (52). However, the main reason for the high concentration in the leaves is that zinc deficiency enhances the uptake rate of phosphorus by the roots and translocation to the shoots (53). This enhancement effect is specific for zinc deficiency and is not observed when other micronutrients are deficient. Enhanced phosphorus uptake in zinc-deficient plants can be part of an expression of higher passive permeability of the plasma membranes of root cells or impaired control of xylem loading. Zinc-deficient plants also have a high phosphorus content because the retranslocation of phosphorus is impaired.

15.8   TRYPTOPHAN AND INDOLE ACETIC ACID SYNTHESIS

The most distinct zinc deficiency symptoms are 'little leaf' and 'rosette' in pecans and peaches (Figure 15.1 and Figure 15.2). These symptoms have long been considered to represent problems in indole acetic acid (IAA, auxin) metabolism. However, the mode of action of zinc in auxin metabolism is unidentified. Retarded stem elongation in zinc-deficient tomato (*Lycopersicon esculentum* Mill.) plants was correlated with a decrease in IAA level, but resumption of stem elongation and IAA content occur after zinc is resupplied. Increased IAA levels preceded elongation growth upon resupply of zinc (54), which would be expected if growth was a response of increased supply of auxin caused by application of zinc. Low levels of IAA in zinc-deficient plants are probably the results of inhibited synthesis of IAA (55). There is an increase in tryptophan content in the dry matter of rice (*Oryza sativa* L.) grains by zinc fertilization of plants grown in calcareous soil (56). The lower IAA content in zinc-deficient leaves may be due to the biosynthesis of IAA tryptophan (57). Lower IAA contents may be the result of enhanced oxidative degradation of IAA caused by superoxide generation enhanced under conditions of zinc deficiency (55).

15.9   ROOT UPTAKE

Zinc absorbed by pecan seedlings was translocated predominately to the youngest, physiologically active tissue, in agreement with the results of Millikan and Hanger (35), who worked with subterranean clover (*Trifolium subterraneum* L.). Autoradiograph and radio assays revealed variation between seedlings of open pollinated pecans with respect to rate of Zn absorption (37). For example, one set of seedlings absorbed extremes from 0.7 to 91 mg Zn kg^{-1} if roots were exposed to ^{65}Zn in a beaker of water for 96 h.

Grauke et al. (58) detected the highest concentration of zinc in pecan seedlings originating from west Texas populations compared to those populations indigenous to east Texas, regardless of whether they were grown in central Texas or Georgia. Selecting hard woodcuttings from the best of the west Texas populations would appear to be an ideal way to use clonal rootstocks as a means of establishing pecan orchards on uniformly zinc-absorbing rootstocks in place of the very heterozygous seedlings used in the last 100 years. McEachern (59) consistently was able to root 40% of the juvenile stem cuttings that he treated, whereas less than 10% of the adult cuttings survived. However, the juvenile growth of a pecan tree is confined to the bottom 3 m of the trunk up from the ground line (60). This portion of the trunk is intermediate in rooting response, and all distal trunk and branches are adult. Heavy pollarding of the trees produce only adult compensatory growth that will not root. Juvenile tissue tends to have a high IAA / low ABA ratio, whereas adult tissue tends to have low IAA / high ABA (59). Only about 12% of juvenile pecan stem cuttings developed viable root systems in greenhouse studies, and none of the adult cuttings initiated roots (59). Only the lower 2 m of the trunk of the original seedling tree of a pecan cultivar is juvenile and eligible to produce cuttings that are capable of rooting (59).

Tissue culture became the popular means of clonal propagation in the 1960s because of the work of Skoog and Miller (61). Smith (62) was unsuccessful after trying most of the known plant growth regulators because of endogenous fungi that defied all sanitation procedures. Pecan tissue culture was plagued with *Alternaria* spp. in another study (63). This contamination is more severe in orchard-grown than in greenhouse-grown pecan seedlings but was still present under the most sterile growing conditions. Knox's attempt to culture pecan was unsuccessful. Knox advanced the theory that *Alternaria* is an endophyte or resident fungus. Knox (63) stated that the host pecan tree does not appear to be disadvantaged or diseased. If the vigor of the tree is essentially unaltered, then the fungus cannot be considered a pathogen and is more appropriately described as an endophyte or resident. The vigor of cultured pecan tissues apparently is enhanced by the fungus, perhaps implying a mutualistic relationship between *Alternaria* and pecan trees. There has been a long precedence for resident fungi in pecan roots because ectomycorrhizal fungi are prominent in native pecan groves and are considered to enhance zinc absorption by pecan roots from leaf mulch. Native pecan trees on fence lines, separating a cultivated field from a native pecan grove that is not tilled, will inevitably be rosetted on the side of the tree where the soil has been disturbed by disking compared to normal healthy growth on the untilled side of the tree.

Pecan tissue finally was cultured successfully by using single-node cuttings obtained from 2-month-old seedlings of pecan (64). Cuttings were induced to break buds and form multiple shoots in liquid, woody plant medium and 2% glucose supplemented with 6-benzylamino purine. *In vitro*-derived shoots soaked in 1 to 3 mg indolebutyric acid (IBA) per liter produced adventitious shoots *in vitro*; when soaked for 8 days in 10 mg IBA per liter, they were rooted successfully in soil and acclimated to greenhouse conditions. Etiolation of stock plants did not improve shoot proliferation or rooting under *in vitro* culture (64).

Absorption of zinc varies with species. For example, Khadr and Wallace (65) reported that rough lemon (*Citrus aurantium* L.) absorbed more ^{65}Zn and ^{59}Fe from the soil than trifoliate orange (*Poncirus trifoliate* Raf.).

15.10   FOLIAR ABSORPTION

Tank mixing urea-ammonium nitrate fertilizer (UAN; 0.5% by weight) with $ZnSO_4$ increased leaflet zinc concentration compared to using $ZnSO_4$ alone in pecan. Zinc nitrate was more efficient than $ZnSO_4$ in increasing leaflet concentration, especially if tank mixed with UAN (0.5%). Zinc concentrations of spray solutions can be reduced by one eighth to one fourth of the current recommended rate as $ZnSO_4$ at 86 g per 100 L of water. Use of the lowest rate of $Zn(NO_3)_2$, 10.8 g per 100 L of water + UAN, increased yield and income over the recommended rate of $ZnSO_4$ (66). This paper plus earlier work that led to the formulation of $Zn(NO_3)_2$ + UAN was patented under the name NZN. (NZN® was patented in 1971 by J. Benton Storey and Allied Chemical Co. under the trade mark registration No. 1041108). The work was documented by Storey and coworkers (34,45,46,66–75).

Grauke (76) followed with research which evaluated and expanded previous work with NZN and considered problems of precipitation of zinc in spray formulations. He noted that precipitation of $ZnSO_4$ occurs from NZN stock solutions with 5% Zn and that use of solutions with 1% Zn avoided precipitation. Earlier, Wallace et al. (77) reported increasing absorption of zinc from $ZnSO_4$ with increasing alkalinity up to pH 8. However, use of high-pH zinc formulations is limited because of low stability of the formulations and the precipitation of zinc when stock solutions of high pH are diluted with water. To avoid precipitation, the $ZnSO_4$ and UAN should be sprinkled into an agitated, full tank of water (76).

Pecan and corn (*Zea mays* L.) leaves absorbed more Zn from NZN than from $ZnSO_4$, and absorption of both formulations was increased at high humidity. Grauke (76) noted that differences in the absorption of the formulations were related to their effective concentrations, calculated by multiplying the molecular concentration of the solution by its activity coefficient. Activity coefficients are factors which, when multiplied by the molar concentrations, yield the active mass or effective concentration. Activity coefficients may be calculated for solutions are less that 0.01 M by using the Debye-Huckel equation

$$\text{Log}\,Y\pm = -0.509\,|Z+|\,|Z-|\,(\mu)^{1/2}$$

where $Y\pm$ is the mean ionic activity coefficient, $|Z+|$ the absolute value of the formal charge on the cation, $|Z-|$ the absolute value of the formal charge on the anion, and μ the ionic strength. The ionic strength is a measure of the electrical environment of ions in solution and is a function of concentration:

$$\mu = \frac{1}{2}\sum C_i Z_i^2$$

where $\sum$ is the sum of the concentrations, C_i, for each ionic species multiplied by the formal charge Z_i on the *i*th ion. For example, a 200 mg L^{-1} solution of $Zn(NO_3)_2$ has an ionic strength (μ) of 0.009. When that figure is used in the above equation, the activity coefficient ($Y\pm$) is equal to 0.597. When each of these factors are multiplied by the mole concentration of the solutions, which is 0.003 for each solution, the active mass of respective solutions is obtained: 0.0024 M (156.9 mg L^{-1}) for $Zn(NO_3)_2$ and 0.0018 M (117.7 mg L^{-1}) for $ZnSO_4$. Therefore, although equal concentrations of the two solutions were applied, the active mass of the $ZnSO_4$ solution was only 75% of that in the $Zn(NO_3)_2$ solution.

Application of a 10-µL drop of a 200 mg L^{-1} solution of $^{65}ZnSO_4$ resulted in sorption of 46% of the applied label. The portion of the applied label absorbed by a leaf treated with a 10-µL drop of 200 mg L^{-1} $^{65}Zn(NO_3)_2$ was 74%. Therefore, sorption from the $ZnSO_4$ solution was 62% of that for the $Zn(NO_3)_2$ solution (76).

The inclusion of NH_4NO_3 and urea to either $Zn(NO_3)_2$ or $ZnSO_4$ resulted in a significant increase in translocation of absorbed zinc. There was no significant difference in movement of absorbed zinc between $ZnSO_4 + NH_4NO_3 +$ urea and $Zn(NO_3)_2 + NH_4NO_3 +$ urea. However, the total amount of zinc available to leaves treated with $Zn(NO_3)_2 + NH_4NO_3 +$ urea would be greater, since much more of the applied zinc was absorbed. These data indicate that the efficiency of a foliar zinc application could be increased by using the $Zn(NO_3)_2 + NH_4NO_3 +$ urea treatment, which increases the amount of total zinc absorbed by the leaf as well as the percentage of absorbed zinc translocated from the treatment site. The latter two ingredients of the triad are contained in a commercial 32% N, liquid UAN fertilizer. Grauke's (76) meticulous evaluation of this triad proved that the presence of NH_4NO_3 + urea did not result in increased sorption of either $Zn(NO_3)_2$ or $ZnSO_4$ as would be expected if urea facilitated cuticular penetration (78). Wadsworth (37) and Grauke (76) showed that $Zn(NO_3)_2$ increased zinc absorption more than $ZnSO_4$ with or without urea. By increasing the total absorption of labeled zinc from $Zn(NO_3)_2$ and by increasing the translocation of absorbed zinc from NH_4NO_3 + urea, these treatment showed increased efficiency for foliar zinc fertilization.

A 1975 article in *California Farmer* (California Farmer was a trade journal that featured new products but was not given a publication number) reported positive response with NZN on almonds, cherries, peaches, apples, walnuts, grapes, tomatoes, and head lettuce. The NZN provides the leaf with zinc that is available for synthesis of IAA, which stimulates shoot growth and leaf expansion. The necessity of applying zinc when the cuticles are less formidable dictates application when the leaves are first developing. Most leaf expansion of bearing pecan shoots occurs in the first 2 months of growth, so zinc foliar sprays should be applied at first sign of the green tip emerging through the terminal bud scales. Subsequent foliar Zn sprays should be applied 1, 3, 5, and 8 weeks after green tip (74,79). These early season Zn sprays were based on the work by Wadsworth (37) with pecans and are also supported by the conclusion of Franke (80) that immature leaves with thinner cuticles were more absorptive than mature leaves and that the lower leaf surfaces, which also had thinner cuticles, were slightly more absorptive than the upper leaf surfaces. Labelled ^{65}Zn absorbed by the immature leaves moved primarily acropetally and was deposited in the midrib and lateral veins of the treated leaf.

Small amounts of ^{65}Zn were transported basipetally within the leaf from the treatment spot down the petiole into the transport system of the stem. Acropetal movement of ^{65}Zn was consistently dramatic when $73\,\mu g$ of Zn as $ZnSO_4$, which contained $3.4\,\mu Ci$ ^{65}Zn, was applied to the stem of pecan seedlings by insertion under a phloem patch, thus proving that once zinc negotiates the cuticle there is no problem of rapid acropetal transport (37).

An important unique feature of NZN is its ability to transport zinc absorbed from a $10\,\mu L$ droplet of $200\,mg\,Zn\,L^{-1}$ labeled with $0.3\,\mu Ci$ ^{65}Zn. The percentage of absorbed zinc detected away from the treatment site was greater in leaves treated with NZN (81).

Landscape maintenance firms in the Southwest have long had problems with $ZnSO_4$-induced defoliation of woody ornamentals and fruit trees during spraying of the large ubiquitous pecan trees in landscapes because of drift to landscape species that are susceptible to $ZnSO_4$-induced defoliation. Foliar treatment of 18 species of container-grown woody ornamentals with NZN resulted in no spray damage (82). Zinc concentrations were increased in 13 species compared to untreated plants. Quality was improved in three species without a related increase in zinc content. The ornamentals in this study were not expected to benefit from zinc because they were growing in acid media.

Peach trees are notoriously susceptible to $ZnSO_4$-induced defoliation (83). However, trees suffering from zinc deficiency may develop 'little leaf' if not supplied with zinc. In early practices, use of $ZnSO_4$ was recommended commonly for control of bacterial leaf spot (*Phytomonas pruni*) (84,85). $ZnSO_4$ was considered effective in controlling bacterial leaf spot on peaches in the 1940s, but the spray solution had to include hydrated lime to prevent defoliation (79). Storey Orchards was established on upland sand in 1932 in Red River county, Texas, and grew to 70 acres in the early 1940s. All of the labor, with the exception of harvest, was supplied by the three family members. My remembrance of childhood was spraying the 'Burbank July Elberta' trees with $ZnSO_4$ for the control of bacterial leaf spot and use of hydrated lime to prevent $ZnSO_4$ spray burn. Similarly, Sherbakoff and Andes (86) and Kadow and Anderson (84,85) reported that hydrated lime was used with lead arsenate ($PbHAsO_4$) to prevent leaf burn. Lead arsenate was used for plum curculio control (85). It is interesting to note that $PbHAsO_4$, $ZnSO_4$, and $Ca(OH)_2$ were last reported in a peach spray guide (87) in which DDT was mentioned first. DDT was far more effective in plum curculio control than $PbHAsO_4$, but its use diminished the amounts of zinc applied. Johnson et al. (88) published a spray guide that recommended a copper fungicide and eliminated the need for $ZnSO_4$ in pest control. This recommendation also overlooked the value of $ZnSO_4$ to supply zinc for tree vigor. Today, NZN is used to supply zinc without the danger of spray burn.

Some sandy soils where peaches are grown, such as in Hidalgo county in South Texas and the ridge in Florida (89), are zinc-deficient. In both areas the typical symptom of 'little leaf' was common. Arce (19,20) used three different zinc fertilizers in a Hidalgo county peach orchard. All three fertilizers gave excellent response in preventing little leaf.

15.10.1 INFLUENCE OF HUMIDITY ON FOLIAR ABSORPTION

The method of zinc application is critical. Growers are tempted to use custom-fixed-wing aircraft instead of investing in hydraulic or air-mist ground sprayers. An application of $ZnSO_4$ at 11.2 kg Zn ha^{-1} produced leaves containing 117 mg Zn kg^{-1} on ground-sprayed trees compared with 34 mg Zn kg^{-1} in aerially sprayed trees (34). A typical airplane application is 52 L ha^{-1} (5 gal per acre), whereas a ground application is typically 1728 L ha^{-1} (200 gal per acre). The limited spray volume of water from air application evaporates before adequate absorption occurs, particularly in arid climates.

Pecan leaves treated either with $ZnSO_4$ or NZN at 80% relative humidity showed increased zinc absorption relative to those treated at 40% RH (76). This result is consistent with observations made by Rossi and Beauchamp (90) of increased absorption of $ZnSO_4$ and $ZnCl_2$ at high humidity. Leaves treated under high humidity conditions maintained substantial amounts of surface moisture for 24 h. The increase in sorption is a reflection of the increased hydration, which permitted a longer period of uptake. The inclusion of humectants in foliar soybeans increased leaf nitrogen contents (91). Stein and Storey (91) evaluated 46 different adjuvants in a variety of classes, including alcohols, amines, carbohydrates, esters, ethoxylated hydrocarbons, phosphates, polyethylene glycols, proteins, silicones, sulfates, sulfonates, and alcohol alkoxylates. Glycerol was the only adjuvant that increased the percentage of nitrogen and phosphorus in leaves over the foliar fertilizer controls, which had no adjuvant.

A simple demonstration often used in classroom lectures utilizes a Petri dish of dry $ZnSO_4$ that remains dry throughout a 50-min class period, whereas a Petri dish containing dry $Zn(NO_3)_2$ will contain large drops of water at the end of the class period. The facts that $ZnSO_4$ is hydrophobic and $Zn(NO_3)_2$ is hydrophilic makes the latter more appropriate for arid climates. Relative humidity normally rises to 30% within 30 min after sunrise and rapidly falls to as low as 5% in the El Paso and Mesilla Valleys of Texas and New Mexico (34).

Addition of surfactants reduced hydration time of aerially applied zinc solutions to one third of those without surfactant. The hydration time of a chelated zinc fertilizer alone was 34 min and that of the fertilizer with surfactant was only 12 min in the arid climate of the El Paso Valley (37). With aerial application at 4 kg Zn ha^{-1} (in 76 L of water), foliar zinc content was significantly different at 43 mg kg^{-1} without surfactant and 31 mg kg^{-1} with surfactant. In another experiment, zinc absorption from chelated zinc was reduced from 43 mg kg^{-1} without surfactant to 31 mg kg^{-1} with surfactant. Likewise, zinc accumulation from $ZnSO_4$ treatments containing no surfactant was reduced from 59 to 38 mg kg^{-1} with surfactant. Accelerated evaporation rate was probably due to the surfactants reducing the surface tension of the solution droplets, thus allowing the droplets to spread more evenly over the leaf and thus accelerated loss of spray solution. With the treatment solutions devoid of surfactants, the droplets stood higher thereby decreasing the evaporative surface, allowing additional time for Zn absorption (80). Likewise, pecan trees treated with $ZnSO_4$, via a ground sprayer, at the rate of 5.6 kg Zn per acre in 1892 L of water, at 40% RH, produced leaves containing 189 mg Zn kg^{-1} with a surfactant and 301 mg kg^{-1} without a surfactant (37).

Fully expanded mature pecan leaves were inefficient in foliar absorption of $ZnSO_4$. Abaxial pecan leaf surfaces are only slightly more absorptive than adaxial surfaces (37). The differences were much greater than those reported by Malavolta et al. (92) but were similar to those reported by Heymann-Herschberg (93) for citrus, who further concluded that absorption through the stomata was unimportant. Franke (80) pointed to the cuticular leaf surface as the controller of ion absorption. Wadsworth (37) noted that the immature leaves with thinner cuticles absorbed more zinc than mature leaves. He also found that abaxial surfaces with thinner cuticles were more absorptive than adaxial surfaces. Acropetal transport of zinc was the primary direction of movement. Fourteen percent of the zinc was translocated from auxiliary buds compared with 1% from zinc applied to leaf midribs. This difference suggests that the tender buds had less cuticle than a fully expanded leaf.

Zinc accumulates in the young, expanding leaves. Translocated ^{65}Zn was found predominately in the stem, midrib, and lateral veins with relatively small amounts in the mesophyll (37). Resistance of movement was in the abscission zone. Millikan and Hanger (36) determined that ^{65}Zn accumulated in the nodes. Histological studies would probably confirm a concentration of small cells in the abscission zone, thus accounting for the accumulation of zinc.

15.11   ROLE OF ZINC IN DNA AND RNA METABOLISM AND PROTEIN SYNTHESIS

The role of zinc in cell division and protein synthesis has been known for a long time, but recently a new class of zinc-dependent protein molecules (zinc metalloproteins) has been identified in DNA replication and transcription, thus regulating gene expression (10,11). Zinc is required for binding of specific genes with tetrahedral bonds that result in transcription. By this means the polypeptide chain forms a loop of usually 11 to 13 amino acid residues, which bind the specific DNA sequences. Zinc is therefore directly involved in the translation step of gene expression of DNA elements in these DNA-binding metalloproteins.

Amino acids accumulate in zinc-deficient plants as protein content decreases (54). Protein synthesis resumes when zinc is resupplied because zinc is a structural component of the ribosomes and responsible for their structural integrity. Ribosomes disintegrate in the absence of zinc, but reconstitution reoccurs with the resupply of zinc.

15.12   ZINC TRANSPORTERS AND ZINC EFFICIENCY

The goal of improving Zn utilization efficiency in grafted tree crops is complicated by a complex genetic system involving scion and rootstock, each of which may contribute to the zinc uptake mechanism via systems that are only poorly understood. In pecan (research at Texas A&M University by Storey and colleagues), the genetic adaptations related to nutrient uptake in general vary across the geographic distribution of the species. Leaves were analyzed from ungrafted pecan seedlings grown from seed collected from native pecan populations representing the range of the species. Differences in leaf structure and composition were related to seed origin, with highest specific leaf weights and lowest leaflet area in seedlings originating from Western populations on alkaline soils. These populations were also characterized by higher leaf zinc concentration (58). Pecan cultivars grafted to a common rootstock in a replicated test orchard manifested dramatically different levels of apparent zinc deficiency. Leaves were analyzed for zinc concentrations, which were determined to be quite variable, with the most severe deficiency symptoms on the cultivar with the lowest leaf zinc concentration. However, leaf Zn was correlated poorly to visual deficiency symptoms. Some cultivars with no visual deficiency symptoms had leaf levels in the lowest range, whereas some of these had high leaf Zn concentration.

In an effort to develop a molecular understanding for these zinc nutritional observations, efforts have been initiated to identify zinc transporter genes in this species. Zinc transport across cellular and intracellular membranes is facilitated by several types of membrane-localized proteins, especially the recently characterized Zip transporter family. The name Zip stands for *zrt*-like, *irt*-like protein, with *zrt* (zinc-regulated transporter) and *irt* (iron-regulated transporter) referring to metal transporter genes identified in yeast (94). Several plant genes from various species (e.g., *Arabidopsis thaliana*, pea, tomato, soybean) have now been identified whose translation products demonstrate high homology with the Zip family (95). Functional analysis of several of these proteins has demonstrated them to be divalent metal transporters, with some having high selectivity for Zn2+ (96). Recent work in Grusak's laboratory (M.A. Grusak, USDA-ARS Baylor College of Medicine, Weslaco, TX, U.S.A., personal communication) has led to the identification of six new *Zip* genes in the model legume, annual or barrel medic (*Medicago truncatula* Gaertn.), with some of the genes showing differential expression in leaves versus roots, or in response to Zn-replete versus Zn-deficient conditions (Grusak, personal communication). With the assistance of Grauke (USDA-ARS, Somerville, TX, U.S.A.), Grusak's group has used polymerase chain reaction (PCR) approaches to attempt to clone *Zip* genes in pecan. Primers developed from the *Medicago truncatula Zip* sequences were used to perform PCRs with mRNA isolated from pecan leaves. Leaf samples were collected from a cultivar with low leaf zinc concentration and severe deficiency from a cultivar with low leaf zinc and no apparent deficiency, and from a cultivar with high leaf Zn and no apparent deficiency. Current results have yielded at least three different PCR products from the pecans, whose predicted translations indicate high amino acid sequence homology to *Zip* proteins from *M. truncatula* and other species (see (97,98) and López-Millán, Grusak, and Grauke, unpublished results). Preliminary qualitative PCR analysis also suggests that a putative pecan *Zip* shows higher levels of mRNA expression in the pecan cultivars with no apparent leaf Zn deficiency (i.e., those with either high or low leaf Zn concentration). This *Zip* could be localized to a subcellular membrane and might influence or improve the intracellular partitioning of zinc. These results are exciting because they suggest that whole-plant zinc efficiency may be influenced by scion characteristics. For maximum benefit to cultivated pecan, therefore, appropriate root-mediated uptake mechanisms (e.g., root vigor) may need to be compatibly combined with scion-mediated uptake mechanisms (e.g., the expression or regulation of Zn transport proteins). Further characterization of the pecan *Zip* genes, including analysis of possible polymorphisms between genotypes of diverse geographic origin, should enhance our understanding of zinc nutrition in this crop, and possibly provide tools for breeding new zinc-efficient cultivars.

15.13 SUMMARY

Twentieth century zinc research has discovered that a lack of zinc is expressed in plants as rosettes, low vigor, poor leaf development, and eventual death progressing from the terminal branches. Zinc is unavailable in alkaline soils because of formation of insoluble $ZnCO_3$ and in acid soil where zinc is in competition with nickel. Foliar application has proven difficult because of cuticular barriers as leaves become mature. Frequent zinc foliar applications are more successful than occasional treatments. Traditional $ZnSO_4$ foliar treatments have proven inadequate compared to a nitrate-based zinc spray. The new formula is NZN consisting of $Zn(NO_3)_2 + NH_4NO_3 +$ urea. Nitrogen is superior to sulfur for many reasons in enhancing zinc absorption. Nitrogen is an integral part of all amino acids, whereas sulfur is found in only a few. Sulfur accumulates on the surface of treated crops and can cause spray burn in many. Nitrates are hydrophilic and sulfates are hydrophobic which influence their ability to enter cuticles of treated crops in arid environments.

The increase from 200,000 to 12 million pounds of pecan production in the 30 year span from 1967 to 1997 of the zinc research in the Trans Pecos area of Texas is more than a coincidence (USDA Agricultural Statistics, Texas Department of Agriculture, 1997). This comparison is more justified than in other areas because lack of zinc was the limiting factor in that area. The zinc nutrition problem that confronted the industry in 1965 has been solved. Obviously, the efforts of a number of hard-working pecan growers and horticulturists were instrumental in securing this massive production increase.

There has been a long, unsuccessful struggle to develop a rootstock that will facilitate zinc root absorption. A small percentage of pecan seedlings will absorb and transport zinc. Zinc-regulated transporter proteins have been found in some pecan seedlings that promise to revolutionize the pecan industry and other species. This development is the future to which we can all look, for all of our zinc-deficient species. The preceding horticulturist and agronomists cited in this chapter have discovered the problem. Now the next generation, using advanced technology like zinc-regulated transporter proteins, will eliminate the expense of foliar sprays and soil treatments.

REFERENCES

1. A.L. Sommer, C.B. Lipman. Evidence on the indispensable nature of zinc and boron for higher green plants. *Plant Physiol.* 1:231–249, 1926.
2. W.H. Chandler, D.R. Hoagland, P.F. Hibbard. Little leaf or rosette of fruit trees. II. *Proc. Am. Soc. Hortic. Sci.* 29:255–263, 1932.
3. J.J. Skinner, J.B. Demaree. Relation of soil conditions and orchard management to the rosette of pecan trees. *US Dept. Agric. Bull.* 1378, 1926.
4. A.O. Alben, J.R. Cole, R.D. Lewis. Chemical treatment of pecan rosette. *Phytopathology* 22:595–601, 1932.
5. A.O. Alben, J.R. Cole, R.D. Lewis. New developments in treating pecan rosette with chemicals. *Phytopathology* 22:979–981, 1932.
6. J.W. Roberts, J.C. John, C. Dunegan. Peach brown rot and scab. *US Dept. Agric. Farmers Bull.* 1527, 1927.
7. H. Marschner. *Mineral Nutrition of Higher Plants*, 2nd ed. New York: Academic Press, 1995.
8. L.V. Kochian. Mechanism of micronutrient uptake and translocation in plants. In: J.J. Mortvedt, ed. *Micronutrients in Agriculture.* Madison, WI: Soil Science Society of America Book Series No.4, 1991, pp. 229–296.
9. B.L. Vallee, D.S. Auld. Zinc coordination, function, and structure of zinc enzymes and other proteins. *Biochemistry* 29:5647–5659, 1990.
10. B.L. Vallee, K.H. Falchuk.. The biochemical basis of zinc physiology. *Physiol. Rev.* 73:79–118, 1993.
11. J.E.Coleman. Zinc proteins: Enzymes, storage proteins, transcription factors, and replication proteins. *Annu. Rev. Biochem.* 61:897–946, 1992.
12. P.A. Moore, W.H. Patrick. Effect of zinc deficiency on alcohol dehydrogenase activity and nutrient uptake in rice. *Agron. J.* 80:882–885, 1988.
13. K. Ohki. Effect of zinc nutrition on photosynthesis and carbonic anhydrase activity in cotton. *Physiol. Plant* 38:300–304, 1976.
14. J.N. Burnell, M.D. Hatch. Low bundle sheath carbonic anhydrase is apparently essential for effective C_4 pathway operation. *Plant Physiol.* 86:1252–1256, 1988.
15. C.R. Slack, M.D. Hatch, D.J. Goodchild. Distribution of enzymes in mesophyll and parencyma-sheath chloroplasts of maize leaves in relation to the C_4-dicarboxylic acid pathway of photosynthesis. *Biochem. J.* 114:489–498, 1969.
16. S.P. Trehan, G.S.S. Sekhon. Effect of clay, organic matter and $CaCO_3$ content on zinc absorption by soils. *Plant Soil* 46:329–336, 1977.
17. W.E. Ballinger, H.K. Bell, N.F. Childers. Peach nutrition. In: N.F. Childers, ed. *Temperate to Tropical Fruit Nutrition.* New Brunswick, NJ: Rutgers—The State University, 1966, Chapter 12.
18. J.B. Storey. Peach fertilization, *Proceedings of the Texas Peach and Plum Growers Association.* Texas Agricultural Experiment Station, College Station, TX, 1957.
19. J.P. Arce. Effectiveness of Three Different Zn Fertilizers and Two Methods of Application for the Control of 'Little-Leaf' in Peach Trees in South Texas. Master of Science Thesis, Texas A&M University, College Station, TX:1991.
20. J.P. Arce, J.B. Storey, C.G. Lyons. Effectiveness of three different Zn fertilizers and two methods of application for the control of 'Little-Leaf' in peach trees in south Texas. *Commun. Soil Sci. Plant Anal.* 23:1945–1962, 1992.
21. R. Boardman, D.O. McGuire. The role of zinc in forestry. I. Zinc in forest environments, ecosystems, and tree nutrition. *Forest Ecol. Manag.* 37:167–205, 1990.

22. A.M.C. Furlani, R.B. Clark, C.Y. Sullivan, J.W. Maranville. Sorghum genotype differences to leaf "red speckling" induced by phosphorous. *J. Plant Nutr.* 9:1435–1451, 1986.

23. I.P. Cumbus. Development of wheat roots under zinc deficiency. *Plant Soil* 83:313–316, 1985.

24. J.A. Qureshi, D.A. Thurman, K. Hardwick, H.A. Collin. Uptake and accumulation of zinc, lead and copper in zinc and lead tolerant *Anthoxanthum odoratum* L. *New Phytol.* 100:429–434, 1985.

25. M.D. Vazquez, J. Barcelo, C. Poschenreider, J. Madico, P. Hatton, A.J.M. Baker, G.H. Cope. Localization of zinc and cadmium in *Thlaspi caerulescens*, a metallophyte that can hyperaccumulate both metals. *J. Plant Physiol.* 140:350–355, 1992.

26. J.V. Colpaert, J.A. van Assche, K. Luijtens. The growth of the extrametrical mycelium of ectomycorrhizal fungi and the growth response of *Pinus sylvestris* L. *New Phytol.* 120:127–135, 1992.

27. D.H. Marx. Pecan mycorrhizae—a partnership between fungi and pecan roots. *The Pecan Quart.* 5:4–7, 1971.

28. A. Brookes, J.C. Collins, D.A. Thurman. The mechanism of zinc tolerance in grasses.. *J. Plant Nutr.* 3:695–705, 1981.

29. D.L. Godbold, W.J. Horst, H. Marschner, J.C. Collins, D.A. Thurman. Root growth and Zn uptake by two ecotypes of *Deschampsia caespitosa* as affected by high Zn concentrations. *Z. Pflanzenphysiol* 112:315–324, 1983.

30. D.L. Godbold, W.J. Horst, J.C. Collins, D.A. Thurman, H. Marschner. Accumulation of Zn and organic acids in roots of Zn tolerant and non-tolerant ecotypes of *Deschampsia caespitosa*. *J. Plant Physiol.* 116:59–69, 1984.

31. N. Smirnoff, G.R. Stewart. Nitrogen assimilation and zinc toxicity to zinc tolerant and non-tolerant clones of *Deschampsia caespitosa* (L.) Beauv. *New Phytol.* 107:671–680, 1987.

32. R.E. Worley, S.A. Hammon, R.L. Carter. Effect of zinc sources and methods of application on yield and mineral concentration of pecan, *Carya illinoensis*, Koch. *J. Am. Soc. Hortic. Sci.* 97:364–359, 1972.

33. K. Mengel, E.A. Kirkby. *Principles of Plant Nutrition*. Bern: International Potash Institute, 1987, pp. 527–539.

34. J.B. Storey. The Zn story from beginning to the present. *Proceedings of the Texas Pecan Growers Assn 81st and 82nd Annual Conference*, 2003, pp. 55–58.

35. C.R. Millikan, B.C. Hanger. Effects of chelation and of various cations on the mobility of foliar applied ^{65}Zn in subterranean clover. *Aust. J. Agric. Res.* 18:953–957, 1965.

36. C.R. Millikan, B.C. Hanger. Distribution of ^{65}Zn in pear trees following bark injection. *Aust. J. Agric. Res.* 18:85–93, 1967.

37. G.L. Wadsworth. Absorption and translocation of zinc in pecan trees *Carya illinoensis* (Wang) K. Koch. Master's Thesis, Texas A&M University, College Station, TX, 1970.

38. S.R. Taylor. Abundance of chemical elements in the continental crust: A new table. *Cosmochim. Acta* 28:1273–1286, 1964.

39. K.K. Turekian, K.H. Wedepohl. Distribution of the elements in some major units of the earth's crust. *Geol. Soc. Am. Bull.* 72:175–192, 1961.

40. A. Kabata-Pendias, H. Pendias. *Trace Elements in Soils and Plants*, 2nd ed. Boca Raton, FL: CRC Press, 1992.

41. L.M. Shuman. Chemical forms of micronutrients in soils. In: J.J. Mortvedt, F.R. Cox, L.M. Shuman, R.M. Welch, eds. *Micronutrients in Agriculture*, 2nd ed. Madison, WI: Soil Science Society of America, 1991, pp. 113–144.

42. W.L. Lindsay. Iron oxide solutes solubilization by organic matter and its effects on iron availability. In: Y. Hadar, ed. *Iron Nutrition and Interaction in Plants*. Dordrecht: Kluwer Academic, 1991, pp. 29–36.

43. Q.Y. Ma, W.L. Lindsay. Measurements of free Zn^{2+} activity in uncontaminated and contaminated soils using chelation. *Soil Sci. Soc. Am. J.* 57:963–967, 1993.

44. E.A. Jenny. Controls on Mn, Fe, Co, Ni, Cu, and Zn concentrations in soils and water: The significance role of hydrous Mn and Fe oxides. *Adv.Chem.* 73:377–387, 1968.

45. J.B. Storey, W.B. Anderson. Pecan Zinc Nutrition Research. Texas Agric Expt Station Progress Report 2710, 1970.

46. J.B. Storey, P.W. Westfall, M. Smith. Why do pecans need zinc? *The Pecan Quarterly* 13:3–8, 1979.

47. W.L. Hoover. Retention of Zinc by Soils as Related to Mineralogy and Extraction Methods. Ph. D. Dissertation, Texas A&M University, College Station, TX, 1966.

48. W.L. Lott. The relation of hydrogen ion concentration to the availability of zinc in soil. *Soil Sci. Soc. Am. Proc.* 3:115–121, 1938.

49. M.W. Smith, J.B. Storey, P.N. Westfall, W.B. Anderson. Zinc and sulfur content in pecan leaflets as affected by application of sulfur and zinc to calcareous soils. *HortScience* 15:77–78, 1980.

50. B.W. Wood, C.C. Reilly, A.N. Nyczepir. Nickel corrects mouse-ear. *The Pecan Grower* XV: 7–8, 2003.

51. L.E. Sommers, W.L. Lindsay. Effect of pH and redox on predicted heavy metal-chelate equilibria in soils. *Soil Sci. Soc. Am. J.* 43:39–47, 1979.

52. M.J. Webb, J.F. Loneragan. Effect of zinc deficiency on growth, phosphorous concentration and phosphorous toxicity of wheat plants. *Soil Sci. Soc. Am. J.* 52:1676–1680, 1988.

53. I.H. Cakmak, H. Marschner. Mechanism of phosphorus-induced zinc deficiency in cotton. I. Zinc deficiency-enhanced uptake rate of phosphorus. *Physiol. Plant.* 68:483–490, 1986.

54. C. Tsui. The role of zinc in auxin synthesis in the tomato plant. *Am. J. Bot.* 35:172–179, 1948.

55. I.H. Cakmak, H. Marschner, F. Bangerth. Effect of zinc nutritional status on growth, protein metabolism and levels of indole-3-acetic acid and other phytohormones in bean (*Phaseolus vulgaris* L.). *J. Exp. Bot.* 40:405–412, 1989.

56. M. Singh. Effect of zinc, phosphorous, and nitrogen on tryptophan concentration in rice grains grown on limed and un-limed soils. *Plant Soil* 62:305–308, 1981.

57. A.U. Salimi, D.G. Kenfick. Stimulation of growth in zinc-deficient corn seedlings by the addition of tryptophan. *Crop Sci.* 10:291–294, 1970.

58. L.J. Grauke, B.W. Wood, T.E. Thompson, J.B. Storey. Population of origin affects leaf structure and nutrient concentration of pecan seedlings. *HortScience* 38:663, 2003.

59. G.R. McEachern. The Influence of Propagation Techniques, the Rest Phenomenon, and Juvenility on the Propagation of Pecan, *Carya illinoensis* (Wang) K. Koch. Stem Cuttings. Ph.D. Dissertation, Texas A&M University, College Station, TX, 1973.

60. L.D. Romberg. Some characteristics of the juvenile and the bearing pecan tree. *Proc. Am. Soc. Hortic. Sci.* 44:255–259, 1944.

61. F. Skoog, C.O. Miller. Chemical regulation of growth and organ formation in tissue culture *in vitro*. *Symp. Soc. Exp. Biol.* 11:118–131, 1957.

62. M.W. Smith. Shoot Meristems and Callus Tissue Culture of Pecans *Carya illinoensis* (Wang) K. Koch. Ph.D. Dissertation, Texas A&M University, College Station, TX, 1979.

63. C.A. Knox. Histological and Physiological Aspects of Growth Responses and Differentiation of Pecan, *Carya illinoensis* (Wang) Koch. *In Vitro*. Ph.D. Dissertation, Texas A&M University, College Station, TX, 1980.

64. K.C Hansen, J.E. Lazarte. *In vitro* propagation of pecan seedlings. *HortScience* 19:237–239, 1984.

65. A. Khadr, A. Wallace. Uptake and translocation of radioactive iron and zinc by trifoliate orange and rough lemon. *Proc. Am. Soc. Hortic. Sci.* 85:189–200, 1964.

66. M. Smith, J.B. Storey. Zinc concentration of pecan leaflets as influenced by zinc source and adjuvants. *J. Am. Soc. Hortic. Sci.* 104:474–477, 1979.

67. J.B. Storey, G. Wadsworth, M. Smith, P. Westfall, J.D. Hanna. Pecan zinc nutrition. *Proc. Southeastern Pecan Growers Assn.* 64:87–91, 1971.

68. J.B. Storey, M. Smith, P.W. Westfall, J.D. Hanna, W. Gass, W.C. Henderson. A new method to increase zinc absorption by pecan leaves. *The Pecan Quart.* 7:10–11, 1973.

69. J.B. Storey, M. Smith, P. Westfall. Zinc nitrate opens new frontiers of rosette control. *The Pecan Quart.* 8:9–10, 1974.

70. J.B. Storey. New NZN foliar spray available to pecan growers. *The Pecan Quart.* 9:26, 1975.

71. J.B. Storey, P.W. Westfall, M. Smith. Why do pecans need zinc? *The Pecan Quart.* 13:3–8, 1979.

72. J.B. Storey. Pecan foliar nutrition. *The Pecan Press* 5:4,6, 1985.

73. J.B. Storey. Zinc fertilization for pecans. *The Pecan Press* 5:10, 1986.

74. J.B. Storey. Pecan foliar zinc research at Texas A&M University. *Pecan South* 30:24–25, 1997.

75. J.B. Storey. Zinc nutrition. In: G.R. McEachern, L.A. Stein, eds. *The Texas Pecan Handbook*, Texas A&M University, College Station TX, Texas Horticultural Handbook 105 VI, Extension Horticulture, 1997, pp. 6–7.

76. L.J. Grauke. The Influence of Zinc Carriers on the Foliar Absorption of Zinc by Pecan and Corn. Ph. D. Dissertation, Texas A&M University, College Station, TX, 1982.

77. A. Wallace, V.Q. Hale, C.B. Joven. DTPA and pH effects on leaf uptake of ^{59}Fe, ^{65}Zn, ^{137}Cs, ^{241}Am, and ^{210}Pb. *J. Am. Soc. Hortic. Sci.* 94:684–686, 1969.

78. Y. Yamada, W.H. Jyung, S. Wittwer, M.J. Bukovac. The effects of urea on ion penetration through isolated cuticular membranes and ion uptake by leaf cells. *Proc. Am. Soc. Hortic. Sci.* 87:429–432, 1955.

79. T.P. Cooper. Peach and plum spray schedule. Extension Division, University of Kentucky, Circular 356, 1941.

80. W. Franke. Mechanisms of foliar penetration of solutions. *Annu Rev Plant Physiol* 18:281–300, 1967.

81. L.J. Grauke, J.B. Storey, E.R. Emino, D.W. Reed. The influence of leaf surface, leaf age, and humidity on the foliar absorption of zinc from two zinc sources by pecan. *HortScience* 17:474, 1982.

82. E.R. Emino, J.B. Storey, M.W. Smith. Enhanced zinc uptake by container-grown shrubs with applications of nitrogen zinc nitrate solution. *HortScience* 15:93–94, 1980.

83. H.F. Morris, T.E. Denman, U.A. Randolph, J.B. Storey, H.B. Sorensen, F.R. Brison, E.E. Burns, B.G. Hancock. Production and Marketing Practices for Texas Peaches. Texas Agricultural Experiment Station, Texas Agricultural Extension Service, Bulletin B-986, 1961

84. K.J. Kadow, H.W. Anderson. The Role of Zinc Sulfate in Peach Sprays. Illinois Agricultural Experiment Station Bulletin 414, 1935

85. K.J. Kadow, H.W. Anderson. Further Studies on Zinc Sulfate in Peach Sprays with Limited Tests in Apple Sprays. Illinois Agricultural Experiment Station Bulletin 424, 1936.

86. C.D. Sherbakoff, J.O. Andes. Peach Diseases and Their Control in Tennessee. Tennessee Agricultural Experiment Station Bulletin 157, 1936.

87. D.R. King, H.F. Morris. Control of Insects and Diseases Attacking Peaches in East Texas. Texas Agricultural Experiment Station Progress Report 1656, 1954.

88. J.D. Johnson, M. McWhorter, J.G. Thomas. Suggestions for Controlling Insects and Diseases on Commercial Peaches and Plums. Texas Agricultural Extension Series Leaflet 1329, 1975.

89. R.D. Dickey, G.H. Blackmon. A Preliminary Report on Little-Leaf of the Peach in Florida—A Zinc Deficiency. Florida Agriculture Experiment Station Bulletin 344, 1940.

90. N. Rossi, E.G. Beauchamp. Influence of relative humidity and associated anion on the absorption of Mn and Zn by soybean leaves. *Agron. J.* 63:860–863, 1971.

91. L.A. Stein, J.B. Storey. Influence of adjuvants on foliar absorption of nitrogen and phosphorus by soybeans. *J. Am. Soc. Hortic. Sci.* 111:829–832, 1986.

92. E. Malavolta, J.P. Arzolla, H.P. Haag. Preliminary note on the absorption of radiozinc by young coffee plants grown in a nutrient solution. *Pyton* 6:1–6, 1956.

93. L. Heymann-Herschberg. Effects of combined zinc and sulfur applications on zinc deficiency in orange trees. *Ktavim* 6:83–89, 1956.

94. D.R. Eide. The molecular biology of metal ion transport in *Saccharomyces cerevisiae*. *Annu. Rev. Nutr.* 18:441–469, 1998.

95. M.L. Guerinot. The Zip family of metal transporters. *Biochim. Biophys. Acta* 1465:190–198, 2000.

96. S. Moreau, R.M. Thomson, B.N. Kaiser, B. Trevaskis, M.L. Guerinot, M.K. Udvardi, A. Puppo, D.A. Day. GmZIP1 encodes a symbiosis specific zinc transporter in soybean. *J. Biol. Chem.* 277:4738–4746, 2002.

97. D.R. Ellis, A.F. Lopez-Millan, M.A. Grusak. Metal physiology and accumulation in a Medicago truncatula mutant exhibiting an elevated requirement for zinc. *New Phytol.* 158:207–218, 2003.

98. A.F. Lopez-Millan, D.R. Ellis, M.A. Grusak. Identification and characterization of several new members of the ZIP family of metal ion transporters in Medicago truncatula. *Plant Molec. Biol.* 54:583–596, 2004.

99. A. Bauer. Considerations in the development of soil test for 'available zinc'. *Commun. Soil Sci. Plant Anal.* 2:161–193, 1971.

100. J.W. Mastalerz. *The Greenhouse Environment.* New York: Wiley, 1977, pp. 510–516.

101. W.C. Dahnke, W.C. Barley. In: D.L. Pluncknett, H.B. Sprague, eds. *Detecting Mineral Nutrient Deficiencies in Tropical and Temperate Crops.* Boulder, CO: Westview Press, 1989, pp. 81–90.

102. C.R. Millikan. Relative effects of zinc and copper deficiencies on lucerne and subterranean clover. *Austr. J. Biol. Sci.* 6:164–177, 1953.

103. J. Beutel, K. Uriu, O. Litteland. Leaf analysis of California deciduous fruits. In: H.M. Reisenauer, ed. *Soil and Plant Tissue Testing in California.* Berkeley: Division of Agricultural Sciences, University of California, 1976, pp. 11–14.

104. N.F. Childers, J.R. Morris, G.S. Sibbett. *Modern Fruit Science.* Gainesville, FL: Horticultural Publications, 1995.

105. W.H. Chandler, D.R. Hoagland, P.L. Hibbard. Little leaf or rosette of fruit trees. II. *Proc. Am. Soc. Hortic. Sci.* 29:255–263, 1932.

106. E. Lahav, A. Kadman. Fertilizing for High Yielding Avocado. *Int. Potash Inst. Bull.* 6, 1980.

107. F.G. Viets, Jr., L.C. Boawn, C.L. Crawford. Zinc content of bean plants in relation to deficiency symptoms and yield. *Plant Physiol.* 29:76–79, 1954.

108. N.H. Peck, D.L. Grunes, R.M. Welch, G.E. MacDonald. Nutritional quality of vegetable crops as affected by phosphorus and zinc fertilizers. *Agron. J.* 74:583–585, 1982.

109. G.M. Geraldson, G.R. Klacan, O.A. Lorenz. Plant analysis as an aid in fertilizing vegetable crops. In: L.M. Walsh, J.D. Beaton, eds. *Soil Testing and Plant Analysis*. Madison, WI: Soil Science Society of America, 1973, pp. 365–379.

110. C.C. Doughty, E.B. Adams, L.W. Martin. High Bush Blueberry Production in Washington and Oregon. Co-operative extension Washington and Oregon State Universities and University of Idaho Bulletin No. PNW 215, 1981.

111. G.H. Gilliam, C.E. Evans, R.L. Schumack, C.O. Plank. Foliar sampling of Boston fern. *J. Am. Soc. Hortic. Sci.* 108:90–93, 1983.

112. J.A. Cutcliffe. Effects of lime and gypsum on yields and nutrition of two cultivars of Brussels sprouts. *Can. J. Soil Sci.* 68:611–615, 1988.

113. H.A. Mills, J.B. Jones. *Plant Analysis Handbook II*. Athens, GA: MicroMacro Publishing, 1996.

114. U.C. Gupta, E.W. Chapman. Influence of iron and pH on the yield and iron, manganese, zinc and sulphur concentrations of carrots grown on sphagnum peat soil. *Plant Soil* 44:559–566, 1976.

115. D.G. Edwards, C.J. Asher. Tolerance of Crop and Pasture Species to Manganese Toxicity. *Proceedings of the 9th Int. Colloq. Plant Anal. Fert. Probl.*, Warwich, U.K., 1982, pp. 145–150.

116. R.H. Howeler. The Mineral Nutrition and Fertilization of Cassava. Cassava Production Course. Colombia, CA: CIAT, 1978, pp. 247–292.

117. D.R. Leece. Diagnostic leaf analysis for stone fruit. 5. Sweet cherry. *Aust. J. Exp. Agric. Anim. Husb.* 15:118–122, 1978.

118. O.R. Lunt, A.M. Kofranek, J. Oertli. Some critical nutrient levels in *Chrysanthemum morithlium* cv. 'Good News'. *Plant Anal. Fert. Prob.* 4:398–413, 1964.

119. T.W. Embleton, W.W. Jones, C. Pallares, R.G. Platt. Effect of fertilization of citrus on fruit quality and ground water nitrate-pollution potential. *Proceedings of the Int Soc Citriculture, Griffin, NSW, Australia, 1978*, pp. 280–285, 1980.

120. M. St. J. Clowes, R.H.K. Hill, eds. *Coffee Growers Handbook*. Salisbury, Zimbabwe: Coffee Growers Association, 1981, pp. 62–63.

121. F.G. Viets, Jr., L.C. Boawn, C.L. Crawford, C.E. Nelson. Zinc deficiency in corn in central Washington. *Agron. J.* 45:559–565, 1953.

122. S.W. Melsted, H.L. Motto, T.R. Peck. Critical plant nutrient composition values useful in interpreting plant analysis data. *Agron. J.* 61:17–20, 1969.

123. I.S. Cornforth. Plant analysis. In: I.S. Cornforth, A.G. Sinclair, eds. *Fertilizer and Lime Recommendations for Pastures and Crops in New Zealand*. Wellington: Ministry of Agriculture and Fisheries, 1982, pp. 34–36.

124. K. Ohki. Effect of zinc nutrition on photosynthesis and carbonic anhydrase activity in cotton. *Physiol. Plant* 38:300–304, 1976.

125. D.B. Marsh, L. Waters, Jr. Critical deficiencies and toxicity levels of tissue zinc in relation to cow pea growth and N_2 fixation. *J. Am. Soc. Hortic. Sci.* 110:365–370, 1985.

126. W.F. Bennett. ed. *Nutrient Deficiencies and Toxicities in Crop Plants*. St. Paul, ME: American Phytopathological Society, 1993.

127. J.N. Joiner, W.E. Waters. Influence of cultural conditions on the chemical composition of six tropical foliage plants. *Proc. Trop. Reg. Am. Soc. Hortic. Sci.* 14:254–267, 1969.

128. R.G. Weir, G.C. Cresswell. *Plant Nutrient Disorders 1. Temperate and Subtropical Fruit and Nut Crops*. Melbourne: Inkata Press, 1993.

129. J.F. Loneragan. The effect of applied phosphate on the uptake of zinc by flax. *Aust. J. Sci. Res.* 4:108–114, 1951.

130. J.A. Cook. 1966. Grape nutrition. In: N.F. Childers, ed. *Nutrition of Fruit Crops*, 2nd ed.. New Brunswick, NJ: Horticultural Publications, 1966, pp. 777–812.

131. J.B. Robinson, M.G. McCarthy. Use of petiole analysis for assessment of vineyard nutrient status in the Barossa district of South Australia. *Aust. J. Exp. Agric.* 25:231–240, 1985.

132. G.S. Smith, B.F. Asher, C.J. Clark. *Kiwifruit Nutrition, Diagnosis of Nutritional Disorders.* Wellington: Agpress Communication Ltd., 1985.

133. W.L. Berry, D.T. Krizek, D.P. Ormrod, J.C. McFarlane, R.W. Langhans, T.W. Tibbets. Variation in elemental contents of lettuce grown in base line conditions in five controlled-environment facilities. *J. Am. Soc. Hortic. Sci.* 106:661–666, 1981.

134. R.L. Aitken, P.W. Moody, B.L. Compton, E.C. Gallagher. Plant and soil diagnostic tests for accessing phosphorous status of seedling *Macadamia integrifolia. Aust. J. Agric. Res.* 43:191–201, 1992.

135. T.W. Young, R.C.J. Koo. Increasing yield of 'Parwin' and 'Kent' mangos on Lakeland sand by increased nitrogen and potassium fertilization. *Proc. Florida State Hortic. Soc.* 87:380–384, 1974.

136. R.C. Ward, D.A. Whitney, D.G. Westfall. Plant analysis as an aid in fertilizing small grains. In: L.M. Walsh, J.D. Beaton, eds. *Soil Testing and Plant Analysis.* Madison, WI: Soil Science Society of America, Inc., 1973, pp. 329–348.

137. W. Reuther, P.F. Smith. Leaf analysis of citrus. In: N.F. Childers, ed. *Mineral Nutrition of Fruit Crops.* Sommerville, NJ: Somerset Press, 1954, pp. 257–294.

138. E.A. Rosenquist. Manuring oil palms. In: *The Oil Palm of Malaya.* Kuala Lumpur: Ministry of Agriculture and Co-operatives, 1966, pp. 167–194.

139. R.G. Weir, G.C. Cresswell. *Plant Nutrient Disorders 3. Vegetable Crop.* Melbourne: Inkata Press, 1993.

140. L.C. Boawn, P.E. Rasmussen. Crop response to excessive zinc fertilization of alkaline soil. *Agron. J.* 63:874–876, 1991.

141. P.R. Nicholas, J.B. Robinson. The nutritional status of in the Murray irrigation areas of South Australia. *Agric. Rec. (S. Aust.)* 4:18–21, 1977.

142. D. Sparks. Nutrient concentrations of pecan leaves with deficiency symptoms and normal growth. *HortScience* 13:256–257, 1978.

143. K. Uriu, J.C. Crane. Mineral element changes in pistachio leaves. *J. Am. Soc. Hortic. Sci.* 102:155–158, 1977.

144. D.R. Leece. Diagnostic leaf analysis for stone fruit. 4. Plum. *Aust. J. Exp. Agric. Anim. Husb.* 15:112–117, 1975.

145. M.H. Chaplin, L.W. Martin. The effect of nitrogen and boron fertilizer applications on leaf levels, yield and fruit size of red raspberry. *Commun. Soil Sci. Plant Anal.* 11:547–566, 1980.

146. R.K. Rattan, L.M. Shukla. Critical limits of deficiency and toxicity of zinc in paddy in a typic Ustipsamment. *Commun. Soil Sci. Plant Anal.* 15:1041–1050, 1984.

147. M. Prasad, R.E. Widmer, R.R. Marshall. Soil testing of horticultural substrates for cyclamen substrates and poinsettia. *Commun. Soil Sci. Plant Anal.* 14:553–573, 1983.

148. R.B. Lockman. Mineral composition of grain sorghum plant samples. III. Suggested nutrient sufficiency limits at various stages of growth. *Commun. Soil Sci. Plant Anal.* 3:295–303, 1972.

149. W.E. Sabbe, J.L. Keogh, R. Maples, L.H. Hileman. Nutrient analysis of Arkansas cotton and soybean leaf tissue. *Arkansas Farm Res.* 21:2, 1972.

150. A. Ulrich, M.A.E. Mostafa, W.W. Allen. Strawberry Deficiency Symptoms: A Visual and Plant Analysis Guide to Fertilization. University of California, Division of Agriculture Science Publication No. 4098, 1980.

151. A. Ulrich, F.J. Hills. Sugar Beet Nutrient Deficiency Symptoms. A Colour Atlas and Chemical Guide. University of California Division of Agriculture Science Publication (unnumbered), 1969.

152. W.R. Schmehl, R.P. Humbert. Nutrient deficiencies in sugar crops. In: H.W. Sprague, ed. *Hunger Signs in Crops.* New York: David McKay, 1964, pp. 415–450.

153. N. Khurana, C. Chatterjee. Influence of variable zinc on yield, oil content, and physiology of sunflower. *Commun. Soil Sci. Plant Anal.* 32:3023–3030, 2001.

154. F.N. Fahmy Soil and Leaf Analyses in Relation to Tree Crop Nutrition in Papua New Guinea. Conference on Classification and Management of Tropical Soils. Int. Soc. Soil Commission IV and V, Kuala Lumpur, 1977.

155. J.B. Jones, Jr. Plant analysis handbook for Georgia. *Univ. Georgia Coll. Agric. Bull.* 735, 1974.

156. H.D. Chapman. Zinc. In: H.D. Chapman, ed. *Diagnostic Criteria for Plants and Soils.* Riverside, CA: H. D. Chapman, 1966, pp. 484–499.

Section IV

Beneficial Elements

16 Aluminum

*Susan C. Miyasaka, N.V. Hue, and
Michael A. Dunn*
University of Hawaii-Manoa, Honolulu, Hawaii

CONTENTS

16.1 Introduction 441
16.2 Aluminum-Accumulating Plants 441
16.3 Beneficial Effects of Aluminum in Plants 442
    16.3.1 Growth Stimulation 442
    16.3.2 Inhibition of Plant Pathogens 442
16.4 Aluminum Absorption and Transport within Plants 442
    16.4.1 Phytotoxic Species 442
    16.4.2 Absorption 443
    16.4.3 Aluminum Speciation in Symplasm 443
    16.4.4 Radial Transport 444
    16.4.5 Mucilage 444
16.5 Aluminum Toxicity Symptoms in Plants 444
    16.5.1 Short-Term Effects 444
        16.5.1.1 Inhibition of Root Elongation 444
        16.5.1.2 Disruption of Root Cap Processes 444
        16.5.1.3 Callose Formation 445
        16.5.1.4 Lignin Deposition 445
        16.5.1.5 Decline in Cell Division 445
    16.5.2 Long-Term Effects 445
        16.5.2.1 Suppressed Root and Shoot Biomass 445
        16.5.2.2 Abnormal Root Morphology 446
        16.5.2.3 Suppressed Nutrient Uptake and Translocation 446
        16.5.2.4 Restricted Water Uptake and Transport 446
        16.5.2.5 Suppressed Photosynthesis 446
        16.5.2.6 Inhibition of Symbiosis with Rhizobia 447
16.6 Mechanisms of Aluminum Toxicity in Plants 447
    16.6.1 Cell Wall 447
        16.6.1.1 Modification of Synthesis or Deposition of Polysaccharides 448
    16.6.2 Plasma Membrane 448
        16.6.2.1 Binding to Phospholipids 448
        16.6.2.2 Interference with Proteins Involved in Transport 449
            16.6.2.2.1 H^+-ATPases 449
            16.6.2.2.2 Potassium Channels 449
            16.6.2.2.3 Calcium Channels 450
            16.6.2.2.4 Magnesium Transporters 450

| | | 16.6.2.2.5 | Nitrate Uptake .... | 450 |

            16.6.2.2.5   Nitrate Uptake 450
            16.6.2.2.6   Iron Uptake 450
            16.6.2.2.7   Water Channels 450
      16.6.2.3   Signal Transduction 451
            16.6.2.3.1   Interference with Phosphoinositide Signal
                         Transduction 451
            16.6.2.3.2   Transduction of Aluminum Signal 451
   16.6.3   Symplasm 451
      16.6.3.1   Disruption of the Cytoskeleton 451
      16.6.3.2   Disturbance of Calcium Homeostasis 452
      16.6.3.3   Interaction with Phytohormones 452
            16.6.3.3.1   Auxin 452
            16.6.3.3.2   Cytokinin 452
      16.6.3.4   Oxidative Stress 452
      16.6.3.5   Binding to Internal Membranes in Chloroplasts 453
      16.6.3.6   Binding to Nuclei 453
16.7   Genotypic Differences in Aluminum Response of Plants 453
   16.7.1   Screening Tests 454
   16.7.2   Genetics 454
16.8   Plant Mechanisms of Aluminum Avoidance or Tolerance 454
   16.8.1   Plant Mechanisms of Aluminum Avoidance 454
      16.8.1.1   Avoidance Response of Roots 455
      16.8.1.2   Organic Acid Release 455
      16.8.1.3   Exudation of Phosphate 457
      16.8.1.4   Exudation of Polypeptides 457
      16.8.1.5   Exudation of Phenolics 457
      16.8.1.6   Alkalinization of Rhizosphere 457
      16.8.1.7   Binding to Mucilage 458
      16.8.1.8   Binding to Cell Walls 458
      16.8.1.9   Binding to External Face of Plasma Membrane 458
      16.8.1.10   Interactions with Mycorrhizal Fungi 459
   16.8.2   Plant Mechanisms of Aluminum Tolerance 460
      16.8.2.1   Complexation with Organic Acids 460
      16.8.2.2   Complexation with Phenolics 460
      16.8.2.3   Complexation with Silicon 460
      16.8.2.4   Sequestration in Vacuole or in Other Organelles 460
      16.8.2.5   Trapping of Aluminum in Cells 461
16.9   Aluminum in Soils 461
   16.9.1   Locations of Aluminum-Rich Soils 461
   16.9.2   Forms of Aluminum in Soils 461
   16.9.3   Detection or Diagnosis of Excess Aluminum in Soils 465
      16.9.3.1   Extractable and Exchangeable Aluminum 466
      16.9.3.2   Soil-Solution Aluminum 467
   16.9.4   Indicator Plants 468
16.10   Aluminum in Human and Animal Nutrition 468
   16.10.1   Aluminum as an Essential Nutrient 468
   16.10.2   Beneficial Effects of Aluminum 469
      16.10.2.1   Beneficial Effects of Aluminum in Animal Agriculture 469
      16.10.2.2   Beneficial Uses of Aluminum in Environmental
                   Management and Water Treatment 470
   16.10.3   Toxicity of Aluminum to Animals and Humans 471
      16.10.3.1   Toxicity to Wildlife 471

		16.10.3.2	Toxicity to Agricultural Animals	472
			16.10.3.2.1 Toxicity to Ruminants (Cattle and Sheep)	473
			16.10.3.2.2 Toxicity to Poultry	474
	16.10.3.3	Toxicity to Humans	474
			16.10.3.3.1 Overview of Aluminum Metabolism	474
			16.10.3.3.2 Overview of the Biochemical Mechanisms of Aluminum Toxicity	475

16.11   Aluminum Concentrations 476
    16.11.1   In Plant Tissues 476
        16.11.1.1   Aluminum in Roots 476
        16.11.1.2   Aluminum in Shoots 476
    16.11.2   Soil Analysis 479
References 481

16.1 INTRODUCTION

Soils contain an average of 7% total aluminum (Al), and under acidic conditions, aluminum is solubilized (1), increasing availability to plants and aquatic animals. Soil acidification due to application of fertilizers, growing of legumes, or acid rain is an increasing problem in agricultural and natural ecosystems (2–4).

No conclusive evidence suggests that aluminum is an essential nutrient for either plants (5) or animals (6,7), although there are a few instances of beneficial effects. Aluminum is toxic to plants and animals, interfering with cytoskeleton structure and function, disrupting calcium homeostasis, interfering with phosphorus metabolism, and causing oxidative stress (discussed in later sections).

16.2 ALUMINUM-ACCUMULATING PLANTS

Relative to aluminum accumulation, there appears to be two groups of plant species: aluminum excluders and aluminum accumulators (8). Most plant species, particularly crop plants, are aluminum excluders. Aluminum contents in most herbaceous plants averaged $200 \, mg \, kg^{-1}$ in leaves (Hutchinson, cited in [9]). Chenery (10,11) analyzed leaves of various species of monocots and dicots for aluminum content, and defined aluminum accumulators as those plants with $1000 \, mg \, Al \, kg^{-1}$ or greater in leaves. Aluminum accumulation appears to be a primitive character, found frequently among perennial, woody species in tropical rain forests (9,12).

Masunaga et al. (13) studied 65 tree species and 12 unidentified species considered to be aluminum accumulators in a tropical rain forest in West Sumatra and suggested that aluminum accumulators be divided further into two groups: (a) those with aluminum concentrations lower than $3000 \, mg \, kg^{-1}$; and (b) those with higher aluminum concentrations. For trees with foliar aluminum concentrations greater than $3000 \, mg \, kg^{-1}$, positive correlations were noted between aluminum concentrations and phosphorus or silicon concentrations in leaves.

Although Chenery (11) did not consider gymnosperms to be aluminum accumulators, Truman et al. (14) proposed that most *Pinus* species are facultative aluminum accumulators. In Australia, values of foliar aluminum ranged from 321 to $1412 \, mg \, kg^{-1}$ for Monterey pine (*Pinus radiata* D. Don), 51 to $1251 \, mg \, kg^{-1}$ for slash pine (*Pinus elliotii* Engelm.), and 643 to $2173 \, mg \, kg^{-1}$ for loblolly pine (*Pinus taeda* L.) (15). In addition, foliar aluminum concentrations $\geq 1000 \, mg \, kg^{-1}$ were reported in Monterey pine and black pine (*Pinus nigra* J.F. Arnold) grown in nutrient solutions containing aluminum (14,16,17).

Tea (*Camellia sinensis* Kuntze) is one crop plant considered to be an aluminum accumulator, with aluminum concentrations of $30,700 \, mg \, kg^{-1}$ in mature leaves, but much lower concentrations of only $600 \, mg \, kg^{-1}$ in young leaves (18). Most of the aluminum was localized in the cell walls of the epidermis of mature leaves (18).

Another well-known aluminum-accumulating plant is hydrangea (*Hydrangea macrophylla* Ser.), which has blue-colored sepals when the plant is grown in acidic soils and red-colored sepals when grown in alkaline soils. The blue color of hydrangea sepals is due to aluminum complexing with the anthocyanin, delphinidin 3-glucoside, and the copigment, 3-caffeoylquinic acid (19).

Two excellent reviews of aluminum accumulators are by Jansen et al. (9) and Watanabe and Osaki (8). Possible mechanisms of aluminum tolerance will be discussed in later sections.

16.3 BENEFICIAL EFFECTS OF ALUMINUM IN PLANTS

16.3.1 GROWTH STIMULATION

Not surprisingly, aluminum addition has a growth stimulatory effect on aluminum accumulators. In tea, addition of aluminum and phosphorus increased phosphorus absorption and translocation as well as root and shoot growth (20,21). Similarly, the aluminum-accumulating shrub, *Melastoma malabathricum* L., exhibited increased growth of leaf, stem, and roots as well as increased phosphorus accumulation when aluminum was added to culture solutions (22).

Low levels of aluminum sometimes stimulate root and shoot growth of nonaccumulators. Turnip (*Brassica rapa* L. subsp. *campestris* A.R. Clapham) root lengths were increased by increasing aluminum levels up to $1.2 \mu M$ at pH 4.6 (23). Soybean (*Glycine max* Merr.) root elongation and $^{15}NO_3^-$ uptake increased with increasing aluminum concentrations up to $10 \mu M$, but were reduced when aluminum levels increased further to $44 \mu M$ (24). Shoot and root growth of Douglas fir (*Pseudotsuga menziesii* Franco) seedlings were stimulated by increasing aluminum levels up to $150 \mu M$ but were reduced at higher aluminum levels (25). Root elongation of an aluminum-tolerant race of silver birch (*Betula pendula* Roth) increased as solution aluminum increased up to $930 \mu M$ Al but then decreased at $1300 \mu M$ Al (26). Several researchers (23–25,27,28) have hypothesized that low levels of Al^{3+} ameliorated the toxic effects of H^+ on cell walls, membranes, or nutrient transport, but aluminum-toxic effects predominated at higher aluminum levels.

16.3.2 INHIBITION OF PLANT PATHOGENS

Aluminum can be toxic to pathogenic microorganisms, thus helping plants to avoid disease. Spore germination and vegetative growth of the black root rot pathogen, *Thielaviopsis basicola* Ferraris, were inhibited by $350 \mu M$ Al at pH 5 (29). Similarly, mycelial growth and sporangial germination of potato late blight pathogen, *Phytophthora infestans*, were inhibited by $185 \mu M$ Al, and Andrivon (30) speculated that amendment of soils with aluminum might be used as a means of disease control.

16.4 ALUMINUM ABSORPTION AND TRANSPORT WITHIN PLANTS

16.4.1 PHYTOTOXIC SPECIES

The most phytotoxic form of aluminum is Al^{3+} (more correctly, $Al(H_2O)_6^{3+}$), which predominates in solutions below pH 4.5 (31–33) (Figure 16.1). Possibly, hydroxyl-aluminum ($AlOH^{2+}$ and $Al(OH)_2^+$) ions are also phytotoxic, particularly to dicotyledonous plants (31,34). However, as pointed out by many researchers (35,36), these aluminum species are interrelated along with the pH variable, so it is difficult to rank their relative toxicity.

In contrast, Al-F, Al-SO_4, and Al-P species are much less toxic or even nontoxic to plants (34,37). Barley (*Hordeum vulgare* L.) roots were unaffected by aluminum when 2.5 to $10 \mu M$ F^- was added to nutrient solution containing up to $8 \mu M$ total soluble aluminum (37). Also using nutrient solution, Kinraide and Parker (38) positively demonstrated the nontoxic nature of Al-SO_4 complexes ($AlSO_4^+$ and $Al(SO_4)_2^-$) for wheat (*Triticum aestivum* L.) and red clover (*Trifolium pratense* L.). Soybean had longer root growth when increasing amounts of phosphorus were added to nutrient solutions having constant total aluminum concentrations (39).

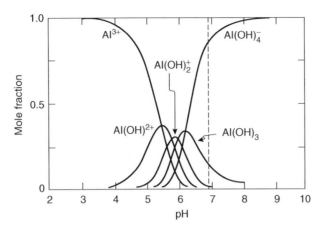

FIGURE 16.1 Speciation of aluminum as affected by solution pH. (From R.B. Martin. Fe^{3+} and Al^{3+} hydrolysis equilibria. Cooperativity in Al^{3+} hydrolysis reactions. *J. Inorg. Biochem.* 44:141–147, 1991.)

16.4.2 ABSORPTION

Since aluminum is a trivalent cation in its phytotoxic form in the external medium, it does not easily cross the plasma membrane. Akeson and Munns (40) calculated that the endocytosis of Al^{3+} could contribute to its absorption. Alternatively, it is possible that Al^{3+} could be absorbed through calcium channels (41) or nonspecific cation channels.

Our understanding of aluminum absorption across plant membranes has been limited by the complex speciation of Al, its binding to cell walls, lack of an affordable and available isotope, and lack of sensitive analytical techniques to measure low levels of aluminum in subcellular compartments (42). Aluminum absorption by excised roots of wheat, cabbage (*Brassica oleracea* L.), lettuce (*Lactuca sativa* L.), and kikuyu grass (*Pennisetum clandestinum* Hochst. ex Chiov.), and by cell suspensions of snapbean (*Phaseolus vulgaris* L.) followed biphasic kinetics (43–45). A rapid, nonlinear, nonmetabolic phase of uptake occurred during the first 20 to 30 min. This nonsaturable phase was thought to be accumulation in the apoplastic compartment due to polymerization or precipitation of aluminum or binding to exchange sites in cell walls (44). A linear, metabolic phase of uptake was superimposed over the nonlinear phase and thought to be accumulation in the symplasmic compartment (i.e., within the plasma membrane).

Using the rare ^{26}Al isotope and accelerator mass spectrometry on giant algal cells of *Chara corallina* Klein ex Willd., Taylor et al. (42) provided the first unequivocal evidence that aluminum rapidly crosses the plasma membrane into the symplasm. Accumulation of ^{26}Al in the cell wall was nonsaturable during 3 h of aluminum exposure and accounted for most of aluminum uptake. Absorption of aluminum into the protoplasm occurred immediately but accounted for less than 0.05% of the total accumulation (42). Accumulation in the vacuole occurred after a 30-min lag period (42).

16.4.3 ALUMINUM SPECIATION IN SYMPLASM

The pH of the cytoplasmic compartment generally ranges from 7.3 to 7.6 (5). Once aluminum enters the symplasm, the aluminate ion, $Al(OH)_4^-$ or insoluble $Al(OH)_3$ could form (Figure 16.1) (46). Alternatively, Al^{3+} could precipitate with phosphate as variscite, $Al(OH)_2H_2PO_4$ (47). Based on higher stability constants, it is likely that Al^{3+} would be complexed by organic ligands, such as adenosine triphosphate (ATP) or citrate (47,48). Martin (47) hypothesized that based on their similar effective ionic radii and affinity for oxygen donor ligands, Al^{3+} would compete with Mg^{2+} rather than Ca^{2+} in metabolic processes.

16.4.4 RADIAL TRANSPORT

The main barrier to radial transport of aluminum across the root into the stele appears to be the endodermis. Rasmussen (49) used electron microprobe x-ray analysis to show little penetration of aluminum past the endodermis of corn (*Zea mays* L.) roots. Similarly, in Norway spruce (*Picea abies* H. Karst.) roots, a large aluminum concentration was detected outside the endodermis, but very low aluminum concentrations on the inner tangential wall (3,50). Using secondary-ion mass spectrometry, Lazof et al. (51) confirmed that the highest aluminum accumulation occurred at the root periphery of soybean root tips, with substantial aluminum in cortical cells, but very low aluminum in stellar tissues. Similar to calcium, aluminum is thought to bypass the endodermis, entering the xylem in maturing tissues where the endodermis is not fully suberized.

16.4.5 MUCILAGE

Aluminum must cross the root mucilage before it can penetrate to the root apical meristem. Mucilage is produced by the root cap and is a complex mixture of high-molecular-weight polysaccharides, a population of several thousand border cells, and an array of cell wall fragments (52). Archambault et al. (53) showed that aluminum binds tightly to wheat mucilage, with 25 to 35% of total aluminum remaining after citrate desorption.

16.5 ALUMINUM TOXICITY SYMPTOMS IN PLANTS

16.5.1 SHORT-TERM EFFECTS

Owing to the numerous biochemical processes with which aluminum can interfere, researchers have attempted to determine the primary phytotoxic event by searching for the earliest responses to aluminum. Symptoms of aluminum toxicity that occur within a few hours of aluminum exposure are inhibition of root elongation, disruption of root cap processes, callose formation, lignin deposition, and decline in cell division.

16.5.1.1 Inhibition of Root Elongation

The first, easily observable symptom of aluminum toxicity is inhibition of root elongation. Elongation of adventitious onion (*Allium cepa* L.) roots (54), and primary roots of soybean (55,56), corn (57,58), and wheat (59–61) were suppressed within 1 to 3 h of aluminum exposure. The shortest time of aluminum exposure required to inhibit elongation rates was observed in seminal roots of an aluminum-sensitive corn cultivar BR 201F after 30 min (62).

Application of aluminum to the terminal 0 to 3 mm of corn root must occur for inhibition of root elongation to occur; however, the presence of the root cap was not necessary for aluminum-induced growth depression (63). Using further refinement of techniques, Sivaguru and Horst (58) determined that the most aluminum-sensitive site in corn was between 1 and 2 mm from the root apex, or the distal transition zone (DTZ), where cells are switching from cell division to cell elongation.

Lateral root growth of soybean was inhibited by aluminum-containing solutions to a greater extent than that of the taproot (64,65). Interestingly, Rasmussen (49) observed greater aluminum accumulation in lateral roots that emerged from the root surface, breaking through the endodermal layer. Similarly, root hair formation was more sensitive to aluminum toxicity than root elongation in white clover (*Trifolium repens* L.) (66).

16.5.1.2 Disruption of Root Cap Processes

The Golgi apparatus is the site of synthesis of noncellulosic polysaccharides targeted to the cell wall (67). Activity of the Golgi apparatus in the peripheral cap cells of corn was disrupted at 18 μM Al, a concentration below that necessary to inhibit root growth (68). In wheat, mucilage from the root cap disappeared within 1 h of aluminum exposure, and dictyosome volume and presence of endoplasmic reticulum decreased within 4 h (69). Death of root border cells (a component of root mucilage) occurred within 1 h of exposure to aluminum in snapbean roots (70).

16.5.1.3 Callose Formation

Callose is a polysaccharide consisting of 1,3-β-glucan chains, which are formed naturally by cells at a specific stage of wall development or in response to wounding (67). An early symptom of aluminum toxicity is formation of callose in roots. Using fluorescence spectrometry, callose could be quantified in soybean root tips (0 to 3 cm from root apex) after 2 h of exposure to 50 μM Al (55). In root cells surrounding the meristem of Norway spruce roots, distinct callose deposits were observed after 3 h of exposure to 170 μM Al (71). Zhang et al. (72) showed that callose accumulated in roots of aluminum-sensitive wheat cultivars exposed to 75 μM Al and they proposed using callose synthesis as a rapid, sensitive marker for aluminum-induced injury. However, callose was not accumulated in two aluminum-sensitive arabidopsis (*Arabidopsis thaliana* Heynh.) mutants exposed to aluminum, indicating no obligatory relationship between callose deposition and aluminum-induced inhibition of root growth (73). Sivaguru et al. (74) showed that aluminum-induced callose deposition in plasmodesmata of epidermal and cortical cells of aluminum-sensitive wheat roots reduced movement of micro-injected fluorescent dyes between cells.

16.5.1.4 Lignin Deposition

Lignins are complex networks of aromatic compounds that are the distinguishing feature of secondary walls (67). Deposition of lignin in response to aluminum was found in wheat cortical cells located 1.4 to 4.5 mm from the root tip (elongating zone [EZ]) after 3 h of exposure to 50 μM Al (75). Lignin occurred in cells with damaged plasma membranes as indicated by staining with propidium iodide, and Sasaki et al. (61) proposed that aluminum-induced lignification was a marker of aluminum injury and was closely associated with inhibition of root elongation. Interestingly, Snowden and Gardner (76) showed that a cDNA induced by aluminum treatment in wheat exhibited high homology with the gene for phenylalanine ammonia-lyase, a key enzyme in the pathway for biosynthesis of lignin.

16.5.1.5 Decline in Cell Division

A decrease in abundance of mitotic figures was observed in adventitious roots of onion after 5 h of exposure to 1 mM Al (54). Similarly, a decrease in the mitotic index of barley root tips was found within 1 to 4 hours of exposure to 5 to 20 μM AI (pH 4.2) (77).

16.5.2 LONG-TERM EFFECTS

Although they may not be indicative of initial, primary phytotoxic events, long-term effects of aluminum are important for plants growing in aluminum-toxic soils or subsoils. Long-term exposure to aluminum over several days or weeks results in suppressed root and shoot biomass, abnormal root morphology, suppressed nutrient uptake and translocation, restricted water uptake and transport, suppressed photosynthesis, and inhibition of symbiosis with rhizobia.

16.5.2.1 Suppressed Root and Shoot Biomass

Increasing aluminum concentrations in solution, sand, or soil decreased fine root biomass of red spruce (*Picea rubens* Sarg.) (78). Typically, aluminum reduces root biomass to a greater degree than shoot biomass, resulting in a decreased root/shoot ratio (78–80). In contrast, in 3-year-old Scots pine (*Pinus sylvestris* L.), increasing solution of aluminum up to 5.6 mM produced no obvious aluminum toxicity symptoms on roots but decreased needle length and whole shoot length, resulting in increased needle density (81).

16.5.2.2 Abnormal Root Morphology

Often, one symptom of aluminum toxicity is 'coralloid' root morphology with inhibited lateral root formation and thickened primary roots (54). Cells in the elongation zone of primary wheat roots exposed to aluminum had decreased length and increased diameter, resulting in appearance of lateral swelling (61). This abnormal root morphology combined with reduced root length could result in decreased nutrient uptake and multiple deficiencies.

16.5.2.3 Suppressed Nutrient Uptake and Translocation

Increasing aluminum levels in the medium have been reported to decrease uptake and translocation of calcium, magnesium, and potassium (78,82). Forest declines in North America and Europe have been proposed to be due to aluminum-induced reductions in calcium and magnesium concentrations of tree roots and needles (3). Excess aluminum reduced magnesium concentration of Norway spruce needles to a level considered to be critical for magnesium deficiency (3). Also, aluminum toxicity reduced calcium and magnesium leaf concentrations in beech (*Fagus sylvatica* L.) (83). In sorghum (*Sorghum bicolor* Moench), magnesium deficiency was a source of acid-soil stress (84).

In the case of phosphorus, concentrations increased in roots but typically decreased in shoots. In roots of red spruce, ^{32}P accumulation increased but ^{32}P translocation to shoots decreased (85). Clarkson (86) proposed that there were two interactions between aluminum and phosphorus: (a) an adsorption–precipitation reaction in the apoplast; and (b) reaction with various organic phosphorus compounds within the symplasm of the cell. Aluminum and phosphorus were shown to be coprecipitated in the apoplast of corn roots, using x-ray microprobe analysis (49). Excised corn roots exposed to 20 h of 0.1 to 0.5 mM Al had decreased mobile inorganic phosphate (40%), ATP (65%), and uridine diphosphate glucose (UDGP) (65%) as shown by ^{31}P-NMR (nuclear magnetic resonance), indicating aluminum interference with phosphorus metabolism within the symplasm (87,88).

16.5.2.4 Restricted Water Uptake and Transport

Typically, aluminum toxicity decreases water uptake and movement in plants. Stomatal closure of arabidopsis occurred after 9 h of exposure to 100 μM Al at pH 4.0 (89). In wheat, transpiration decreased after 28 days of exposure to 148 μM Al (90). Treatment of 1-year-old black spruce (*Picea mariana* Britton) with 290 μM Al resulted in wilting and reduced water uptake within 7 days (91). Hydraulic conductivity of red oak roots was reduced after 48 to 63 days of exposure to aluminum, although no effect was observed after only 4 days (92). In contrast, transpiration in sorghum increased after 28 days of aluminum treatment (90).

16.5.2.5 Suppressed Photosynthesis

Net photosynthesis is reported to decrease with excess aluminum relative to normal rates. Exposure to 250 μM Al for 6 to 8 weeks reduced the photosynthetic rate of red spruce, and McCanny et al. (79) attributed this effect to an aluminum-induced decrease in root/shoot ratio. Similarly, exposure of beech seedlings to 0.37 mM Al for 2 months significantly decreased net CO_2 assimilation rates (83).

16.5.2.6 Inhibition of Symbiosis with Rhizobia

Biological nitrogen fixation results in release of H^+, acidification of legume pastures, and increased solubilization of aluminum (2). Excess aluminum has an inhibitory effect on rhizobial symbiosis. In an Australian pasture, the percentage of plant nitrogen derived from the atmosphere declined in subterranean clover (*Trifolium subterraneum* L.) as foliar concentration of aluminum increased (93). In four tropical pasture legumes, aluminum at $>25\,\mu M$ for 28 days delayed appearance of nodules, decreased percentage of plants that nodulated, and decreased number and dry weight of nodules (94). In phasey-bean (*Macroptilium lathyroides* Urb.) and centro (*Centrosema pubescens* Benth.), nodulation was more sensitive to aluminum toxicity than host plant growth (94).

Aluminum also inhibited the multiplication and nodulating ability of the symbiotic bacterium, *Rhizobium leguminosarum* bv. *trifolii* Frank (66). Recent research efforts have focused on identifying aluminum-tolerant rhizobial strains. For example, strains of *Bradyrhizobium* spp. that were isolated from acid soils were found to more tolerant of $50\,\mu M$ Al at pH 4.5 than commercial strains (95).

16.6 MECHANISMS OF ALUMINUM TOXICITY IN PLANTS

Controversy exists over mechanisms of aluminum phytotoxic effects (96–99). Researchers long have debated whether the primary toxic effect of aluminum is on inhibition of cell elongation or inhibition of cell division. Lazof and Holland (28) demonstrated in soybean, pea (*Pisum sativum* L.), and bean (*Phaseolus vulgaris* L.) that both effects occur, with rapid, largely reversible responses to aluminum toxicity due to cell extension effects and irreversible responses due to cell division effects.

Another question puzzling researchers is whether the primary injury due to aluminum in plants is symplasmic or apoplastic. Horst (100) and Horst et al. (101) reviewed the evidence supporting the apoplast as the site of the primary aluminum-toxic event. However, dividing aluminum effects into symplasmic or apoplastic can be arbitrary, because aluminum could enter the symplasm to produce effects in the cell wall or outer face of the plasma membrane.

Since cell walls occur in plants and not animals, aluminum injuries at this site are unique to plants. Possible mechanisms of aluminum injury in cell walls include: (a) aluminum binding to pectin; or (b) modification of synthesis or deposition of polysaccharides. Jones and Kochian (102) proposed that the plasma membrane is the most likely site of aluminum toxicity in plants. Possible mechanisms of toxicity in the plasma membrane are: (a) aluminum binding to phospholipids; (b) interference with proteins involved in transport; or (c) signal transduction. Once aluminum enters the symplasm, there are many possible interactions with molecules containing oxygen donor ligands (47,48). Probable mechanisms of aluminum toxicity within plant cells include: (a) disruption of the cytoskeleton, (b) disturbance of calcium homeostasis, (c) interaction with phytohormones, (d) oxidative stress, (e) binding to internal membranes in chloroplasts, or (f) binding to nuclei.

16.6.1 CELL WALL

Pectins are a mixture of heterogenous polysaccharides rich in D-galacturonic acid; one major function is to provide charged structures for ion exchange in cell walls (67). Under acidic conditions, aluminum binds strongly to negatively charged sites in the root apoplast, sites consisting mostly of free carboxyl groups on pectins. Klimashevskii and Dedov (103) isolated cell walls from pea roots, exposed them to aluminum, and found that aluminum decreased plasticity and elasticity of cell walls. Blamey et al. (104) demonstrated in vitro a rapid sorption of aluminum by calcium pectate and proposed that aluminum phytotoxicity is due to strong binding between aluminum and calcium pectate in cell walls. Reid et al. (105) proposed that aluminum could disrupt normal cell wall growth either by reducing Ca^{2+} concentration below that required for cross-linking of pectic residues or through formation of aluminum cross-linkages that alter normal cell wall structure. Using x-ray microanalysis, Godbold and

Jentschke (106) showed that aluminum displaced calcium and magnesium from root cortical cell walls of Norway spruce. Using a vibrating calcium-selective microelectrode, Ryan and Kochian (107) observed that addition of aluminum commonly resulted in an initial efflux of calcium from wheat roots, probably due to displacement of calcium from cell walls.

Pectin is secreted in a highly esterified form from the symplast to the apoplast, where demethylation takes place by pectin methylesterase (PME), resulting in free carboxylic groups available to bind aluminum (108). Transgenic potato (*Solanum tuberosum* L.) overexpressing PME is more sensitive to aluminum based on inhibition of root elongation relative to unmodified control plants, indicating that increased binding sites for aluminum in the apoplast are associated with increased aluminum sensitivity (108).

16.6.1.1 Modification of Synthesis or Deposition of Polysaccharides

In addition to external binding to cell wall components, aluminum also could interfere with the internal synthesis or deposition of cell wall polysaccharides. Exposure of wheat seedlings to $10 \mu M$ Al for 6 h decreased mechanical extensibility of subsequently isolated cell walls (109). Tabuchi and Matsumoto (109) showed that aluminum treatment modified cell wall components, increasing the molecular mass of hemicellulosic polysaccharides, thus decreasing the viscosity of cell walls, and perhaps restricting cell wall extensibility.

Uridine diphosphate glucose (UDGP) is the substrate for cellulose synthesis. Using ^{31}P-NMR, Pfeffer et al. (87) demonstrated that a 20-h exposure of excised corn roots to 0.1 mM Al decreased UDGP by 65%, and they speculated that such suppression could limit production of cell wall polysaccharides. In barley, one of the most aluminum-sensitive cereals, callose was excreted from the junction between the root cap and the root epidermis after 38 min of exposure to $37 \mu M$ Al, and Kaneko et al. (110) proposed that aluminum-induced inhibition of root elongation could be due to reduced cell wall synthesis caused by a shortage of substrate to form polysaccharides.

16.6.2 PLASMA MEMBRANE

16.6.2.1 Binding to Phospholipids

Biological membranes are composed of phospholipids that contain a phosphate group (67), and aluminum can bind to this negatively charged group. Using electron paramagnetic resonance spectroscopy, Vierstra and Haug (111) demonstrated that 100 mM Al at pH 4 decreased fluidity in membrane lipids of a thermophilic microorganism (*Thermoplasma acidophilum* Darland, Brock, Samsonoff and Conti). Using physiologically significant concentrations of aluminum, Deleers et al. (112) showed that $25 \mu M$ Al increased rigidity of membrane vesicles as indicated by the increased temperature required to maintain a specific polarization value. In addition, aluminum at $< 30 \mu M$ could induce phase separation of phosphatidylserine (PS; a negatively charged phospholipid) vesicles, as shown by leakage of a fluorescent compound (113).

Phosphatidylcholine (PC) is the most abundant phospholipid in plasma membranes of eukaryotes, and Akeson et al. (114) showed that in vitro, Al^{3+} has a 560-fold greater affinity for the surface of PC than Ca^{2+}. Further, Jones and Kochian (102) found that lipids with net negatively charged head groups such as phosphatidyl inositol (PI) had a much greater affinity for aluminum than PC with its net neutral head group. Interestingly, Delhaize et al. (115) found that expression of a wheat cDNA (TaPSS1) encoding for phosphatidylserine synthase (PSS) increased in response to excess aluminum in roots. Overexpression of this cDNA conferred aluminum resistance in one strain of yeast (*Saccharomyces cerevisiae*) but not in another. In addition, a disruption mutant of the endogenous yeast *CHO1* gene that encodes for PSS was sensitive to aluminum (115).

Aluminum reduced membrane permeability to water as shown by a plasmometric method on root disks of red oak (116). To remove the confounding effect of aluminum binding to cell walls, Lee et al. (117) used protoplasts of red beet (*Beta vulgaris* L.). Within 1 min of exposure to 0.5 mM

Al, volumetric expansion of red beet cells was reduced under hypotonic conditions, and Lee et al. (117) hypothesized that aluminum could bridge neighboring negatively charged sites on the plasma membrane, stabilizing the membrane.

Binding of Al^{3+} to the exterior of phospholipids reduces the surface negative charge of membranes. Kinraide et al. (27) proposed that accumulation of aluminum at the negatively charged cell surface plays a role in rhizotoxicity and that amelioration of aluminum toxicity by cations is due to reduced negativity of the cell-surface electrical potential by charge screening or cation binding. Kinraide et al. (27) found a good correlation between the reduction in relative root length of an aluminum-sensitive wheat cultivar with aluminum activity as calculated at the membrane surface, but not in the bulk external solution. Ahn et al. (118) measured the zeta potential (an estimate of surface potential) of plasma membrane vesicles from squash (*Cucurbita pepo* L.) roots and showed that aluminum exposure resulted in a less negative surface potential. Measuring uptake of radioisotopes by barley roots, Nichol et al. (119) showed that influx of cations (K^+, NH_4^+, and Ca^{2+}) decreased whereas influx of anions (NO_3^-, HPO_4^{2-}) increased in the presence of aluminum. They speculated that binding of Al^{3+} to the exterior of a plasma membrane forms a positively charged layer that retards movement of cations to the membrane surface and increases movement of anions to the surface.

In contrast, Silva et al. (120) demonstrated that Mg^{2+} was 100-fold more effective than Ca^{2+} in alleviating aluminum-induced inhibition of soybean taproot elongation. They (120) suggested that such an effect could not be explained by changes in membrane surface potential and proposed that the protective effects of Mg could be due to alleviation of aluminum binding to G-protein.

16.6.2.2 Interference with Proteins Involved in Transport

In addition to phospholipids, biological membranes are composed of proteins, many of which are involved in transport functions across the membrane (5,67). Aluminum is reported to interfere with the uptake of many nutrients, perhaps through interactions with cross-membrane transporters or channels.

16.6.2.2.1 H^+-ATPases
Transmembrane electric potential (V_m) is the difference in electric potential between the external environment and the symplasm; typically, the interior of the cell is negatively charged with respect to the outside (67). The potential depends on transient fluxes of H^+ through membrane-bound H^+-ATPases, as well as fluxes of K^+ and other cations through membrane transporters. Measurements of net H^+ flux using either a microelectrode or vibrating probe demonstrated that net inward currents of H^+ occurred between 0 to 3 mm from root tips of wheat (60,121). Exposure of roots of an aluminum-sensitive wheat cultivar to 10 µM Al for 1 to 3 h inhibited H^+ influx; however, there was no obligatory association between inhibition of H^+ influx and inhibition of root elongation (60). Ryan et al. (60) speculated that the H^+ influx near the root apex could be due to cotransport of H^+ with unloaded sugars and amino acids into the cytoplasm, or a membrane more permeable to H^+.

Conducting an in vitro enzyme test, Jones and Kochian (102) found little effect of aluminum on H^+-ATPase activity. Similarly, Tu and Brouillette (122) found no effect of aluminum on plasma membrane-bound ATPase activity in the presence of free ATP; however, exposure of Mg^{2+}-ATP to 18 µM Al competitively inhibited hydrolysis of ATP. Based on immunolocalization, H^+-ATPases in epidermal and cortical cells (2 to 3 mm from tip) of squash roots decreased after 3 h of exposure to 50 µM Al (118). Similarly, 2 days of exposure to $\geq$ 75 µM Al decreased activity of plasma membrane-bound ATPases in 1-cm root tips of five wheat cultivars (123). Since H^+-ATPases generate the proton motive force that drives secondary transporters and channels (5,67), a decrease in activity of this membrane-bound enzyme could result in an overall decrease in nutrient uptake.

16.6.2.2.2 Potassium Channels
Uptake of K^+ by pea roots was depressed by aluminum (124). Similarly, exposure of mature root cells ($\geq$ 10 mm from root tip) of an aluminum-sensitive wheat cultivar to 5 µM Al inhibited K^+ influx (121). In addition, Reid et al. (105) showed partial inhibition of Rb^+ (analog for K^+) uptake by > 50 µM Al

in giant algal (*Chara corallina*) cells, and they attributed this effect to partial blocking by aluminum of K^+ channels. Using the patch-clamp technique on isolated plasma membranes or whole cells from an aluminum-tolerant corn cultivar, Pineros and Kochian (125) showed that instantaneous outward K^+ channels were blocked by $12\,\mu M$ Al, whereas inward K^+ channels were inhibited by $400\,\mu M$ Al.

A strong dysfunction in K^+ fluxes between guard cells and epidermal cells was observed in beech (*Betula* spp.) seedlings exposed to excess aluminum for 2 months (83). Measuring currents of inside-out membrane patches from fava bean (*Vicia faba* L.) guard cells, Liu and Luan (41) demonstrated that the K^+ inward rectifying channel (KIRC) was inhibited by $50\,\mu M$ Al when exposed on the inward-facing side of the membrane. They (41) proposed that calcium channels conduct Al^{3+} across the plasma membrane because, verapamil, a Ca^{2+} channel blocker, prevented aluminum-induced inhibition of KIRC in the whole cell configuration. In addition, Liu and Luan (41) expressed the gene, *KAT1*, which encodes for a KIRC, in *Xenopus* oocytes, injected aluminum into the cytoplasm, and observed inhibition of the KAT1 current.

16.6.2.2.3 Calcium Channels

Uptake by roots and translocation of ^{45}Ca to shoots was decreased in wheat by $100\,\mu M$ Al (126). Similar results occurred with 4-week-old Norway spruce seedlings, in which uptake of ^{45}Ca was reduced by 77 to 92% by 100 to $800\,\mu M$ Al (3). Net Ca^{2+} influx was highest between 0 and 2 mm from the root apex of wheat, based on a calcium-selective vibrating microelectrode (127). Addition of $20\,\mu M$ Al to roots of an aluminum-sensitive wheat cultivar resulted in a dramatic decrease in Ca^{2+} influx, and this effect was attributed to blockage by aluminum of a putative calcium channel (128). However, Ryan and Kochian (107) did not find an obligatory relationship between inhibition of calcium uptake and reduction of root growth in wheat. Similarly, in *Chara corallina* cells, aluminum inhibited calcium influx by less than 50% at $100\,\mu M$ Al, and Reid et al. (105) thought it unlikely that such a small degree of inhibition would be sufficient to inhibit growth so rapidly.

16.6.2.2.4 Magnesium Transporters

Exposure of annual ryegrass (*Lolium multiflorum* Lam.) to $6.6\,\mu M$ Al competitively inhibited net Mg^{2+} uptake (129). Interestingly, McDiarmid and Gardner (130) isolated two yeast genes, ALR1 and ALR2, that encode proteins homologous to bacterial Mg^{2+} and Co^{2+} transport systems. Overexpression of these genes conferred increased tolerance to Al^{3+}, indicating that aluminum toxicity in yeast is related to reduced Mg^{2+} influx (130).

16.6.2.2.5 Nitrate Uptake

In white clover, 3 weeks of exposure to $50\,\mu M$ Al inhibited nitrate uptake as measured by nitrogen content in plants (131). In all regions of soybean roots, $^{15}NO_3^-$ influxes were reduced within 30 min of exposure to $80\,\mu M$ Al (132). In corn, 30 min of exposure to $100\,\mu M$ Al decreased NO_3^- uptake as measured by NO_3-N depletion in solution, but aluminum-induced inhibition of root elongation was not attributed to inhibition of nitrate uptake (133). Aluminum treatment for 3 days followed by measurement of $^{15}NO_3^-$ uptake in the final hour decreased $^{15}NO_3^-$ uptake in soybean at $\geq 44\,\mu M$ Al but increased $^{15}NO_3^-$ uptake at aluminum levels below $10\,\mu M$, probably as a result of Al^{3+} amelioration of H^+ toxicity (24).

16.6.2.2.6 Iron Uptake

Iron acquisition in Strategy II plants (gramineous plants) involves secretion of mugineic acids (MA) and uptake of MA–Fe^{3+} complexes (67). Chang et al. (134) demonstrated that exposure to 100 mM Al for 21 h depressed biosynthesis and secretion of 2′-deoxymugineic acid in wheat.

16.6.2.2.7 Water Channels

Aluminum is reported to reduce permeability of the plasma membrane to water, perhaps through reduced aquaporin (water channel) activity. Milla et al. (135) found that expression of a rye (*Secale cereale* L.) gene encoding for aquaporin (water channel) was decreased by aluminum.

16.6.2.3 Signal Transduction

16.6.2.3.1   Interference with Phosphoinositide Signal Transduction

Under in vitro conditions, aluminum interacted strongly with the phosphoinositide signal transduction element, the plasma-membrane-bound phosphatidylinositol-4,5-bisphosphate (PIP_2) (136). In animals, cleavage of the plasma membrane lipid, PIP_2, by phospholipase C (PLC) releases inositol 1,4,5-triphosphate (IP_3) into the cytoplasm. Then, IP_3 could produce a signaling cascade by binding to a Ca^{2+} channel and releasing Ca^{2+} into the cytosol. In microsomal membranes of wheat roots, aluminum $\geq 20\,\mu M$ dramatically inhibited PLC activity (136). Under in vitro conditions, aluminum was shown to block the PLC-activated cleavage of PIP_2 to IP_3 (136).

16.6.2.3.2   Transduction of Aluminum Signal

Cell wall-associated kinases could serve as a connecting molecule between the cell wall and the cytoplasmic cytoskeleton. These kinases span the plasma membrane, with the extracellular portion covalently bound to pectin in the cell wall and the cytoplasmic portion containing kinase activity. Recently, expression of a cell wall associated kinase (WAK1) in arabidopsis was induced within 3 h of exposure to aluminum (89). Sivaguru et al. (89) hypothesized that WAK1 could be involved in the aluminum signal transduction pathway.

16.6.3 SYMPLASM

16.6.3.1 Disruption of the Cytoskeleton

The cytoskeleton is a network of filamentous protein polymers that permeates the cytoplasm, providing structural stability and motility for macromolecules and organelles (67). In plants, there are two major families of proteins: actin and tubulin (67). Actin binds and hydrolyzes the nucleotide, ATP, during polymerization to form microfilaments. Proteins α- and β-tubulin bind and hydrolyze guanosine triphosphate (GTP) during polymerization to form microtubules.

Actin filaments are important in cytoplasmic streaming in giant algal cells. With an alga (*Vaucheria longicaulis* Hoppaugh), Alessa and Oliveira (137) demonstrated that cytoplasmic streaming of chloroplasts and mitochondria (mediated by microfilaments) decreased within 30 s of aluminum exposure and completely ceased within 3 min. Using suspension-cultured soybean cells, Grabski and Schindler (138) demonstrated that aluminum rapidly increased rigidity of the transvacuolar actin network, and they proposed that the cytoskeleton is the primary target of aluminum toxicity in plants. Grabski et al. (139) hypothesized that phosphorylated sites on myosin or other actin-binding proteins could bind aluminum, preventing access to phosphatases and resulting in a stabilized actin network. Alternatively, they hypothesized that a calcium-dependent phosphatase could be inhibited directly by aluminum. Interestingly, aluminum toxicity in wheat causes increased expression of a gene encoding for a fimbrin-like (actin-binding) protein involved in maintenance of cytoskeletal function (140). They speculated that the increased tension of cytoskeletal actin by aluminum (138) could involve cross-linking of actin filaments by fimbrins, leading to increased fimbrin gene expression.

Aluminum could disrupt microtubule assembly and disassembly through inhibition of GTP hydrolysis and reduced sensitivity to regulatory signals from Ca^{2+}. When magnesium concentrations were below 1.0 mM, MacDonald et al. (141) demonstrated in vitro that 4×10^{-10} M Al could replace Mg^{2+} in polymerization of tubulin. Disappearance of microtubules was observed sometimes in cells of the EZ of aluminum-treated (3 h, $50\,\mu M$ Al) wheat roots (61). In outer cortical cells of the DTZ of aluminum-sensitive corn roots, microtubules disappeared within 1 h of exposure to $90\,\mu M$ Al (142). Treatment of corn roots with $50\,\mu M$ Al for 3 h resulted in random or obliquely oriented microtubules in inner cortical cells compared to the transverse orientation of those from control roots (57). In addition, a 1 h pretreatment with aluminum prevented auxin-induced reorientation of microtubules in inner cortical cells of corn, and Blancafor et al. (57) proposed that aluminum induced greater stabilization of microtubules. Microfilaments seemed to be less sensitive to aluminum toxicity, with random arrays detectable in the inner cortical cells after 6 h (57).

16.6.3.2 Disturbance of Calcium Homeostasis

Siegel and Haug (143) proposed that the primary biochemical injury due to aluminum was caused by aluminum complexes with calmodulin (a calcium-dependent, regulatory protein). Similarly, Rengel (144) proposed that aluminum is the primary environmental signal, with Ca^{2+} as the secondary messenger that triggers aluminum-toxic events in plant cells. Using a fluorescent calcium-binding dye, Fura 2, Lindberg and Strid (145) showed that exposure of wheat root protoplasts to 50 μM Al caused a transient and oscillating increase in cytoplasmic Ca^{2+} concentration. Similarly, using a cytosolic calcium indicator dye, Fluo-3, in intact wheat apical cells, Zhang and Rengel (146) showed an increase in cytoplasmic Ca^{2+} after 1 h treatment with 50 μM Al. Using Fluo-3 and an indicator of membrane-bound Ca^{2+}, chlorotetracycline (CTC), Nichol and Oliveira (147) found increased calcium concentration in the zone of elongation of an aluminum-sensitive barley cultivar. Since aluminum is known to block calcium channels that allow calcium to move into the cytoplasm, Nichol and Oliveira (147) suggested that Ca^{2+} was released from intracellular storage sites. Interestingly, aluminum-induced callose formation, a rapid marker of aluminum toxicity, is always preceded by elevated cytoplasmic Ca^{2+} (67).

In contrast, Jones et al. (148) used the fluorescent dye, Indo-1, and showed a rapid reduction in cytosolic Ca^{2+} in suspension cultures of tobacco (*Nicotiana tabacum* L.) cells. They (148) attributed this effect to blockage of calcium channels in the plasma membrane by aluminum.

16.6.3.3 Interaction with Phytohormones

The spatial separation between the most aluminum-sensitive site, the DTZ, and the root region that exhibits reduced cell elongation, the EZ, indicates that a signaling pathway is involved. Perhaps, the phytohormones, auxin (IAA) or cytokinin, are involved in the transduction of an aluminum-stress signal.

16.6.3.3.1 Auxin

Corn roots were observed to curve away from unilaterally applied aluminum (149). Similar results were found for snapbean roots that curved away from an agar surface containing aluminum (52). Hasenstein and Evans (150) showed that aluminum inhibited basipetal transport of indoleacetic acid (IAA), perhaps resulting in the tropic root response. Kollmeier et al. (151) confirmed this result, showing that exogenous ^{3}H-IAA application to the meristematic zone of corn roots with aluminum application to the DTZ resulted in decreased basipetal transport of auxin to the EZ. They also showed that exogenous IAA application to the EZ partially ameliorated the aluminum-induced (Al applied to DTZ) inhibition of root elongation. Kollmeier et al. (151) hypothesized that aluminum inhibition of auxin transport mediated the aluminum signal between the DTZ and EZ. Sivaguru et al. (74) speculated that aluminum-induced callose in plasmodesmata could be a primary factor in aluminum inhibition of root growth through disturbance of auxin transport.

16.6.3.3.2 Cytokinin

Bean root elongation was inhibited after 360 min of exposure to 6.5 μM Al (152). Ethylene evolution as well as the level of zeatin (a cytokinin) from root tips increased after 5 min of aluminum exposure. Massot et al. (152) suggested a role for cytokinin and ethylene in transduction of aluminum-induced stress signal.

16.6.3.4 Oxidative Stress

Aluminum is redox inactive and is not able to initiate oxidation of lipids or proteins on its own. Yet, lipid peroxidation has been observed in barley roots after 3 h incubation with aluminum (100 μM $AICl_3$, pH, 4.3) (153). Similarly, in pea roots, increase of lipid peroxidation and inhibition of root elongation occurred after 4 h of exposure to 10 μM aluminium (154). Sakihama and Yamasaki (153) proposal that aluminum stabilizes the oxidized form of phenolics (normally unstable), resulting in phenoxyl radicals that initiate lipid peroxidation. Alternatively, aluminum could increase formation

of reactive oxygen species (ROS). Cell defense against ROS includes the enzymes, superoxide dismutase (SOD) and glutathione peroxidase (PX), which reduce ROS (153). If levels of these enzymes
are not sufficient, then ROS could lead to oxidation of lipids, proteins, and DNA, and even cell death.
In corn, 24 h of exposure to aluminum increased activities of SOD and PX, and increased protein
oxidation in the aluminum-sensitive genotype (155).

Another possibility proposed by Ikegawa et al. (156) is aluminum-enhanced, Fe(II)-medicated
peroxidation of lipids as a cause of cell death. Exposure of tobacco suspension cultures to aluminum
alone for 24 h resulted in aluminum accumulation but no significant cell death (156). Addition of
Fe(II) (a redox active metal) to cells with accumulated aluminum after 12 h resulted in enhanced
lipid peroxidation and cell death. Lipid peroxidation does not appear to be the mechanism involved
in reduction of root elongation (154). In pea roots, treatment with an antioxidant prevented
aluminum-enhanced lipid peroxidation, reduced callose formation, but did not prevent aluminum-
induced inhibition of root elongation (154).

Interestingly, three of four cDNA up-regulated by aluminum stress in *Arabidopsis thaliana* encoded
genes were induced also by oxidative stress (157). Similarly, the vast majority of isolated cDNAs,
whose expression increased in response to aluminum toxicity in sugarcane (*Saccharum officinarum* L.),
showed greater expression in response to oxidative stress (158). These results indicate that oxidative
stress is an important component of the plant's response to aluminum toxicity. Overexpression of a
tobacco gene encoding for glutathione S-transferase (*parB*) in *Arabidopsis thaliana* conferred a degree
of aluminum resistance as well as resistance to oxidative stress induced by diamide, providing genetic
evidence of a linkage between aluminum stress and oxidative stress in plants (159).

16.6.3.5 Binding to Internal Membranes in Chloroplasts

As discussed earlier, one long-term effect of aluminum toxicity is the suppression of photosynthetic
activity (79,90). Photosynthetic $^{14}CO_2$ fixation of isolated spinach (*Spinacia oleracea* L.) chloroplasts
was inhibited by 10 μM Al at pH 7 (160). Hampp and Schnabel (160) attributed this effect to damage
of the membrane system. Aluminum exposure of wheat for 14 days decreased the maximum photochemical yield F_v/F_m of photosystem II, (ratio of variable fluorescence over maximum fluorescence,
as measured by a fluorometer) (161). Moustakas and Ouzounidou (161) attributed this effect to loss
of Ca^{2+}, Mg^{2+}, and K^+ from chloroplasts. Seventy days of aluminum exposure decreased F_v/F_0, or the
ratio of variable fluorescence over initial fluorescence (162). Pereira et al. (162) speculated that this
decrease was an indicator of aluminum-induced structural damage in the thylakoids. In the cyanobacterium, *Anabaena cylindrica* Lemm., aluminum was found to degrade thylakoid membranes (163).

16.6.3.6 Binding to Nuclei

Aluminum entered soybean root cells and was associated with nuclei only after 30 min of exposure
to 1.45 μM Al (164). In corn root tips, high chromatin fragmentation and loss of plasma membrane
integrity occurred after 48 h exposure to 36 μM Al (155). However, Al^{3+} binding to DNA is very
weak and cannot compete with phosphate, ATP, or other organic ligands such as citrate (47,48).
Martin (47) stated that the observed association of aluminum with nuclear chromatin must be due
to its complexation to other ligands and not to DNA.

16.7 GENOTYPIC DIFFERENCES IN ALUMINUM RESPONSE OF PLANTS

Comparative studies of aluminum effects in 22 species in seven plant families have established that
some species or genotypes within species can resist aluminum toxicity (82). Foy (165) proposed
'tailoring the plant to fit the soil; in other words, he suggested that it was more economical to
develop mineral-stress-resistant plants than to correct the soil for nutrient deficiencies or toxicities.
This statement is particularly true for acid subsoils, where it is not economically feasible to lime at
such depths, or for developing countries, where farmers cannot afford the high-input costs of lime.

16.7.1 Screening Tests

Screening for genotypic differences in response to aluminum toxicity can be conducted in pots or in fields with aluminum-toxic soil. A more rapid screening test for differences in aluminum tolerance among species or genotypes within species utilizes the aluminum-induced inhibition of root elongation as a measure of aluminum sensitivity (166). These tests are conducted with varying levels of aluminum in solution at an acid pH (≤ 4.5) to maintain a high activity of Al^{3+}, the phytotoxic ion. Some researchers have found a poor correlation between plant responses in soil with those in nutrient solution (167). Others have found a good correlation (168–171).

Hematoxylin stains extracellular aluminum phosphate compounds that result from aluminum damage to root cells (172). Another quick screening test is to stain roots grown in an aluminum-containing solution with hematoxylin and to assess the intensity of staining (173). With wheat, Scott et al. (174) found a good agreement between root elongation results and those using hematoxylin. However, Bennet (175) warned that many aspects of hematoxylin staining are not well understood and that aluminum-treated roots do not always respond to hematoxylin even when symptoms of aluminum toxicity occurred. Further, sometimes roots will stain in the absence of aluminum (175).

Moore et al. (176) proposed that recovery of root elongation after 48 h of exposure to aluminum is a better measure of irreversible damage to the root apical meristem. Hecht-Buchholz (177) reported that aluminum toxicity in barley caused stunted roots, destruction of root cap cells, swelling, and destruction of both root epidermal and cortical cells. She found large differences between cultivars and proposed that aluminum resistance could be attributed to greater resistance of the root meristem of the aluminum-tolerant genotype to irreversible destruction. Lazof and Holland (28) suggested that root recovery experiments in soybean, pea, and snapbean allowed separation of H^+ toxicity effects from Al^{3+} toxicity effects. Zhang et al. (178) showed that root regrowth after aluminum stress could be used to improve aluminum tolerance in triticale (*Triticosecale* spp.).

16.7.2 Genetics

Aluminum tolerance is a heritable trait in sorghum (179), barley (180), wheat (181,182), rice (*Oryza sativa* L.) (183), soybean (184), and *Arabidopsis thaliana* (185). With sorghum, Magalhaes (cited in 179) has found a pattern of inheritance of aluminum tolerance that is consistent with a single locus. With barley, Tang et al. (180) confirmed that aluminum tolerance segregation in F_2 genotypes was due to a single gene, *Alp*, and they proposed the use of molecular markers in selection of aluminum tolerance in barley genotypes without the need for field trials, soil bioassays, or solution culture tests. In wheat, controversy exists over the number and location of genes that are involved in aluminum tolerance (181,182). In rice, nine different genomic regions on eight chromosomes have been associated with genetic control of plant response to aluminum, indicating that aluminum tolerance is a multigenic trait (183). Similarly, with soybean, aluminum tolerance is likely to be governed by 3 to 5 genes (184). In *Arabidopsis*, two quantitative trait loci occurring on two chromosomes could account for 43% of total variability in aluminum tolerance among a recombinant inbred population (185). A recent review of genetic analysis of aluminum tolerance in plants is found in Kochian et al. (179).

16.8 PLANT MECHANISMS OF ALUMINUM AVOIDANCE OR TOLERANCE

There are two types of mechanisms whereby a plant can avoid or tolerate aluminum toxicity: (a) exclusion of aluminum from the symplasm, or (b) internal tolerance of aluminum in the symplasm. Good reviews on this subject are in Taylor (186,187), Matsumoto (99), Kochian et al. (179, 188), and Barcelo and Poschenrieder (96).

16.8.1 Plant Mechanisms of Aluminum Avoidance

Based on chemical analysis of aluminum in root sections, Horst et al. (189) showed that the root tips of an aluminum-tolerant cultivar of cowpea (*Vigna unguiculata* Walp.) had a lower aluminum concentration than those of an aluminum-sensitive cultivar, suggesting that reduced aluminum absorption into the root tip was responsible for its higher aluminum tolerance. Using direct measurement of aluminum with atomic absorption spectrophotometry or ion chromatography, Rincon and Gonzales (190) showed that aluminum content was 9 to 13 times greater in the 0-to-2-mm root tips of an aluminum-sensitive wheat cultivar than in an aluminum-tolerant cultivar. Similar results were reported by Delhaize et al. (191), who showed using x-ray microanalysis that aluminum-sensitive wheat root apices accumulated 5 to 10 times greater aluminum than aluminum-tolerant root apices.

These results indicate that aluminum exclusion occurs in several plant species. Possible mechanisms of aluminum avoidance include: (a) root avoidance response, (b) organic acid release, (c) exudation of phosphate, (d) exudation of polypeptides, (e) exudation of phenolics, (f) alkalinization of rhizosphere pH, (g) binding to mucilage, (h) binding to cell walls, (i) binding to external face of membrane, and (j) interactions with mycorrhizal fungi.

16.8.1.1 Avoidance Response of Roots

Classic avoidance response of roots to aluminum toxicity was shown by research (149) in which corn roots curved away from aluminum applied to one side of root. Also, aluminum toxicity killed cells in the corn root apical meristem, and Boscolo et al. (155) speculated that this phenomenon would result in loss of apical dominance and greater lateral root growth into environments with lower aluminum levels. Interestingly, taproots of corn cv. SA-6 and soybean cv. Perry did not penetrate much into an aluminum-toxic subsoil layer, although lateral root lengths increased in the nontoxic top soil layer (192). However, although increased lateral root growth in topsoil layers could help to maintain crop yields in areas with acid subsoils, under drought conditions, lack of root growth into deeper layers could limit water uptake.

16.8.1.2 Organic Acid Release

Considerable evidence supports organic acid release as a mechanism of aluminum avoidance in plants (179,188,193,194). Hue et al. (195) used elongation of cotton (*Gossypium hirsutum* L.) taproots as a measure of aluminum toxicity to document the aluminum detoxification effect of several low-molecular-weight organic acids or anions. The relative ameliorative capacity of the organic acids followed closely the stability constants of the aluminum–organic acid complexes in the order:

$$\text{Citric} > \text{Oxalic} > \text{Tartaric} > \text{Malic} > \text{Acetic}$$

The formation of stable rings (5-, 6-, and to a lesser extent 7-membered structures) between aluminum and organic anions or molecules seems to be responsible for the detoxification (195). Structure of an aluminum–citrate complex is shown below.

$Al(H_2O)_6{}^{3+}$ Citric acid   Al-Citrate

The first evidence of aluminum-induced root exudation of an organic acid was identified in snapbean, in which an aluminum-tolerant cultivar exuded ten times as much citrate as an aluminum-sensitive cultivar in the presence of aluminum (196). Aluminum-induced root release of malate was characterized thoroughly in wheat by Delhaize and co-workers (197–200). They showed that exposure of an aluminum-tolerant genotype to 10 μM Al induced malate exudation from roots within 15 min. Wheat root apices contained sufficient malate for excretion for over 4 h (198). After 24 h of exposure to 100 μM Al, de novo synthesis of malate was demonstrated by measuring ^{14}C incorporation into malate (199). The efflux of malate from root apices was electroneutral, because it was accompanied by an efflux of K^+ (198). Evaluating 36 wheat cultivars, Ryan et al. (200) showed a significant correlation between relative tolerance of wheat genotypes to aluminum and amount of malate released from root apices. Other researchers have argued against the effectiveness of malate exudation on alleviating aluminum toxicity because of rapid degradation by soil microorganisms (201) and the low concentrations and relatively weak chelating ability of malate for aluminum (202).

Other plant species have been shown to exude organic acids in response to aluminum stress. Aluminum-tolerant corn genotypes exuded higher concentrations of citrate (203). An aluminum-tolerant tree species, *Senna tora* Roxb. (formerly *Cassia tora*), exuded citric acid after 4 h of exposure to 50 μM Al (204). In rye, after 10 h of exposure to 10 μM Al, increased activity of citrate synthase (CS) occurred along with increased citrate secretion (205). In all soybean genotypes, citrate exudation increased within 6 h of aluminum exposure; however, only citrate efflux in aluminum-tolerant genotypes was sustained for an extended time period (206). A positive correlation was found between citrate in root tips of soybean and aluminum tolerance (206). The aluminum-accumulating plant, buckwheat (*Fagopyrum esculentum* Moench), was found to exude oxalate, a strong aluminum chelator (207). Taro (*Colocasia esculenta* Schott), a tropical root crop that is not an aluminum accumulator, also exuded oxalate from roots in response to aluminum (208). Aluminum-resistant mutants of *Arabidopsis thaliana* constitutively released higher concentrations of citrate or malate compared to the wild type (209). A mutant carrot (*Daucus carota* L.) cell line that solubilized phosphate from aluminum phosphate exuded citrate from roots (210). This cell line had a greater activity of mitochondrial CS and a lower activity of a cytoplasmic enzyme, NADP-specific isocitrate dehydrogenase (NADP-ICDH), involved in citrate degradation (211,212).

Anion channels are involved in the aluminum-activated exudation of organic anions. Using electrophysiology to measure current passing across whole apical cells of wheat roots, Ryan et al. (213) showed that 20 to 50 μM Al activated an anion channel. Genotypic differences were found with the aluminum-induced currents across protoplasts from the aluminum-tolerant wheat genotype occurring more frequently and being sustained for a longer period of time than those from the aluminum-sensitive genotype (214). Using subtractive hybridization of cDNAs from near-isogenic lines of aluminum-sensitive and aluminum-tolerant wheat, Sasaki et al. (215) found greater expression of a gene that cosegregated with aluminum tolerance. Heterologous expression of this gene, named ALMT1 (aluminum-activated malate transporter), in *Xenopus* oocytes, rice, and cultured tobacco cells conferred an aluminum-activated malate efflux, and enhanced the ability of tobacco cells to recover from 18 h of exposure to 100 μM Al (215). Transgenic barley cultivars with the ALMT1 transgene showed increased malate effux and increased root grwoth at concentrations up to 12 μM Al (216).

Another means of increasing aluminum tolerance in plants is to increase synthesis as well as exudation of organic acids. De la Fuente et al. (217) overexpressed a CS gene from the bacterium, *Pseudomonas aeruginosa* Migula, in the cytoplasm of transgenic tobacco and found increased citrate levels within roots, increased citrate efflux, and increased root elongation in the presence of ≥ 100 μM Al. However, Delhaize et al. (218) were unsuccessful in repeating this work (217), and they suggested that the activity of *P. aeruginosa* cytoplasmic CS in transgenic tobacco is either sensitive to environmental conditions, or that the improved aluminum tolerance observed by de la Fuente et al. (217) was due to other factors. Koyama et al. (219) overexpressed a mitochondrial CS gene, isolated from carrot, in *Arabidopsis thaliana* and found increased CS activity, increased excretion of citrate, and slightly increased amelioration of aluminum toxicity based on root elongation at pH 5.

Tesfaye et al. (220) overexpressed genes for nodule-enhanced forms of the enzymes that catalyze malate synthesis, phosphoenolpyruvate carboxylase and malate dehydrogenase in alfalfa (*Medicago sativa* L.). They found increased enzyme activities, increased root exudation of organic acids (citrate, oxalate, malate, succinate, and acetate), and increased root elongation in the presence of 50 to 100 μM Al. However, such root exudation represented a drain of plant resources, and transgenic lines had reduced biomass compared to untransformed control plants when grown at soil pH 7.25. In acid soils, however, transgenic alfalfa had 1.6 times greater biomass than untransformed control plants.

Although abundant evidence exists for aluminum-induced organic acid excretion as a mechanism of aluminum tolerance, other mechanisms probably exist. Ishikawa et al. (221) found no correlation between species or within species for organic acid exudation and aluminum tolerance. Similarly, Wenzl et al. (222) reported that the greater aluminum tolerance of signalgrass (*Urochloa decumbens* R.D. Webster, formerly *Brachiaria decumbens*) relative to ruzigrass (*Urochloa ruziziensis* Crins, formerly *Brachiaria ruziziensis*) was not due to greater exudation of organic acids.

16.8.1.3 Exudation of Phosphate

Root apices of an aluminum-tolerant genotype of wheat exuded phosphate as well as citrate in response to aluminum exposure (223). Pellet et al. (223) speculated that phosphate release contributed to aluminum tolerance in wheat. In contrast, no major differences in phosphate release were found among near-isogenic lines of wheat that differed in aluminum tolerance (224).

16.8.1.4 Exudation of Polypeptides

Aluminum-resistant lines of wheat exuded an aluminum-induced 23 kDa polypeptide (225). This polypeptide, synthesized de novo in response to aluminum, binds aluminum, and cosegregates with the aluminum-resistant phenotype in F_2 populations (225,226). The gene encoding this polypeptide still needs to be isolated.

16.8.1.5 Exudation of Phenolics

Phenolics are aromatic secondary metabolites of plants (e.g., quercetin, catechin, morin, or chlorogenic acid) that can bind aluminum (67,227). Silicon ameliorates aluminum toxicity in some plants (228, 229). In an aluminum-resistant corn cultivar, silicon and aluminum triggered the release of phenolic compounds (e.g., catechol, catechin, and quercetin) up to 15 times the release by plants not pretreated with silicon (230). However, the binding capacity of many of these phenolic compounds for aluminum is greater at pH 7 than at pH 4.5 (227).

16.8.1.6 Alkalinization of Rhizosphere

The solubility of aluminum is dependent on pH; as pH rises above 5.0, precipitation of aluminum as $Al(OH)_3$ increases (Figure 16.1). An aluminum-tolerant wheat cultivar grown in a nutrient solution increased the pH, whereas an aluminum-sensitive cultivar lowered the solution pH (231). Foy et al. (231) proposed that aluminum tolerance is associated with plant-induced alkalinization of pH. However, rhizosphere pH associated with apical root tissues did not appear to be a primary mechanism of differential aluminum tolerance in wheat. The root apex of an aluminum-tolerant wheat genotype had only a slightly higher rhizosphere pH in the presence of aluminum than an aluminum-sensitive genotype, resulting in a 6% decrease in free Al^{3+} activity (121). Yet the aluminum-tolerant wheat genotype had 140% greater relative root elongation compared to the aluminum-sensitive genotype, indicating that rhizosphere pH did not play a major role in differential aluminum tolerance (121). In contrast, Degenhardt et al. (232) reported that aluminum exposure induced a doubling in net H^+ influx at the root tip of an aluminum-resistant *Arabidopsis* mutant relative to the wild-type, increasing pH by 0.15 units. Although the pH difference was small, solution pH maintained at 4.5 was shown to increase *Arabidopsis* root growth relative to that at pH 4.4.

16.8.1.7   Binding to Mucilage

Horst et al. (233) reported that mucilage from root tips of cowpea had a high binding capacity for aluminum and that removal of this mucilage resulted in greater inhibition of root elongation by aluminum. They proposed that mucilage served to protect the apical meristem against aluminum injury. Similarly, Brigham et al. (234) showed that removal of snapbean mucilage (including root border cells) resulted in reduced root elongation and greater aluminum accumulation in root tips as shown by lumogallion staining. Pan et al. (777) demonstrated that the presence of mucilage and border cells in wheat reduced aluminum injury to root meristems, as shown by a greater mitotic index. In contrast, Li et al. (235) found that although mucilage from corn root apices binds strongly to aluminum, the presence or absence of mucilage did not affect aluminum-induced inhibition of root elongation.

16.8.1.8   Binding to Cell Walls

Some researchers observed that root cation exchange capacity (CEC) of Al-tolerant genotypes were lower than that of aluminum-sensitive ones (236); however, other researchers found no such correlation (237,238). Interestingly, a transgenic potato overexpressing PME exhibited greater activity of PME (which should result in more free carboxylic groups in cell walls), greater aluminum accumulation in root tips, and greater sensitivity to aluminum as shown by aluminum-induced callose formation and inhibition of root elongation (108). These results suggest that genotypic differences in number of negatively charged binding sites in the cell wall could result in differential aluminum tolerance.

Interestingly, overexpression of WAK1 in arabidopsis conferred increased aluminum tolerance as shown by increased root elongation in the presence of aluminum (89). Sivaguru et al. (89) speculated that WAKs could interact with cell wall components such as callose or pectins, alleviating aluminum toxicity. Alternatively, they speculated that the cytoplasmic kinase domain could be cleaved off from WAKs and participate in cytoplasmic aluminum response pathways.

16.8.1.9   Binding to External Face of Plasma Membrane

Among five plant species differing in aluminum tolerance, the zeta potential (i.e., an estimate of plasma membrane surface potential) was higher (membrane surface less negative) in aluminum-resistant plant species than in sensitive ones (239). Wagatsuma and Akiba (239) hypothesized that aluminum-sensitive plant species had more negative charges on the plasma membrane, resulting in greater aluminum-binding to its surface. Similarly, Ishikawa and Wagatsuma (240) pretreated protoplasts of four plant species with aluminum for 10 min followed by a hypotonic aluminum-free solution. They found that protoplasts from aluminum-sensitive species exhibited greater leakage of K^+ and proposed that aluminum binding to plasma membrane induced greater rigidity, reduced extensibility, and increased leakage under hypotonic conditions.

In contrast, Yermiyahu et al. (241) found that the surface-charge density of vesicles isolated from an aluminum-sensitive wheat cultivar was 26% more negative than those from an aluminum-tolerant wheat cultivar. However, they (241) argued that this small difference in surface-charge density did not account for the large difference in sensitivity to aluminum (50%).

16.8.1.10   Interactions with Mycorrhizal Fungi

Conflicting reports occur in the literature with a few researchers finding negative or no effect of mycorrhizal colonization on host-response to aluminum toxicity (242–245) and a greater number showing a beneficial effect of colonization with either ectomycorrhizal (ECT) (246,247) or arbuscular mycorrhizal fungi (AMF) (248–250). Host response to aluminum toxicity depended on the species of ECT (242) or AMF (243). Scots pine (*Pinus sylvestris* L.) colonized by an aluminum-sensitive ECT fungus (*Hebeloma* cf. *longicaudum* Kumm. ss. Lange) exhibited decreased shoot and root biomass compared to nonmycorrhizal plants in the presence of 2500 µM Al (242). In contrast, Scots pine colonized by an aluminum-tolerant ECT fungus (*Laccaria bicolor* Orton) had greater shoot and root biomass, greater shoot P, and lower shoot aluminum compared to nonmycorrhizal plants in the presence of 740 µM Al (242). Similarly, only five of eight isolates of AMF increased growth of switchgrass and reduced foliar Al concentrations in an acid soil (243).

Pitch pine (*Pinus rigida* Mill.) colonized with the ECT fungus, *Pisolithus tinctorius* Coker and Couch, had greater shoot and root biomass at 50 to 200 µM Al than noninoculated plants (246). Colonization of white pine (*Pinus strobus* L.) with the ECT fungus, *P. tinctorius*, resulted in greater shoot dry weight, height, and needle length relative to nonmycorrhizal seedlings at aluminum levels ≥ 460 µM (247). Schier and McQuattie (247) attributed the beneficial effects of ECT fungi to reduced aluminum concentrations and higher phosphorus concentrations in needles.

Colonization of switchgrass (*Panicum virgatum* L.) with the AMF, *Glomus occultum* Walker, resulted in higher total shoot biomass at 500 µM Al as well as lower tissue aluminum and higher calcium concentrations (248). In an aluminum-sensitive barley cultivar, colonization with the AMF, *Glomus etunicatum* Becker and Gerdemann, resulted in greater shoot biomass and greater P concentrations in shoots and roots at 600 µM Al (249). Colonization of tissue-cultured banana (*Musa acuminata* Colla) with the AMF, *Glomus intraradices* N.C. Schenck & S.S. Sm., increased shoot dry weight, water uptake, and nutrient uptake and decreased aluminum content in roots and shoots (250). Apparently, one of the benefits of either ecto- or endomycorrhizal colonization is to ameliorate the detrimental effects of aluminum toxicity on root growth and nutrient or water uptake.

Aluminum has toxic effects also on mycorrhizal fungi, adversely affecting the quality and quantity of mycorrhizal colonization (243,251). Differences in response to aluminum have been found between ECT fungal species (243). Also, genotypic differences within an ECT fungal species have been found in response to aluminum. For example, isolates of ECT fungus, *P. tinctorius*, from old coal-mining sites (pH 4.3, 12.1 mM Al) exhibited greater aluminum tolerance based on mycelial mass at ≥ 440 µM Al than isolates from rehabilitated mine sites (pH 4.9, 800 µM Al) and those from forest sites (pH 4.3, 220 µM Al) (252). Strains of the ECT fungus, *Suillus luteus* Gray, that differed in aluminum sensitivity were inoculated on Scots pine, and the extramatrical mycelia developed by the aluminum-resistant strain were more abundant in the presence of aluminum compared to those of the aluminum-sensitive strain (251). Scots pine seedlings colonized by this aluminum-tolerant ECT strain in the presence of aluminum had greater shoot heights compared to noninoculated seedlings (251).

Cuenca et al. (253) showed that the tropical woody species, *Clusia multiflora* Knuth., inoculated with AMF accumulated less aluminum in roots; instead aluminum was bound to the cell walls of the fungal mycelium and in vesicles. Using ^{27}Al-NMR, aluminum was found to be taken up and accumulated into polyphosphate complexes in the vacuole of the ECT fungus, *Laccaria bicolor* Orton (254). Martin et al. (254) suggested that sequestration of aluminum in polyphosphate complexes could help to protect mycorrhizal plants against aluminum toxicity. An aluminum-adapted strain of an ECT fungus, *Suillus bovines* Kuntze, had a shorter average chain length of mobile polyphosphates and greater terminal phosphate groups (255). Gerlitz (255) proposed that this change increased binding and detoxification of polyphosphates to aluminum. A good review of possible aluminum tolerance mechanisms in ECT is found in Jentschke and Godbold (256).

16.8.2 Plant Mechanisms of Aluminum Tolerance

Mechanisms of internal tolerance of aluminum involve: (a) complexation with organic acids, (b) complexation with phenolics, (c) complexation with silicon, (d) sequestration in the vacuole or other storage organs, and (e) trapping of aluminum in cells.

16.8.2.1 Complexation with Organic Acids

In the leaves of aluminum-accumulating hydrangea, Ma et al. (257) used molecular sieve chromatography to determine that citrate eluted at the same time as aluminum and that the molar ratio of aluminum to citric acid was approximately 1:1. In the aluminum accumulator, buckwheat, aluminum was complexed with citrate in the xylem (258), but with oxalic acid in vacuoles of leaf cells (259,260). In the aluminum accumulator, *Melastoma malabathricum* L., aluminum citrate occurred in the xylem sap and was then transformed into aluminum oxalate for storage in leaves (261,262).

16.8.2.2 Complexation with Phenolics

In aluminum-accumulating tea, Nagata et al. (263) used ^{27}Al-NMR to demonstrate that aluminum was bound to catechin in young leaves and buds; in mature leaves, aluminum–phenolic acid and aluminum–organic acid complexes were found. Interestingly, Ofei-Manu et al. (227) showed that at pH 7 (cytoplasmic pH), aluminum binding capacity is in the order: quercetin > catechin, chlorogenic acid, morin > organic acids. Among ten woody plant species and two marker crop species, a positive linear correlation was found between root phenolic compounds and aluminum tolerance, based on aluminum-inhibited root elongation (227).

16.8.2.3 Complexation with Silicon

Cocker et al. (229) proposed that amelioration of aluminum toxicity by silicon is due to formation of an aluminosilicate compound in the root apoplast. Hodson and Sangster (264) proposed that codeposition of aluminum and silicon in needles of conifers is responsible for aluminum detoxification by silicon. Hodson and Evans (228) reviewed the evidence in support of various mechanisms of silicon amelioration of aluminum toxicity, and they divided plants into four groups: (a) aluminum accumulators in arborescent dicots, (b) silicon accumulators in grasses, (c) gymnosperms and arborescent dicots with moderate amounts of aluminum and silicon, and (d) herbaceous dicots that exclude aluminum and silicon. Obviously, aluminum can codeposit with silicon only in plants that accumulate both elements. Aluminum was deposited in phytoliths (hydrated silica deposits) of conifers, graminaceous plants, and dicots in the Ericaceae family (265,266). Using x-ray microanalysis, Hodson and Sangster (267) found codeposition of aluminum and silicon in the outer tangential wall of the endodermis of sorghum. In *Faramea marginata* Cham., a woody member of the Rubiaceae family that is known to accumulate aluminum and silicon in leaves, colocalization of aluminum and silicon in a molar ratio of 1:2 occurred in the cortex of stem sections and throughout leaves (268). A good review of aluminum and silicon interactions can be found in Hodson and Evans (228), Cocker et al. (229), and Hodson and Sangster (264).

16.8.2.4 Sequestration in the Vacuole or in Other Organelles

Aluminum ions could be sequestered in vacuoles or other storage organelles where they would not affect metabolism in the cytoplasm adversely. The presence of 50 μM Al increased pyrophosphate-dependent and ATP-dependent H$^+$ pump activity in tonoplast membrane vesicles isolated from barley roots, and Kasai et al. (269) hypothesized that Al^{3+} was sequestered in the vacuole perhaps by an Al/nH$^+$ exchange reaction. Interestingly, expression of two 51 kDa proteins is strongly induced in an aluminum-tolerant wheat cultivar, and only weakly expressed in an aluminum-sensitive wheat cultivar (270). Sequence analysis of the purified peptides showed that one is homologous to the B subunit of the vacuolar H^+-ATPase (V-ATPase) (270).

In an aluminum-tolerant unicellular red alga (*Cyanidium caldarium* Geitler), aluminum accumulated in spherical electron-dense bodies in the cytoplasm near the nucleus (271). These bodies contained high levels of iron and phosphorus, and the researchers speculated that they might be iron-storage sites under normal culture conditions. Interestingly, transferrin, an iron carrier, is the main protein that binds Al^{3+} in the blood plasma of animals (47).

16.8.2.5 Trapping of Aluminum in Cells

Fiskesjo (272) proposed that aluminum could be trapped in root border cells, which were then detached and sloughed away from roots. Consistent with this hypothesis, detached root border cells of snap bean were killed by aluminum within 2 h of aluminum exposure (70).

A punctated pattern of cell death was observed in aluminum-tolerant wheat roots after 8 h of exposure to aluminum, with an increase in oxalate oxidase activity and H_2O_2 production after 24 h (273). Delisle et al. (273) speculated that cell death could be a means for root tip cells to trap or exclude aluminum from live tissues. Interestingly, a hypersensitive cell death response is a common means for plants to trap pathogens, not allowing them to spread to other cells. Many genes up-regulated by aluminum in wheat are similar to pathogenesis-related genes (274).

16.9 ALUMINUM IN SOILS

Aluminum in soil forms the structure of primary and secondary minerals, especially aluminosilicates, such as feldspars, micas, kaolins, smectites, and vermiculites (275). As the soils continue to weather (especially under conditions of high rainfall and warm climates), silicon is leached away, usually as $Si(OH)_4$ in solution, leaving aluminum behind in the solid forms of aluminum oxyhydroxides, such as boehmite and gibbsite, as shown below (276):

$$Al_2Si_2O_5(OH)_4 \text{ (kaolinite)} + 5H_2O \rightleftharpoons 2Al(OH)_3 \text{ (gibbsite)} + 2Si(OH)_4$$

The soils themselves become 'older,' more acidic, and more aluminum toxic and would be classified as Oxisols or Ultisols.

16.9.1 Locations of Aluminum-Rich Soils

According to FAO/UNESCO recent maps (277), most Oxisols and Ultisols are located in the Tropics and Subtropics (Figure 16.2 and Figure 16.3). More specifically, about one third of the Tropics (1.5 billion ha) has sufficiently strong soil acidity for soluble aluminum to be toxic to most crops (278). Geographically, Latin America has 821 million ha, Africa 479 million ha, South and Southeast Asia 236 million ha (278). In the United States (Figure 16.4), a major portion of acid Ultisols is in the Southeast (88 million ha), from Alabama, Arkansas to Virginia (279). Other states, such as California, New York, Oregon, Pennsylvania, and Washington, also have acid Ultisols, but to a much smaller extent (280). In contrast, only Hawaii and Puerto Rico have Oxisols (Figure 16.5). A detailed review of global distribution of acid soils was given by Sumner and Noble (281).

16.9.2 Forms of Aluminum in Soils

To be bioavailable, soil aluminum must first be in solution (279). Soluble aluminum, however, is controlled by several processes (Figure 16.6). For example, aluminum-containing minerals, such as gibbsite and kaolinite, can dissolve under acidic conditions, release aluminum into solution, and

Distribution of FERRALSOLS
Based on WRB and the FAO/Unesco Soil Map of the World

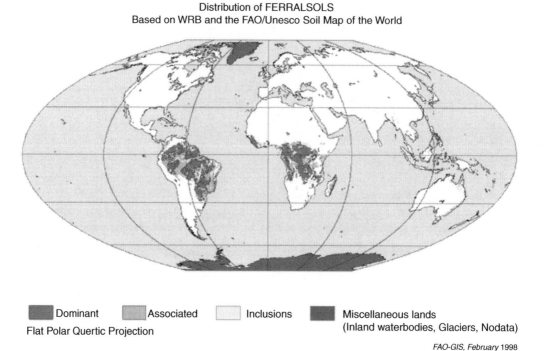

■ Dominant   ■ Associated   □ Inclusions   ■ Miscellaneous lands
Flat Polar Quertic Projection   (Inland waterbodies, Glaciers, Nodata)

FAO-GIS, February 1998

FIGURE 16.2 Oxisols distribution in the world. (From FAO/UNESCO. http://www.fao.org/ag/agl/agll/ wrb/mapindex.stm, 1998. Accessed March 2003.) (For a color presentation of this figure, see the accompanying compact disc.)

Distribution of ACRISOLS
Based on WRB and the FAO/Unesco Soil Map of the World

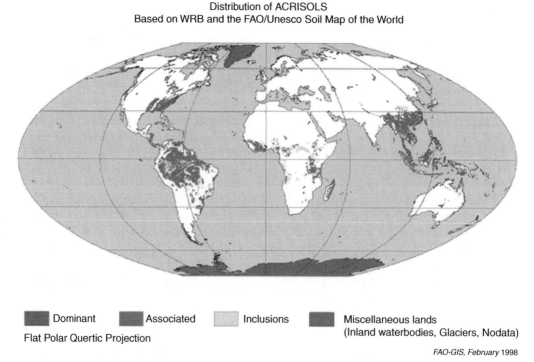

■ Dominant   ■ Associated   □ Inclusions   ■ Miscellaneous lands
Flat Polar Quertic Projection   (Inland waterbodies, Glaciers, Nodata)

FAO-GIS, February 1998

FIGURE 16.3 Ultisols distribution in the world. (From FAO/UNESCO. http://www.fao.org/ag/agl/agll/ wrb/mapindex.stm, 1998. Accessed March 2003.) (For a color presentation of this figure, see the accompanying compact disc.)

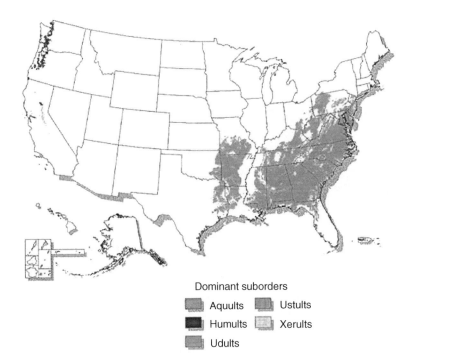

FIGURE 16.4 Ultisols distribution in the United States. (From NRCS (Natural Resources Conservation Service). http://soils.usda.gov/classification/orders/main.htm, 2002. Accessed March 2003.) (For a color presentation of this figure, see the accompanying compact disc.)

FIGURE 16.5 Oxisols distribution in the United States. (From NRCS (Natural Resources Conservation Service). http://soils.usda.gov/classification/orders/main.htm, 2002. Accessed March 2003.) (For a color presentation of this figure, see the accompanying compact disc.)

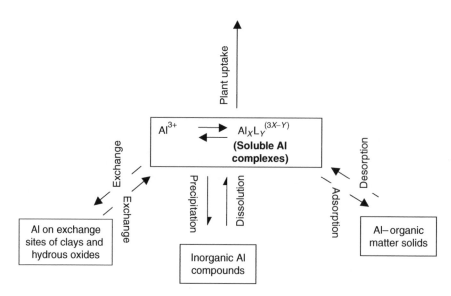

FIGURE 16.6 Processes controlling forms, solubility, and availability of Al in soils. (Adapted from G.S.P. Ritchie, in *Soil Acidity and Plant Growth*, Academic Press Australia, Marrickville, Australia, 1989, pp. 1–60.)

thus, control soluble aluminum concentration and activity (282). The dissolution of gibbsite is expressed by

$$\gamma\text{-}Al(OH)_3(\text{gibbsite}) + 3H^+ \rightleftharpoons Al^{3+}(\text{aqueous}) + 3H_2O$$

On the other hand, clay minerals with negative charges on their surface, resulting from isomorphic substitution (permanent charge) or from hydrolysis of hydroxyl (OH^-) groups at broken edges (variable charge), can take aluminum from solution by electrostatic attraction in cation exchange. Allophane and imogolite, which are amorphous aluminosilicates with large surface areas and high variable charges, can retain large quantities of aluminum (283). So can solid organic matter (OM) with many negative charges from carboxyl ($-COO^-$) functional groups. Solid OM also can retain aluminum strongly by another process called specific adsorption or complexation. Bloom et al. (284) proposed that aluminum–solid OM interactions were central to the exponential decreases of soluble aluminum at pH $<$ 5. They reported a 40% reduction in soluble aluminum after adding 2% of a decomposed leafy material to an acid B horizon of an inceptisol.

Aluminous minerals in soils are numerous (275). Besides the aluminosilicates and aluminum oxyhydroxides mentioned previously, aluminum can form sparingly soluble compounds with common soil anions, such as phosphates and sulfates (1). Alunite [$KAl_3(OH)_6(SO_4)_2$], basaluminite [$Al_4(OH)_{10}SO_4$] and jurbanite [$Al(OH)SO_4 \cdot 5H_2O$] have been found in soils where concentration of SO_4^{2-} was high from fertilization with gypsum or by acid sulfate natural occurrence (282,285,286). With prolonged phosphorus fertilization, soluble phosphorus concentration was increased with time, and Al-P minerals, such as variscite, could be formed (287).

The concentration and activity of Al^{3+} in soil solutions not only depend on the processes by which aluminum is distributed between the solid and liquid phases, but also on its many reactions in solution. The extent of these aqueous reactions depends on (a) solution pH, (b) ionic strength, (c) kind and concentration of complexing ligands, and (d) kind and concentration of competing cations (288). Important among these reactions are hydrolysis, polymerization, and complexation with inorganic (e.g., SO_4^{2-}, F^-) and organic anions (e.g., citrate, malate, fulvates) (Table 16.1) (289).

Thus, there are several different species of aluminum in the soil solution, with widely different bioavailability or toxicity (35,37,195). Another implication is that Al^{3+} concentration (activity) makes up only a relatively small fraction of the total soluble aluminum. Wolt (285) found that free

TABLE 16.1
Possible Reactions of Al^{3+} in the Soil Solution

	log K^a (at 25°C)
1. Hydrolysis reactions	
$Al^{3+} + H_2O = Al(OH)^{2+} + H^+$	−5.0
$Al^{3+} + 2H_2O = Al(OH)_2^+ + 2H^+$	−10.1
$Al^{3+} + 3H_2O = Al(OH)_3^0 + 3H^+$	−16.8
$Al^{3+} + 4H_2O = Al(OH)_4^- + 4H^+$	−22.99
2. Polymerization	
$2Al^{3+} + 2OH^- = Al_2(OH)_2^{4+}$	
$13Al^{3+} + 28OH^- = Al_{13}O_4(OH)_{24}^{7+} + 4H^+$	
3. Complexation with inorganic anions	
$Al^{3+} + SO_4^{2-} = Al(SO_4)^+$	3.5
$Al^{3+} + F^- = AlF^{2+}$	7.0
$Al^{3+} + H_2PO_4^- = Al(H_2PO_4)^{2+}$	3.1
4. Complexation with organic anions	
$Al^{3+} + oxalate^{2-} = (Al\text{-}oxalate)^+$	6.0
$Al^{3+} + citrate^{3-} = (Al\text{-}citrate)^0$	8.1
$Al^{3+} + fulvate^{n-} = (Al\text{-}fulvate)^{(n-3)-}$	

[a]From D.K. Nordstrom, H.M. May, in *The Environmental Chemistry of Aluminum*, CRC Press, Boca Raton, FL, 1996, pp. 39–80.

Al^{3+} comprised 2 to 61% of total aluminum in soil solutions of acid Ultisols where SO_4^{2-} was the dominant ligand. Similarly, Hue et al. (195) reported that 76 to 93% of total soil solution aluminum of two acid Ultisols in Alabama was complexed with low-molecular-weight organic acids.

As discussed earlier, it is generally accepted that Al^{3+} and monomeric Al-hydroxy species are more toxic to plants than other forms (35,37,195). Several lines of evidence have shown the nontoxic nature of organically complexed aluminum (195,207,217,290–292). In addition, ionic strength of the soil solution also plays an important role in modifying aluminum toxicity (293). Expressing aluminum species in terms of activity instead of concentration significantly improved the correlation between plant growth and aluminum toxicity across many soils and soil horizons (293,294).

In addition to monomeric aluminum species, polymeric aluminum species have recently been studied intensively perhaps because of their reportedly acute phyto/rhizo-toxicities (31,35,295,296). The 'Al$_{13}$' polymer [$AlO_4Al_{12}(OH)_{24} (H_2O)_{12}^{7+}$] was identified using ^{27}Al NMR spectroscopy, where 'clean' solutions containing relatively high aluminum ($> 10\,mM$) were partially neutralized (297). However, this polymeric aluminum species (Al$_{13}$) could not be detected in soil solutions containing SO_4^{2-} or silicates (298).

16.9.3 DETECTION OR DIAGNOSIS OF EXCESS ALUMINUM IN SOILS

As discussed earlier, soil aluminum can exist in many different pools, and its reactions within the soil solution are also quite intricate. It is generally accepted that the activity of monomeric hydroxyaluminum species should be a good predictor of aluminum toxicity for a given plant species if (a) the aluminum absorption by plants is small relative to the quantity of toxic aluminum species in the soil solution such that the solution activity remains virtually constant as the plant grows (steady-state condition) or (b) any decrease in the activity of toxic aluminum species is readily compensated for by solid phase aluminum or nontoxic aluminum in solution (equilibrium condition). In reality, these conditions are hardly met, thus solution activity (intensity factor) and an estimate of the aluminum-buffering capacity (capacity factor) are required to evaluate or predict the toxicity of soil aluminum.

16.9.3.1 Extractable and Exchangeable Aluminum

Different methods have been used to extract solid-phase aluminum, which presumably correlates well with aluminum phytotoxicity (299). An unbuffered solution of 1 M KCl is commonly used to extract the fraction of aluminum (often referred to as 'exchangeable'), which is presumably held by negative charges on the soil surface. When exchangeable aluminum is expressed as a percentage of the effective cation exchange capacity (ECEC), it is referred to as the aluminum saturation percentage. Table 16.2

TABLE 16.2
Selected Chemical Properties of Some Acid Soils from Latin America

Horizon (cm)	pH (H₂O)	Org. C (g kg⁻¹)	Al	Ca	Mg	K	ECEC	Al Sat. (%)
			Exchangeable (cmol$_c$ kg⁻¹)					
Florencia, Colombia. Typic Tropudult								
0–16	4.8	20	3.60	0.95	0.80	0.23	5.58	64
16–85	4.7	5	7.76	0.22	0.43	0.03	8.44	92
Napo, Ecuador. Orthoxic Tropudult								
0–13	4.7	10	0.30	2.06	0.50	2.15	5.01	6
13–25	4.3	6	1.97	0.20	0.09	0.64	2.90	68
25–40	4.0	5	2.07	0.20	0.06	0.18	2.51	82
40–60	4.2	2	2.27	0.22	0.17	0.04	2.70	84
Yurimaguas, Peru. Typic Paleudult								
0–10	4.4	17	1.29	1.13	0.60	0.28	3.30	39
10–30	4.4	5	3.31	0.29	0.14	0.08	3.82	87
30–50	4.5	3	4.26	0.29	0.22	0.07	4.45	87
Iquitos, Peru. Typic Paleudult								
0–16	4.0	24	5.9	1.0	0.2	0.20	7.30	81
16–35	4.5	10	6.7	0.4	0.1	0.08	7.28	92
35–70	4.3	5	9.5	0.2	0.1	0.08	9.88	96
Manaus – AM, Brazil. Typic Acrorthox								
0–8	4.6	30	1.1	1.7	0.3	0.19	3.29	33
8–22	4.4	9	1.1	0.2	—	0.09	1.39	79
22–50	4.3	7	1.2	0.2	—	0.07	1.47	82
Paragominas – PA, Brazil. Typic Acrorthox								
0–6	4.2	28	1.45	2.08	0.88	0.14	4.55	32
6–23	4.1	9	1.86	0.64	0.56	0.07	3.13	59
23–60	4.7	7	1.03	0.48	0.48	0.04	2.03	51
Barrolandia – BA, Brazil. Typic Paleudult								
0–30	4.7	13	0.7	0.8	1.3	0.07	2.87	24
10–23	4.7	10	0.9	0.0	0.6	0.06	1.56	58
23–49	4.8	5	1.0	0.0	0.6	0.04	1.64	61
Porto Velho – RO, Brazil. Orthoxic Palehumult								
0–5	4.5	31	2.2	0.6	—	0.20	3.00	73
5–20	4.2	13	1.4	0.1	—	0.08	1.58	93
20–40	4.4	10	1.1	0.1	—	0.05	1.25	88
40–60	4.7	7	1.0	0.1	—	0.04	1.14	88

Source: From P.A. Sanchez, in *Management of Acid Tropical Soils for Sustainable Agriculture*. IBSRAM Proceedings No. 2, 1987, pp. 63–107.

lists values of exchangeable aluminum and aluminum saturation percentage for some acid soils from Latin America (300). The amount of aluminum extracted by neutral salts, such as 1 M KCl or 0.01 M CaCl$_2$, however, varies with extraction time, concentration of the extracting solution (301), and with the number of successive extractions (302).

Other solutions such as 1 M NH$_4$Cl, 0.01 M CaCl$_2$, or 0.01 M Ca(NO$_3$)$_2$ have also been used to extract aluminum. There are indications that aluminum extracted with 0.01 M CaCl$_2$, an extractant that mimics the ionic strength (and composition) of highly weathered acid soils, correlates well with the free Al^{3+} activity in soil solution and with aluminum phytotoxicity (303–304).

Also, 0.5 M CuCl$_2$ and 0.33 M LaCl$_3$ have been used to extract organically bound aluminum (284,305). Copper reacts strongly with carboxylate sites that bind aluminum and can readily replace aluminum bound to the solid organic matter. Lanthanum is less effective than copper, but more effective than potassium, in displacing organically bound aluminum (306).

Despite potential difficulties in extracting toxic forms of aluminum with neutral salt solutions, exchangeable aluminum and aluminum saturation percentage have been used extensively as an indicator of aluminum toxicity in acid soils and in estimating the lime requirement (307). Growth of many plants in acid soils was reduced by 50% or more compared to growth in limed soil when the soil aluminum saturation was > 60% (307). As for lime requirement, it is generally accepted that the amount of CaCO$_3$ required to neutralize toxic aluminum can be estimated as follows:

The CaCO$_3$ requirement (t ha^{-1}) = $K \times$ exchangeable aluminum (cmol$_c$ kg^{-1})

where K ranges from 1.5 to 3.0 and averages 2.0 (307). Often K is > 1 to partly account for the fraction of aluminum that is not extracted by KCl. On the other hand, as pointed out by Adams (279), the critical aluminum-saturation percentage, above which relative plant growth would be restricted by 10% or more, varies markedly with soils and crops. For example, the critical aluminum saturation for soybean was about 20 to 25% for Ultisols in Alabama and North Carolina (308–310). It was about 6% for an Ultisol in South Carolina (308), 5% for a Spodosol in Florida (311), and 30% for an Oxisol in Brazilian Amazon (312). As for different crops, the critical aluminum saturation was 4 to 5% for alfalfa, white clover, tall fescue (*Festuca arundinacea* Schreb.), and sericea lespedeza (*Kummerowia striata* Schindl., formerly *Lespedeza striata*) (313,314). It was 40 to 50% for corn grown on three Ultisols in North Carolina (315), 1 to 8% for six Ultisols in Georgia (316) and 30% for an Oxisol in Brazil (312). Similarly, Adams and Moore (317), using the elongation rate of cotton taproot as an indicator of aluminum toxicity, found that the critical aluminum saturation was 2% in the Bt2 horizon of one soil but more than 56% in the Bt1 of another soil in Alabama. For peanut (*Arachis hypogaea* L.), the critical aluminum saturation was 60% (312). Evidently, additional and perhaps better methods for identifying the toxic aluminum forms are needed.

16.9.3.2 Soil-Solution Aluminum

Soil solution can be collected by several techniques, such as zero-tension lysimeters (in situ field sampling), column displacement with a miscible liquid, or high-speed centrifugation with or without a heavy liquid that is immiscible with water (laboratory sampling) (299,318). These techniques, however, are time consuming and often require high skills and care (in terms of pH changes due to CO$_2$ loss, and contamination) especially when aluminum concentrations are at micromolar levels.

Once in solution, be it soil solution or dilute neutral salt extracts, soluble aluminum can be quantified readily using atomic absorption (preferably flameless) spectroscopy or inductively coupled plasma emission spectroscopy. Alternatively, total soluble aluminum can be measured colorimetrically after forming a colored complex with an organic agent (319).

The separation of total soluble aluminum into different forms (speciation) is more involved, and many techniques have been proposed, which can be grouped into three main categories: (a) analytical separation of various aluminum fractions based on differential reaction kinetics with complexing agents or the physico-chemical separation of aluminum fractions based on size and charge; (b) computational differentiation of aluminum species from an analytically determined 'total' aluminum fraction, using a thermodynamically based geochemical speciation model with mass balance constraints (320); and (c) combination of one or more analytical techniques with a geochemical speciation model (321).

The most common timed spectrophotometric methods for aluminum determination include 8-hydroxyquinoline (HQ) and pyrocatechol violet (PCV) (322–325). James et al. (322) used a 15 s reaction with HQ buffered at pH 5.2, followed by extraction into butyl acetate, as a method for measuring monomeric aluminum species; a 30-min reaction would measure the total soluble aluminum. The PCV method requires a longer reaction time (approximately 20 min as suggested by Menzies et al. (325)) to complex completely with monomeric aluminum; thus, it is more suitable for an automated procedure.

Aluminum fractionation methods based on size or charge include dialysis, ultrafiltration, size-exclusion chromatography, ion chromatography, capillary zone electrophoresis, and C-18 reverse-phase chromatography (299). Soluble aluminum can also be measured indirectly by reacting it with F^-, then measuring the unreacted free F^- with an ion-elective electrode (326). A quantitative ^{27}Al NMR method is often preferred for the measurement of the 'Al_{13}' polymer (327).

The use of solution Al^{3+} activities to predict or characterize aluminum phytotoxicity are discussed in the later section on soil analysis.

16.9.4   INDICATOR PLANTS

Baker (328) proposed that there are three types of plant responses to increasing heavy metal contents in soil: (a) accumulators, where heavy metals are concentrated in above-ground plant parts; (b) indicators, where internal concentrations reflect external levels; and (c) excluders, where metal concentrations in shoots are low and constant over a wide range of soil concentrations up to a critical soil level above which unrestricted transport occurs. It might be expected that aluminum accumulators would be good indicator plant species; however, this relationship has not been found to be true. Truman et al. (14) reported that only a weak linear relationship was found between foliage aluminum concentration of *Pinus* spp. and exchangeable aluminum in soil. Even in controlled nutrient solution culture, foliar aluminum levels of red spruce varied almost fivefold at a similar solution of aluminum concentration (78).

An alternate method of determining the status of soil aluminum is to grow pairs of aluminum-tolerant and sensitive genotypes of some common crops, such as barley or snapbean, then observe their differential responses. For example, shoots of the aluminum-sensitive 'Romano' snapbean showed a significant response to liming of an acid (pH 5.1) soil from Beltsville, Maryland, but those of the aluminum-tolerant 'Dade' did not; this dry weight difference indicated that aluminum toxicity was the main factor limiting growth (329). Sanchez (300) reported that there was a high degree of tolerance to acid (mostly Al) soil in many varieties of upland rice and cowpea. Such knowledge would be very useful in identifying and managing aluminum-toxic soils.

16.10   ALUMINUM IN HUMAN AND ANIMAL NUTRITION

16.10.1   ALUMINUM AS AN ESSENTIAL NUTRIENT

Speculation that aluminum is an essential nutrient has persisted for at least 70 years (330); yet to date, there is no conclusive evidence for its essentiality in the diets of animals or humans (6,7). One of the earliest speculations about the essentiality was by E. E. Smith, president of the New York Academy of Sciences in the early 1900s. In his 1928 book on aluminum, he described the effects of adding different elements to milk on the growth and fertility of rats consuming only a milk diet (330,331). Aluminum was one of the added elements that appeared to be necessary for normal fertility and survival of offspring. On this basis, and the fact that aluminum was present in tissues of the rat, Smith concluded that aluminum 'exercises a true and essential biological function.' This early research with milk diets must be considered equivocal, however, and has never been repeated.

Since this early work, few studies have directly addressed the question of aluminum's essentiality. In 1980, the National Academy of Sciences reviewed the existing research and stated that 'aluminum has not been proven to be essential to animals, but indirect evidence suggests it may be' (332). The indirect evidence included accumulation of aluminum in regenerating bone, stimulation of certain enzyme systems, effective use as an adjuvant, and a report that aluminum stimulated growth in poultry.

Despite this optimism, recent reviews conclude that the evidence for the essentiality of aluminum remains quite limited (6,7). The reports of aluminum accumulation in regenerating bone, stimulation of certain enzymes, and the often-cited ability of aluminum to combine with fluoride and activate the guanine nucleotide (GTP) binding regulatory element of adenylate cyclase (333) are actions of aluminum that have never been proven to be required for normal biological function in any organism. This leaves, then, two isolated studies indicating that a deficiency of aluminum in the diet may modestly inhibit the growth of goats and chickens as the only support for essentiality (6,7). These studies, however, have yet to be validated by others. If aluminum is ever shown to be essential, it appears that the levels required in the diet are so low (less than $200\,\mu g\;kg^{-1}$ diet in the goat study) that dietary deficiency would be very rare.

16.10.2   BENEFICIAL EFFECTS OF ALUMINUM

Although the essentiality of aluminum as a nutrient is questionable, aluminum compounds have been used for many years in animal agriculture, environmental management, and the food and pharmaceutical industries for beneficial purposes. In animals and humans, the beneficial effects usually occur at levels of aluminum intake far above that found in typical diets and, as such, in pharmacological treatments that may carry some risk of aluminum toxicity.

16.10.2.1   Beneficial Effects of Aluminum in Animal Agriculture

Aluminum is generally not added to animal diets because of the lack of any known nutritional function, and no evidence suggests beneficial effects occur in livestock grazing high-aluminum pastures. Rather, aluminum toxicity is of concern as some forages contain over $2000\,mg\;Al\;kg^{-1}$ (334). For a variety of useful reasons, however, aluminum compounds have been added to animal diets.

One of the oldest uses of aluminum compounds in agriculture is the use of bentonite clay (Al silicates of sodium, calcium, or other cations) as a binder for pelleted feeds. Studies in the 1950s with poultry indicated no detrimental effects of ingesting bentonite, and some indicated a beneficial effect on growth rate. Benefits were attributed to an increase in feed intake and a delay in the passage of feed through the digestive tract resulting in better absorption of nutrients (335). More recently, bentonite and other aluminosilicates have been investigated for their ability to ameliorate the toxic effects of aflotoxin-contaminated feeds on growth and feed intake in poultry and swine (336,337). Feeding hydrated sodium calcium aluminosilicates has also been shown to reduce the passage of aflatoxins into milk (338). The mechanism of action appears to be adsorption of aflatoxins by the aluminosilicates, reducing aflatoxin bioavailability.

The addition of aluminosilicates to poultry diets has also been reported to enhance eggshell quality (339). Feeding sodium zeolite A, a synthetic aluminosilicate with a 1:1 ratio of aluminum to silicon, increased the levels of silicon and aluminum in the blood. The authors suggested that the increase in blood silicon stimulated calcium use for eggshell formation. Wisser et al. (340), however, were able to show small increases in eggshell quality by adding aluminum sulfate to poultry diets, suggesting that aluminum had an effect independent of silicon. With aluminum sulfate, however, aluminum accumulated in the bones of the hens and reduced fertility. Similar, but less severe toxic effects were reported with sodium zeolite A, suggesting that zeolites may be a safer way to stimulate eggshell formation (341).

Sodium zeolite A has also been shown to prevent a condition referred to as milk fever (parturient hypocalcemia) in dairy cows, a relatively common problem in the dairy industry (342). Around the time of calving, the metabolic demand for calcium to support gestational growth and milk production is large. This demand for calcium can result in hypocalcemia leading to muscle tremors, weakness, and eventually death if not treated. Sodium zeolite A added to the ration for 3 weeks prior to calving was found to stimulate calcium mobilization from bone and enhance the efficiency of calcium absorption, preventing hypocalcemia (342). The stimulus for these changes in calcium metabolism appeared to come from an aluminum-induced reduction in phosphate availability, since treated cows had significantly lower plasma inorganic phosphate levels.

Similar to the above concept of using aluminum to inhibit phosphate absorption, aluminum has been shown to inhibit fluoride absorption and protect against fluoride toxicity in poultry (343). Aluminum fluoride complexes may be formed in the body, however, and may have detrimental effects of their own (344). Aluminum has also been studied for its beneficial effects on reducing lead toxicity (345).

Some of the beneficial roles of aluminum compounds in animal agriculture are unrelated to aluminum ingestion. Aluminum sulfate has been used to acidify poultry litter to reduce the growth and transmission of bacterial infections caused by *Campylobacter. Campylobacter* is a common cause of diarrhea in humans, and undercooked poultry is a potential source. In a recent study, litter contaminated with this bacterium was treated with aluminum sulfate, then, newly hatched chicks were raised on the treated litter (346). No transmission of *Campylobacter* to the chicks was observed. Unfortunately, the treatment was not effective against *Salmonella*. Aluminum compounds have also been used to treat animal manure prior to land applications to reduce environmental impacts. This practice will be discussed in the next section.

16.10.2.2 Beneficial Uses of Aluminum in Environmental Management and Water Treatment

The use of animal manures as fertilizers can increase water pollution problems due to runoff of soluble phosphorus. Several aluminum-containing compounds have been shown to reduce phosphate runoff if applied to manure. Applications of aluminum sulfate or aluminum chloride to swine manure reduced soluble phosphate in runoff by 84%, presumably by forming insoluble phosphate complexes (347). In a large scale, on-farm trial, aluminum sulfate was applied over a 16-month period to litter in 97 poultry houses on the Delmarva Peninsula. Compared to litter from untreated houses, treated litter had decreased soluble phosphates, a lower pH, and higher total nitrogen and sulfur concentrations, thereby increasing its value as a fertilizer (348). Zeolite and aluminum sulfate were evaluated in amending slurries of dairy manure (349). Aluminum sulfate eliminated soluble phosphorus, and zeolite reduced it by over half. Both aluminum compounds reduced ammonia emissions by 50%, presumably by reducing the pH or by adsorbing ammonium cations. Peak et al. (350) used x-ray absorption near edge spectroscopy to determine the chemical species of aluminum and phosphorus in treated manures. No evidence of aluminum phosphate precipitation was found. Therefore, the mechanism of action is not clear and brings up the possibility that soluble forms of aluminum may be present in the treated manures and, hence, in the runoff, especially if excess aluminum is used in the treatment process.

Aluminum sulfate also has been used to treat algal-rich, eutrophied lakes. Welch and Cooke (351) reported the effectiveness and longevity of treatments in 21 lakes across the United States. They concluded that aluminum sulfate effectively reduced total soluble phosphate levels (and the algae that depend on this nutrient) for 8 years on average, especially in lakes without large external inputs of phosphorus. Aluminum is thought to form insoluble aggregates of aluminum phosphate, hydroxide, and organic material that settle to the bottom of the lake and remain in the sediment unless solubilized by acidic conditions. Acid conditions release soluble forms of aluminum that can be toxic to fish, prompting guidelines that lake pH should remain between 5.5 and 9.0.

Very little evidence suggests that aluminum is beneficial to aquatic species under normal circumstances. Short-term protective effects of aluminum against acid (H^+) toxicity have been shown in some studies (352). Uptake of protons from acidic water can fatally disrupt electrolyte regulation in fish. However, under acidic conditions, monomeric aluminum (Al^{+3}) may bind to gill surfaces blocking the binding and systemic uptake of H^+, thereby improving survival. This protective effect may only last a few hours and has been reported only under laboratory conditions. Aluminum in acidic water (pH 5.2 to 5.9) was also shown to eliminate ectoparasites on Atlantic salmon better than acidic water alone (353).

Municipal water treatment facilities often use aluminum sulfate as a water-clarifying agent in a process similar to that described above for treating eutrophied lakes. The basic process is ancient, originating in China thousands of years ago. When aluminum sulfate is added to turbid water at pH 6.5 to 8, aluminum hydroxide forms as a gel-like precipitate (floc). Suspended particles and oils are trapped in the floc, which is then removed by various methods. Some aluminum, however, can remain in solution. Concentrations of aluminum in treated drinking water have ranged from undetectable to 2.7 mg L^{-1}, with a median of 0.1 mg L^{-1} (354). The Environmental Protection Agency has suggested a maximum contamination level for aluminum in drinking water at a concentration range of 0.05 to 0.2 mg L^{-1}. Recently, other types of aluminum-based clarifying agents such as polyaluminum chloride have been used that may result in less residual aluminum and different chemical species of residual aluminum in treated water compared to current methods (355,356). Clarification of water by aluminum compounds has been investigated for its potential to reduce drinking water fluoride concentrations in regions where fluoride toxicity is a concern (357).

16.10.3 TOXICITY OF ALUMINUM TO ANIMALS AND HUMANS

The ubiquitous presence of aluminum in soil, water, food, and pharmaceuticals makes exposure to this metal unavoidable for most species. The potential toxicity to humans has been debated since at least the 1920s with the advent of commercially available aluminum-containing baking powders (330). In natural habitats, concern about toxicity increased in the 1970s with the knowledge that acidification of natural waters from acid rain, mine drainage, and deforestation increased the mobilization and bioavailability of soil aluminum (352). The growing awareness of increased exposure to aluminum and the clear demonstration of its potential toxicity to animals and humans (discussed below), combined with its possible association with Alzheimer's disease has given rise to an exponential increase in research related to the metabolism and toxicity of this metal. In the decade from 1970 to 1980, only 140 publications are listed by a bibliographic search using the keywords 'aluminum toxicity,' compared to 1035 publications in the decade from 1990 to 2000. For this reason, a detailed review of aluminum toxicity and metabolism in animals and humans is outside the scope of this chapter and the reader is referred to several recent reviews for this purpose (358–360). The focus of this section will be on the consequences of aluminum exposure from common sources in the food chain with reference, when possible, to potential toxic mechanisms.

16.10.3.1 Toxicity to Wildlife

Much of the concern about aluminum toxicity to wildlife stems from the fact that many lakes and streams have been acidified by natural or industrial causes resulting in increased concentrations of aluminum in their waters. Sparling and Lowe (352) presented a comprehensive review of the environmental toxicity of aluminum and discuss its toxicity in invertebrates, fish, and other wildlife.

Aquatic species, especially freshwater fish, have been studied the most, and it is clear that their survival can be reduced greatly as aluminum concentrations increase in acidic water (361). In fact, aluminum toxicity is thought to be the most common cause of fish die-offs. Levels of aluminum above 100 to 500 µg L^{-1} are usually needed to cause death depending on fish species and water conditions such as the amount of dissolved organic matter and pH. Acidity is also toxic and is additive to the effects of aluminum.

The mechanisms of aluminum toxicity fall into two categories based on water pH: asphyxiation in the pH range of 6.5 to 5.5, and loss of electrolytes from the blood in the pH range of 5.5 to 4.5. At the more acidic pH range, soluble cationic species of aluminum are thought to bind to negatively charged sites on the gill surface, displacing bound calcium ions that regulate electrolyte fluxes. This displacement results in the diffusion of sodium and chloride out of the body. In the less acidic pH range of 5.5 to 6.5, the formation of uncharged Al(OH)$_3$ is more likely. These uncharged species form colloids and precipitates that collect on the gill surface, stimulating excess mucus formation. The excess mucus inhibits oxygen and CO$_2$ diffusion leading to asphyxiation (362). Aluminum appears to be relatively nontoxic to fish at basic pHs where anionic species would predominate.

Dissolved organic matter, such as humic acid, can chelate positively charged aluminum species preventing aluminum from interacting with the gill, thereby reducing aluminum stress (352). Birchall (363) has proposed that silicon can also ameliorate aluminum toxicity by forming colloidal hydroxyaluminosilicates that limit the availability of aluminum for binding to gill surfaces.

Much less is known about aluminum toxicity to other aquatic species such as crustaceans, mollusks, and insect larvae. In general, these invertebrate species are more tolerant to aluminum than fish, but toxic mechanisms appear to be similar in those that have gills, i.e., related to alterations in calcium and electrolyte balance or respiration rates. In contrast to fish, however, invertebrates may accumulate large amounts of aluminum on or within their bodies reaching concentrations as high as 1000 mg kg^{-1} (352,363,364).

There has been some concern about transfer of aluminum up the food chain. Nyholm (365) postulated that elevated levels of aluminum in invertebrates could affect wild birds feeding in or near aluminum-laden waters. In studies with flycatchers, it was reported that female birds had elevated bone aluminum levels and laid deformed eggs with soft shells leading to dehydration and reduced hatchability. Other concerns were with bone growth and body weight gain in growing chicks since aluminum in the diet at a level of 1000 mg kg^{-1} has been shown to inhibit phosphate absorption, reduce feed intake, and accumulate in bone (366). Not all studies, however, have found significant toxic effects on wild birds (352).

Although the ecological impacts of aluminum mobilization into acidified water has been an important concern, recent studies by Palmer and Driscoll (367) indicate, at least in northern hardwood forests in the United States, that stream water aluminum concentrations are declining. They suggested that within 10 years, at the current rate of decline, aluminum toxicity would no longer pose a threat to fish. Remediation of acidic aluminum-laden water also is being accomplished by adding powered limestone (CaCO$_3$) to increase pH and reduce levels of soluble aluminum and, in some cases, total aluminum (352).

16.10.3.2 Toxicity to Agricultural Animals

Generally, aluminum toxicity has not been a serious problem in livestock production (cattle, swine, sheep, and poultry). Levels of aluminum in most common feedstuffs, forages, pastures, and water supplies usually are not high enough to cause problems in animal performance or in the safety of food derived from animals, i.e., they result in diets that contain less than the maximum tolerable levels listed by the National Research Council: 1000 mg kg^{-1} dry feed for cattle and sheep and 200 mg kg^{-1} for swine, poultry, horses, and rabbits (332). These values are for highly soluble forms of aluminum, and higher levels of less soluble forms may be tolerated.

Nevertheless, there has been concern about the toxic levels of intake in cattle and sheep foraging on plants that either accumulate high levels of aluminum or are contaminated with large amounts of soil, and in poultry consuming diets that contain aluminum from contaminated feed ingredients or from added zeolites. Toxicity symptoms are rather consistent across species. Symptoms include decreased feed intake, reduced efficiency in converting feed to body weight gain, disturbances in mineral metabolism including reduced phosphate absorption, hypercalcemia, reduced bone mineralization, and accumulation of aluminum in body tissues. Large intakes of soluble forms of aluminum (above 3000 to 4000 mg kg^{-1} diet) can be fatal, especially in young animals, or when dietary calcium or phosphorus is low (332).

Storer and Nelson (368) were one of the first to compare the toxicity of different chemical forms of aluminum using young chickens as an animal model. They showed that compounds that were not soluble in dilute acid or water, such as aluminum oxide, did not produce symptoms of toxicity even at dietary levels up to 16,000 mg kg^{-1} diet. Compounds that were soluble such as aluminum chloride, sulfate, acetate and nitrate produced severe toxicity at the 5000 mg kg^{-1} level. Interestingly, aluminum phosphate, which is soluble in dilute acid but not in water, did not produce toxicity apparently due to precipitation in the alkaline environment of the small intestine and its inability to reduce the bioavailability of other forms of dietary phosphate.

16.10.3.2.1 Toxicity to Ruminants (Cattle and Sheep)

Aluminum toxicity to ruminants has not been reported under most livestock production systems. But, some concern has been expressed about the risks of inducing either a phosphorus deficiency or a condition known as grass tetany when ruminants consume large amounts of aluminum from soil or aluminum-rich forages. In general, soil does not appear to be toxic, but the more soluble forms of aluminum in plants may pose some risk.

Ruminants can consume large amounts of soil under some pasture conditions and, therefore, may consume large amounts of aluminum (up to 1.5% of the diet dry matter) (369). Since phosphorus is the mineral most likely to be deficient in the diet of grazing cattle, studies have looked at the effects of soil intake on phosphorus nutrition. Most have shown that soil intake has a minimal effect on phosphorus balance and animals are able to maintain normal serum phosphate levels (370,371). Apparently, the aluminum species in soil are not soluble enough in the intestinal tract of the ruminant to cause significant precipitation of available phosphate.

It is clear, however, that soluble forms of aluminum can induce toxicity. Crowe et al. (369) fed diets that contained soluble aluminum chloride hexahydrate at 2000 mg Al kg^{-1} diet to Holstein dairy calves for 7 weeks. The results are typical of studies in ruminants using soluble forms of aluminum (370). Feed intake decreased by 17%, average daily weight gain decreased by 47%, and the amount of feed needed to produce a kilogram of weight gain increased by 50%. Fecal phosphorus excretion increased by 79% and plasma inorganic phosphate concentrations dropped to levels found in phosphorus-deficient animals. Aluminum accumulated in bone thereby causing demineralization, serum calcium concentrations rose, and urinary and fecal calcium excretion increased. To what extent natural aluminum species in forages can cause these symptoms is not known.

Grass tetany is a serious, often fatal metabolic disorder, characterized by low magnesium levels in the blood. Grass tetany occurs most often in female ruminants in the early stages of lactation while grazing on succulent, immature, magnesium-deficient grasses in springtime. Symptoms include poor coordination, convulsions, and death, presumably related to a metabolic deficiency of magnesium. Several outbreaks of grass tetany have been associated with pastures and forages containing high aluminum concentrations such as wheat and tall fescue containing 1000 to 2000 mg Al kg^{-1} (372). Although most studies looking at soil aluminum intake have not shown significant effects on serum magnesium levels, some studies using soluble aluminum (such as aluminum citrate) have shown small decreases (370,372). It was suggested that the decrease in serum magnesium was not caused by reduced magnesium absorption. Rather, aluminum can cause hypercalcemia, which induces the loss of magnesium in urine. This loss may contribute to the appearance of grass tetany.

16.10.3.2.2 Toxicity to Poultry

Aluminum toxicity has not been reported as a significant problem in poultry production, but concerns have been raised due to the possible intake of soluble aluminum compounds from feed ingredients such as aluminum-flocculated algae, aluminum-contaminated mineral mixes, or the intentional use of zeolites to improve eggshell quality.

Sodium zeolite A ($Na_{12}[(AlO_2)_{12} (SiO_2)_{12}]$- $27H_2O$) is a synthetic aluminosilicate with cation exchange properties that has been shown to improve eggshell quality when added to the diet at 0.75 to 1.5%, as mentioned earlier under beneficial effects. When added to the diets of young chicks, however, it caused reductions in feed intake, growth, bone ash, and serum phosphate, and increased serum calcium and bone aluminum content (373–375).

The soluble forms of aluminum are relatively more toxic, but generally show the same biological effects as sodium zeolite A (340,366,376). Interestingly, however, soluble forms tend to inhibit calcium absorption from low calcium diets, whereas, zeolites seem to enhance it. No studies have been done to evaluate the effects of including natural, aluminum-loaded plant or animal products in the diet.

The fact that consuming high levels of aluminum usually decreases food intake makes it difficult to identify toxic effects of aluminum that are independent of reduced nutrient intakes. Wisser et al. (340), however, showed that adding aluminum sulfate to the diet of laying hens decreased egg production and fertility, and increased serum calcium without causing significant decreases in food intake or plasma phosphate. This implies that systemic aluminum can have direct toxic effects on metabolism.

16.10.3.3 Toxicity to Humans

There is no doubt that aluminum intake can be toxic to humans under certain conditions. Regular intake of large doses of aluminum hydroxide can cause bone disease, anemia, and neurological problems in patients with poor renal function that cannot adequately excrete aluminum from the body. Similar effects can occur in healthy individuals if aluminum intake is high enough, over a long enough period. There are questions about the relationship of aluminum to Alzheimer's disease and the health consequences of long-term, low-level exposures that remain unanswered. The reader is referred to several recent reviews for detailed discussions of these topics (358–360).

16.10.3.3.1 Overview of Aluminum Metabolism

The intestine is viewed as a protective barrier against aluminum toxicity as only a small fraction (0 to 0.5%) of ingested aluminum is absorbed from any source. However, of the small amount absorbed, about half is retained in tissues and the other half is excreted, primarily in urine. Elimination from tissues is not rapid so, in the face of constant intake, tissues accumulate aluminum over time.

Drueke (377), and Yokel and McNamara (359) provide recent reviews of the absorption and metabolism of aluminum. A number of factors influence the efficiency of absorption. Most are dietary factors that affect solubility; hence, phosphate reduces absorption as does ingesting insoluble forms of aluminum such as aluminum oxide. Silicon has shown conflicting results, but does not appear to reduce absorption except when given as insoluble, oligomeric forms. The soluble aluminum salts have higher absorption efficiencies, although the hydroxide appears to be less bioavailable than more soluble forms. Citrate, as well as other organic acids including ascorbic, oxalic, lactic, and tartaric acids can greatly enhance absorption possibly by increasing solubility or charge neutralization when complexed species are formed. The mechanism, however, is not yet understood. Aluminum-accumulating plants which store aluminum bound to organic acids would be expected to contain bioavailable aluminum, but this concept has never been tested. Polyphenolic acids have recently been shown to increase tissue uptake of aluminum from food, suggesting increased absorption (378). Fluoride may also enhance absorption.

The mechanism of aluminum absorption is not well understood but appears to involve active transport through the intestinal cells as well as passive diffusion. High iron diets inhibit transport whereas low iron diets enhance it, suggesting that aluminum can follow iron transport pathways.

In the blood, about 80% of aluminum is bound to iron binding sites on transferrin, the major iron transport protein in plasma (47,48). The remainder is bound to low-molecular-weight molecules, possibly citrate. Since most tissues take up transferrin to acquire iron, this process provides a mechanism for aluminum to enter cells, including the brain. Tissue uptake from the citrate-bound form is also possible. In fact, increased dietary citrate appears to enhance tissue accumulation of aluminum as well as urinary excretion. In renal-failure patients, citrate greatly enhances risk of toxicity.

Bone is the major tissue deposition site with aluminum accumulating at areas of active mineralization, possibly as aluminum citrate. Aluminum also enters and is toxic to the bone forming cells (osteoblasts). Other tissues accumulate lesser amounts of aluminum, usually in the order: bone > liver > kidney > spleen > brain. Contrary to other tissues, the brain has not always been found to accumulate aluminum in association with increased dietary intake. Nevertheless, aluminum is routinely found in the brain in measurable amounts. Elimination of aluminum from tissues is relatively slow compared with its rapid uptake, with half-lives estimated in terms of months or years. Elimination from bone is the most rapid, and that from brain is the slowest. Body loads are typically low, 30 to 50 mg in healthy individuals on usual diets.

The intracellular metabolism of aluminum is poorly understood. Presumably, it initially follows the pathways of iron metabolism being incorporated with transferrin-bound iron into endosomes. Its subsequent fate, or the fate of citrate bound aluminum are unknown.

16.10.3.3.2 Overview of the Biochemical Mechanisms of Aluminum Toxicity

The biochemical mechanisms of aluminum toxicity leading to neurodegeneration, bone loss, and anemia are not understood and an explanation for these symptoms cannot be made at this time. At its most fundamental level, the systemic toxicity of aluminum is probably related to its strong binding affinity for three-oxygen-donor ligands, especially negatively charged oxygen donors found in organic phosphates and proteins with carboxylic acid or phosphorylated residues (379). This strong binding can displace magnesium ions, alter the structure and function of substrates, enzymes, regulatory and structural proteins, and in poorly understood ways interfere with iron metabolism. The biochemical aspects of aluminum toxicity in animals and man have recently been reviewed (360). It is likely that the basic biochemical effects of aluminum are similar in plant and animal cells.

Before systemic toxicity is discussed, it should be remembered that dietary aluminum toxicity often induces a phosphate deficiency. Appetite and growth are depressed. Bone mineral is dissolved in an attempt to raise serum phosphate levels and hypercalcemia may result. Skeletal muscle may also lose intracellular phosphorus and magnesium to the blood, resulting in lowered ATP synthesis and a general lack of phosphate for metabolic use within the muscle. Intracellular calcium levels become elevated. Bone pain, muscle weakness, and neurological symptoms including confusion, seizures, and coma can occur (380).

Once aluminum gains entry into the body and enters cells, it is thought to bind to phosphate ligands, particularly ATP. It also binds to proteins. Bound aluminum may alter enzyme activity by displacing cofactors such as Mg^{++}, by affecting the binding of substrates such as ATP, or by inducing conformational changes. For example, aluminum has been shown to inhibit ATP dependent enzymes such as hexokinase. The mechanism is thought to involve formation of Al-ATP that is much more stable and binds much tighter to proteins than Mg-ATP, inhibiting enzyme action. More than 20 other enzymes are reportedly inhibited or stimulated by aluminum (379).

Aluminum may also influence protein–protein interactions (381). For example, aluminum may bind to calmodulin, a calcium-activated regulatory protein that controls the activity of more than 40 different enzymes by binding to them via hydrophobic interactions resulting in the induction or inhibition of activity. Aluminum binding does not affect calcium binding to calmodulin; rather, aluminum induces conformational changes that inhibit the ability of calmodulin to bind target proteins.

Aluminum also can cross-link proteins by forming intermolecular bridges between binding sites on amino acid side chains. The binding of aluminum to proteins may also affect their turnover, either stabilizing them, such as in insoluble aggregates, or enhancing degradation via conformational changes.

Since many signal transduction processes involve phosphate group transfers, this is another likely site for aluminum toxicity (382). The phosphatidylinositol 4,5-bisphosphate (PIP_2) signaling pathway has been inhibited by aluminum. Aluminum apparently binds to phosphate groups of PIP_2 in membrane phospholipids inhibiting PIP_2 hydrolysis by phospholipase C. An alteration in signal transduction pathways may help explain the altered pattern of gene expression seen in tissues exposed to aluminum (383). G-proteins and protein kinases are also reportedly affected by aluminum (383).

Aluminum has been shown to interfere with iron metabolism. It blocks the incorporation of iron into heme resulting in poor hemoglobin production and anemia (384). Aluminum also appears to disrupt the mechanisms that control intracellular iron homeostasis. The result may be altered iron distribution in the cell leading to increased levels of reactive or "free" iron and iron-induced oxidative stress (384–386). Normally, increasing intracellular "free" iron concentrations coordinately stimulate the synthesis of the iron storage protein ferritin, and inhibit the synthesis of transferrin receptors that control iron uptake. Studies suggest that aluminum antagonizes the ability of intracellular iron to regulate the translation of mRNAs for both ferritin and the transferrin receptor. Under these conditions, the amount of "free" iron in the cell becomes elevated relative to the amount of its storage and detoxification by ferritin, thus increasing the risk for iron-induced oxidative stress. Aluminium has also been shown to inhibit the ATP-dependent proton pump on endosomes, resulting in the trapping of transferrin-bound iron inside these vesicles. The trapping of iron would limit its ability to stimulate ferritin synthesis. Aluminium may also inhibit the incorporation of iron into ferritin, further increasing the levels of reactive "free" iron in the cell.

Recent studies have shown that aluminum can induce oxidative stress even though it is not a redox metal, and that antioxidants can attenuate this effect supporting the concept that aluminum toxicity involves oxidative damage (387,388). Oxidative stress could result from altered membrane structure, a reduction in antioxidant defense systems, or the induction of free radical generating systems such as increased levels of reactive "free" iron.

16.11 ALUMINUM CONCENTRATIONS

16.11.1 IN PLANT TISSUES

16.11.1.1 Aluminum in Roots

Increasing aluminum levels in the medium tended to result in increasing aluminum concentrations in roots of aluminum accumulators or aluminum excluders (Table 16.3). Concentrations of aluminum in roots were 2- to 250-fold higher than those in shoots (Table 16.3). In red spruce, root aluminum concentrations associated with a 20% decrease in root biomass ranged from 1700 to 6000 mg Al kg^{-1} (78). Aluminum in roots is present mostly as precipitated hydroxy or phosphate compounds outside the root cells (86). As a result, it is difficult to use aluminum concentrations in roots as a measure of aluminum toxicity unless an effort is made to remove or prevent extracellularly precipitated and adsorbed aluminum. Alternatively, it might be possible to analyze aluminum concentrations in root apices alone as a measure of toxicity (189–191).

16.11.1.2 Aluminum in Shoots

In accumulators, foliar aluminum concentrations of 65 tree species and 12 unidentified trees from an Indonesian rain forest ranged from 1 g kg^{-1} in delta tree (*Aporusa* spp. Blume, Euphorbiaceae) to 37 g kg^{-1} in *Maschalocorymbosus corymbosus* Bremek. (Rubiaceae) (13). Aluminum accumulators (*Melastoma malabathricum* L., *Hydrangea macrophylla* Ser., and *Fagopyrum esculentum* Moench.) exposed to increasing aluminum in solution showed increasing aluminum concentrations in leaves (22) (Figure 16.7). Facultative aluminum accumulators, jack pine (*Pinus banksiana*

TABLE 16.3
Aluminum Concentrations in Roots and Leaves

Species	Al Level (μM)	Effect on Growth[a]	Root Al (mg kg^{-1})	Young Foliar Al (mg kg^{-1})[b]	Reference
Al accumulators					
Jack pine (*Pinus banksiana* Lamb.)	0	0	211	39	390
	185	0	411	85	
	370	0	747	139	
	740	0	849	196	
	1480	−	1227	251	
	2960	−	1744	380	
	5930	−	3654	988	
Black pine	0	0	108[c]	189[d]	16
(*Pinus nigra* Arnold)	100	+	1863	891	
	500	+	1593	999	
	1000	−	5400	999	
Al excluders					
European white birch	0	0	—	—	26
(*Betula pendula*	74	−	1050	70	
Roth race SMM)	185	−	270	160	
	370	+	270	100	
	555	+	260	120	
	930	+	240	40	
	1296	+	310	130	
Tomato (*Lycopersion*	0	0	59	15	397
esculentum Mill.)	10	−	1937	14	
	25	−	5888	51	
	50	−	11,838	48	
Phasey bean	0	0		125	398
(*Macroptilium*	18	+		125	
lathyroides Urb.)	37	+		125	
	74	+		140	
Alfalfa	0	0		70	398
(*Medicago sativa* L.)	18	−		100	
	37	−		150	
	74	−		315	
Red spruce	0	0	243	29	390
(*Picea rubens* Sarg.)	185	−	446	47	
	370	−	739	67	
	740	−	1690	162	
	1480	−	2212	272	
	2960	−	2905	492	
	5930	−	5351	772	
Douglas fir	0	0	304	27[d]	25
[*Pseudotsuga menziesii*	148	+	1350	157	
(Mirb.) Franco]	296	0	1753	369	
	593	0	2375	430	
	1185	0	3591	447	
Northern Red oak	56	0	7560	66	399
(*Quercus rubra* L.)	169	−	6567	168	

Continued

TABLE 16.3 (*Continued*)
Aluminum Concentrations in Roots and Leaves

Species	Al Level (μM)	Effect on Growth[a]	Root Al (mg kg^{-1})	Young Foliar Al (mg kg^{-1})[b]	Reference
	360	−	6422	138	
Stylo	825	−	6982	147	
[*Stylosanthes guianensis*	0	0	180	74	22
(Aubl.) Sw.]	111	−	886	61	
	555	−	890	146	
African marigold	0	0	71	36	396
(*Tagetes erecta* L.)	37	+	650	32	
	148	+	1230	33	
White clover	0	0	1120	<25	131
(*Trifolium repens*[c] L.)	25	−	1621	44	
	50	−	2998	83	
	100	−	4008	66	
Corn	0	0	116	30	400
(*Zea mays* L.)	93	−	2150	38	
	185	−	2470	142	
	370	−	2500	163	
	741	−	2730	282	

[a]Positive (+), negative (−), or no effect (0) on growth relative to control (0 Al).
[b]Foliar concentration in young leaves if young and old leaves were analyzed separately; otherwise, foliar concentration averaged across all leaves.
[c]Al concentrations in coarse roots.
[d]Al concentrations in needles.
[e]Plants supplied with N and no further Al given after pretreatment with Al.

Lamb.) and loblolly pine (*Pinus taeda* L.), also had increasing foliar aluminum concentrations as solution aluminum increased (389) (Table 16.3).

Efforts to establish critical aluminum concentrations for toxicity in plants generally have been unsuccessful (78,82,390). For example, foliar concentrations in red spruce associated with a 20% decrease in foliar biomass ranged from 70 to 250 mg kg^{-1} (78). Similarly, foliar aluminum concentrations in red oak associated with a 20% decrease in leaf biomass ranged from 93 to 188 mg kg^{-1} (391). Within slash pine families, aluminum sensitivity was correlated positively with foliar aluminum concentration; however, no such correlation was found within loblolly pine families (392).

In accumulators, internal complexation of aluminum by organic anions, silicate, or other ligands resulted in poor correlations between foliar aluminum concentrations and restrictions in biomass growth. Raynal et al. (78) reported the absence of any significant correlation between biomass response and foliar aluminum levels in *Pinus* species. In the case of aluminum excluders, aluminum concentrations in shoots do not increase with increasing aluminum levels in the medium until a toxic threshold is exceeded (328), again resulting in poor correlation between foliar aluminum levels and biomass response. For example, in rice and barley, only trace amounts of aluminum were found in leaves at solution aluminum levels up to 111 μM, then foliar aluminum concentrations increased as aluminum levels in solution increased to 555 μM (22) (Figure 16.7). Similarly, increasing solution aluminum levels from 0 to 620 μM had no effect on biomass growth of Western hemlock (*Tsuga heterophylla* Sarg.), then foliar aluminum concentrations decreased from 300 to 250 mg kg^{-1} when biomass was affected adversely by solution aluminum (393). In sugar maple

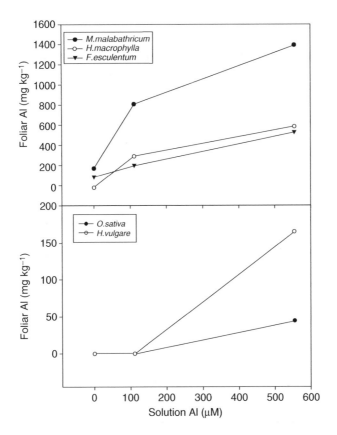

FIGURE 16.7 The pattern of increasing foliar aluminum concentrations with increasing solution aluminum differs in aluminum accumulator species (top) and aluminum excluder species (bottom) (From M. Osaki, T. Watanabe, T. Tadano. Beneficial effect of aluminum on growth of plants adapted to low pH soils. *Soil Sci. Plant Nutr.* 43:551–563, 1997.)

(*Acer saccharum* Marsh.), aluminum concentrations in leaves increased from 50 to 200 mg kg^{-1} as aluminum levels in solution increased from 0 to 600 µM, but then foliar aluminum concentration dropped to 150 mg kg^{-1} when shoot growth was restricted at 1000 µM aluminum in solution (394). Other examples of a lack of correlation between aluminum-induced growth inhibition and foliar aluminum concentrations can be found in Table 16.3 (395–399).

16.11.2 SOIL ANALYSIS

Aluminum bioavailability in soils and toxicity to plants is difficult to quantify because toxic levels vary with species and even with cultivars within a species (82). For example, 1.5 µM Al^{3+} activity was reportedly toxic to cotton roots (294), and 4.0 µM Al^{3+} was toxic for coffee (32). For rice, an aluminum-tolerant crop, the critical Al^{3+} activity was approximately 100 µM (400).

Chemical composition of some soil solutions, including aluminum and its various species, is listed in Table 16.4 (294). Table 16.5 lists critical Al^{3+} activities, as measured by root elongation, for selected plants (401). In general, trees are more tolerant of aluminum than most agronomic crops (Table 16.5). For 2-year-old seedlings of Norway spruce, aluminum toxicity was not evident when Al^{3+} activities in soil solutions ranged from 7.7 to 64.3 µM (402).

Instead of using Al^{3+} activity as the sole indicator of phytotoxicity, Alva et al. (34) used the sum of the activities of monomeric aluminum species (Al^{3+} + AlOH^{2+} + Al(OH)$_2^+$ + Al(OH)$_3^0$ + AlSO$_4^+$). They observed 50% reductions in root elongation, relative to roots of plants not receiving

TABLE 16.4
Range of Values, Means, and Standard Deviations (s.d.) for Attributes of Soil Solutions from 48 Surface and 48 Subsoil Samples from Queensland, Australia

Attribute	Unit	Surface Soil			Subsoil		
		Range	Mean	S.d.	Range	Mean	S.d.
pH		3.73–7.99	5.4	0.85	3.78–6.77	5.28	0.7
EC	dS m^{-1}	0.13–1.92	0.48	0.35	0.03–1.12	0.24	0.28
I	mM	1.2–22.6	5.3	4.2	<0.1–13.1	2.4	3.3
Ca	μM	34–1854	38	339	8–1437	79	206
Mg	μM	82–1366	345	240	14–560	138	134
Na	μM	262–8378	1279	1591	106–6960	1333	1730
K	μM	65–3171	386	481	12–2110	143	304
SO$_4$	μM	63–3858	585	597	14–1369	220	264
Al	μM	2.1–101	23	25	0.05–378	12	54
Al^{3+}	μM	0.05–34	3.4	7	0.05–126	3.6	18
Al(OH)$^{2+}$	μM	0.05–8.3	1.4	1.9	0.05–7.2	0.5	1.1
Al(OH)$_2^+$	μM	0.05–38	0.8	8.9	0.05–11	1.3	1.8
Al(OH)$_3^0$	μM	0.05–22	4.1	4.2	0.05–5.8	0.5	0.9
Al(SO$_4$)$^+$	μM	0.05–30	2.1	5.3	0.05–7.7	0.4	1.2
Σ(Al)	μM	2.1–67	19	18	0.05–143	6.2	21

Source: From R.C. Bruce et al., *Aus. J. Soil Res.*, 27:333–351, 1989.

TABLE 16.5
Threshold of Al Toxicity to Some Plants Where Root Elongation Was the Measure of Response and Where Available Al Was Expressed as Al^{3+} Activity in Solution

Plant	Al^{3+} at Phytotoxic Threshold (μM)	Rooting Medium
Gramineae spp.	0.90	Solution, soil
Cotton	1.5	Solution, soil
Barley	1.5	Solution
Coffee	4.0	Solution, soil
Cotton	6.0	Soil
Wheat	20	Soil
Honey Locust	40	Solution
Red spruce	50	Solution
Hybrid poplar	100	Solution
Red spruce, balsam fir	300	Solution
Autumn-olive	400	Solution
Pine, oak, birch	800	Solution

Source: From J.D. Wolt, in *Soil Solution Chemistry*, Wiley, New York, 1994, pp. 220–245.

any aluminum, as this sum ranged from 12 to 17 μM for soybean, <8 to 16 μM for sunflower (*Helianthus annuus* L.), <7 to 15 μM for subterranean clover, and <5 to 10 μM for alfalfa. Alternatively, Cronan and Grigal (390) proposed the use of calcium/aluminum ratios as indicators of aluminum stress in forest ecosystems.

REFERENCES

1. W.L. Lindsay. *Chemical Equilibria in Soils*. New York: Wiley, 1979, pp. 35–49.
2. F.P.C. Blamey, C.J. Asher, D.G. Edwards. Hydrogen and aluminium tolerance. *Plant Soil* 99:31–37, 1987.
3. D.L. Godbold, E. Fritz, A. Huttermann. Aluminum toxicity and forest decline. *Proc. Natl. Acad. Sci. USA* 85:3888–3892, 1988.
4. N. van Breemen. Acidification and decline of Central European forests. *Nature* 315:16, 1985.
5. H. Marschner. *Mineral Nutrition of Higher Plants*. San Diego: Academic Press, 1986, pp. 433–435.
6. F.H. Nielsen. Ultratrace elements in nutrition: Current knowledge and speculation. *J. Trace Elem. Exp. Med.* 11:251–274, 1998.
7. F.H. Nielsen. Boron, manganese, molybdenum, and other trace elements. In: B.A. Bowman, R.M. Russell, eds. *Present Knowledge in Nutrition*. Washington, DC: ILSI, 2001, pp. 384–400.
8. T. Watanabe, M. Osaki. Mechanisms of adaptation to high aluminum condition in native plant species growing in acid soils: A review. *Commun. Soil Sci. Plant Anal.* 33:1247–1260, 2002.
9. S. Jansen, M.R. Broadley, E. Robbrecht, E. Smets. Aluminum hyperaccumulation in agiosperms: A review of its phylogenetic significance. *Bot. Rev.* 68: 235–269, 2002.
10. E.M. Chenery. Aluminum in the Plant World. Part I, General Survey in Dicotyledons. *Kew Bull.* 2: 173–183, 1948.
11. E.M. Chenery. Aluminium in the plant world. Part II. Monocotyledons and gymnosperms. *Kew Bull.* 4:463–466, 1949.
12. E.M. Chenery, K.R. Sporne. A note on the evolutionary status of aluminium-accumulators among dicotyledons. *New Phytol.* 76: 551–554, 1976.
13. T. Masunaga, D. Kubota, M. Hotta, T. Wakatsuki. Mineral composition of leaves and bark in aluminum accumulators in tropical rain forest in Indonesia. *Soil Sci. Plant Nutr.* 44:347–358, 1998.
14. R.A. Truman, F.R. Humphreys, P.J. Ryan. Effect of varying solution ratios of Al to Ca and Mg on the uptake of phosphorus by *Pinus radiata*. *Plant Soil* 96:109–123, 1986.
15. F.R. Humphreys, R. Truman. Aluminum and phosphorus requirements of *Pinus radiata*. *Plant Soil* 20:131–134, 1964.
16. A.W. Boxman, H. Krabbendam, M.J.S. Bellemakers, J.G.M. Roelofs. Effects of ammonium and aluminium on the development and nutrition of *Pinus nigra* in hydroculture. *Environ. Pollut.* 73:119–136, 1991.
17. J. Huang, E.P. Bachelard. Effects of aluminium on growth and cation uptake in seedlings of *Eucalyptus mannifera* and *Pinus radiata*. *Plant Soil* 149:121–127, 1993.
18. H. Matsumoto, E. Hirasawa, S. Morimura, E. Takahashi. Localization of aluminium in tea leaves. *Plant Cell Physiol.* 17: 627–631, 1976.
19. K. Takeda, M. Kariuda, H. Itoi. Blueing of sepal colour of *Hydrangea macrophylla*. *Phytochemistry* 24:2251–2254, 1985.
20. W. Konishi, S. Miyamoto, T. Taki. Stimulatory effects of aluminum on tea plants grown under low and high phosphorus supply. *Soil Sci. Plant Nutr.* 31:361–368, 1985.
21. S. Konishi. Promotive effects of aluminium on tea plant growth. *JARQ* 26:26–33, 1992.
22. M. Osaki, T. Watanabe, T. Tadano. Beneficial effect of aluminum on growth of plants adapted to low pH soils. *Soil Sci. Plant Nutr.* 43:551–563, 1997.
23. T.B. Kinraide, D.R. Parker. Apparent phytotoxicity of mononuclear hydroxy-aluminum to four dicotyledonous species. *Physiol. Plant* 79:283–288, 1990.
24. T.W. Rufty, Jr., D.T. MacKown, D.B. Lazof, T.E. Carter. Effects of aluminium on nitrate uptake and assimilation. *Plant Cell Environ.* 18:1325–1331, 1995.
25. W.G. Keltjens. Effects of aluminum on growth and nutrient status of Douglas-fir seedlings grown in culture solution. *Tree Physiol.* 6:165–175, 1990.
26. P.S. Kidd, J. Proctor. Effects of aluminium on the growth and mineral composition of *Betula pendula* Roth. *J. Exp. Bot.* 51:1057–1066, 2000.
27. T.B. Kinraide, P.R. Ryan, L.V. Kochian. Interactive effects of Al^{3+}, H^+, and other cations on the root elongation considered in terms of cell-surface electrical potential. *Plant Physiol.* 99:1461–1468, 1992.
28. D.B. Lazof, M.J. Holland. Evaluation of the aluminium-induced root growth inhibiton in isolation from low pH effects in *Glycine max*, *Pisum sativum* and *Phaseolus vulgaris*. *Aust. J. Plant Physiol.* 26:147–157, 1999.
29. J.R. Meyer, H.D. Shew, U.J. Harrison. Inhibition of germination and growth of *Thielaviopsis basicola* by aluminum. *Phytopathology* 84:598–602, 1994.

30. D. Andrivon. Inhibition by aluminum of mycelia growth and of sporangial production and germination in *Phytophthora infestans*. *Eur. J. Plant Pathol.* 101:527–533, 1995.
31. D.R. Parker, T.B. Kinraide, L.W. Zelazny. Aluminum speciation and phytotoxicity in dilute hydroxyl-aluminum solutions. *Soil Sci. Soc. Am. J.* 52:438–444, 1988.
32. M.A. Pavan, F.T. Bingham. Toxicity of aluminum to coffee seedlings grown in nutrient solution. *Soil Sci. Soc. Am. J.* 46:993–997, 1982.
33. A. Tanaka, T. Tadano, K. Yamamoto, N. Kanamura. Comparison of toxicity to plants among Al^{3+}, $AlSO_4^+$, and Al-F complex ions. *Soil Sci. Plant Nutr.* 33:43–55, 1987.
34. A.K. Alva, D.G. Edwards, C.J. Asher, F.P.C. Blamey. Relationships between root length of soybean and calculated activities of aluminum monomers in nutrient solution. *Soil Sci. Soc. Am. J.* 50:959–962, 1986.
35. T.B. Kinraide. Identity of the rhizotoxic aluminium species. *Plant Soil* 134:167–178, 1991.
36. N.W. Menzies. Toxic elements in acid soils: Chemistry and measurement. In: Z Rengel, ed. *Handbook of Soil Acidity*. New York: Marcel Dekker, 2003, pp. 267–296.
37. R.C. Cameron, G.S.P. Ritchie, A.D. Robson. Relative toxicities of inorganic aluminium complexes to barley. *Soil Sci. Soc. Am. J.* 50:1231–1236, 1986.
38. T.B. Kinraide, D.R. Parker. Non-phytotoxicity of the aluminum sulfate ion, $AlSO_4^+$. *Physiol. Plant* 71:207–212, 1987.
39. F.P.C. Blamey, D.G. Edwards, C.J. Asher. Effects of aluminium, OH:Al and P:Al molar ratios, and ionic strength on soybean root elongation in solution culture. *Soil Sci.* 136:197–207, 1983.
40. M. Akeson, D.N. Munns. Uptake of aluminum into root cytoplasm: Predicted rates for important solution complexes. *J. Plant Nutr.* 13:467–484, 1990.
41. K. Liu, S. Luan. Internal aluminum block of plant inward K^+ channels. Plant Cell. 13:1453–1465, 2001.
42. G.J. Taylor, J.L. McDonald-Stephens, D.B. Hunter, P.M. Bertsch, D. Elmore, Z. Rengel, R.J. Reid. Direct measurement of aluminum uptake and distribution in single cells of *Chara corallina*. *Plant Physiol.* 123:987–996, 2000.
43. E.O. Huett, R.C. Menary. Aluminium uptake by excised roots of cabbage, lettuce and kikuyu grass. *Aust. J. Plant Physiol.* 6:643–653, 1979.
44. G. Zhang, G.J. Taylor. Kinetics of aluminum uptake in *Triticum aestivum* L.: Identity of the linear phase of aluminum uptake by excised roots of aluminum-tolerant and aluminum-sensitive cultivars. *Plant Physiol.* 94:577–584, 1990.
45. J.L. McDonald-Stephens G.J. Taylor. Kinetics of aluminum uptake by cell suspensions of *Phaseolus vulgaris* L. *J. Plant Physiol.* 145:327–334, 1995.
46. R.B. Martin. Fe^{3+} and Al^{3+} hydrolysis equilibria. Cooperativity in Al^{3+} hydrolysis reactions. *J. Inorg. Biochem.* 44:141–147, 1991.
47. R.B. Martin. Aluminium speciation in biology. In: D.J. Chadwick, J. Whela, eds. *Aluminum in Biology and Medicine*. New York: Wiley , 1992, pp. 5–25.
48. W.R. Harris, G. Berthon, J.P. Day, C. Exley, T.P. Flaten, W.F. Forbes, T. Kiss, C. Orvig, P.F. Zatta. Speciation of aluminum in biological systems. In: R.A. Yokel, M.S. Golub, eds. *Research Issues in Aluminum Toxicity*. New York: Taylor & Francis, 1997, pp. 91–116.
49. H.P. Rasmussen. Entry and distribution of aluminum in Zea mays: The mode of entry and distribution of aluminum in *Zea mays*: Electron microprobe x-ray analysis. *Planta* 81:28–37, 1968.
50. G. Jentschke, H. Schlegel, D.L. Godbold. The effect of aluminium on uptake and distribution of magnesium and calcium in roots of mycorrhizal Norway spruce seedlings. *Physiol. Plant* 82:266–270, 1991.
51. D.B. Lazof, J.G. Goldsmith, T.W. Rufty, R.W. Linton. Early entry of Al into cells of intact soybean roots: A comparison of three developmental root regions using secondary ion mass spectrometry imaging. *Plant Physiol.* 112:1289–1300, 1996.
52. M.C. Hawes, U. Gunawardena, S. Miyasaka, X. Zhao. The role of root border cells in plant defense. *Trends Plant Sci.* 5:128–133, 2000.
53. D.J Archambault, G. Zhang, G.J. Taylor. Accumulation of Al in root mucilage of an Al-resistant and an Al-sensitive cultivar of wheat. *Plant Physiol.* 112:1741–1748, 1996.
54. D.T. Clarkson. The effect of aluminium and some other trivalent metal cations on cell division in the root apices of *Allium cepa*. *Ann. Bot.* 29:309–315, 1965.
55. W.J. Horst, C.J. Asher, I. Cakmak, P. Szulkiewicz, A.H. Wissemeier. Short-term responses of soybean roots to aluminium. *J. Plant Physiol.* 140:174–178, 1992.

56. D.B. Lazof, J.G. Goldsmith, T.W. Rufty, R.W. Linton. Rapid uptake of aluminum into cells of intact soybean root tips: A microanalytical study using secondary ion mass spectrometry. *Plant Physiol.* 106:1107–1114, 1994.

57. E.B. Blancafor, D.L. Jones, S. Gilroy. Alterations in the cytoskeleton accompany aluminum-induced growth inhibition and morphological changes in primary roots of maize. *Plant Physiol.* 118:159–172, 1998.

58. M. Sivaguru, W.J. Horst. The distal part of the transition zone is the most aluminum-sensitive apical root zone of maize. *Plant Physiol.* 116:155–163, 1998.

59. D.R. Parker. Root growth analysis: An underutilized approach to understanding aluminium rhizotoxicity. *Plant Soil* 171:151–157, 1995.

60. P.R. Ryan, J.E. Shaff, L.V. Kochian. Aluminum toxicity in roots: Correlation between ionic currents, ion fluxes, and root elongation in aluminum-sensitive and aluminum-tolerant wheat cultivars. *Plant Physiol.* 99:1193–1200, 1992.

61. M. Sasaki, Y. Yamamoto, J.F. Ma, H. Matsumoto. Early events induced by aluminum stress in elongating cells of wheat root. *Soil Sci. Plant Nutr.* 43:1009–1014, 1997.

62. M. Llugany, C. Poschenrieder, J. Barcelo. Monitoring of aluminium-induced inhibition of root elongation in four maize cultivars differing in tolerance to aluminium and proton toxicity. *Physiol. Plant* 93:265–271, 1995.

63. P.R. Ryan, J.M. Ditomaso, L.V. Kochian. Aluminum toxicity in roots: An investigation of spatial sensitivity and the role of the root cap. *J. Exp. Bot.* 44:437–446, 1993.

64. A. Ferrufino, T.J. Smyth, D.W. Israel, T.E. Carter, Jr. Root elongation of soybean genotypes in response to acidity constraints in a subsurface solution compartment. *Crop Sci.* 40:413–421, 2000.

65. E.R. Silva, R.J. Smyth, C.D. Raper, T.E. Carter, T.W. Rufty. Differential aluminum tolerance in soybean: An evaluation of the role of organic acids. *Physiol. Plant* 112:200–210, 2001.

66. M. Wood, J.E. Cooper, A.J. Holding. Aluminium toxicity and nodulation of *Trifolium repens*. *Plant Soil* 78:381–391, 1984.

67. B.B. Buchanan, W. Gruiseem, R.L. Jones. *Biochemistry and Molecular Biology of Plants*. Rockville, MD: American Society of Plant Physiology, 2000, pp. 2–50, 52–108, 110–158, 202–258, 930–987, 1204–1249, 1250–1318.

68. R.J. Bennet, C.M. Breen, M.V. Fey. The effects of aluminium on root cap function and root development in *Zea mays* L. *Environ. Exptl* 27:91–104, 1987.

69. V. Puthota, R. Cruz-Ortega, J. Johnson, J Ownby. An ultrastructural study of the inhibition of mucilage secretion in the wheat root cap by aluminium. In: R.J. Wright, V.C. Baligar, R.P. Murrmann, eds. *Plant-Soil Interactions at Low pH*. Dordrecht: Kluwer Academic Publishers, 1991, pp. 779–787.

70. S.C. Miyasaka, M.C. Hawes. Possible role of root border cells in detection and avoidance of aluminum toxicity. *Plant Physiol.* 125:1978–1987, 2001.

71. A.C. Jorns, C. Hecht-Buchholz, A.H. Wissemeier. Aluminium-induced callose formation in root tips of Norway spruce (*Picea abies* (L.) Karst.). *Z. Pflanzenernahr Bodenk* 154:349–353, 1991.

72. G. Zhang, J. Hoddinott, G.J. Taylor. Characterization of 1,3-β-D-Glucan (callose) synthesis in roots of *Triticum aestivum* in response to aluminum toxicity. *J. Plant Physiol* 144:229–234, 1994.

73. P.B. Larsen, C.Y. Tai, L.V. Kochian, S.H. Howell. Arabidopsis mutants with increased sensitivity to aluminum. *Plant Physiol.* 110:743–751, 1996.

74. M. Sivaguru, T. Fujiwara, J. Samaj, F. Baluska, Z. Yang, H. Osawa, T. Maeda, T. Mori, D. Volkmann, H. Matsumoto. Aluminum-induced 1-3-β-D-glucan inhibits cell-to-cell trafficking of molecules through plasmodesmata. A new mechanism of aluminum toxicity in plants. *Plant Physiol.* 124:991–1005, 2000.

75. M. Sasaki, Y. Yamamoto, H. Matsumoto. Lignin deposition induced by aluminum in wheat (*Triticum aestivum*) roots. *Plant Physiol.* 96:193–198, 1996.

76. K.C. Snowden, R.C. Gardner. Five genes induced by aluminum in wheat (*Triticum aestivum* L.) roots. *Plant Physiol.* 103:855–861, 1993.

77. J.W. Pan, D. Ye, L.L. Wang, J. Hua, G.F. Hua, W.H. Pan, N. Han, M.Y. Zhu. Root border cell development is a temperature-insensitive and Al-sensitive process in barley. *Plant Cell Physiol.* 45:751–760, 2004.

78. D.J. Raynal, J.D. Joslin, F.C. Thornton, M. Schaedle, G.S. Henderson. Sensitivity of tree seedlings to aluminum: III. Red spruce and loblolly pine. *J. Environ. Qual.* 19:180–187, 1990.

79. S.J. McCanny, W.H. Hendershot, M.J. Lechowicz, B. Shipley. The effects of aluminum on *Picea rubens*: factorial experiments using sand culture. *Can. J. For. Res.* 25:8–17, 1995.

80. F.C. Thornton, M. Schaedle, D.J. Raynal. Effects of aluminum on red spruce seedlings in solution culture. *Environ. Exptl. Bot.* 27:489–498, 1987.

81. S. Janhunen, V. Palomaki, T. Holopainen. Aluminium causes nutrient imbalance and structural changes in the needles of Scots pine without inducing clear root injuries. *Trees* 9:134–142, 1995.

82. C.D. Foy. Plant adaptation to acid, aluminum-toxic soils. Commun. *Soil Sci. Plant Anal.* 19:959–987, 1988.

83. M. Ridolfi, J.P. Garrec. Consequences of an excess Al and a deficiency in Ca and Mg for stomatal functioning and net carbon assimilation of beech leaves. *Ann. For. Sci.* 57:209–218, 2000.

84. K. Tan, W.G. Keltjens. Analysis of acid-soil stress in sorghum genotypes with emphasis on aluminium and magnesium interactions. *Plant Soil* 171:147–150, 1995.

85. J.R. Cumming, R.T. Eckert, L.S. Evans. Effect of aluminum on ^{32}P uptake and translocation by red spruce seedlings. *Can. J. For. Res.* 16:864–867, 1986.

86. D.T. Clarkson. Effect of aluminum on the uptake and metabolism of phosphorus by barley seedlings. *Plant Physiol.* 41:16–172, 1966.

87. P.E. Pfeffer, S.I. Tu, W.V. Gerasimowicz, J.R. Cavanaugh. *In vivo* ^{31}P NMR studies of corn root tissue and its uptake of toxic metals. *Plant Physiol.* 80:77–84, 1986.

88. P.E. Pfeffer, S.I. Tu, W.V. Gerasimowicz, R.T. Boswell. Effects of aluminum on the release and-or immobilization of soluble phosphate in corn root tissue: A ^{31}P-nuclear magnetic resonance study. *Planta* 172:200–208, 1987.

89. M. Sivaguru, B. Ezaki, Z.-H. He, H. Tong, H. Osawa, F. Baluska, D. Volkmann, H. Matsumoto. Aluminum-induced gene expression and protein localization of a cell wall-associated receptor kinase in Arabidopsis. *Plant Physiol.* 132:2256–2266, 2003.

90. K. Ohki. Photosynthesis, chlorophyll, and transpiration responses in aluminum stressed wheat and sorghum. *Crop Sci.* 26:572–575, 1986.

91. P.A. Arp, I. Strucel. Water uptake by black spruce seedlings from rooting media (solution, sand, peat) treated with inorganic and oxalated aluminum. *Water Air Soil Pollut.* 44:57–70, 1989.

92. E. Kruge,r E. Sucoff. Aluminium and the hydraulic conductivity of *Quercus rubra* L. root systems. *J. Exp. Bot.* 40:659–665, 1989.

93. P Sanford, J.S. Pate, M.J. Unkovich. A survey of proportional dependence of subterranean clover and other pasture legumes on N_2 fixation in South-west Australia utilizing ^{15}N natural abundance. *Aust. J. Agric. Res.* 45:165–181, 1993.

94. H.E. Murphy, D.G. Edwards, C.J. Asher. Effects of aluminium on nodulation and early growth of four tropical pasture legumes. *Aust. J. Agric. Res.* 35:663–673, 1984.

95. G.R. Cline, Z.N. Senwo. Tolerance of Lespedeza *Bradyrhizobium* to acidity, aluminum, and manganese in culture media containing glutamate or ammonium. *Soil Biol. Biochem.* 26:1067–1072, 1994.

96. J Barcelo, C. Poschenrieder. Fast root growth responses, root exudates, and internal detoxification as clues to the mechanisms of aluminium toxicity and resistance: A review. *Environ. Exptl. Bot.* 48:75–92, 2002.

97. L.V. Kochian. Cellular mechanisms of aluminum toxicity and resistance in plants. *Annu. Rev. Plant Physio.l Plant Mol. Biol.* 46:237–260, 1995.

98. L.V. Kochian, D.L. Jones. Aluminum toxicity and resistance in plants. In: R.A. Yokel, M.S. Golub, eds. *Research Issues in Aluminum Toxicity.* New York: Taylor & Francis, 1997, pp. 69–89.

99. H. Matsumoto. Cell biology of aluminum toxicity and tolerance in higher plants. *Int. Rev. Cytol.* 200:1–46, 2000.

100. W.J. Horst. The role of the apoplast in aluminium toxicity and resistance of higher plants: a review. *Z. Pflanzenernahr. Bodenk.* 158:419–428, 1995.

101. W.J. Horst, N. Schmohl, M. Kollmeier, F. Baluska, M Sivaguru. Does aluminium affect root growth of maize through interaction with the cell wall—plasma membrane—cytoskeleton continuum? *Plant Soil* 215:163–174, 1999.

102. D.L. Jones, L.V. Kochian. Aluminum interaction with plasma membrane lipids and enzyme metal binding sites and its potential role in Al cytotoxicity. *FEBS Letters* 400:51–57, 1997.

103. E.F. Klimashevskii, V.M. Dedov. Localization of the mechanism of growth-inhibiting action of Al^{3+} in elongating cell walls. *Soviet Plant Physiol.* 22:1040–1046, 1976.

104. F.P.C. Blamey, C.J. Asher, G.L. Kerven, D.G. Edwards. Factors affecting aluminum sorption by calcium pectate. *Plant Soil* 149:87–94, 1993.

105. R.J. Reid, M.A. Tester, F.A. .Smith. Calcium/aluminium interactions in the cell wall and plasma membrane of *Chara*. *Planta* 195:362–368, 1995.

106. D.L. Godbold, G. Jentschke. Aluminium accumulation in root cell walls coincides with inhibition of root growth but not with inhibition of magnesium uptake in Norway spruce. *Physiol. Plant* 102:553–560, 1998.

107. P.R. Ryan, L.V. Kochian. Interaction between aluminum toxicity and calcium uptake at the root apex in near-isogenic lines of wheat (*Triticum aestivum* L.) differing in aluminum tolerance. *Plant Physiol.* 102:975–982, 1993.

108. N. Schmohl, J. Pilling, J. Fisahn, W.J. Horst. Pectin methylesterase modulates aluminium sensitivity in *Zea mays* and *Solanum tuberosum*. *Physiol. Plant* 109:419–427, 2000.

109. A. Tabuchi, H. Matsumoto. Changes in cell-wall properties of wheat (*Triticum aestivum*) roots during aluminum-induced growth inhibition. *Physiol. Plant* 112:353–358, 2001.

110. M. Kaneko, E. Yoshimura, N.K. Nishizawa, S. Mori. Time course study of aluminum-induced callose formation in barley roots as observed by digital microscopy and low-vacuum scanning electron microscopy. *Soil Sci. Plant Nutr.* 45:710–712, 1999.

111. R. Vierstra, A. Haug. The effect of Al^{3+} on the physical properties of membrane lipids in *Thermoplasma acidophilum*. *Biochem. Biophys. Res. Commun.* 84:138–143, 1978.

112. M. Deleers, J.P. Servais, E. Wulfert. Neurotoxic cations induce membrane rigidificationi and membrane fusion at micromolar concentrations. *Biochim. Biophys. Acta* 855:271–276, 1986.

113. M. Deleers, J.P. Servais, E. Wulfert. Micromolar concentrations of Al^{3+} induce phase separation, aggregation and dye release in phosphatidylserine-containing lipid vesicles. *Biochim. Biophys. Acta.* 813:195–200, 1985.

114. M.A. Akeson, D.N. Munns, R.G. Burau. Adsorption of Al^{3+} to phosphatidylcholine vesicles. *Biochim. Biophys. Acta* 986:33–40, 1989.

115. E. Delhaize, D.M. Hebb, K.D. Richards, J.M. Lin, P.R. Ryan, R.C. Gardner. Cloning and expression of a wheat (*Triticum aestivum* L.) phosphatidylserine synthase cDNA. *J. Biol. Chem.* 274:7082–7088, 1999.

116. J. Chen, E.I. Sucoff, E.J. Stadelmann. Aluminum and temperature alteration of cell membrane permeability of *Quercus rubra*. *Plant Physiol.* 96:644–649, 1991.

117. Y.S. Lee, G. Mitiku, A.G. Endress. Short-term effects of Al^{3+} on osmotic behavior of red beet (*Beta vulgaris* L.) protoplasts. *Plant Soil* 228:223–232, 2001.

118. S.J. Ahn, M. Sivaguru, H. Osawa, G.C. Chung, H. Matsumoto. Aluminum inhibits the H^+-ATPase activity by permanently altering the plasma membrane surface potentials in squash roots. *Plant Physiol.* 126:1381–1390, 2001.

119. B.E. Nichol, L.A. Oliveira, A.D.M. Glass, M.Y. Siddiqi. The effects of aluminum on the influx of calcium, potassium, ammonium, nitrate, and phosphate in an aluminum-sensitive cultivar of barley (*Hordeum vulgare* L.) *Plant Physiol.* 101:1263–1266, 1993.

120. I.R. Silva, T.J. Smyth, D.W. Israel, C.D. Raper, T.W. Rufty. Magnesium is more efficient than calcium in alleviating aluminum rhizotoxicity in soybean and its ameliorative effect is not explained by the Gouy–Chapman-Stern model. *Plant Cell Physiol.* 42:538–545, 2001.

121. S.C. Miyasaka, L.V. Kochian, J.E. Shaff, C.D. Foy. Mechanisms of aluminum tolerance in wheat: An investigation of genotypic differences in rhizosphere pH, K^+, and H^+ transport, and root-cell membrane potentials. *Plant Physiol.* 91:1188–1196, 1989.

122. S.I. Tu, J.N. Brouillette. Metal ion inhibition of corn root plasma membrane ATPase. *Phytochemistry* 26:65–69, 1987.

123. C.A. Hamilton, A.G. Good, G.J. Taylor. Induction of vacuolar ATPase and mitochondrial ATP synthase by aluminum in an aluminum-resistant cultivar of wheat. *Plant Physiol.* 125:1068–1077, 2001.

124. H. Matsumoto, T. Yamaya. Inhibition of potassium uptake and regulation of membrane-associated Mg^{2+}-ATPase activity of pea roots by aluminium. *Soil Sci. Plant Nutr.* 32:179–188, 1986.

125. M.A. Pineros, L.V. Kochian. A patch-clamp study on the physiology of aluminum toxicity and aluminum tolerance in maize. Identification and characterization of Al^{3+}-induced anion channels. *Plant Physiol.* 125:292–305, 2001.

126. R.E. Johnson, W.A. Jackson. Calcium uptake and transport by wheat seedlings as affected by aluminum. *Soil Sci. Soc. Am. Proc.* 28:381–386, 1964.

127. J.W. Huang, D.L. Grunes, L.V. Kochian. Aluminum effects on the kinetics of calcium uptake into cells of the wheat root apex. *Planta* 188:414–421, 1992.

128. J.W. Huang, D.L. Grunes, L.V. Kochian. Aluminium and calcium transport inhibitions in intact roots and root plasmalemma vesicles from aluminium-sensitive and tolerant wheat cultivars. *Plant Soil* 171:131–135, 1995.

129. Z. Rengel, D.L. Robinson. Competitive Al^{3+} inhibition of net Mg^{2+} uptake by intact *Lolium multiflorum* roots. *Plant Physiol.* 91:1407–1413, 1989.

130. C.W. McDiarmid, R.C. Gardner. Overexpression of the Saccharomyces cerevisiae magnesium transport system confers resistance to aluminum ion. *J. Biol. Chem.* 273:1727–1732, 1998.

131. S.C. Jarvis, D.J. Hatch. The effects of low concentrations of aluminium on the growth and uptake of nitrate-N by white clover. *Plant Soil* 95:43–55, 1986.

132. D.B. Lazof, M. Rincon, T.W. Rufty, C.T. Mackown, T.E. Carter. Aluminum accumulation and associated effects on $^{15}NO_3^-$ influx in roots of two soybean genotypes differing in Al tolerance. *Plant Soil* 164:291–297, 1994.

133. R.P. Durieux, R.J. Bartlett, F.R. Magdoff. Separate mechanisms of aluminium toxicity for nitrate uptake and root elongation. *Plant Soil* 172:229–234, 1995.

134. Y.C. Chang, J.F. Ma, H. Matsumoto. Mechanisms of Al-induced iron chlorosis in wheat (*Triticum aestivum*). Al-inhibited biosynthesis and secretion of phytosiderophore. *Plant Physiol.* 102:9–15, 1998.

135. M.A.R. Milla, E. Butler, A.R. Huete, C.F. Wilson, O. Anderson, J.P. Gustafson. Expressed sequence tag-based gene expression analysis under aluminum stress in rye. *Plant Physiol.* 130:1706–1716, 2002.

136. D.L. Jones, L.V. Kochian. Aluminum inhibition of the inositol 1,4,5,-triphosphate signal transduction pathway in wheat roots: A role in aluminum toxicity? *Plant Cell* 7:1913–1922, 1995.

137. L. Alessa, L. Oliveira. Aluminum toxicity studies in *Vaucheria longicaulis* var. macounii (Xanthophyta, Tribophyceae). I. Effects on cytoplasmic organization. *Environ. Exptl. Bot.* 45:205–222, 2001.

138. S. Grabski, M. Schindler. Aluminum induces rigor within the actin network of soybean cells. *Plant Physiol.* 108:897–901, 1995.

139. S. Grabski, E. Arnoys, B. Busch, M. Schindler. Regulation of actin tension in plant cells by kinases and phosphatases. *Plant Physiol.* 116:279–290, 1998.

140. R. Cruz-Ortega, J.C. Cushman, J.D. Ownby. cDNA clones encoding 1,3-β-glucanase and a fimbrin-like cytoskeletal protein are induced by Al toxicity in wheat roots. *Plant Physiol.* 114:1453–1460, 1997.

141. T.L. MacDonald, W.G. Humphries, R.B. Martin. Promotion of tubulin assembly by aluminum ion in vitro. *Science* 236:183–186, 1987.

142. M. Sivaguru, F.Baluska, D. Volkmann, H.H. Felle, W.J. Horst. Impacts of aluminum on the cytoskeleton of the maize root apex. Short-term effects on the distal part of the transition zone. *Plant Physiol.* 119:1073–1082, 1999.

143. N. Siegel, A. Haug. Calmodulin-dependent formation of membrane potential in barley root plasma membrane vesicles: A biochemical model of aluminum toxicity in plants. *Physiol. Plant* 59:285–291, 1983.

144. Z. Rengel. Disturbance of cell Ca^{2+} homeostasis as a primary trigger of Al toxicity syndrome. *Plant Cell. Environ.* 15:931–938, 1992.

145. S. Lindberg, H. Strid. Aluminium induces rapid changes in cytosolic pH and free calcium and potassium concentrations in root protoplasts of wheat (*Triticum aestivum*). *Plant Physiol.* 99:405–414, 1997.

146. W.H. Zhang, Z. Rengel. Aluminium induces an increase in cytoplasmic calcium in intact wheat root apical cells. *Aust. J. Plant Physiol.* 26:401–419, 1999.

147. B.E. Nichol, L.A. Oliveira. Effects of aluminum on the growth and distribution of calcium in roots of an aluminum-sensitive cultivar of barley (*Hordeum vulgare*). *Can. J. Bot.* 73:1849–1858, 1995.

148. D.L. Jones, L.V. Kochian, S Gilroy. Aluminum induces a decrease in cytosolic calcium concentration in BY-2 tobacco cell cultures. *Plant Physiol.* 116:81–89, 1998.

149. K.H. Hasenstein, M. Evans, C.L. Stinemetz, R. Moore, W.M. Fondren, E.C. Koon, M.A. Higby, A.J.M. Smucker. Comparative effectiveness of metal ions in inducing curvature of primary roots of *Zea mays. Plant Physiol.* 86:885–889, 1988.

150. K.H. Hasenstein, M.L. Evans. Effects of cations on hormone transport in primary roots of *Zea mays. Plant Physiol.* 86:890–894, 1988.

151. M. Kollmeier, H.H. Felle, W.J. Horst. Genotypical differences in aluminum resistance of maize are expressed in the distal part of the transition zone. Is reduced basipetal auxin flow involved in inhibition of root elongation by aluminum? *Plant Physiol.* 122:945–956, 2000.

152. N. Massot, B. Nicander, J. Barcelo, Ch. Poschenrieder, E. Tillbert. A rapid increase in cytokinin levels and enhanced ethylene evolution precede Al^{3+}-induced inhibition of root growth in bean seedlings (*Phaseolus vulgaris* L.) *Plant Growth Regulation* 37:105–112, 2002.

153. Y. Sakihama, H. Yamasaki, Lipid peroxidation induced by phenolics in conjunction with aluminum ions. *Biologia Plantarum* 45:249–254, 2002.

154. Y. Yamamoto, Y. Kobayashi, H. Matsumoto. Lipid peroxidation is an early symptom triggered by aluminum, but not the primary cause of elongation inhibition in pea roots. *Plant Physiol.* 125:199–208, 2001.

155. P.R.S. Boscolo, M. Menossi, R.A. Jorge. Aluminum-induced oxidative stress in maize. *Phytochemistry* 62:181–189, 2003.

156. H. Ikegawa, Y. Yamamoto, H. Matsumoto. Responses to aluminium of suspension-cultured tobacco cells in a simple calcium solution. *Soil Sci. Plant Nutr.* 46:503–514, 2000.

157. K.D. Richards, E.J. Schott, Y.K. Sharma, K.R. Davis, R.C. Gardner. Aluminum induces oxidative stress genes in *Arabidopsis thaliana*. *Plant Physiol.* 116:409–418, 1998.

158. D.A. Watt. Aluminum-responsive genes in sugarcane: Identification and analysis of expression under oxidative stress. *J. Exp. Bot.* 54:1163–1174, 2003.

159. B. Ezaki, R.C. Gardner, Y. Ezaki, H. Matsumoto. Expression of aluminum-induced genes in transgenic Arabidopsis plants can ameliorate aluminum stress and/or oxidative stress. *Plant Physiol.* 122:657–665, 2000.

160. R. Hampp, H Schnabl. Effect of aluminium ions on $^{14}CO_2$-fixation and membrane system of isolated spinach chloroplasts. *Z. Pflanzenphysiol Bd* 76:300–306, 1975.

161. M. Moustakas, G. Ouzounidou. Increased non-photochemical quenching in leaves of aluminum-stressed wheat plants is due to Al^{3+}-induced elemental loss. *Plant Physiol. Biochem.* 32:527–532, 1994.

162. W.E. Pereira, D.L. de Siqueira, C.A. Martinez, M. Puiatti. Gas exchange and chlorophyll fluorescence in four citrus rootstocks under aluminium stress. *J. Plant Physiol.* 157:513–520, 2000.

163. A. Petterson, L. Hallbom, B. Bergman. Physiological and structural responses of the cyanobacterium *Anabaena cylindrica* to aluminium. *Physiol. Plant* 63:153–158, 1985.

164. I.R. Silva, T.J. Smyth, D.F. Moxley, T.E. Carter, N.S. Allen, T.W. Rufty. Aluminum accumulation at nuclei of cells in the root tip. Fluorescence detection using lumogallion and confocal laser scanning microscopy. *Plant Physiol.* 123:543–552, 2000.

165. C.D. Foy. Plant adaptation to mineral stress in problem soils. *Iowa State J. Res.* 57:339–354, 1983.

166. P.C. Kerridge, M.D. Dawson, D.P. Moore. Separation of degrees of aluminum tolerance in wheat. *Agron. J.* 63:586–591, 1971.

167. T.A. Campbell, N.J. Nuernberg, C.D. Foy. Differential responses of alfalfa cultivars to aluminum stress. *J. Plant Nutr.* 12:291–305, 1989.

168. A.C. Baier, D.J. Somers, J.P. Gustafson. Aluminum tolerance in wheat: Correlating hydroponic evaluations with field and soil performances. *Plant Breeding* 114:291–296, 1995.

169. J.J. Bilski, C.D. Foy. Differential tolerances of oat cultivars to aluminum in nutrient solutions and in acid soils of Poland. *J. Plant Nutr.* 10:129–141, 1987.

170. R.H. Howeler, L.F. Cadavid. Screening of rice cultivars for tolerance to Al-toxicity in nutrient solutions as compared with a field screening method. *Agron. J.* 68:551–555, 1976.

171. D.A. Reid, A.L. Fleming, C.D. Foy. A method for determining aluminum response of barley in nutrient solutions in comparison to response in Al-toxic soil. *Agron. J.* 63:600–603, 1971.

172. J.D. Ownby. Mechanisms of reaction of hematoxylin with aluminum-treated wheat roots. *Physiol. Plant* 87:371–380, 1993.

173. E. Polle, A.F. Konzak, J.A. Kittrick. Visual detection of aluminum tolerance levels in wheat by hematoxylin staining of seedling roots. *Crop Sci.* 18:823–827, 1978.

174. B.J. Scott, J.A. Fisher, L.J. Spohr. Tolerance of Australian wheat varieties to aluminum toxicity. *Commun. Soil Sci. Plant Anal.* 23:509–526, 1992.

175. R.J. Bennet. The response of lucern and red clover roots to aluminium/hematoxylin: how universal is the hematoxylin test for aluminium? *S. Afr. Tydskr. Plant Grond.* 14:120–126, 1997.

176. D.P. Moore, W.E. Kronstad, R.J. Metzger. Screening wheat for aluminum tolerance. In: M.J. Wright, S.A. Ferrari, eds. *Plant Adaptation to Mineral Stress in Problem Soils*. Ithaca, NY: Cornell University, 1976, pp. 287–295.

177. Ch. Hecht-Buchholz. Light and electron microscopic investigations of the reactions of various genotypes to nutritional disorders. *Plant Soil* 72:151–165, 1983.

178. X.G. Zhang, R.S. Jessop, F. Ellison. Differential responses to selection for aluminium stress tolerance in triticale. *Aus. J. Agric. Res.* 53:1295–1303, 2002.

179. L.V. Kochian, O.A. Hoekenga, M.A. Pineros. How do crop plants tolerate acid soils? Mechanisms of aluminum tolerance and phosphorus efficiency. *Annu. Rev. Plant Biol.* 55:459–493, 2004.

180. Y. Tang, M.E. Sorrells, L.V. Kochian, D.F. Garvin. Identification of RFLP markers linked to the barley aluminum tolerance gene Alp. *Crop Sci.* 40:778–782, 2000.

181. A.M. Aniol. Physiological aspects of aluminium tolerance associated with the long arm of chromosome 2D of the wheat (*Triticum aestivum* L.) genome. *Theor. Appl. Genet.* 91:510–516, 1995.

182. L.G. Campbell, H.N. Lafever. Heritability of aluminum tolerance in wheat. *Cereal Res. Commun.* 9:281–287, 1981.

183. V.T. Nguyen, M.D. Burow, H.T. Nguyen, B.T. Le, T.D. Le, A.H. Paterson. Molecular mapping of genes conferring aluminum tolerance in rice (*Oryza sativa* L.) *Theor. Appl. Genet.* 102:1002–1010, 2001.

184. C.M. Bianchi-Hall, T.E. Carter, Jr., T.W. Rufty, C. Arellano, H.R. Boerma, D.A. Ashley, J.W. Burton. Heritability and resource allocation of aluminum tolerance derived from soybean PI 416937. *Crop Sci.* 38:513–522, 1998.

185. Y. Kobayashi, H. Koyama. QTL analysis of Al tolerance in recombinant inbred lines of *Arabidopsis thaliana*. *Plant Cell Physiol.* 43:1526–1533, 2002.

186. G.J. Taylor. Current views of the aluminum stress response, The physiological basis of tolerance. *Curr. Topics Plant Biochem. Physiol.* 10:57–93, 1991.

187. G.J. Taylor. Overcoming barriers to understanding the cellular basis of aluminium resistance. *Plant Soil* 171:89–103, 1995.

188. L.V. Kochian, N.S. Pence, D.L.D. Letham, M.A. Pineros, J.V. Magalhaes, O.A. Hoenkenga, D.F. Garvin. Mechanisms of metal resistance in plants: Aluminum and heavy metals. *Plant Soil* 247:109–119, 2002.

189. W.J. Horst, A. Wagner, H. Marschner. Effect of aluminium on root growth, cell-division rate and mineral element contents in roots of *Vigna unguiculata* genotypes. *Z. Pflanzenphysiol. Bd* 109:95–103, 1983.

190. M. Rincon, R.A. Gonzales. Aluminum partitioning in intact roots of aluminum-tolerant and aluminum-sensitive wheat (*Triticum aestivum* L.) cultivars. *Plant Physiol.* 99:1021–1028, 1992.

191. E. Delhaize, S. Craig, C.D. BEaton, R.J. Bennet, V.C. Jagadish, P.R. Randall. Aluminum tolerance in wheat (*Triticum aestivum* L.) I. Uptake and distribution of aluminum in root apices. *Plant Physiol.* 103:685–693, 1993.

192. V.N. Bushamuka, R.W. Zobel. Maize and soybean tap, basal, and lateral root growth responses to a stratified acid, aluminum-toxic soil. *Crop Sci.* 38:416–421, 1998.

193. J.F. Ma. Role of organic acids in detoxification of aluminum in higher plants. *Plant Cell Physiol.* 41:383–390, 2000.

194. J.F. Ma, P.R. Ryan, E. Delhaize. Aluminium tolerance in plants and the complexing role of organic acids. *TRENDS Plant Sci.* 6:273–278, 2001.

195. N.V. Hue, G.R. Craddock, F. Adams. Effect of organic acids on aluminum toxicity in subsoils. *Soil Sci. Soc. Am. J.* 50:28–34, 1986.

196. S.C. Miyasaka, J.G. Buta, R.K. Howell, C.D. Foy. Mechanism of aluminum tolerance in snapbeans: Root exudation of citric acid. *Plant Physiol.* 96:737–743, 1991.

197. E. Delhaize, P.R. Ryan, P.J. Randall. Aluminum tolerance in wheat (*Triticum aestivum* L.) II. Aluminum-stimulated excretion of malic acid from root apices. *Plant Physiol.* 103:695–702, 1993.

198. P.R. Ryan, E. Delhaize, P.J. Randall. Characterization of Al-stimulated efflux of malate from the apices of Al-tolerant wheat roots. *Planta* 196:103–110, 1995.

199. U. Basu, D. Godbold, G.J. Taylor. Aluminum resistance in *Triticum aestivum* associated with enhanced exudation of malate. *J. Plant Physiol.* 144:747–753, 1994.

200. P.R. Ryan, E. Delhaize, P.J. Randall. Malate efflux from root apices and tolerance to aluminium are highly correlated in wheat. *Aus. J. Plant Physiol.* 22:531–536, 1995.

201. D.L. Jones, A.M. Prabowo, L.V. Kochian. Kinetics of malate transport and decomposition in acid soils and isolated bacterial populations: The effect of microorganisms on root exudation of malate under Al stress. *Plant Soil* 182:239–247, 1996.

202. D.R. Parker, J.F. Pedler. Probing the "malate hypothesis" of differential aluminum tolerance in wheat by using other rhizotoxic ions as proxies for Al. *Planta* 205:389–396, 1998.

203. D.M. Pellet, D.L. Grunes, L.V. Kochian. Organic acid exudation as an aluminum-tolerance mechanism in maize (*Zea mays* L.). *Planta* 196:788–795, 1995.

204. J.F. Ma, S.J. Zheng, H. Matsumoto. Specific secretion of citric acid induced by Al stress in *Cassia tora* L. *Plant Cell Physiol.* 38:1019–1025, 1997.

205. X.F. Li, J.F. Ma, H. Matsumoto. Pattern of aluminum-induced secretion of organic acids differs between rye and wheat. *Plant Physiol.* 123:1537–1543, 2000.

206. I.R. Silva, T.J. Smyth, D. Raper, T.E. Carter, T.W. Rufty. Differential aluminum tolerance in soybean: An evaluation of the role of organic acids. *Physiol. Plant* 112:200–210, 2001.

207. J.F. Ma, S.J. Zheng, S. Hiradate, H. Matsumoto. Detoxifying aluminum with buckwheat. *Nature* 390:569–570, 1997.

208. Z. Ma, S.C. Miyasaka. Oxalate exudation by taro in response to Al. *Plant Physiol.* 118:861–865, 1998.

209. P.B. Larsen, J. Degenhardt, C.Y. Tai, L.M. Stenzler, S.H. Howell, L.V. Kochian. Aluminum-resistant Arabidopsis mutants that exhibit altered patterns of aluminum accumulation and organic acid release from roots. *Plant Physiol.* 117:9–18, 1998.

210. H. Koyama, R. Okawara, K. Ojima, T. Yamaya. Re-evaluation of characteristics of a carrot cell line previously selected as aluminum-tolerant cells. *Physiol. Plant* 74:683–687, 1988.

211. T. Kihara, T. Ohno, H. Koyama, T. Sawafuji, T. Hara. Characterization of NADP-isocitrate dehydrogenase expression in a carrot mutant cell line with enhanced citrate excretion. *Plant Soil* 248:145–153, 2003.

212. E. Takita, H. Koyama, T. Hara. Organic acid metabolism in aluminum-phosphate utilizing cells of carrot (*Daucus carota* L.). *Plant Cell Physiol.* 40:489–495, 1999.

213. P.R. Ryan, M. Skerrett, G.P. Findlay, E. Delhaize, S.D. Tyerman. Aluminum activates an anion channel in the apical cells of wheat roots. *Proc. Natl. Acad. Sci. USA* 94:6547–6552, 1997.

214. W.H. Zhang, P.R. Ryan, S.D. Tyerman. Malate-permeable channels and cation channels activated by aluminum in the apical cells of wheat roots. *Plant Physiol.* 125:1459–1472, 2001.

215. T. Sasaki, Y. Yamamoto, B. Ezaki, M. Katsuhara, S.J. Ahn, P.R. Ryan, E. Delhaize, H. Matsumoto. A wheat gene encoding an aluminum-activated malate transporter. *Plant J.* 37:645–653, 2004.

216. E. Delhaize, P.R. Ryan, D.M. Hebb, Y. Yamamoto, T. Sasaki, H. Matsumoto. Engineering high-level aluminum tolerance in barley with ALMT1 gene. *PNAS.* 101:15249–15254, 2004.

217. J.M. de la Fuente, V. Ramirez-Rodriguez, J.L. Cabrera-Ponce, L. Herrera-Estrella. Aluminum tolerance in transgenic plants by alteration of citrate synthesis. *Science* 276:1566–1568, 1997.

218. E. Delhaize, D.M. Hebb, P.R. Ryan. Expression of a *Pseudomonas aeruginosa* citrate synthase gene in tobacco is not associated with either enhanced citrate accumulation or efflux. *Plant Physiol.* 125:2059–2067, 2001.

219. H. Koyama, A. Kawamura, T. Kihara, T. Hara, E. Takita, D. Shibata. Over expression of mitochondrial citrate synthase in *Arabidopsis thaliana*: Improved growth on a phosphorus-limited soil. *Plant Cell Physiol.* 41:1030–1037, 2000.

220. M. Tesfaye, S.J. Temple, D.L. Allan, C.P. Vance, S.A. Samac. Overexpression of malate dehydrogenase in transgenic alfalfa enhances organic acid synthesis and confers tolerance to aluminum. *Plant Physiol.* 127:1836–1844, 2001.

221. S. Ishikawa, T. Wagatsuma, R. Sasaki, P. Ofei-Manu. 2000. Comparison of the amount of citric and malic acids in Al media of seven plant species and two cultivars each in five plant species. *Soil Sci. Plant Nutr.* 46:751–758, 2000.

222. P. Wenzl, G.M. Patino, A.L. Chaves, J.E. Mayer, I.M. Rao. The high level of aluminum resistance in Signalgrass is not associated with known mechanisms of external aluminum detoxification in root apices. *Plant Physiol.* 125:1473–1484, 2001.

223. D.M. Pellet, L.A. Papernik, L.V. Kochian. Multiple aluminum-resistance mechanisms in wheat: Roles of root apical phosphate and malate exudation. *Plant Physiol.* 112:591–597, 1996.

224. Y. Tang, D.F. Garvin, L.V. Kochian, M.E. Sorrells, B.F. Carver. Physiological genetics of aluminum tolerance in the wheat cultivar Atlas 66. *Crop Sci.* 42:1541–1546, 2002.

225. U. Basu, J.L. McDonald-Stephens, D.J. Archambault, A.G. Good, K.G. Briggs, T. Aung, G.J. Taylor. Genetic and physiological analysis of doubled-haploid, aluminum-resistant lines of wheat provide evidence for the involvement of a 23 kD, root exudates polypeptide in mediating resistance. *Plant Soil* 196:283–288, 1997.

226. U. Basu, A.G. Good, T. Aung, J.J. Slaski, A. Basu, K.G. Briggs, G.J. Taylor. A 23-kDa, root exudates polypeptide co-segregates with aluminum resistance in *Triticum aestivum. Physiol. Plant* 106:53–61, 1999.

227. P. Ofei-Manu, T. Wagatsuma, S. Ishikawa, K. Tawaraya. The plasma membrane strength of the root-tip cells and root phenolic compounds are correlated with Al tolerance in several common woody plants. *Soil Sci. Plant Nutr.* 47:359–375, 2001.

228. M.J. Hodson, D.E. Evans. Aluminum/silicon interactions in higher plants. *J. Exp. Bot.* 46:161–171, 1995.

229. K.M. Cocker, D.E. Evans, M.J. Hodson. The amelioration of aluminium toxicity by silicon in higher plants: Solution chemistry or an in planta mechanism. *Physiol. Plant* 104:608–614, 1998.

230. P.S. Kidd, M. Llugany, C. Poschenrieder, B. Gunse, J. Barcelo. The role of root exudates in aluminium resistance and silicon-induced amelioration of aluminium toxicity in three varieties of maize (*Zea mays* L.). *J. Exp. Bot.* 52:1339–1352, 2001.

231. C.D. Foy, G.R. Burns, J.C. Brown, A.L. Fleming. Differential aluminum tolerance of two wheat varieties associated with plant-induced pH changes around their roots. *Soil Sci. Soc. Proc.* 29:64–67, 1965.

232. J. Degenhardt, P.B. Larsen, S.H. Howell, L.V. Kochian. Aluminum resistance in the Arabidopsis mutant alr-104 is caused by an aluminum-induced increase in rhizosphere pH. *Plant Physiol.* 117:19–27, 1998.

233. W.J. Horst, A. Wagner, H. Marschner. Mucilage protects root meristems from aluminum injury. *Z Pflanzenphysiol Bd* 105:435–444, 1982.

234. L.A. Brigham, M.C. Hawes, S.C. Miyasaka. Avoidance of aluminum toxicity: Role of root border cells. In: W.J. Horst, M.K. Schenk, A. Burkert, N. Claassen, H. Flessa, W.B. Frommer, H. Goldbach, H.W. Olfs, V. Romheld, eds. *Plant Nutrition: Food Security and Sustainability of Agro-Ecosystems Through Basic and Applied Research.* Boston: Kluwer Academic , 2001, pp. 452–453.

235. X.F. Li, J.F. Ma, S. Hiradate, H. Matsumoto. Mucilage strongly binds aluminum but does not prevent roots from aluminum injury in *Zea mays. Physiol. Plant* 108:152–160, 2000.

236. L.M. Mugwira, S.M. Elgawhary. Aluminum accumulation and tolerance of triticale and wheat in relation to root cation exchange capacity. *Soil Sci. Soc. Am. J.* 43:736–740, 1979.

237. G. Zhang, G.J. Taylor. Effects of biological inhibitors on kinetics of aluminium uptake by excised roots and purified cell wall material of aluminium-tolerant and aluminium-sensitive cultivars of *Triticum aestivum* L. *J. Plant Physiol.* 138:533–539, 1991.

238. F.P.C. Blamey, N.J. Robinson, C.J.Asher. Interspecific differences in aluminium tolerance in relation to root cation exchange capacity. In: P.J. Randall, ed. Genetic Aspects of Plant Mineral Nutrition. New York: Kluwer Academic Press, 1993, pp. 91–96.

239. T. Wagatsuma, R. Akiba. Low surface negativity of root protoplasts from aluminum-tolerant plant species. *Soil Sci. Plant Nutr.* 35:443–452, 1989.

240. S. Ishikawa, T. Wagatsuma. Plasma membrane permeability of root-tip cells following temporary exposure to Al ions is a rapid measure of Al tolerance among plant species. *Plant Cell Physiol.* 39:516–525, 1998.

241. U. Yermiyahu, D.K. Brauer, T.B. Kinraide. Sorption of aluminum to plasma membrane vesicles isolated from roots of Scout 66 and Atlas 66 cultivars of wheat. *Plant Physiol.* 115:1119–1125, 1997.

242. U. Ahonen-Jonnarth, A. Goransson, R.D. Finlay. Growth and nutrient uptake of ectomycorrhizal *Pinus sylvestris* seedlings in a natural substrate treated with elevated Al concentrations. *Tree Physiol.* 23:157–167, 2003.

243. R.B. Clark, R.W. Zobel, S.K. Zeto. Effects of mycorrhizal fungus isolates on mineral acquisition by *Panicum virgatum* in acidic soil. *Mycorrhiza* 9:167–176, 1999.

244. G. Jentschke, D.L. Godbold, A. Huttermann. Culture of mycorrhizal tree seedlings under controlled conditions: Effects of nitrogen and aluminium. *Physiol. Plant* 81:408–416, 1991.

245. E. Hentschel, D.L. Godbold, P. Marschner, H. Schlegel, G. Jentschke. The effect of Paxillus involutus fr. On aluminum sensitivity of Norway spruce seedlings. *Tree Physiol.* 12:379–390, 1993.

246. J.R. Cumming, L.H. Weinstein. Aluminum-mycorrhizal interactions in the physiology of pitch pine seedlings. *Plant Soil* 125:7–18, 1990.

247. G.A. Schier, C.J. McQuattie. Effect of aluminum on the growth, anatomy, and nutrient content of ectomycorrhizal and nonmycorrhizal eastern white pine seedlings. *Can. J. For. Res.* 25:1252–1262, 1995.

248. S.D. Koslowsky, R.E.J. Boerner. Interactive effects of aluminum, phosphorus and mycorrhizae on growth and nutrient uptake of *Panicum virgatum* L. (Poaceae). *Environ. Pollut.* 61:107–125, 1989.

249. J. Mendoza, F. Borie. Effect of *Glomus etunicatum* inoculation on aluminum, phosphorus, calcium, and magnesium uptake of two barley genotypes with different aluminum tolerance. *Commun. Soil Sci. Plant Anal.* 29:681–695, 1998.

250. G. Rufyikiri, S. Declerck, J.E. Dufey, B. Delvaux. Arbuscular mycorrhizal fungi might alleviate aluminium toxicity in banana plants. *New Phytol.* 148:343–352, 2000.

251. M. Rudawska, B. Kieliszewska-Rokicka, T. Leski. Effect of aluminium on *Pinus sylvestris* seedlings mycorrhizal with aluminum-tolerant and aluminium-sensitive strains of *Suillus luteus*. *Dendrobiology* 45:89–96, 2000.

252. L.M. Egerton-Warburton, B.J. Griffin. Differential responses of *Pisolithus tinctorius* isolates to aluminum in vitro. *Can. J. Bot.* 73:1229–1233, 1995.

253. G. Cuenca, Z. De Andrade, E. Meneses. The presence of aluminum in arbuscular mycorrhizas of *Clusia multiflora* exposed to increased acidity. *Plant Soil* 231:233–241, 2001.

254. F. Martin, P. Rubini, R. Cote, I. Kottke. Aluminum polyphosphate complexes in the mycorrhizal basidiomycete Laccaria bicolor: A ^{27}Al-nuclear magnetic resonance study. *Planta* 194:241–246, 1994.

255. T.G.M. Gerlitz. Effects of aluminium on polyphosphate mobilization of the ectomycorrhizal fungus *Suillus bovines*. *Plant Soil* 178:133–140, 1996.

256. G. Jentschke, D.L. Godbold. Metal toxicity and ectomycorrhizas. *Physiol. Plant* 109:107–116, 2000.

257. J.F. Ma, S. Hiradate, K. Nomoto, T. Iwashita, H. Matsumoto. Internal detoxification mechanism of Al in hydrangea. *Plant Physiol.* 113:1033–1039, 1997.

258. J.F. Ma, S. Hiradate. Form of aluminum for uptake and translocation in buckwheat (*Fagopyrum esculentum* Moench). *Planta* 211:355–360, 2000.

259. J.F. Ma, S. Hiradate, H. Matsumoto. High aluminum resistance in buckwheat: II. Oxalic acid detoxifies aluminum internally. *Plant Physiol.* 117:753–759, 1998.

260. R. Shen, J.F. Ma, M. Kyo, T. Iwashita. Compartmentation of aluminium in leaves of an Al-accumulator, *Fagopyrum esculentum* Moench. *Planta* 215:394–398, 2002.

261. T. Watanabe, M. Osaki, T. Yoshihara, T. Tadano. Distribution and chemical speciation of aluminum in the Al accumulator plant, *Melastoma malabathricum* L. *Plant Soil* 201:165–173, 1998.

262. T. Watanabe, M. Osaki. Influence of aluminum and phosphorus on growth and xylem sap composition in *Melastoma malabathricum* L. *Plant Soil* 237:63–70, 2001.

263. T. Nagata, M. Hayatsu, N. Kosuge. Identification of aluminium forms in tea leaves by ^{27}Al-NMR. *Phytochemistry* 31:1215–1218, 1992.

264. M.J. Hodson, A.G. Sangster. Aluminum/silicon interactions in conifers. *J. Inorg. Biochem.* 76:89–98, 1999.

265. F. Bartoli, L.P. Wilding. Dissolution of biogenic opal as a function of its physical and chemical properties. *Soil Sci. Soc. Am. J.* 44:873–878, 1980.

266. A.L. Carnelli, M. Madella, J.P. Theurillat, B. Ammann. Aluminum in the opal silica reticule of phytoliths: A new tool in palaeoecological studies. *Am. J. Bot.* 89:346–351, 2002.

267. M.J. Hodson, A.G. Sangster. Interaction between silicon and aluminum in *Sorghum bicolor* (L.) Moench: Growth analysis and X-ray microanalysis. *Ann. Bot.* 72:389–400, 1993.

268. R.M. Britez, T. Watanabe, S. Jansen, C.B. Reissman, M. Osaki. The relationship between aluminium and silicon accumulation in leaves of *Faramea marginata* (Rubiaceae). *New Phytol.* 156:437–444, 2002.

269. M. Kasai, M. Sasaki, Y. Yamamoto, H. Matsumoto. Aluminum stress increases K^+ efflux and activities of ATP- and PP_i-dependent H^+ pumps of tonoplast-enriched membrane vesicles from barley roots. *Plant Cell Physiol.* 33:1035–1039, 1992.

270. G.J. Taylor, A. Basu, U. Basu, J.J. Slaski, G. Zhang, A. Good. Al-induced, 51-kilodalton, membrane-bound proteins are associated with resistance to Al in a segregating population of wheat. *Plant Physiol.* 114:363–372, 1997.

271. S. Nagasaka, N.K. Nishizawa, T. Negishi, K. Satake, S. Mori, E. Yoshimura. Novel iron-storage particles may play a role in aluminum tolerance of *Cyanidium caldarium*. *Planta* 215:399–404, 2002.

272. G. Fiskesjo. Occurrence and degeneration of "Al structures" in root cap cells of *Allium cepa* L. after Al treatment. *Hereditas* 112:193–202, 1990.

273. G. Delisle, M. Champoux, M. Houde. Characterization of oxalate oxidase and cell death in Al-sensitive and tolerant wheat roots. *Plant Cell Physiol.* 42:324–333, 2001.

274. F. Hamel, C. Breton, M. Houde. Isolation and characterization of wheat aluminum-regulated genes: possible involvement of aluminum as a pathogenesis response elicitor. *Planta* 205:531–538, 1998.

275. B.L. Allen, B.F. Hajek. Mineral occurrence in soil environments. In: J.B. Dixon, S.B. Weed, eds. *Minerals in Soil Environments*, 2nd ed. Madison, WI: Soil Science Society of America, 1989, pp. 199–278.

276. M. Conyers. The control of aluminium solubility in some acidic Australian sSoils. *J. Soil Sci.* 41:147–156, 1990.

277. FAO/UNESCO. http://www.fao.org/ag/agl/agll/wrb/mapindex.stm, 1998. Accessed March 2003.

278. P.A. Sanchez, T.J. Logan. Myths and science about the chemistry and fertility of soils in the tropics. In: R. Lal, P.A. Sanchez, eds. *Myths and Science of Soils of the Tropics*. Madison, WI: Soil Science Society of America, 1992, pp. 35–46.

279. F. Adams. Crop response to lime in the southern United States. In: F. Adams, ed. *Soil Acidity and Liming*, 2nd ed. Madison, WI: Soil Science Society of America, 1984, pp. 211–265.

280. NRCS (Natural Resources Conservation Service). http://soils.usda.gov/technical/classification/orders/, 2002. Accessed May 2006.

281. M.E. Sumner, A.D. Noble. Soil acidification: The World Story. In: Z. Rengel, ed. *Handbook of Soil Acidity*. New York, NY: 2003, pp. 1–28.

282. N.W. Menzies, L.C. Bell, D.G. Edwards. Exchange and solution phase chemistry of acid, highly weathered soils: II. Investigation of mechanisms controlling Al release into solution. *Aust. J. Soil Res.* 32:269–283, 1994.

283. K. Wada. Allophane and imogolite. In: J.B. Dixon, S.B. Weed, eds. *Minerals in Soil Environments*, 2nd ed. Madison, WI: Soil Science Society of America, 1989, pp. 1051–1087.

284. P.R. Bloom, M.B. McBride, R.M. Weaver. Aluminum organic matter in acid soils: Buffering and solution aluminum activity. *Soil Sci. Soc. Am. J.* 43:488–493, 1979.

285. J.D. Wolt. Sulfate retention by acid sulfate-polluted soils in the copper basin area of Tennessee. *Soil Sci. Soc. Am. J.* 45:283–287, 1981.

286. N.V. Hue, F. Adams, C.E. Evans. Sulfate retention by an acid BE horizon of an Ultisol. *Soil Sci. Soc. Am. J.* 49:1196–1200, 1985.

287. R.W. Blanchard, G.K. Stearman. Ion products and solid-phase activity to describe phosphate sorption by soils. *Soil Sci. Soc. Am. J.* 48:1253–1258, 1984.

288. G.S.P. Ritchie. The chemical behaviour of aluminum, hydrogen and manganese in acid soils. In: A.D. Robson, ed. *Soil Acidity and Plant Growth*. Marrickville, Australia: Academic Press Australia, 1989, pp. 1–60.

289. D.K. Nordstrom, H.M. May. Aqueous equilibrium data for mononuclear aluminum species. In: G. Sposito, ed. *The Environmental Chemistry of Aluminum*. Boca Raton, FL: CRC Press, 1996, pp. 39–80.

290. R.J. Bartlett, D.C. Riego. Effect of chelation on the toxicity of aluminum. *Plant Soil* 37:419–423, 1972.

291. N.V. Hue, I. Amien. Aluminum detoxification with green manures. *Commun. Soil Sci. Plant Anal.* 20:1499–1511, 1989.

292. N.V. Hue. Correcting soil acidity of a highly weathered Ultisol with chicken manure and sewage sludge. *Commun. Soil Sci. Plant Anal.* 23:241–264, 1992.

293. F. Adams, Z.F. Lund. Effect of chemical activity of soil solution aluminum in cotton root penetration of acid subsoils. *Soil Sci.* 101:193–198, 1966.

294. R.C. Bruce, L.A. Warrell, L.C. Bell, D.G. Edwards. Chemical attributes of some Queensland acid soils. I. Solid and solution phase compositions. *Aust. J. Soil Res.* 27:333–351, 1989.

295. P.M. Bertsch, D.R. Parker. Aqueous polynuclear aluminum species. In: G. Sposito, ed. *The Environmental Chemistry of Aluminum*. Boca Raton, FL: Lewis Publisher, 1996, pp. 117–168.

296. D.R. Parker, T.B. Kinraide, L.W. Zelazny. On the phytotoxicity of polynuclear hydroxy aluminum complexes. *Soil Sci. Soc. Am. J.* 53:789–796, 1989.

297. P.M. Bertsch. Conditions for Al_{13} polymer formation in partially neutralized aluminum solutions. *Soil Sci. Soc. Am. J.* 51:825–828, 1987.

298. P.L. Larsen. Dynamics of Amelioration of Aluminium Toxicity and Base Deficiency by Organic Materials in Highly Weathered Acid Soils. PhD dissertation, University of Queensland, Queensland, Australia, 2002.

299. P.M. Bertsch, P.R. Bloom. Aluminum. In: D.L. Sparks, ed. *Methods of Soil Analysis, Part 3: Chemical Methods*. Madison, WI: Soil Science Society of America, 1996, pp. 517–574.

300. P.A. Sanchez. Management of acid soils in the humid tropics of Latin America. In: *Management of Acid Tropical Soils for Sustainable Agriculture. IBSRAM Proceedings No. 2*, 1987, pp. 63–107.

301. P.R. Bloom, M.B. McBride, R.M. Weaver. Aluminum organic matter in acid soils. Salt-extractable aluminum. *Soil Sci. Soc. Am. J.* 43:813–815, 1979.

302. G. Amedee, M. Peech. The significance of KCl extractable Al (III) as an index to lime requirement of soils of the humid tropics. *Soil Sci.* 121:227–233, 1976.

303. R.J. Wright, V.C. Baligar, J.L. Ahlrichs. The influence of extractable and soil solution aluminum on root growth of wheat seedlings. *Soil Sci.* 148:293–302, 1989.

304. L.M. Shuman. Comparison of exchangeable Al, extractable Al, and Al in soil fractions. *Can. J. Soil Sci.* 70:263–275, 1990.

305. W.L. Hargrove, G.W. Thomas. Extraction of aluminum from aluminum-organic matter in relation to titratable acidity. *Soil Sci. Soc. Am. J.* 48:1458–1460, 1984.

306. K.M. Oates, E.J. Kamprath. Soil acidity and liming: I. Effect of the extracting solution cation and pH on the removal of aluminum from acid soils. *Soil Sci. Soc. Am. J.* 47:686–689, 1983.

307. E.J. Kamprath. Crop response to lime on soils in the tropics. In: F. Adams, ed. *Soil Acidity and Liming*, 2nd ed. Madison, WI: Soil Science Society of America, 1984, pp. 349–368.

308. R.W. Pearson, R. Perez-Escobar, F. Abruna, Z.F. Lund, E.J. Brenes. Comparative responses of three crop species to liming several soils of the southeastern United States and of Puerto Rico. *J. Agric. Univ. PR* 61:361–382, 1977.

309. C.E. Evans, E.J. Kamprath. Lime response as related to percent Al saturation, solution Al, and organic matter content. *Soil Sci. Soc. Am. Proc.* 34:893–896, 1970.

310. J.B. Sartain, E.J. Kamprath. Effect of liming a high Al-saturated soil on the top and root growth and soybean nodulation. *Agron. J.* 67:507–510, 1975.

311. Z.Z. Zakaria, V.N. Schroder, K.J. Boote. Soybean response to calcium and phosphorus under aluminum saturation. *Proc. Soil Crop Sci. Soc. Fla.* 36:178–181, 1977.

312. T.J. Smyth, M.S. Cravo. Aluminum and calcium constraints to continuous crop production in a Brazilian Amazon Oxisol. *Agron. J.* 84:843–850, 1992.

313. W.W. Moschler, G.D. Jones, G.W. Thomas. Lime and soil acidity effects on alfalfa growth in a Red-Yellow Podzolic soil. *Soil Sci. Soc. Am. Proc.* 24:507–509, 1960.

314. G.J. Shoop, C.R. Brooks, R.E. Blaser, G.W. Thomas. Differential responses of grasses and legumes to liming and phosphorus fertilization. *Agron. J.* 53:111–115, 1961.

315. E.J. Kamprath. Exchangeable aluminum as a criterion for liming leached mineral soils. *Soil Sci. Soc. Am. Proc.* 34:252–254, 1970.

316. M.P.W. Farina, M.E. Sumner, C.O. Plank, W.S. Letzsch. Exchangeable aluminum and pH as indicators of lime requirement for corn. *Soil Sci. Soc. Am. J.* 44:1036–1041, 1980.

317. F. Adams, B.L. Moore. Chemical factors affecting root growth in subsoil horizons of Coastal Plain soils. *Soil Sci. Soc. Am. J.* 47:99–102, 1983.

318. F. Adams, C. Burmester, N.V. Hue, F.L. Long. Comparison of column-displacement and centrifuge methods for obtaining soil solution. *Soil Sci. Soc. Am. J.* 44:733–735, 1980.

319. P.R. Bloom, M.S. Erich. The quantitation of aqueous aluminum. In: G. Sposito, ed. *The Environmental Chemistry of Aluminum*, 2nd ed. Boca Raton, FL: Lewis Publisher, 1996, pp. 1–38.

320. D.R. Parker, R.L. Chaney, W.A. Norvel. Chemical equilibrium models: Applications to plant nutrition research. In: R.H. Loeppert, ed. *Chemical Equilibrium and Reaction Models*. Madison, WI: Soil Science Society of America Spec Publ 42, 1995, pp. 253–269.

321. J.K. Jallah, T.J. Smyth. Assessment of rhizotoxic aluminum in soil solutions by computer and chromogenic speciation. *Commun. Soil Sci. Plant Anal.* 29:37–50, 1998.

322. B.R. James, C.J. Clark, S.J. Riha. An 8-hydroxyquinoline method for labile and total aluminum in soil extracts. *Soil Sci. Soc. Am. J.* 47:893–897, 1983.

323. W.K. Dougan, A.L. Wilson. The absorptiometric determination of aluminum in water: A comparison of some chromogenic reagents and the development of an improved method. *Analyst* 99:413–430, 1974.

324. D.C. McAvoy, R.C. Santore, J.D. Shosa, C.T. Driscoll. Comparison between pyrocatechol and 8-hydroxyquinoline procedures for determining aluminum fractions. *Soil Sci. Soc. Am. J.* 56:449–455, 1992.

325. N.W. Menzies, G.L. Kerven, L.C. Bell, D.G. Edwards. Determination of total soluble aluminium in soil solution using pyrocatechol violet, lanthanum and iron to discriminate against micro-particulates and organic ligands. *Commun. Soil. Sci. Plant Anal.* 23:2525–2545, 1992.

326. S.C. Hodges. Aluminum speciation: A comparison of five methods. *Soil Sci. Soc. Am. J.* 51:57–64, 1987.

327. P.M. Bertsch, W.J. Layton, R.I. Barnhisel. Speciation of hydroxy-Al solutions by wet chemical and Al-27 NMR methods. *Soil Sci. Soc. Am. J.* 50:1449–1454, 1986.

328. A.J.M. Baker. Accumulators and excluders — Strategies in response of plants to heavy metals. *J. Plant Nutr.* 3:643–654, 1981.

329. C.D. Foy, A.M. Sadeghi, J.C. Ritchie, D.T. Krizek, J.R. Davis, W.D. Kemper. Aluminum toxicity and high bulk density: Role in limiting shoot and root growth of selected aluminum indicator plants and eastern gammagrass in an acid soil. *J. Plant Nutr.* 22:1551–1566, 1999.

330. E.E. Smith. *Aluminum Compounds in Food*. New York: Hoeber, 1928.

331. A.L. Daniels, M.K. Hutton. Mineral deficiencies of milk as shown by growth and fertility of white rats. *J. Biol. Chem.* 63:143–150, 1925.

332. NRC (National Research Council). *Mineral Tolerances of Domestic Animals*. Washington DC: National Academy of Sciences, 1980.

333. P.C. Sternweis, A.G. Gilman. Aluminum: A requirement for activation of the regulatory component of adenylate cyclase by fluoride. *Proc. Natl. Acad. Sci. USA* 79:4888–4891, 1982.

334. L.R. McDowell. *Minerals in Animal and Human Nutrition*. San Diego: Academic Press, 1992, pp. 355–357.

335. W.R. Ewing. *Poultry Nutrition*, 5th ed. Pasadena:Hoffman-La Roche, 1963, pp. 691–693.

336. C.A. Rosa, R. Miazzo, C. Magnoli, M. Salvano, S.M. Chiacchiera, S. Ferrero, M. Saenz, E.C. Carvallo, A. Dalcero. Evaluation of the efficacy of bentonite from the south of Argentina to ameliorate the toxic effects of aflatoxin in broilers. *Poult. Sci.* 80:139–144, 2001.

337. T.C. Schell, M.D. Lindemann, E.T. Kornegay, D.J. Blodgett. Effects of feeding aflatoxin-contaminated diets with and without clay to weanling and growing pigs on performance, liver function and mineral metabolism. *J. Anim. Sci.* 71:1209–1218, 1993.

338. E.E. Smith, T.D. Phillips, J.A. Ellis, R.B. Harvey, L.F. Kubena, J. Thompson, G. Newton. Dietary hydrated sodium calcium aluminosilicate reduction of aflatoxin M1 residue in dairy goat milk and effect on milk production and components. *J. Anim. Sci.* 72:677–682, 1994.

339. H.W. Rabon, Jr., D.A. Roland, Sr., M.M. Bryant, R.C. Smith, D.G. Barnes, S.M. Laurent. Absorption of silicon and aluminum by hens fed sodium zeolite A with various levels of dietary cholicalciferol. *Poult. Sci.* 74:352–369, 1995.

340. L.A. Wisser, B.S. Heinrichs, R.M. Leach. Effect of aluminum on performance and mineral metabolism in young chicks and laying hens. *J. Nutr.* 120:493–498, 1990.

341. J. Moshtaghian, C.M. Prsons, R.W. Leeper, P.C. Harrison, K.W. Koelkebeck. Effect of sodium aluminosilicate on phosphorus utilization by chicks and laying hens. *Poult. Sci.* 70:955–962, 1991.

342. T. Thilsing-Hansen, R.J. Jorgensen, J.M. Enemark, T. Larsen. The effect of zeolite a supplementation in the dry period on periparturient calcium, phosphorus, and magnesium homeostasis. *J. Dairy Sci.* 85:1855–1862, 2002.

343. P.H.B. Hahn, W. Guenter. Effect of dietary fluoride and aluminum on laying hen performance and fluoride concentration in blood, soft tissue, bone and egg. *Poult. Sci.* 65:1343–1349, 1986.

344. L. Li. The biochemistry and physiology of metallic fluoride: action, mechanism and implications. *Crit. Rev. Oral Biol. Med.* 14:100–114, 2003.

345. A. Shakoor, P.K. Gupta, Y.P. Singh, M. Kataria. Beneficial effects of aluminum on the progression of lead-induced nephropathy in rats. *Pharmacol. Toxicol.* 87:258–260, 2000.

346. J.E. Line. Campylobacter and Salmonella populations associated with chickens raised on acidified litter. *Poult. Sci.* 81:1473–1477, 2002.

347. D.R. Smith, P.A. Moore, Jr., C.L. Griffis, T.C. Daniel, D.R. Edwards, D. Boothe. Effect of alum and aluminum chloride on phosphorous runoff from swine manure. *J. Environ. Qual.* 30:992–1008, 2001.

348. J.T. Sims, N.J. Luka-McCaffetry. On-farm evaluation of aluminum sulfate (alum) as poultry litter amendment: effect on litter properties. *J. Environ. Qual.* 31:2066–2073, 2002.

349. A.M. Lefcourt, J.J. Meisinger. Effect of adding alum or zeolite to dairy slurry on ammonia volatilization composition. *J. Daily Sci.* 8:1814–1821, 2001.

350. D. Peak, J.T. sims, D.L. Sparks. Solid-state speciation of natural and alum-amended poultry litter using XANES spectroscopy. *Environ. Sci. Technol.* 36:4253–4261, 2002.

351. E.B. Welch, G.D. Cooke. Effectiveness and longevity of phosphorous inactivation with alum. *Lake and Reservoir Management* 15:5–27, 1999.

352. D.W. Sparling, T.P. Lowe. Environmental hazards of aluminum to plants, invertebrates, fish, and wildlife. In: G. Ware, ed. *Reviews of Environmental Contamination and Toxicology.* New York: Springer, Vol. 145, 1996. pp. 1–127.

353. A. Soleng, A.B. Poleo, N.E. Alstad, T.A. Bakke. Aqueous aluminum eliminates *Gyrodactylus salaries* (Platyhelminthes, Monogenea) infections in atlantic salmon. *Parasitology* 119:19–25, 1999.

354. ATSDR (Agency for Toxic Substances and Disease Registry). *Toxicology Profile for Aluminum.* Atlanta: Public Health Service, U.S. Department of Health and Human Services, 1999, pp. 1–368.

355. M. Schintu, P. Meloni, A. Contu. Aluminum fractions in drinking water from reservoirs. *Ecotoxicol. Environ. Safety* 46:29–33, 2000.

356. K.N. Exall, G.W. vanLoon. Effect of raw water conditions on solution-state aluminum speciation during coagulant dilution. *Water Res.* 37:3341–3350, 2003.

357. S. Malhotra, D.N. Kulkarni, S.P. Pande. Effectiveness of poly aluminum chloride (PAC) vis-à-vis alum in the removal of fluorides and heavy metals. *J. Environ. Sci. Health. Part A Environ. Sci. Eng. Toxic Hazardous Sub. Con.* 32:2563–2574, 1997.

358. P. Nayak. Aluminum: Impacts and disease. *Environ. Res. Sec.* A 89:101–115, 2002.

359. R.A. Yokel. Aluminum. In: E. Merian, M. Anke, M. Inhat, M. Stoeppler, eds. *Elements and Their Compounds in the Environment*, 2nd ed. Weinheim, Germany: Wiley-VCH Verlag, 2004, pp. 635–658.

360. C. Exley, ed. *Aluminum and Alzheimer's Disease: The Science that Describes the Link.* Amsterdam: Elsevier, 2001.

361. C.M. Neville, P.G.C. Campbell. Possible mechanism of aluminum toxicity in a dilute acidic environment to fingerlings and older life stages of salmonids. *Water, Air and Soil Pollut.* 42:311–327, 1998.

362. R.C. Playle, C.M. Wood. Mechanism of aluminum extraction and accumulation at the gills of rainbow trout, *Oncorhynchus mykiss* (Walbaum) in acidic softwater. *J. Fish Biol.* 38:731–805, 1991.

363. J.D. Birchall. The role of silicon in biology. *Chemistry in Britain* 26:141–144, 1990.

364. D. Kadar, J. Slanki, R. Jugdaohsingh, J.J. Powell, C.R. McCrohan, K.N. White. Avoidance responses to aluminum in the freshwater bivalve *Anodonta cygnea. Aquat. Toxicol.* 55:137–148, 2001.

365. N.E.I. Nyholm. Evidence of involvement of aluminum in causation of defective formation of eggshells and of impaired breeding in wild passerine birds. *Environ. Res.* 26:363–371, 1981.

366. M.C. Capdevielle, L.E. Hart, J. Goff, C.G. Scanes. Aluminum and acid effects on calcium and phosphorus metabolism in young growing chickens (*Gallus gallus domesticus*) and mallard ducks (*Anas playtrhynchos*). *Arch. Environ. Contam. Toxicol.* 35:82–88, 1998.

367. S.M. Palmer, C.T. Driscoll. Acidic deposition: Decline in mobilization of toxic aluminum. *Nature* 417:242–243, 2002.

368. N.L. Storer, T.S. Nelson. The effect of various aluminum compounds on chick performance. *Poult. Sci.* 47:244–247, 1968.

369. N.A. Crowe, M.N. Neathery, W.J. Miller, L.A. Muse, C.T. Crowe, J.L. Varnadoe, D.M. Blackmon. Influence of high dietary aluminum on performance and phosphorus bioavailability in dairy calves. *J. Dairy Sci.* 73:808–818, 1990.

370. V.G. Allen, J.P. Fontenot, S.H. Rahnema. Influence of aluminum citrate and citric acid on mineral metabolism in wether sheep. *J. Anim. Sci.* 68:2496–2505, 1990.

371. C.M. Garcia-Bojalil, G.B. Ammerman, P.R. Henry, R.C. Littell, W.G. Blue. Effects of dietary phosphorus, soil ingestion and dietary intake level on performance, phosphorus utilization and serum and alimentary tract mineral concentrations in lambs. *J. Anim. Sci.* 66:1508–1519, 1998.

372. J.P. Fontenot, V.G. Allen, G.E. Bunce, J.P. Goff. Factors influencing magnesium absorption and metabolism in ruminants. *J. Anim. Sci.* 67:3445–3455, 1989.

373. R.M. Leach, Jr., B.S. Heinrichs, J. Burdette. Broiler chicks fed low calcium diets. 1. Influence of zeolite on growth rate and parameters of bone metabolism. *Poult. Sci.* 69:1539–1543, 1990.

374. K.L. Watkins, L.L. Southern. Effect of dietary sodium zeolite A and graded levels of calcium and phosphorus on growth, plasma, and tibia characteristics of chicks. *Poult. Sci.* 71:1048–1058, 1992.

375. H.M. Edwards, Jr., M.A. Elliot, S. Sooncharernying. Effect of dietary calcium on tibiasl dyschondrophasia. Interaction with light, cholecalciferol, 1,25-dihydroxycholecalciferol, protein, and synthetic zeolite. *Poult. Sci.* 71:2041–2055, 1992.

376. A.S. Hussein, A.H. Cantor, A.J. Pescatore, T.H. Johnson. Effect of dietary aluminum and vitamin D interaction on growth and calcium and phosphorus metabolism on broiler chicks. *Poult. Sci.* 72:306–309, 1993.

377. T.B. Drueke. Intestinal absorption of aluminum in renal failure. *Nephrol. Dial. Transplant* 17 (suppl. 2):13–16, 2002.

378. Z. Deng, C. Coudray, L. Gouzoux, A. Mazur, Y. Rayssiguier, D. Pepin. Effects of acute and chronic coingestion of AlCl3 with citrate or polyphenolic acids on tissue retention and distribution of aluminum in rats. *Biol. Trace Elem. Res.* 76:245–256, 2000.

379. J.P. Knochel. Phophorus. In: M. Shils, J.A. Olson, M. Shike, A.C. Ross. *Modern Nutrition in Health and Disease*. Baltimore:Williams & Wilkins, 1999, pp. 157–168.

380. T. Kiss, M. Hollosi. The interaction of aluminum with peptides and proteins. In: C. Exley, ed. *Aluminum and Alzheimer's Disease: The Science that Describes the Link*. Amsterdam: Elsevier, 2001, pp. 361–392.

381. B. Solomon. Calmodulin, Aluminum and alzheimer's disease. In: C. Exley, ed. *Aluminum and Alzheimer's Disease: The Science that Describes the Link*. Amsterdam: Elsevier, 2001, pp. 393–410.

382. W.R. Mundy, T.J. Shafer. Aluminum-induced alteration of phosphoinositide and calcium signaling. In: C. Exley, ed. *Aluminum and Alzheimer's Disease: The Science that Describes the Link*. Amsterdam: Elsevier, 2001, pp. 345–360.

383. W.J. Lukiw. Aluminum and gene transcription in the mammalian central nervous system — implications for Alzheimer's disease. In: C. Exley, ed. *Aluminum and Alzheimer's Disease: The Science that Describes the Link*. Amsterdam: Elsevier, 2001, pp. 147–168.

384. A. Nesse, G. Garbossa. Aluminum toxicity in erythropoiesis. Mechanisms related to cellular dysfunction in Alzheimer's disease. In: C. Exley, ed. *Aluminum and Alzheimer's Disease: The Science that Describes the Link*. Amsterdam: Elsevier, 2001, pp. 261–278.

385. C. Exley. The pro-oxidant activity of aluminum. *Free Radic. Biol. Med.* 36:380–387, 2004.

386. R.J. Ward, R.R. Crichton. Iron homeostasis and aluminum toxicity. In: C. Exley, ed. *Aluminum and Alzheimer's Disease: The Science that Describes the Link*. Amsterdam: Elsevier, 2001, pp. 293–310.

387. D. Pratico, K. Uryu, S. Sung, S. Tang, J.Q. Trojanowski, V.M. Lee. Aluminum modulates brain amyloidosis through oxidative stress in APP transgenic mice. *J. FASEB* 16:1138–1140, 2002.

388. M.G. Abubakar, A. Taylor, G.A. Ferns. Aluminum administration is associated with enhanced hepatic oxidant stress that may be offset by dietary vitamin E in the rat. *Int. J. Exp. Pathol.* 84:49–54, 2003.

389. T.C. Hutchinson, L. Bozic, G. Munoz-Vega. Responses of five species of conifer seedlings to aluminum stress. *Water Air Soil Pollut.* 31:283–294, 1986.

390. C.S. Cronan, D.F. Grigal. Use of calcium/aluminum ratios as indicators of stress in forest ecosystems. *J. Environ. Qual.* 24:209–226, 1995.

391. J.M. Kelly, M. Schaedle, F.C. Thornton, J.D. Joslin. Sensitivity of tree seedlings to aluminum: II. Red oak, sugar maple, and European beech. *J. Environ. Qual.* 19:172–179, 1990.

392. J. Nowak, A.L. Friend. Aluminum sensitivity of loblolly pine and slash pine seedlings grown in solution culture. *Tree Physiol.* 15:605–609, 1995.

393. P.J. Ryan, S.P. Gessel, R.J. Zasoski. Acid tolerance of Pacific Northwest conifers in solution culture. II. Effect of varying aluminum concentration at constant pH. *Plant Soil* 96:259–272, 1986.

394. F.C. Thornton, M. Schaedle, D.J. Raynal. Effect of aluminum on the growth of sugar maple in solution culture. *Can. J. For. Res.* 16: 892–896, 1986.

395. T.J. Smalley, F.T. Lasseigne, H.A. Mills, G.G. Hussey. Effect of aluminum on growth and chemical composition of marigolds. *J. Plant Nutr.* 16:1375–1384, 1993.

396. L. Simon, T.J. Smalley, J. Benton Jones, Jr., F.T. Lasseigne. Aluminum toxicity in tomato. Part 1. Growth and mineral nutrition. *J. Plant Nutr.* 17:293–306, 1994.

397. C.S. Andrew, A.D. Johnson, R.L. Sandland. Effect of aluminium on the growth and chemical composition of some tropical and temperate pasture legumes. *Aust. J. Agric. Res.* 24:325–339, 1973.

398. L.E. DeWald, E.I. Sucoff, T. Ohno, C.A. Buschena. Response of northern red oak (*Quercus rubra*) seedlings to soil solution aluminum. *Can. J. For. Res.* 20:331–336, 1990.

399. R.B. Clark. Effect of aluminum on growth and mineral elements of Al-tolerant and Al-intolerant corn. *Plant Soil* 47:653–662, 1977.

400. T.V. Hai, T.T. Nga, H. Laudelout. Effect of aluminum on the mineral nutrition of rice. *Plant Soil* 114:173–185, 1989.
401. J.D. Wolt. *Soil Solution Chemistry*. New York: Wiley, 1994, pp. 220–245.
402. H.J. Van Praag, F. Weissen. Aluminum effects on spruce and beech seedlings. I. Preliminary observations on plant and soil. *Plant Soil* 83:331–338, 1985.

17 Cobalt

Geeta Talukder
Vivekananda Institute of Medical Sciences, Kolkata, India

Archana Sharma
University of Calcutta, Kolkata, India

CONTENTS

17.1 Introduction 500
17.2 Distribution 500
    17.2.1 Microorganisms and Lower Plants 500
        17.2.1.1 Algae 500
        17.2.1.2 Fungi 501
        17.2.1.3 Moss 501
    17.2.2 Higher Plants 501
17.3 Absorption...502
17.4 Uptake and Transport 502
    17.4.1 Absorption as Related to Properties of Plants 502
    17.4.2 Absorption as Related to Properties of Soil 503
    17.4.3 Accumulation as Related to the Rhizosphere 503
17.5 Cobalt Metabolism in Plants 504
17.6 Effect of Cobalt in Plants on Animals 505
17.7 Interaction of Cobalt with Metals and Other Chemicals in Mineral Metabolism 505
    17.7.1 Iron 506
    17.7.2 Zinc 506
    17.7.3 Cadmium 506
    17.7.4 Copper 506
    17.7.5 Manganese 507
    17.7.6 Chromium and Tin 507
    17.7.7 Magnesium 507
    17.7.8 Sulfur 507
    17.7.9 Nickel 507
    17.7.10 Cyanide 507
17.8 Beneficial Effects of Cobalt on Plants 507
    17.8.1 Senescence 507
    17.8.2 Drought Resistance 507
    17.8.3 Alkaloid Accumulation 507
    17.8.4 Vase Life 508
    17.8.5 Biocidal and Antifungal Activity 508
    17.8.6 Ethylene Biosynthesis 508

17.8.7   Nitrogen Fixation 508
17.9 Cobalt Tolerance by Plants 508
    17.9.1   Algae 508
    17.9.2   Fungi 509
    17.9.3   Higher Plants 509
References 509

17.1   INTRODUCTION

Cobalt has long been known to be a micronutrient for animals, including human beings, where it is a constituent of vitamin B_{12} (1). However, its presence and function has not been recorded to the same extent in higher plants as in animals, leading to the suggestion that vegetarians and herbivorous animals need to ingest extra cobalt or vitamin B_{12} in diets to prevent deficiency. Vitamin B_{12} is synthesized in some bacteria, but not in animals and plants (1). Intestinal absorption and subsequent plasma transport of vitamin B_{12} are mediated by specific vitamin B_{12} proteins and their receptors in mammals. Vitamin B_{12}, taken up by the cells, is converted enzymatically into methyl and adenosyl vitamin B_{12}, which function as coenzymes. Feeding trials of cattle (*Bos taurus* L.), which also suffer from vitamin B_{12} deficiency, show that the normal diet is deficient in cobalt to the extent that supplemental provision of the element can improve their performance, something that could also be achieved by feeding them feedstuffs grown in cobalt-rich soil (2).

The only physiological role so far definitely attributed to cobalt in higher plants has been in nitrogen fixation by leguminous plants (3).

17.2   DISTRIBUTION

17.2.1   MICROORGANISMS AND LOWER PLANTS

17.2.1.1   Algae

Cobalt is essential for many microorganisms including cyanobacteria (blue–green algae). It forms part of cobalamin, a component of several enzymes in nitrogen-fixing microorganisms, whether free-living or in symbiosis. It is required for symbiotic nitrogen fixation by the root nodule bacteria of legumes (3). Soybeans grown with $0.1 \mu g L^{-1}$ cobalt with atmospheric nitrogen and no mineral nitrogen showed rapid nitrogen fixation and growth (4). Cobalt is distributed widely in algae, including microalgae, *Chlorella*, *Spirulina*, *Cytseira barbera*, and *Ascophyllum nodosum*. Alginates, such as fucoiden, in the cell wall play an important role in binding cobalt in the cell-wall structure (5,6).

Bioaccumulation of heavy metals in aquatic macrophytes growing in streams and ponds around slag dumps has led to high levels of cobalt (7). Certain marine species such as diatoms (*Septifer virgatus* Wiegman) and brown algae *Sargassum horneri* (Turner) and *S. thunbergii* (Kuntze) from the Japanese coast act as bioindicators of cobalt (8). Accumulation has been shown to be controlled by salinity of the medium with bladder wrack (brown alga, *Fucus vesiculosus* L.) (9).

The cell walls of plants, including those of algae, have the capacity to bind metals at negatively charged sites. The wild type of *Chlamydomonas reinhardtii* Dangeard, owing to the presence of its cell wall, was more tolerant to metals such as cobalt, copper, cadmium, and nickel, than the wall-less variant (10). When exposed to metals, singly in solutions for 24 h, cells of both strains accumulated the metals. Absorbed metals not removed by chelation with EDTA–$CaCl_2$ wash were considered strongly bound. Cobalt and nickel were present in significantly higher amounts loosely bound to the walled organism than in the wall-less ones. It was concluded that metal ions were affected by the chelating molecules in walled algae, which limited the capacity of the metal to penetrate the cell. Thus, algae appear to contain a complex mechanism involving internal and external detoxification of metal ions (10).

In a flow-through wetland treatment system to treat coal combustion leachates from an electrical power system using cattails (*Typha latifolia* L.), cobalt and nickel in water decreased by an average of 39 and 47% in the first year and 98 and 63% in the second year, respectively. Plants took up 0.19% of the cobalt salts per year. Submerged *Chara* (a freshwater microalga), however, took up 2.75% of the salts, and considerably higher concentrations of metals were associated with cattail roots than shoots (11).

17.2.1.2 Fungi

In fungi, cobalt accumulates by two processes. The essential process is a metabolically independent one presumably involving the cell surface. Accumulation may reach 400 mg g^{-1} of yeast and is rapid in *Neurospora crassa* Shear & BO Dodge (12,13).

In the next step, which is metabolism dependent, progressive uptake of large amounts of cations takes place. Two potassium ions are released for each Co^{2+} ion taken up in freshly prepared yeast-cell suspensions. The Co^{2+} appears to accumulate via a cation-uptake system. Its uptake is specifically related to the ionic radius of the cation (14). Accumulated cobalt is transported (at the rate of 40 μg h^{-1} 100 mg^{-1} dry weight of *N. crassa*) mainly into the intercellular space and vacuoles (13,15). Acidity and temperature of media are factors involved in Co^{2+} uptake and transport. In *N. crassa,* Mg^{2+} inhibits Co^{2+} uptake and transport, suggesting that the processes of the two cations are interrelated. In yeast cells exposed to elevated concentrations of cobalt, uptake is suppressed, and intercellular distribution is altered (15).

Yeast mitochondria passively accumulate Co^{2+} in levels linearly proportional to its concentration in the medium. The density of mitochondria is slightly increased and their appearance is altered, based on observations with electron microscopy (16). The more dense mitochondria are exchanged by hyphal fusion in the fully compatible common A and common AB matings of tetrapolar basidiomycetes *Schizophyllum commune* Fries, but not in the common B matings (17). Toxicity and the barrier effect of the cell wall inhibit surface binding of Co^{2+}. As a result, isolated protoplasts from yeast-like cells of hyphae and chlamydospores of *Aureobasidium pollulans* were more sensitive to intracellular cobalt uptake than intact cells and chlamydospores (18).

17.2.1.3 Moss

The absorption and retention of heavy metals in the woodland moss *Hylocomium splendens* Hedw followed the order of Cu, Pb>Ni>Co>Zn, and Mn within a wide range of concentrations and was independent of the addition of the ions (19).

17.2.2 Higher Plants

Cobalt is not known to be definitely essential for higher plants. Vitamin B$_{12}$ is neither produced nor absorbed by higher plants. It is synthesized by soil bacteria, intestinal microbes, and algae. In naturally cobalt-rich areas, cobalt accumulates in plants in a species-specific manner. Plants such as astragalus (*Astragalus* spp. L.) may accumulate from 2 or 3 to 100 mg kg^{-1} dried plant mass. Cobalt occurs in a high concentration in the style and stigma of *Lilium longifolium* Thunb. It was not detected in the flowers of green beans (*Phaseolus sativus* L.) and radishes (*Raphanus sativus* L.) though the leaves of the latter contain it. It was shown to occur in high amounts in leafy plants such as lettuce (*Lactuca sativa* L.), cabbage (*Brassica oleracea* var. *capitata* L.), and spinach (*Spinacea oleracea* L.) (above 0.6 ppm) by Kloke (20). Forage plants contain 0.6 to 3.5 mg Co kg^{-1} and cereals 2.2 mg kg^{-1} (21). Rice (*Oryza sativa* L.) contains 0.02 to 0.150 mg kg^{-1} plant mass (22).

Cobalt chloride markedly increases elongation of etiolated pea stems when supplied with indole acetic acid (IAA) and sucrose, but elongation is inhibited by cobalt acetate. Cobalt in the form of vitamin B$_{12}$ is necessary for the growth of excised tumor tissue from spruce (*Picea glauca* Voss.) cultured *in vitro*. It increases the apparent rate of synthesis of peroxides and prevents the peroxidative destruction of IAA. It counteracts the inhibition by dinitrophenol (DNP) in oxidative phosphoryla-tion and reduces activity of ATPase and is known to be an activator of plant enzymes such as car-boxylases and peptidases (4). The Co^{2+} ion is also an inhibitor of the ethylene biosynthesis pathway, blocking the conversion of 1-amino-cyclopropane-l-carboxylic acid (ACC) (23).

17.3 ABSORPTION

Kinetic studies of cobalt absorption by excised roots of barley (*Hordeum vulgare* L.) exhibited a Q_{10} of 2.2 in a concentration range of 1 to100 μM $CoCl_2$. It has been suggested that a number of car-rier sites are available, which are concentration dependent (24). Entry of divalent cations in the roots of maize is accompanied by a decrease in the pH of the incubation media and of the cell sap and also a decrease in the malate content (25). The uptake by different species probably depends on the various physiological and biological needs of the species (26,27).

  Accumulation of cobalt by forage plants has been studied in wetlands, grasslands, and forests close to landfills and mines (11,28,29). Irrigation with cobalt-rich water in meadows has shown high intake of cobalt, which was also demonstrated in the blood serum and plasma of bulls fed on the hay grown in the field (29). African buffalos *(Syncerus caffer* Sparrman*)* in the Kruger National Park (KNP) downwind of mining and refining of cobalt, copper, and manganese showed the pres-ence of the metals in liver in amounts related significantly to age and gender differences (30).

17.4 UPTAKE AND TRANSPORT

17.4.1 ABSORPTION AS RELATED TO PROPERTIES OF PLANTS

The molecular basis of metal transport through membranes has been studied by several workers. Korshunova et al. (31) reported that IRT 1, an *Arabidopsis thaliana* Heynh (mouse-ear cress) metal-ion transporter, could facilitate manganese absorption by a yeast mutant *Saccharomyces cerevisiae* Meyen ex E.C. Hansen strain defective in manganese uptake (smfl delta). The IRT 1 protein has been identified as a transporter for iron and manganese and is inhibited by cadmium and zinc. The IRT 1 cDNA also complements a Zn-uptake-deficient yeast mutant. It is therefore suggested that IRT 1 protein is a broad-range metal-ion transporter in plants (31).

  Macfie and Welbourn (10) reviewed the function of cell wall as a barrier to the uptake of several metal ions in unicellular green algae. The cell walls of plants, including those of algae, have the capacity to bind metal ions in negatively charged sites. As mentioned above, the wild-type (walled) strain of the unicellular green alga *Chlamydomonas reinhardtii* Dangeard was more tolerant to cobalt than a wall-less mutant of the same species. In a study to determine if tolerance to metals was asso-ciated with an increased absorption, absorbed metal was defined as that fraction that could be removed with a solution of Na-EDTA and $CaCl_2$. The fraction that remained after the EDTA–$CaCl_2$ wash was considered strongly bound in the cell. When exposed to metals, singly, in solution for 24 h, cells of both strains accumulated the metals. Significantly higher concentrations of cobalt were in the loosely bound fraction of the walled strain than in the wall-less strain.

  Passive diffusion and active transport are involved in the passage of Co^{2+} through cortical cells. A comparison of concentration of Co^{2+} in the cytoplasm and vacuoles indicates that active trans-port occurs outward from the cytoplasm at the plasmalemma and also into the vacuoles at the tono-plast. Light–dark cycles play an important role in transport through the cortical cells of wheat (*Triticum aestivum* L.) (32). A small amount of absorption at a linear rate takes place in the water-free space, Donnan-free space, and cytoplasm in continuous light, whereas a complete inhibition of absorption occurs during the dark periods (32). In ryegrass (*Lolium perenne* L.), 15% of the Co^{2+} absorbed was transported to the shoot after 72 h. Absorption and transport of Co^{2+} markedly increased with increasing pH of the solution, but were not affected by water flux through the plants. With 0.1 μM Co^{2+} treatment, concentration of cobalt in the cytoplasm was regulated by an efflux pump at the plasmalemma and by an influx pump at the tonoplast. Stored cobalt in the vacuole was not available for transport (33).

Cobalt tends to accumulate in roots, but free Co^{2+} inhibited hydrolysis of Mg-ATP and protein transport in corn-root tonoplast vesicles (34). ATP complexes of Co^{2+} inhibited proton pumping, and the effect was modulated by free Co^{2+}. Free cations affected the structure of the lipid phase in the tonoplast membrane, possibly by interaction with a protogenic domain of the membrane through an indirect link mechanism (34).

Upward transport of cobalt is principally by the transpirational flow in the xylem (35). Usually, the shoot receives about 10% of the cobalt absorbed by the roots, most of which is stored in the cortical cell vacuoles and removed from the transport pathway (32). Distribution along the axis of the shoot decreases acropetally (36). Cobalt is bound to an organic compound of negative overall charge and molecular weight in the range of 1000 to 5000 and is transported through the sieve tubes of castor bean (*Ricinus communis* L.) (37). Excess cobalt leads to thick callose deposits on sieve plates of the phloem in white bean (*Phaseolus vulgaris* L.) seedlings, possibly reducing the transport of ^{14}C assimilates significantly (38).

The distribution of cobalt in specific organs indicates a decreasing concentration gradient from the root to the stem and from the leaf to the fruit. This gradient decreases from the root to the stem and leaves in bush beans (*Phaseolus vulgaris* L.) and *Chrysanthemum* (39,40). No strong gradient occurs from the stem to the leaves because of the low mobility of cobalt in plants, leading to its transport to leaves in only small amounts (41,42). In seeds of lupin (*Lupinus angustifolius* L.), concentrations of cobalt are higher in cotyledons and embryo than in seed coats (43). The distribution depends on the phase of development of the plant. At the early phase of growth of potatoes (*Solanum tuberosum* L.) on lixiviated (washed) black earth, large quantities of cobalt are accumulated in the leaves and stalks (44), whereas before flowering and during the ripening of beans (*Phaseolus vulgaris* L.), the largest amount is in the nodules. Plant organs contain cobalt in the following increasing order: root, leaves, seed, and stems (44). During flowering, a large amount shifts to the tuber of potato and, in the case of beans, to flowers, followed by nodules, roots, leaves, and stems. Movement is more rapid in a descending direction than in an ascending one (36). The cobalt content was observed to be higher in pickled cucumber (*Cucumis sativus* L.) than in young fresh fruit (45). In grains of lupins (*Lupinus* spp. L.) and wheat, the concentration varied with the amount of rainfall and soil types (46).

17.4.2 ABSORPTION AS RELATED TO PROPERTIES OF SOIL

Soil pH has a major effect on the uptake of cobalt, manganese, and nickel, which become more available to plants as the pH decreases. Increase in soil pH reduces the cobalt content of ryegrass (*Lolium* spp.) (47). Reducing conditions in poorly drained soils enhance the rate of weathering of ferromagnesian minerals, releasing cobalt, nickel, and vanadium (48). Liming decreased cobalt mobility in soil (49). The presence of humus facilitates cobalt accumulation in soil, but lowers its absorption by plants. Five percent humus has been shown to decrease cobalt content by one-half or two-thirds in cultures (50).

High manganese levels in soil inhibit accumulation of cobalt by plants (51). Manganese dioxides in soil have a high sorption capacity and accumulate a large amount of cobalt from the soil solution. Much of the cobalt in the soil is fixed in this way and is thus not available to plants (52). Water logging of the soil increases cobalt uptake in French bean (*Phaseolus vulgaris* L.) and maize (*Zea mays* L.) (53).

17.4.3 ACCUMULATION AS RELATED TO THE RHIZOSPHERE

Cobalt may be absorbed through the leaf in coniferous forests, but the majority is through the soil, especially in wetlands. The physicochemical status of transition metals such as cobalt in the rhizosphere is entirely different from that in the bulk soil. A microenvironment is created around the root system (e.g., wheat and maize), characterized by an accumulation of root-derived organic material with a gradual shift from ionic metal to higher-molecular weight forms such as cobalt, manganese, and zinc. These three metals are increasingly complexed throughout the growth period. Fallow soil has been shown to complex lower amounts (6.4%) of tracers (^{57}Co) than cropped soil, 61% for maize and 31% for wheat (54). Cobalt has a stimulatory effect on the microflora of tobacco (*Nicotiana tabacum* L.) rhizosphere, shown by an intensification of the immobilization of nitrogen and mineralization of phosphorus (55). Cobalt status in moist soil from the root zone of field-grown barley shows seasonal variation, being low in late winter and higher in spring and early summer. Discrete maxima are achieved frequently between May and early July, depending on the extent of the development of the growing crop and on seasonal influences. Increased concentration may result from the mobilization of the micronutrient from insoluble forms by biologically produced chelating ligands.

17.5 COBALT METABOLISM IN PLANTS

Interactions between cobalt and several essential enzymes have been demonstrated in plants and animals. Two metal-bound intermediates formed by Co^{2+} activate ribulose-1,5 bisphosphate carboxylase/oxygenase (EC 4.1.1.39). Studies by electron paramagnetic resonance (EPR) spectroscopy have shown the activity to be dependent on the concentration of ribulose 1,5 bisphosphate (23). This finding suggested that the enzyme–metal coordinated ribulose 1,5 bisphosphate and an enzyme–metal coordinated enediolate anion of it, where bound ribulose 1,5 bisphosphate appears first, constitute the two EPR detectable intermediates, respectively.

Ganson and Jensen (56) showed that the prime molecular target of glyphosate (*N*-[phosphonomethyl]glycine), a potent herbicide and antimicrobial agent, is known to be the shikimate-pathway enzyme 5-enol-pyruvylshikimate-3-phosphate synthetase. Inhibition by glyphosate of an earlier pathway enzyme that is located in the cytosol of higher plants, 3-deoxy-D-arabino-heptulosonate-7 phosphate synthase (DS-Co), has raised the possibility of dual enzyme targets *in vivo*. Since the observation that magnesium or manganese can replace cobalt as the divalent-metal activator of DS-Co, it has now been possible to show that the sensitivity of DS-Co to inhibition by glyphosate is obligately dependent on the presence of cobalt. Evidence for a cobalt(II):glyphosate complex with octahedral coordination was obtained through examination of the effect of glyphosate on the visible electronic spectrum of aqueous solutions of $CoCl_2$.

Two inhibition targets of cobalt and nickel were studied on oxidation–reduction enzymes of spinach (*Spinacia oleracea* L.) thylakoids. Compounds of complex ions and coordination compounds of cobalt and chromium were synthesized and characterized (57). Their chemical structures and the oxidation states of their metal centers remained unchanged in solution. Neither chromium(III) chloride ($CrCl_3$) nor hexamminecobalt(III) chloride [$Co(NH_3)_6Cl_3$] inhibited photosynthesis. Some other coordination compounds inhibited ATP synthesis and electron flow (basal phosphorylating, and uncoupled) behaving as Hill-reaction inhibitors, with the compounds targeting electron transport from photosystem II (P680 to plastoquinones, QA and QB, and cytochrome).

The final step in hydrocarbon biosynthesis involves the loss of cobalt from a fatty aldehyde (58). This decarbonylation is catalyzed by microsomes from *Botyrococcus braunii*. The purified enzyme releases nearly one mole of cobalt for each mole of hydrocarbon. Electron microprobe analysis revealed that the enzyme contains cobalt. Purification of the decarbonylase from *B. braunii* grown in $^{57}CoCl_2$ showed that ^{57}Co co-eluted with the decarbonylase. These results indicate that the enzyme contains cobalt that might be part of a Co-porphyrin, although a corrin structure (as in vitamin B_{12}) cannot be ruled out. These results strongly suggest that biosynthesis of hydrocarbons is effected by a microsomal Co-porphyrin-containing enzyme that catalyzes decarbonylation of aldehydes and, thus, reveals a biological function for cobalt in plants (58).

The role of hydrogen bonding in soybean (*Glycine max* Merr.) leghemoglobin was studied (59,60). Two spectroscopically distinct forms of oxycobaltous soybean leghemoglobin (oxyCoLb), acid and neutral, were identified by electron spin echo envelope modulation. In the acid form, a coupling to 2H was noted, indicating the presence of a hydrogen bond to bound oxygen. No coupled 2H occurred in the neutral form (60). The oxidation–reduction enzymes of spinach thylakoids are also affected by chromium and cobalt (23,57).

The copper chaperone for the superoxide dismutase (CCS) gene encodes a protein that is believed to deliver copper to Cu–Zn superoxide dismutase (CuZnSOD). The CCS proteins from different organisms share high sequence homology and consist of three distinct domains, a CuZnSOD-like central domain flanked by two domains, which contain putative metal-binding motifs. The Co^{2+}-binding properties of proteins from arabidopsis and tomato (*Lycopersicon esculentum* Mill.) were characterized by UV–visible and circular dichroism spectroscopies and were shown to bind one or two cobalt ions depending on the type of protein. The cobalt-binding site that was common in both proteins displayed spectroscopic characteristics of Co^{2+} bound to cysteine ligands (61).

The inhibition of photoreduction reactions by exogenous manganese chloride ($MnCl_2$) in Tris-treated photosystem II (PSII) membrane fragments has been used to probe for amino acids on the PSII reaction-center proteins, including the ones that provide ligands for binding manganese (62,63). Inhibition of photooxidation may involve two different types of high-affinity, manganese-binding components: (a) one that is specific for manganese, and (b) others that bind manganese, but may also bind additional divalent cations such as zinc and cobalt that are not photooxidized by PSII. Roles for cobalt or zinc in PSII have not been proposed, however.

17.6 EFFECT OF COBALT IN PLANTS ON ANIMALS

Cobalt uptake by plants allows its access to animals. Kosla (29) demonstrated the effect of irrigation of meadows with the water of the river Ner in Poland on the levels of iron, manganese, and cobalt in the soil and vegetation. Experiments were also carried out on young bulls (*Bos taurus* L.) fed with the hay grown on these meadows. The levels of iron and cobalt were determined in the blood plasma, and manganese level in the hair of the bulls. The irrigation caused an increase of the cobalt content in the soil, but had no effect on cobalt content in the plants or in the blood plasma of the bulls. Webb et al. (30) stated that animals may act as bioindicators for the pollution of soil, air, and water. To monitor changes over time, a baseline status should be established for a particular species in a particular area. The concentration of minerals in soil is a poor indicator of mineral accumulation by plants and availability to animals.

The chemical composition of the body tissue, particularly the liver, is a better reflection of the dietary status of domestic and wild animals. Normal values for copper, manganese, and cobalt in the liver have been established for cattle, but not for African buffalo. As part of the bovine-tuberculosis (BTB) monitoring program in the KNP in South Africa, 660 buffalo were culled. Livers were randomly sampled in buffered formalin for mineral analysis. The highest concentrations of copper in livers were measured in the northern and central parts of the KNP, which is downwind of mining and refining activities. Manganese, cobalt, and selenium levels in the liver samples indicated neither excess nor deficiency although there were some significant area, age, and gender differences. It was felt that these data could serve as a baseline reference for monitoring variations in the level and extent of mineral pollution on natural pastures close to mines and refineries. Cobalt is routinely added to cattle feed, and deficiency diseases are known. Of interest also are the possible effects of minor and trace elements in Indian herbal and medicinal preparations (64).

17.7 INTERACTION OF COBALT WITH METALS AND OTHER CHEMICALS IN MINERAL METABOLISM

The interaction of cobalt with other metals depends to a major extent on the concentration of the metals used. The cytotoxic and phytotoxic responses of a single metal or combinations are considered in terms of common periodic relations and physicochemical properties, including electronic structure, ion parameters (charge–size relations), and coordination. But, the relationships among toxicity, positions, and properties of these elements are very specific and complex (65). The mineral elements in plants as ions or as constituents or organic molecules are of importance in plant metabolism. Iron, copper, and zinc are prosthetic groups in certain plant enzymes. Magnesium, manganese, and cobalt may act as inhibitors or as activators. Cobalt may compete with ions in the biochemical reactions of several plants (66,67).

17.7.1 IRON

Many trace elements in high doses induce iron deficiency in plants (68). Combinations of increased cobalt and zinc in bush beans have led to iron deficiency (69). Excess metals accumulated in shoots, and especially in roots, reduce ion absorption and distribution in these organs, followed by the induction of chlorosis, decrease in catalase activity, and increase in nonreducing sugar concentration in barley (70,71). Supplying chelated iron ethylenediamine di(o-hydroxyphenyl) acetic acid [Fe-(EDDHA),] could not overcome these toxic effects in *Phaseolus* spp. L. (72). Simultaneous addition of cobalt and zinc to iron-stressed sugar beet (*Beta vulgaris* L.) resulted in preferential transport of cobalt into leaves followed by ready transport of both metals into the leaf symplasts within 48 h (73). A binuclear binding site for iron, zinc, and cobalt has been observed (74).

17.7.2 ZINC

Competitive absorption and mutual activation between zinc and cobalt during transport of one or the other element toward the part above the ground were recorded in pea (*Pisum sativum* L.) and wheat seedlings (75). Enrichment of fodder beet (*Beta vulgaris* L.) seeds before sowing with one of these cations lowers the content of the other in certain organs and tissues. It is apparently not the result of a simple antagonism of the given cations in the process of redistribution in certain organs and tissue, but is explained by a similar effect of cobalt and zinc as seen when the aldolase and carbonic anhydrase activities and intensity of the assimilators' transport are determined (76).

Cobalt tends to interact with zinc, especially in high doses, to affect nutrient accumulation (77). The antagonism is sometimes related to induced nutrient deficiency (69). In bush beans, however, cobalt suppressed to some extent the ability of high concentration of zinc to depress accumulation of potassium, calcium, and magnesium. The protective effect was stated to be the result of zinc depressing the leaf concentration of cobalt rather than the other nutrients (69). Substitution of Zn^{2+} by Co^{2+} reduces specificity of Zn^{2+} metalloenzyme acylamino-acid-amido hydrolase in *Aspergillus oryzae* Cohn (78).

17.7.3 CADMIUM

Combinations of elements may be toxic in plants when the individual ones are not (72). Trace elements usually give protective effects at low concentrations because some trace elements antagonize the uptake of others at relatively low levels. For example, trace elements in various combinations (Cu–Ni–Zn, Ni–Co–Zn–Cd, Cu–Ni–Co–Cd, Cu–Co–Zn–Cd, Cu–Ni–Zn–Cd, and Cu–Ni–Co–Zn–Cd) on growth of bush beans protected against the toxicity of cadmium. It was suggested that part of the protection could be due to cobalt suppressing the uptake of cadmium by roots. Other trace elements in turn suppressed the uptake of cobalt by roots (69). These five trace elements illustrated differential partitioning between roots and shoots (40). The binding of toxic concentration of cobalt in the cell wall of the filamentous fungus (*Cunninghamella blackesleeana* Lender) was totally inhibited and suppressed by trace elements (79).

17.7.4 COPPER

The biphasic mechanism involved in the uptake of copper by barley roots after 2 h was increased with 16 μM Co^{2+}, but after 24 h, a monophasic pattern developed with lower values of copper absorption, indicating an influence of Co^{2+} on the uptake site (80).

17.7.5   MANGANESE

Cobalt and zinc increased the accumulation of manganese in the shoots of bush beans grown for 3 weeks in a stimulated calcareous soil containing Yolo loam and 2% $CaCO_3$ (40).

17.7.6   CHROMIUM AND TIN

The inhibitory effects of chromium and tin on growth, uptake of NO_3^- and NH_4^+, nitrate reductase, and glutamine synthetase activity of the cyanobacterium (*Anabaena doliolum* Bharadwaja) was enhanced when nickel, cobalt, and zinc were used in combination with test metals in the growth medium in the following degree: Ni>Co>Zn (81).

17.7.7   MAGNESIUM

The activating effect of cobalt on Mg^{2+}-dependent activity of glutamine synthetase by the blue–green alga *Spirulina platensis* Geitler may be considered as an important effect. Its effect in maintaining the activity of the enzyme *in vivo* is independent of ATP (82).

17.7.8   SULFUR

The mold *Cunninghamella blackesleeana* Lendner, grown in the presence of toxic concentration of cobalt, showed elevated content of sulfur in the mycelia. Its cell wall contained higher concentrations of phosphate and chitosan, citrulline, and cystothionine as the main cell wall proteins (79).

17.7.9   NICKEL

In moss (*Timmiella anomala* Limpricht), nickel overcomes the inhibitory effect of cobalt on protonemal growth whereas cobalt reduces the same effect of nickel on bud number (83).

17.7.10   CYANIDE

Cyanide in soil was toxic to bush beans and also resulted in the increased uptake of the toxic elements such as copper, cobalt, nickel, aluminum, titanium, and, to a slight extent, iron. The phytotoxicity from cyanide or the metals led to increased transfer of sodium to the leaves and roots (40).

17.8   BENEFICIAL EFFECTS OF COBALT ON PLANTS

17.8.1   SENESCENCE

Senescence in lettuce leaf in the dark is retarded by cobalt, which acts by arresting the decline of chlorophyll, protein, RNA and, to a lesser extent, DNA. The activities of RNAase and protease, and tissue permeability were decreased, while the activity of catalase increased (84). Cobalt delays ageing and is used for keeping leaves fresh in vetch (*Vicia* spp.) (85). It is also used in keeping fruits such as apple fresh (86).

17.8.2   DROUGHT RESISTANCE

Presowing treatment of seeds with cobalt nitrate increased drought resistance of horse chestnut (*Aesculus hippocastanum* L.) from the Donets Basin in southeastern Europe (87).

17.8.3   ALKALOID ACCUMULATION

Alkaloid accumulation in medicinal plants such as downy thorn apple *Datura innoxia* Mill. (88), *Atropa caucasica* (89), belladonna *A. belladonna* L. (90), and horned poppy *Glaucium flavum* Crantz (91) is regulated by cobalt. It also increased rutin (11.6%) and cyanide (67%) levels in different species of buckwheat (*Fagopyrum sagittatum* Gilib., *F. tataricum* Gaertn., and *F. emargitatum*) (89,92).

17.8.4   VASE LIFE

Shelf and vase life of marigold (*Tagetes patula* L.), chrysanthemum (*Chrysanthemum* spp.), rose (*Rosa* spp.), and maidenhair fern (*Adiantum* spp.) is increased by cobalt. Cobalt also has a long-lasting effect in preserving apple (*Malus domestica* Borkh.). The fruits are kept fresh by cobalt application after picking (86,93–96).

17.8.5   BIOCIDAL AND ANTIFUNGAL ACTIVITY

Cobalt acts as a chelator of salicylidine-*o*-aminothiophenol (SATP) and salicylidine-*o*-aminopyridine (SAP) and exerts biocidal activity against the molds *Aspergillus nidulans* Winter and *A. niger* Tiegh and the yeast *Candida albicans* (97). Antifungal activities of Co^{2+} with acetone salicyloyl hydrazone (ASH) and ethyl methyl ketone salicyloyl hydrazone (ESH) against *A. niger* and *A. flavus* have been established by Johari et al. (98).

17.8.6   ETHYLENE BIOSYNTHESIS

Cobalt inhibits IAA-induced ethylene production in gametophores of the ferns *Pteridium aquilinum* Kuhn and sporophytes of ferns *Matteneuccia struthiopteris* Tod. and *Polystichum munitum* K. Presl (99); in pollen embryo culture of horse nettle (*Solanum carolinense* L.) (100); in discs of apple peel (101); in winter wheat and beans (102); in kiwifruit (*Actinidia chinensis* Planch) (103); and in wheat seedlings under water stress (104). Cobalt also inhibits ethylene production and increases the apparent rate of synthesis of peroxides and prevents the peroxidative destruction of IAA. Other effects include counteraction of the uncoupling of oxidative phosphorylation by dinitrophenol (4).

Cobalt acts mainly through arresting the conversion of methionine to ethylene (105) and thus inhibits ethylene-induced physiological processes. It also causes prevention of cotyledonary prickling-induced inhibition of hypocotyls in beggar tick (*Bidens pilosa* L.) (106), promotion of hypocotyl elongation (107), opening of the hypocotyl hook (bean seedlings) either in darkness or in red light, and the petiolar hook (*Dentaria diphylla* Michx.) (108,109). Cobalt has also been noted to cause reduction of RNAase activity in the storage tissues of potato (110), repression of developmental distortion such as leaf malformation and accumulation of low-molecular-weight polypeptides in velvet plant (*Gynura aurantiaca* DC) (111), delayed gravitropic response in cocklebur (*Xanthium* spp.), tomato and castor bean stems (112), and prevention of 3,6-dichloro-*o*-anisic acid-induced chlorophyll degradation in tobacco leaves (73). Prevention of auxin-induced stomatal opening in detached leaf epidermis has been observed (85). The effects of ethylene on the kinetics of curvature and auxin redistribution in the gravistimulated roots of maize are known (113). ^{60}Co γ-rays and EMS influence antioxidase activity and ODAP content of grass pea (*Lathyrus sativus* L.) (114).

17.8.7   NITROGEN FIXATION

Cobalt is essential for nitrogen-fixing microorganisms, including the cyanobacteria. Its importance in nitrogen fixation by symbiosis in Leguminosae (Fabaceae) has been established (115–119). For example, soybeans grown with only atmospheric nitrogen and no mineral nitrogen have rapid nitrogen fixation and growth with 1.0 or 0.1 µg Co ml^{-1}, but have minimal growth without cobalt additions (4).

17.9   COBALT TOLERANCE BY PLANTS

17.9.1   ALGAE

Stonewort (*Chara vulgaris* L.) resistant to metal pollution, when cultivated in a natural medium containing $CoCl_2$ showed high level of cobalt in dry matter as insoluble compounds (120). On the other hand, a copper-tolerant population of a marine brown alga (*Ectocarpus siliculosus* Lyng.) had an increased tolerance to cobalt. The copper-tolerance mechanism of other physiological processes may be the basis of this cotolerance (121).

17.9.2 FUNGI

A genetically stable cobalt-resistant strain, Co^R, of *Neurospora crassa* Shear & Dodge, exhibited an approximately ten-fold higher resistance to Co^{2+} than the parent strain. The Co^{2+} toxicity was reversed by Mg^{2+}, but not by Fe^{3+}, indicating that the Co^{2+} did not affect iron metabolism. Alternatively, the mechanism of resistance probably involves an alteration in the pattern of iron metabolism so that the toxic concentration of cobalt could not affect the process (122). Magnesium (Mg^{2+}) may reverse the toxicity of Co^{2+}, either by increasing the tolerance to high intracellular concentration of heavy metal ions or by controlling the process of uptake and accumulation of ions (123). In several mutants of *Aspergillus niger* growing in toxic concentrations of Zn^{2+}, Co^{2+}, Ba^{2+}, Ni^{2+}, Fe^{3+}, Sn^{2+}, and Mn^{2+}, the resistance is due to an intracellular detoxification rather than defective transport. Each mutation was due to a single gene located in its corresponding linkage group. Toxicity of metals is reversed in the wild-type strain by definite amounts of K^+, NH_4^+, Mg^{2+}, and Ca^{2+}. These competitions between pairs of cations indicate a general system responsible for the transport of cations (124). In *Aspergillus fumigatus*, cobalt increased thermophily at 45°C and fungal tolerance at 55°C (125).

17.9.3 HIGHER PLANTS

In higher plants, cobalt tolerance has been mainly reported in members of 'advanced' families such as the Labiatae and Scrophulariaceae growing in the copper-field belt of Shaba (Zaire) (126). Among these plants, *Haumaniastrum robertii*, a copper-tolerant species, is also a cobalt-accumulating plant. The plant contains abnormally high cobalt (about 4304 $\mu g\ g^{-1}$ dry weight), far exceeding the concentration of copper. This species has the highest cobalt content of any phanerogam (127). *Haumaniastrum katangense* and *H. robertii* grow on substrates containing 0 to 10,000 μg Co g^{-1}. Although they can accumulate high concentrations of cobalt, an exclusion mechanism operates in these species at lower concentrations of the element in the soil. Uptake of cobalt was not linked to a physiological requirement of the element. The plant–soil relationship for Co was significantly high enough for these species to be useful in the biogeochemical prospecting for cobalt (128).

Tolerance and accumulation of copper and cobalt were investigated in three members of phylogenetic series of taxa within the genus *Silene* (Caryophyllaceae) from Zaire, which were regarded as representing a progression of increasing adaptation to metalliferous soils. Effects of both metals (singly and in combination) on seed germination, seedling and plant performances, yield, and metal uptake from soil culture confirmed the ecotypic status of *S. burchelli*, which is a more tolerant variant of the nontolerant *S. burchelli* var. *angustifolia*. But both the ecotype and metallophyte variants of *S. cobalticola* are relatively more tolerant to copper than to cobalt.

REFERENCES

1. F. Watanabe, Y. Nakano. Vitamin B12. *Nippon Rinsho* 57:2205–2210, 1999.
2. P.A. Olson, D.R. Brink, D.T. Hickok, M.P. Carlson, N.R. Schneider, G.H. Deutscher, D.C. Adams, D.J. Colburn, A.B. Johnson. Effects of supplementation of organic and inorganic combinations of copper, cobalt, manganese and zinc above nutrient requirement levels on postpartum two-year-old cows. *J. Anim. Sci.* 77:522–532, 1999.
3. F.B. Salisbury, C.W. Ross. *Plant Physiology*, 4th edition. Belmont, CA: Wadsworth Publishers, 1992, pp. 124–125.

4. S. Ahmed, H.J. Evans. Cobalt: a micronutrient for the growth of soybean plants under symbiotic conditions. *Soil Sci.* 90, 205–210, 1960.

5. N. Kuyacak, B. Volesky. The mechanism of cobalt biosorption. *Biotechnol. Bioeng.* 33:823–831, 1989.

6. D.D. Ryrdina, G.G. Polikarpov. Distribution of certain chemical elements in biochemical fractions of the black sea alga *Cystoseira barbata. Gidrobiol. Zh.* 19:79–84, 1983.

7. A. Samecka-Cymerman, A.J. Kempers. Bioaccumulation of heavy metals by aquatic macrophytes around Wroclaw, Poland. *Ecotoxicol. Environ. Saf.* 35:242–247, 1996.

8. Y. Tateda, J. Misonolu. Marine indicator organisms of cobalt, strontium, cesium. *Denryoku Chvo Kenkyusho Hokoku* 9U88007:1–19, 1988.

9. I.M. Munda, V. Hudnik. The effects of zinc, manganese and cobalt accumulation on growth and chemical composition of *Fucus vesiculosus* L.under different temperature and salinity conditions. *Mar. Ecol.* 9:213–216, 1988.

10. S.M. Macfie, P.M. Welbourn. The cell wall as a barrier to uptake of metal ions in the unicellular green alga *Chlamydomonas reinhardtii* (Chlorophyceae). *Arch. Environ. Contam. Toxicol.* 39:413–419, 2000.

11. Z.H. Ye, S.N. Whiting, Z.Q. Lin, C.M. Lytle, J.H. Qian, N. Terry. Removal and distribution of iron, manganese, cobalt and nickel within a Pennsylvania-constructed wetland treating coal combustion byproduct leachate. *J. Environ. Qual.* 30:1464–1473, 2001.

12. E.G. Davidova, A.P. Belov, V.V. Zachinskii. The accumulation of labelled cobalt in yeast cells. *Izh. Timiryazev S-KH Akad.* Jul-Aug (4), pp. 109–114, 1986.

13. G. Venkateswerlu, K. Sivaramasastry. The mechanism of uptake of cobalt ions by *Neurospora crassa. Biochem. J.* 118:497–503, 1970.

14. P.R. Norris, D.P. Kelly. Accumulation of cadmium and cobalt by *Saccharomyces cerevisiae. J. Gen. Microbiol.* 99:317–324, 1977.

15. C. White, G.M. Gadd. Uptake and cellular distribution of copper, cobalt and cadmium in strains of *Saccharomyces cerevisiae* cultured on elevated concentrations of these metals. *FEMS Microbiol. Ecol.* 38:277–284, 1986.

16. H. Tuppy, W. Sieghart. Effects of Co^{2+} on yeast mitochondria. *Monatsh Chem.* 104:1433–1443, 1973.

17. L.S. Watrud, A.H. Ellingboe. Cobalt as a mitochondrial density marker in a study of cytoplasmic exchange during mating of *Schizophyllum commune. J. Cell Biol.* 59:127–133, 1973.

18. G.M. Gadd, C. White, J.L. Mowel. Heavy metal uptake by intact cells and protoplast of *Aureobasidium pollulans. FEBS Microbiol. Ecol.* 45:261–268, 1987.

19. A. Ruhling, G. Tyler. Sorption and retention of heavy metals in woodland moss *Hylocomium splendens* (Hedw.) Br. *et* Sch. *Oikos* 21:92–97, 1970.

20. A. Kloke. Effects of heavy metals on soil fertility, flora and fauna as well as its importance for the nutrition and fodder-chain (in German). In: *Proceedings of the Utilization of Refuse and Sewage sludge in Agriculture*, Zurich, 1980, pp. 58–87.

21. A.K. Roy, L.L. Srivastava. Removal of some micronutrients by forage crops in soils. *J. Indian Soc. Soil Sci.* 36:133–137, 1988.

22. S. Palit, A. Sharma, G. Talukder. Effects of cobalt on plants. *Bot. Rev.* 60:149–181, 1994.

23. R. Branden, K. Jamson, P. Nilsson, R. Aasa. Intermediates formed by the Co^{2+} activated ribulose-1-5, biphosphate carboxylase/oxygenase from spinach studied by EPR spectroscopy. *Biochem. Biophys. Acta* 916:293–303, 1987.

24. L.G. Craig, W.E. Schmid. Absorption of cobalt by excised barley roots. *Plant Cell Physiol.* 15:273–279, 1974.

25. S.M. Cocucci, S. Morgutti. Stimulation of proton extrusion by potassium ion and divalent cations (nickel, cobalt, zinc) in maize (*Zea mays* cultivar Dekalab XL85) root segments. *Physiol. Plant* 68:497–501, 1986.

26. I.T. Platash, L.I. Dyeryuhina, V.S. Art'omchenko. *Astragalus* micro-element. *Farm Zh.* 27:64–65, 1972.

27. G.N. Schrauzer. Cobalt. In: E. Merian, ed. *Metals and Their Compounds in the Environment.* Weinheim, Germany: VCH Verlag, 1991, pp. 879–892.

28. U. Boikat, A. Fink, J. Bleck-Neuhaus. Cesium and cobalt transfer from soil to vegetation on permanent pastures. *Radiat. Environ. Biophys.* 24:287–301, 1985.

29. T. Kosla. Iron, manganese and cobalt levels in soil, pasture grass and bodies of young bulls after irrigation of the soil with waste water. *Pol. Arch. Weter* 24:587–596, 1987.

30. E.C. Webb, J.B. van Ryssen, M.E. Eramus, C.M. McCrindle. Copper, manganese, cobalt and selenium concentrations in liver samples from African buffalo (*Syncerus caffer*) in the Kruger National Park. *J. Environ. Monit.* 3:583–585, 2001.

31. Y.O. Korshunova, D. Eide, W.G. Clarke, M.L. Guirinot, H.B. Pakrasi. The IRT 1 protein from *Arabidopsis thaliana* is a metal transporter with a broad substrate range. *Plant Mol. Biol.* 40:37–44, 1999.

32. A.E.S. Macklon, A. Sim. Cellular cobalt fluxes in roots and transport to the shoots of wheat seedlings. *J. Exp. Bot.* 38:1663–1677, 1987.

33. A.E.S. Macklon, A. Sim. Cortical cell fluxes in roots and transport to the shoots of rye grass seedlings. *Plant Physiol.* 80:409–416, 1990.

34. S. Tu, E. Nungesser, D. Braver. Characterization of the effects of divalent cations on the coupled activities of the proton-ATPase in tonoplast vesicles. *Plant Physiol.* 90:1636–1643, 1989.

35. J. Jarosick, P. Z. Vara, J. Koneeny, M. Obdrzalek. Dynamics of cobalt 60 uptake by roots of pea plants. *Sci. Total Environ.* 71:225–229, 1988.

36. T.A. Danilova, I.V. Tishchenko, E.N. Demikina. Distribution and translocation of cobalt in legumes. *Agrokhimiya* 2:100–104, 1970.

37. D. Wiersma, B.J.V. Goor. Chemical forms of nickel and cobalt in phloem of *Ricinus communis*. *Plant Physiol.* 45:440–442, 1979.

38. C.A. Peterson, W.E. Rauser. Callose deposition and photo assimilate export in *Phaseolus vulgaris* exposed to excess cobalt, nickel and zinc. *Plant Physiol.* 63:1170–1174, 1979.

39. P.M. Patel, A. Wallace, R.T. Mueller. Some effects of copper, cobalt, cadmium, zinc, nickel and chromium on growth and mineral element concentration in *Chrysanthemum. J. Am. Soc. Hortic. Sci.* 101:553–556, 1976.

40. A. Wallace, R.T. Muller, C.V. Alexander. High levels of four heavy metals on the iron status of plants. *Commun. Soil Sci. Plant Anal.* 7:43–46, 1976.

41. F.A. Austenfeld. Effects of Ni, Co and Cr on net photosynthesis in *Phaseolus vulgaris*. *Photosynthetica* 13:434–438, 1979.

42. F.I. Rehab, A. Wallace. Excess trace metal elements on cotton: 2. Copper, zinc, cobalt and manganese in yolo loam soil. *Commun. Soil Sci. Plant Anal.* 9:519–528, 1978.

43. A.D. Robson, G.R. Mead. Seed cobalt in *Lupinus angustifolius*. *Aust. J. Agric. Res.* 31:109–116, 1980.

44. N.A. Kenesarina. The effects of mineral fertilizers on cobalt content in potato plants. *IZV Akad Nauk. Kaz SSR Ser. Biol.* 6:31–35, 1972.

45. J. Jorgovic-Kremzer, M. Duricic, V. Bjelic. Levels of copper, manganese, zinc, cobalt, iron and lead in cucumber before and after pickling. *Agrohemija* (5/6), pp. 165–171, 1980.

46. C. White, G.M. Gadd, A.D. Robson, H.M. Fisher. Variation in nitrogen, sulfur, selenium, cobalt, manganese, copper and zinc contents of grain from wheat (*Triticum aestivum*) and 2 lupine (*Lupinus*) species grown in a range of Mediterranean environment. *Aust. J. Agric. Res.* 32:47–60, 1981.

47. M. Coppenet, E. More, L.L. Corre, M.L. Mao. Variations in rye grass cobalt content: investigating enriching methods. *Annu. Agron.* 23:165–192, 1972.

48. M.L. Berrow, R.L. Mitchell. Location of trace elements in soil profile: total and extractable contents of individual horizons. *Trans. R Soc. Earth Sci.* 71:103–122, 1980.

49. D. Afusaic, T. Muraru. A study of trace element mobility in acid limed soils. *Agrochim. Agrotech. Pasuni Finete* 35:45–56, 1967.

50. G.Y. Freiberg. Absorption of trace elements—Cu and Co by some field cultivars in relation to the content of organic matter in soil. *Izv. Akad. Nauk. Latvijsk SSR* 2:116–121, 1970.

51. D.B.R. Poole, L. Moore, T.F. Finch, M.J. Gardiner, G.A. Fleming. An unexpected occurrence of cobalt pine in lambs in North Leinster. *IR J. Agric. Res.* 13:119–122, 1974.

52. R.M. McKenzie. The manganese oxides in soils: a review. *Zh. Boden Pflanz.* 131:221–242, 1972.

53. K.L. Iu, I.D. Pulford, H.J. Duncan. Influence of soil waterlogging on subsequent plant growth and trace element content. *Plant Soil* 66:423–428, 1982.

54. R. Mercky, J.H. Van Grinkel, J. Sinnaeve, A. Cremera. Plant-induced changes in the rhizosphere of maize and wheat: II Complexion of Co, Zn and Mn in the rhizosphere of maize and wheat. *Plant Soil* 96:95–101, 1986.

55. G.K. Kasimova, P.B. Zamanov, R.A. Abushev, M.G. Safarov. The effect of certain trace elements molybdenum, boron, manganese and cobalt in the background of mineral fertilizers on the biological activity of tobacco rhizosphere. *Ref. Zh. Biol.* 3:7–9, 1971.

56. R.J. Ganson, R.A. Jensen. The essential role of cobalt in the inhibition of the cytosolic isozyme of 3-deoxy-D-arabino-heptulosonate-7-phosphate synthase from *Nicotiana silvestris* by glyphosate. *Arch. Biochem. Biophys.* 260:85–93, 1988.

57. A.E. Ceniceros-Gomez, B. King-Diaz, N. Barba-Behrens, B. Lotina-Hennsen, S.E. Castillo-Blum. Two inhibition targets by [Cr(2gb)(3)](3+) and [Co(2gb)(3)](3+) on redox enzymes of spinach thylakoids. *J. Agric. Food Chem.* 47:3075–3080, 1999.

58. M. Dennis, P.E. Kolattukudy. A cobalt porphyrin enzyme converts a fatty aldehyde to a hydrocarbon and Co. *Proc. Natl. Acad. Sci. USA* 12:5204–5210, 1992.

59. C. Preston, M. Seibert. The carboxyl modifier 1-ethyl-3-[3-(Dimethylamino)propyl] carbodiimide (EDC) inhibits half of the high-affinity Mn-binding site in photosystem II membrane fragments. *Biochemistry* 30:9615–9633, 1991.

60. H.C. Lee, J.B. Wittenberg, J. Peisach. Role of hydrogen bonding to bound dioxygen in soybean leg haemoglobin. *Biochemistry* 32:11500–11506, 1993.

61. H. Zhu, E. Shipp, R.J. Sanchez, A. Liba, J.E. Stine, P.J. Hart, E.B. Gralla, A.M. Nersissian, J.S. Valentine. Cobalt (2+) binding to tomato copper chaperone for superoxide dismutase: implications for the metal ion transfer mechanism. *Biochemistry* 39:5413–5421, 2000.

62. C. Preston, M. Seibert. Partial identification of the high-affinity Mn-binding site in *Scenedesmus obliquus* Photosystem II. In: M Baltcheffsky, ed. *Current Research in Photosynthesis*, Vol. 1. Dordrecht, Netherlands: Kluwer, 1990, pp. 925–928.

63. M.L. Ghirardi, T.W. Lutton, M. Seibert. Interactions between diphenylcarbazide, zinc, cobalt, and manganese on the oxidizing side of photosystem II. *Biochemistry* 35:1820–1828, 1996.

64. L. Samudralwar, A.N. Garg. Minor and trace elemental determination of Indian herbal and other medicinal preparations. *Biol. Trace Element Res.* 54:113–121, 1996.

65. S.M. Siegel. The cytotoxic response of *Nicotiana* protoplast to metal ions: a survey of the chemical elements. *Water Air Soil Pollut.* 8:292–304, 1977.

66. H.E. Christensen, L.S. Conrad, J. Ulstrup. Redox potential and electrostatic effects in competitive inhibition of dual-path electron transfer reactions of spinach plastocyanin. *Arch. Biochem. Biophys.* 301:385–390, 1993.

67. S.Y. Hachar-Hill, R.G. Shulman. Co^{2+} as a shift reagent for 35Cl NMR of chloride with vesicles and cells. *Biochemistry* 31:6267–6268, 1992.

68. J.G. Hunter, O. Verghano. Trace-element toxicities in oat plants. *Annu. Appl. Biol.* 40:761–777, 1953.

69. A. Wallace, A.M. Abou-Zamzam. Low levels but excesses of five different trace elements, singly and in combination, on interaction in bush beans grown in solution culture. *Soil Sci.* 147:439–441, 1989.

70. M. Agarwala, H.D. Kumar. Cobalt toxicity and its possible mode of action in blue–green alga *Anacystis nidulans*. *Beitr. Biol. Pflanz.* 53:157–164, 1977.

71. C.P. Sharma, S.S. Bisht, S.C. Agrawala. Effect of excess supply of heavy metals on the absorption and translocation of iron (59 Fe) in barley. *J. Nucl. Agric. Biol.* 7:12–14, 1978.

72. A. Wallace. Additive, protective and synergistic effects of plants with excess trace elements. *Soil Sci.* 133:319–323, 1982.

73. L.A. Young, E.C. Sisler. Interaction of dicamba (3,6-dichloro-*o*-anisic-acid) and ethylene on tobacco leaves. *Tob. Sci.* 34:34–35, 1990.

74. G. Battistuzzi, M. Dietrich, R. Löcke, H. Witzel. Evidence for a conserved binding motif of the dinuclear metal site in mammalian and plant acid phosphatases: 1H NMR studies of the di-iron derivative of the Fe(III)Zn(II) enzyme from kidney bean. *Biochem. J.* 323:593–596, 1997.

75. F.M. Chaudhury, J.F. Loneragan. Zinc absorption by wheat seedlings: II. Inhibition by hydrogen ions and by micronutrient cations. *Soil Sci. Soc. Am. Proc.* 36:327–331, 1972.

76. A.A. Anisimov, O.P. Ganicheva. Possible interchangeability between Co and Zn in plants. *Fiziol. Biokhim. Kul't Rast.* 10:613–617, 1978.

77. W.L. Berry, A. Wallace. Toxicity: The concept and relationship to the dose response curve. *J. Plant Nutr.* 3:13–19, 1981.

78. I. Gilles, H.G. Loeffler, F. Schneider. Cobalt-substituted acylamino-acid amido-hydrolase from *Aspergillus oryzae*. *Zh. Naturf. Sect. C Biosci.* 36:751–754, 1981.

79. G. Venkateswerlu, G. Stotzky. Binding of metals by cell wall of *Cunninghamella blakesleeana* grown in the presence of copper or cobalt. *Appl. Microbiol. Biotechnol.* 31:619–625, 1986.

80. V. Werner. Effect of nickel, cadmium and cobalt on the uptake of copper by intact barley (*Hordeum distichon*) roots. *Zh. Pflanz. Physiol.* 93:1–10, 1979.

81. L.C. Rai, S.K. Dubey. Impact of chromium and tin on the nitrogen-fixing cyanobacterium *Anabaena doliolum*: interaction with bivalent cations. *Environ. Saf.* 17:94–104, 1989.

82. H.F.G. Dang, N.A. Solovieva, Z.G. Evstigneeva, V.L. Kretovich. Purification, physicochemical properties, and kinetics of *Spirulina platensis* glutamine synthetase. *Applied biochemistry and microbiology* 23(6):621–626, 1988.

83. A. Kapur, R.H. Chopra. Effects of some metal ions on protonemal growth and bud formation in the moss *Timiella anomala* grown in aseptic cultures. *J. Hattori Bot. Lab.* 10:283–298, 1989.

84. S. Tosh, M.A. Choudhuri, S.K. Chatterjee. Retardation of lettuce (*Lactuca sativa*) leaf senescence by cobalt ions. *Indian J. Exp. Biol.* 17:1134–1136, 1979.

85. F. Merritt, A. Kemper, G. Tallman. Inhibitors of ethylene synthesis inhibit auxin-induced stomatal opening in epidermis detached from leaves of *Vicia faba* L. *Plant Cell Physiol.* 42:223–230, 2001.

86. E.A. Bulantseva, E.M. Glinka, M.A. Protsenko, E.G. Sal'kova. A protein inhibitor of polygalacturonase in apple fruits treated with aminoethoxyvinylglycine and cobalt chloride. *Prikl. Biokhim. Mikrobiol.* 37:100–104, 2001.

87. V.P. Tarabrin, T.R. Teteneva. Presowing treatment of seeds and its effect on the resistance of wood plant seedlings against drought. *Sov. J. Ecol.* 10:204–211, 1979.

88. B.N. Yadrov, S.E. Donitruk, V.G. Baturin. The effects of copper, manganese and cobalt on the productivity of a culture of isolated tissue of *Datura innoxia* Mill. *Rastit. Resursy* 14:408–411, 1978.

89. V.V. Koval'Skii, N.I. Grinkevich, I.F. Gribovskaya, L.S. Dinevech, A.N. Shandova. Cobalt in medicinal plants and its effect on the accumulation of biologically active compounds. *Rastit. Resursy* 7:503–510, 1971.

90. I.A. Petrishek, M. Ya. Lovkova, N.I. Grinkevich, L.P. Orlova, L.V. Poludennyi, Effects of cobalt and copper on the accumulation of alkaloids in Atropa belladonna. *Biology bulletin of the Academy of Sciences of the USSR*, 10(6):509–516, 1984.

91. M.Y. Lovkova, G.N. Buzuk, N.S. Sabirova, N.I. Kliment'eva, N.I. Grinkevich. Pharmacognostic examination of *Glaucium flavum* Cr. *Farmatsiya* 37:31–34, 1988.

92. N.L. Grinkevich, L.F. Gribovskaya, A.N. Shandova, L.S. Dinevich. Concentration of cobalt in some medicinal plants and its effect on the accumulation of flavonoids in buckwheat. *Biol. Nauk.* 14:88–91, 1971.

93. I. Barbat, M. Tomsa, T. Suciu. Influence of foliar nutrition with microelements on some physiological processes in apple tree. *Bull. Inst. Agron. Cluj-Napora Ser. Agric.* 33:69–74, 1979.

94. G. Chandra, K.S. Reddy, H.Y. Mohan Ram. Extension of vase life of cut marigold and chrysanthemum flowers by the use of cobalt chloride. *Indian J. Exp. Biol.* 19:150–154, 1981.

95. D.W. Fujino, M.S. Reid. Factors affecting the vase life of fronds of maidenhair fern (*Adiantum raddianum*). *Sci. Hortic.* 21:181–188, 1983.

96. T. Venkatarayappa, M.J.Tsujita, D.P. Murr. Influence of cobaltous ion on the post-harvest behaviour of roses (*Rosa hybrida* Samantha). *J. Am. Soc. Hort. Sci.* 105:148–151, 1980.

97. R.K. Parashar, R.S. Sharma, R. Nagar, R.C. Sharma. Biological studies of ONS and ONN donor Schiff bases and their copper (II), nickel (II) and manganese (II) complexes. *Curr. Sci.* 56:518–521, 1987.

98. R.B. Johari, R. Nagar, R.C. Sharma. Studies on copper (II), nickel (II), cobalt (II) and zinc (II) complexes of acetone salicyloyl hydrazone and ethyl methyl ketone salicyloyl hydrazone. *Indian J. Chem. Soc. Inorg. Phys. Theor. Anal.* 26:962–963, 1987.

99. F.L. Tittle. Auxin-stimulated ethylene production in fern gametophytes and sporophytes. *Plant Physiol.* 70:499–502, 1987.

100. T.L. Reynolds. A possible role of ethylene during IAA-induced pollen embryogenesis in anther cultures of *Solanum carolinense* L. *Amer. J. Bot.* 74:967–969, 1987.

101. E.G. Sal'kova, E.A. Bulantseva. Effect of cobalt and silver ions on ethylene evolution by discs from the peel. *Prikl. Biokhim. Mikrobiol.* 24:698–702, 1988.

102. O.I. Romanovskaya, V.V. Ilin, O.Z. Kreitsbreg. Ethylene biosynthesis in winter wheat and kidney beans upon growth inhibition with chlorocholine chloride. *Fiziol. Rast.* 35:893–898, 1988.

103. H. Hyodo, R. Fukasawa. Ethylene production in kiwi fruit (*Actinidia chinensis* cultivar. Hayward). *J. Jpn. Soc. Hortic. Sci.* 54:209–215, 1985.

104. I. Gaal, H. Ariunaa, M. Gyuris. Influence of various stress on ethylene production in wheat seedlings. *Acta Univ. Szeged. Acta Biol.* 34:35–44, 1988.

105. O.L. Lau, S.F. Yang. Inhibition of ethylene production by cobaltous ion. *Plant Physiol.* 58:114–117, 1976.

106. D. Crouzillat, M.O. Desbiel, C. Penel, T. Gasper. Lithium, aminoethoxy-vinylglycine and cobalt reversal of the cotyledonary prickling-induced growth inhibition in the hypocotyl of *Bidens pilosus* in relation to ethylene and peroxidase. *Plant Sci.* 40:7–12, 1985.

107. S. Grover, W.K. Purves. Cobalt and plant development: interaction with ethylene in hypocotyl growth. *Plant Physiol.* 57:886–889, 1976.

108. B.G. Kang. Effects of inhibitors of RNA and protein synthesis on bean hypocotyl hook opening and their implication regarding phytochrome action. *Planta* 87:217–226, 1969.

109. J.M. Yopp. The role of light and growth regulators in the opening of the *Dentaria* petiolar hook. *Plant Physiol.* 54:7141–7147, 1973.

110. M.C. Isola, L. Franzoni. Effect of ethylene on the increase in RNAase activity in potato tuber tissue. *Plant Physiol. Biochem.* 27:245–250, 1989.

111. J.M. Belles, V. Conejero. Ethylene mediation of the viroid-like syndrome induced by silver ions in *Gynura aurantiaca* DC. *Phytopathology* 124:275–284, 1989.

112. R.M. Wheeler, F.B. Salisbury. Gravitropism in higher plants shoots: 1. A role for ethylene. *Plant Physiol.* 67:686–690, 1981.

113. J.S. Lee, W.K. Chang, M.L. Evans. Effects of ethylene on the kinetics of curvature and auxin redistribution in gravistimulated roots of *Zea mays*. *Plant Physiol.* 94:1770–1775, 1990.

114. X. Qin, F. Wang, X. Wang, G. Zhou, Z. Li. Effect of combined treatment of 60 Co γ-rays and EMS on antioxidase activity and ODAP content in *Lathyrus sativus*. *Ying Yong Sheng Tai Xue Bao* 11:957–958, 2000.

115. E.G. Hallsworth, S.B. Wilson, E.A.N. Greenwood. Copper and cobalt in nitrogen fixation. *Nature* 187:79–80, 1960.

116. H.M. Reisenauer. Cobalt in nitrogen fixation by a legume. *Nature* 186:375–376, 1960.

117. C.C. Delwiche, C.M. Johnson, R.M. Reisenauer. Influence of cobalt on nitrogen fixation by *Medicago*. *Plant Physiol.* 36:73, 1961.

118. A.D. Robson, M.J. Dilworth, D.L. Chatel. Cobalt and nitrogen fixation in *Lupinus angustifolius* L. 1. Growth, nitrogen concentratins and cobalt distribution. *New Phytol.* 83:53–62, 1979.

119. M.J. Dilworth, A.D. Robson, D.L. Chatel. Cobalt and nitrogen fixation in *Lupinus angustifolius* L. 2. Nodule formation and function. *New Phytol.* 83:63–79, 1979.

120. R. Strauss. Nickel and cobalt accumulation by Characeae. *Hydrobiologia* 14:263–268, 1986.

121. A. Hall. Heavy metal co-tolerance in a copper-tolerant population of the marine alga, *Ectocarpus siliculosus*. *New Physiol.* 85:73–78, 1980.

122. G. Venkateswerlu, K. Sivaramasastry. Interrelationship in trace-element metabolism in metal toxicities in a cobalt resistant strain of *Neurospora crassa*. *Biochem. J.* 132:673–680, 1973.

123. P.M. Mohan, K. Sivaramasastry. Interelationship in trace-element metabolism in metal toxicities in nickel-resistant strains of *Neurospora crassa*. *Biochem. J.* 212:205–210, 1983.

124. V. Florza. Toxicity of metal ions for *Aspergillus nidulans*. *Microbiol. Espan.* 22:131–138, 1969.

125. S. Ramada, A.A. Razak, A.M. Hamed. Partial dependence of *Aspergillus fumigatus* thermophilism on additive nutritional requirements. *Mikrobiologia* 25:57–66, 1988.

126. R.S. Brooks, R.D. Reeves, R.S. Morrison, F. Malaisee. Hyperaccumulation of copper and cobalt: a review. *Bull. Soc. Roy. Bot. Belg.* 113:166–172, 1980.

127. R.S. Brooks. Copper, magnesium and cobalt uptake by *Haumanistrum* species. *Plant Soil* 48:541–544, 1977.

128. R.S. Morrison, R.R. Brooks, R.D. Reeves, F. Malaisse. Accumulation of copper and cobalt by metallophytes from Zaire. *Plant Soil* 53:535–540, 1979.

18 Selenium

Dean A. Kopsell
University of Tennessee, Knoxville, Tennessee

David E. Kopsell
University of Wisconsin-Platteville, Platteville, Wisconsin

CONTENTS

18.1 The Element Selenium 515
    18.1.1 Introduction 515
    18.1.2 Selenium Chemistry 516
18.2 Selenium in Plants...517
    18.2.1 Introduction 517
    18.2.2 Uptake 517
    18.2.3 Metabolism 518
    18.2.4 Volatilization 520
    18.2.5 Phytoremediation 520
18.3 Selenium Toxicity to Plants 521
18.4 Selenium in the Soil 521
    18.4.1 Introduction 521
    18.4.2 Geological Distribution..522
    18.4.3 Selenium Availability in Soils...523
18.5 Selenium in Human and Animal Nutrition 524
    18.5.1 Introduction 524
    18.5.2 Dietary Forms 524
    18.5.3 Metabolism and Form of Selenium...525
18.6 Selenium and Human Health 525
    18.6.1 Introduction 525
    18.6.2 Selenium Deficiency and Toxicity in Humans 525
    18.6.3 Anticarcinogenic Effects of Selenium...526
    18.6.4 Importance of Selenium Methylation in Chemopreventive Activity 526
18.7 Selenium Enrichment of Plants..526
18.8 Selenium Tissue Analysis Values of Various Plant Species 543
References 543

18.1 THE ELEMENT SELENIUM

18.1.1 INTRODUCTION

Selenium (Se), a beneficial element, is one of the most widely distributed elements on Earth, having an average soil abundance of 0.09 mg kg^{-1} (1). It is classified as a Group VI A metalloid, having metallic and nonmetallic properties. Selenium was identified in 1818 by the Swedish chemist Jöns Jacob Berzelius as an elemental residue during the oxidation of sulfur dioxide from copper pyrites in the production of sulfuric acid (2). The name selenium originates through its chemical similarities to tellurium (Te), discovered 35 years earlier. Tellurium had been named after the Earth (*tellus* in Latin), so selenium was named for the moon (*selene* in Greek) (3). Although selenium is not considered as an essential plant micronutrient (4), it is essential for maintaining mammalian health (5). Selenium deficiency or toxicity in humans and livestock is rare, but can occur in localized areas (5,6) owing to low selenium contents in soils and locally produced crops (7). Recently, much attention has been given to the role of selenium in reducing certain types of cancers and diseases. Efforts in plant improvement have begun to enhance the selenium content of dietary food sources.

18.1.2 SELENIUM CHEMISTRY

Selenium has an atomic number of 34 and an atomic mass of 78.96. The atomic radius of Se is 1.40 Å, the covalent radius is 1.16 Å, and the ionic radius is 1.98 Å. The ionization potential is 9.74 eV, the electron affinity is -4.21 eV, and the electronegativity is 2.55 on the Pauling Scale (8). The chemical and physical properties of selenium are very similar to those of sulfur (S). Both have similar atomic size, outer valence-shell electronic configurations, bond energies, ionization potentials, electron affinities, electronegativities, and polarizabilities (8). Selenium can exist as elemental selenium (Se^0), selenide (Se^{2-}), selenite (SeO_3^{2-}), and selenate (SeO_4^{2-}). There are six stable isotopes of selenium in nature: ^{74}Se (0.87%), ^{76}Se (9.02%), ^{77}Se (7.58%), ^{78}Se (23.52%), ^{80}Se (49.82%), and ^{82}Se (9.19%) (8). Some of the commercially available forms of selenium are H_2Se, metallic selenides, SeO_2, H_2SeO_3, SeF_4, $SeCl_2$, selenic acid (H_2SeO_4), Na_2SeO_3, Na_2SeO_4, and various organic Se compounds (9).

In the elemental form, selenium exists in either an amorphous state or in one of three crystalline states. The amorphous form of selenium is a hard, brittle glass at 31°C, vitreous at 31 to 230°C, and liquid at temperatures above 230°C (10). The first of three crystalline states takes the form of flat hexagonal and polygonal crystals called α-monoclinic or red selenium. The second form is the prismatic or needle-like crystal called β-monoclinic or dark-red selenium. The third crystalline state is made up of spiral polyatomic chains of Se_n, often referred to as hexagonal or black selenium. The black forms of crystalline Se are the most stable. At temperatures above 110°C, the monoclinic amorphous forms convert into this stable black form. Conversion of the amorphous form into the black form occurs readily at temperatures of 70 to 210°C. When Se^0 is heated above 400°C in air, it becomes the very pungent and highly toxic gas H_2Se. This gas decomposes in air back to Se^0 and water (10).

Reduction or oxidation of elemental selenium can be to the -2-oxidation state (Se^{2-}), the $+4$-oxidation state (SeO_3^{2-}), or the $+6$-oxidation state (SeO_4^{2-}). The Se^{2-} ion is water-soluble (270 ml per 100 ml H_2O at 22.5°C) and will react with most metals to form sparingly soluble metal selenides. Selenium in the $+4$-oxidation state can occur as selenium dioxide (SeO_2), SeO_3^{2-}, or selenious acid (H_2SeO_3). Selenium dioxide is water-soluble (38.4 g per 100 ml H_2O at 14°C) and is produced when Se^0 is burned or reacts with nitric acid. Reduction back to Se^0 can be carried out in the presence of ammonium, hydroxylamine, or sulfur dioxide. In hot water, SeO_2 will dissolve to H_2SeO_3, which is weakly dibasic. Organic selenides, which are electron donors, will oxidize readily to the higher oxidation states of selenium. Selenites are electron acceptors. At low pH, SeO_3^{2-} is reduced to Se^0 by ascorbic acid or sulfur dioxide. In the soil, SeO_3^{2-} is bound strongly by hydrous oxides of iron and is sparingly soluble at pH 4 to 8.5 (10).

In the $+6$-oxidation state, selenium is in the form of selenic acid (H_2SeO_4) or SeO_4^{2-} salts. Selenic acid is formed by the oxidation of H_2SeO_3 and is a strong, highly soluble acid. Selenate salts are soluble, whereas SeO_3^{2-} salts and metal Se^{2-} salts are sparingly soluble. Their solubilities and stabilities are the greatest in alkaline environments. Conversion of SeO_4^{2-} to the less-stable SeO_3^{2-} and to Se^0 occurs very slowly (10).

18.2 SELENIUM IN PLANTS

18.2.1 INTRODUCTION

The question of whether or not selenium is a micronutrient for plants is still considered unresolved (3). Selenium has not been classified as an essential element for plants, but its role as a beneficial element in plants that are able to accumulate large amounts of it has been considered (11). Uptake and accumulation of selenium by plants is determined by the form and concentration of selenium, the presence and identity of competing ions, and affinity of a plant species to absorb and metabolize selenium (10). Variation in selenium contents of plants seems to exceed that of nearly every other element (12). Nonconcentrator or nonaccumulator plant species will accumulate >25 mg Se kg^{-1} dry weight. Most crops species such as grains, grasses, fruits, vegetables, and many weed species are considered nonconcentrators (8,13). Secondary absorbers normally grow in areas with low to medium soil-selenium concentrations and can accumulate from 25 to 100 mg Se kg^{-1} dry weight. They belong to a number of different genera, including *Aster*, *Atriplex*, *Castelleja*, *Grindelia*, *Gutierrezia*, *Machaeranthera*, and *Mentzelia*. The primary indicator or selenium-accumulator species can accumulate from 100 to 10,000 mg Se kg^{-1} dry weight. This group includes species of *Astragalus*, *Machaeranthera*, *Haplopappus*, and *Stanleya* (14). These plant species are suspects for causing acute selenosis, or selenium toxicity, of range animals that consume the plants as forages (10,15). Selenium-accumulator plants can contain 100 times more selenium than nonaccumulator plants when grown on the same soil (16). Surveys of selenium concentrations in crops reveal that areas producing low-selenium crops (<0.1 mg Se kg^{-1}) are more common than those producing crops with toxic selenium levels (>2 mg Se kg^{-1}) (16).

18.2.2 UPTAKE

Selenium can be absorbed by plants as inorganic SeO$_4^{2-}$ or SeO$_3^{2-}$ or as organic selenium compounds such as the selenoamino acid, selenomethionine (Se-Met) (10). Selenate and organic selenium forms are taken up actively by plant roots, but there is no evidence that SeO$_3^{2-}$ uptake is mediated by the same process (3). Because of the close chemical and physical similarities between selenium and sulfur, their uptake by plants is very similar. Sulfur is absorbed actively by plants, mainly as SO$_4^{2-}$. The controlling enzymes for sulfur uptake are sulfur catabolic enzymes such as aryl sulfatase, choline sulfatase, and various S permeases (3,17,18). Uptake of SO$_4^{2-}$ and SeO$_4^{2-}$ was shown to be controlled by the same carrier with a similar affinity for both ions (19). This action demonstrated competition between SO$_4^{2-}$ and SeO$_4^{2-}$ for the same binding sites on these permeases (20,21).

Many studies have demonstrated an antagonistic relationship for uptake between SeO$_4^{2-}$ and SO$_4^{2-}$ (10,19,22–25). When SeO$_4^{2-}$ is present in high concentrations, it can competitively inhibit SO$_4^{2-}$ uptake. Adding SeO$_4^{2-}$ lowered SO$_4^{2-}$ absorption and transport in excised barley (*Hordeum vulgare* L.) roots. Conversely, adding SO$_4^{2-}$ lowered SeO$_4^{2-}$ absorption and transport (19,26). These studies involved an SeO$_4^{2-}$/SO$_4^{2-}$ ratio of 1:1. In a preliminary solution culture experiment, an SeO$_4^{2-}$/SO$_4^{2-}$ ratio of 1:3 resulted in the death of onion (*Allium cepa* L.) plant within 6 weeks (D.A. Kopsell and W.M. Randle, University of Georgia, unpublished results, 1994). When the SeO$_4^{2-}$/SO$_4^{2-}$ ratio was lowered to 1:500 or 1:125 in solution culture, Kopsell and Randle (27) reported significant increases in SO$_4^{2-}$ uptake by whole onion plants. Increasing SO$_4^{-2}$ levels from 0.25 to 10 mM in solution culture inhibited SeO$_4^{2-}$ uptake of broccoli (*Brassica oleracea* var. *botrytis* L.), Indian mustard (*Brassica juncea* Czern.), sugarbeet (*Beta vulgaris* L.), and rice (*Oryza sativa* L.) by 90% (22). Applications of gypsum (CaSO$_4 \cdot 2H_2O$) at the rates of 5.6 to 16.8 t ha^{-1} reduced selenium uptake in alfalfa (*Medicago sativa* L.) and oats (*Avena sativa* L.) grown on fly-ash landfill soils (28).

Although phosphate (H$_2$PO$_4^-$) is not expected to affect SeO$_4^{2-}$ uptake because of the chemical dissimilarities of the two radicals, the relationship between phosphate additions and selenium levels in plants has been inconsistent (9,10,29). Hopper and Parker (29) reported that a 10-fold increase (up to 200 µM) in phosphate solution culture decreased the selenium content of ryegrass (*Lolium perenne* L.) shoots and roots by 30 to 50% if selenium was supplied as SeO_3. In contrast, Carter et al. (30) reported that applying up to 160 kg P ha^{-1} either as H_3PO_4 or concentrated superphosphate to Gooding sandy loam increased selenium concentrations in alfalfa.

Selenate can accumulate in plants to concentrations much greater than that of selenium in the surrounding medium. In contrast, SeO_3^{2-} did not accumulate to levels surpassing the selenium levels of the external environment (31). When broccoli, Indian mustard, and rice were grown in the presence of SeO_4^{2-}, SeO_3^{2-}, or selenomethionine (Se-Met), plants accumulated the greatest amount of shoot selenium when selenium was supplied as SeO_4^{2-}, followed by those provided with Se-Met (22). In the same study, sugarbeet (*Beta vulgaris* L.) accumulated the most shoot-Se when treated with Se-Met (22). Broccoli, swiss chard (*Beta vulgaris* var. *cicla* L.), collards (*Brassica oleracea* var. *acephala* D.C.), and cabbage (*Brassica oleracea* var. *capitata* L.) grown in soil treated with 4.5 mg SeO_3^{2-} kg^{-1} or 4.5 mg SeO_4^{2-} kg^{-1} had a tissue concentration of Se in the range from 0.013 to 1.382 g Se kg^{-1} dry weight and absorbed 10 times the amount of selenium if treated with SeO_4^{2-} than with SeO_3^{2-} (32). When roots of bean (*Phaseolus vulgaris* L.) were incubated in 5 mmol m^{-3} Na_2SeO_3 or 5 mmol m^{-3} Na_2SeO_4 for 3 h, there was no significant difference in selenium accumulation, but distribution within the plant was different (33). In contrast, time-dependent kinetic studies showed that Indian mustard absorbed SeO_4^{2-} up to 2-fold faster than SeO_3^{2-} (34).

Increasing levels of selenium in plants may act to suppress the tissue concentrations of nitrogen, phosphorus, and sulfur. It can also inhibit the absorption of several heavy metals, especially manganese, zinc, copper, iron, and cadmium (35). This detoxifying effect of selenium has been demonstrated as reducing cadmium effects on garlic (*Allium sativum* L.) cell division (36). In contrast, the application of nitrogen, phosphorus, or sulfur is known to detoxify selenium. This effect may be due to either lowering of selenium uptake by the roots or to establishment of a safe ratio of selenium to other nutrient elements (35).

Selenomethionine was readily taken up by wheat (*Triticum aestivum* L.) seedlings, and the uptake followed a linear pattern in response to increasing selenium solution concentrations up to 1.0 µM (37). Western wheatgrass (*Pascopyrum smithii* Löve) also showed linear selenium uptake with Se-Met solution concentrations up to 0.6 mg Se L^{-1} (38). Results from Bañuelos et al. (39) showed that alfalfa accumulated selenium in plant tissues when selenium-laden mustard plant tissue was added to the soil. These studies provide evidence that organic selenium compounds in the soils may become available sources of selenium (40).

Genetic differences for selenium uptake and accumulation within species have also been reported. In 1939, Trelease and Trelease reported that cream milkvetch (cream locoweed, *Astragalus racemosus* Pursh.), a selenium-accumulator, produced 3.81 g dry weight in solution culture with 9 mg Na_2SeO_3 L^{-1}, whereas ground plum (*A. crassicarpus* Nutt.), a nonaccumulator, produced only 0.20 g dry weight (41). Shoots of different land races of Indian mustard grown hydroponically in the presence of 2.0 mg Na_2SeO_4 L^{-1} ranged from 501 to 1092 mg Se kg^{-1} dry matter, whereas shoots grown in soil culture at 2.0 mg Na_2SeO_4 kg^{-1} concentration ranged from 407 to 769 mg Se kg^{-1} dry matter (42). Total accumulation of selenium in onion bulb tissue ranged from 60 to 113 µg Se g^{-1} dry weight among 16 different cultivars responding to 2.0 mg Na_2SeO_4 L^{-1} nutrient solution (43).

18.2.3 METABOLISM

The incorporation of SeO_4^{2-} into organic compounds in plants occurs in the leaves (44). In a similar manner, SO_4^{2-} is reduced to sulfide (S^{2-}) in the leaves before being assimilated into the S-containing amino acid, cysteine (45). After SO_4^{2-} enters the cell it can be bound covalently in different secondary metabolites or immediately reduced and assimilated (46). Selenate is assimilated in the same metabolic pathways as SO_4^{2-}. Discrimination between SO_4^{2-} and SeO_4^{2-} was

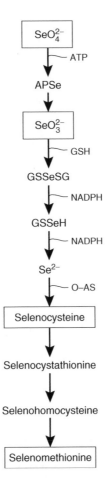

FIGURE 18.1   Proposed pathway for formation of the two Se-amino acids, Se-cysteine and Se-methionine in plants. (Abbreviations: APSe, adenosine 5′-selenophosphate; GSH, reduced glutathione; GSSeSG, selenotrisulphide; GSSeH, selenoglutathione; O-AS, acetylserine.) From A. Läuchli. *Bot. Acta* 106:455–468, 1993.

noted to occur at the level of amino acid incorporation into proteins. Uptake ratios between SO_4^{2-} and SeO_4^{2-} remained constant over a 60-h period for excised barley roots, but the ratio of S/Se decreased for free amino acid content and increased for proteins during assimilation (24).

In a series of solution-culture experiments with corn (*Zea mays* L.), Gissel-Neilsen (47) reported immediate selenium uptake and translocation to the leaves. Xylem sap contained 80 to 90% of ^{75}Se supplied as SeO_3 in amino-acid form, whereas 90% of ^{75}Se supplied as SeO_4 was recovered unchanged (47). In the leaves, selenate is converted into adenosine phosphoselenate (APSe) by ATP sulfurylase (Figure 18.1). In a similar fashion, SO_4^{2-} is first activated by ATP sulfurylase to form adenosine phosphosulfate (48). It has been suggested that ATP sulfurylase is not only the rate-limiting enzyme controlling the reduction of SO_4^{2-} (46), but it also appears to be the rate-limiting step in reduction of SeO_4^{2-} to SeO_3^{2-} (34,49). Overexpression of ATP sulfurylase in Indian mustard increased reduction of supplied SeO_4^{2-} (49). Following reduction of SeO_4^{2-}, APSe is converted into SeO_3^{2-}. Selenite is coupled to reduced glutathione (GSH), a sulfur-containing tripeptide to form a selenotrisulfide. Selenotrisulfide is reduced first to selenoglutathione and then to Se^{2-}. Selenide reacts with O-acetylserine to form selenocysteine (Se-Cys), which is further converted into Se-Met via selenocystathionine and selenohomocysteine (40). Ng and Anderson (50) reported that cysteine synthase enzymes extracted from selenium accumulator and nonaccumulator plants utilize Se^{2-} as an alternative substrate to S^{2-} to form Se-Cys in lieu of cysteine and that the affinity for Se^{2-} was substantially greater than for S^{2-}.

18.2.4 VOLATILIZATION

Biological methylation of selenium to produce volatile compounds occurs in plants, animals, fungi, bacteria, and microorganisms (9). The predominant volatile selenium species is dimethylselenide, which is less toxic (1/500 to 1/700) than the inorganic selenium species (51). Plant species differ in their rates of selenium volatilization, and these rates are correlated with tissue selenium concentrations (52). The ability of plants to accumulate selenium is a good indicator of their potential volatilization rate. It was reported that selenium was more readily transported to the shoots of an accumulator plant (*Astragalus bisulcatus* A. Gray), whereas a barrier to selenium movement to the shoots was seen in the nonaccumulator plant, western wheatgrass (*Pascopyrum smithii* A. Löve) (38). However, in broccoli, the roots were shown to be the primary site for selenium volatilization (53). In an earlier experiment with broccoli, Zayed and Terry (54) revealed that a decrease in selenium volatilization was observed with increased application of SO_4^{2-} fertilizer.

Volatilization of selenium is also influenced by the chemical form of selenium in the growing medium. The rate of selenium volatilization of a hybrid poplar (*Populus tremula* × *alba*) was 230-fold higher in sand culture if $20\,\mu M$ Se was supplied as Se-Met than as SeO_3^{2-}, and volatilization from SeO_3^{2-} was 1.5-fold that from SeO_4^{2-} (49). Selenium volatilization by shoots of broccoli, Indian mustard, sugarbeet, or rice supplied with Se-Met was also many folds higher than that from plants supplied with SeO_3^{2-} (22). In Indian mustard, Se-volatilization rates were doubled or tripled in sand culture amended with $20\,\mu M$ SeO_3^{2-} relative to rates with $20\,\mu M$ SeO_4^{2-} (34). These data indicate that selenium volatilization from SeO_4^{2-} is limited by the rate of SeO_4^{2-} reduction as well as by the form of selenium available (22,34).

18.2.5 PHYTOREMEDIATION

An increasing problem with irrigation agriculture in arid and semi-arid regions is the appearance of selenium in soils, ground water, and drainage effluents (12,55,56). The greatest concerns for selenium contamination come in areas where water systems drain seleniferous soils. One area of the United States that has come under close investigation because of elevated levels of selenium in the water is the San Joaquin Valley in California (57,58). Selenium enters the groundwater as soluble selenites and selenates and as suspended particles of sparingly soluble and organic forms of the element (8). The mobility of selenium in groundwater is related to its speciation in the aqueous solution, sorption properties of the substrate, and solubility of the solid phases (59). The ability of certain plants to take up, accumulate, and volatilize selenium has an important application in phytoremediation of selenium from the environment (3). Phytoremediation of selenium from contaminated soils is more practical and economical than its physical removal (60). Bioaccumulation of selenium in wetland habitats is also a problem and results in selenium toxicity to wildlife (61). There is a danger of selenium re-entering the local ecosystem if plant tissues that have accumulated selenium are consumed by wildlife or allowed to degrade (62).

The search for germplasm with the potential for effective phytoremediation has begun (63). The most ideal plant species for selenium phytoremediation should have the ability for rapid establishment and growth, ability to accumulate or volatilize large amounts of selenium, tolerate salinity and elevated soil boron, and develop large amounts of biomass on high-selenium soils (3,62–64). Indian mustard was more efficient at accumulating selenium than milkvetch (*Astragalus incanus* L.), Australian saltbush (*Atriplex semibaccata* R. Br.), old man saltbush (*Atriplex nummularia* Lindl.), or tall fescue (*Festuca arundinacea* Schreb.) when grown in potting soil amended with $3.5\,mg\ Se^{6+}$ kg^{-1} or $3.5\,mg\ Se^{4+}\ kg^{-1}$ as selenate or selenite (60).

Two of the options available once selenium is phytoextracted from contaminated soils are volatilization of methylated Se forms or harvest and removal of selenium-enriched plant biomass.

Plant species with a high affinity for phytovolatilization could remove selenium from the environment by releasing it into the atmosphere, where it is dispersed and diluted by air currents (3,11,62). Most of the selenium in the air comes from windblown dusts, volcanic activity, and discharges from human activities such as the combustion of fossil fuels, smelting and refining of nonferrous metals, and the manufacturing of glass and ceramics (8). The large particulate and aerosol forms of selenium generally are not readily available for intake by plants or animals. When 15 crop species were grown in solution culture with $20\,\mu M$ SeO_4^{2-}, rice, broccoli, or cabbage volatized 200 to $350\,\mu g$ Se m^{-2} leaf area day^{-1}, whereas sugar beet, bean, lettuce (*Lactuca sativa* L.), or onion volatized less than $15\,\mu g$ Se m^{-2} leaf area day^{-1} (52). One of the proposed disposal schemes for selenized plants from phytoremediation is as a source of forage for selenium-deficient livestock (3,60) Accurate determination of selenium levels as well as other trace elements in plant tissues and the use of other forages in a blended mixture would be needed to ensure proper dietary selenium levels in animal feeds (60,62).

18.3 SELENIUM TOXICITY TO PLANTS

Selenium toxicity is influenced by plant type, form of selenium in the growth medium, and presence of competing ions such as sulfate and phosphate (9). Interestingly, there are no written reports of selenium toxicity under cultivated conditions (9,12). This result may be because most crop plants show no injury or yield suppression until they accumulate at least 300 mg Se kg^{-1}, which is usually more than they contain even on seleniferous soils (9,14). In nonaccumulator plants, the threshold selenium concentration in shoot tissue that resulted in a 10% restriction in yield ranged from 2 mg Se kg^{-1} in rice to 330 mg Se kg^{-1} in white clover (*Trifolium repens* L.) (10). Wild-plant species growing in areas of elevated soil selenium tend to be adapted to those regions. Indicator plants can hyperaccumulate selenium to levels above 10,000 mg Se kg^{-1}, but possess biochemical means to avoid toxicity.

Descriptions for toxicity symptoms come only from solution-culture experiments. Stunting of growth, slight chlorosis, decreases in protein synthesis and dry matter production, and withering and drying of leaves are most often associated with selenium toxicity (4). Toxicity of selenium appears as chlorotic spots on older leaves that also exhibit bleaching symptoms. A pinkish, translucent color appearing on roots can also occur (65). Onions grown under extremely toxic Se concentrations showed sulfur-deficiency symptoms just before plant death (D.A. Kopsell and W.M. Randle, unpublished data, 1994).

The toxic effect of selenium to plants results mainly from interferences of selenium with sulfur metabolism (10). In most plant species, selenoamino acids replace the corresponding S-amino acids and are incorporated into proteins. Nuehierl and Böck (66) reported on a proposed mechanism of selenium tolerance in plants. In nonaccumulator plant species, Se-cys would either be incorporated into proteins or function as a substrate for downstream-sulfur pathways, which would allow selenium to interfere with sulfur metabolism. Replacing cysteine (Cys) with Se-Cys in S-proteins will alter the tertiary structure and negatively affect their catalytic activity (31). In contrast, accumulator plant species would instantly and specifically methylate Se-cys using Se-Cys methyltransferase, thereby avoiding Se-induced phytotoxicity (31). This action would remove selenium from the pool of substrates for cysteine metabolism. Thus, Se-Cys methyltransferase may be a critical enzyme conferring selenium tolerance in selenium-accumulating plants. Alternatively, tolerance may be achieved by sequestering selenium as selenate or other nonprotein Se-amino acids in the vacuole in accumulator plant cells (3).

18.4 SELENIUM IN THE SOIL

18.4.1 INTRODUCTION

The two forms of selenium that predominate in cultivated soils are SeO_4^{2-} and SeO_3^{2-} (8). Soils also contain organic selenium compounds such as Se-Met (67). Selenium occurs in the highest concentration in the surface layers of soils, where there is an abundance of organic matter (9).

Selenium in soils is generally considered to be controlled by an adsorption mechanism rather than by precipitation–dissolution reactions (68). In acid soils, sesquioxides control the sorption of selenium. Absorption controls the co-precipitation of SeO_3^{2-} by $Fe(OH)_3$. In mineral soils, SeO_4^{2-} was absorbed by soil solids. Adsorption is also believed to control the distribution of selenium in the soil under oxidizing conditions (68).

Transformation of SeO_3^{2-} to SeO_4^{2-} and vice versa occurs very slowly. The transformation of SeO_3^{2-} to Se^0 was found to be even slower (9). After Se^0 is added to soil, it oxidizes rapidly to SeO_3^{2-}. But, after the initial oxidation, the remaining selenium in the soil becomes inert, and any further oxidation proceeds very slowly. The rate of oxidation will vary in different soil types (68).

18.4.2 GEOLOGICAL DISTRIBUTION

Selenium attracts interest because the amount in which it is present in soils is not evenly distributed geographically. Seleniferous soils and vegetation in North America extend from Alberta, Saskatchewan, and Manitoba south along the west coast into Mexico (12). The mean total selenium in soils of the United States is reported to be 0.26 mg kg^{-1} (69). Considerable variability exists from one location to another, and high Se concentrations occur in a few localized regions. In the United States, seleniferous soils occur in the northern Great Plains states of North Dakota, South Dakota, Wyoming, Montana, Nebraska, Kansas, and Colorado and in the Southwest states of Utah, Arizona, and New Mexico. These soils average 4 to 5 mg Se kg^{-1} and can reach levels as high as 80 mg kg^{-1} in some areas (8). The primary selenium sources are the western shales of the Cretaceous Age and the carbonic debris of sandstone ores of the Colorado Plateau (9).

In the other parts of the world, selenium occurs in high amounts only in the semi-arid and arid regions derived from cretaceous soils (14). Seleniferous soils occur in Mexico, Columbia, Hawaii, and China. Toxic soil selenium levels (>300 mg kg^{-1}) in Europe are limited to a few locations in Wales and Ireland (16). High-selenium soils also occur in Iceland, probably because of the volcanic activity on the island (16). In contrast, soils in Denmark, the Netherlands, Switzerland, Australia, and New Zealand, and Finland are naturally low in selenium (16). In humid climates, or in irrigated areas, most of the selenium is leached from soils (9). The most severe selenium-deficient area in the world is the Keshan region in southeastern China (16), where many children have died owing to insufficient dietary selenium. Variations in soil selenium can give rise to differences of selenium in the food chain (70).

Selenium can enter the soil through weathering of selenium-containing rocks, volcanic activity, phosphate fertilizers, and water movement. The selenium content in the soil reflects the concentration in the parent material, secondary deposition or redistribution of selenium in the soil profile, accumulation and deposition by selenium-accumulating plant materials, and erosion from selenium-containing rocks (71). The highest amounts of selenium are in igneous rock formations, existing as Se^{2-} or sulfoselenides with copper, silver, lead, mercury, and nickel (8). Selenium also occurs under sedimentary rock formations. The weathering of selenium-containing rocks under alkaline and well-aerated conditions releases selenium into the soil, which oxidizes it into the SeO_4^{2-} form. Selenium released from rocks under acidic, poorly aerated conditions will form insoluble Se^{2-} and SeO_3^{2-}. These forms of selenium develop stable adsorption complexes with ferric hydroxide and are less available to plants (8). The level of selenium in a phosphate fertilizer is governed by the concentration of selenium in the phosphatic rock (9). Fifteen different rock-phosphate fertilizers from sources in Canada and the United States ranged in selenium concentration from 0.07 to 178 mg kg^{-1} (72). Ordinary and concentrated super phosphate can be expected to contain between 40 and 60% more selenium than the phosphate rock from which it was made (72).

The distribution of selenium in the soil profile is determined by factors such as soil type, amount of organic matter, soil pH, and to some extent, leaching caused by rainfall. Organic matter helps to retain selenium in the surface horizon and has a greater SeO_3-fixation capacity than clay minerals do (9,16). Soil pH, aeration, water levels, and oxidation–reduction conditions have an effect on the form of selenium in the soil and its availability to plants. Selenates are highly soluble in water and do not have stable adsorption complexes, thereby making them highly leachable (8).

Metal selenides occur in metal sulfide ores of iron, copper, and lead. Selenium occurs in small quantities in pyrite and in the minerals clausthalite (PbSe), naumannite ((Ag,Pb)Se), and tiemannite (HgSe). The similarity of the ionic radii of Se^{2-} (0.191 nm) and S^{2-} (0.184 nm) results in substitution of Se^{2-} for S^{2-}. Soil pH will affect the capacity of clays and ferric oxides to adsorb selenium (10). Selenite has a strong affinity for sorption, especially by iron oxides like geothite, amorphous iron hydroxide, and aluminum sesquioxides. Adsorption of SeO_3^{2-} is also a function of soil-particle concentration and composition, SeO_3^{2-} concentration, and the concentration of competing anions such as phosphate (10). Being stable in reducing environments, Se^0 can be oxidized to SeO_3^{2-} and to trace amounts of SeO_4^{2-} by some microorganisms.

18.4.3 SELENIUM AVAILABILITY IN SOILS

Soil texture can affect selenium availability and uptake by plants. Because of the adsorption of SeO_3^{2-} to clay fractions in the soil, plants grown on sandy soils take up twice as much selenium as those grown on loamy soils (10). Organic matter has the ability to draw selenium from the soil solution (10). In general, selenium concentrations in plants will increase as the level of soil selenium increases, but will decrease with the addition of SO_4^{2-} (10). Extraction of selenium from soils is increased when SO_4^{2-} is used in the leaching process (9). The presence of low-molecular-weight organic acids in the soil–root interface resulted in the loss of SeO_3^{2-} sorption sites on aluminum hydroxides (73). A decrease in total selenium accumulation from soils supplied with sodium selenate (Na_2SeO_4) resulted under conditions of increasing levels of sodium (NaCl) and calcium (CaCl) salinity for canola (*Brassica napus* L.), kenaf (*Hibiscus cannibinus* L.), and tall fescue (74).

The chemical form of selenium in the soil is determined mainly by soil pH and redox potential (Figure 18.2). In alkaline soils, selenium is in the available SeO_4^{2-} form. When soil conditions become neutral to acidic, sparingly soluble ferric oxide–selenite complexes develop. Since sparingly soluble forms dominate at low pH, liming of the soil to raise the pH also has an effect by increasing the availability of selenium to plants (9). This response to addition of lime is probably

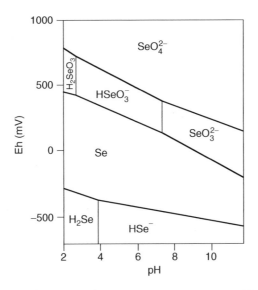

FIGURE 18.2 Selenium speciation in an aqueous system: effect of pH and oxidation–reduction potential E_h. From R.L. Mikkelsen, et al., *Selenium in Agriculture and the Environment*. Madison, WI: American Society of Agronomy, Soil Science Society of America, 1989, pp. 65–94.

caused by the reduced absorption to clays and iron oxides, resulting from increases in the soil pH (75). In the soil solution, the pH can change the speciation of selenium present. Below pH 4.5, soluble selenium speciation was 71% SeO_4^{2-} and 8% SeO_3^{2-}. When the pH was 7.0, the percentages were 51% for SeO_4^{2-} and 23% for SeO_3^{2-}. After 105 days, SeO_4^{2-} accounted for 22% and SeO_3^{2-} for 20% at pH 4.5, and were 12 and 22%, respectively, at pH 7.0 (76).

Selenium can be supplied to plants by application to soil, by foliar sprays, and by seed treatments (16). Slow-release selenium fertilizers were effective over a 4-year period in maintaining selenium levels in subterranean clover (*Trifolium subterraneum* L.) to prevent selenium deficiency in sheep in Australia (77). Use of selenium-enriched $Ca(NO_3)_2$ significantly increased selenium in wheat (*Triticum aestivum* L.) (78). Coal fly ash has been used as a source of soil-applied selenium as well as many heavy metals (9). One should be careful when using phosphate fertilizers as soil amendments, since they may contain substantial amounts of selenium (10). Selenium incorporation into fertilizers is becoming common in some countries with low soil-Se levels. Spraying SeO_4^{2-} onto pumice has been used for the production of selenium prills in New Zealand (16,77).

18.5 SELENIUM IN HUMAN AND ANIMAL NUTRITION

18.5.1 INTRODUCTION

After its discovery, selenium was most noted for its harmful effects. Selenium was the first element identified to occur in native vegetation at levels toxic to animals. Poisoning of animals can occur through consumption of plants containing toxic levels of selenium (79). Livestock consuming excessive amounts of selenized forages are afflicted with 'alkali disease' and 'blind staggers.' Typical symptoms of these diseases include loss of hair, deformed hooves, blindness, colic, diarrhea, lethargy, increased heart and respiration rates, and eventually death. On the other hand, selenium deficiency in animal feeds can cause 'white muscle disease,' a degenerative disease of the cardiac and skeletal muscles (9). Perceptions of selenium changed when Schwarz and Foltz (80) reported that additions of selenium prevented liver necrosis in rats (*Rattus* spp.) deficient in vitamin E. Its role in human health was established in 1973 when selenium, the last of 40 nutrients proven to be essential, was shown to be a component of glutathione peroxidase (GSHx), an enzyme that protects against oxidative cell damage (81). The United States' recommended daily allowance for selenium is 50 to 70 µg in human diets (5). Currently, all of the known functions of selenium as an essential nutrient in humans and other animals have been associated with selenoproteins (82).

18.5.2 DIETARY FORMS

Organic forms of selenium appear to be more bioavailable than the inorganic ones because the organic forms are more easily absorbed, have the ability to be stored in seleno- and other nonspecific proteins, and have lower renal clearance (83). The organic-selenium compounds identified in plants include Se-Cys, Se-methylselenocysteine, selenohomocystine, Se-Met, Se-methyl-selenomethionine, selenomethionine selenoxide, selenocystathionine, and di-methyl diselenide, selenoethionine, and Se-allyl selenocysteine (41,84,85). The majority of selenium in seleniferous wheat was shown to be Se-Met (86). The effect of consumption of seleniferous wheat on urinary excretion and retention in the body was similar to that of Se-Met supplementation (87). The form of selenium in nuts is selenocystathionine (88). The high-selenium-accumulating species of milkvetch (*Astragalus* spp. L.) contain Se-methylselenocysteine and selenocystathionine (89). Most fruits and vegetables contain >0.1 mg Se kg^{-1}, (13) but some have the potential to be enriched. Marine fish such as tuna are high in selenium, but bioactivity is much lower than selenium from other foods (90). Inorganic SeO_3^{2-}, SeO_4^{2-}, and Se^{2-} have been identified in plants at low levels (91). Selenate and SeO_3^{2-} are not regarded as naturally occurring forms of selenium in foods, but they have high biological activity, and animals can metabolize them into more active forms such

as Se-Cys (90). Selenocysteine is a component of glutathione peroxidase and constitutes the majority of selenium in animal proteins.

18.5.3 METABOLISM AND FORM OF SELENIUM

The bioavailability and metabolism of selenium and its distribution in an organism depend on the
form of selenium ingested (83). The chemical form of selenium in foods and supplements determines absorption, speciation, and metabolism within the body, bioavailability for selenoproteins,
and toxicity (87). Inorganic forms of selenium are absorbed rapidly, but are equally rapidly excreted
in the urine, in contrast to Se-Met, which is retained in the body. Total recovery of inorganic forms
of selenium in urine and feces of human subjects was 82 to 95% of the total dose, whereas only
26% of the total Se-Met was recovered after being ingested (87). Prolonged consumption of any one
single form of selenium can produce side effects such as exaggerated accumulation in body tissues
(Se-Met) and changes in cellular glutathione homeostasis (selenite) (92). When high levels of inorganic SeO_3^{2-} or organic Se-Met were fed to rats, higher selenium concentrations in body tissues
were found for Se-Met than for SeO_3^{2-}. Selenium levels in erythrocytes, testes, kidney, and lungs
were not significantly different between rats fed with 0.2 mg kg^{-1} Se as SeO_3^{2-} and those fed with
Se as Se-Met, but higher levels of selenium were found in liver, muscle, and brain tissues for rats
fed with Se-Met (93). There was an increase of up to 26-fold in the concentration of selenium localized in muscle tissues for rats fed with high levels of selenium as Se-Met when compared with those
fed with SeO_3^{2-}. Selenium from Se-Met and seleno yeast showed higher accumulation in liver and
muscle tissues than that from SeO_3^{2-} for channel catfish (94).

18.6 SELENIUM AND HUMAN HEALTH

18.6.1 INTRODUCTION

Immune system enhancement, cancer suppression, and cardiovascular disease reduction are all
associated with increased dietary selenium (95–97). The chief biological function of selenium is as
an essential cofactor to the enzyme GSHx (81). The antioxidant enzyme GSHx protects against
oxidative stress by removing DNA-damaging hydrogen peroxide and lipid hydroperoxides. The
chemopreventive action of selenium may come from its role in GSHx (98). Other protective qualities attributed to selenium, independent of GSHx activity, include repair of damaged DNA (99),
reduction in DNA binding of carcinogens (100), and suppressing genetic mutations (101).

18.6.2 SELENIUM DEFICIENCY AND TOXICITY IN HUMANS

The average selenium intake by humans in most countries is sufficient to meet the United States'
recommended daily allowances, and selenium deficiency in healthy humans is relatively rare (5,6).
Selenium status in a population correlates highly with the selenium content of locally produced
crops (7). In areas of the world with low soil selenium, addition of selenium in normal fertility
regimes is practiced to avoid selenium deficiencies in humans and livestock (16). A significant
inverse relationship between low-selenium status and increased risk of cancer mortality has been
established for some rural counties of the United States (102).

The link between selenium deficiency and disease is associated with more than 40 different
health conditions (103). The first reports of diseases linked to selenium status came from regions of
China having extremely low soil selenium. Keshan disease, an endemic cardiomyopathy, and
Kashin-Beck disease, a chronic and deforming arthritis, have been linked to selenium deficiency
(104). Selenium deficiency also depresses the effectiveness of immune cells. Selenium deficiency
was found to be an independent predictor of survival rates among patients infected with HIV
(human immunodeficiency virus) (105). Increasing selenium intake in animals and human beings

increases antitumorigenic activities (106), and selenium-dietary supplementation decreases severity of several viral diseases (107).

The United States National Academy of Sciences has identified selenium intake of up to 200 µg day^{-1} as safe (108). However, sustained consumption of selenium levels exceeding 750 µg day^{-1} can cause selenium poisoning or selenosis (109). Signs of human selenosis include morphological changes in fingernails and hair loss, with an accompanied garlicky breath odor. Human selenosis reports have come from regions in China, where extremely high levels of soil selenium caused human-dietary intake to be >900 µg day^{-1} (110).

18.6.3 ANTICARCINOGENIC EFFECTS OF SELENIUM

There is perhaps no more extensive body of evidence for the cancer preventive potential of a normal dietary component than there is for selenium (106). Evidence for inverse associations between nutritional selenium status and cancer risk exist from epidemiological studies (111,112), experimental animal models (92,113), and most recently, clinical trials (5). Selenium supplementation resulted in a 63% reduction in the incidence of prostate cancer over a 10-year period in an at-risk group of men given 200 µg Se day^{-1} (5). Experimental antitumorigenic effects of selenium are associated with supranutritional levels of at least 10 times those required to prevent clinical signs of selenium deficiency (106). These levels are higher than those experienced by most people, an amount which tends to be <150 to 200 µg Se day^{-1}. Anticarcinogenic activity of selenium may not involve its usual role as a nutrient because selenium-dependent enzyme activities are already at a maximum at levels of selenium below effective anticarcinogenic level and the forms of selenium that lack nutritional activity (not synthesized by Se-dependent enzymes) show good cancer-preventing activity (82). Therefore, for anticarcinogenic effects to be seen, supplementation of selenium in the diet is usually needed. Inorganic SeO_3^{2-} and yeast-derived Se-Met are the most common selenium supplements for human consumption.

18.6.4 IMPORTANCE OF SELENIUM METHYLATION IN CHEMOPREVENTIVE ACTIVITY

Methylation is the best-known fate of selenium, and fully methylated metabolites are regarded as detoxified forms of selenium. Selenium methylselenocysteine has very high chemopreventive activity. This form of selenium is naturally occurring in plants enriched with selenium and does not get incorporated into proteins, thus minimizing excessive accumulation in body tissues. The metabolism of Se-methylselenocysteine produced monomethylated forms of selenium as excretory products (82). The potential activity of selenium can be enhanced in the course of being metabolized in plants, especially in those having specialized alkyl-group capabilities. Some plants such as alliums can transfer allyl groups to sulfur, or possibly, selenium. These allyl groups can undergo methylation to form highly chemopreventive alkylated derivatives (82). Selenium-enriched garlic (*Allium sativum* L.) had higher chemopreventive activity than regular garlic alone in animal models (113). Natural selenium products formed in plants are very active chemopreventive metabolites. They show higher activity in animals than the selenium compounds metabolized from inorganic selenium sources (82).

18.7 SELENIUM ENRICHMENT OF PLANTS

Substantial genetic variation in plants has been reported for mineral (43,114,115), vitamin (116), and phytochemical content (117). Breeding plants that are enriched with mineral nutrients and vitamins could substantially reduce the recurrent costs associated with fortification (118,119). Successful programs are now in place for improving zinc (120) and iron (119) contents of wheat. Selenium fertilizer has been used in Finland on vegetable crops to increase the uptake levels of dietary Se in both humans and other animals (121). However, there is very little information on the

TABLE 18.1
Selenium Tissue Analysis Values of Various Plant Species

Plant Common and Scientific Name	Variety	Type of Culture[a]	Type of Tissue Sampled	Age, Stage, Condition, or Date of Sample	Selenium Treatment	Selenium Concentration in Dry Matter (mg kg⁻¹ unless otherwise noted)			Reference
						Low	Medium	High	
Alfalfa (*Medicago sativa* L.)		Sand	Shoot	Three cuttings	No Se treatment; pH 4.5	0.10	—	0.20	
					0.25 mg L⁻¹ Na₂SeO₃; pH 4.5	14.1	—	28.7	
					0.50 mg L⁻¹ Na₂SeO₃; pH 4.5	27.6	—	28.9	
					1.0 mg L⁻¹ Na₂SeO₃; pH 4.5	32.7	—	49.9	
					0.25 mg L⁻¹ Na₂SeO₃; pH 4.5	21.6	—	24.3	
					0.50 mg L⁻¹ Na₂SeO₃; pH 4.5	38.3	—	52.6	
					1.0 mg L⁻¹ Na₂SeO₃; pH 4.5	73.8	—	165.4	
					3.0 mg L⁻¹ Na₂SeO₃; pH 4.5	478.2	—	912.7	133
					No Se treatment; pH 7.0	0.10	—	0.50	
					0.25 mg L⁻¹ Na₂SeO₃; pH 7.0	19.2	—	60.1	
					0.50 mg L⁻¹ Na₂SeO₃; pH 7.0	52.7	—	63.5	
					1.0 mg L⁻¹ Na₂SeO₃; pH 7.0	92.4	—	131.4	
					3.0 mg L⁻¹ Na₂SeO₃; pH 7.0	183.3	—	382.4	

Continued

TABLE 18.1 (*Continued*)

Plant Common and Scientific Name	Variety	Type of Culture[a]	Type of Tissue Sampled	Age, Stage, Condition, or Date of Sample	Selenium Treatment	Selenium Concentration in Dry Matter (mg kg^{-1} unless otherwise noted)			Reference
						Low	Medium	High	
					0.25 mg L^{-1} Na$_2$SeO$_3$; pH 7.0	28.4	—	65.1	
					0.50 mg L^{-1} Na$_2$SeO$_3$; pH 7.0	61.5	—	169.0	
					1.0 mg L^{-1} Na$_2$SeO$_3$; pH 7.0	174.4	—	503.30	
					3.0 mg L^{-1} Na$_2$SeO$_3$; pH 7.0	722.3	—	1581.60	
	'Germain WL 512'	Sand	Shoot	First harvest	No Se treatment	—	<0.05 mg kg^{-1}	—	134
				Second harvest	No Se treatment	—	<0.05 mg kg^{-1}	—	
				First harvest	0.25 mg L^{-1} Na$_2$SeO$_4$	—	44.3 mg kg^{-1}	—	
				Second harvest	0.25 mg L^{-1} Na$_2$SeO$_4$	—	30.1 mg kg^{-1}	—	
				First harvest	0.5 mg L^{-1} Na$_2$SeO$_4$	—	133.3 mg kg^{-1}	—	
				Second harvest	0.5 mg L^{-1} Na$_2$SeO$_4$	—	45.5 mg kg^{-1}	—	
				First harvest	1.0 mg L^{-1} Na$_2$SeO$_4$	—	620 mg kg^{-1}	—	
				Second harvest	1.0 mg L^{-1} Na$_2$SeO$_4$	—	98.6 mg kg^{-1}	—	
	'Honey-oye'	Soil	Shoot		50 ton A^{-1} Se as fly ash (16.8 ppm Se)	—	0.13 mg kg^{-1}	—	135

Species	Tissue	Medium	Treatment		Concentration		Ref.
Astragalus, (Two-grooved milkvetch, *Astragalus bisulcatus* A. Gray) See entry under milkvetch.		Solution	No Se treatment	—	44 µg kg⁻¹	—	136
			0.25 µg L⁻¹ SeO₃	—	272 µg kg⁻¹	—	
			5 µg L⁻¹ SeO₃	—	6200 µg kg⁻¹	—	
			10 µg L⁻¹ SeO₃	—	10,700 µg kg⁻¹	—	
	Roots		No Se treatment	—	27 µg kg⁻¹	—	
			0.25 µg L⁻¹ SeO₃	—	252 µg kg⁻¹	—	
			5 µg L⁻¹ SeO₃	—	3480 µg kg⁻¹	—	
			10 µg L⁻¹ SeO₃	—	6650 µg kg⁻¹	—	
Astragalus crotalariae A. Gray	Tops	Solution	No Se treatment	—	0.97 µg kg⁻¹	—	
			0.25 µg L⁻¹ SeO₃	—	238 µg kg⁻¹	—	
			1 µg L⁻¹ SeO₃	—	452 µg kg⁻¹	—	
			2.5 µg L⁻¹ SeO₃	—	1530 µg kg⁻¹	—	
			10 µg L⁻¹ SeO₃	—	4960 µg kg⁻¹	—	
			50 µg L⁻¹ SeO₃	—	26,900 µg kg⁻¹	—	
			100 µg L⁻¹ SeO₃	—	30,300 µg kg⁻¹	—	
	Roots		No Se treatment	—	22 µg kg⁻¹	—	
			0.25 µg L⁻¹ SeO₃	—	151 µg kg⁻¹	—	
			1 µg L⁻¹ SeO₃	—	363 µg kg⁻¹	—	
			2.5 µg L⁻¹ SeO₃	—	750 µg kg⁻¹	—	

Continued

TABLE 18.1 (*Continued*)

Plant Common and Scientific Name	Variety	Type of Culture[a]	Type of Tissue Sampled	Age, Stage, Condition, or Date of Sample	Selenium Treatment	Selenium Concentration in Dry Matter (mg kg⁻¹ unless otherwise noted) Low	Medium	High	Reference
					$10\,\mu g\,L^{-1}$ SeO_3	—	$2400\,\mu g\,kg^{-1}$	—	
					$50\,\mu g\,L^{-1}$ SeO_3	—	$10{,}200\,\mu g\,kg^{-1}$	—	
					$100\,\mu g\,L^{-1}$ SeO_3	—	$20{,}800\,\mu g\,kg^{-1}$	—	
Barley (*Hordeum vulgare* L.)		Native soil[a]	Grain		No Se treatment	—	0.09	—	
					$1.12\,kg\,ha^{-1}$ Na_2SeO_3; pH 6.6	—	1.24	—	137
					$2.24\,kg\,ha^{-1}$ Na_2SeO_3 pH 6.6	—	—	—	
	'Iona'	Foliar application	Grain		$10\,g\,ha^{-1}$ Na_2SeO_4	—	2.00	—	
					$20\,g\,ha^{-1}$ Na_2SeO_4	—	0.51	—	138
			Straw		$10\,g\,ha^{-1}$ Na_2SeO_4	—	1.13	—	
					$20\,g\,ha^{-1}$ Na_2SeO_4	—	0.50	—	
					Na_2SeO_4	—	0.79	—	
Bean (*Phaseolus vulgaris* L.)	'Tender-crop'	Soil	Pods		$100\,ton\,A^{-1}$ Se as fly ash (16.8 ppm Se)	—	0.47	—	139
					$50\,ton\,A^{-1}$ Se as fly ash (16.8 ppm Se)	—	0.07	—	135

	RCBP							
Rapid-growing brassica (*Brassica oleracea* L.)	Solution	Leaves	No Se treatment	—	ND	—		140
			3.0 mg L⁻¹ Na₂SeO₄	—	522	—		
			6.0 mg L⁻¹ Na₂SeO₄	—	1275	—		
			9.0 mg L⁻¹ Na₂SeO₄	—	1916	—		
		Stem	No Se treatment	—	ND	—		
			3.0 mg L⁻¹ Na₂SeO₄	—	267	—		
			6.0 mg L⁻¹ Na₂SeO₄	—	721	—		
			9.0 mg L⁻¹ Na₂SeO₄	—	1165	—		
		Root	No Se treatment	—	ND	—		
			3.0 mg L⁻¹ Na₂SeO₄	—	338	—		
			6.0 mg L⁻¹ Na₂SeO₄	—	857	—		
			9.0 mg L⁻¹ Na₂SeO₄	—	1636	—		
Broccoli (*Brassica oleracea* var. *botrytis* L.)	Soil	Floret	5 mg kg⁻¹ Na₂SeO₃	—	155	—		32
			5 mg kg⁻¹ Na₂SeO₄	—	1382	—		
		Composite leaves	5 mg kg⁻¹ Na₂SeO₃	—	49	—		
			5 mg kg⁻¹ Na₂SeO₄	—	377	—		
Cabbage (*Brassica oleracea* var. *capitata* L.)	Soil	Young leaves	5 mg kg⁻¹ Na₂SeO₃	—	52	—		32
			5 mg kg⁻¹ Na₂SeO₄	—	479	—		

Continued

TABLE 18.1 (*Continued*)

Plant Common and Scientific Name	Variety	Type of Culture[a]	Type of Tissue Sampled	Age, Stage, Condition, or Date of Sample	Selenium Treatment	Selenium Concentration in Dry Matter (mg kg^{-1} unless otherwise noted)			Reference
						Low	Medium	High	
			Old leaves		5 mg kg^{-1} Na$_2$SeO$_3$	—	38	—	—
			Composite leaves		5 mg kg^{-1} Na$_2$SeO$_4$	—	275	—	—
					5 mg kg^{-1} Na$_2$SeO$_3$	—	41	—	—
					5 mg kg^{-1} Na$_2$SeO$_4$	—	316	—	—
	'Scandic'	Native soil	Leaves		No Se treatment	11.00 µg kg^{-1} fresh weight	45.00 µg kg^{-1} fresh weight	100.00 µg kg^{-1} fresh weight	141
	'Golden Acre'	Soil	Leaves		100 ton A^{-1} Se as fly ash (16.8 ppm Se)	—	0.95	—	139
					50 ton A^{-1} Se as fly ash (16.8 ppm Se)	—	0.20	—	135
Canola (*Brassica napus* L.)	'Wester'	Soil	Leaves	First harvest	1.5 mg kg^{-1} as SeO$_4^{2-}$ or Se organic materials	1.60	—	283	142
				Second harvest	1.5 mg kg^{-1} as SeO$_4^{2-}$ or Se organic materials	0.80	—	7.70	
			Stems	First harvest	1.5 mg kg^{-1} as SeO$_4^{2-}$ or Se organic materials	0.50	—	57.00	

Plant	Growth medium	Plant part	Harvest	Treatment				
			Second harvest	1.5 mg kg^{-1} as SeO_4^{2-} or Se organic materials	0.30	—	5.60	
		Roots	First harvest	1.5 mg kg^{-1} as SeO_4^{2-} or Se organic materials	0.60	—	87.50	
			Second harvest	1.5 mg kg^{-1} as SeO_4^{2-} or Se organic materials	0.80	—	5.80	
	Native soil	Shoots	—	40 mg kg^{-1} Se in soil	280	—	470.0	32
			—	0.1 mg kg^{-1} Se in soil	0.20	—	0.60	
		Roots	—	40 mg kg^{-1} Se in soil	25	—	44	
			—	0.1 mg kg^{-1} Se in soil	0.10	—	0.20	
Carrot (*Daucus carota* L.) 'Scarlet Nantes'	Soil	Root		100 ton A^{-1} Se as fly ash (16.8 ppm Se)	—	0.19	—	139
				50 ton A^{-1} Se as fly ash (16.8 ppm Se)		0.06	—	135
Celery (*Apium graveolens* L.) 'Seoul'	Solution	Leaves, petioles		6 mg L^{-1} Na_2SeO_4		57.3	—	143
Collards (*Brassica oleracea* var *acephala* DC.)	Soil	Leaf		5 mg kg^{-1} Na_2SeO_3		36	—	32
				5 mg kg^{-1} Na_2SeO_4		398	—	
		Mid-rib/petiole		5 mg kg^{-1} Na_2SeO_3		23		
				5 mg kg^{-1} Na_2SeO_4		240		

Continued

TABLE 18.1 (*Continued*)

Plant Common and Scientific Name	Variety	Type of Culture[a]	Type of Tissue Sampled	Age, Stage, Condition, or Date of Sample	Selenium Treatment	Selenium Concentration in Dry Matter (mg kg⁻¹ unless otherwise noted)			Reference
						Low	Medium	High	
			Composite leaves		5 mg kg⁻¹ Na₂SeO₃	—	33	—	
					5 mg kg⁻¹ Na₂SeO₄	—	455	—	
			Seeds		5 mg kg⁻¹ Na₂SeO₃	—	18	—	
					5 mg kg⁻¹ Na₂SeO₄	—	491	—	
Tall fescue (*Festuca arundinacea* L.)	'Fawn'	Soil	Shoots	First harvest	1.5 mg kg⁻¹ as SeO₄²⁻ or Se organic materials	0.40	—	75.2	142
				Second harvest	1.5 mg kg⁻¹ as SeO₄²⁻ or Se organic materials	0.80	—	74.6	142
		Native soil	Shoots	First clipping (60 days)	0.46 mg kg⁻¹ Se in soil	—	310	—	55
				Second clipping (115 days)	0.46 mg kg⁻¹ Se in soil	—	630	—	
				First clipping (60 days)	0.65 mg kg⁻¹ Se in soil	—	170	—	
				Second clipping (85 days)	0.65 mg kg⁻¹ Se in soil	—	200	—	
				Third clipping (115 days)	0.65 mg kg⁻¹ Se in soil	—	270	—	
	'Alta'	Native soil	Shoots		40 mg kg⁻¹ Se in soil	10	—	50	62

Species	Soil	Part	Treatment				Ref.
Fourwing Saltbush [Atriplex canescens Nutt.]	Native soil	Shoots	0.1 mg kg⁻¹ Se in soil	0.01	—	0.14	
	Soil	Shoots	1.8 mg Se kg⁻¹ in soil	—	2.10	—	144
			4.8 mg Se kg⁻¹ (3.0 mg Na₂SeO₄ kg⁻¹)	—	172	—	
Grape (Vitis vinifera L.) 'Cabernet Sauvignon'	Soil	Leaves	0 to 1.5 kg Se ha⁻¹ as Na₂SeO₄	0.02 to 0.12 µg g⁻¹	—	10.41	145
	Native soil	Fruit	0.15 ± 0.02 µg Se g⁻¹	—	0.02	—	146
			0.31 ± 0.06 µg Se g⁻¹ in soil	—	0.04	—	
			0.49 ± 0.03 µg Se g⁻¹ in soil	—	0.06	—	
Kanef (Hibiscus cannabinus L.) 'Indian'	Native soil	Shoots	40 mg kg⁻¹ Se in soil	36	—	45	42
			0.1 mg kg⁻¹ Se in soil	0.75	—	1.10	
		Roots	40 mg kg⁻¹ Se in soil	36	—	62	
			0.1 mg kg⁻¹ Se in soil	0.86	—	1.10	
	Native soil	Shoots	0.75 mg kg⁻¹ Se in soil	—	520	—	55
		Roots	0.75 mg kg⁻¹ Se in soil	—	420	—	
Lettuce (Lactuca sativa L.)	Soil	Leaves	No Se treatment	—	0.05	—	147
			0.1 mg kg⁻¹ H₂SeO₄	—	6.40	—	
			1.0 mg kg⁻¹ H₂SeO₄	—	270.0	—	

Continued

TABLE 18.1 (*Continued*)

Plant Common and Scientific Name	Variety	Type of Culture[a]	Type of Tissue Sampled	Selenium Treatment	Age, Stage, Condition, or Date of Sample	Selenium Concentration in Dry Matter (mg kg⁻¹ unless otherwise noted) Low	Medium	High	Reference
Milkvetch, two-grooved (*Astragalus bisulcatus* A. Gray)		Solution	Tops	1.0 mg L⁻¹ Na₂SeO₃		—	243	—	38
				2.0 mg L⁻¹ Na₂SeO₃		—	510	—	
				0.3 mg L⁻¹ Se-Met		—	283	—	
				0.6 mg L⁻¹ Se-Met		—	274	—	
				0.3 mg L⁻¹ Se-Cys		—	46.8	—	
				0.6 mg L⁻¹ Se-Cys		—	95.2	—	
See entry under *Astragalus*.			Roots	1.0 mg L⁻¹ Na₂SeO₃		—	202	—	
				2.0 mg L⁻¹ Na₂SeO₃		—	407	—	
				0.3 mg L⁻¹ Se-Met		—	350	—	
				0.6 mg L⁻¹ Se-Met		—	428	—	
				0.3 mg L⁻¹ Se-Cys		—	124	—	
				0.6 mg L⁻¹ Se-Cys		—	222	—	
Millet, Japanese; barnyardgrass (*Echinochloa crusgalli* var. *frumentacea* Wight)		Soil	Grain	100 ton A⁻¹ Se as fly ash (16.8 ppm Se)			0.90		139
				50 ton A⁻¹ Se as fly ash (16.8 ppm Se)		—	0.16	—	135
Indian mustard (*Brassica juncea* L.)	Land races	Solution	Shoots	2.0 mg L⁻¹ Na₂SeO₄		501.00 ± 26.00 mg kg⁻¹	—	1092	62
			Roots	2.0 mg kg⁻¹ Na₂SeO₄		197.00 ± 16.00 mg kg⁻¹	—	470	

Continued

		Treatment				
	Soil	Shoots	2.0 mg L⁻¹ Na₂SeO₄	407.00 ± 26.00 mg kg⁻¹	—	769
		Roots	2.0 mg kg⁻¹ Na₂SeO₄	152.00 ± 38.00 mg kg⁻¹	—	332
	Native soil	Shoots	0.50 mg kg⁻¹ Se in soil	—	950	—	55
			0.86 mg kg⁻¹ Se in soil	—	1050	—
Onion (*Allium cepa* L.)	'Granex 33' Solution	Leaves	No Se treatment	—	ND	—	27
			0.5 mg L⁻¹ Na₂SeO₄	—	47.3	—
			1.0 mg L⁻¹ Na₂SeO₄	—	109.3	—
			1.5 mg L⁻¹ Na₂SeO₄	—	140	—
			2.0 mg L⁻¹ Na₂SeO₄	—	208	—
		Bulb	No Se treatment	—	ND	—
			0.5 mg L⁻¹ Na₂SeO₄	—	18.9	—
			1.0 mg L⁻¹ Na₂SeO₄	—	41.4	—
			1.5 mg L⁻¹ Na₂SeO₄	—	56.5	—
			2.0 mg L⁻¹ Na₂SeO₄	—	70.9	—
		Root	No Se treatment	—	ᵃND	—
			0.5 mg L⁻¹ Na₂SeO₄	—	37.7	—
			1.0 mg L⁻¹ Na₂SeO₄	—	78.9	—
			1.5 mg L⁻¹ Na₂SeO₄	—	104.3	—
			2.0 mg L⁻¹ Na₂SeO₄	—	148.5	—

TABLE 18.1 (*Continued*)

Plant Common and Scientific Name	Variety	Type of Culture[a]	Type of Tissue Sampled	Age, Stage, Condition, or Date of Sample	Selenium Treatment	Selenium Concentration in Dry Matter (mg kg⁻¹ unless otherwise noted)			Reference
						Low	Medium	High	
		Solution	Bulb		$2.0\,\text{mg L}^{-1}$ Na_2SeO_4	65.7	—	156.2	148
	'Downing Yellow Sweet Spanish'	Soil	Bulb		100 ton A⁻¹ Se as fly ash (16.8 ppm Se)	—	0.30	—	139
	'1620 Pedro'	Soil	Bulb		50 ton A⁻¹ Se as fly ash (16.8 ppm Se)	—	0.21	—	135
	'Stuttgart'	Soilless media	Bulb		7.59% Se as coal fly ash (13.3 ppm Se)	—	0.25	—	149
					10% Se as coal fly ash (10.1 ppm Se)	—	0.22	—	
Orach (*Atriplex patula* L.)		Native soil	Shoots		45.20 ± 19.79 mg kg⁻¹ Se in soil	—	20.79	—	150
					75.78 ± 28.78 mg kg⁻¹ Se in soil	—	79.96	—	
Potato (*Solanum tuberosum* L.)	'Katahdin'	Soil	Tuber		100 ton A⁻¹ Se as fly ash (16.8 ppm Se)	—	0.49	—	139
		Soil			50 ton A⁻¹ Se as fly ash (16.8 ppm Se)	—	0.03	—	135
Raspberry (*Rubus idaeus* L.)		Soil	Roots		0 to 1.5 kg Se ha⁻¹ as Na_2SeO_4	0.02	—	0.21	151

Species	Soil	Plant part	Subcategory	Treatment				Reference
		Floricanes		0 to 1.5 kg Se ha^{-1} as Na_2SeO_4	—	—	0.32	
		Primicanes		0 to 1.5 kg Se ha^{-1} as Na_2SeO_4	—	—	1.10	
		Leaves		0 to 1.5 kg Se ha^{-1} as Na_2SeO_4	—	—	1.81	
		Brambles		0 to 1.5 kg Se ha^{-1} as Na_2SeO_4	—	—	0.65	
Rice (*Oryza sativa* L.)	Native soil	Grain	First year	2.4 mg Se kg^{-1} in soil	—	9.9	—	152
			Second year	2.4 mg Se kg^{-1} in soil	—	8.9	—	
		Straw	First year	2.4 mg Se kg^{-1} in soil	—	18.0	—	
			Second year	2.4 mg Se kg^{-1} in soil	—	16.6	—	
'M101'	Native soil	Grain		No Se treatment (0 to 5 g kg^{-1} OM)	0.10	—	0.40	153
				1.5 mg kg^{-1} Na_2SeO_4 (0 to 5g kg^{-1} OM)	6.0	—	213	
				3.0 mg kg^{-1} Na_2SeO_4 (0 to 5 g kg^{-1} OM)	6.8	—	215	
				6.0 mg kg^{-1} Na_2SeO_4 (0 to 5 g kg^{-1} OM)	13.9	—	455	
		Shoots		No Se treatment (0 to 5 g kg^{-1} OM)	0.10	—	2.00	
				1.5 mg kg^{-1} Na_2SeO_4 (0 to 5 g kg^{-1} OM)	5.60	—	360	
				3.0 mg kg^{-1} Na_2SeO_4 (0 to 5 g kg^{-1} OM)	10.2	—	668	

Continued

TABLE 18.1 (*Continued*)

Common and Scientific Name	Variety	Type of Culture[a]	Type of Tissue Sampled	Age, Stage, Condition, or Date of Sample	Selenium Treatment	Selenium Concentration in Dry Matter (mg kg^{-1} unless otherwise noted)			Reference
						Low	Medium	High	
Ryegrass (*Lolium perenne* L.)		Soil	Shoots		$6.0\ \text{mg kg}^{-1}$ Na_2SeO_4 (0 to 5 g kg^{-1} OM)	20.9	—	1233	147
					No Se treatment	—	0.05	—	
					$0.1\ \text{mg kg}^{-1}$ H_2SeO_4	—	5.70	—	
					$1.0\ \text{mg kg}^{-1}$ H_2SeO_4	—	72.0	—	
Sprouts, Brussels (*Brassica oleracea* var. *gemmifera* Zenker)	'Explorer'	Native soil	Leaves		No Se treatment	38.00 µg kg^{-1} fresh weight	66.00 µg kg^{-1} fresh weight	220.00 µg kg^{-1} fresh weight	141
Sweet clover, Yellow [*Melilotus officinalis* Pallas]		Native soil	Shoots		1.8 mg Se kg^{-1} in soil	—	2.75	—	144
		Soil			4.8 mg Se kg^{-1} (3.0 mg Na_2SeO_4 kg^{-1})	—	216	—	
Sweet clover (*Melilotus indica* L.)		Native soil	Shoots		75.78 ± 29.78 mg Se kg^{-1} in soil		183.01		150
Swiss chard (*Beta vulgaris* L.)		Soil	Leaf		$5\ \text{mg kg}^{-1}$ Na_2SeO_3		29	—	32
					$5\ \text{mg kg}^{-1}$ Na_2SeO_4		735	—	
			Mid-rib/petiole		$5\ \text{mg kg}^{-1}$ Na_2SeO_3		13	—	

Species	Cultivar	Soil	Plant part	Growth stage	Treatment		Concentration		Ref.
			Composite leaves		5 mg kg⁻¹ Na₂SeO₄	—	120	—	
					5 mg kg⁻¹ Na₂SeO₃	—	30	—	
					5 mg kg⁻¹ Na₂SeO₄	—	449	—	
Timothy (*Phleum pratense* L.)		Natural soil	Shoots	First cutting	No Se treatment	—	0.10	—	137
					1.12 kg ha⁻¹ Na₂SeO₃; pH 6.6	—	0.68	—	
					2.24 kg ha⁻¹ Na₂SeO₃ pH 6.6	—	1.18	—	
Tomato (*Lycopersicon esculentum* Mill.)	'Vendor'	Soil	Fruit		100 ton A⁻¹ Se as fly ash (16.8 ppm Se)	—	0.33	—	139
	'Super-sonic'	Soil	Fruit		50 ton A⁻¹ Se as fly ash (16.8 ppm Se)	—	0.02	—	135
Trefoil, birdsfoot (*Lotus corniculatus* L.)		Native soil	Shoots	First clipping (60 days)	0.39 mg kg⁻¹ Se in soil	—	440 µg kg⁻¹	—	63
				Second clipping (115 days)	0.39 mg kg⁻¹ Se in soil	—	870 µg kg⁻¹	—	
				First clipping (60 days)	0.82 mg kg⁻¹ Se in soil	—	360 µg kg⁻¹	—	
				Second clipping (85 days)	0.82 mg kg⁻¹ Se in soil	—	290 µg kg⁻¹	—	
				Third clipping (115 days)	0.82 mg kg⁻¹ Se in soil	—	220 µg kg⁻¹	—	
Wheat (*Triticum aestivum* L.)		Native soil	Grain	First year	2.4 mg Se kg⁻¹ in soil	—	19.6	—	152
				Second year	2.4 mg Se kg⁻¹ in soil	—	12.4	—	

Continued

TABLE 18.1 (*Continued*)

Plant Common and Scientific Name	Variety	Type of Culture[a]	Type of Tissue Sampled	Age, Stage, Condition, or Date of Sample	Selenium Treatment	Selenium Concentration in Dry Matter (mg kg^{-1} unless otherwise noted) Low	Medium	High	Reference
Western wheatgrass (*Pascopyrum smithii* Löve)		Solution	Straw	First year	2.4 mg Se kg^{-1} in soil	—	16.6	—	38
				Second year	2.4 mg Se kg^{-1} in soil	—	11.1	—	
			Tops		1.0 mg L^{-1} Na$_2$SeO$_3$	—	20.2	—	
					2.0 mg L^{-1} Na$_2$SeO$_3$	—	55.1	—	
					0.3 mg L^{-1} Se-Met	—	31.5	—	
					0.6 mg L^{-1} Se-Met	—	92.8	—	
					0.3 mg L^{-1} Se-Cys	—	17.4	—	
					0.6 mg L^{-1} Se-Cys	—	28.6	—	
			Roots		1.0 mg L^{-1} Na$_2$SeO$_3$	—	187	—	
					2.0 mg L^{-1} Na$_2$SeO$_3$	—	647	—	
					0.3 mg L^{-1} Se-Met	—	81	—	
					0.6 mg L^{-1} Se-Met	—	161	—	
					0.3 mg L^{-1} Se-Cys	—	158	—	
					0.6 mg L^{-1} Se-Cys	—	220	—	

Note: ND = not determined.

[a]Native soil denotes experiments or studies where crops were harvested from untreated soil and the selenium level was determined from a soil sample to estimate selenium fertility.

inheritance of Se uptake and accumulation in plants. Investigation into the genetic variation for Se content in tall fescue revealed that progress from selection for selenium content is possible and that the trait was heritable (122). Narrow-sense heritability estimates for selenium accumulation in a rapid-cycling *Brassica oleracea* L. population were moderate (0.55), and gains from selection were 4.8 and 4.0% per selection cycle for high and low selenium accumulation, respectively (114). Knowledge of the genetic variances for selenium accumulation will be useful in selecting efficient strategies designed to enhance food crops. Further research is needed to identify the form and dosage of selenium delivered by selenium-enriched plants (92).

18.8 SELENIUM TISSUE ANALYSIS VALUES OF VARIOUS PLANT SPECIES

Selenium is unevenly distributed within plant tissues. Actively growing tissues usually contain the highest amounts of Se (35), and many plant species accumulate higher amounts of selenium in shoot or leaf tissues than in root tissues. Plant species differ greatly in their ability to accumulate seed selenium. Nelson and Johnson (123) reported seed selenium levels up to $3750\,\mu g$ Se g^{-1} dry weight in native milkvetch (*Astragalus* L.) species. Selenium accumulation in a rapid-cycling *Brassica oleracea* L. population increased linearly with increasing Na_2SeO_4 treatment concentrations in nutrient solution culture, ranging from nondetectable at $0\,mg\,Na_2SeO_4\,L^{-1}$ to $753\,\mu g$ Se g^{-1} dry weight at $7.0\,mg\,Na_2SeO_4\,L^{-1}$ (124). Selenium is also unevenly distributed within seeds. In dried grains of barley, the husk and pericarp accumulated selenium up to $0.6\,\mu g$ Se g^{-1}, the scutellum $0.4\,\mu g$ Se g^{-1}, the embryo $0.3\,\mu g$ Se g^{-1}, and the aleurone layer, embryonic leaves, and root initials $0.2\,\mu g$ Se g^{-1} (125).

Selenium treatment and selenium-enriched media will affect seed germination in a number of species. Soybeans (*Glycine max* Merr.) pretreated with 10 to $100\,g$ Se ha^{-1} as either seed or foliar treatments were grown on a nonseleniferous sandy loam soil and subsequently produced seeds accumulating 0.78 to $38.5\,mg$ Se kg^{-1}. When these seeds were planted without application of selenium fertilizer, the concentration of harvested seeds decreased to 0.11 to $1.02\,mg$ Se kg^{-1} (126). Seed germination was reduced if wheat (*Triticum aestivum* L.) was grown in soils with $>16.0\,mg$ Se kg^{-1} (127). Weight of fresh Alfalfa seedling was suppressed in response to $>10.0\,mg$ Se L^{-1} in solution culture (128). Turnip (*Brassica campestris* L.) seed germination was $>98\%$ when seeds were incubated in $<484\,mg$ $NaSeO_3\,L^{-1}$, but decreased to 51% if the concentration of $NaSeO_4$ was increased to $4.84\,g$ $NaSeO_3\,L^{-1}$. In response to $NaSeO_3$, turnip seed germination was 97% at Se levels $<95\,mg$ $NaSeO_3\,L^{-1}$, 53% at $484\,mg$ $NaSeO_3\,L^{-1}$, 17% at $951\,mg$ $NaSeO_3\,L^{-1}$, and 0% at $4.84\,g$ $NaSeO_3\,L^{-1}$ (129). Interestingly, several studies report that seed germination was enhanced in response to $<1.0\,mg$ Se L^{-1} in nutrient solutions (127,130,131). Activity of β-galactosidase, an enzyme important in the hydrolysis of complex carbohydrates during seed germination, in fenugreek (*Trigonella foenum-graecum* L.) was enhanced by 40% when exposed to $0.5\,mg\,L^{-1}$ Na_2SeO_3-seed treatment, but decreased by 60 to 65% if Na_2SeO_3-seed treatment was increased to $1\,mg\,L^{-1}$ (132). Seed germination was $>96\%$ after 72 h in a rapid-cycling *Brassica oleracea* population when the content of selenium in the seed was $<700\,\mu g$ Se g^{-1} dry weight (124).

REFERENCES

1. H.W. Lakin. Selenium accumulation in soils and its absorption by plants and animals. *Geological Soc. Am. Bull.* 83:181, 1972.
2. H.F. McNeal, L.S. Balistrieri. Geochemistry and occurrence of selenium: An overview. In: L.W. Jacobs, ed. *Selenium in Agriculture and the Environment.* Madison, WI: American Society of Agronomy, Soil Science Society of America, 1989, pp. 1–13.
3. N. Terry, A.M. Zayed, M.P. deSouza, A.S. Tarun. Selenium in higher plants. *Annu. Rev. Plant Physiol. Plant Mol. Biol.* 51:401–432, 2000.

4. K. Mengel, E.A. Kirkby. *Principles of Plant Nutrition*. Bern, Switzerland: International Potash Institute, 1987, pp. 589–606.

5. L.C. Clark, B. Dalkin, A. Krongrad, G.F. Combs, B.W. Trunbull, E.H. Slate, R. Witherington, J.H. Herlong, E. Janosko, D. Carpenter, C. Borosso, S. Falk, J. Rounder. Decreased incidence of prostate cancer with selenium supplementation: results of a double-blind cancer prevention trial. *Br. J. Urol.* 81:730–734, 1998.

6. G. Lockitch. Selenium: clinical significance and analytical concepts. *Crit. Rev. Clin. Lab. Sci.* 27:483–541, 1989.

7. G.F. Combs Jr. Selenium. In: T.E. Moon, M.S. Micozzi, eds. *Nutrition and Cancer Prevention: Investigating the Role of Micronutrients*. New York: Marcel Dekker, 1989, pp. 389–419.

8. G.F. Combs Jr., S.B. Combs. *The Role of Selenium in Nutrition*. Orlando, FL: Academic Press, 1986.

9. D.C. Adriano. *Trace Elements in the Terrestrial Environment*. New York: Springer-Verlag, 1986, pp. 390–420.

10. R.L. Mikkelsen, A.L. Page, F.T. Bingham. Factors affecting selenium accumulation by agricultural crops. In: L.W. Jacobs, ed. *Selenium in Agriculture and the Environment*. Madison, WI: American Society of Agronomy, Soil Science Society of America, 1989, pp. 65–94.

11. N. Terry, A.M. Zayed. Selenium volatilization by plants. In: W.T. Frankenberger Jr., S. Benson, eds. *Selenium in the Environment*. New York: Marcel Dekker, 1994, pp. 343–368.

12. T.J. Ganje. Selenium. In: H.D. Chapman, ed. *Diagnostic Criteria for Plants and Soils*. Berkeley, CA: University of California Division of Agricultural Sciences Press, 1966, pp. 394–404.

13. V.C. Morris, O.A. Lavander. Selenium content of foods. *J. Nutr.* 100:1383–1388, 1970.

14. I. Rosenfeld, O.A. Beath. *Selenium, Geobotany, Biochemistry, Toxicity, and Nutrition*. New York: Academic Press, 1964.

15. WHO Ernst. Selenpflanzen (Selenophyten). In: H. Kinzel, ed. *Pflanzenökologie und Mineralstoffwechsel*. Stuttgart, Germany: Verlag Eugen Ulmer, 1982, pp. 511–519.

16. G. Gissel-Nielsen, U.C. Gupta, M. Lamand, T. Westermarck. Selenium in soils and plants and its importance in livestock and human nutrition. *Adv. Agron.* 37:397–461, 1984.

17. D.T. Clarkson, U. Lüttge. II. Mineral nutrition: inducible and repressible nutrient transport systems. *Progr. Bot.* 52:61–83, 1991.

18. J.S. Ketter, G. Jarai, Y.H. Fu, G.A. Marzluf. Nucleotide sequence, messenger RNA stability, and DNA recognition elements of cys-14, the structural gene for sulfate permease II in *Neurospora crassa*. *Biochemistry* 30:1780–1787, 1991.

19. J.E. Leggett, E. Epstein. Kinetics of sulfate absorption by barley roots. *Plant Physiol.* 31:222–226, 1956.

20. T.C. Stadtman. Selenium biochemistry. *Science* 183:915–922, 1974.

21. R.D. Bryant, E.J. Laishley. Evidence for two transporters of sulfur and selenium oxyanions in *Clostridium pasteurianum*. *Can. J. Microbiol.* 34:700–703, 1988.

22. A.M. Zayed, C.M. Lytle, N. Terry. Accumulation and volatilization of different chemical species of selenium by plants. *Planta* 206:284–292, 1998.

23. P. Barak, I.L. Goldman. Antagonistic relationship between selenate and sulfate uptake in onion (*Allium cepa*): implications for the production of organosulfur and organoselenium compounds in plants. *J. Agr. Food Chem.* 45:1290–1294, 1997.

24. G. Ferrari, F. Renosto. Regulation of sulfate uptake by excised barley roots in the presence of selenate. *Plant Physiol.* 49:114–116, 1972.

25. S.F. Trelease, H.M. Trelease. Selenium as a stimulation and possibly essential element for indicator plants. *Am. J. Bot.* 25:372–380, 1938.

26. M.J. Hawkesford, J.-C. Davidian, C. Grignon. Sulfate/proton cotransport on plasma-membrane vesicles isolated from roots of *Brassica napus* L.: increased transport in membranes isolated from sulfur-starved plants. *Planta* 190:297–304, 1993.

27. D.A. Kopsell, W.M. Randle. Selenate concentration affects selenium and sulfur uptake and accumulation by 'Granex 33' onions. *J. Am. Soc. Hortic. Sci.* 122:721–726, 1997.

28. M.A. Arthur, G. Rubin, P.B. Woodbury, R.E. Schneider, L.H. Weinstein. Uptake and accumulation of selenium by terrestrial plants growing on a cola fly ash landfill. Part 2. Forage and roots crops. *Environ. Toxicol. Chem.* 11:1298–1299, 1992.

29. J.L. Hopper, D.R. Parker. Plant availability of selenite and selenate as influenced by the competing ions phosphate and sulfate. *Plant Soil* 210:199–207, 1999.

30. D.L. Carter, C.W. Robbins, M.J. Brown. Effect of phosphorous on the selenium concentration in alfalfa (*Medicago sativa*). *Soil Sci. Soc. Am. Proc.* 36:624–628, 1972.

31. T.A. Brown, A. Shrift. Selenium: toxicity and tolerance in higher plants. *Biol. Rev.* 57:59–84, 1982.

32. G.S. Bañuelos, D.W. Meek. Selenium accumulation in selected vegetables. *J. Plant Nutr.* 12:1255–1272, 1989.

33. M.P. Arvy. Selenate and selenite uptake and translocation in bean plants (*Phaseolus vulgaris*). *J. Exp. Bot.* 44:1083–1087, 1993.

34. M.P. de Souza, E.A.H. Pilon-Smits, C.M. Lytle, S. Hwang, J. Tai, T.S.U. Honma, L. Yeh, N. Terry. Rate-limiting steps in selenium assimilation and volatilization by Indian mustard. *Plant Physiol.* 117:1487–1494, 1998.

35. A. Kabata-Pendias, H. Pendias. *Trace Elements in Soils and Plants.* Boca Raton, FL: CRC Press, 1992.

36. A. Mukherjee, A. Sharma. Effects of cadmium and selenium on cell division and chromosomal aberrations in *Allium sativum* L. *Water Air Soil Pollut.* 37:433–438, 1988.

37. M.M. Abrams, C. Shennan, J. Zasoski, R.G. Burau. Selenomethionine uptake by wheat seedlings. *Agron. J.* 82:1127–1130, 1990.

38. M.C. Williams, H.F. Mayland. Selenium absorption by two grooved milkvetch and western wheatgrass from selenomethionine, selenocysteine, and selenite. *J. Range Manage.* 45:374–378, 1992.

39. G.S. Bañuelos, R. Mead, S. Akohoue. Adding selenium-enriched plant tissue to soil causes the accumulation of selenium in alfalfa. *J. Plant Nutr.* 14:701–713, 1991.

40. A. Läuchli. Selenium in plants: uptake, functions, and environmental toxicity. *Bot. Acta* 106:455–468, 1993.

41. A. Shrift. Aspects of selenium metabolism in higher plants. *Annu. Rev. Plant Physiol.* 20:475, 1969.

42. G.S. Bañuelos, H.A. Ajwa, B. Mackey, L. Wu, C. Cook, S. Akohoue, S. Zambruzuski. Evaluation of different plant species used for phytoremediation of high soil selenium. *J. Environ. Qual.* 26:639–646, 1997.

43. D.A. Kopsell, W.M. Randle. Short-day onion cultivars differ in bulb selenium and sulfur accumulation which can affect bulb pungency. *Euphytica* 96:385–390, 1997.

44. T.A. Brown, A. Shrift. Exclusion of selenium from proteins of selenium-tolerant *Astragalus* species. *Plant Physiol.* 67:1051–1053, 1981.

45. J.E. Lancaster, M.J. Boland. Flavor biochemistry. In: H.D. Rabinowitch, J.L. Brewster, eds. *Onions and Allied Crops*, Vol. 3. Boca Raton, FL: CRC Press, 1990, pp. 33–72.

46. T. Leustek. Molecular genetics of sulfate assimilation in plants. *Plant Physiol.* 97:411–419, 1996.

47. G. Gissel-Nielsen. Uptake and translocation of selenium-75 in *Zea mays* L. In: *Proceedings of Symposium on Isotopes and radiation in Soil–Plant Relationships Including Forestry.* December 13–17, 1971, Vienna, Austria. Vienna: International Atomic Energy Agency, 1972, pp. 427–436.

48. A. Setya, M. Murillo, T. Leustek. Sulfate reduction in higher plants: Molecular evidence for a novel 5′-adenylsulfate reductase. *Proc. Natl. Acad. Sci. USA* 93:13383–13388, 1996.

49. E.A.H. Pilon-Smits, S. Hwang, C.M. Lytle, Y. Zhu, J.C. Tai, R.C. Bravo, Y. Chen, T. Leustek, N. Terry. Overexpression of ATP sulfurylase in Indian mustard leads to increased selenate uptake, reduction, and tolerance. *Plant Physiol.* 119:123–132, 1998.

50. B.H. Ng, J.W. Anderson. Synthesis of selenocysteine by cysteine synthases from selenium accumulator and non-accumulator plants. *Phytochemistry* 17:2069–2074, 1978.

51. H.E. Ganther, O.A. Levander, C.A. Saumann. Dietary control of selenium volatilization in the rat. *J. Nutr.* 88:55–60, 1966.

52. N. Terry, C. Carlson, T.K. Raab, A.M. Zayed. Rates of selenium volatilization among crop species. *J. Environ. Qual.* 21:341–344, 1992.

53. A.M. Zayed, N. Terry, Selenium volatilization in roots and shoots: effects of shoot removal and sulfate level. *J. Plant Physiol.* 143:8–14, 1994.

54. A.M. Zayed, N. Terry. Selenium volatilization in broccoli as influenced by sulfate supply. *J. Plant Physiol.* 140:646–652, 1992.

55. G.S. Bañuelos, D. Dyer, R. Ahmad, S. Ismail, R.N. Raut, J.C. Dagar. In search of *Brassica* germplasm in saline semi-arid and arid regions of India and Pakistan for reclamation of selenium-laden soils in the U.S. *J. Soil Water Conserv.* 48:530–534, 1993.

56. T.S. Presser, M.A. Sylvester, W.H. Law. Bioaccumulation of Se in the west. *Environ. Manage.* 18:423–436, 1994.

57. R. Fujii, S.J. Deverel. Mobility and distribution of selenium and salinity in groundwater and soil of drained agricultural fields, western San Joaquin Valley of California. In: L.W. Jacobs, ed. *Selenium in Agriculture and the Environment*. Madison, WI: American Society of Agronomy, Soil Science Society of America, 1989, pp. 195–212.

58. S.J. Deverel, J.L. Fio, N.M. Dubrovsky. Distribution and mobility of selenium in groundwater in the western San Joaquin Valley of California. In: W.T. Frankenberger Jr., S. Benson, eds. *Selenium in the Environment*. New York: Marcel Dekker, 1994, pp. 157–184.

59. A.F. White, N.M. Dubrovsky. Chemical oxidation–reduction controls on selenium mobility in groundwater systems. In: W.T. Frankenberger Jr., S. Benson, eds. *Selenium in the Environment*. New York: Marcel Dekker, 1994, pp. 185–222.

60. G.S. Bañuelos, D.W. Meek. Accumulation of selenium in plants grown on selenium treated soils. *J. Environ. Qual.* 19:772–777, 1990.

61. D.R. Clark Jr. Selenium accumulation in mammals exposed to contaminated California irrigation drainwater. *Sci. Total Environ.* 66:147–168, 1987.

62. G.S. Bañuelos, H.A. Ajwa, L. Wu, X. Guo, S. Akohoue, S. Zambrzuski. Selenium-induced growth reduction in *Brassica* land races considered for phytoremediation. *Ecotoxic Environ. Safety* 36:282–287, 1997.

63. G.S. Bañuelos, G. Cardon, B. Mackey, J. Ben-Asher, L. Wu, P. Beuselinck, S. Akohoue, S. Zambrzuski. Boron and selenium removal in boron-laden soils by four sprinkler-irrigated plant species. *J. Environ. Qual.* 22:786–792, 1993.

64. D.R. Parker, A.L. Page, D.N. Thomason. Salinity and boron tolerances of candidate plants for the removal of selenium from soils. *J. Environ. Qual.* 20:157–164, 1991.

65. L. Wu. Selenium accumulation and colonization of plants in soils with elevated selenium and salinity. In: W.T. Frankenberger Jr., S. Benson, eds. *Selenium in the Environment*. New York: Marcel Dekker, 1994, pp. 279–326.

66. B. Neuhierl, A. Böck. On the mechanism of selenium tolerance in selenium-accumulating plants, Purification and characterization of a specific selenocysteine methyltransferase from cultured cells of *Astragalus bisculatus*. *Eur. J. Biochem.* 239:235–238, 1996.

67. M.M. Abrams, R.G. Burau, R.J. Zasoski. Organic selenium distribution in selected California soils. *Soil Sci. Soc. Am. J.* 54:979–982, 1990.

68. M.A. Elrashidi, D.C. Adriano, W.L. Lindsay. Solubility, speciation, and transformation of selenium in soils. In: L.W. Jacobs, ed. *Selenium in Agriculture and the Environment*. Madison, WI: American Society of Agronomy, Soil Science Society of America, 1989, pp. 51–64.

69. H.T. Shacklette, J.G. Boerngen. Elemental concentrations in soils and other surface materials of the conterminous United States. *Geological Surv. Prof. Pap.* 1270, 1984.

70. O.A. Levander, M.A. Beck. The role of selenium as an essential nutrient in humans and its role as a determinant of viral virulence. In: J. Mass, ed. *Selenium in the Environment: Essential Nutrient or Potential Toxicant*. Oakland, CA: University of California Agricultural Natural Resources Cooperative Extension, 1995, pp. 7–10.

71. D.Y. Boon. Potential selenium problems in Great Plains soils. In: L.W. Jacobs, ed. *Selenium in Agriculture and the Environment*. Madison, WI: American Society of Agronomy, Soil Science Society of America, 1989, pp. 107–122.

72. C.W. Robbins, D.L. Carter. Selenium concentrations in phosphorus fertilizer materials and associated uptake by plants. *Soil Sci. Soc. Am. Proc.* 34:506–509, 1970.

73. J.J. Dynes, P.M. Huang. Influence of organic acids on selenite sorption by poorly ordered aluminum hydroxides. *Soil Sci. Soc. Am. J.* 61:772–783, 1997.

74. G.S. Bañuelos, A. Zayed, N. Terry, L. Wu, A. Akohoue, S. Zambrzuski. Accumulation of selenium by different plant species grown under increasing sodium and calcium chloride salinity. *Plant Soil* 183:49–59, 1996.

75. R.H. Neal, G. Sposito, K.M. Holtzclaw, S.J. Traina. Selenite adsorption on alluvial soil. I. Soil composition and pH effects. *J. Soil Sci. Soc. Am.* 51:1161–1165, 1987.

76. S.H. van Dorst, P.J. Peterson. Selenium speciation in the soil solution and its relevance to plant uptake. *J. Sci. Food Agric.* 35:601–605, 1984.

77. B.R. Whelan, N.J. Barrow. Slow-release selenium fertilizers to correct selenium deficiency in grazing sheep in western Australia. *Fert. Res.* 38:183–188, 1994.

78. B.R. Singh. Effects of selenium-enriched calcium nitrate, top-dressed at different growth stages, on the selenium concentration in wheat. *Fert. Res.* 38:199–203, 1994.

79. L.F. James, K.E. Panter, H.F. Mayland, M.R. Miller, D.C. Baker. Selenium poisoning in livestock: A review and progress. In: L.W. Jacobs, ed. *Selenium in Agriculture and the Environment*. Madison, WI: American Society of Agronomy, Soil Science Society of America, 1989, pp. 123–132.

80. K. Schwarz, C.M. Foltz. Selenium as an integral part of factor 3 against dietary necrotic liver degeneration. *J. Am. Chem. Soc.* 70:3292–3293, 1957.

81. U.C. Gupta, S.C. Gupta. Selenium in soils and crops, its deficiencies in livestock and humans: implications for management. *Commun. Soil Sci. Plant Anal.* 31:1791–1807, 2000.

82. H.E. Ganther, J.R. Lawrence. Chemical transformations of selenium in living organisms. Improved forms of selenium and cancer prevention. *Tetrahedron* 53:12299–12310, 1997.

83. L.A. Daniels. Selenium metabolism and bioavailability. *Biol. Trace Element Res.* 54:185–199, 1996.

84. H. Ge, X. Cai, J.F. Tyson, P.C. Uden, E.R. Denoyer, E. Block. Identification of selenium species in selenium-enriched garlic, onion, and broccoli using high-performance ion chromatography with inductively coupled plasma mass spectrometry detection. *Anal. Commun.* 33:279–281, 1996.

85. E. Block, X.J. Cai, P.C. Uden, X. Zhang, B.D. Quimby, J.J. Sullivan. *Allium* chemistry: natural abundance of organoselenium compounds from garlic, onion, and related plants and in human garlic breath. *Pure Appl. Chem.* 68:937–944, 1996.

86. O.E. Olson, E.J. Novacek, E.I. Whitehead, I.S. Palmer. Investigations on selenium in wheat. *Phytochemistry* 9:1181–1188, 1970.

87. C.D. Thomson. Selenium speciation in human body fluids. *Analyst* 123:827–831, 1998.

88. L. Aronow, F. Kerdel-Vegas. Selenocystathionine, a pharmacologically active factor in the seeds of *Lecythis ollaria. Nature* 205:1185–1186, 1965.

89. A. Shrift. Metabolism of selenium by plants and microorganisms. In: D.L. Klayman, W.H.H. Günther, eds. *Organic Selenium Compounds: Their Chemistry and Biology*. New York: Wiley-Interscience, 1973, pp. 763–814.

90. H.E. Ganther. Pathways of selenium metabolism including respiratory excretory products. *J. Am. Coll. Toxicol.* 5:1–5, 1986.

91. D.A. Martens, D.L. Suarez. Selenium in water management wetlands in the semi-arid west. *HortScience* 34:34–39, 1999.

92. C. Ip, D.J. Lisk. Characterization of tissue selenium profiles and anticarcinogenic responses in rats fed natural sources of selenium-enriched products. *Carcinogenesis* 15:573–576, 1994a.

93. P.D. Whanger, J.A. Butler. Effects of various dietary levels of selenium as selenite or selenomethionine on tissue selenium levels and glutathione peroxidase activity in rats. *J. Nutr.* 118:846–852, 1988.

94. C. Wang, R.T. Lovell. Organic selenium sources, selenomethionine and selenoyeast, have higher bioavailability than an inorganic selenium source, sodium selenite, in diets for channel catfish (*Ictalurus punctatus*). *Aquaculture* 152:223–234, 1997.

95. G.N. Schrauzer, J. Sacher. Selenium in the maintenance and therapy of HIV-infected patients. *Chem. Biol. Interactions* 91:199–205, 1994.

96. P.A. Vandenbrandt, R.A. Goldbohlm, P. Vantveer, P. Bode, E. Dorant, R.J.J. Hermun, F. Sturmans. A prospective cohort study on selenium status and the risk of lung cancer. *Cancer Res.* 53:4860–4865, 1993.

97. O.A. Levander. Selenium: Biochemical actions, interactions, and some human health implications. In: A.S. Prasad, ed. *Clinical Biochemical, and Nutritional Aspects of Trace Elements*. New York: Alan R. Liss, 1982, pp. 345–368.

98. C.K. Chow. Nutritional influences on cellular antioxidant defense systems. *Am. J. Clin. Nutr.* 32:1066–1081, 1979.

99. T. Lawson, D.F. Birt. Enhancement of the repair of carcinogen-induced DNA damage in the hamster pancreas by dietary selenium. *Chem. Biol. Interaction* 45:95–104, 1983.

100. T. Chen, M.P. Goelchius, G.F. Combs, T.C. Campbell. Effects of dietary selenium and vitamin E on covalent binding of aflatoxin to chick liver cell macromolecules. *J. Nutr.* 112:350–355, 1982.

101. M.P. Rosin. Effects of sodium selenite on the frequency of spontaneous mutation of yeast mutator strains. *Proc. Am. Assoc. Cancer Res.* 22:55, 1981.

102. L.C. Clark, K.P. Cantor, W.H. Allaway. Selenium in forage crops and cancer mortality in U.S. counties. *Arch. Environ. Health* 46:37–42, 1991.

103. C. Reilly. Selenium: a new entrant into the functional food arena. *Trends Food Sci. Technol.* 9: 114–118, 1998.
104. K. Ge, G. Yang. The epidemiology of selenium deficiency in the etiology of endemic diseases in China. *Am. J. Clinical Nutr. Suppl.* 57:259S–263S, 1993.
105. M.K. Baum, G. Shor-Posner, S. Lai, G. Zhang, H. Lai, M. Fletcher, H. Sauberlich, J.B. Page. High risk of HIV-related mortality is associated with selenium deficiency. *J. Acquired Immune Deficiency Syndromes Human Retrovirol.* 15:370-374, 1997.
106. G.F. Combs Jr., W.P. Gray. Chemopreventative agents: selenium. *Pharmacol. Therapeut.* 79:179–192, 1998.
107. R.C. McKenzie, T.S. Rafferty, G.J. Beckett. Selenium: an essential element for immune function. *Immunol. Today* 19:342–345, 1998.
108. National Research Council. *Selenium in Nutrition*, revised ed. Washington, DC: National Academic Press, 1983.
109. Department of Health. *Dietary Reference Values for Food Energy and Nutrients in the United Kingdom*. London: Her Majesties Stationery Office, 1991.
110. L.H. Foster, S. Sumar. Selenium in health and disease: a review. *Critical Rev. Food Sci. Nutr.* 37:211–228, 1997.
111. G.F. Combs Jr. Selenium and cancer. In: H. Garewal, ed. *Antioxidants and Disease Prevention*. New York: CRC Press, 1997, pp. 97–113.
112. L.C. Clark, G.F. Combs Jr., L. Hixon, D.R. Deal, J. Moore, J.S. Rice, M. Dellasega, A. Rogers, J. Woodard, B. Schurman, D. Curtis, B.W. Turnbull. Low plasma selenium predicts the prevalence of colorectal adenomatous polyps in a cancer prevention trial. *FASEB J.* 7:A65, 1993.
113. C. Ip, D.J. Lisk, G.S. Stoewsand. Mammary cancer prevention by regular garlic and selenium-enriched garlic. *Nutr. Cancer* 17:279–286, 1992.
114. D.A. Kopsell, W.M. Randle. Genetic variances and selection potential for selenium accumulation in a rapid-cycling *Brassica oleracea* population. *J. Am. Soc. Hortic. Sci.* 126:329–335, 2001.
115. J.M. Quintana, H.C. Harrison, J. Nienhuis, J.P. Palta, M.A. Grusak. Variation in calcium concentration among sixty S_1 families and four cultivars of snap bean (*Phaseolus vulgaris* L.). *J. Am. Soc. Hortic. Sci.* 121:789–793, 1996.
116. E.C. Tigchelaar. Tomato breeding. In: M. Basset, ed. *Breeding Vegetable Crops*. Westport, CT: AVI Publishing, 1986, pp. 135–171.
117. M. Wang, I.L. Goldman. Phenotypic variation in free folic acid content among F_1 hybrids and open-pollinated cultivars of red beet. *J. Am. Soc. Hortic. Sci.* 121:1040–1042, 1996.
118. R. Graham, D. Senadhira, S. Beebe, C. Iglesias, I. Monasterio. Breeding for micronutrient density in edible portions of staple food crops: conventional approaches. *Field Crops Res.* 60:57–80, 1999.
119. H.E. Bouis. Enrichment of food staples though plant breeding: a new strategy for fighting micronutrient malnutrition. *Nutr. Rev.* 54:131–137, 1996.
120. M.T. Ruel, H.E. Bouis. Plant breeding: a long-term strategy for the control of zinc deficiency in vulnerable populations. *Am. J. Clin. Nutr.* 68:488S–494S, 1998.
121. M. Eurola, P. Ekholm, M. Ylinen, P. Koivistoinen, P. Varo. Effects of selenium fertilization on the selenium content of selected Finnish fruits and vegetables. *Acta Agric. Scand.* 39:345–350, 1989.
122. S.D. McQuinn, D.A. Sleper, H.F. Mayland, G.F. Krause. Genetic variation for selenium content in tall fescue. *Crop Sci.* 31:617–620, 1991.
123. D.M. Nelson, C.D. Johnson. Selenium in seeds of *Astragalus* (Leguminosae) and its effects on host preferences of Bruchid beetles. *J. Kansas Entomol. Soc.* 56:267–272, 1983.
124. D.A. Kopsell, D.E. Kopsell, W.M. Randle. Seed germination response of rapid-cycling *Brassica oleracea* grown under increasing sodium selenate. *J. Plant Nutr.* 26:1355–1366, 2003.
125. K.X. Huang, J. Clausen. Uptake, distribution, and turnover rates of selenium in barley. *Biol. Trace Element Res.* 40:213–223, 1994.
126. U.C. Gupta, J.A. MacLeod. Relationship between soybean seed selenium and harvested grain selenium. *Can. J. Soil Sci.* 79:221–223, 1999.
127. A.T. Perkins, H.H. King. Selenium and Tenmarq wheat. *J. Am. Soc. Agron.* 30:664–667, 1938.
128. M. Hu, J.E. Spallholz. Toxicological evaluations of selenium compounds assessed by alfalfa seed germination. *Proc. Third Intl. Symp. Selenium Biol. Med.* 1:530–533, 1987.
129. N.E. Spencer, S.M. Siegel. Effects of sulfur and selenium oxyanions on Hg-toxicity in turnip seed germination. *Water Air Soil Pollut.* 9:423–427, 1978.

130. Z. Wang, Y. Xu, A. Peng. Influences of fulvic acid on bioavailability and toxicity of selenate for wheat seedling growth. *Biol. Trace Element Res.* 55:147–162, 1996.

131. K. Lalitha, K. Easwari. Kinetic analysis of ^{75}selenium uptake by mitochondria of germinating *Vigna radiata* of different selenium status. *Biol. Trace Element Res.* 48:67–89, 1995.

132. M. Shreekala, K. Lalitha. Selenium-mediated differential response of β-glucosidase and β-galactosidase of germinating *Trigonella foenum-graecum. Biol. Trace Element Res.* 64:247–258. 1998.

133. R.L. Mikkelsen, G.H. Haghnia, A.L. Page. Effects of pH and selenium oxidation state on the selenium accumulation and yield of alfalfa. *J. Plant Nutr.* 10:937–950, 1987.

134. R.L. Mikkelsen, G.H. Haghnia, A.L. Page, F.T. Bingham. The influence of selenium, salinity, and boron on alfalfa tissue composition and yield. *J. Environ. Qual.* 17:85–88, 1988.

135. A.K. Furr, T.F. Parkinson, W.H. Gutenmann, I.S. Pakkala, D.J. Lisk. Elemental content of vegetables, grains, and forages field-grown on fly ash amended soil. *J. Agric. Food Chem.* 26:357–359, 1978.

136. T.C. Broyer, C.M. Johnson, R.P. Huston. Selenium and nutrition of astragalus. I. Effects of selenite or selenate supply on growth and selenium content. *Plant Soil* 36:635–649, 1972.

137. U.C. Gupta, K.A. Winter. Long-term residual effects of applied selenium on the selenium uptake by plants. *J. Plant Nutr.* 3:493–502, 1981.

138. J.A. MacLeod, U.C. Gupta, P. Milburn, J.B. Sanderson. Selenium concentration in plant material, drainage and surface water as influenced by Se applied to barley foliage in barley–red clover–potato rotation. *Can. J. Soil Sci.* 78:685–688, 1998.

139. A.K. Furr, W.C. Kelly, C.A. Bache, W.H. Gutenmann, D.J. Lisk. Multielemental uptake by vegetables and millet grown in pots on fly ash amended. *J. Agric. Food Chem.* 26:885–888, 1976.

140. D.A. Kopsell, W.M. Randle. Selenium accumulation in rapid-cycling *Brassica oleracea* population responds to increasing sodium selenate concentrations. *J. Plant Nutr.* 22:927–937, 1999.

141. A. Bibak, S. Stürup, L. Knudsen, V. Gundersen. Concentrations of 63 elements in cabbage and sprouts in Denmark. *Commum. Soil Sci. Plant Anal.* 30:2409–2418, 1999.

142. H.A. Ajwa, G.S. Bañuelos, H.F. Mayland. Selenium uptake by plants from soils amended with inorganic and organic materials. *J. Environ. Qual.* 27:1218–1227, 1998.

143. G. Lee, K. Park. Quality improvement of 'Seoul' celery by selenium in nutrient solution culture. *Acta Hortic.* 483:185–192, 1999.

144. P.L. Wanek, G.F. Vance, P.D. Stahl. Selenium uptake by plants: Effects of soil steaming, root addition, and selenium augmentation. *Commun. Soil Sci. Plant Anal.* 30:265–278, 1999.

145. V. Licina, M. Jakovljevic, S. Antic-Mladenovic. An effect of specific selenium nutrition of grapevine. *Acta Hortic.* 526:225–228, 2000.

146. H. Pinochet, I. DeGregori, M.G. Lobos, E. Fuentes. Selenium and copper in vegetables and fruits grown on long-term impacted soils from Valparaiso Region, Chile. *Bull. Environ. Contam. Toxicol.* 63:327–334, 1999.

147. H. Hartikainen, T. Xue. The promotive effect of selenium on plant growth as triggered by ultraviolet irradiation. *J. Environ. Qual.* 28:1372–1375, 1999.

148. D.A. Kopsell, W.M. Randle. Selenium affects the S-alk(en)yl cysteine sulfoxides among short-day onion cultivars. *J. Am. Soc. Hortic. Sci.* 124:307–311, 1999b.

149. W.H. Gutenmann, G.J. Doss, D.J. Lisk. Selenium in onions grown in media amended with coal fly ashes collected with differing efficiencies. *Chemosphere* 37:398–390, 1998.

150. L. Wu, A. Enberg, K.K. Tanji. Natural establishment and selenium accumulation of herbaceous plant species in soils with elevated concentrations of selenium and salinity under irrigation and tillage practices. *Ecotoxic Environ. Safety* 25:127–140, 1993.

151. V. Licina, M. Jakovljevic, S. Antic-Mladenovic, C. Oparnica. The content of selenium in raspberry plant and its improvement by Se-fertilization. *Acta Hortic.* 477:167–172, 1998.

152. K.S. Dhillon, S.K. Dhillon. Selenium accumulation by sequentially grown wheat and rice as influenced by gypsum application in a seleniferous soil. *Plant Soil* 227:243–248, 2000.

153. R.L. Mikkelsen, D.S. Mikkelsen, A. Abshahi. Effects of soil flooding on selenium transformations and accumulation by rice. *Soil Sci. Soc. Am. J.* 53:122–127, 1989.

19 Silicon

George H. Snyder
University of Florida/IFAS,
Belle Glade, Florida

Vladimir V. Matichenkov
Russian Academy of Sciences, Pushchino, Russia

Lawrence E. Datnoff
University of Florida/IFAS,
Gainesville, Florida

CONTENTS

19.1   Introduction 551
19.2   Historical Perspectives 552
19.3   Silicon in Plants 553
    19.3.1   Plant Absorption of Silicon 553
    19.3.2   Forms of Silicon in Plants 553
    19.3.3   Biochemical Reactions with Silicon 553
19.4   Beneficial Effects of Silicon in Plant Nutrition 554
    19.4.1   Effect of Silicon on Biotic Stresses 554
    19.4.2   Effect of Silicon on Abiotic Stresses 557
19.5   Effect of Silicon on Plant Growth and Development 557
    19.5.1   Effect of Silicon on Root Development 557
    19.5.2   Effect of Silicon on Fruit Formation 557
    19.5.3   Effect of Silicon on Crop Yield 557
19.6   Silicon in Soil 561
    19.6.1   Forms of Silicon in Soil...561
    19.6.2   Soil Tests 561
19.7   Silicon Fertilizers 562
19.8   Silicon in Animal Nutrition..562
References 562

19.1 INTRODUCTION

Silicon (Si) is the second-most abundant element of the Earth's surface. Beginning in 1840, numerous laboratory, greenhouse, and field experiments have shown benefits of application of silicon fertilizer for rice (*Oryza sativa* L.), corn (*Zea mays* L.), wheat (*Triticum aestivum* L.), barley

(*Hordeum vulgare* L.), and sugar cane (*Saccharum officinarum* L.). Silicon fertilizer has a double effect on the soil–plant system. First, improved plant-silicon nutrition reinforces plant-protective properties against diseases, insect attack, and unfavorable climatic conditions. Second, soil treatment with biogeochemically active silicon substances optimizes soil fertility through improved water, physical and chemical soil properties, and maintenance of nutrients in plant-available forms.

19.2 HISTORICAL PERSPECTIVES

In 1819, Sir Humphrey Davy wrote:

> The siliceous epidermis of plants serves as support, protects the bark from the action of insects, and seems to perform a part in the economy of these feeble vegetable tribes (Grasses and Equisetables) similar to that performed in the animal kingdom by the shell of crustaceous insects (1)

In the nineteenth and twentieth centuries, many naturalists measured the elemental composition of plants. Their data demonstrated that plants usually contain silicon in amounts exceeding those of other elements (2) (Figure 19.1). In 1840, Justius von Leibig suggested using sodium silicate as a silicon fertilizer and conducted the first greenhouse experiments on this subject with sugar beets (3). Starting in 1856, and being continued at present, a field experiment at the Rothamsted Station (England) has demonstrated a marked effect of sodium silicate on grass productivity (4).

The first patents on using silicon slag as a fertilizer were obtained in 1881 by Zippicotte and Zippicotte (5). The first soil test for plant-available silicon was conducted in the Hawaiian Islands by Professor Maxwell in 1898 (6).

Japanese agricultural scientists appear to have been the most advanced regarding the practical use of silicon fertilizers, having developed a complete technology for using silicon fertilizers for rice in the 1950s and 1960s. Other investigations of the effect of silicon on plants were conducted in France, Germany, Russia, the United States, and in other countries.

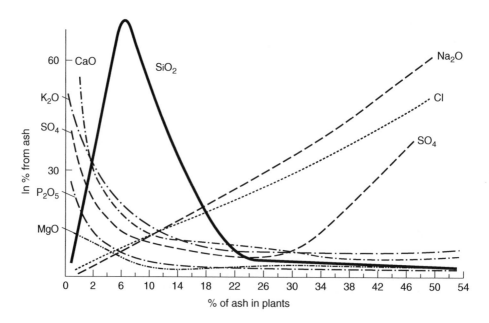

FIGURE 19.1   Silicon in ash of cultivated plants. (From V.A. Kovda, *Pochvovedenie* 1:6–38, 1956.)

19.3 SILICON IN PLANTS

19.3.1 PLANT ABSORPTION OF SILICON

Tissue analyses from a wide variety of plants showed that silicon concentrations range from 1 to 100 g Si kg^{-1} of dry weight, depending on plant species (7). Comparison of these values with those for elements such as phosphorus, nitrogen, calcium, and others shows silicon to be present in amounts equivalent to those of macronutrients (Figure 19.1).

Plants absorb silicon from the soil solution in the form of monosilicic acid, also called orthosilicic acid [H$_4$SiO$_4$] (8,9). The largest amounts of silicon are adsorbed by sugarcane (300–700 kg of Si ha^{-1}), rice (150–300 kg of Si ha^{-1}), and wheat (50–150 kg of Si ha^{-1}) (10). On an average, plants absorb from 50 to 200 kg of Si ha^{-1}. Such values of silicon absorbed cannot be fully explained by passive absorption (such as diffusion or mass flow) because the upper 20 cm soil layer contains only an average of 0.1 to 1.6 kg Si ha^{-1} as monosilicic acid (11–13). Some results have shown that rice roots possess specific ability to concentrate silicon from the external solution (14).

19.3.2 FORMS OF SILICON IN PLANTS

Basically, silicon is absorbed by plants as monosilicic acid or its anion (9). In the plant, silicon is transported from the root to the shoot by means of the transportation stream in the xylem. Soluble monosilicic acid may penetrate through cell membranes passively (15). Active transport of mono-silicic acid in plants has received little study.

After root adsorption, monosilicic acid is translocated rapidly into the leaves of the plant in the transpiration stream (16). Silicon is concentrated in the epidermal tissue as a fine layer of silicon–cellulose membrane and is associated with pectin and calcium ions (17). By this means, the double-cuticular layer can protect and mechanically strengthen plant structures (18).

With increasing silicon concentration in the plant sap, monosilicic acid is polymerized (8). The chemical nature of polymerized silicon has been identified as silicon gel or biogenic opal, amorphous SiO$_2$, which is hydrated with various numbers of water molecules (9,19). Monosilicic acid polymerization is assigned to the type of condensable polymerization with gradual dehydration of monosilicic acid and then polysilicic acid (20,21):

$$n(Si(OH)_4) \rightarrow (SiO_2) + 2n(H_2O)$$

Plants synthesize silicon-rich structures of nanometric (molecular), microscopic (ultrastructural), and macroscopic (bulk) dimensions (22). Ninety percent of absorbed silicon is transformed into various types of phytoliths or silicon–cellulose structures, represented by amorphous silica (18). Partly biogenic silica is generated as unique cell or inter-cell structures at the nanometer level (23). The chemical composition of oat (*Avena sativa* L.) phytoliths (solid particles of SiO$_2$) was shown to be amorphous silica (82–86%) and varying amounts of sodium, potassium, calcium, and iron (24). Phytoliths are highly diversified, and one plant can synthesize several forms (25,26). A change in plant-silicon nutrition has an influence on phytolith forms (27).

19.3.3 BIOCHEMICAL REACTIONS WITH SILICON

Soluble silicon compounds, such as monosilicic acid and polysilicic acid, affect many chemical and physical-chemical soil properties. Monosilicic acid possesses high chemical activity (21,28). Monosilicic acid can react with aluminum, iron, and manganese with the formation of slightly soluble silicates (29,30):

$$Al_2Si_2O_5 + 2H^+ + 3H_2O = 2Al^{3+} + 2H_4SiO_4, \quad \log K^\circ = 15.12$$

$$Al_2Si_2O_5(OH)_4 + 6H^+ = 2Al^{3+} + 2H_4SiO_4 + H_2O, \quad \log K^\circ = 5.45$$

$$Fe_2SiO_4 + 4H^+ = 2Fe^{2+} + 2H_4SiO_4, \quad \log K^\circ = 19.76$$

$$MnSiO_3 + 2H^+ + H_2O = Mn^{2+} + 2H_4SiO_4, \quad \log K^\circ = 10.25$$

$$Mn_2SiO_4 + 4H^+ = 2Mn^{2+} + H_4SiO_4, \quad \log K^\circ = 24.45$$

Monosilicic acid under different concentrations is able to combine with heavy metals (Cd, Pb, Zn, Hg, and others), forming soluble complex compounds if monosilicic acid concentration is less (31), and slightly soluble heavy metal silicates when the concentration of monosilicic acid is greater in the system (28,32).

$$ZnSiO_4 + 4H^+ = 2Zn^{2+} + H_4SiO_4, \quad \log K^\circ = 13.15$$

$$PbSiO_4 + 4H^+ = 2Pb^{2+} + H_4SiO_4, \quad \log K^\circ = 18.45$$

Silicon may play a prominent part in the effects of aluminum on biological systems (33). Significant amelioration of aluminum toxicity by silicon has been noted by different groups and in different species (34). The main mechanism of the effect of silicon on aluminum toxicity is probably connected with the formation of nontoxic hydroxyaluminosilicate complexes (35).

The anion of monosilicic acid $[Si(OH)_3]^-$ can replace the phosphate anion $[HPO_4]^{2-}$ from calcium, magnesium, aluminum, and iron phosphates (12). Silicon may replace phosphate from DNA and RNA molecules. As a result, proper silicon nutrition is responsible for increasing the stability of DNA and RNA molecules (36–38).

Silicon has also been shown to result in higher concentrations of chlorophyll per unit area of leaf tissue (39). This action may mean that a plant can tolerate either low or high light levels by using light more efficiently. Moreover, supplemental levels of soluble silicon are responsible for producing higher concentrations of the enzyme ribulose bisphosphate carboxylase in leaf tissue (39). This enzyme regulates the metabolism of CO_2 and promotes more efficient use of CO_2 by plants.

The increase in the content of sugar in sugar beets (*Beta vulgaris* L.) (3,40) and sugar cane (41,42) as a result of silicon fertilizer application may be assessed as a biochemical influence of silicon as well. The optimization of silicon nutrition for orange resulted in a significant increase in fruit sugar (brix) (43).

There have been few investigations of the role and functions of polysilicic acid and phytoliths in higher plants.

In spite of numerous investigations and observed effects of silicon on plants and the considerable uptake and accumulation of silicon by plants, no evidence yet shows that silicon takes part directly in the metabolism of higher plants.

19.4 BENEFICIAL EFFECTS OF SILICON IN PLANT NUTRITION

19.4.1 EFFECT OF SILICON ON BIOTIC STRESSES

Silicon has been found to suppress many plant diseases (Table 19.1) and insect attacks (Table 19.2). The effect of silicon on plant resistance to pests is considered to be due either to accumulation of absorbed silicon in the epidermal tissue or expression of pathogensis-induced host-defense responses. Accumulated monosilicic acid polymerizes into polysilicic acid and then transforms to amorphous silica, which forms a thickened silicon–cellulose membrane (44,45), and, which can be associated with pectin and calcium ions (46). By this means, a double-cuticular layer protects and mechanically strengthens plants (9) (Figure 19.2). Silicon might also form complexes with organic compounds in the cell walls of epidermal cells, therefore increasing their resistance to degradation by enzymes released by the rice blast fungus (*Magnaporthe grisea* M.E. Barr) (47). Indeed, silicon can be associated with lignin–carbohydrate complexes in the cell wall of rice epidermal cells (48).

TABLE 19.1
Plant Diseases Suppressed by Silicon

Plant	Disease	Pathogen	Reference
Barley (*Hordeum vulgare* L.)	Powdery mildew	*Erysiphe graminis*	87–89
Creeping bent grass	Dollar spot	*Sclerotinia homoeocarpa*	90
Cucumber (*Cucumis sativus* L.)	Root disease	*Pythium aphanidermatum*	91
Cucumber	Root disease	*Pythium ultimum*	92
Cucumber	Stem rotting	*Didymella bryoniae*	93
Cucumber	Stem lesions	*Botrytis cineria*	93
Cucumber, muskmelon (*C. melo* L.)	Powdery mildew	*Sphaerotheca fuliginea*	39, 94, 95
Grape (*Vitis vinifera* L.)	Powdery mildew	*Oidium tuckeri*	96
Grape	Powdery mildew	*Uncinula necator*	97
Pea (*Pisum sativum* L.)	*Mycosphaerella* leaf spot	*Mycosphaerella pinodes*	50
Rice (*Oryza sativa* L.)	Brown leaf spot	*Helminthosporium oryzae*	98
Rice	Brown spot (husk discoloration)	*Cochiobolus miyabeanus* (*Bipolaris oryzae*)	99–105
Rice	Grain discoloration	*Bipolaris, Fusarium, Epicoccum,* etc.	101, 106–109
Rice	Leaf and neck blast	*Magnaportha grisea* (*Pyricularia grisea*) (*Pyricularia oryzae*)	47, 101–103, 106, 107, 110–116
Rice	Leaf scald	*Gerlachia oryzae*	101, 106, 107, 117
Rice	Sheath blight	*Thanatephorus cucumeris* (*Rhizoctonia solani*)	52, 117–119
Rice	Sheath blight	*Corticum saskii* (*Shiriai*)	120
Rice	Stem rot	*Magnaporthe salvanii* (*Sclerotium oryzae*)	117
St. Augustine grass (*Stenotaphrum secundatum* Kuntze)	Gray leaf spot	*Magnaporthe grisea*	121
Sugarcane (*Saccharum officinarum* L.)	Leaf freckle	Probably a nutrient disorder	122
Sugarcane	Sugarcane rust	*Puccinia melanocephala*	123
Sugarcane	Sugarcane ring spot	*Leptosphaeria sacchari*	124
Tomato (*Lycopersicon esculentum* Mill.)	Fungal infection	*Sphaerotheca fuliginea*	39
Wheat (*Triticum aestivum* L.)	Powdery mildew	*Septoria nodorum*	89
Wild rice (*Zizania aquatica* L.)	Fungal brown spot	*Bipolaris oryzae*	125
Zoysia grass (*Zoysia japonica* Steud.)	Brown patch	*Rhizoctania solani*	126
Zucchini squash (*Cucurbita pepo* L.)	Powdery mildew	*Erysiphe cichoracearum*	95

Research also points to the role of silicon in plants as being active and suggests that the element might be a signal for inducing defense reactions to plant diseases. Silicon has been demonstrated to stimulate chitinase activity and rapid activation of peroxidases and polyphenoxidases after fungal infection (49). Glycosidically bound phenolics extracted from amended plants when subjected to acid or β-glucosidase hydrolysis displayed strong fungistatic activity. Dann and Muir (50) reported

TABLE 19.2
Plant Insects and Other Pests Suppressed by Silicon

Plant	Pest	Insect	Reference
Grape (*Vitis vinifera* L.)	Fruit cracking[a]		127
Italian ryegrass (*Lolium multiforum* Lam.)	Stem borer	*Oscinella frut*	128
Maize (*Zea mays* L.)	Borer	*Sesamia calamistis*	129
Rice (*Oryza sativa* L.)	Stem borer	*Chilo suppressalis*	9, 130–134
		Scirpophaga incertulas	
Rice	Stem maggot	*Chlorops oryzae*	135
Rice	Green leaf hopper	*Nephotettix bip nctatus cinticeps*	135
Rice	Brown plant hopper	*Nalaparrata lugens*	136
Rice	White-back plant hopper	*Sogetella furcifera*	137
Rice	Leaf spider[a]	*Tetranychus* spp.	9
Rice	Mites[a]	—	138
Rice	Grey garden slug[a]	*Deroceras reticulatum*	139
Rice	Lepidopteran (Pyralidae)	*Chilo zacconius*	140
Sargent crabapple (*Malus sylvestris* Mill.)	Japanese beetle	*Papilla japonica*	141
Sorghum (*Sorghum bicolor* Moench.)	Root striga, parasitic angiosperm	Scrophulariaceae; *Striga asiatica* Kuntze	142
Sugarcane (*Saccharum officinarum* L.)	Stem borer	*Diatraea succharira*	143
Sugarcane	Stalk borer	*Eldana saccharira*	144
Wheat (*Triticum aestivum* L.)	Red flour beetle	*Tribotium castaneum*	129
Zoysia grass (*Zoysia japonica* Steud.)	Fall army worm	*Spodoptera depravata*	126

[a]Noninsect pests.

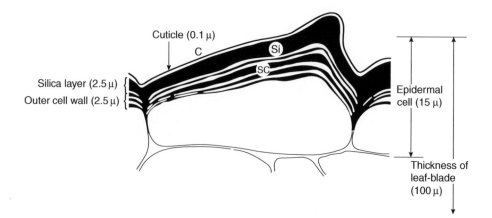

FIGURE 19.2   Schematic representation of the rice (*Oryza sativa* L.) leaf epidermal cell. (From S. Yoshida, Technical bulletin, no. 25, Food and Fertilizer Technology Center, Taipei, Taiwan, 1975.)

that pea (*Pisum sativum* L.) seedlings amended with potassium silicate showed an increase in the activity of chitinase and β-1,3-glucanase prior to being challenged by the fungal blight caused by *Mycosphaerella pinodes* Berk. et Blox. In addition, fewer lesions were observed on leaves from silicon-treated pea seedlings than on leaves from pea seedlings not amended with silicon. More recently, flavonoids and momilactone phytoalexins were found to be produced in both dicots and monocots, respectively, and these antifungal compounds appear to be playing an active role in plant disease suppression (51,52).

19.4.2 EFFECT OF SILICON ON ABIOTIC STRESSES

Silicon deposits in cell walls of xylem vessels prevent compression of the vessels under conditions of high transpiration caused by drought or heat stress. The silicon–cellulose membrane in epidermal tissue also protects plants against excessive loss of water by transpiration (53). This action occurs owing to a reduction in the diameter of stomatal pores (54) and, consequently, a reduction in leaf transpiration (15).

The interaction between monosilicic acid and heavy metals, aluminum, and manganese in soil (discussed below) helps clarify the mechanism by which heavy metal toxicity of plants is reduced (55,56).

Silicon may alleviate salt stress in higher plants (57,58). There are several hypotheses for this effect. They are (a) improved photosynthetic activity, (b) enhanced K/Na selectivity ratio, (c) increased enzyme activity, and (d) increased concentration of soluble substances in the xylem, resulting in limited sodium adsorption by plants (58–61).

Proper silicon nutrition can increase frost resistance by plants (58,62). However, this mechanism remains poorly understood.

19.5 EFFECT OF SILICON ON PLANT GROWTH AND DEVELOPMENT

19.5.1 EFFECT OF SILICON ON ROOT DEVELOPMENT

Optimization of silicon nutrition results in increased mass and volume of roots, giving increased total and adsorbing surfaces (39,63–66). As a result of application of silicon fertilizer, the dry weight of barley increased by 21 and 54% over 20 and 30 days of growth, respectively, relative to plants receiving no supplemental silicon (67). Silicon fertilizer increases root respiration (68).

A germination experiment with citrus (*Citrus* spp.) has demonstrated that with increasing monosilicic acid concentration in irrigation water, the weight of roots increased more than that of shoots (69). The same effect was observed for bahia grass (*Paspalum notatum* Flügge) (70).

19.5.2 EFFECT OF SILICON ON FRUIT FORMATION

Silicon plays an important role in hull formation in rice, and, in turn, seems to influence grain quality (71). The hulls of poor-quality, milky-white grains (kernels) are generally low in silicon content, which is directly proportional to the silicon concentration in the rice straw (72).

Barley grains that were harvested from a silicon-fertilized area had better capacity for germination than seeds from a soil poor in plant-available silicon (37). Poor silicon nutrition had a negative effect on tomato (*Lycopersicon esculentum* Mill.) flowering (73). It is important to note that the application of silicon fertilizer accelerated citrus growth by 30 to 80%, speeded up fruit maturation by 2 to 4 weeks, and increased fruit quantity (74). A similar acceleration in plant maturation with silicon fertilizer application was observed for corn (37).

19.5.3 EFFECT OF SILICON ON CROP YIELD

Numerous field experiments under different soil and climatic conditions and with various plants clearly demonstrated the benefits of application of silicon fertilizer for crop productivity and crop quality (Table 19.3).

TABLE 19.3
Effect of Silicon Fertilizers on Crop Production

No.	Soil, Country	Silicon Fertilizer	Dose (kg ha^{-1})	Regime	Plant	Crop, Grain, Mg ha^{-1}	Straw Mg ha^{-1}	Reference
1	Clay-with-flints chalk, Rothamsted Station, England	Sodium silicate	0	Control	Barley (*Hordeum vulgare* L.)	2.02	1.13	145
			0	N		3.03	2.32	
			448	N		**5.04**	**4.32**	
			0	N, P		6.32	5.04	
			448	N, P		**6.52**	**5.04**	
			0	N, K, Na, and Mg		3.82	3.70	
			448	N, K, Na, and Mg		**5.22**	**4.49**	
			0	N, P, K, Na, and Mg		6.42	5.08	
			448	N, P, K, Na, and Mg		**7.31**	**5.76**	
2	Clay-with-flints chalk, Rothamsted Station, England	Sodium silicate	0	N, P, K, Na, and Mg	Hay		5.98	146
			448	N, P, K, Na, and Mg			**7.78**	
3	Soddy podzolic soil	Amorphous silica	0	N, K	Barley	2.47	3.47	147
			870	N, K		**2.88**	**3.57**	
			0	N, P, K		2.74	3.72	
			870	N, P, K		**3.17**	**4.00**	
4	Soddy podzolic soil, Russia	Amorphous silica	0		Barley	4.6		37
			100			**5.26**		
			500			**6.84**		
5	Soddy podzolic soil, Russia	Amorphous silica	0		Corn (*Zea mays* L.)	0	7.68	37
			30			**4.2**	**11.44**	
			100			**6.3**	**13.68**	
6	Soddy podzolic soil, Russia	Zeolite	0	N, P, K	Strawberry (*Fragaria vesca* L.)	8.9		148
			10%	N, P, K		**9.8**		
			0			10.6		
			10%			**15.3**		
7	Acid podzolic soil, Sweden	Si-Mn slag	0	Lime 2000	Oats (*Avena sativa* L.)	0.6		149
			0			0.93		
			2000			**1.48**		
8	Alluvial soil, Russia	Slag	0		Hay		1.85	150
			1000				**2.33**	

No.	Soil	Si source	Rate	Fertilizer	Crop			Ref.
9	Chernozem, Russia (mollisol)	Slag	0	N, P, and K	Beet (*Beta vulgaris* L.)	37.5	7.37	40
			0	N, P, and H+lime		40.2	7.72	
			18,000	N, P, and K		**4.10**	**7.98**	
10	Chernozem, Russia (mollisol)	Zeolite	0	Manure (120 t ha^{-1})	Corn forage		160	151
			120,000	Manure (120 t ha^{-1})			202	
							280.4	
11	Chernozem, Russia (mollisol)	Sodium silicate	0	N	Wheat (*Triticum aestivum* L.)	2.6		152
			10	N		**2.9**		
12	Chestnut soil, Russia	Zeolite	0		Sorghum (*Sorghum bicolor* Moench.)	3.72	10.5	153
			20,000			**4.3**	**14.7**	
13	Chestnut soil, Russia	Zeolite	0		Barley	2.36		154
			10,000			**2.66**		
14	Chestnut soil, Russia	Amorphous silica	0		Barley	3.48	5.56	155
			3000			**3.85**	**6.16**	
			0	N, P, and K		3.66	5.85	
			3000	N, P, and K		**4.08**	**6.52**	
15	Histosol acid, Norway	Iron slag	0		Hay	9.09		156
			3600			**9.97**		
16	Muck soil, Russia	Dunite	0	N, P, and K	Potato (*Solanum tuberosum* L.)	7.26		157
			1500	N, P, and K		**13.05**		
17	Muck acid soil, Russia	Amorphous silica	0		Barley	3.7		158
			8000			**5.2**		
18	Alluvial-swamp with salt, Russia	Rice straw	0		Rice (*Oryza sativa* L.)	2.77		159
			6000			**4.78**		
19	Alluvial-swamp Chernozem, Russia	Sodium silicate	0		Rice	5.09		160
			310			**5.9**		
20	Dark chestnut soil, Russia	Sodium silicate	0		Rice	3.52		161
			310			**4.01**		
			0	Manure		3.98		
			310	Manure		**4.28**		
21	Sandy loam, Sri Lanka	Rice straw ash	0		Rice	3.9		162
			1000			**4.6**		
			0	K		4.3		
			1000	K		**5.0**		
22	Ultisol, Nigeria	Sodium silicate	0	N, P, and K	Rice	2.4		101
			0	N, P, and K		6.3		
			4.7			**9.3**		

Continued

TABLE 19.3 *(Continued)*

No.	Soil, Country	Silicon Fertilizer	Dose (kg ha^{-1})	Regime	Plant	Crop, Grain, Mg ha^{-1}	Straw Mg ha^{-1}	Reference
23	Hydromorphe organic Gley, Madagascar		0	N, P, K + Mg	Rice	8.1		163
			4.7	N, P, K + Mg		**14.7**		
			0	Mg		2.34	4.96	
			4.7	Mg		**2.48**	**4.86**	
			0			2.04	4.58	
			4.7			**3.14**	**6.02**	
24	Mineral semi-tropic Gley, Madagascar	Amorphous silica	0	N, P, and K	Rice	3.876		163
			0			5.571		
			1500	N, P, and K		**6.186**		
			1600	K_{120}		3.520		
			0			**5.172**		
			1600	K_{120}		6.1775		
			0			**6.920**		
25	Humic latosol, Hawaii	Calcium silicate	0	P	Sugarcane (*Saccharum officinarum* L.)		141	164
			830	P			**157**	
26	Humic latosol, Hawaii	Calcium silicate	0	pH 5.8	Sugarcane		124	165
			830				**147**	
			1660				**151**	
27	Humic latosol, Hawaii	Calcium silicate	0	pH −6.2	Sugarcane		131	165
			830				151	
			1660				166	
28	Humic ferriginous latosol, Hawaii	TVA slag	0	P 280	Sugarcane	23.4	253	42
			4500	P 280		**31.6**	**327**	
			0	$CaCO_3$ (4.5 Mg ha^{-1}) + P (1120 kg ha^{-1})		20.7	262	
			4500	P (1120 kg ha^{-1})		**32.7**	**338**	
29	Aluminos humic, ferruginous latosol, Mauritius	Electric furnace slag	0	N, P, and K	Sugarcane	27.4	266.7	41
			0	N, P, and K + $CaCO_3$ (4.5 t ha^{-1})		26.67	256.8	
30	Histosol, Florida	Calcium silicate slag	6177	N, P, and K	Sugarcane	**33.84**	**313.7**	124
			0			18.1	150	
			6700			23.8	194	

Note: Response to application of silicon fertilizer is shown in bold type in the columns.

19.6 SILICON IN SOIL

19.6.1 FORMS OF SILICON IN SOIL

Soils generally contain from 50 to 400 g Si kg^{-1} of soil. Soil-silicon compounds usually are present as SiO_2 and various aluminosilicates. Quartz, together with crystalline forms of silicates (plagioclase, orthoclase, and feldspars), secondary or clay- and silicon-rich minerals (kaolin, vermiculite, and smectite), and amorphous silica are major constituents of most soils (75). These silicon forms are only sparingly soluble and usually biogeochemically inert. Monosilicic and polysilicic acids are the principal soluble forms of silicon in soil (76).

For the most part, monosilicic acid occurs in a weakly adsorbed state in the soil (13,37). Monosilicic acid has a low capacity for migration down the soil profile (77). The chemical similarity between the silicate anion and the phosphate anion results in a competitive reaction between the various phosphates and monosilicic acid in the soil. Increasing monosilicic acid concentration in the soil solution causes transformation of the plant-unavailable phosphates into the plant-available ones (12). Monosilicic acid can interact with aluminum, iron, manganese, and heavy metals to form slightly soluble silicates (29,30).

Polysilicic acids are an integral component of the soil solution. They mainly affect soil physical properties. The mechanism of polysilicic acid formation is not clearly understood. Unlike monosilicic acid, polysilicic acid is chemically inert and basically acts as an adsorbent, forming colloidal particles (34). Polysilicic acids are readily sorbed by minerals and form siloxane bridges (78). Since polysilicic acids are highly water saturated, they may have an effect on the soil water-holding capacity. Polysilicic acids have been found to be important for the formation of soil structure (79). There is a pressing need to obtain additional information about biogeochemically active silicon-rich substances involved in soil-formation processes.

19.6.2 SOIL TESTS

Silicon forms may be defined as total, extractable, and soluble. Total silicon comprises all existing forms of soil silicon that can be dissolved by strong alkali-fusion or acid-digestion methods (80). This parameter does not provide information about plant-available and chemically active silicon because silicon in soil is in the form of relatively inert minerals (62).

Usually for determination of soil plant-available silicon, different extracts are used. Extracts remove silicon of intermediate stability that is often associated with crystalline or amorphous soil components. The most common chemical extracts used are 0.5 M ammonium acetate (pH 4.8), 0.1 or 0.2 M HCl, water, sodium acetate buffer (pH 4.0), and ammonium oxalate (pH 3.0) among others (71,81–83). Unfortunately, soil drying is a component of all these extraction methods. During drying, all monosilicic acid (plant-available form of Si) is dehydrated and transformed into amorphous silica (21). Concern has been expressed that data obtained on dried soil may not adequately describe plant-available soil silicon and may be unsatisfactory for evaluating soil previously amended with silicon fertilizer (71). Nevertheless, extractable silicon has been correlated with the plant yield (84).

To overcome problems associated with soil drying, soluble monosilicic acid can be determined in water extracted from field-moist soil samples. After 1 h of shaking and filtration, the clean extract is analyzed for soluble monosilicic acid. This method also facilitates the testing for polysilicic acid in the soil (13). It should be noted that a change in the soil-water concentration from 5 to 50% of the field capacity had no effect on the sensitivity of the method (12,13).

To fully characterize soil plant-available silicon, it appears that more than one parameter of measurement is required. The combination of data on soluble monosilicic acid, polysilicic acid, and silicon in some extracts could give more complete information about the soil-silicon status.

19.7   SILICON FERTILIZERS

Although silicon is a very abundant element, for a material to be useful as a fertilizer, it must have a relatively high content of silicon, provide sufficient water-soluble silicon to meet the needs of the plant, be cost effective, have a physical nature that facilitates storage and application, and not contain substances that will contaminate the soil (85). Many potential sources meet the first requirement; however, only a few meet all of these requirements. Crop residues, especially of silicon-accumulating plants such as rice, are used as silicon sources either intentionally or unintentionally. When available, they should not be overlooked as sources of silicon. However, the crop demand for application of silicon fertilizer generally exceeds that which can be supplied by crop residues.

Inorganic materials such as quartz, clays, micas, and feldspars, although rich in silicon, are poor silicon-fertilizer sources because of the low solubility of the silicon. Calcium silicate, generally obtained as a byproduct of an industrial procedure (steel and phosphorus production, for example) is one of the most widely used silicon fertilizers. Potassium silicate, though expensive, is highly soluble and can be used in hydroponic culture. Other sources that have been used commercially are calcium silicate hydrate, silica gel, and thermo-phosphate (85).

19.8   SILICON IN ANIMAL NUTRITION

In the last 30 years, a few studies on silicon effects on mammals, fish, and birds were conducted (33,38,86). Data have shown that active silicon (fine amorphous silica) increased the weight and quality of animals. Chicken (*Gallus gallus domesticus*), pig (*Sus scrofa*), and sheep (*Ovis aries*) with silicon-rich diets were healthier and stronger than animals without silicon supplements (33,38).

REFERENCES

1. H. Davy. *The Elements of Agricultural Chemistry*. Hartford: Hudson and Co., 1819.
2. V.A. Kovda. The mineral composition and soil formation. *Pochvovedenie* 1:6–38, 1956.
3. J. Leibig. *Organic Chemistry in Its Application to Agriculture and Physiology*. From the manuscript of the author by Lyon Playfair. London: Taylor & Walton, 1840.
4. Rothamsted Experimental Station. *Guide the Classical Experiment*. Watton, Norfolk: Lawes Agricultural Trust, Rapide Printing, 1991.
5. J. Zippicotte. Fertilizer. U.S. Patent no. 238240. Official Gazette of the United States Patent Office, 1881, pp. 19, 9, and 496.
6. W. Maxwell. *Lavas and Soils of the Hawaiian Islands*. Honolulu, Hawaii: Hawaiian Sugar Planters' Association, 1898, p. 189.
7. E. Epstein. Silicon. *Ann. Rev. Plant Physiol. Plant Mol. Biol.* 50:641–664, 1999.
8. C.J. Lewin, B.E. Reimann. Silicon and plant growth. *Annu. Rev. Plant Physiol.* 20:289–304, 1969.
9. S. Yoshida. The physiology of silicon in rice. Technical bulletin, no. 25, Food and Fertilizer Technology Center, Taipei, Taiwan, 1975.
10. N.I. Bazilevich. *The Biological Productivity of North Eurasian Ecosystems*. RAS Institute of Geography, Moscow: Nayka, 1993.
11. V.V. Matichenkov, E.A. Bocharnikova, D.V. Calvert, G.H. Snyder. Comparison study of soil silicon status in sandy soils of south Florida. *Soil Crop Sci. Soc. Florida Proc.* 59:132–137, 2000.
12. V.V. Matichenkov, M.Y. Ammosova. Effect of amorphous silica on soil properties of a sod-podzolic soil. *Euras. Soil Sci.* 28:87–99, 1996.
13. V.V. Matichenkov, Y.M. Ammosova, E.A. Bocharnikova. The method for determination of plant-available silica in soil. *Agrochemistry* 1:76–84, 1997.
14. E. Takahashi. Uptake mode and physiological functions of silica. *Japan J. Soil Sci. Plant Nutr.* 49:357–360, 1995.
15. M.J. Aston, M.M. Jones. A study of the transpiration surfaces of *Avena sterilis* L. var. algerian leaves using monosilicic acid as a tracer for water movement. *Planta* 130:121–129, 1976.

16. J.F. Ma. Function of silicon in higher plants. *Prog. Mol. Subcell. Biol.* 33:127–147, 2003.
17. L. Waterkeyn, A. Bientait, A. Peters. Callose et silice epidermiques rapports avec la transpiration culticulaire. *La Cellule* 73:263–287, 1982.
18. S. Yoshida. Chemical aspects of the role of silicon in physiology of the rice plant. *Bull. Nat. Inst. Agric. Sci. Series B* 15:1–58, 1965.
19. F.C. Lanning. Plant constituents, silicon in rice. *J. Agric. Food Chem.* 11:435–437, 1963.
20. L.V. Dracheva. The study of silicic acid condition in model and technological solutions and surface waters. Autoref. Diss. Cand., MITHT, Moscow, 1975.
21. R.K. Iler. *The Chemistry of Silica*. New York: Wiley, 1979.
22. S. Mann, C.C. Perry. Structural aspects of biogenic silica. In: *Silicon Biochemistry. Ciba Foundation Symposium 121*. New York: Wiley, 1986, pp. 40–53.
23. S. Mann, G.A. Ozin. Synthesis of inorganic materials with complex form. *Nature* 382:313–318, 1996.
24. L.H.P. Jones, K.A. Handreck. Silica in soil, plants and animals. *Adv. Agron.* 19:107–149, 1967.
25. P.N. Balabko, V.E. Prikhod'ko, J.M. Ammosova. The biolites of silica in plants and some forest soils. *Biol. Sci.* 12:92–96, 1980.
26. V.I. Mica. Rezidua sodiku a zevera ve frakcioxidu kremiciteho pri rozforech rostein. *Agrochemia* 26:270–272, 1986.
27. G.V. Dobrovolsky, A.A. Bobrov, A.A. Gol'eva, S.A. Shoba. The opal phytoliths in taiga biogeocenose of media taiga. *Biol. Sci.* 2:96–101, 1988.
28. W.L. Lindsay. *Chemical Equilibria in Soil*. New York: Wiley, 1979.
29. T. Horiguchi. Mechanism of manganese toxicity and tolerance of plant. IV. Effect of silicon on alleviation of manganese toxicity of rice plants. *Soil Sci. Plant Nutr.* 34:65–73, 1988.
30. D.G. Lumsdon, V.C. Farmer. Solubility characteristics of proto-imogolite sols: how silicic acid can de-toxify aluminium solutions. *Eur. Soil Sci.* 46:179–186, 1995.
31. P.W. Schindler, B. Furst, R. Dick, P.O. Wolf. Ligand properties of surface silanol groups. I. Surface complex formation with Fe^{3+}, Cu^{2+}, Cd^{3+}, and Pb^{2+}. *J. Colloid Interface Sci.* 55:469–475, 1976.
32. K.A. Cherepanov, G.I. Chernish, V.M. Dinelt, J.I. Suharev. *The Utilization of Secondary Material Resources in Metallurgy*. Moscow: Metallurgy, 1994.
33. J.D. Birchall, C. Exley, J.S. Chappell. Acute toxicity of aluminum to fish eliminated in silicon-rich acid waters. *Nature* 338:146–148, 1989.
34. M.J. Hodson, D.E. Evans. Aluminium/silicon interactions in higher plants. *J. Exp. Bot.* 46: 161–171, 1995.
35. C. Exley, J.D. Birchall. A mechanism of hydroxyaluminosilicate formation. *Polyhedron* 12:1007–1017, 1993.
36. N.E. Aleshin. The content of silicon in DNA of rice. *Doklady VASHNIL* 6:6–7, 1982.
37. V.V. Matichenkov. Amorphous oxide of silicon in soddy podzolic soil and its influence on plants. Autoref. Diss. Cand., Moscow State University, Moscow, 1990.
38. M.G. Voronkov G.I. Zelchan, A.Y. Lykevic *Silicon and Lie*. Riga: Zinatne, 1978.
39. M.H. Adatia, R.T. Besford. The effects of silicon on cucumber plants grown in recirculating nutrient solution. *Ann. Bot.* 58:343–351, 1986.
40. V.M. Klechkovsky, A.V. Vladimirov. New fertilizer. *Chem. Soc. Agric.* 7:55, 1934.
41. A.S. Ayres. Calcium silicate slag as a growth stimulator for sugarcane on low-silicon soils. *Soil Sci.* 101:216–227, 1966.
42. R.L. Fox, J.A. Silva, O.R. Younge, D.L. Plucknett, G.D. Sherman. Soil and plant silicon and silicate response by sugar cane. *Soil Sci. Soc. Am.* 31:775–779, 1967.
43. V.V. Matichenkov, E.A. Bocharnikova, D.V. Calvert. Response of citrus to silicon soil amendments. *Proc. Florida State Hortic. Soc.* 114:94–97, 2002.
44. N.E. Aleshin. About the biological role of silicon in rice. *Vestnik Agric. Sci.* 10:77–85, 1988.
45. M.J. Hodson, A.G. Sangster. Silica deposition in the influence bracts of wheat (*Triticum aestivum*). 1 Scanning electron microscopy and light microscopy. *Can. J. Bot.* 66:829–837, 1988.
46. L. Waterkeyn, A. Bientait, A. Peeters. Callose et silice epidermiques rapports avec la transpiration culticulaire. *La Cellule* 73:263–287, 1982.
47. R.J. Volk. Silicon content of the rice plant as a factor influencing its resistance to infection by the blast fungus *Piricularia oryzae. Phytopathology* 48:179–184, 1958.
48. S. Inanaga, A. Okasaka, S. Tanaka. Does silicon exist in association with organic compounds in rice plant? *Japan J. Soil Sci. Plant Nutr.* 11:111–117, 1995.

49. M. Cherf, J.G. Menzies, D.L. Ehret, C. Bopgdanoff, R.R. Belanger. Yield of cucumber infected with *Pythium aphanidermatum* when grown with soluble silicon. *HortScience* 29:896–897, 1994.

50. E.K. Dann, S. Muir. Peas grown in media with elevated plant-available silicon levels have higher activities of chitinases and β-1,3-glucanase, are less susceptible to a fungal leaf spot pathogen and accumulate more foliar silicon. *Aust. Plant Pathol.* 31:9–13, 2002.

51. A. Fawe, M. Abou-Zaid, J.G. Menzies, R.R. Bélanger. Silicon-mediated accumulation of flavonoid phytoalexins in cucumber. *Phytopathology* 88:396–401, 1998.

52. F.A. Rodrigues, D. McNally, L.E. Datnoff, J.B. Jones, C. Labbé, N. Benhamou, J.M. Menzies, R. Bélanger. Silicon enhances the accumulation of diterpenoid phytoalexins in rice: a potential mechanism for blast resistance. *Phytopathology* 93(Suppl.):S74, 2003.

53. S.F. Emadian, R.J. Newton. Growth enhancement of loblolly pine (*Pinus taeda* L.) seedlings by silicon. *J. Plant Physiol.* 134:98–103, 1989.

54. G.V. Efimova, S.A. Dokynchan. Anatomo-morphological construction of epidermal tissue of rice leaves and increasing of its protection function under silicon effect. *Agric. Biol.* 3:57–61,1986.

55. J. Barcelo, P. Guevara, C.H. Poschenrieder. Silicon amelioration of aluminum toxicity in teosinte (*Zea mays* L. ssp. mexicana). *Plant Soil* 154:249–255, 1993.

56. C.D. Foy. Soil chemical factors limiting plant root growth. *Adv. Soil Sci.* 19:97–149, 1992.

57. Y. Liang, Z. Shen. Interaction of silicon and boron in oilseed rape plants. *J. Plant Nutr.* 17:415–425, 1994.

58. V.V. Matichenkov, E.A. Bocharnikova. The relationship between silicon and soil physical and chemical properties. In: L.E. Datnoff, G.H. Snyder, H. Korndorfer, eds. *Silicon in Agriculture*. Amsterdam: Elsevier, 2001, pp. 209–219.

59. R. Ahmad, S. Zaheer, S. Ismail. Role of silicon in salt tolerance of wheat (*Triticum aestivum* L.). *Plant Sci.* 85:43–50, 1992.

60. M. Bradbury, R. Ahmad. The effect of silicon on the growth of *Prosopis juliflora* growing in saline soil. *Plant Soil* 125:71–74, 1990.

61. Y. Liang. Effects of silicon on enzyme activity and sodium, potassium and calcium concentration in barley under salt stress. *Plant Soil* 209:217–224, 1999.

62. V.V. Matichenkov, E.A. Bocharnikova, D.V. Calvert, G.H. Snyder. Comparison study of soil silicon status in sandy soils of South Florida. *Proc. Soil Crop Sci. Florida* 59:132–137, 1999.

63. E.A. Bocharnikova. The study of direct silicon effect on root demographics of some cereals. In: *Proceedings of the Fifth Symposium of the International Society of Root Research. Root Demographics and Their Efficiencies in Sustainable Agriculture, Grasslands, and Forest Ecosystems*, Madrea Conference Conter-Clenson, South Carolina, 14–18 July, 1996.

64. H.M. Kim. The influence of nitrogen and soil conditioners on root development, root activity and yield of rice. 3. The effects of soil conditioners on development of root, rooting zone and rice yield. *Research Rep. Rural Development Admin., Plant Environ., Mycol. & Farm Prod Util, Korea Republic* 29:12–29, 1987.

65. L.I. Kudinova. The effect of silicon on growth, size of leaf area and sorbed surface of plant roots. *Agrochemistry* 10:117–120, 1975.

66. V.V. Matichenkov. The silicon fertilizer effect of root cell growth of barley. Abstr. in *The fifth Symposium of the International Society of Root Research*, Clemson, SC, USA, 1996, p. 110.

67. L.I. Kudinova The effect of silicon on weight of plant barley. *Sov. Soil Sci.* 6:39–41, 1974.

68. T. Yamaguchi, Y. Tsuno, J. Nakano, P. Mano. Relationship between root respiration and silica:calcium ratio and ammonium concentration in bleeding sap from stem in rice plants during the ripening stage. *Jpn. J. Crop Sci.* 64:529–536, 1995.

69. V.V. Matichenkov, D.V. Calvert, G.H. Snyder. Silicon fertilizers for citrus in Florida. *Proc. Florida State Hortic. Soc.* 112:5–8, 1999.

70. V.V. Matichenkov, D.V. Calvert, G.H. Snyder. Effect of Si fertilization on growth and P nutrition of bahiagrass. *Proc. Soil Crop Sci. Soc. Florida* 60:30–36, 2000.

71. N.K. Savant, G.H. Snyder, L.E. Datnoff. Silicon management and sustainable rice production. *Adv. Agron.* 58:151–199, 1997.

72. E.P. Aleshin, N.E. Aleshin, A.R. Avakian. The effect of various nutrition and gibberillins on SiO_2 content in hulls of rice. *Agrochemistry* 7:64–68, 1978.

73. Y. Miyake. On the environmental condition and nitrogen source to appearance of silicon deficiency of the tomato plant. *Sci. Rep. of the faculty of Agriculture Okayama Univ., Japan* 81:27–35, 1993.

74. V.G. Taranovskaia. The silicication of subtropic greenhouse and plantations. *Sov. Subtropics* 7:32–37, 1939.

75. D.S. Orlov. *Soil Chemistry*. Moscow: Moscow State University, 1985.

76. V.V. Matichenkov, G.H. Snyder. The mobile silicon compounds in some South Florida soils. *Euras. Soil Sci.* 24:1165–1173, 1996.

77. R.A. Khalid, J.A. Silva. Residual effect of calcium silicate on pH, phosphorus and aluminum in tropical soil profile. *Soil Sci. Plant Nutr.* 26:87–98, 1980.

78. O.A. Chadwick, D.M. Hendriks, W.D. Nettleton. Silica in durick soil. *Soil Sci. Soc. Am. J.* 51:975–982, 1987.

79. V.V. Matichenkov, D.L. Pinsky, E.A. Bocharnikova. Influence of mechanical compaction of soils on the state and form of available silicon. *Euras. Soil Sci.* 27:58–67, 1995.

80. G.H. Snyder. Methods for silicon analysis in plants, soils, and fertilizers. In: L.E. Datnoff , G.H. Snyder, G.H. Korndorfer, eds. *Silicon in Agriculture*. Amsterdam: Elsevier, 2001, pp. 185–196.

81. P.K. Nayer, A.K. Mistra, S. Patnaik. Evaluation of silica-supplying power of soils for growing rice. *Plant Soil* 47:487–494, 1977.

82. K. Imaizumi, S. Yoshida. Edaphological studies on silicon supplying power of paddy soils. *Bull. Natl. Inst. Agric. Sci. (Jpn.) B* 8:261–304, 1958.

83. M.B.C. Haysom, L.S. Chapman. Some aspects of the calcium silicate trials at Mackay. *Proc. Queens Soc. Sugar Cane Tech.* 42, 177–222, 1975.

84. G.H. Korndorfer, G.H. Snyder, M. Ulloa, G. Powell, L.E. Datnoff. Calibration of soil and plant silicon analysis for rice production. *J. Plant Nutr.* 24:1071–1084, 2001.

85. G.J. Gascho. Silicon sources for agriculture. In: L.E. Datnoff, G.H. Snyder, G.H. Korndorfer, eds. *Silicon in Agriculture*. Amsterdam: Elsevier, 2001, pp. 197–207.

86. E.M. Carlisle. Silicon as an essential trace element in animal nutrition. In: D. Evered, M. O'Connor, eds. *Silicon Biochemistry*. Ciba Foundation Symposium, Chichester, UK: Wiley, 1986, pp. 121, 123–139.

87. T.L.W. Carver, R.J. Zeyen, G.G. Ahlstrand. The relation between insoluble silicon and success or failure of attempted penetration by powdery mildew (*Erysiphe graminis*) germlings on barley. *Physiol. Plant Pathol.* 31:133–148, 1987.

88. D. Jiang, R.J. Zeyen, V. Russo. Silicon enhances resistance of barley to powdery mildew (*Erusiphe graminis* f. sp. hordei). *Phytopathology* 79:1198, 1989.

89. H.J. Leusch, H. Buchenaner, Effect of soil treatments with silica-rich lime fertilizers and sodium silicate on the incidence of wheat by *Erysiphe graminis* and *Septoria nodorum* depending on the form of N-fertilizer. *J. Plant Dis. Prot.* 96:154–172, 1989.

90. R.E. Schmidt, X. Zhang. Antioxidant response to hormone-containing product in Kentucky bluegrass subjected to drought. *Crop Sci.* 39:545–551, 1999.

91. M. Cherif, J.G. Menzies, D.L. Ehret, C. Bogdanoff, R.R. Belanger. Yield of cucumber infected with *Pythium aphanidermatum* when grown with soluble silicon. *HortScience* 29:896–897, 1994.

92. M. Cherif, R.R. Belanger. Use of potassium silicate amendments in recirculating nutrient solution to suppress *Pythium ultimum* on long English cucumber. *Plant Dis.* 76:1008–1011, 1992.

93. T.M. O'Neill. Investigation of glasshouse structure, growing medium and silicon nutrition as factors affecting disease incidence in cucumber crops. *Med. Fac. Landbouw Rijksuniv Gent.* 56:359–367, 1991.

94. R.R. Belanger, P.A. Bowen, D.L. Ehret, J.G. Menzies. Soluble silicon: its role in crop and disease management of greenhouse crops. *Plant Dis.* 79:329–336, 1995.

95. J.G. Menzies, D.L. Ehret, A.D.M. Glass, T. Helmer, C. Koch, F. Seywerd. Effects of soluble silicon on the parasitic fitness of *Sphaerotheca fuliginea* on *Cucumus sativus*. *Phytopathology* 81:84–88, 1991.

96. H. Grundnofer. Eifluss von silikataufnahme und einlagerung auf den befall der rebe mit echtem mehltau. *Diss* 114:102–114, 1994.

97. P. Bowen, J. Menzies, D. Ehret, L. Samuel, A.D.M. Glass. Soluble silicon sprays inhibit powdery mildew development on grape leaves. *J. Am. Soc. Hortic. Sci.* 117:906–912, 1992.

98. M.F. Hegazi, D.I. Harfoush, M.H. Mostafa, I.K. Ibrahim, M.F.Hegazi, D.I. Harfoush, M.H. Mostafa, I.K. Ibrahim. Changes in some metabolites and oxidative enzymes associated with brown leaf spot of rice. *Ann. Agric. Sci.* 38:291–299, 1993.

99. E. Takahashi. Nutritional studies on development of *Helminthosporium* leaf spot. In: *Proceedings of the Symposium on Rice Diseases and Their Control by Growing Resistance Varieties and Other Measures*. Tokyo, Japan: Forestry and Fisheries Research Council, 1967, pp. 157–170.

100. K. Ohata, C. Kubo, K. Kitani. Relationship between susceptibility of rice plants to *Helmithosporium* blight and physiological changes in plants. *Bull. Shikoku Agric. Exp. Stn.* 25:1–19, 1972.

101. M. Yamaguchi, M.D. Winslow. Effect of silica and magnesium on yield of upland rice in humid tropics. *Plant Soil* 113:265–269, 1987.

102. L.E. Datnoff, R.N. Raid, G.H. Snyder, D.B. Jones. Effect of calcium silicate slag on blast and brown spot intensities and yields of rice. *Plant Dis.* 75:729–732, 1991.

103. L.E. Datnoff, G.H. Snyder, C.W. Deren. Influence of silicon fertilizer grades on blast and brown spot development and on rice yields. *Plant Dis.* 76:1011–1013, 1992.

104. T.S. Lee, L.S. Hsu, C.C. Wang, Y.H. Jeng. Amelioration of soil fertility for reducing brown spot incidence in the paddy fields of Taiwan. *J. Agric. Res. China* 30:35–49, 1981.

105. H.P. Nanda, R.S. Gangopadhyay. Role of silicated cells of rice leaf on brown spot disease incidence by *Bipolaris oryzae*. *Int. J. Trop. Plant Dis.* 2:89–98, 1984.

106. F.J. Correa-Victoria, L.E. Datnoff, M.D. Winslow, K. Okada, D.K. Friesen, J.I. Danz, G.H. Snyder. Silicon deficiency of upland rice on highly weathered Savanna soil in Columbia. In: *II. Diseases and grain quality. IX Conf. Int. de arroz para a America Latina e para o Caribe, V Reuniao Nacional de Ressquisa de Arroz*, Castro's Park Hotel, Goiania, Goias, Brazil, 21–25 March 1994.

107. M.D. Winslow. Silicon, disease resistance and yield of rice genotypes under upland cultural conditions. *Crop Sci.* 32:1208–1213, 1992.

108. G.K. Korndorfer, L.E. Datnoff, G.F. Correa. Influence of silicon on grain discoloration and upland rice grown on four savanna soils of Brazil. *J. Plant Nutr.* 22:93–102, 1999.

109. A.S. Prabhu, M.P. Barbosa Filho, M.C. Filippi, L.E. Datnoff, G.H. Snyder. Silicon from disease control perspective in Brazil. In: L.E. Datnoff, G.H. Snyder, G.H. Korndorfer, eds. *Silicon in Agriculture*. Amsterdam:Elsevier, 2001, pp. 293–311.

110. N.E. Aleshin, E.R. Avakyan, S.A. Dyakunchan, E.P. Aleshin, V.P. Barushok, M.G. Voronkov. Role of silicon in resistance of rice to blast. *Dokl. Acad. Nauk SSSR* 291:217–219, 1987.

111. C.W. Deren, L.E. Datnoff, G.H. Snyder, F.G. Martin. Silicon concentration, disease response and yield components of rice genotypes grown on flooded organic histosols. *Crop Sci.* 34:733–737, 1994.

112. C.K. Kim, S. Lee. Reduction of the incidence of rice neck blast by integrated soil improvement practice. *Kor. J. Plant Prot.* 21:15–18, 1982.

113. C.K. Kim, M.C. Rush, D.R. MacKenzie. Food-mediated resistance to the rice blast disease. In: A.S.R. Juo, J.A. Lowe, eds. *The Wetlands and Rice in Subsaharan Africa*. Ibadan, Nigeria: IITA, 1986, pp. 15–169.

114. T. Kozaka. Control of rice blast by cultivation practices in Japan. In: *The Rice Blast Disease. Proceedings of the Symposium of the International Rice Research Institute*, Los Banos, Philippines, July 1993. Baltimore: John Hopkins, 1965, pp. 421–438.

115. C.T. Kumbhar, A.G. Nevase, N.K. Savant. Rice hull ash applied to soil reduces leaf blast incidence. *Internat. Rice Res. Newslett.* 20:16–21, 1995.

116. F.J. Osuna-Canizales, S.K. DeDatta, J.M. Bonman. Nitrogen form and silicon nutrition effects on resistance to blast disease of rice. *Plant Soil* 135:223–231, 1991.

117. S.H. Elawad, V.E. Green. Silicon and the rice plant environment: a review of recent research. *Il Riso* 28:235–253, 1979.

118. L.E. Datnoff, R.N. Raid, G.H. Snyder, D.B. Jones. Evaluation of calcium silicate slag and nitrogen on brown spot, neck, and sheath blight development on rice. *Biol. Cult.Test Cont.Plant Dis.* 5:65, 1990.

119. H. Kunoh. Ultrastructure and mobilization of ions near infection sites. *Annu. Rev. Phytopathol.* 28:93–111, 1990.

120. G. Mathai, P.V. Paily, M.R. Menon. Effect of fungicides and silica in the control of sheath blight disease of rice caused by *Corticumsaskii* (*Shiriai*). *Agr. Res. J. Kerala* 19:79–83, 1978.

121. L.E. Datnoff, R.T. Nagata. Influence of silicon on gray leaf spot development in St. Augustine grass. *Phytopathology* 89(Suppl.):S10, 1999.

122. R.L. Fox, A. James, J.A. Silva, D.Y. Teranishi, M.H. Matsuda, P.C. Ching. Silicon in soils, irrigation water, and sugarcane of Hawaii. *Hawaii Farm Sci.* 16:1–4, 1967.

123. J.L. Dean, E.H. Todd. Sugarcane rust in Florida. *Sugar J.* 42:10, 1979.

124. R.N. Raid, D.L. Anderson, M.F. Ulloa. Influence of cultivar and amendment of soil with calcium silicate slag on foliar disease development and yield of sugarcane. *Crop Prot.* 11:84–87, 1992.

125. D.K. Malvick, J.A. Percich. Hydroponic culture of wild rice (*Zizania palustris* L.) and its application to studies of silicon nutrition and fungal brown spot disease. *Can. J. Plant Sci.* 73:969–975, 1993.

126. M.K. Saigusa, K. Onozawa. Effects of porous hydrate calcium silicate on the silica nutrition of turf grasses. *Grassland Sci.* Jan. 45(4):411–415, 2000.

127. N. Sang-Young, M.K. Kyong, C.L. Sang, C.P. Jong. Effects of lime and silica fertilizer application on grape cracking. *J. Agric. Sci. Soil Fert.* 38:410–415, 1996.

128. D. Moore. The role of silica in protecting Italian ryegrass (*Lolium multiflorum*) from attack by dipterous stem-boring larvae (*Oscinella* fruit and other related species). *Ann. Appl. Biol.* 104:161–166, 1984.

129. M. Setamou, F. Schulthess, N.A. Bosque-Perez, A. Thomas-Odjo. Effect of plant nitrogen and silicon on the biomocs of *Sesamia calamistics. Bull. Ent. Res.* 83:405–411, 1993.

130. A.S. Djamin, M.D. Pathak. Role of silica in resistance to asiatic rice borer, *Chilo suppresalis* (Walker), in rice. *J. Econ. Ent.* 60:347–351, 1967.

131. M. Ota, H. Kobayashi, Y. Kawaguchi. Effect of slag on paddy rice. 2. Influence of different nitrogen and slag levels on growth and composition of rice plant. *Soil Plant Food* 3:104–107, 1957.

132. N. Panda, B. Pradhan, A.P. Samalo, P.S.P. Rao. Note on the relationship of some biochemical factors with the resistance in rice varieties to rice borer. *Indian J. Agric. Sci.* 45:499–501, 1975.

133. A.S. Savant, V.H. Patit, N.K. Savant. Rice hull ash applied to seedbed reduces deadhearts in transported rice. *Internat. Rice Res. Notes* 19:21–22, 1994.

134. S. Yoshida, S.A. Javasero, E.A. Ramirez. Effect of silica and nitrogen supply on some leaf characters of the rice plant. *Plant Soil* 31:48–56, 1969.

135. F.G. Maxwell, J.N. Jenkons, W.L. Parrott. Resistance of plants to insects. *Adv. Agron.* 24:187–265, 1972.

136. G. Sujathata, G.P.V. Reddy, M.M.K. Murthy. Effect of certain biochemical factors on expression of resistance of rice varieties to brown plantthopper (*Nilaparvata lugens* Stal). *J. Res. Andra Pradesh Agric. Univ.* 15:124–128, 1987.

137. M. Salim, R.C. Saxena. Iron, silica, and aluminum stresses and varietal resistance in rice: Effects on whitebacked planthopper. *Crop Sci.* 32:212–219, 1992.

138. A. Tanaka, Y.D. Park. Significance of the absorption and distribution of silica in the rice plant. *Soil Sci.* 12:191–195, 1966.

139. M.D. Wadham, P.W. Parry. The silicon content of *Oryza sativa* L. and its effect on the grazing behaviour of *Agriolimax reticulatus* Muller. *Ann. Bot.* 48:399–402, 1981.

140. M.N. Ukwungwu. Effect of silica content of rice plants on damage caused by the larvae of *Chilo zacconius* (Lepidoptera: Pyralidae). *WARDA Tech. Newslett.* 5:20–21, 1984.

141. A.M. Shirazi, F.D. Miller. Pre-treatment of Sargent crabapple leaf discs with potassium silicate reduces feeding damage by adult Japanese beetle. *Proceedings of the Second Conference on Silicon in Agriculture*, Japan, 2002, p. 41.

142. M.J. Hodson, A.G. Sangster. X-ray microanalysis of the seminal root of *Sorghum bicolor* with particular reference to silicon. *Ann. Bot.* 64:659–675, 1989.

143. S.H. Elawad, L.H. Allen Jr., G.J. Gascho. Influence of UV-B radiation and soluble silicates on the growth and nutrient concentration of sugarcane. *Proc. Soil Crop Sci. Soc. Florida* 44:134–141, 1985.

144. J.H. Meyer, M.G. Keeping. Past, present and future research of the role of silicon for sugarcane in southern Africa. In: L.E. Datnoff, G.H. Snyder, G.H. Korndorfer, eds. *Silicon in Agriculture.* Amsterdam: Elsevier, 2001, pp. 257–275.

145. R.G. Warren, A.E. Johnton. Hoosfield continuous barley. Rep. Rothamsted Exp. Stn. 320–338, 1966.

146. J.M. Thurston, E.D. Williams, A.E. Johnston. Modern developments in an experiment on permanent grassland started in 1856: effects of fertilizers and lime on botanical composition and crop and soil analyses. *Ann. Agron.* 27:1043–1082, 1976.

147. I.P. Derygin, J.K. Chyprikov, M.V. Vasil'eva. Barley crop production and quality under improving of plant Si nutrition. *Izv. TSHA* 2:52–56, 1988.

148. V.K. Cily. The efficiency of natural zeolites on strawberry for increasing of the yield and reduction of heavy metal contamination. Autoref. Diss. Cand., Moscow, 1992.

149. E. Haak, G. Siman. Field experiments with Oyeslag (Faltlorsok med Oyeslag). Report 185, Uppsala, 1992.

150. J.A. Shugarov, Z.I. Gosudareva, S.N. Shukov. The agrochemical efficiency of steel slags. *Chem. Agric.* 8:41–43, 1986.

151. A.Z. Malahidze, V.R. Kvaliashvili, G.V. Cicishvili, T.G. Andronikashvili, G.V. Maisuradze, T.I. Hulicishvili, Z.U. Kvrivishvili. Technology for manufacturing of organo-mineral fertilizer based on the manure and zeolite. USSR Patent 1240757, 1985.

152. I.D. Komisarov, L.A. Panfilova. The method for production of slowly soluble fertilizer. USSR patent 1353767, 1984.
153. A.P. Carev, S.P. Kojda, V.N. Chishenkov. The effect of zeolites on crops. *Corn Sorghum* 4:15–16, 1995.
154. J.X. Mustafaev. The performance of mineral fertilization under zeolite application for barley on eroded Gray–Brown mountain soils on south-east slope of Great Caucus. Autoref. Diss. Cand., Acad. Sci. Aserbadjan SSR, Inst. Soil Sci. and Agrochim., Baky, Azerbadjan, 1990.
155. A.G. Barsykova, V.A. Rochev. Effect of silica gel-based fertilizer on the silicic acid mobility in the soil and availability for plant. In: *The Control and Management of the Content of the Macro- and the Microelements on Media in Ural Region. Proc. Sverdlovsky ACI* 54:84–88, 1979.
156. K. Myhr, K. Erstad. Converter slag as a liming material on organic soils. *Norwegian J. Agric. Sci.* 10:81–93, 1996.
157. E.I. Ratner. Using natural silicon-rich wastes of mining-rock industry and some steel slags as fertilizers. *The New Fertilizers*, ser. 2, Moscow: Selhozgiz, pp. 110–128, 1937.
158. V.K. Bahnov. The silicon is a deficient nutrient on histosol. *Agrochemistry* 11, 119–124, 1979.
159. A.A. Kurmanbaev, A.K. Sadanov, A. Sadikov. The meliorative effect of rice straw on salt-affected soil and their effect on biological soil activity. *Izv. Nat. Acad. Sci. Rep. Kazahstan, Ser. biol.* 5:87–89, 1994.
160. N.E. Aleshin. The peculiarity of rice crop formation independent from silicon nutrition. Autoref. Diss. Cand., Moscow, 1982.
161. M.G.F. Kintanal'ia. The effect of single application of silicon-rich slag on rice dark-chestnut soil properties on south Ukraine. Autoref. Diss. Cand., Univ. Friendship Peoples, Moscow, 1987.
162. S.L. Amarasiri, K. Wickramasingke. Use of rice straw as a fertilizer material. *Trop. Agric.* 133:39–49, 1977.
163. J. Velly. La fertilisation in silice du riza Madagascar. *Agron. Tropic.* 30:305–324, 1975.
164. J.A. Silva. Possible mechanisms for crop response to silicate application. In: *Proceedings of the International Symposium on Soil Fertility Evaluation*, New Delhi, 1971, pp. 805–814.
165. J.A. Silva. The role of research in sugar production. Hawaiian Sugar Technologies: 1969 Report, 1969.

20 Sodium

John Gorham
Tottori University, Tottori, Japan
University of Wales, Bangor, United Kingdom

CONTENTS

20.1 Sodium in Soils and Water 569
    20.1.1 Salinity 570
    20.1.2 Sodicity 570
20.2 Sodium as an Essential Element 571
20.3 Beneficial Effects..571
    20.3.1 Growth Stimulation 571
    20.3.2 Interaction with Other Nutrients 572
20.4 Sodium in Fertilizers..573
20.5 Sodium Metabolism in Plants 573
    20.5.1 Effects on C_4 Species 573
    20.5.2 Toxicity of Sodium 573
20.6 Intracellular and Intercellular Compartmentation...574
20.7 Sodium in Various Plant Species 574
References 575

20.1 SODIUM IN SOILS AND WATER

Sodium and potassium, being adjacent elements in Group 1 of the Periodic Table, have similar chemical properties. In the biology of higher organisms, however, these two elements have very different roles and are treated very differently by mechanisms involved in short- and long-range transport. Estimates of the percentages of sodium and potassium in the Earth's crust vary between 2.5 and 3% (by weight), with slightly more sodium than potassium (1), and these concentrations are similar to the percentages of calcium and magnesium. Much of the sodium is in seawater, to the extent of 30.6% by weight compared with only 1.1% for potassium and 1.2% for calcium. Chloride, although present at only 0.05% in the Earth's crust, makes up 55% of the mass of seawater salts. For humans and most animals, physiological solutions are dominated by sodium (around 0.8% [w/v] compared with about 0.02% for potassium, calcium, and magnesium) and chloride (0.9%), and both elements are essential for animals. Thus, when we think of sodium, we think first of common salt—sodium chloride. In soils, the situation is more complex than in bulk solutions, and concentrations of cations (as experienced by the plant root) are influenced by ion exchange, diffusion, and mass-flow processes. The osmotic effects of excessive salts are also influenced by the exact amounts and proportions of anions and cations.

Some sodium occurs in most soils, but in temperate climates, the concentrations are often similar to, or lower than, those of potassium. Excessive amounts of sodium may be present in the soil in arid and semi-arid areas, and where evapotranspiration is similar to or greater than precipitation. The excess may be in the form of high concentrations of sodium ions in solution, usually accompanied by chloride and sulfate (saline soils), or where sodium is the main cation associated with cation-exchange sites (sodic soils). There is no absolute division of salt-affected soils into these two categories, saline or sodic, as there is a range from purely saline to purely sodic, with most salt-affected soils falling somewhere between the two extremes. The FAO estimated that in 2000, 3.1% of the Earth's land area was affected by salinity and a further 3.4% had sodic soils (2). These figures include 19.5% of irrigated land and 2.1% of land under dry-land agriculture. Detailed properties of these soils are presented in a number of monographs (3–9). A brief summary is given below.

20.1.1 SALINITY

A widely accepted definition of a saline soil is one that gives a saturated paste extract with an electrical conductivity (EC_e) of >4 dS m^{-1} (mmho cm^{-1}). Seawater is about 55 dS m^{-1}. These saline soils will also have an exchangeable sodium percentage (ESP) of <15 and a pH of <8.5. Saline soils are a problem for most plants because of the high concentrations of soluble salts in the soil solution. Soil salinity usually involves other ions in addition to those of sodium and chloride, particularly calcium, magnesium, and sulfate. The proportions of these ions depend on the chemistry and hydrology of the soil, but all saline soils have high concentrations of salts that may be harmful in three ways. First, the high concentrations result not only in higher electrical conductivity, but also in high osmotic pressures (more negative osmotic potentials). This action makes it more difficult for plants to establish a continuous gradient of water potential between the soil solution and the atmosphere—the driving force for transpiration and water uptake by osmosis. Plants must make their own tissue solutions more concentrated (higher osmotic pressure) in order to draw water into their tissues. This response is called osmotic adjustment, and in a strict sense, it refers to an increase in solutes on a dry weight basis (a higher osmotic pressure can also be achieved to some extent by a reduction in the amount of water). The simplest and energetically the cheapest way to achieve osmotic adjustment is by the accumulation of inorganic ions (10). This action can lead to the second problem—the toxicity of high concentrations of inorganic ions in plant tissues (11). Toxicity, in this context, can result from direct interference with cellular metabolism or from an osmotic imbalance caused by the accumulation of salts in the leaf apoplast, known as the Oertli effect (12,13). The third problem is that high concentrations of salts can inhibit the uptake of other nutrients such as potassium and nitrate (see below).

20.1.2 SODICITY

In contrast, soils with little soluble sodium, and hence a low EC_e (<4 dS m^{-1}), but with a substantial proportion of the exchangeable cations in the form of sodium (ESP>15) and a pH of >8.5, are called sodic soils. In purely sodic soils, a substantial osmotic problem does not occur, since the concentrations of free ions in the soil solution are low. Nutrition is a problem because of the replacement of nutrient cations (K^+, Ca^{2+}, and Mg^{2+}) at ion-exchange sites in the soil by sodium (Na^+) and because of the high pH. Sodic soils have poor physical structure and may be impermeable to water and to plant roots, so that there are often secondary problems such as waterlogging and hypoxia.

Primary salinization is the result of geological processes such as the deposition of salt from drying lakes and seas. The large areas of salt-affected soil in parts of Hungary, Australia, and the western United States of America are the result of such natural events. Secondary salinization refers to the impact of man, mainly resulting from unsustainable irrigation for agriculture and rising water tables. Secondary salinization has played a role in the decline of several civilizations. The Sumerian civilization in Mesopotamia is probably the best known. This civilization was initially based on irrigated wheat farming, but lack of adequate drainage and excessive use of irrigation water with an appreciable salt content led to accumulation of salts in the irrigated lands. Wheat (*Triticum aestivum* L.) was replaced gradually by the more tolerant cereal barley (*Hordeum vulgare* L.), until it was abandoned completely in about 1700 BC (6). Eventually, the salinity reached levels at which not even barley would grow. Clearly, this presentation is a simplification of a complex series of events, but the pattern of irrigation without adequate drainage or control of salt fluxes in the soil has been repeated in other civilizations such as the Hohokam of the Sonoran Desert and the Indus civilization of Pakistan. The mistakes of ancient civilizations have, unfortunately, been repeated in more modern times. Examples are the vast irrigation systems in the Indian subcontinent and central Asia. In the former case, remedial civil engineering is tackling the problem (6). In the former Soviet Union, large-scale irrigation schemes built in the 1950s abstracted water from the Amu Darya and Syr Darya rivers for the cultivation of cotton (*Gossypium hirsutum* L.) and other crops. These rivers flow into the Aral Sea, and with the reduction in river flows, the level of the sea dropped by more than 10 m; and its area decreased by over 40% in the latter half of the 20th century and is still decreasing. Even the United States of America, with all of its technological and financial resources, is not immune to the impact of secondary salinization, as in the San Joachim valley and the Salton Sea.

Secondary salinization is most severe in arid and semi-arid regions, where potential evapotranspiration rates are high, as in parts of the United States, the Indian subcontinent, Australia, the Middle East, and South America.

20.2 SODIUM AS AN ESSENTIAL ELEMENT

Some uncertainty exists about the status of sodium as a nutrient, partly arising from the semantics of 'essentiality'. The original criteria of Arnon and Stout (14) were that an essential element should be necessary for completion of the life cycle, should not be replaceable by other elements, and should be involved directly in plant metabolism. Sodium fails to meet all the three criteria for most plants and is generally regarded as a beneficial nutrient (see below). Only a few plants have any difficulty completing their life cycles in the absence of sodium, and these include some euhalophytes and some C_4 species. The osmotic functions of cations in the vacuoles of plants growing at low salinity can be performed to some extent by any of the common cations. In particular, the monovalent alkali metals can perform similar functions in generating solute osmotic pressures and turgor (1,15–18).

The term 'functional nutrient' has been suggested for sodium, and, perhaps also for silicon and selenium (19,20). It might equally be applied to some of the rare earth elements that promote plant growth in certain circumstances (21). As Tyler (21) has pointed out for the latter group, research on essentiality, even of sodium, has examined only a small proportion of the total number of species in the Plant Kingdom. Even so, it is clear that for most species, sodium is not essential in any sense.

20.3 BENEFICIAL EFFECTS

20.3.1 Growth Stimulation

Halophytes. The responses of halophytes and glycophytes to salinity have been reviewed many times (4,7,22–28). One feature of the response of halophytes, and, particularly the succulent halophytes predominantly from the family Chenopodiaceae, is that maximum biomass is achieved at moderate-to-high salinity (29–33). In other species, growth can be stimulated at low salinity, compared with the absence of salt (34), but this effect may depend on the overall nutritional status of the plant and the purity of the sodium chloride.

A part of the biomass of halophytes is the inorganic ions that they accumulate, especially in the shoots (23,26,27,30). It has been argued that, for a better assessment of plant productivity, only the organic portion of the biomass should be considered—that is, the ash-free dry weight (35–37). This consideration certainly reduces the apparent stimulation of 'growth' by sodium in the salt-accumulating, succulent euhalophytes, but a positive effect on ash-free dry weight is still apparent.

20.3.2 INTERACTION WITH OTHER NUTRIENTS

The role of potassium in generating turgor can be fulfilled by sodium and to some extent, by calcium and magnesium, particularly at low concentrations of potassium (38–41). The estimated extent to which potassium can be replaced by sodium in the edible portions of crops varies from 1% in wheat (*Triticum aestivum* L.) and rice (*Oryza sativa* L.) to 90% in red beet (*Beta vulgaris* L.) (42). The interactions among cations in terms of uptake and accumulation rates are complex. The ability of low concentrations (<500 μM) of sodium to stimulate potassium uptake when potassium concentrations are low does not appear to be of importance outside the laboratory (43). The extensive literature on the physiology and genetics of potassium–sodium interactions, especially related to membrane transport, is beyond the scope of this chapter and has been reviewed comprehensively by other researchers (44–50). Some evidence suggests that shoot sodium concentrations (altered by spraying sodium onto leaves) affects the transport of potassium to the shoots, or at least leaf potassium concentrations (51).

Interactions between sodium and other nutrients have been observed (52–54). Excessive sodium inhibits the uptake of potassium (43,55), calcium (56–67), and magnesium (53). A deficiency of calcium, or a high sodium/calcium ratio, results in enhanced sodium uptake. For most species, this calcium requirement is satisfied at a few moles per cubic meter of calcium in solution and is rarely detected in soils. It can become a problem in hydroponics if the calcium concentration in the nutrient solution is low, and no extra calcium is added. Maintaining low sodium/calcium ratios (as a general rule, not >10:1 for dicots and 20:1 for monocots) will prevent this problem. Similar considerations apply to silicon (68–75).

Nitrogen nutrition modifies the effects of sodium on Chenopodiaceae such as goosefoot (*Suaeda salsa* L.) (76). Plants of this family accumulate large amounts of nitrogen in the form of nitrate and glycinebetaine (30,77–80). The interactions among salinity, nitrogen, and sulfur nutrition have been investigated in relation to the accumulation of different organic solutes in the halophytic grasses of the genus *Spartina* (81–83). Generally, adequate nitrogen nutrition is necessary to minimize the inhibition of growth caused by excess salt, but with some differences between the ammonium- and nitrate-fed plants (84–94).

Salinity may interfere with nitrogen metabolism in a number of ways, starting with the uptake of nitrate and ammonium (87,95). Under nonsaline conditions, nitrate is an important vacuolar solute in many plants, including members of the Chenopodiaceae and Gramineae. Under saline conditions, much of the vacuolar nitrate may be replaced by chloride, possibly releasing some nitrate-nitrogen for plant growth and metabolism. On the other hand, salinity can result in the synthesis of large amounts of nitrogen-containing compatible solutes such as glycinebetaine (and in a few cases, proline) and lead to the accumulation of amides and polyamines. Changes may occur at the site of nitrate reduction from the leaves to the roots, and hence changes in nitrate transport to the shoots. Since the latter is linked to potassium recirculation (96,97) and long-range signaling mechanisms controlling growth and resource allocation (98), the implications of such changes are wide ranging. The activity of nitrate reductase may also be affected by salinity. Although toxic ions can affect all aspects of nitrogen metabolism, little evidence suggests that nitrogen supply directly limits the growth of plants under conditions of moderate salinities (99).

In comparison with the other nutrients, the interactions between salinity and phosphorus have received relatively little attention (100) and depend to a large extent on the substrate (52,53). When investigating interactions between salinity and nutrients, one has to be aware of the effects of the substrate, the environment, the genotype–nutrient balances, the nutrient and salt concentrations, the time of exposure to salinity, and the phenology of the plant. These interactions are complex and cannot be comprehended adequately from one or two experiments.

20.4 SODIUM IN FERTILIZERS

Application of sodium to many crops has been reported to stimulate growth, particularly when potassium is deficient (15,101–107). This phenomenon has been documented repeatedly with *Beta* species (red beet, fodder beet, and sugar beet) (108–126), and in a range of other crops including asparagus (*Asparagus officinalis* L.), Italian ryegrass (*Lolium multiflorum* Lam.), tomato (*Lycopersicon esculentum* Mill.), potato (*Solanum tuberosum* L.), carrots (*Daucus carota* L.), celery (*Apium graveolens* L.), and flax (*Linum usitatissimum* L.) (15,74,101,103,104,107,127,128).

There is particular interest in sodium fertilizer application to forage crops, since animals require substantial amounts of sodium (129,130). Lactating dairy cows need a concentration of about 2 g Na kg^{-1} in forage (131). The problem is particularly evident on soils that are intensively managed and deficient in nutrients (132–134), although there are exceptions (135). Application of sodium fertilizer improves the quality of fodder crops and makes them more acceptable to animals (136–140).

20.5 SODIUM METABOLISM IN PLANTS

20.5.1 EFFECTS ON C$_4$ SPECIES

Sodium was reported to be necessary for the growth of some halophyte species (32,141–143); notably, bladder saltbush (*Atriplex vesicaria* Heward, Chenopodiaceae). Sodium specifically stimulates the growth of Joseph's coat (*Amaranthus tricolor* L., Amaranthaceae) (144), possibly by an effect on nitrate uptake and assimilation (145,146). Sodium appears to be essential for the C$_4$ grasses such as proso millet (*Panicum miliaceum* L.), kleingrass (*P. coloratum* L.) and saltgrass (*Distichlis spicata* Greene) (20,147,148) and has been found to stimulate the growth of grasses such as marsh grass (*Sporobolus virginicus* Kunth) and alkali sacaton (*S. airoides* Torr.) in some studies (149–151). Subsequent work showed that this requirement was linked with the C$_4$ pathway of photosynthesis (141,142,152–157) and specifically with pyruvate–Na$^+$ co-transport into mesophyll chloroplasts (158–163), a step that is necessary for the regeneration of phosphoenolpyruvate and the fixation of CO$_2$. Not all C$_4$ plants require sodium for photosynthesis or grow better when it is present (161). The C$_4$ species of the NADP$^+$-malic enzyme (ME) type have a different co-transport system for pyruvate that uses protons rather than sodium ions.

In sorghum species (*Sorghum* L.), there is a specific effect of higher concentrations of sodium (and low concentrations of lithium) on the kinase that regulates the activity of phosphoenolpyruvate (PEP) carboxylase, the primary carbon-fixing enzyme in C$_4$ and crassulacean acid metabolism (CAM) plants (164). The kinase also seems to be linked to the responses of PEP carboxylase to nitrate in C$_3$ and C$_4$ *Alternanthera* Forssk. species (165). There was a report that sodium was required for CAM in Chandlier plant (*Kalanchoe tubiflora* Hamet) (166), but little further work has been published on this aspect, and no relationship occurs between CAM and halophytism (167). On the other hand, salinity and other stresses are known to induce CAM photosynthesis in the facultative CAM species, ice plant (*Mesembryanthemum crystallinum* L., Aizoaceae) (168,169).

20.5.2 TOXICITY OF SODIUM

Application of sodium to recently transplanted seedlings or cuttings runs the risk of uncontrolled bypass flow of water and sodium to the shoots through damaged roots. Hence sodium is often applied in the laboratory, greenhouse, or growth-chamber experiments after the plants have become established in the growing medium. For such situations, Munns (24,25,33) has described a series of events that occurs in most plants. At its simplest, these effects start with the initial osmotic stress caused by making the external (medium) water potential more negative. Subsequently, external inorganic ions are taken up and organic solutes synthesized for osmotic adjustment of the plant cells. Failure to properly control the influx of inorganic salts results in the direct toxicity of high intracellular (particularly cytoplasmic) concentrations of ions or to osmotic imbalances within tissues such as the accumulation of salts in the apoplast of species like rice (12,13). Although this description has been challenged in detail regarding the implications for stress-resistance breeding (11) and the point at which specific ion effects become evident (170), it is still the best model of physiological responses to applied salinity. The same concepts, with modifications of timescale and phenology, can be useful in the crop field and in natural environments, although in both cases the severity of salinity (and other stresses) is subject to fluctuations that the laboratory experiment is designed to avoid.

Important questions are what, when, and why salts are toxic to plants. The question of whether sodium or chloride is a toxic ion is still difficult to answer in most plants, though of course, this action is not important if the problem is primarily osmotic. The question of when inorganic salts (mainly sodium chloride) become toxic is a little easier to answer, at least in theory. Accumulation of salts is required for osmotic adjustment, as cellular dehydration may make a contribution, but generally perturbs metabolism by changing the concentrations of critical intermediates and signaling molecules in the cytoplasm. If salts accumulate much in excess of the concentrations needed for osmotic adjustment of plant cells, it is likely that they will become inhibitory to metabolism and growth, although this may depend on the intracellular location of the salts (see below). The cytoplasm of eukaryotic cells has evolved to work best within a limited range of concentrations of solutes, and particularly of certain ions. Exceeding these ranges for inorganic (and some organic) ions (including potassium) creates problems for macromolecular structures, and hence enzyme activities and nucleic acid metabolism (171,172).

20.6   INTRACELLULAR AND INTERCELLULAR COMPARTMENTATION

From the above, it follows that plants growing in saline environments and accumulating high concentrations of salts must have a mechanism that facilitates high rates of metabolic activity in the cytoplasm. Enzymes from halophytes were shown not to have any enhanced capacity to work at high salt concentrations compared with those from glycophytes (1,171–176). This observation led to the hypothesis that toxic inorganic salts might be preferentially accumulated in vacuoles, where they could still have an osmotic role. In this intracellular-compartmentation model (17,177–179), the osmotic potential of the cytoplasm is adjusted by the accumulation of 'compatible' organic solutes such as glycinebetaine, proline, and cyclitols (27,171,173,177,180–184). For the interpretation of plant-sodium contents in saline environments, it is not therefore sufficient to know how much sodium a plant tissue contains. It is also necessary to consider the relative and absolute concentrations within different parts of the tissue, both at the inter and intracellular levels (178).

20.7   SODIUM IN VARIOUS PLANT SPECIES

One has to be cautious about interpreting concentrations expressed on the basis of different units (30,185). A tissue dry weight basis is often used in the agricultural literature, but conveys no information about the osmotic effects of solutes such as sodium ions or about changes in other dry weight components such as chloride in euhalophytes. Thus, ash-free dry weight might be a more appropriate basis for measuring concentrations. Using a fresh-weight basis does not facilitate the proper assessment of osmotic contributions of solutes, nor does it provide information about changes in the amount of solute independent of the amount of solvent (water). Expressing concentrations on a plant-water basis, or as measured concentrations in cell sap, does convey information about the osmotic effects of solutes, but does not allow a distinction to be made between osmotic adjustment *sensu stricto* and changes in the water content of the tissue. An example is given in Reference (185), where sodium concentrations in the roots and shoots of mammoth wildrye (*Leymus sabulosus* Tzvel.) are compared as concentrations in sap or as concentrations per kilogram dry weight. The conclusion

TABLE 20.1

Sodium Concentrations in a Variety of Plants under Saline and Nonsaline Conditions

Species	Conditions	Sodium Concentration	Units	Reference	Notes and Additional References
Phragmites communis	Inland saline lake, Austria	11	mol m^{-3} water	186	
Scirpus maritimus	Estuarine salt marsh, U.K.	144	mol m^{-3} water	187	Middle of the marsh
Spartina anglica	Estuarine salt marsh, U.K.	346	mol m^{-3} water	187	Seaward end of marsh
Salicornia europaea	Estuarine salt marsh, U.K.	820	mol m^{-3} water	187	Seaward end of marsh
Avicennia marina	Mangrove swamp, Australia	520	mol m^{-3} water	188	Sodium concentrations close to, or below, that of seawater have been reported in some mangrove species by others (189–193)
Triticum aestivum	Hydroponics, 0 mol Na m^{-3}	1	mol m^{-3} plant sap	194	cv. SARC1
Triticum aestivum	Hydroponics, 100 mol Na m^{-3}	44	mol m^{-3} plant sap	194	cv. SARC1
Triticum aestivum	Hydroponics, 100 mol Na m^{-3}, hypoxic	143	mol m^{-3} plant sap	194	cv. SARC1
Eragrostis tef	Hydroponics, 100 mol Na m^{-3}	176	mol m^{-3} plant sap	195	Salt-sensitive glycophyte

Note: Seawater has about 480 mol Na m^{-3}.

about whether there are higher concentrations of sodium in the roots or shoots is reversible depending on which units are used.

Table 20.1 shows the concentrations of sodium in the healthy shoots of different species. Under nonsaline conditions, the sodium concentrations in most plant tissues are a few moles per cubic meter plant water at most. As external salinity is increased, the amount of sodium within the plant increases, but the rate at which this increase occurs varies from slow in wheat to very rapid in tef, a salt-sensitive glycophyte with little ability to control the influx of sodium. Halophytes accumulate substantial amounts of sodium, but are able to tightly control this accumulation at salinities close to or below that of seawater.

In conclusion, sodium is essential only for some C$_4$ species, but is undoubtedly beneficial to the growth of euhalophytes. It may stimulate the growth of some species with an evolutionary history in saline environments, and even of apparently totally glycophytic species under certain conditions. Whether there is a need to reclassify sodium as a 'functional' nutrient is open to debate. These considerations are, however, of minor importance compared with the problems caused by the secondary salinization of agricultural land.

REFERENCES

1. T.J. Flowers, A. Läuchli. Sodium versus potassium: substitution and compartmentation. In: A. Läuchli, R.L. Bielski, eds. *Inorganic Plant Nutrition*. Heidelberg, Berlin: Springer, 1983, pp. 651–681.
2. FAO Land and Plant Nutrition Management Services. Table 1. Regional distribution of salt-affected soils in million ha. http://www.fao.org/ag/agl/agll/spush/topic2.htm, 2000.

3. L.A. Richards. *The Diagnosis and Improvement of Saline and Alkaline Soils.* Davis, CA: USDA, 1954, pp. 1–160.

4. K.K. Tanji. *Agricultural Salinity Assessment and Management.* New York: American Society of Civil Engineers, 1990, pp. 1–619.

5. B. Hanson, S.R. Grattan, A. Fulton. *Agricultural Salinity and Drainage.* Davis, CA: USDA, 1993, pp. 1–156.

6. F. Ghassemi, A.J. Jakeman, H.A. Nix. *Salinisation of Land and Water Resources; Human Causes, Extent, Management and Case Studies.* Sydney, Australia; Wallingford, UK: UNSW Press; CAB International, 1995.

7. A. Läuchli, U. Lüttge. *Salinity: Environment—Plants—Molecules.* Dordrecht: Kluwer Academic Publishers, 2002.

8. M. Pessarakli. *Handbook of Plant and Crop Stress.* New York: Marcel Dekker, 1999.

9. I. Szabolcs. *Salt-Affected Soils.* Boca Raton: CRC Press, 1989.

10. A.R. Yeo. Salinity resistance: physiologies and prices. *Physiologia Plantarum* 58:214–222, 1983.

11. P. Neumann. Salinity resistance and plant growth revisited. *Plant, Cell Environ.* 20:1193–1198, 1997.

12. J.J. Oertli. Extracellular salt accumulation, a possible mechanism of salt injury. *Agrochimica* 12:461–469, 1968.

13. T.J. Flowers, M.A. Hajibagheri, A.R. Yeo. Ion accumulation in the cell walls of rice growing under saline conditions: evidence for the Oertli hypothesis. *Plant, Cell Environ.* 14:319–325, 1991.

14. D.I. Arnon, P.R. Stout. The essentiality of certain elements in minute quantity for plants with special reference to copper. *Plant Physiol.* 14:371–375, 1939.

15. A. Montasir, H. Sharoubeem, G. Sidrack. Partial substitution of sodium for potassium in water cultures. *Plant Soil* 25:181–194, 1966.

16. H. Marschner. Why can sodium replace potassium in plants? In: *Potassium in Biochemistry and Physiology. Eighth Colloquium of the International Potash Institute,* Bern, 1971, pp. 50–63.

17. R.A. Leigh, R.G. Wyn Jones. Cellular compartmentation in plant nutrition: the selective cytoplasm and the promiscuous vacuole. In: P.B. Tinker, A. Läuchli, eds. *Advances in Plant Nutrition,* Vol. 2. New York: Praeger, 1986, pp. 249–279.

18. M.G. Lindhauer, H.E. Haeder, H. Beringer. Osmotic potentials and solute concentrations in sugar beet plants cultivated with varying potassium/sodium ratios. *Zh. Pflanz. Boden* 153:25–32, 1990.

19. D. Nicholas. Minor mineral nutrients. *Annu. Rev. Plant Physiol.* 12:63–90, 1961.

20. G.V. Subbarao, O. Ito, W.L. Berry, R.M. Wheeler. Sodium—a functional plant nutrient. *Crit. Rev. Plant Sci.* 22:391–416, 2003.

21. G. Tyler. Rare earth elements in soil and plant systems—a review. *Plant Soil* 267:191–206, 2004.

22. H. Greenway, R. Munns. Mechanisms of salt tolerance in non-halophytes. *Annu. Rev. Plant Physiol.* 31:149–190, 1980.

23. T.J. Flowers, M.A. Hajibagheri, N. Clipson. Halophytes. *Q. Rev. Biol.* 61:313–337, 1986.

24. R. Munns. Physiological processes limiting plant growth in saline soils—some dogmas and hypotheses. *Plant Cell Environ.* 16:15–24, 1993.

25. R. Munns. Comparative physiology of salt and water stress. *Plant Cell Environ.* 25:239–250, 2002.

26. I.A. Ungar. *Ecophysiology of Vascular Halophytes.* Boca Raton: CRC Press, 1991, pp. 1–209.

27. T.J. Flowers, P.F. Troke, A.R. Yeo. The mechanism of salt tolerance in halophytes. *Annu. Rev. Plant Physiol.* 28:89–121, 1977.

28. E.P. Glenn, J.J. Brown, E. Blumwald. Salt tolerance and crop potential of halophytes. *Crit. Rev. Plant Sci.* 18:227–255, 1999.

29. K. Kreeb. Plants in saline habitats. *Naturwissenschaften* 61:337–343, 1974.

30. M.A. Khan, I.A. Ungar, A.M. Showalter. The effect of salinity on the growth, water status, and ion content of a leaf succulent perennial halophyte, *Suaeda fruticosa* (L.) *Forssk. J. Arid Environ.* 45:73–84, 2000.

31. M.D. Williams, I.A. Ungar. The effect of environmental parameters on the germination, growth, and development of *Suaeda depressa* (Pursh) Wats. *Am. J. Bot.* 59:912–918, 1972.

32. P.F. Brownell. Sodium as an essential micronutrient for a higher plant (*Atriplex vesicaria*). *Plant Physiol.* 40:460–468, 1965.

33. R. Munns. Salinity, growth and phytohormones. In: A. Läuchli, U. Lüttge, eds. *Salinity: Environment—Plants—Molecules.* Dordrecht, Berlin, London: Kluwer Academic Publishers, 2002, pp. 271–290.

34. J. Gorham, J. Bridges. Effects of calcium on growth and leaf ion concentrations of *Gossypium hirsutum* grown in saline hydroponic culture. *Plant Soil* 176:219–227, 1995.

35. T. Flowers. Salt tolerance in *Suaeda maritima* (L.) Dum.—effects of sodium chloride on growth, respiration, and soluble enzymes in a comparative study with *Pisum sativum* L. *J. Exp. Bot.* 23:310–321, 1972.

36. M.A. Khan, I.A. Ungar, A.M. Showalter. Effects of salinity on growth, water relations, and ion accumulation of the subtropical perennial halophyte, *Atriplex griffithii* var. stocksii. *Ann. Bot.* 85:225–232, 2000.

37. S.W. Breckle. Salinity, halophytes and salt-affected natural ecosystems. In: A. Läuchli, U. Lüttge, eds. *Salinity: Environment—Plants—Molecules*. Dordrecht, Boston, London: Kluwer Academic Publishers, 2002, pp. 53–77.

38. J. Khan, J. Gorham. Salinity effects on 4D recombinant tetraploid wheat genotypes. *Pak. J. Arid Agric.* 4:37–43, 2001.

39. J. Gorham, J. Bridges, J. Dubcovsky, J. Dvorak, P.A. Hollington, M.C. Luo, J.A. Khan. Genetic analysis and physiology of a trait for enhanced K^+/Na^+ discrimination in wheat. *New Phytol.* 137:109–116, 1997.

40. G.V. Subbarao, R.M. Wheeler, G.W. Stutte, L.H. Levine. How far can sodium substitute for potassium in red beet? *J. Plant Nutr.* 22:1745–1761, 1999.

41. G.V. Subbarao, R.M. Wheeler, G.W. Stutte, L.H. Levine. Low potassium enhances sodium uptake in red beet under moderate saline conditions. *J. Plant Nutr.* 23:1449–1470, 2000.

42. G.V. Subbarao, R.M. Wheeler, G.W. Stutte. Feasibility of substituting sodium for potassium in crop plants for advanced life support systems. *Life Support Biosphere Sci.* 7:225–232, 2000.

43. S. Box, D.P. Schachtman. The effect of low concentrations of sodium on potassium uptake and growth of wheat. *Aust. J. Plant Physiol.* 27:175–182, 2000.

44. F.J.M. Maathuis, A. Amtmann. K^+ nutrition and Na^+ toxicity: the basis of cellular K^+/Na^+ ratios. *Ann. Bot.* 84:123–133, 1999.

45. M. Tester, R.J. Davenport. Na^+ tolerance and Na^+ transport in higher plants. *Ann. Bot.* 91:503–527, 2003.

46. L. Reinhold, M. Guy. Function of membrane transport systems under salinity: plasma membrane. In: A. Läuchli, U. Lüttge, eds. *Salinity: Environment—Plants—Molecules*. Dordrecht, Berlin, London: Kluwer Academic Publishers, 2002, pp. 397–421.

47. M. Binzel, W. Ratajczak. Function of membrane transport systems under salinity: tonoplast. In: A. Läuchli, U. Lüttge, eds. *Salinity: Enviroinment—Plants—Molecules*. Dordrecht, Berlin, London: Kluwer Scientific Publishers, 2002, pp. 423–449.

48. P. Maser, M. Gierth, J.I. Schroeder. Molecular mechanisms of potassium and sodium uptake in plants. *Plant Soil* 247:43–54, 2002.

49. J. Schroeder, P. Buschmann, B. Eckelman, E. Kim, M. Sussman, N. Uozumi, P. Maser. Molecular mechanisms of potassium and sodium transport in plants. In: W.J. Horst, M.K. Schenk, A. Burkert, N. Claassen, H. Flessa, W.B. Frommer, H. Goldbach, H.W. Olfs, V. Römheld, eds. *Developments in Plant and Soil Sciences*, Vol. 92. Dordrecht: Kluwer Academic Publishers, 2001, pp. 10–11.

50. J. Song, H. Fujiyama. Ameliorative effect of potassium on rice and tomato subjected to sodium salinization. *Soil Sci Plant Nutr.* 42:493–501, 1996.

51. J. Song, H. Fujiyama. Importance of Na content and water status for growth in Na-salinized rice and tomato plants. *Soil Sci. Plant Nutr.* 44:197–208, 1998.

52. S.R. Grattan, C.M. Grieve. Mineral element acquisition and growth response of plants grown in saline environments. *Agric. Ecosystems Environ.* 38:275–300, 1992.

53. S.R. Grattan, C.M. Grieve. Salinity–mineral nutrient relations in horticultural crops. *Sci. Hortic.* 78:127–157, 1999.

54. P.C. Chiy, C.J.C. Phillips. Sodium in forage crops. In: P.C. Chiy, C.J.C. Phillips, eds. *Sodium in Agriculture*. Canterbury: Chalcombe Publications, 1995, pp. 43–69.

55. G.N. Al Karaki. Growth, sodium, and potassium uptake and translocation in salt-stressed tomato. *J. Plant Nutr.* 23:369–379, 2000.

56. G.R. Cramer, A. Läuchli, E. Epstein. Effects of NaCl and $CaCl_2$ on ion activities in complex nutrient solutions and root growth of cotton. *Plant Physiol.* 81:792–797, 1986.

57. E. Epstein. The essential role of calcium in selective cation transport by plant cells. *Plant Physiol.* 36:437–444, 1961.

58. G.R. Cramer. Sodium–calcium interactions under salinity stress. In: A. Läuchli, U. Lüttge, eds. *Salinity: Environment—Plants—Molecules.* Dordrecht, Berlin, London: Kluwer Academic Publishers, 2002, pp. 205–227.

59. G.R. Cramer, A. Läuchli, V.S. Polito. Displacement of Ca^{2+} by Na^+ from the plasma lemma of root cells. A primary response to salt stress? *Plant Physiol.* 79:207–211, 1985.

60. R.J. Davenport, R.J. Reid, F.A. Smith. Sodium–calcium interactions in two wheat species differing in salinity tolerance. *Physiol. Plant* 99:323–327, 1997.

61. O.E. Elzam. Interactions between sodium, potassium and calcium in their absorption by intact barley plants. *Recent Advances in Plant Nutrition*, Vol. 2. New York: Gordon and Breach Scientific Publishers, 1971, pp. 491–507.

62. L.M. Kent, A. Läuchli. Germination and seedling growth of cotton: salinity–calcium interactions. *Plant Cell Environ.* 8:155–159, 1985.

63. J. Lynch, G.R. Cramer, A. Läuchli. Salinity reduces membrane-associated calcium in corn root protoplasts. *Plant Physiol.* 83:390–394, 1987.

64. J. Song, H. Fujiyama. Difference in response of rice and tomato subjected to sodium salinization to the addition of calcium. *Soil Sci. Plant Nutr.* 42:503–510, 1996.

65. Z. Rengel. The role of calcium in salt toxicity. *Plant Cell Environ.* 15:625–632, 1992.

66. T.B. Kinraide. Three mechanisms for the calcium alleviation of mineral toxicities. *Plant Physiol.* 118:513–520, 1998.

67. T.B. Kinraide. Interactions among Ca^{2+}, Na^{2+}, and K^+ in salinity toxicity: quantitative resolution of multiple toxic and ameliorative effects. *J. Exp. Bot.* 50:1495–1505, 1999.

68. D. Chandramony, M.K. George. Nutritional effects of calcium, magnesium, silica, and sodium chloride on certain anatomical characters of rice plant related to lodging. *Agric. Res. J. Kerala* 13:39–42, 1975.

69. Y. Liang, R. Ding, Q. Liu. Effects of silicon on salt tolerance of barley and its mechanism. *Sci. Agric. Sin.* 32:75–83, 1999.

70. Y. Liang, Q. Shen, Z. Shen, T. Ma. Effects of silicon on salinity tolerance of two barley cultivars. *J. Plant Nutr.* 19:173–183, 1996.

71. Y. Liang, Q. Shen, A. Zhang, Z. Shen. Effect of calcium and silicon on growth of and nutrient uptake by wheat under stress of acid rain. *Chin. J. Appl. Ecol.* 10:589–592, 1999.

72. O.F. Lima Filho, M.T.G. Lima, S. Tsai. Silicon in Agriculture. Informacoes Agronomicas Technical Supplement, 1–7, 1999.

73. T. Matoh, P. Kairusmee, E. Takahashi. Salt-induced damage to rice plants and alleviation effect of silicate. *Soil Sci. Plant Nutr.* 32:295–304, 1986.

74. J. Wooley. Sodium and silicon as nutrients for the tomato plant. *Plant Physiol.* 1:317–321, 1957.

75. A.R. Yeo, S.A. Flowers, G. Rao, K. Welfare, N. Senanayake, T.J. Flowers. Silicon reduces sodium uptake in rice (*Oryza sativa* L.) in saline conditions and this is accounted for by a reduction in the transpirational bypass flow. *Plant Cell Environ.* 22:559–565, 1999.

76. X. Liu, Y. Yang, W. Li, D. Duan, T. Tadano. Interactive effects of sodium chloride and nitrogen on growth and ion accumulation of a halophyte. *Commun. Soil Sci. Plant Anal.* 35:2111–2123, 2004.

77. J. Gorham, R.G. Wyn Jones. Solute distribution in *Suaeda maritima. Planta* 157:344–349, 1983.

78. R. Storey, R.G. Wyn Jones. Quaternary ammonium compounds in plants in relation to salt resistance. *Phytochemistry* 16:447–453, 1977.

79. R. Storey, N. Ahmad, R.G. Wyn Jones. Taxonomic and ecological aspects of the distribution of glycinebetaine and related compounds in plants. *Oecologia* 27:319–332, 1977.

80. J. Guil, I. Rodriguez-Garcia, E. Torija. Nutritional and toxic factors in selected wild edible plants. *Plant Foods Hum. Nutr.* 51:99–107, 1997.

81. A. Cavalieri. Proline and glycinebetaine accumulation by *Spartina alterniflora* Loisel. in response to NaCl and nitrogen in a controlled environment. *Oecologia* 57:24–1983.

82. A. Cavalieri, A. Huang. Accumulation of proline and glycinebetaine in *Spartina alterniflora* Loisel. in response to NaCl and nitrogen in the marsh. *Oecologia* 49:224–228, 1981.

83. T.D. Colmer, T.W.M. Fan, A. Läuchli, R.M. Higashi. Interactive effects of salinity, nitrogen, and sulphur on the organic solutes in *Spartina alterniflora* leaf blades. *J. Exp. Bot.* 47:369–375, 1996.

84. M.G. Khan, M. Silberbush, S.H. Lips. Effect of nitrogen nutrition on growth and mineral status of alfalfa plants in saline conditions. *Indian J. Plant Physiol.* 2:279–283, 1997.

85. M.G. Khan, M. Silberbush, S.H. Lips. Responses of alfalfa to potassium, calcium, and nitrogen under stress induced by sodium chloride. *Biologia Plantarum* 40:251–259, 1998.

86. M.G. Khan, H.S. Srivastava. Nitrate application improves plant growth and nitrate reductase activity in maize under saline conditions. *Indian J. Plant Physiol.* 5:154–158, 2000.

87. H. Jaenicke, H.S. Lips, W.R. Ullrich. Growth, ion distribution, potassium and nitrate uptake of *Leucaena leucocephala*, and effects of NaCl. *Plant Physiol. Biochem.* 34:743–751, 1996.

88. E.O. Leidi, R. Nogales, S.H. Lips. Effect of salinity on cotton plants grown under nitrate or ammonium nutrition at different calcium levels. *Field Crops Res.* 26:35–44, 1991.

89. E.O. Leidi, M. Silberbush, S.H. Lips. Wheat growth as affected by nitrogen type, pH and salinity. I. Biomass production and mineral composition. *J. Plant Nutr.* 14:235–246, 1991.

90. E.O. Leidi, M. Silberbush, S.H. Lips. Wheat growth as affected by nitrogen type, pH, and salinity. II. Photosynthesis and transpiration. *J. Plant Nutr.* 14:247–256, 1991.

91. E.O. Leidi, M. Silberbush, M.I.M. Soares, S.H. Lips. Salinity and nitrogen nutrition studies on peanut and cotton plants. *J. Plant Nutr.* 15:591–604, 1992.

92. M. Sagi, A. Dovrat, T. Kipnis, H. Lips. Nitrate reductase, phosphoenolpyruvate carboxylase, and glutamine synthetase in annual ryegrass as affected by salinity and nitrogen. *J. Plant Nutr.* 21:707–723, 1998.

93. M. Sagi, A. Dovrat, T. Kipnis, H. Lips. Ionic balance, biomass production, and organic nitrogen as affected by salinity and nitrogen source in annual ryegrass. *J. Plant Nutr.* 20:1291–1316, 1997.

94. M. Silberbush, S.H. Lips. Potassium, nitrogen, ammonium/nitrate ratio, and sodium chloride effects on wheat growth. I. Shoot and root growth and mineral composition. *J. Plant Nutr.* 14:751–764, 1991.

95. W.R. Ullrich. Salinity and nitrogen nutrition. In: A. Läuchli, U. Lüttge, eds. *Salinity: Environment—Plants—Molecules.* Dordrecht, Boston, London: Kluwer Academic Publishers, 2002, pp. 229–248.

96. C. Engels, E.A. Kirkby. Cycling of nitrogen and potassium between shoot and roots in maize as affected by shoot and root growth. *J. Plant Nutr. Soil Sci.* 164:183–191, 2001.

97. H. Marschner, E.A. Kirkby, C. Engels. Importance of cycling and recycling of mineral nutrients within plants for growth and development. *Botanica Acta* 110:265–273, 1997.

98. B.G. Forde. Local and long-range signalling pathways regulating plant responses to nitrate. *Annu. Rev. Plant Biol.* 53:203–224, 2002.

99. J.A. Memon. Interaction Between Salinity and Nutrients in Cotton. Ph.D. dissertation, University of Wales, Bangor, U.K., 1999.

100. G.N. Al Karaki. Barley response to salt stress at varied levels of phosphorus. *J. Plant Nutr.* 20:1635–1643, 1997.

101. P. Harmer, E. Benne. Sodium as a crop nutrient. *Soil Sci. Soc. Am. J.* 60:137–148, 1945.

102. W.E. Larson, W. Pierre. Interaction of sodium and potassium on yield and cation composition of selected crops. *Soil Sci. Soc. Am. J.* 76:51–64, 1953.

103. J. Lehr. Sodium as a plant nutrient. *J. Sci. Food Agric.* 4:460–471, 1953.

104. J. Lehr. The importance of sodium for plant nutrition. *Soil Sci.* 63:479, 1947.

105. C.R.S. Devi, P. Padmaja. Effects of partial substitution of muriate of potash by common salt on the tuber quality parameters of cassava. *J. Root Crops* 22:23–27, 1996.

106. C.R.S. Devi, P. Padmaja. Effect of K and Na applied in different proportions on the growth, yield, and nutrient content of cassava (*Manihot esculenta* Crantz.). *J. Indian Soc. Soil Sci.* 47:84–89, 1999.

107. E. Troug, K. Berger, O. Attoe. Response of nine economic plants to fertilization with sodium. *Soil Sci. Soc. Am. J.* 76:41–50, 1953.

108. A.P. Draycott. *Sugar-Beet Nutrition.* London, UK: Applied Science Publishers Ltd., 1972, 250 pp.

109. A.P. Draycott, S. Bugg. Response by sugar beet to various amounts and times of application of sodium chloride fertilizer in relation to soil types. *J. Agric. Sci.* 98:579–592, 1982.

110. A.P. Draycott, M.J. Durrant. Response by sugar beet to potassium and sodium fertilizers, particularly in relation to soils containing little exchangeable potassium. *J. Agric. Sci.* 87:105–112, 1976.

111. A.P. Draycott, M.J. Durrant, A.B. Messem. Effects of plant density, irrigation, and potassium and sodium fertilizers on sugar beet. 2. Influence of soil moisture and weather. *J. Agric. Sci.* 82:261–268, 1974.

112. A.P. Draycott, J. Marsh, P.B. Tinker. Sodium and potassium relationships in sugar beet. *J. Agric. Sci.* 74:568–573, 1970.

113. M.J. Durrant, A.P. Draycott, G.F.J. Milford. Effect of sodium fertilizer on water status and yield of sugar beet. *Ann. Appl. Biol.* 88:321–328, 1978.

114. A. El-Sheikh, A. Ulrich, T. Broyer. Sodium and rubidium as possible nutrients for sugar beet plants. *Plant Physiol.* 42:1202–1208, 1967.

115. S. Haneklaus, L. Knudsen, E. Schnug. Relationship between potassium and sodium in sugar beet. *Commun. Soil Sci. Plant Anal.* 29:1793–1798, 1998.

116. S. Haneklaus, E. Schnug. Evaluation of critical values of soil and plant nutrient concentrations of sugar beet by means of boundary lines applied to a large dataset from production fields. *Aspects Appl. Biol.* 52:87–93, 1998.

117. S. Haneklaus, E. Schnug, L. Knudsen. Minimum factors for the mineral nutrition of field-grown sugar beet in northern Germany and eastern Denmark. *Aspects Appl. Biol.* 52:57–64, 1998.

118. G. Judel, H. Kuhn. Uber die Wirkung einer Natriumdungung zu Zuckerruben bei guter Versorgung mit Kalium in Gefabversuchen. *Zuckerindustrie* 28:68–71, 1975.

119. S.S. Magat, K.M. Goh. Effect of chloride fertilizers on ionic composition and cation–anion balance and ratio of fodder beet (*Beta vulgaris* L.) grown under field conditions. *N. Z. J. Agric. Res.* 33:29–40, 1990.

120. C.I. Bell, J. Jones, G.F.J. Milford, R.A. Leigh. The effects of crop nutrition on sugar beet quality. *Aspects Appl. Biol* 32:19–26, 1992.

121. D.W. Lawlor, G.F.J. Milford. The effect of sodium on growth of water-stressed sugar beet. *Ann. Bot.* 37:597–604, 1973.

122. G.F.J. Milford, W.F. Cormack, M.J. Durrant. Effects of sodium chloride on water status and growth of sugar beet. *J. Exp. Bot.* 28:1380–1388, 1977.

123. M. Nunes, M. Dias, M. Correia, M. Oliveira. Further studies on growth and osmoregulation of sugar beet leaves under low-salinity conditions. *J. Exp. Bot.* 35:322–331, 1984.

124. N.H. Peck, J.P. Van Burren, G.E. MacDonald, M. Hemmat, R.F. Becker. Table beet plant and canned root responses to Na, K, and Cl from soils and from application of NaCl and KCl. *J. Am. Soc. Hortic. Sci.* 112:188–194, 1987.

125. P.B. Tinker. The effects of nitrogen, potassium, and sodium fertilizers on sugar beet. *J. Agric. Sci.* 65:207–212, 1965.

126. M.C. Williams. Effect of sodium and potassium salts on growth and oxalate content of *Halogeton*. *Plant Physiol.* 35:500–509, 1960.

127. J. Lehr, J. Wybenga. Exploratory pot experiments on sensitiveness of different crops to sodium. *Plant Soil* 3:251–261, 1955.

128. D.D. Warncke, T.C. Reid, M.K. Hausbeck. Sodium chloride and lime effects on soil cations and elemental composition of asparagus fern. *Commun. Soil Sci. Plant Anal.* 33:3075–3084, 2002.

129. D.C. Edmeades, M.B. O'Connor. Sodium requirements for temperate pastures in New Zealand: a review. *N. Z. J. Agric. Res.* 46:37–47, 2003.

130. C.J.C. Phillips, P.C. Chiy. *Sodium in Agriculture.* Canterbury: Chalcombe Publications, 1995, pp. 1–217.

131. G.S. Smith, K.R. Middleton. Sodium and potassium content of top-dressed pastures in New Zealand in relation to plant and animal nutrition. *N. Z. J. Exp. Agric.* 6:217–225, 1978.

132. G.N. Mundy. Effects of potassium and sodium application to soil on growth and cation accumulation of herbage. *Aust. J. Agric. Res.* 35:85–97, 1984.

133. G.N. Mundy. Effects of potassium and sodium concentrations on growth and cation accumulation in pasture species grown in sand culture. *Aust. J. Agric. Res.* 34:469–481, 1983.

134. A.H. Sinclair, D.C. Macdonald, R. Ferrier, A.C. Edwards. The use of minor nutrients for grassland. In: *Forward with Grass into Europe. Proceedings of the British Grassland Society Winter meeting, 1992.* Reading: British Grassland Society, 1993, pp. 73–84.

135. A. Cushnahan, J.S. Bailey, F.J. Gordon. Some effects of sodium application on the yield and chemical composition of pasture grown under differing conditions of potassium and moisture supply. *Plant Soil* 176:117–127, 1995.

136. W.O. Boberfeld, M. Schlosser, H. Laser. Effect of Na amounts on forage quality and feed consumption on *Lolium perenne* depending on fertilizer and nutrient ratio. *Agribiol. Res.* 52:261–270, 1999.

137. P.C. Chiy, A. Al-Tulihan, M.H. Hassan, C.J.C. Phillips. Effects of sodium and potassium fertilizers on the composition of herbage and its acceptability to dairy cows. *J. Sci. Food Agric.* 76:289–297, 1998.

138. P.C. Chiy, C.J.C. Phillips. Sodium fertilizer application to pasture. 8. Turnover and defoliation of leaf tissue. *Grass Forage Sci.* 54:297–311, 1999.

139. P.C. Chiy, C.J.C. Phillips. Sodium fertilizer application to parture. 6. Effects of combined application with sulfur on herbage production and chemical composition in the season of application. *Grass Forage Sci.* 53:1–10, 1998.

140. P.C. Chiy, C.J.C. Phillips. Effects of sodium fertiliser on the chemical composition of grass and clover leaves, stems, and inflorescences. *J. Sci. Food Agric.* 72:501–510, 1996.

141. T.S. Boag, P.F. Brownell. C$_4$ photosynthesis in sodium-deficient plants. *Aust. J. Plant Physiol.* 6:431–434, 1979.

142. P.F. Brownell. Sodium as an essential micronutrient element for plants and its possible role in metabolism. *Adv. Bot. Res.* 7:117–224, 1979.

143. P.F. Brownell, J. Wood. Sodium as an essential micronutrient element for *Atriplex vesicaria*. *Nature* 179:365–366, 1957.

144. T. Matoh, D. Ohta, E. Takahashi. Effect of sodium application on growth of *Amaranthus tricolor* L. *Plant Cell Physiol.* 27:187–192, 1986.

145. D. Ohta, T. Matoh, E. Takahashi. Early responses of sodium-deficient *Amaranthus tricolor* L. plants to sodium application. *Plant Physiol.* 84:112–117, 1987.

146. D. Ohta, S. Yasuoka, T. Matoh, E. Takahashi. Sodium stimulates growth of *Amaranthus tricolor* L. plants through enhanced nitrate assimilation. *Plant Physiol.* 89:1102–1105, 1989.

147. T. Matoh, S. Murata. Sodium stimulates growth of *Panicum coloratum* through enhanced photosynthesis. *Plant Physiol.* 92:1169–1173, 1990.

148. S. Murata, J. Sekiya. Effects of sodium on photosynthesis in *Panicum coloratum*. *Plant Cell Physiol.* 33:1239–1242, 1992.

149. H. Ball, J. O'Leary. Effects of salinity on growth and cation accumulation of *Sporobolus virginicus* (Poaceae). *Am. J. Bot.* 90:1416–1424, 2003.

150. K.B. Marcum, C.L. Murdoch. Salt tolerance of the coastal salt marsh grass, *Sporobolus virginicus* (L.) Kunth. *New Phytol.* 120:281–288, 1992.

151. K. Hunter, L. Wu. Morphological and physiological response of five California native grass species to moderate salt spray: implications for landscape irrigation with reclaimed water. *J. Plant Nutr.* 28:247–270, 2005.

152. P.F. Brownell, L.M. Bielig, C.P.L. Grof. Increased carbonic anhydrase activity in leaves of sodium-deficient C$_4$ plants. *Aust. J. Plant Physiol.* 18:589–592, 1991.

153. P.F. Brownell, C.J. Crossland. The requirement for sodium as a micronutrient by species having the C4 dicarboxylic photosynthetic pathway. *Plant Physiol.* 49:794–797, 1972.

154. P.F. Brownell, M. Jackman. Changes during recovery from sodium deficiency in *Atriplex*. *Plant Physiol.* 41:617–622, 2005.

155. C.P.L. Grof, M. Johnston, P.F. Brownell. Free amino acid concentrations in leaves of sodium-deficient C$_4$ plants. *Aust. J. Plant Physiol.* 13:343–346, 1986.

156. C.P.L. Grof, M. Johnston, P.F. Brownell. Effect of sodium nutrition on the ultrastructure of chloroplasts of C$_4$ plants. *Plant Physiol.* 89:539–543, 1989.

157. C.P.L. Grof, D.B.C. Richards, M. Johnston, P.F. Brownell. Characterisation of leaf fluorescence of sodium-deficient C$_4$ plants: kinetics of emissions from whole leaves and fluorescence properties of isolated thylakoids. *Aust. J. Plant Physiol.* 16:459–468, 1989.

158. P.F. Brownell, L.M. Bielig. The role of sodium in the conversion of pyruvate to phosphoenolpyruvate in mesophyll chloroplasts of C$_4$ plants. *Aust. J. Plant Physiol.* 23:171–177, 1996.

159. M. Johnston, C.P.L. Grof, P.F. Brownell. The effect of sodium nutrition on the pool sizes of intermediates of the C$_4$ photosynthetic pathway. *Aust. J. Plant Physiol.* 15:749–760, 1988.

160. N. Aoki, J. Ohnishi, R. Kanai. Two different mechanisms for transport of pyruvate into mesophyll chloroplasts of C$_4$ plants—a comparative study. *Plant Cell Physiol.* 33:805–809, 1992.

161. J. Ohnishi, U.I. Flugge, H.W. Heldt, R. Kanai. Involvement of Na$^+$ in active uptake of pyruvate in mesophyll chloroplasts of some C$_4$ plants. Na$^+$/pyruvate cotransport. *Plant Physiol.* 94:950–959, 1990.

162. J. Ohnishi, R. Kanai. Na$^+$-induced uptake of pyruvate into mesophyll chloroplasts of a C$_4$ plant, *Panicum miliaceum*. *FEBS Lett.* 219:347–350, 1987.

163. N. Aoki, R. Kanai. Reappraisal of the role of sodium in the light-dependent active transport of pyruvate into mesophyll chloroplasts of C$_4$ plants. *Plant Cell Physiol.* 38:1217–1225, 1997.

164. J. Monreal, R. Alvarez, J. Vidal, C. Echevarria. Characterization of salt stress-enhanced phosphoenolpyruvate carboxylase kinase activity in leaves of *Sorghum vulgare*: independence from osmotic

stress, involvement of ion toxicity and significance of dark phosphorylation. *Planta* 216:648–655, 2003.

165. A. Rajagopalan, R. Agarwal, A. Raghavendra. Modulation *in vivo* by nitrate salts of the activity and properties of phosphoenolpyruvate carboxylase in leaves of *Alternanthera pungens* (C$_4$ plant) and *A. sessilis* (C$_3$ species). *Photosynthetica* 42:345–349, 2004.

166. P.F. Brownell, C.J. Crossland. Growth responses to sodium by *Bryophyllum tubiflorum* under conditions inducing crassulacean acid metabolism. *Plant Physiol.* 54:416–417, 1974.

167. U. Lüttge. Peformance of plants with C$_4$-carboxylation modes of photosynthesis under salinity. In: A. Läuchli, U. Lüttge, eds. *Salinity: Environment—Plants—Molecules*. Dordrecht, Berlin, London: Kluwer Academic Publishers, 2002, pp. 341–360.

168. A.J. Bloom. Salt requirement for crassulacean acid metabolism in the annual succulent *Mesembryanthemum crystallinum*. *Plant Physiol.* 63:749–753, 1979.

169. J. Cushman, H.J. Bohnert. Induction of crassulacean acid metbolism by salinity—molecular aspects. In: A. Läuchli, U. Lüttge, eds. *Salinity: Enviroment—Plants—Molecules*. Dordrecht, Berlin, London: Kluwer Academic Publishers, 2002, pp. 361–393.

170. A. Sumer, C. Zorb, F. Yan, S. Schubert. Evidence of sodium toxicity for the vegetative growth of maize (*Zea mays* L.) during the first phase of salt stress. *J. Appl. Bot. Food Qual.* 78:135–139, 2004.

171. R.G. Wyn Jones, C. Brady, J. Speirs. Ionic and osmotic relations in plant cells. In: D. Laidman, R.G. Wyn Jones, eds. *Recent Advances in the Biochemistry of Cereals*. New York: Academic Press, 1979, pp. 63–103.

172. R.G. Wyn Jones, A. Pollard. Proteins, enzymes and inorganic ions. In: A. Läuchli, R.L. Bielski, eds. *Encyclopedia of Plant Physiology*, New Series 15B. Berlin: Springer, 1983, pp. 528–562.

173. A. Pollard, R.G. Wyn Jones. Enzyme activities in concentrated solutions of glycinebetaine and other solutes. *Planta* 144:291–298, 1979.

174. T.J. Flowers. The effect of sodium chloride on enzyme activities from four halophyte species of Chenopodiaceae. *Phytochemistry* 11:1881–1886, 1972.

175. T.J. Flowers, J.L. Hall, M.E. Ward. Salt tolerance in the halophyte *Suaeda maritima*. Further properties of the enzyme malate dehydrogenase. *Phytochemistry* 15:1231–1234, 1976.

176. H. Greenway, C.B. Osmond. Salt responses of enzymes from species differing in salt tolerance. *Plant Physiol.* 49:256–259, 1972.

177. R.G. Wyn Jones, R. Storey, R.A. Leigh, N. Ahmad, A. Pollard. A hypothesis on cytoplasmic osmoregulation. In: E. Marre, O. Ciferri, eds. *Regulation of Cell Membrane Activities in Higher Plants*. Amsterdam: Elsevier/North Holland, 1977, pp. 121–136.

178. R.G. Wyn Jones, J. Gorham. Intra- and inter-cellular compartmentation of ions: a study in specificity and plasticity. In: A. Läuchli, U. Lüttge, eds. *Salinity: Environment—Plants—Molecules*. Dordrecht, Boston, London: Kluwer Academic Publishers, 2002, pp. 159–180.

179. R.G. Wyn Jones. Cytoplasmic potassium homeostasis: review of the evidence and its implications. In: D.M. Oosterhuis, G. Berkowitz, eds. *Frontiers in Potassium Nutrition: New Perspectives on the Effects of Potassium on Physiology of Plants*. Saskatoon: Potash and Phosphate Institute of Canada, 1999, pp. 13–22.

180. R. Storey, R.G. Wyn Jones. Response of *Atriplez spongiosa* and *Suaeda monoica* to salinity. *Plant Physiol.* 63:156–162, 1979.

181. R.G. Wyn Jones, R. Storey. Betaines. In: L. Paleg, D. Aspinall, eds. *Physiology and Biochemistry of Drought Resistance in Plants*. Sydney: Academic Press, 1981, pp. 171–204.

182. R.G. Wyn Jones, J. Gorham. Osmoregulation. In: O. Lange, P.S. Nobel, C.B. Osmond, H. Zeigler, eds. *Encyclopedia of Plant Physiology*, New Series 12C. Berlin: Springer, 1983, pp. 35–58.

183. D. Rhodes, A.D. Hanson. Quaternary ammonium and tertiary sulfonium compounds in higher plants. *Annu. Rev. Plant Physiol. Plant Mol. Biol.* 44:357–384, 1993.

184. D. Rhodes, A. Nadolska-Orczyk, P. Rich. Salinity, osmolytes and compatible solutes. In: A. Läuchli, U. Lüttge, eds. *Salinity: Environment—Plants—Molecules*. Dordrecht, Boston, London: Kluwer Academic Publishers, 2002, pp. 181–204.

185. J. Gorham. Sodium content of agricultural crops. In: C. Phillips, P.C. Chiy, eds. *Sodium in Agriculture*. Canterbury: Chalcombe Publications, 1995, pp. 17–32.

186. R. Albert, H. Kinzel. Unterscheidung von Physiotypen bei Halophyten des Neusiedlerseegebietes (Österreich). *Z. Pflanzenphysiol.* 70:138–157, 1973.

187. J. Gorham, Ll. Hughes, R.G. Wyn Jones. Chemical composition of salt-marsh plants from Ynys Môn (Anglesey): the concept of physiotypes. *Plant Cell Environ.* 3:309–318, 1980.

188. M. Popp. Chemical composition of Australian mangroves. I. Inorganic ions and organic acids. *Z Pflanzenphysiol.* 113:395–409, 1984.

189. F. Anjum. Leaf cuticular and epidermal traits and elemental status in *Rhizophora* species in a coastal wetland ecosystem. *Phytomorphology* 50:317–325, 2000.

190. B.F. Clough. Growth and salt balance of the mangroves *Avicennia marina* (Forsk.) Vierh. and *Rhizophora stylosa* Griff. in relation to salinity. *Aust. J. Plant Physiol.* 11:419–430, 1984.

191. W.J.S. Downton. Growth and osmotic relations of the mangrove *Avicennia marina*, as influenced by salinity. *Aust. J. Plant Physiol.* 9:519–528, 1982.

192. E. Medina, A.E. Lugo, A. Novelo. Mineral content of foliar tissue of mangrove species of the Sontecomapan lagoon (Veracruz, Mexico) and its relation with salinity. *Biotropica* 27:317–323, 1995.

193. M.A. Khan, I. Aziz. Salinity tolerance in some mangrove species from Pakistan. *Wetlands Ecol. Manage.* 9:219–223, 2001.

194. J. Akhtar, J. Gorham, R.H. Qureshi, M. Aslam. Does tolerance of wheat to salinity and hypoxia correlate with root dehydrogenase activities or aerenchyma formation? *Plant Soil* 201:275–284, 1998.

195. J. Gorham, C. Hardy. Response of *Eragrostis tef* to salinity and acute water shortage. *J. Plant Physiol.* 135:641–645, 1990.

21 Vanadium

David J. Pilbeam and Khaled Drihem
University of Leeds, Leeds, United Kingdom

CONTENTS

21.1 Historical 585
21.2 Growth Effects 586
    21.2.1 Growth Stimulation 586
    21.2.2 Toxicity 587
21.3 Metabolism 588
21.4 Vanadium in Plant Species 589
Acknowledgment 594
References 594

21.1 HISTORICAL

The transition element vanadium exists mostly in the $+3$, $+4$, and $+5$ oxidation states (Table 21.1), with the $+4$ and $+5$ states predominating under oxidizing conditions in the normal soil acidity of below pH 8 (1,2). Vanadium, with many other heavy metals, is released by anthropogenic activity, and its concentration has been steadily increasing in the environment. A study on peat dating back 12,370 years from a bog in Switzerland indicated a large increase in inputs of vanadium since the industrial revolution (3). Analysis of herbarium specimens of 24 species of vascular plants and 3 bryophytes collected over many years in Spain has shown a large increase in leaf vanadium concentrations, particularly since the 1960s (4).

In soils, the main source of vanadium is from the burning of coal, and the subsequent addition of fly ash and bottom ash. In 1988, this ash contributed 11 to 67×10^6 kg V yr^{-1} to soils, 25% of the total vanadium deposited (5). Agricultural and food wastes contributed 3 to 22×10^6 kg yr^{-1}, and atmospheric fallout added 3.2 to 21×10^6 kg yr^{-1}.

TABLE 21.1
Oxidation States of Some Important Species of Vanadium

Species	Formula	Oxidation State
Vanadous	V^{2+}	$+2$
Vanadic	V^{3+}	$+3$
Vanadyl	VO^{2+}	$+4$
Pervanadyl	VO^{3+}; $V(OH)_4{}^+$	$+5$
Metavanadate	$VO_3{}^-$	$+5$

Total atmospheric fallout in a typical year in recent times (1983) resulted mainly from the burning of oil in electricity generation (estimated to be 6960 to $52,200 \times 10^3$ kg) and from industrial and domestic combustion of oil (30,150 to $141,860 \times 10^3$ kg) (5). Of the 15 heavy metals considered in that study, vanadium was the highest to be emitted during oil combustion (5), and its presence is often taken as an indicator of oil pollution (4).

In a study of microelements in the needles of white fir (*Abies alba* Mill.) in the Carpathian mountains of Eastern Europe, vanadium was found in high concentrations in the vicinity of ferrous metal plants (6), and it is emitted into the atmosphere during the production of copper, nickel, iron, and steel, and during the incineration of sewage sludge (5). With the discontinuation of sewage sludge incineration in many countries, it might be expected that direct addition of vanadium to soils in sewage sludge could increase worldwide.

The natural vanadium, occurring at approximately 110 to 150 mg kg^{-1} (1,7) in the crust of the Earth, is found particularly in roscoelite ($KV_3Si_3O_{10}(OH)_2$), vanadinite ($Pb_5(VO_4)_3Cl$), and patronite (VS_4) (1). During weathering of these rocks, vanadium is oxidized to the vanadate ion, which because of its solubility in water across a range of pH values makes vanadium readily available to plants. However, in practice, vanadium is not very mobile in soil, and in a study on a loamy sand, only a very small proportion of vanadium added to the top 7.5 cm of soil migrated down within 18 or 30 months; 81% remained in the top of the soil where it was added (2). The amount of vanadium that was removed by HCl–H$_2$SO$_4$ extraction of the top 7.5 cm of soil decreased by 81% during 18 months; hence, vanadium must have been transformed to an immobile form with time. Vanadium is known to adsorb to iron and aluminum oxides in the clay fraction (2). Some vanadium may be precipitated as Fe(VO$_3$)$_2$, and some may be immobilized by anion exchange (2).

The correlation is good between soil organic matter content and the oxidizable (immobile) fraction of vanadium (8). Insoluble humic acid is known to reduce mobile metavanadate (VO$_3^-$) anions to vanadyl (VO$_2^+$) cations, which probably bind to the humic acid by cation exchange (1). In an industrial area of Poland, most of the vanadium was bound to soil organic matter in a recent study of a soil that was rich in the element. The next largest fraction was the residual fraction followed, in order, by a fraction bound to iron-manganese oxides, a fraction in exchangeable form, and finally a fraction bound to carbonates in amounts too small to measure. The much lower amounts of vanadium in soil from an agricultural area occurred in the order of exchangeable fraction, residual fraction, the fraction bound to iron-manganese oxides, and the fraction bound to organic matter, with the fraction bound to carbonates being again too small to measure (9).

Uptake and accumulation are influenced by soil type, as soil composition affects the availability of vanadium. Vanadium generally is accumulated in plants in very small amounts in comparison to the total vanadium content of the soil (1). In a comparison of soybeans (*Glycine max* Merr.) grown in a fluvo-aquic soil and an Oxisol, an increase in shoot vanadium concentration occurred when concentrations of more than 30 mg V kg^{-1} were added to the fluvo-aquic soil, but no increase occurred at concentrations of up to 75 mg V kg^{-1} added to the Oxisol (10). Plant growth was inhibited when the concentration of vanadium supplied exceeded 30 mg kg^{-1} in the fluvo-aquic soil but was not inhibited in the Oxisol. In a study on bush bean (*Phaseolus vulgaris* L.), the accumulation of vanadium from a loamy sand was more than double the accumulation of cadmium and more than 300 times the accumulation of thallium (2). Concentrations of vanadium in plants are typically 0.27 to 4.2 mg kg^{-1} dry weight (11). At low rates of supply, vanadium appears to stimulate plant growth, but at higher rates of supply it appears to be toxic to many plants (7).

21.2   GROWTH EFFECTS

21.2.1   GROWTH STIMULATION

Vanadium was considered to be a micronutrient for the green alga *Scenedesmus obliquus* Kützing during experiments in which impure iron salts were being used to assess the iron requirement of the species (12). It was difficult to confirm a similar requirement in higher plants (13). First, it is difficult to eliminate vanadium entirely from nutrient cultures (13). Also, although vanadate is a well-known inhibitor of plasma membrane proton-pumping ATPases, trace concentrations have been reported to benefit plant growth. In an experiment on sand-grown corn (*Zea mays* L.), a supply of vanadium increased grain yield, probably because leaf area was increased but also possibly due to physiological effects (14). Supply of vanadium to tomato (*Lycopersicon esculentum* Mill.) at 0.25 mg L^{-1} of nutrient solution gave greater plant height, more leaves, more flowers, and greater plant mass than supplying no vanadium (15).

Hewitt, working with data from Welch and Huffman (16), calculated that the concentrations of vanadium in tomato plant cells are less than 1% of the concentration of vanadium in vanadium-deficient *Chlorella* cells, suggesting that vanadium is not an essential element for the growth of higher plants (13). In the paper on which Hewitt's calculations were based, lettuce (*Lactuca sativa* L.) and tomato plants were grown to maturity in nutrient solutions containing less than 0.04 mg V L^{-1} and with tissue concentrations of <2 to 18 mg V kg^{-1} dry weight (16). Plant growth in this low concentration of vanadium was comparable to that in nutrient solutions containing 50 mg V L^{-1}, with tissue concentrations of 117 to 419 mg kg^{-1} dry weight, whereas it might have been expected that the low concentration of vanadium should have had a beneficial effect on growth. However, iron was supplied as the citrate salt, and in work on *Chlorella pyrenoidosa*, vanadium stimulated growth when iron was supplied as $FeCl_3$ but had only negligible effect when iron was supplied as citrate or iron EDTA (17). Therefore, part of its requirement as an essential element in algae, at least, is as a replacement for unavailable iron, and supply of iron in a readily available form removes this requirement. If vanadium is a beneficial element for higher plants it may be so only when iron or other metals are limiting.

21.2.2 TOXICITY

If some doubt exists about the role of vanadium as a beneficial element, there is no doubt that at high rates of supply (10 to 20 mg L^{-1}) it is harmful to plants (12). Sorghum (*Sorghum bicolor* Moench.) seedlings supplied with vanadium as ammonium metavanadate at 1, 10, or 100 mg L^{-1} in nutrient solutions showed no toxic effects in the 1 mg L^{-1} solution, but showed a noticeable reddening of the lower stems, and later the leaf tips, in the 10 mg L^{-1} or higher solution (7). In an experiment on bush beans planted 15 months after application of 5.6 kg $VOSO_4$ H_2O ha^{-1} on the surface and harvested 3 months later, growth of shoots and roots was significantly less than in unfertilized plants (2).

In the experiments in which soybeans were grown in a fluvo-aquic soil or in an Oxisol, plant growth was inhibited when the concentration of vanadium supplied exceeded 30 mg kg^{-1} in the fluvo-aquic soil, a rate of supply that gave a shoot concentration of approximately 1 mg V kg^{-1} dry matter (10). With a supply of 75 mg V kg^{-1} soil, the shoot concentration was approximately 4 mg kg^{-1} dry matter, and plant growth was even more depressed than with the lower supply of vanadium (10).

One of the reasons for the harmful effects of vanadium is that it induces iron deficiency. Noticeably decreased concentrations of iron were measured in leaves of a manganese-sensitive bush bean cultivar supplied with vanadate (18). Cereals, strawberries (*Fragaria* X *ananassa* Duchesne), and flax (*Linum usitatissimum* L.) are noted as being very sensitive species (19). Wheat (*Triticum aestivum* L.) and barley (*Hordeum vulgare* L.) are more sensitive than rice (*Oryza sativa* L.) or soybean (20). In addition to causing chlorosis from iron deficiency, vanadium has been shown to lower the concentration of iron in roots of soybeans (21) and to lower root concentrations of magnesium and potassium in soybean (22,23) and lettuce (23). Vanadium also decreased root and hypocotyl accumulation of molybdenum in white mustard (*Sinapis alba* L.) (25) and decreased calcium concentrations in leaves of soybean (23,24). Root and hypocotyl concentrations of manganese, copper, and nickel were increased in *Sinapis alba* (25), and leaf concentration of manganese was increased to toxic levels in bush bean (18). Some evidence indicates that vanadium may increase aluminum concentrations in soybeans (22).

In a field experiment with soybean, seed yields decreased with an increase in vanadium concentration in the soil, or more precisely with an increase in the V:(V+P) ratio (26). Seed yield decreased by approximately 20% as the resin-extractable V:(V+P) ratio increased to 0.15 mol mol^{-1} (26), although a decrease also occurred in relation to vanadium alone (27). The negative relationship between vanadium and phosphorus is not surprising given that the inhibition of ATPases by vanadate is brought about by competitive inhibition of phosphate-binding on the enzymes.

If the harmful effects of vanadium become more important with time as anthropogenic sources increase, it would be helpful to be able to alleviate them. The effects of vanadium in the soil can be reduced by adding a chelating agent, such as γ-irradiated chitosan, to the soil (20). Furthermore, it might be expected that since vanadium induces iron deficiency in plants, increased iron supply might alleviate vanadium toxicity, and this effect has been shown to be the case (28).

21.3   METABOLISM

Vanadium has been shown to enhance chlorophyll formation and iron metabolism of tomato plants and to enhance the Hill reaction of isolated chloroplasts (15). Corn plants that had higher grain yield with a supply of vanadium in sand culture had increased concentrations of chlorophyll *a* and chlorophyll *b* (14). Supply of vanadium increased the synthesis of chlorophyll through enhanced synthesis of the porphyrin precursor δ-aminolevulinic acid in the green alga *Chlorella pyrenoidosa* Chick. (29), although the pH optimum for the enhancement of chlorophyll synthesis by vanadium was slightly different from the pH optimum for enhancement of algal cell growth (30). The substitution of vanadium for iron in green algae highlights the involvement of both ions in chlorophyll synthesis.

No clear evidence is available for the role of vanadium in chlorophyll synthesis in higher plants, but iron deficiency gives rise to lower amounts of chlorophyll per chloroplast (31), and the requirement for iron in chlorophyll synthesis has been narrowed down to a specific step (32) rather than to secondary effects. The requirement for iron is clear, and vanadium may possibly influence chlorophyll synthesis only through an effect on iron metabolism. At one stage it was proposed that green algae may have a pathway of synthesis of δ-aminolevulinic acid that is vanadium-dependent but differs from the pathway in higher plants (13); however, such a pathway has not been identified. In recent years, genes coding for the enzymes involved in this synthesis have been identified in higher plants and in algae, so differences in the pathway, if they exist, appear to be at the level of control rather than in the pathway itself. It is possible that vanadium is an essential cofactor for one of the enzymes of chlorophyll biosynthesis in green algae, but in higher plants this role is normally taken on by another metal for which vanadium can substitute.

Vanadate (but not vanadyl) promoted the evolution of oxygen from intact cells of *Chlorella fusca* at the same concentrations that gave maximum promotion of algal growth (1 to 2 μM) (33). Vanadium was thought to work in the chain of electron transport between photosystems 2 and 1 by virtue of the ability of the vanadium to change reversibly between its tetravalent and pentavalent states (33). Vanadium also increased photosystem 1 activity (but not photosystem 2 activity) in isolated chloroplasts of spinach (*Spinacia oleracea* L.), with an optimum at approximately 20 μM V (33).

Corn plants that showed enhanced grain yield with supply of vanadium had more nitrogen, phosphorus, potassium, calcium, and magnesium in the leaves, although high concentrations of vanadium decreased the concentrations of these elements (14). Vanadium was shown to increase foliar concentrations of calcium and iron in lettuce, although in these plants, yield was actually depressed by the vanadium supplied (23).

The presence of vanadium certainly affects the metabolism of plants. Addition of vanadium at 1 mg L^{-1} to solution reduced nicotine concentrations in tobacco (*Nicotiana tabacum* L.) by 25%

(34). In lupin (*Lupinus polyphyllus* Lindl.), a negative correlation between alkaloid and vanadium concentrations in the leaves has been observed (35).

Given the inhibitory effects of vanadate on plasma membrane ATPases, it is not surprising that vanadium should affect metabolism. Changes in concentrations of other ions in plants supplied with vanadium could in part be due to the effects on proton-pumping APTases, although uptake of phosphate into isolated corn root tips was inhibited less than the activity of ATPase in the tips at the same amount of sodium vanadate supplied (36). Nevertheless, heavy exposure of these enzymes to vanadium might be expected to stop plant transport completely. Some evidence indicates that vanadium may also inhibit the absorption of water (37).

Absorption of vanadium appears to be a passive process as it is a linear function of external vanadium concentration and is not affected by putting excised roots into anaerobic conditions (38). Absorption is highly pH-dependent, being fastest at pH 4 and dropping to a very slow rate by pH 10, although being relatively constant between pH 5 and 8 (38). This effect of pH on absorption appears to be due to the ionic form in which vanadium is present, with VO_2^+ predominating at pH 4, HVO_3 predominating between pH 4 and 5, VO_3^- predominating between pH 5 and 8, and HVO_4^{2-} predominating at pH 9 to 10 (38). The VO_2^+ form that predominates in acid soil is taken up by plants far more readily than the other forms that predominate in neutral and alkaline soils (11).

Absorption of vanadium appears to occur at the expense of calcium uptake, there being a linear decrease in calcium accumulation into sorghum cultivars with log concentration of vanadate supplied (39). This result is probably due to an effect on calcium channels that more than compensates for the inhibition by vanadate of the H^+-translocating ATPase responsible for calcium flux. The presence of calcium is required for absorption of vanadium, and this effect, together with the fact that vanadium concentrates in the roots at up to twice the concentration in the external medium, indicates that the passive absorption cannot be purely by diffusion. A concentration gradient from outside to inside the root could be maintained by the vanadium changing form inside the root, with up to 10% of VO_3^- taken up being reduced to VO_2^+ (40), or it could be chelated (38).

Indeed, various complexes of vanadium have been detected in plants. At low rates of vanadium supply, plants form low-molecular-weight complexes thought to be vanadyl amino compounds, and at high rates of supply, plants form high molecular weight complexes, probably vanadyl cellulose compounds (41). It seems that following absorption, vanadium is partially immobilized on the root cell walls. It then develops soluble complexes outside the plasmalemma and finally is absorbed into the vacuoles within the cells (41). Concentrations in roots are usually higher than in leaves.

Calcium seems to accumulate in roots along with vanadium. In soybeans supplied with vanadium, both elements were concentrated in the roots, and very high concentrations of calcium have been detected in the roots of vanadium-accumulating species. Perhaps, calcium may work to detoxify the vanadium (7,24). It is possible that the vanadium occurs as insoluble calcium vanadate (1). This action may be only a partially successful detoxification as it has been suggested that the accumulation of calcium might give rise to the imbalance in other cations associated with vanadium toxicity (24).

There does not appear to be much inhibition of absorption of vanadium by molybdate, borate, chloride, selenate, chromate, or nitrate (38). However, in *Sinapis alba* nickel, manganese, and copper inhibited the accumulation of vanadium in roots and hypocotyls, whereas molybdate decreased its accumulation in the hypocotyls and enhanced its accumulation in the roots (25).

21.4 VANADIUM IN PLANT SPECIES

In general, lower plants contain more vanadium than seed-bearing plants, and older parts contain more than younger parts (7). Despite this overall trend, some angiosperms seem to be accumulator plants (Table 21.2). In an experiment where sorghum seedlings showed noticeable harmful effects

TABLE 21.2
A List of Concentrations of Vanadium in Various Plant Species

Plant Species	Plant Part	Type of Culture	Concentration in Dry Matter (mg kg^{-1})	Reference	Comments
Allium macropetalum Rydb. (onion)	Root	Wild	133	7	Accumulator species
Anethum graveolens L. (dill)	Shoot	Field	0.84	44	
Astragalus confertiflorus Gray (yellow milkvetch)	Shoot	Wild	144	7	Accumulator species
Astragalus preussi A. Gray (milkvetch)	Shoot	Wild	67	7	Accumulator species
Avena sativa L. (oat)	Seed	Nutrient solution	0.055 0.151	45	No added V 0.25 mg V L^{-1}
Brassica napus L. (rape)	Seed	Nutrient solution	0.018 0.132	45	No added V 0.25 mg V L^{-1}
Brassica oleracea var. *botrytis* L. (cauliflower)	Florets	Field	1.09×10^{-3}	44	
Carthamus tinctorius L. (safflower)	Seed	Nutrient solution	0.019–0.021 0.173–0.184	45	No added V 0.25 mg V L^{-1}
Castilleja angustifolia G. Don. (desert paintbrush)	Shoot	Wild	22	7	Accumulator species
Chrysothamnus viscidiflorus Nutt. (rabbitbrush)	Shoot	Wild	37	7	Accumulator species
Conifers (unidentified species)	Leaves	Soil	0.69	7	
Cowania mexicana D.Don var. *stansburiana* (cliff rose)	Shoot	Wild	7.4	7	Accumulator species
Cucumis sativus L. (cucumber)	Fruit	Field or glasshouse	5.6×10^{-2}	44	
Deciduous shrubs (unidentified species)	Leaves	Soil	2.7	7	
Deciduous trees (unidentified species)	Leaves	Soil	1.65	7	
Equisetum sp. (horsetail)		Soil	2.4	7	
Eriogonum inflatum Torr. & Frém. (desert trumpet)	Shoot	Wild	15	7	Accumulator species
Ferns (unidentified species)	Fronds	Soil	1.28	7	
Forbs (unidentified species excluding legumes)	Leaves	Soil	1.20	7	
Fragaria X *ananassa* Duchesne (strawberry)	Fruit	Field	3.1×10^{-2}	44	
Fragaria vesca L. (wild strawberry)	Fruit	Wild	4.1×10^{-2}	44	

Continued

TABLE 21.2 (*Continued*)

Plant Species	Plant Part	Type of Culture	Concentration in Dry Matter (mg kg^{-1})	Reference	Comments
Glycine max Merr. (soybean)	Shoot	Nutrient solution	2.3	28	No V, no Fe, then low Fe + V
			3.9		V, no Fe, then low Fe
			0.7		No V, no Fe, then high Fe + V
			0.8		High Fe + V
	Root		170		No V, no Fe, then low Fe + V
			129		V, no Fe, then low Fe
			41		No V, no Fe, then high Fe + V
			115		High Fe + V
	Pods	Soil in rhizotron	27/29	24	Control/plus extra metals (including V) (Control is no metals added)
	Upper leaves		22/33		Control/plus extra metals (including V) (Control is no metals added)
	Lower leaves		20/30		Control/plus extra metals (including V) (Control is no metals added)
	Roots		28/77		Control/plus extra metals (including V) (Control is no metals added)
	Upper leaves	Nutrient solution	0/0	24	3.0/6.0 mg V L^{-1}
	Lower leaves		1/1		3.0/6.0 mg V L^{-1}
	Roots		18/20		3.0/6.0 mg V L^{-1}
	Shoot	Soil	1.0	10	30 mg V kg^{-1} fluvo-aquic soil
			4.0		75 mg V kg^{-1} fluvo-aquic soil
			0.5		75 mg V kg^{-1} Oxisol
	Youngest leaf	Vermiculite and nutrient solution	53.6	21	104-day-old plants, 100 μmol V L^{-1}
	Oldest leaf		45.6		104-day-old plants, 100 μmol V L^{-1}
	Oldest part of stem		98.7		104-day-old plants, 100 μmol V L^{-1}
	Root		5680		104-day-old plants, 100 μmol V L^{-1}
	Root		9.16		104-day-old plants, no added V
Grasses (unidentified species)	Leaves	Soil	1.4	7	

Continued

TABLE 21.2   (*Continued*)

Plant Species	Plant Part	Type of Culture	Concentration in Dry Matter (mg kg^{-1})	Reference	Comments
Gutierezzia divaricata (snakeweed)	Shoots	Soil	9.3	7	Accumulator species
Hordeum vulgare L. (barley)	Seeds	Nutrient solution	0.028	45	No added V
			0.175		0.25 mg V L^{-1}
Lactuca sativa L. (lettuce)	Shoots	Field	0.58	44	
	Shoots	Nutrient solution	6	16	0.04 mg V L^{-1}
			283		50 mg V L^{-1}
	Roots		73		0.04 mg V L^{-1}
	Shoots		0.165	45	No added V
			0.780		0.25 mg V L^{-1}
Larrea tridentata Cov. (creosote bush)	Leaf	Wild	1.8–3.4	46	Plants in geothermal area
Legumes (unidentified species)	Leaves	Soil	0.84	7	
Lichens (unidentified species)	Thallus	Soil	8.6	7	
Linum usitatissimum L. (flax)	Seed	Nutrient solution	0.018	45	No added V
			0.102		0.25 mg V L^{-1}
Lycopersicon esculentum Mill. (tomato)	Fruit	Field or glasshouse	0.53×0^{-3}	44	
	Shoots	Nutrient solution	11	16	0.04 mg V L^{-1}
			278		50 mg V L^{-1}
	Roots		61		0.04 mg V L^{-1}
	Shoots		0.15	45	No added V
			0.84		0.25 mg V L^{-1}
	Shoots	Sand and nutrient solution	0.18	15	No added V
			0.39		0.25 mg V L^{-1}
	Roots		0.25		No added V
			0.96		0.25 mg V L^{-1}
	Fruit	Rock-wool and nutrient solution	0.126×10^{-3} (fresh mass)	47	Normal EC
			0.090×10^{-3} (fresh mass)		High EC
	Fruit	Soil and nutrient solution	0.124×10^{-3} (fresh mass)		Normal EC
Malus pumila Mill. [*M. domestica* Borkh.] (apple)	Fruit	Field	0.86×0^{-2}	44	
Medicago sativa L. (alfalfa)	Shoots	Field	0.115	48	
Mosses (unidentified species)		By stream	108	7	
Oryza sativa L. (rice)	Shoots	Nutrient solution	530	20	10 mg V L^{-1}
	Roots		1730		
Oryzopsis hymenoides Ricker (ricegrass)	Shoot	Soil	10	7	Accumulator species
Petroselinum crispum Nyman ex. A.W. Hill (parsley)	Shoots	Field	4.52	44	

Continued

TABLE 21.2 (*Continued*)

Plant Species	Plant Part	Type of Culture	Concentration in Dry Matter (mg kg^{-1})	Reference	Comments
Phaseolus vulgaris L.	Primary	Nutrient	2.6	18	0.05 mg V L^{-1}
(bush bean)	leaf	solution	8.3		2.0 mg V L^{-1}
Mn-sensitive cultivar	Oldest		4.7		0.05 mg V L^{-1}
	trifoliate		2.8		2.0 mg V L^{-1}
	leaf				
	Second		3.1		0.05 mg V L^{-1}
	trifoliate		0.6		2.0 mg V L^{-1}
	leaf				
	Stem		0.6		0.05 mg V L^{-1}
			7.0		2.0 mg V L^{-1}
	Roots		34.3		0.05 mg V L^{-1}
			425.0		2.0 mg V L^{-1}
Mn-tolerant cultivar	Primary leaf		4.7		0.05 mg V L^{-1}
			8.6		2.0 mg V L^{-1}
	Oldest		5.9		0.05 mg V L^{-1}
	trifoliate		3.4		2.0 mg V L^{-1}
	leaf				
	Second		2.0		0.05 mg V L^{-1}
	trifoliate		0.8		2.0 mg V L^{-1}
	leaf				
	Stem		2.1		0.05 mg V L^{-1}
			5.9		2.0 mg V L^{-1}
	Roots		44.0		0.05 mg V L^{-1}
			518.9		2.0 mg V L^{-1}
Pisum sativum L. (pea)	Shoot	Nutrient solution	15.0	28	No V, no Fe, then low Fe + V
			17.0		V, no Fe, then low Fe
			2.8		No V, no Fe, then high Fe + V
			7.2		High Fe + V
			28.0		High Fe, then add V
	Root		186		No V, no Fe, then low Fe + V
			510		V, no Fe, then low Fe
			66		No V, no Fe, then high Fe + V
			163		High Fe + V
			540		High Fe, then add V
	Seed	Nutrient	0.054	45	No added V
		solution	0.075		0.25 mg V L^{-1}
Plantago insularis Eastw. (common plantain)	Leaf	Wild	1.9–3.2	46	Plants in geothermal area
Raphanus sativus L. (radish)	Roots	Field	1.26	44	
Solanum tuberosum L. (potato)	Tuber	Field	0.64×10^{-2}	44	

Continued

TABLE 21.2   (*Continued*)

Plant Species	Plant Part	Type of Culture	Concentration in Dry Matter (mg kg^{-1})	Reference	Comments
Triticum aestivum L. (wheat)	Seed	Nutrient solution	0.046	45	No added V
			0.137		0.25 mg V L^{-1}
	Shoot		1	20	No added V
			560		10 mg V L^{-1}
	Root		10		No added V
			3820		10 mg V L^{-1}
Zea mays L. (corn)	Leaves	Field	0.244	48	

when grown in 10 mg V L^{-1} in the nutrient solution, the selenium-accumulator *Astragalus preussi* A. Gray was not affected by 100 mg V L^{-1} and accumulated vanadium in the tissues (7).

Chicory (*Cichorium intybus* L.) and dogfennel (*Eupatorium capillifolium* Small) have been suggested to have potential as indicators of vanadium bioavailability (42). Since 1981, the Bavarian State Office for Environmental Protection has been analyzing samples of the moss *Hypnum cupressiforme* L. as indicators of emission-derived metals, including vanadium (43).

Even in crop species that are sensitive to vanadium, there are genotypes that are less affected by the element. In a study in which soybean was found to be sensitive to the V:(V+P) ratio, one cultivar showed very little sensitivity to either element (27). Although concentrations of 10 to 20 mg V L^{-1} vanadium in nutrient solutions are generally regarded as harmful to plants, some bush bean and lettuce genotypes have been affected adversely by concentrations as low as 0.20 mg V L^{-1} (18,23).

ACKNOWLEDGMENT

We thank Dr. P. A. Millner for stimulating conversation on the role of vanadium in plant biochemistry.

REFERENCES

1. P.J. Peterson, C.A. Girling. Other trace metals. In: N.W. Lepp, ed. *Effect of Heavy Metal Pollution on Plants*. London: Applied Science Publishers, 1981, pp. 213–278.
2. H.W. Martin, D.I. Kaplan. Temporal changes in cadmium, thallium, and vanadium mobility in soil and phytoavailability under field conditions. *Water Air Soil Pollut.* 101:399–410, 1998.
3. M. Krachler, C. Mohl, H. Emons, W. Shotyk. Atmospheric deposition of V, Cr, and Ni since the late glacial: Effects of climatic cycles, human impacts, and comparison with crustal abundances. *Environ. Sci. Technol.* 37:2658–2667, 2003.
4. J. Peñuelas, I. Filella. Metal pollution in Spanish terrestrial ecosystems during the twentieth century. *Chemosphere* 46:501–505, 2002.
5. J.O. Nriagu, J.M. Pacyna. Quantitative assessment of worldwide contamination of air, water and soils by trace metals. *Nature* 333:134–139, 1988.
6. B. Mankovska. Concentrations of nutrient elements and microelements in the needles of *Abies alba* Mill. as an environmental indicator in the Carpathian Mountains. *J. Forest Sci.* 47:229–240, 2001.
7. H.L. Cannon. The biogeochemistry of vanadium. *Soil Sci.* 96:196–204, 1963.
8. M. Ovari, M. Csukas, G. Zaray. Speciation of beryllium, nickel, and vanadium in soil samples from Csepel Island, Hungary. *Fresenius J. Anal. Chem.* 370:768–775, 2001.
9. J. Poledniok, F. Buhl. Speciation of vanadium in soil. *Talanta* 59:1–8, 2003.
10. J.F. Wang, Z. Liu. Effect of vanadium on the growth of soybean seedlings. *Plant Soil* 216:47–51, 1999.
11. A.J. Aller, J.L. Bernal, M.J. del Nozal, L. Deban. Effect of selected trace elements on plant growth. *J. Sci. Fd. Agric.* 51:447–479, 1990.

12. D.I. Arnon, G. Wessel. Vanadium as an essential element for green plants. *Nature* 172:1039–1040, 1953.

13. E.J. Hewitt. A perspective of mineral nutrition: Essential and functional metals in plants. In: D.A. Robb, W.S. Pierpoint, eds. *Metals and Micronutrients. Uptake and Utilization by Plants*. London: Academic Press, 1983, pp. 277–323.

14. B.B. Singh. Effect of vanadium on the growth, yield and chemical composition of maize (*Zea mays* L.). *Plant Soil* 34:209–212, 1971.

15. F.M. Basiouny. Distribution of vanadium and its influence on chlorophyll formation and iron metabolism in tomato plants. *J. Plant Nutr.* 7:1059–1073, 1984.

16. R.M. Welch, E.W.D. Huffman, Jr. Vanadium and plant nutrition. *Plant Physiol.* 52:183–185, 1973.

17. H.-U. Meisch, H.-J. Bielig. Effect of vanadium on growth, chlorophyll formation and iron metabolism in unicellular green algae. *Arch. Microbiol.* 105:77–82, 1975.

18. Y. Kohno. Vanadium induced manganese toxicity in bush bean plants grown in solution culture. *J. Plant Nutr.* 9:1261–1272, 1986.

19. C. Bould, E.J. Hewitt, P. Needham. *Diagnosis of Mineral Disorders in Plants. Volume 1 Principles*. London: Her Majesty's Stationery Office, 1983, pp. 40–41.

20. L.X. Tham, N. Nagasawa, S. Matsuhashi, N.S. Ishioka, T. Ito, T. Kume. Effect of radiation-degraded chitosan on plants stressed with vanadium. *Radiat. Phys. Chem.* 61:171–175, 2001.

21. S. Ueoka, J. Furukawa, T.M. Nakanishi. Multi-elemental profile patterns during the life cycle stage of the soybean. Part 2. Determination of transition elements, V, Cr, Mn, Fe and Zn. *J. Radioanal. Nucl. Chem.* 249:475–480, 2001.

22. S. Ueoka, J. Furukawa, T.M. Nakanishi. Multi-elemental profile patterns during the life cycle stage of the soybean. Part 1. Determination of representative elements, Na, Mg, Al, K and Ca. *J. Radioanal. Nucl. Chem.* 249:469–473, 2001.

23. J. Gil, C.E. Alvarez, C.M. Martinez, N. Perez. Effect of vanadium on lettuce growth, cationic nutrition, and yield. *J. Environ. Sci. Health Part A Environ. Sci. Eng.* 30:73–87, 1995.

24. D.I. Kaplan, D.C. Adriano, C.L. Carlson, K.S. Sajwan. Vanadium: Toxicity and accumulation by beans. *Water Air Soil Pollut.* 49:81–91, 1990.

25. A. Fargasova, E. Beinrohr. Metal–metal interactions in accumulation of V^{5+}, Ni^{2+}, Mo^{6+}, Mn^{2+} and Cu^{2+} in under- and above-ground parts of *Sinapis alba*. *Chemosphere* 36:1305–1317, 1998.

26. A. Olness, T. Nelsen, J. Rinke, W.B. Voorhees. Ionic ratios and crop performance: I. Vanadate and phosphate on soybean. *J. Agron. Crop Sci.* 185:145–151, 2000.

27. A. Olness, D. Palmquist, J. Rinke. Ionic ratios and crop performance: II. Effects of interactions amongst vanadium, phosphorus, magnesium and calcium on soybean yield. *J. Agron. Crop Sci.* 187:47–52, 2001.

28. K. Warington. Investigations regarding the nature of the interaction between iron and molybdenum or vanadium in nutrient solutions with and without a growing plant. *Ann. App. Biol.* 44:535–546, 1956.

29. H.-U. Meisch, J. Bauer. The role of vanadium in green plants. IV. Influence on the formation of δ-aminolevulinic acid in *Chlorella*. *Archs. Microbiol.* 117:49–52, 1978.

30. H.-U. Meisch, H. Benzschawel, H.-J. Bielig. The role of vanadium in green plants. II. Vanadium in green algae – two sites of action. *Archs. Microbiol.* 114:67–70, 1977.

31. N. Terry, J. Abadia. Function of iron in chloroplasts. *J. Plant Nutr.* 9:609–646, 1986.

32. B.M. Chereskin, P.A. Castelfranco. Effects of iron and oxygen on chlorophyll biosynthesis. II. Observations on the biosynthetic pathway in isolated etiochloroplasts. *Plant Physiol.* 69:112–116, 1982.

33. H.-U. Meisch, L.J.M. Becker. Vanadium in photosynthesis in *Chlorella fusca* and higher plants. *Biochim. Biophys. Acta* 636:119–125, 1981.

34. T.S. Tso, T.P. Sorokin, M.E. Engelhaupt. Effects of some rare elements on nicotine content of the tobacco plant. *Plant Physiol.* 51:805–806, 1973.

35. G.N. Buzuk, M. Ya Lovkova, S.M. Sokolova, Yu V. Tyutekin. Evaluation of the correlation between the content of alkaloids and chemical elements in Washington lupin (*Lupinus polyphyllus* Lindl.) by means of mathematical simulation. *Dokl. Biol. Sci.* 384:227–229, 2002.

36. J. Sklenar, G.G. Fox, B.C. Loughman, A.D.B. Pannifer, R.G. Ratcliffe. Effects of vanadate on the ATP content, ATPase activity and phosphate absorption capacity of maize roots. *Plant Soil* 167:57–62, 1994.

37. J. Furukawa, H. Yokota, K. Tanoi, S. Ueoka, S. Matsuhashi, N.S. Ishioka, S Watanabe, H. Uchida, A. Tsuji, T. Ito, T. Mizuniwa, A. Osa, T. Sekine, S. Hashimoto, T.M. Nakanishi. Vanadium uptake and an

effect of vanadium treatment on F-18-labeled water movement in a cowpea plant by positron emitting tracer imaging system (PETIS). *J. Radioanal. Nucl. Chem.* 249:495–498, 2001.

38. R.M. Welch. Vanadium uptake by plants. . Absorption kinetics and the effects of pH, metabolic inhibitors, and other anions and cations. *Plant Physiol.* 51:828–832, 1973.

39. R.E. Wilkinson, R.R. Duncan. Vanadate influence on calcium (Ca) absorption by sorghum root tips. *J. Plant Nutr.* 16:1991–1994, 1993.

40. S. Deiana, A. Dessi, C. Gessna, A. Premoli. Selective determination of vanadium (IV) and vanadium (V) in excised plant roots. *Commun. Soil Sci. Plant Anal.* 19:355–366, 1988.

41. D.B. McPhail, D.J. Linehan, B.A. Goodman. An electron paramagnetic resonance (EPR) study of the uptake of vanadyl by wheat plants. *New Phytol.* 91:615–620, 1982.

42. H.W. Martin, T.R. Young, D.I. Kaplan, L. Simon, D.C. Adriano. Evaluation of three herbaceous index plant species for bioavailability of soil cadmium, chromium, nickel and vanadium. *Plant Soil* 182:199–207, 1996.

43. T. Faus-Kessler, C. Dietl, J. Tritschler, L. Peichl. Correlation patterns of metals in the epiphytic moss *Hypnum cupressiforme* in Bavaria. *Atmos. Environ.* 35:427–439, 2001.

44. R. Söremark. Vanadium in some biological specimens. *J. Nutr.* 92:183–190, 1967.

45. R.M. Welch, E.E. Cary. Concentrations of chromium, nickel and vanadium in plant materials. *J. Agric. Fd. Chem.* 23:479–482, 1975.

46. E.M. Romney, A. Wallace, J. Kinnear, G.V. Alexander. Baseline mineral analysis of leaves from populations of two native plant species from geothermal areas of Imperial Valley, California. *Soil Sci.* 134:2–12, 1982.

47. V. Gundersen, D. McCall, I.E. Bechmann. Comparison of major and trace element concentrations in Danish greenhouse tomatoes (*Lycopersicon esculentum* cv. Aromata F1) cultivated in different substrates. *J. Agric. Fd. Chem.* 49:3808–3815, 2001.

48. R.M. Welch, W.H. Allaway. Vanadium determination in biological materials at nanogram levels by a catalytic method. *Anal. Chem.* 44:1644–1647, 1972.

Section V

Conclusion

22 Conclusion

Allen V. Barker
University of Massachusetts, Amherst, Massachusetts

David J. Pilbeam
University of Leeds, Leeds, United Kingdom

CONTENTS

22.1 Status of Current Knowledge and Research 599
22.2 Soil Testing and Plant Analysis and Nutrient Availability 599
22.3 Accumulation of Elements by Plants 600
22.4 Genetics of Plant Nutrition 601
22.5 General Remarks..602
References 603

22.1 STATUS OF CURRENT KNOWLEDGE AND RESEARCH

Chapters in this handbook summarize research for each of the plant nutrients and several beneficial elements, and readers should refer to the individual chapters for information on past, current, and future research on these elements. However, some conclusions can be drawn about the kinds of current research that are being carried out in plant nutrition, and literature that addresses this research in a general way can be identified and will be presented in this summary.

Traditionally, research in soil fertility and plant nutrition has addressed soil testing and plant analyses and nutrient availability for plants, nutrient requirements of different crops, fertilizer use, and crop utilization of nutrients in materials applied to soil. Interest in these traditional fields continues, but topics including accumulation and transport of nutrients and nonessential elements have received recent attention. Research in genetics of plant nutrition has risen with the growth in the field of molecular biology.

22.2 SOIL TESTING AND PLANT ANALYSIS AND NUTRIENT AVAILABILITY

Consideration of the environmental and economic consequences of soil fertility practices is an essential component of research in plant nutrition. Soil tests are developed to assess the availability of plant nutrients in soils, and these tests are calibrated for the major field and vegetable crops, and provide the basis for lime and fertilizer recommendations. Recommendations for amounts and application of fertilizers are continually modified to optimize economics of production as the costs of fertilizer application, the value of crop yields, and subsidy regimes change. Criteria for interpreting the results of soil testing and plant analyses are developed through field and glasshouse research that relates test results and plant composition to crop yields. Research in soil fertility and plant nutrition also covers application to the land of agricultural, municipal, and industrial wastes and by-products (1), atmospheric contributions to plant nutrients in soils, short- and long-term availability of plant nutrients, especially nitrogen and phosphorus, and many other factors as well as soil testing and plant analyses.

Work on soil fertility and plant nutrition often involves multidisciplinary research in other areas of soil science and plant physiology. Basic and applied information in such areas as soil–plant relations, nutritional physiology, and plant nutrition technology have been summarized in books and monographs (2–4). Regular meetings of scientists working on plant nutrition occur, leading to continual developments in the subject. For example, 11 symposia on iron nutrition and interactions in plants have been held, with the most recent one covering topics that include the genetics of iron effciency in plants and molecular biology of iron absorption (5).

Some plant nutrients, such as potassium and sodium, are involved in plant responses to salt and water stress (6,7), giving rise to further studies on comparative physiology. Research on nutritional stresses include studying the physiological and biochemical detail of the absorption and transport of nutrients (8–11), and also studying plant composition with respect to factors such as organic acid biosynthesis in relation to nutrient accumulation or deficiency (12).

The complexity of the relations between plants and soils, and the complexity of the assimilatory pathways and cycling of nutrients within plants, has caused some workers to develop models to aid our understanding of the acquisition and uptake of nutrients by plants (13). Some of these models, such as those developed by Warwick HRI for nitrogen, potassium, and phosphorus for a variety of crops in different geographical locations (http://www.qpais.co.uk/nable/nitrogen.htm) are freely available on the internet.

Interest in nutrient absorption and accumulation is derived from the need to increase crop productivity by better nutrition and also to improve the nutritional quality of plants as foods and feeds. Investigations occurring in many different research locations are determining and helping to understand factors that affect nutrient absorption and accumulation in plants. The U.S. Plant, Soil and Nutrition Laboratory at Cornell University, Ithaca, New York (http://www.uspsnl.cornell.edu/index.html) conducts studies in the chemistry and movement of nutritionally important elements in the soil and the absorption of the elements by plant roots. Scientists at the laboratory also investigate factors that affect the concentration and bioavailability of nutrients in plant foods and feeds, and are developing methods to evaluate soil contamination of foods derived from plants. The laboratory is conducting research on identifying and investigating genes that facilitate and regulate plant nutrient uptake and transport. The Plant Physiology Laboratory of the Children's Nutrition Research Center at Baylor University, Waco, Texas (www.bcm.tmc.edu/cnrc), is a unique cooperative venture between a college of medicine (Baylor) and an agricultural research agency (USDA/ARS). This laboratory is dedicated to understanding the nutrient transport systems of plants as a means of improving food crops.

22.3 ACCUMULATION OF ELEMENTS BY PLANTS

Understanding how plants accumulate and store metallic elements are research topics of current interest, and the direct toxicity of elements to plants has been a long-standing topic of interest in plant nutrition research. Meharg and Hartley-Whitaker (14) reviewed literature on the accumulation and metabolism of arsenic in plants. Nable et al. (15) discussed research on the toxicity of boron in soils, noting amelioration methods of soil amendments, selection of plant genotypes that are tolerant of boron, and breeding of boron-tolerant crops.

The mechanisms of toxicity of trace elements are complex, and plants vary considerably in their responses to trace elements in soils. To understand and manage the risks to plant and animal life posed by toxic elements in soils, it is essential to know how these elements are absorbed, translocated, and accumulated in plants. A special issue of *New Phytologist* was dedicated to metal accumulation, metabolism, and detoxification in plants and in the use of plants in remediation of contaminated soils (16). Cobbett and Goldsbrough (17) considered the roles of metal-binding ligands in plants in metal detoxification, and there has been considerable interest in engineering plants for metal accumulation for purposes of phytoremediation of soils (or for providing better nutrition in diets and rations of humans and livestock) (18). Accordingly, the genetics of plants with regard to metal accumulation is a major topic of interest.

Babaoglu et al. (19) noted that *Gypsophila sphaerocephala* Fenzl ex Tchihat. has the potential to accumulate boron (over 3000 mg B/kg in leaves) from soils in which boron is phytotoxic and that the boron-rich plant material may be transported to areas where boron is deficient. Selenium, although often regarded as an element that is dangerous when it accumulates in plants that are ingested by animals, has received considerable attention in programs such as that at Cornell University, as selenium is now seen as being deficient in the human diet worldwide. The fact that its uptake by plants can be enhanced by supply of more selenium to the plants is important in this context (20). These issues are addressed in a chapter on selenium in this handbook. Terry et al. (21) also reviewed the literature on the physiology of plants with regards to selenium absorption and transport, pathways of assimilation, and mechanisms of toxicity and tolerance of plants to selenium. Aluminum toxicity is a long-standing issue for research in plant physiology, and a chapter in this handbook addresses aluminum as a factor in plant and animal nutrition. Rout et al. (22) also reviewed the physiology and biochemistry of aluminum toxicity in plants and discussed ways of increasing the tolerance of plants to aluminum.

The use of organic materials in metal detoxification or in the increase in nutrient availability in soils is also a topic for study (23). Similarly, the role of mycorrhizal associations in alleviating metal toxicity in plants is a topic of current research. Jentschke and Godbold (24) discussed the possibilities of a role of fungal activities in immobilization of metals or otherwise restricting the effects of soil-borne metals on plant growth.

22.4 GENETICS OF PLANT NUTRITION

The genetic and molecular background for plant nutrition is an area in which interest in research is expanding (5,16,25). A special section of *Journal of Experimental Botany* contains six invited papers from a session held at the Society for Experimental Biology Annual Meeting in April 2003, addressing the genetics of plant mineral nutrition. A preface to this section mentions the topics covered (26). The topics include a review of the genes that affect nitrogen absorption, assimilation, utilization, and metabolism in corn (*Zea mays* L.), and how manipulation of these genes might improve grain production. Another article describes the physiological and biochemical characteristics that allow plants to survive in environments containing little available phosphorus. The article explains the genetic events that occur when plants lack phosphorus and how knowledge of these events might be used to improve the efficiency of phosphorus acquisition and utilization by crops. The genetics of control of K^+ transport across plant cell membranes is the topic of another article. Another discussion is of the generation of salt-tolerant plants through transgenic approaches and through conventional plant breeding. Another article surveys the accumulation of nutrients in the shoots of angiosperms under lavish nutrition in hydroponics and under natural environmental conditions. In another article, the micronutrient requirements of humans and the supply of micronutrients from plants to populations at risk from mineral deficiencies is discussed in relation to the varying micronutrient contents in plants. These papers illustrate basic research in plant nutrition and describe how the application of modern genetic techniques contribute to solutions for plant and animal mineral nutrition.

Research in the genetics of plant nutrition covers major and minor nutrients, metals, plant stress, symbioses, and plant breeding. Several publications cover research in this area. A book by Reynolds et al. (27) has several chapters that address genotypic variation in wheat with respect to zinc and other nutrient efficiencies. A review article by Fox and Guerinot (28) summarizes knowledge about genes that influence the transport of cationic nutrients and addresses how genes encode for transporter proteins. These proteins can be divided into three main types, primary ion pumps, ion channels and cotransporters (29), and the genes that code for transporter proteins for all the macronutrients and some micronutrients that have been cloned from plants (29–31). This research studies how genetics affect plant responses to nutrient availability and may allow for creation of food crops with enhanced nutrient levels or with the ability to exclude toxic metals. Smith (10) describes how the expression of genes encoding for high-affinity phosphate transporters may improve phosphate utilization by plants growing under regimes of low phosphate availability in soils. However, it is probably the case that the influx of nutrient ions is not the limiting step in nutrient acquisition, so 'improving' the performance of transporters in plants by breeding may not achieve big increases in plant yield if not accompanied by other changes (29). In terms of improving yields of plants through improving the uptake and assimilation of nitrogen, expression of genes for cytosolic glutamine synthetase could have as large an impact on nutrient use efficiency as expression of genes for transporters (32).

Keeping phosphate, or other nutrients, available at the root surface is a major problem in nutrient-deficient soils; consequently, some research addresses mobilization of nutrients in the soil as well as internal mobilization within plants. Hinsinger (33) reviewed changes in the rhizosphere that can occur with plant species, plant nutrient availability, and soil conditions that can affect the acquisition of phosphorus by plants. Root exudates that are important in the acquisition of nutrients through modifications of the soil environment are topics of research (34), so they are studied for their composition and their effects on the development of mycorrhizal fungi, chelation of nutrients, solubilization of sparingly soluble compounds, and effects on soil acidity, among other actions. Breeding for improved soil–plant–microorganism interactions, especially under suboptimal environmental conditions, may lead to genotypes that are improved for nitrogen fixation and promotion of mycorrhizal symbiosis may bring about increased crop yields under a wide range of environmental conditions.

Bassirirad (35) considered factors of global change, such as increased atmospheric carbon dioxide concentrations, higher soil temperatures, and increased atmospheric nitrogen deposition, that may affect the kinetics of nutrient absorption by roots, noting that the information on the subject was scanty and that rigorous research was needed on the topic. Processes such as transpiration-driven mass flow, root growth, root exudation, biological nitrogen fixation, and tissue dilution are all likely to be affected by climate change (36).

Ionomics has been coined as the study of how genes regulate all the ions in a cell (37). This research is stated to hold promise leading to mineral-efficient plants that might need little fertilizer, to crops with better nutritional value for humans, and to plants that may remove contamination from the soil. Possibly, a simple genetic change can increase nutrient absorption by green plants and allow crop production under conditions of limited nutrient availability or allow plants to be efficient in recovery of fertilizer-borne nutrients. Yanagisawa et al. (38) suggested that utilization of transcription factors might lead to modification of metabolism of crops, because a single transcription factor frequently regulates coordinated expression of a set of key genes for several pathways. They applied the plant-specific transcription factor (*Dof1*) to improve nitrogen assimilation, including the primary assimilation of ammonia to biosynthesize amino acids and other organic compounds containing nitrogen. The authors proposed that similar genetic modifications could reduce dependence on nitrogen fertilizers.

22.5 GENERAL REMARKS

Current research on plant nutrition is extensive, and only a few topics can be mentioned here. Some of the topics mentioned on http://www.plantstress.com, which is sponsored by the Rockefeller Foundation, are noted. With the world population increasing fast, and many people suffering from deficiencies of essential nutrients, there will be continuing pressure to improve our understanding of plant mineral nutrition so that we can grow crops that utilize mineral nutrients as efficiently as possible.

REFERENCES

1. J. Power, W.A. Dick, eds. Land application of agricultural, industrial, and municipal by-products. *Soil Science Society of America Book Series No. 6*, Madison, Wis: Soil Science Society of America, Inc., 2000.
2. T. Ando, K. Fujita, T. Mae, H. Matsumoto, S. Mori, J. Sekiya., eds. *Plant Nutrition for Sustainable Food Production and Environment*. Kluwer Academic Publishers: Dordrecht, 1997.
3. H.S. Srivastava, R.P. Singh, eds. *Nitrogen Nutrition and Plant Growth*. Enfield, New Hampshire: Science Publishers, 1999.
4. H. Hirt, K. Shinozaki, eds. *Plant Responses to Abiotic Stress*. Berlin: Springer, 2004.
5. H.A. Mills., Exec. ed. Proceedings of the Eleventh International Symposium on Iron Nutrition and Interactions in Plants, June 2002, Udine, Italy. *J. Plant Nutr.* 26: 1889–2319, 2003.
6. S.R. Grattan, C.M. Grieve. Salinity-mineral nutrient relations in horticultural crops. *Sci. Hortic.* 78: 127–157, 1999.
7. R. Munns. Comparative physiology of salt and water stress. *Plant Cell Environ.* 25: 239–250, 2002.
8. D.P. Schachtman, R.J. Reid, S.M. Ayling. Phosphorus uptake by plants: From soil to cell. *Plant Physiol.* 116: 447–453, 1998.
9. F. Gastal, G.N. Lemaire. N uptake and distribution in crops: An agronomical and ecophysiological perspective. *J. Exp. Bot.*, Inorganic Nitrogen Assimilation Special Issue, no. 370: 789–799, 2002.
10. F.W. Smith The phosphate uptake mechanism. *Plant Soil* 245: 105–114, 2002.
11. F.W. Smith, S.R. Mudge, A.L. RaeM D Glassop. Phosphate transport in plants. *Plant Soil* 248: 71–83, 2003.
12. J. Abadía, A.-F.L. Millán, A. Rombolà, A. Abadía. Organic acids and Fe deficiency: A review. *Plant Soil* 241: 75–86, 2002.
13. J. Le Bot, S. Adamowicz. P. Robin. Modelling plant nutrition of horticultural crops: A review. *Sci Hortic.* 74: 47–82, 1998.
14. A.A. Meharg, J. Hartley-Whitaker. Arsenic uptake and metabolism in arsenic resistant and nonresistant plant species. *New Phytol.* 154: 29–43, 2002.
15. R.O. Nable, G.S. Bañuelos, J.G. Paull. Boron toxicity. *Plant Soil* 193: 181–198, 1997.
16. C. Cobbett. Heavy metals and plants — model systems and hyperaccumulators. *New Phytol.* 159: 289–293, 2003.
17. C. Cobbett, P. Goldsbrough. Phytochelatins and metallothioneins: Roles in heavy metal detoxification and homeostasis. *Annu. Rev. Plant Biol.* 53: 159–182, 2002.
18. S. Clemens, M.G. Palmgren, U. Krämer. A long way ahead: Understanding and engineering plant metal accumulation. *Trends Plant Sci.* 7: 1360–1385, 2002.
19. M. Babaoglu, S. Gelzin, A. Topal, B. Sade, H. Dural. *Gypsophila sphaerocephala* Fenzl ex Tchihat.: A boron hyperaccumulator plant species that may phytoremediate soils with toxic boron levels. *Turk J. Bot.* 28: 273–278, 2004.
20. R.M. Welch. The impact of mineral nutrients in food crops on global human health. *Plant Soil* 247: 83–90, 2002.
21. N. Terry,.A.M. Zayed, M.P. de Souza,.A.S. Tarun. Selenium in higher plants. *Annu. Rev. Plant Physiol. Plant Mol. Biol.* 51: 401–432, 2002.
22. G.R. Rout, S. Samantaray, P. Das. Aluminium toxicity in plants: A review. *Agronomie* 21: 3–21, 2001.
23. R.J. Haynes, M.S. Mokolobate. Amelioration of Al toxicity and P deficiency in acid soils by additions of organic residues: A critical review of the phenomenon and the mechanisms involved. *Nutr. Cycling Agroecosystems* 59: 47–63, 2001.
24. G. Jentschke, D.L. Godbold. Metal toxicity and ectomycorrhizas. *Physiol. Plant* 109: 107–116, 2001.
25. G.A. Gissel-Nielsen, A. Jensen, eds. *Plant Nutrition — Molecular Biology and Genetics. Proceedings of the 6th International Symposium on Genetics and Molecular Biology of Plant Nutrition*, Kluwer Academic Publishers: Dordrecht, 1999.
26. P.J. White, M.R. Broadley. Preface to genetics of plant mineral nutrition. *J. Exp. Bot.* 55: i–iv, 2004.
27. M.P. Reynolds, J.I. Ortiz-Monasterio, A. McNab, eds. *Application of Physiology in Wheat Breeding* CIMMYT: Mexico, 2001.
28. T.C. Fox, M.L. Guerinot. Molecular biology of cation transport in plants. *Annu. Rev. Plant Physiol. Plant Mol. Biol.* 49: 669–696, 1998.
29. J. Dunlop, T. Phung. Transporter genes to enhance nutrient uptake: Opportunities and challenges. *Plant Soil* 245: 115–122, 2002.

30. M.J. Chrispeels, N.M. Crawford, J.I. Schroeder. Proteins for transport of water and mineral nutrients across the membrane of plant cells. *Plant Cell* 11: 661–675, 1999.

31. J.L. Hall, L.E. Williams. Transition metal transporters in plants. *J. Exp. Bot.* 54: 2601–2613, 2003.

32. A.G. Good, A.K. Shrawat, D.G. Muench. Can less be more? Is reducing nutrient input into the environment compatible with maintaining crop production? *Trends Plant Sci.* 9: 597–605, 2004.

33. P. Hinsinger. Bioavailability of soil inorganic P in the rhizosphere as affected by root-induced chemical changes: A review. *Plant Soil* 237: 173–195, 2001.

34. F.D. Dakora, D.A. Phillips. Root exudates as mediators of mineral acquisition in low-nutrient environments. *Plant Soil* 245: 35–47, 2002.

35. H. Bassirirad. Kinetics of nutrient uptake by roots: Responses to global change. *New Phytol.* 147: 155–169, 2000.

36. J.P. Lynch, S.B. St. Clair. Mineral stress: The missing link in understanding how global climate change will affect plants in real world soils. *Field Crops Res.* 90: 101–115, 2004.

37. B. Lahner, J.M. Gong, M. Mahmoudian, E.L. Smith, K.B. Abid, E.E. Rogers, M.L. Guerinot, J.F. Harper, J.M. Ward, L. McIntyre, J.I. Schroeder, D.E. Salt. Genomic scale profiling of nutrient and trace elements in Arabidopsis thaliana. *Nature Biotechnol.* 21: 1215–1221, 2003.

38. S. Yanagisawa, A. Akiyama, H. Kisaka, H. Uchimiya, T. Miwa. Metabolic engineering with Dof1 transcription factor in plants: Improved nitrogen assimilation and growth under low-nitrogen conditions. *Proc Natl Acad Sci* 101: 7833–7838, 2004.

Index

A

Acetyl coenzyme A synthase and nickel, 397, 398
O-acetylserine, 186, 519
Abscisic acid (ABA), 125, 424
Accumulator plants, 586, 589,
  for aluminum, 441–442, 478–479
  for cobalt, 500, 509
  for copper, 313–314
  for iron, 335
  for nickel, 406
  for selenium, 517, 520–521, 594
  for vanadium 589, 594
Actin, 451
Agmatine, 99–100
Akagare, 332
Alcohol dehydrogenase and zinc deficiency, 11, 412
Aldehyde oxidase and molybdenum, 376, 378
Alfisols, 115, 319
  boron concentration, 245
  calcium concentration, 138
  cation exchange capacity, 113, 138
  copper concentration, 317
  potassium concentration, 106, 110
Aluminum
  and boron, 243
  and calcium, 443, 446, 447–448, 449, 450, 452, 459
  and cell walls, 443, 447–448, 458
  and copper, 311
  and iron, 450
  and magnesium, 153–154, 446, 449, 450
  and membranes, 447, 448–449, 453, 458
  and molybdenum, 385, 389
  and nitrate, 442, 446, 449, 450
  and phosphorus, 442, 446, 459
  and plant disease, 442
  and potassium, 446, 449–450
  and silicon, 460, 554
  and vanadium, 588
  and water uptake, 446, 450, 459
  effect on calcium homeostasis, 452
  effect on cell division, 445, 447
  effect on lignification, 445
  effect on photosynthesis, 446
  effect on root elongation, 444–445, 449, 454, 458, 479–480
  inhibition of symbiosis with Rhizobium, 447
Aluminum citrate, 460
Aluminum oxalate, 460
Aluminum oxides
  and boron sorption, 262
  and copper sorption, 318
  and molybdenum sorption, 385, 389

  and phosphorus sorption, 54
  and vanadium sorption, 586
Aluminum sulfate in water treatment, 470–471
Aluminum toxicity, 154–155, 442, 444–453, 468, 476–479, 601
Alunite, 461
Amidation, 24–25
δ-aminolevulinic acid, 588
δ-aminolevulinic acid synthetase, 330
Amino sugars in soil, 34, 38–39
Ammoniated superphosphate fertilizer, 42
Ammonium
  accumulation in plant tissues, 10, 92
  accumulation in soil, 35, 36, 92
  assimilation, 23–25
  toxicity, 35
Ammonium chloride as fertilizer, 287
Ammonium metavanadate, 587
Ammonium molybdate as fertilizer, 387, 388
Ammonium nitrate fertilizer, 41
Ammonium nitrate sulfate fertilizer, 42
Ammonium phosphate nitrate fertilizer, 42
Ammonium polyphosphate fertilizer, 42, 82
Ammonium sulfate fertilizer, 39, 41
Anhydrous ammonia fertilizer, 40
Anthocyanin accumulation, 5, 7, 199
Apatite, 52, 137, 139
APS reductase, 185–186
Aqua ammonia fertilizer, 40
Aridisols, 138
Arsenic
  accumulation, 600
  competition with sulfur, 197
  metabolism, 600
Ascorbic acid oxidase and copper deficiency, 11, 314
Atmospheric emissions, 600, 602
  of sulfur dioxide, 183–184, 187
  of vanadium, 585, 586, 594
ATPase
  activity limited by boron deficiency, 244
  in photophosphorylation, 147
  inhibition by aluminum, 449
  inhibition by cobalt, 502
  inhibition by copper, 316
  inhibition by vanadate, 587, 588, 589
  role in acidification of rhizosphere, 338
  role in calcium transport, 124, 131
  role in potassium uptake and transport, 94, 95, 96–97, 98
  stimulation by chloride, 280
ATP sulfurlyase, 185–186
Augite, 137, 166

Auxins, 125, 244, 245, 423, 452
Available plant nutrients, 11–12
Azurite, 317

B

Band placement
    of boron fertilizer, 267–268
    of phosphorus fertilizer, 79–80
Basaluminite, 464
Beneficial element, definition, 4, 571
Biological nitrogen fixation, 33, 35; see also Nitrogen
        fixation
Biotite, 105–106, 166
Bitter pit
    and calcium, 126, 136, 139
    and potassium, 100
Blossom end rot
    and calcium, 126–127, 131–132, 136, 139
    and potassium, 100, 132
Borax as fertilizer, 266, 267
Boric acid fertilizer, 267
Boron
    adsorption in soil, 263
    and aluminium, 243
    and calcium, 245, 260–261
    and chloride, 244
    and lignification, 244
    and magnesium, 260–261
    and nitrate concentration, 243
    and nitrate reductase, 243
    and nitrogen, 261–262
    and phosphorus, 244, 262
    and potassium, 245, 262
    and protein synthesis, 243
    and rubidium, 244
    and sugar synthesis, 243
    and sulfate, 244
    and zinc, 246, 262
Boron deficiency, 243–245, 246–249, 261, 262, 264, 266
Boron frits fertilizer, 267
Boron toxicity, 246, 249–251, 262, 263, 264–265, 600,
        601
Boundary Line Development System (BOLIDES),
        215–217
Brown-heart and boron, 242, 248

C

Cadmium, 586
Caffeic acid, 244
Calcareous soil, phosphorus sorption, 54, 133, 138
Calcicole, 122, 132–133, 343
Calcifuge, 122, 132–133
Calcite, 54, 135, 137
Calcium
    accumulation with vanadium, 588, 589
    and aluminum, 443, 446, 447–448, 449, 450, 452, 459
    and boron, 245, 260–261
    and copper, 310, 311
    and enzymes, 124
    and fruit firmness, 124, 127–128, 139
    and magnesium competition, 124, 132, 149, 150, 151
    and nickel, 403

and phosphorus sorption, 132–133, 138
    and potassium competition, 100–101, 132
    and sodium competition, 165, 572
    and strontium, 125
    and vanadate, 589
    channels, 128, 443, 589
    competition with vanadium, 587, 589
    deficiency, 7, 245
    role in pollen tube growth, 125
    transport, 129–131
    uptake, 128–129
Calcium carbonate equivalent (CCE), 140
Calcium chloride and fruit, 139
Calcium chloride as fertilizer, 287
Calcium magnesium phosphate as fertilizer, 171
Calcium nitrate and fruit, 139
Calcium nitrate urea fertilizer, 41
Calcium oxalacetate, 128
Calcium oxalate, 128
Calcium silicate fertilizer, 562
Calcium sulfate fertilizer, 139; see also Gypsum
Calmodulin, 124
Cambisols, 263
Canonical discriminant analysis, 9
Carbamylputrescine, 100
Carbonic anhydrase and zinc deficiency, 11, 412
Carbon monoxide dehydrogenase and nickel, 397
Catalase
    and cobalt, 507
    and iron deficiency, 10–11, 330
    iron as a component, 330
Cation competition, see Ion antagonism
Cation exchange in soil, 113, 137, 138, 140, 331, 586
    in sodic soil, 570
Cation exchange in plant cell walls, 129, 131, 133, 447, 458
Cellular pH, maintenance of, 52
Cell-to-cell adhesion, 124
Cell wall structure, 122–124, 447–448, 554, 556
Chalcocite, 312, 317
    chalcocite as fertilizer, 312
Chalcopyrite, 317
    as fertilizer, 312
Chenopodiaceae as halophytes, 571–573
Chernozems, 317
Children's nutrition, 600
Chitosan, 588
Chlorapatite, 137
Chloride
    and magnesium, 154
    and manganese, 282
    osmotic effect 112, 280, 284
    role in maintenance of electroneutrality, 280–281
    role in stomatal opening, 280
Chlorine deficiency, 279, 280, 281–282, 283–284, 285
Chlorine toxicity, 283
Chlorite, 107, 166
Chlorophyll,
    copper substitution for magnesium, 316
    magnesium as a constituent, 4, 146, 147, 148, 149, 151
Chlorophyll a, 588
Chlorophyll b, 588
Chlorophyll biosynthesis

and iron, 330
and magnesium, 5
and nitrogen, 5, 27
enhanced by vanadium, 588
Chlorophyll meter, 10
Citric acid test for phosphorus, 71–72
Clausthalite, 523
Climate change, 602
Coal, 585
Cobalt
    and cadmium, 505
    and cell walls, 500, 502
    and chromium, 507
    and copper, 505, 509
    and cyanide, 507
    and iron, 506, 509
    and magnesium, 504, 505, 507, 509
    and manganese, 504, 505, 506, 507, 509
    and nickel, 507
    and sulfur, 507
    and tin, 507
    and zinc, 506
Cobalt toxicity, 506
Cobalt uptake, 501, 502–503
Colemanite as fertilizer, 267
Copper
    and aluminum, 311
    and calcium, 310, 311
    and iron, 310, 311
    and magnesium, 154
    and manganese, 311
    and molybdenum, 311
    and nickel, 403, 404
    and nitrogen, 310, 311
    and phosphorus, 310, 311
    and potassium, 316
    and selenium, 310
    and zinc, 310, 311
Copper chelate as fertilizer, 312
Copper chloride as fertilizer, 312
Copper chlorosis, 335
Copper deficiency, 11, 313, 314–315, 320, 379
    and nickel, 403
Copper frits as fertilizer, 312
Copper oxalate as fertilizer, 312
Copper sulfate as fertilizer, 312, 313
Copper toxicity, 294, 314, 315–316
    and magnesium, 154
Copper-induced chlorosis, 315
Copper uptake, 294, 310
Corn stalk test, 10
Covellite, 317
Crassulacean acid metabolism (CAM), 573
Crease, 126
Critical concentration, 9
Cupric nitrate as fertilizer, 312
Cupric oxide as fertilizer, 312, 313
Cuprite, 317
Cuprous oxide as fertilizer, 312
Cytochrome oxidase, 314
Cytokinins, 452
Cytoplasmic potassium homeostasis, 95

D
Dairy cows, sodium requirement, 573
Diagnosis and Recommendation Integrated System
    (DRIS), 9, 32
Diamine oxidase, 314
Diammonium phosphate fertilizer, 42, 79
Dicalcium phosphate dihydrate (DCPD) fertilizer, 54
Dicyandiamide fertilizer/nitrification inhibitor, 41
Diet and minerals, 155, 601, 602
    and aluminum, 468–469, 474–476
    and cobalt, 500
    and copper, 321–323
    and iron, 344
    and manganese, 353
    and selenium, 524–526
Dimethylselenide, 520
Dof1, 602
Dolomite, 113, 135, 138, 167, 170, 317
Dolomite as fertilizer, 151, 170, 171, 172
Dumas method, 33–34

E
Effective calcium carbonate equivalent, 140
Electro-ultra-filtration (EUF), 108–112
Elovich function, 108, 109
Entisols, 106, 110, 138
Epsom salts, 170–172, 221; *see also* Magnesium sulfate
Essential element, definition, 3–4, 396, 571
Exchangeable sodium percentage, 570

F
Facilitated diffusion, 94, 101
FeDTPA, 345
FeEDDHA, 344–345
FeEDTA, 344–345
FeHEDTA, 345
Feldspars, 105–107, 108, 137, 561, 562
Ferrasols, 317
Ferric chelate reductase (Fe(III) chelate reductase), 310,
    336–338, 339, 344
Ferric citrate (Fe(III) citrate) as fertilizer, 344
Ferric dihydroxamate (Fe(III) dihydroxamate) as fertilizer,
    345
Ferric monodihydroxamate (Fe(III) monodihydroxamate)
    as fertilizer, 345
Ferric reductase (Fe(III) reductase) and copper, 310
Ferrimolybdite, 385
Ferrous sulfate (Fe(II) sulfate) as fertilizer, 344–345
Fertigation and phosphorus supply, 81–82
Fluorapatite, 137
Fluvisols, 317, 383
Fluvo-aquic soils, 586, 587
Fly ash, 219, 524, 585
Foliar application
    of boron, 268
    of calcium, 139
    of copper, 312
    of iron, 344, 345
    of molybdenum, 387, 388
    of potassium, 112
    of sulfur, 221
    of zinc, 424–428, 429

Foliar uptake
  of chlorine 285
  of iron, 337
  of phosphorus, 81
  of sulfur, 187–188
  of zinc, 424–428
Forest decline and magnesium, 146
Franklinite, 422
Frost resistance in plants, 557
Functional elements, 571

G
Geographic information system, 13
Geothite, 523
Gibbsite, 461, 464
Gleysols, 383
Glucosinolates
  metabolism, 193–195
  synthesis, 193–195, 207
Glutamate-oxalacetate aminotransferase, 11
Glutamate synthase, 12, 24
Glutamic acid dehydrogenase, 24
Glutamine synthetase, 24, 602
Glycinebetaine, 572
Glyoxylase and nickel, 397, 398
Glyphosate, 504
Golden rice, 344
Goldspot, 126
Grass tetany, *see* Hypomagnesia
Greenback and potassium, 99
Gypsum, 137, 138, 464
Gypsum as fertilizer, 139; *see also* Calcium sulfate as
  fertilizer

H
Haber–Bosch process, 22, 39
Halophytes, 152, 571–575
Heart rot and boron, 242, 248
Heme proteins, 330
Hidden hunger, 7, 9
Histosols
  and copper concentration, 317
  and molybdenum concentration, 383
  and phosphorus concentration, 73, 79
  cation exchange capacity, 113
Hohokan civilization, 571
Homogalacturonan, 122–124
Hornblende, 137, 167
Hubbard Brook Experimental Forest, 137
Humic acid, 34–35, 53, 58
Hydrogen sulfate
  emissions by plants, 217–219
  uptake by plants, 187–188
Hydrogenase and nickel, 397, 398, 399
Hydroxyapatite, 137
Hydroxyferulic acid, 244
Hypomagnesia, 146, 155

I
IAA oxidase, 244
Illite, 105–107, 263

Imogolite, 319
Inceptisols, 113, 138, 245, 317
Indole-3-acetic acid (IAA), 281, 424
  inhibition of breakdown by cobalt, 501, 502, 508
  inhibition of synthesis, or increased degradation, with
    zinc, 423
  inhibition of transport by aluminum, 452
  role of calcium in action of IAA, 125
Indus civilization, 571
Inositol phosphate, 53; *see also myo*-Inositol phosphate
Ion antagonism, 100
Ionomics, 602
Iron
  and aluminum, 450
  and cobalt, 506, 509
  and copper, 310, 311
  and magnesium, 152
  and molybdenum, 379, 385, 389
  and nickel, 403
  and phosphorus, 332
  and potassium, 332
Iron deficiency, 6, 10–11, 330, 332–334, 335–336, 339,
    342
  deficiency with magnesium, 152, 156
  deficiency with vanadium, 587–588
Iron deficiency chlorosis, 133, 335, 337, 342, 343, 355
Iron deficiency chlorosis paradox, 336
Iron EDTA, 587
Iron efficiency, iron-efficient plants, 336, 343, 600
Iron oxides in plants, 335
Iron oxides in soil, 331–332
  and sorption of boron, 262
  and sorption of molybdenum, 389
  and sorption of phosphorus, 54
  and sorption of selenium, 523
  and sorption of vanadium, 586
Iron toxicity, 332, 334
Iron uptake, 336–338, 600
Irrigation
  and boron, 265–266
  and cobalt, 502
  and copper, 317
  and iron, 339, 341
  and magnesium, 172
  and nickel, 404
  and phosphorus, 77, 81–82
  and salinity, 570–571
  and sulfur, 205
Isobutylidene diurea (IBDU) fertilizer, 41

J
Jarosite, 335
Jurbanite, 464

K
Kaolinite, 108, 109, 263, 461, 561
Kastanozems, 317
Kieserite as fertilizer, 171
K$^+$ fixation, 106
K$^+$-fixing soils, 92, 114
Kjeldahl, 33–34, 36

L

Labile phosphorus, 54
Law of diminishing returns, 12
Law of the minimum, 12
Leaf area
    increase with vanadium, 587
    decrease with copper, 311
Leaf canopy reflectance 10, *see also* Spectral reflectance,
    13
Lignification
    and aluminum, 445
    and boron, 244
    and iron, 330
    and manganese, 353, 365
Lime application, 137–139, 139–140, 151, 170–172
    and aluminum tolerance, 468
    and cobalt uptake, 503
    and boron, 260–261
    and mineralization of nitrogen, 139
    and molybdenum, 384, 389
    and selenium, 523–524
    and zinc, 412
Lime-induced chlorosis
    and iron, 332, 335, 339, 342, 343, 345
    and magnesium, 152, 157
Lime requirement, 140, 467
Livestock
    and aluminum, 469–470, 473–474
    and cobalt, 500, 505
    and copper, 321–322
    and molybdenum, 389
    and nickel, 398
    and silicon, 562
    and sodium, 573
Luvisols, 107, 111, 132

M

Magnesite as fertilizer, 171
Magnesium
    accumulation with vanadium, 588
    and aluminum ,153–154, 446, 449, 450
    and boron, 260–261
    and calcium competition, 124, 132, 149, 150, 151,
        165
    and chloride, 154
    and cobalt, 504, 505, 507, 509
    and copper, 154
    and fruit quality, 147–148
    and iron, 152
    and leaf stomatal conductance, 147
    and manganese, 153
    and nickel, 403
    and nitrogen, 151–152
    and phosphorus, 153–154
    and potassium competition, 100–101, 147–148, 149,
        150–151
    and sodium competition, 152, 165, 572
    and water relations, 147
    and zinc, 153
    deficiency, 8, 148–149, 151, 154
    toxicity, 149
Magnesium ammonium phosphate as fertilizer, 171
Magnesium chloride as fertilizer, 287
Magnesium nitrate as fertilizer, 171
Magnesium oxide as fertilizer, 170, 172
Magnesium sulfate as fertilizer, 170–172; *see also* Epsom
    salts
Magnesium uptake and mycorrhizas, 150
Malate dehydrogenase, 457
Manganese
    and chloride, 282
    and cobalt, 504, 505, 506, 507, 509
    and copper, 311
    and lignification, 353, 365
    and magnesium, 153
    and nickel, 403
    and photosynthesis, 353, 365
    and plant diseases, 357
    deficiency, 10, 11, 353, 354–355, 357
    toxicity, 356
Manganese oxide in soil, 353, 586
Manganese toxicity, 153, 356–357
Manganese uptake, 353
Malachite, 317
Mesopotamia, 570
Metal
    accumulation, 600–601
    detoxification, 600–601
    metabolism, 600–601
Metallothioneins, 192, 313
Methyl coenzyme M reductase and nickel, 397
Methylene urea fertilizer, 41
Micas, 105–107, 461, 562
Mineralization
    of nitrogen, 32, 34, 35, 36, 37, 42–43
    of phosphorus, 53–54, 504
Mollisols, 106, 113, 138
Molybdenite, 385
Molybdenum
    and aluminum, 385, 389
    and iron, 379, 385, 389
    and phosphorus, 385, 389
    and sulfate, 379, 385, 389
    deficiency, 11, 378–379, 388
Molybdenum frits as fertilizer, 387
Molybdenum sulfide as fertilizer, 387
Molybdenum toxicity, 379
Molybdenum trioxide as fertilizer, 387, 388
Molybdic acid as fertilizer, 387
Molybdopterin, 376–378
Monoammonium phosphate fertilizer, 42, 79, 81, 82
Monocalcium phosphate fertilizer, 81
Montmorillonite, 168, 263, 318, 319
Muscovite, 105–106
Mycorrhizas, 331, 404, 602
    aluminum toxicity to, 459
    and magnesium uptake, 150
    and zinc uptake, 415, 424
    in alleviating metal toxicity, 601
myo-Inositol phosphate, 52

N

NADP$^+$-malic enzyme, 573
Naumannite, 523

Nickel
and calcium, 403
and cobalt, 507
and copper, 403, 404
and iron, 403
and magnesium, 403
and manganese, 403
and zinc, 403, 404
deficiency, 395, 399, 400, 401–403, 404–405, 406
toxicity, 401–402, 405
uptake, 404
Nickel permeases, 397
Nickel sulfate as fertilizer, 406
Nicotianamine, 339
Nicotine, 588
Nitrate
and aluminum, 442, 446, 449, 450
and boron, 243
and root growth, 12
assimilation, 23
in plant tissues, 9–10, 30, 221, 243, 282
in soil, 35, 36, 37–38
Nitrate reductase, 23
and boron, 243
and iron deficiency, 11
and molybdenum requirement, 11, 377, 381, 388
and nitrogen deficiency, 11
and sodium toxicity, 572
and sulfur deficiency, 221
Nitrification, 35, 37–38, 40, 41; see also Mineralization of nitrogen
and chloride, 282
inhibition by copper, 319
Nitrification inhibitor, 36, 39, 41
Nitrite reductase, 23, 330
Nitritetoxicity, 35
Nitrogen
absorption, 601
accumulation with vanadium, 588
and boron, 261–262
and copper, 310, 311
and magnesium, 151–152
and molybdenum, 184, 188–189, 195–197, 207–208, 213–214, 220–221, 378
and selenium, 518
assimilation, 23–26, 601, 602
availability index, 36
deficiency, 5–6, 11, 26–27
fertilizers, 39–43
fixation, 33, 376–377, 378–378, 388–389, 447, 500, 508
metabolism, 601
uptake, 600
Nitrogenase and molybdenum, 376–377, 381, 388
NRT1, 12
NRT2, 12

O
Oertli effect, 570
Orthoclase, 561
Orthophosphate (orthophosphoric acid) fertilizer, 81, 82
Osmotic adjustment in plants, 147, 570, 573–574

Oxisols, 113, 132, 586, 587
aluminum saturation, 467
calcium concentration and cation exchange capacity, 138
distribution, 461, 462, 463
potassium-binding capacity, 107

P
Patronite, 586
Pectin, 122–124, 128, 447–448
Peroxidase
and cobalt, 501
and iron deficiency, 10–11, 330
and manganese deficiency, 10
and silicon, 555
iron as a component, 330
Peteca, 126–127
Phenolase, 314
Phosphate, high affinity transporters, 602
Phosphoenolpyruvate, sodium requirement for regeneration, 573
Phosphoenolpyruvate carboxylase, 124, 336, 412, 457, 573
Phosphoinositide, 451
Phosphorus
accumulation with vanadium, 588
acquisition, 601
and aluminum, 442, 446, 459
and boron, 244, 262
and copper, 310, 311
and iron, 332
and magnesium, 152
and molybdenum, 385, 389
and selenium, 517–518
and silicon, 554
and sulfur, 197–198
and zinc, 423
cycle, 53–54
deficiency, 7, 8, 11, 54–55
nutrition, 601
sorption in soil, 54, 132–133, 138
uptake, 78, 600
Photosynthesis
inhibition by aluminum, 446
inhibition by manganese, 353, 365
oxygen evolution, 588
photosystem I, 330, 588
photosystem II, 315–316, 453, 504, 505, 588
Phytoalexins, 219
Phytochelatins, 192, 313
Phytoextraction, phytoremediation, 13, 313–314, 406, 520–521, 600–601
Phytoferritin, 335
Phytosiderophores, 336–339, 343
Plagioclase, 137, 561
Plant analysis, see Tissue analysis
Plant disease
and aluminum, 442
and chloride, 282–283
and manganese, 357
and nickel, 399, 400

and potassium, 99
and silicon, 554–557
and sulfur, 184, 217–219
Podzols, 170, 317, 332, 383
Pollen tube growth, 125
Potassium
    accumulation with vanadium, 588
    and aluminum, 446, 449–450
    and boron, 245, 262
    and calcium, 100–101, 132
    and copper, 316
    and fungal infection, 99
    and magnesium, 100–101, 147–148, 149, 150–151
    and sodium competition, 93–94, 100–101, 115–116, 557, 572
    and water use efficiency, 99
    as osmoticum, 95–97, 98–99, 101
    cytoplasmic homeostasis, 95
    deficiency, 6, 10, 11, 99–100
    fixation, 106–107
    in phloem, 97–99
    in xylem, 97–99
    role in enzyme activation, 92–93
    role in protein synthesis, 93–94
    sodium substitution, 101
    transport, 97–99, 601
    uptake, 94–95, 600
Potassium chloride fertilizer, 112, 113, 285, 286, 287
Potassium-fixing soils, 92, 114
Potassium magnesium sulfate as fertilizer, 170–171
Potassium metaphosphate fertilizer, 112, 113
Potassium nitrate fertilizer, 112, 113
Potassium silicate fertilizer, 112, 562
Potassium sulfate fertilizer, 112, 113
Powellite, 385
Precision agriculture, 13
Preplant nitrate test, 37
Pre-sidedress soil nitrate test (PSNT), 37–38
Principal component analysis, 9
Proline, 572
Protein synthesis
    and magnesium, 5
    and nitrogen, 5, 25, 188–190
    and potassium, 93–94
    and sulfur, 188–190
Putrescine, 99–100
Pyrite, 523
Pyrolusite, 352
Pyrophosphatase, 95
Pyruvate-Na⁺ cotransport, 573
Pyruvic kinase, 11, 12

Q
Quantity/Intensity Relationship and potassium, 110–111

R
Rectifying channels, 94
Rhamnogalacturonan, 122
Ribulose bisphosphate carboxylase, 92, 412
Rockefeller Foundation, 602
Rock phosphate, 79, 139, 522

Roscoelite, 168, 586
Rosetting 5–6, 99, 248, 411, 423, 424
Rothamsted Experimental Station, 52

S
Saline soil, 570–571
Salinity
    and boron, 263–264
    and magnesium, 151, 152, 154, 572
    and nitrogen nutrition, 572
    and phosphorus nutrition, 572
    and potassium nutrition, 115–116, 572
    and silicon alleviation, 557
    visual symptoms, 8
Salinization, 570–571
Selenium
    and copper, 310
    and nitrogen, 518
    and phosphorus, 517–518
    and sulfur, 191, 197, 517, 518–519, 521, 526
Selenium deficiency in human diet, 524–526, 601
Selenium toxicity in animals, see Selenosis
Selenium toxicity in plants, 521
Selenium uptake, 517–518
Selenosis, 517, 524, 526
Sewage sludge, sewage effluent, 382, 404, 405, 406, 586
Silicate chrysocolla, 317
Silicon
    and aluminum, 460, 554
    and cell walls, 554–557
    and pests and diseases, 554–557
    and phosphorus, 554
    and salinity, 557
Silicon uptake, 553
Smectites, 107, 108, 109, 461, 561
Sodicity, sodic soil, 570–571
Sodium
    and inhibition of protein synthesis, 93–94
    and inhibition of uptake of calcium, 165, 572
    and inhibition of uptake of magnesium, 152, 165, 572
    and inhibition of uptake of potassium, 93–94, 100–101, 557, 572
    and nitrate assimilation, 572
    and nitrate uptake, 572
Sodium absorption ratio, 165, 263
Sodium bicarbonate soil test for phosphorus, 73, 75
Sodium borates, 246
Sodium-calcium borates, 246
Sodium chloride as fertilizer, 287
Sodium copper EDTA, 312
Sodium molybdate as fertilizer, 387, 388
Sodium nitrate fertilizer, 39
Sodium/potassium replacement in plants, 572
Sodium toxicity, 573–574
Soil quality index, 12
Soil test, 11–12, 599
    for aluminum, 465–468
    for ammonium, 36
    for boron, 257–260
    for calcium, 137
    for chlorine, 286

for copper, 320
for iron, 344
for lime requirement, 139–140
for magnesium, 170
for molybdenum, 386–387
for nickel, 405
for nitrate, 37–38
for nitrogen, 35–38
for phosphorus, 71–75
for potassium, 107–112
for silicon, 561
for sulfur, 202–206
Solubor fertilizer, 266, 267
Sonoran Desert, 571
Spectral reflectance, 13, *see also* Leaf canopy reflectance,
    10
Spodosols, 106, 113, 138, 467
Strontium, 125
Sulfate
    assimilation, 185–187
    reduction, 185–186, 191
    uptake, 185, 219–220, 221
Sulfite oxidase, 197, 376, 378
Sulfite reductase, 185–186, 330
Sulfur
    and antimony, 197
    and arsenic, 197
    and baking quality, 188–189
    and boron, 197, 244
    and bromine, 197
    and cadmium, 192
    and molybdenum, 197, 379, 385, 389
    and nitrogen, 184, 188–189, 195–197, 207–208,
        213–214, 220–221, 378
    and pests/diseases, 184, 217–219
    and phosphorus, 197–198
    and selenium, 191, 197, 517, 518–519, 521, 526
Sulfur cycle, 204
Sulfur deficiency, 184, 198–202, 218
Sulfur dioxide uptake, 187–188
Sumerian civilization, 570
Superoxide dismutase
    and aluminum, 453
    and cobalt, 505
    and manganese deficiency, 11
    and nickel, 397, 398, 399
Superphosphate, 139, 389, 522

T
Tenorite, 317
Tetrapolyphosphate fertilizer, 81
Thallium accumulation, 586
Tiemannite, 523
Tissue analysis, 8–11, 599
    for aluminum, 476–479
    for boron, 251–257
    for calcium, 133–135
    for chlorine, 283–285
    for cobalt, 501
    for copper, 294–312

for iron, 335–336, 340–34
for magnesium, 156–165
for manganese, 358–365
for molybdenum, 379–382, 384
for nickel, 403–404
for nitrogen, 28–32
for phosphorus, 55–71
for potassium, 101–105
for selenium, 518, 527–542, 543
for silicon, 558–560
for sodium 574–575
for sulfur, 206–217
for vanadium 586, 587, 589–594
for zinc, 416–421
Tourmaline, 246
Transamination, 24
Transporter protein genetics, 602
2,3,5-triiodobenzoic acid (TIBA), 125
Triple superphosphate, 82, 139, 389, 522
Tripolyphosphate fertilizer, 81
Trunk injection of zinc, 422

U
Ulexite fertilizer, 267
Ultisols, 113, 138, 461, 465, 467
    distribution, 462, 463
Urea as fertilizer, 39, 39–41
Urea ammonium phosphate fertilizer, 42
Urea formaldehyde fertilizer, 41
Urea phosphate, 82
Urease and nickel, 396, 397, 399, 400, 401

V
Vanadinite, 586
Vanadium
    and aluminum accumulation, 588
    and borate, 589
    and calcium, 587
    and chloride, 589
    and chromate, 589
    and copper, 587, 589
    and increase in leaf area, 587
    and iron deficiency, 587–588
    and magnesium, 587
    and manganese, 587–588, 589
    and molybdate, 589
    and molybdenum, 587
    and nickel, 587, 589
    and potassium, 587
    and selenium, 589
Vanadium bioavailability, 594
Vanadium oxidation states, 585
Vanadyl amino compounds, 589
Vanadyl cellulose compounds, 589
Variscite, 464
Vermiculite, 107, 319, 461, 561
Vertisols, 110, 113, 138, 317
Viets effect, 94
Vitamin B_{12} deficiency and cobalt, 500
Vivianite, 345

W

Water use and silicon, 557
Water use efficiency and potassium, 99
Wulfenite, 385

X

Xanthine dehydrogenase and molybdenum, 376,
    377–378

Z

Zinc
    and boron, 246
    and cobalt, 506
    and copper, 310, 311

and flooding, 412
and magnesium, 153
and nickel, 402, 403, 404
and phosphorus, 423
and protein synthesis, 428
deficiency, 11, 246, 402, 403, 412–415, 428, 429
uptake, 412, 423–424, 428–429
Zinc nitrate as fertilizer, 424, 425–427
Zinc nitrate-ammonium nitrate-urea (NZM™) fertilizer,
    415, 424–427, 429
Zinc nutrition, 601
Zinc sulfate-induced defoliation, 426–427
Zinc sulfate as fertilizer, 422–423, 424, 425–428, 429
ZnEDTA, 422